To Dear Graeme

Enjoy! It may not be my dream but I know it's yours

Love

Shen x

A
Cruising Guide
to the
MAINE
COAST

"...*A Cruising Guide to the Maine Coast*...is by far the best of the cruising guides." This book should be the bible for anyone cruising Maine for the first or the 50th time."
—Joe Brown, *Cruising World* magazine

"What a great book you have produced! Invaluable!"
—John Howarth, *Rain*

"The definitive sailor's guide to the waters of Maine."
—*Down East* magazine

"Wonderful resource!"
—Michael Hatton, *Sweetwater*

"Unsurpassed in completeness, accuracy, and accessibility, this book covers not only the fabled harbors and passages from the sandy beaches of the southern coast to Saint John in Canada, but it also leads you to remote coves where you can enjoy a sunset all by yourself."
—International Marine Publishing

"Greatly appreciate the Guide. Thanks."
—Marc Miller, *Thriller*

"We love your guide and thank you and the Tafts for making the boating waters of Maine so comfortable, even if the bottom is so hard."
—Randy Phelps, *Redfeather* or *Catspaw*

"We hardly made a move without consulting the Tafts."
—*Sail* magazine

"Thank you for the outstanding cruising guide!"
—John and Connie Cermak, *Mirabar*

"Enticing, readable, and easy to use. It gives away almost all the gunkholing secrets from Pepperrell Cove to Calais."
—*WoodenBoat* magazine

"Wish we had purchased this book three years ago when we purchased our boat."
—Randy Overbey, *Finally*

"Great information, thoroughly enjoyable!"
—Wesley Faust, *Osprey*

"Your Cruising Guide is the best in the business! Thanks for the great work."
—Allen Ryan, *Summer Cottage II*

"Your book is terrific. It is quite dog-eared after a wonderful one-month trip from Scituate to Mt. Desert and back in my sweet little 27-foot Albin Vega..."
—Dave Pomerantz, *WishCraft*

"Still the best guide there is."
—John Melchner, *Jocar*

"The Cruising Guide is my bible when I'm Down East!"
—Charles Webb, *Fancy Free*

"This is the best guide we've ever used!"
—Tom and Val Zicko, *Wine and Roses*

"It stands out like Monhegan Island Light in a smoky sou'wester. If ever there was a place where sailors need an indispensable cruising guide, it is the fascinating 3,500-mile coast of Maine. This book is delightful reading, often more like a good Downeast yarn than a guide. You will want to set sail immediately."
—*Sailing* magazine

"You bet we use the Guide—it's the greatest publication there is. Thanks again for a great job."
—Al & Claudia Ryan

"I don't know what we'd do without it! Thank you for writing this book."
—Donald Skeffington, *Anaconda*

"Bluewater's best-selling Maine guide. Fine sketch charts, hands-on navigational information, and good advice."
—Bluewater Books and Charts

"We're cruising now using your book. Terrific!"
—Susan Richmann, *Blue*

"A Cruising Guide is the best--I wouldn't leave home without it."
—Steve Waruick. *The Griswold*

"I just have to say that your Guide is absolutely the best cruising guide to Maine and adds a whole dimension to cruising in Maine that one could not obtain elsewhere. We have made two two-week trips to Maine in the last three years from our home port in Jamestown, Rhode Island, with the two children and my wife aboard, and as we approach the next port of call and want to refresh the details of the approach, I just say to the kids, 'Would you please hand up the Bible' and they know just which book I mean, even though we have all the other Maine cruising guides too. I just discovered your website. Please try to keep it up, it is very informative."
—Steve McInnis

"I would never sail without your guide. The new edition is bigger and better."
—John A. Nelson, *Damsel*

"This book has been an enormous help to us, especially as it lists many non-direct-to-the-east but charming anchorages."
—J. Linzee Coolidge, *Sea Lion*

"Thank you and the Tafts....Could not have done it without your very good cruising guide."
—Bob and Lorelei Diamond

"A standard for cruising guide excellence."
—Armchair Sailor

"A beautiful, well-written, indispensable part of my ship's library. Keep up the good work!"
—David Conley, *Indra*

"Don't even attempt a cruise here without it. We use it like a bible."
—Joel Gleason, *Points East* magazine

"If you take only one guide Down East, this is the one you should take. Its 465 pages are up-to-date, and packed with info about every harbor and practically every anchorage on the coast...If you are going to Maine, buy this book!"
—Landfall Navigation

"Without a doubt, it is the best guide to the waters of Maine, written clearly, carefully and—best of all—with heart."
—Donna Gould, *Maine Times*

"Helpful directions, especially the confusing topic of the Reversing Falls at Saint John. The clearest written explanation of that phenomenon that I've read yet."
—Paul Monfredo, *Sigi*

"Well organized, well researched, and eminently readable. This is a book I could read all winter when the boat is stored away."
—David Platt, *Maine Times*

"Great book! This is our second edition and we use it faithfully while cruising the coast."
—Frank Lingard, Jr., *Second Wind*

"By far the best guide to Maine!"
—John Drozdal, *Ariana*

"Whether it's your first trip cruising the Maine Coast or you've sailed here for years, you'll find this your second most valuable resource (nothing beats a good set of charts), because when you need to decide where you are going to drop the hook for the night, the charts won't be nearly as helpful as the Tafts' years of experience. Where to go, where not to go, how to approach the anchorage, what to see ashore are all combined with slices of history and dry-witted anecdotes."
—Stephen Kosacz, on amazon.com

"Really like the new edition. Indispensable!"
—William Sweeney, *Ariadne*

"Love the guide!"
—Jon Steiner, *Circe*

"In my opinion, *A Cruising Guide to the Maine Coast* is a paradigm of accuracy, accessibility, and functional beauty."
—Herb Payson, *Sail* magazine

"This book is more than a cruising guide. It is a compendium of the Maine coast that no boat venturing east of the Isles of Shoals should be without."
—*Ocean Navigator* magazine

"Thanks for your hard work. It's not often a cruising guide will update..."
—Dan Gilmore

"Best guide yet that I've seen, and the updating makes it all the more worthwhile. Keep up the good work!"
—Dick Winchell

"We like your updates—new harbor charts, etc.. Thank you for continuing in such a superb fashion the wonderful work done by the Tafts."
—Dottie and John Farrar, *Patience*

"The Cruising Guide has been invaluable on my last three charters to Maine."
—Stephen Mayotte

"The new edition of the Cruising Guide to the Maine Coast is the best cruising guide I have ever seen! Thanks for doing it."
—Kris Greene

"I am a longtime devotee of 'Duncan' and first used your guide last September enroute from Christmas Cove to Southwest Harbor. I'm sold."
—Jack St. John, *Figment*

"Detailed and precise information relevant to the cruising sailor is stuffed onto every page, providing both necessary logistical details as well as insight into both the history and the natural world of Maine's seemingly unending coast."
—*Ocean Navigator* magazine

"This guide has enhanced our trip Down East. I only wish I had purchased it sooner. Thank you. It's the best of its class."
—Rob Baker, *Imagine*

Rather than sift through coves and harbors looking for nuggets, the busy cruiser would prefer to go to the vault and withdraw the 24-carat stuff. *A Cruising Guide to the Maine Coast* is such a vault."
—*Sail* magazine

"My summer bible!"
—Robert Ridolfi, *Low Key*

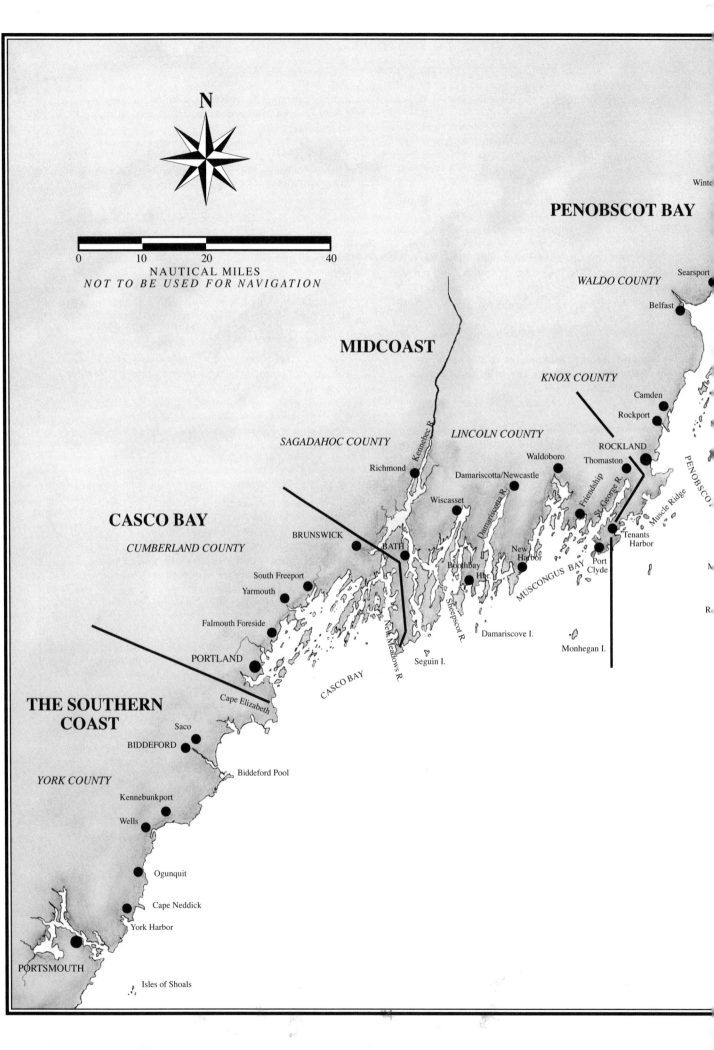

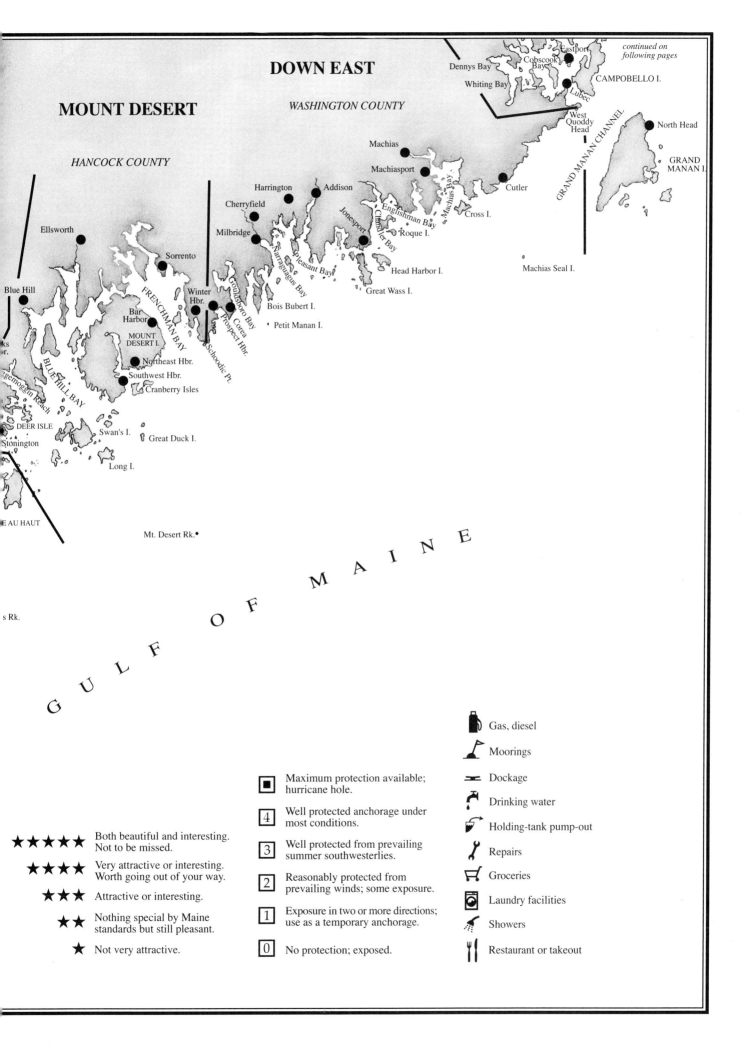

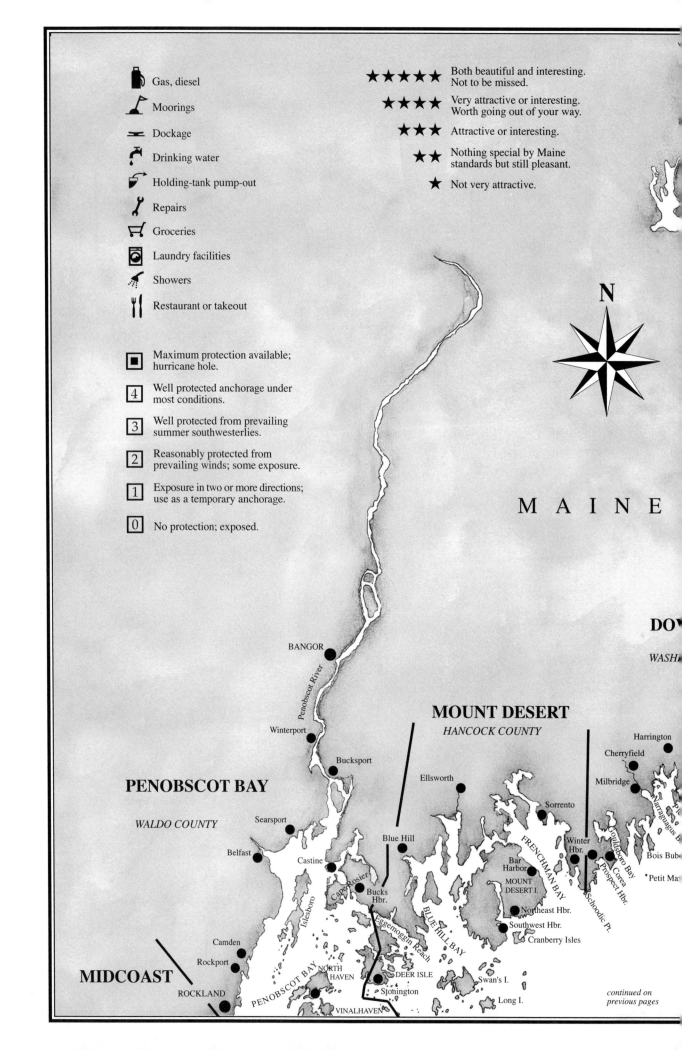

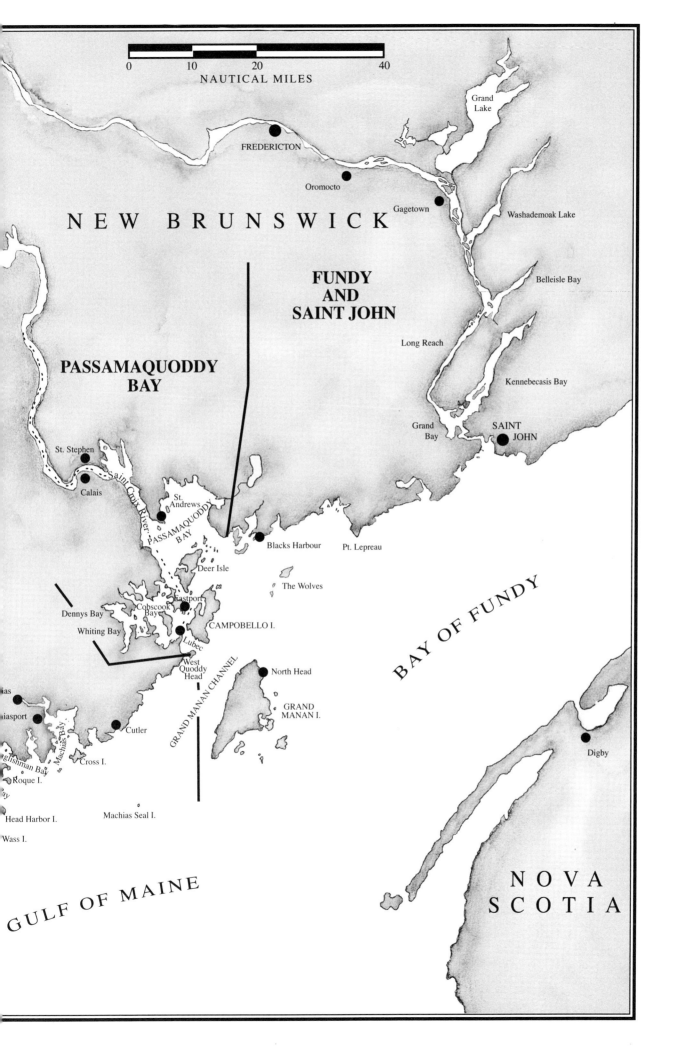

A Cruising Guide
to the

MAINE

COAST

FIFTH EDITION

Hank and Jan Taft
Curtis Rindlaub

> While every effort has been made to assure the accuracy of this book, it is not a substitute for careful judgment, prudent seamanship, and common sense. Neither the publisher nor the authors assume any liability for any errors or omissions or for how this book or its contents are used or interpreted or for any consequences resulting directly or indirectly from the use of this book.

Published by Diamond Pass Publishing, Inc.
Printed in the United States of America

Copyright © 2008 by Curtis C. Rindlaub

All rights reserved. No part of this book may be reproduced or transmitted in any form or by any means, electronic or mechanical, including photocopying, recording, or any information and storage retrieval system without prior permission in writing from the publisher.

    Fourth Edition 2002
    Third Edition 1996
    Second Edition 1991
    First Edition 1988

Versions: This edition is produced in both hardcover and softcover.

    ISBN:
    978-0-9649246-6-6 (hardcover)
    978-0-9649246-7-3 (softcover)

Please visit our website for important updates or corrections.
Questions, comments, and suggestions are greatly appreciated. Please address them to:

    Diamond Pass Publishing, Inc.
    19 Brook Lane
    Peaks Island, Maine 04108
    207-766-2337
    mail@mainecoastguide.com
    www.mainecoastguide.com

It may be it was not my chance to see the best;
but least others may be deceived as I was,
or through dangerous ignorance haphazard themselves as I did,
I have drawn a Map from Point to Point, Ile to Ile, and Harbour to Harbour,
with the Soundings, Sands, Rocks, and Landmarks
as I passed close aboard the Shore in a little Boat....

—Captain John Smith,
The Description of New England, 1616.

It was Captain John Smith's lyric descriptions of the beauty, lushness, and potential of New England which fired the English imagination and spurred the growth of the new colonies. And it was Smith who realized that the real treasure of Maine was not gold or silver, but codfish. (Library of Congress)

CONTENTS

PREFACE xiv

HOW TO USE THIS BOOK 2

INTRODUCTION 5
    Navigating Down East 14
    Approaches to the Coast of Maine 24

THE SOUTHERN COAST   ISLES OF SHOALS TO CAPE ELIZABETH   27
    Isles of Shoals 30
    Piscataqua River, Portsmouth Harbor and vicinity 33
    York River 42
    Kennebunk River 47
    Biddeford Pool and Saco River 54

CASCO BAY   CAPE ELIZABETH TO CAPE SMALL   61
    Portland Harbor and the city islands 65
    Broad Sound 81
    Royal River 83
    Harraseeket River 84
    Merriconeag and Harpswell Sound 89
    Quahog Bay and vicinity 94
    New Meadows River 96
    The Eastern Shore 101

THE MIDCOAST   CAPE SMALL TO MARSHALL POINT   105
    Kennebec River 107
    Sheepscot River 120
    Boothbay Harbor Region 136
    Damariscotta River and Johns Bay 140
    Muscongus Bay and the Saint George River 154
    Monhegan Island 166

PENOBSCOT BAY   MARSHALL POINT TO ISLE AU HAUT   179
    Matinicus and the Outlying Islands 183
    Tenants Harbor and Muscle Ridge Channel 190
    Rockland 198
    Camden 206
    Fox Islands Thorofare and vicinity 213
    Hurricane Sound and Vinalhaven's west coast 221
    Carvers Harbor and vicinity 227
    Islesboro 230
    West Penobscot Bay 236
    Penobscot River 242
    Castine and East Penobscot Bay 247
    Deer Isle, west coast 259
    Isle au Haut 261

MOUNT DESERT ISLE AU HAUT TO SCHOODIC POINT 267
- Merchants Row, Stonington, and Deer Island Thorofare 270
- Deer Isle, east coast 280
- Eggemoggin Reach 282
- Swan's Island 286
- Blue Hill Bay 293
- Mount Desert Island 300
- Somes Sound 309
- Cranberry Islands 316
- Frenchman Bay 322

DOWN EAST SCHOODIC POINT TO WEST QUODDY HEAD 333
- Eastern Schoodic Peninsula to Petit Manan and Cape Split 336
- Moosabec Reach, Jonesport, Western and Eastern Bays 343
- Roque Island Archipelago 349
- Machias Bay to Cutler and West Quoddy Head 355

PASSAMAQUODDY BAY WEST QUODDY HEAD TO CALAIS 363
- Entering the United States and Canada 368
- Grand Manan 369
- Campobello Island 378
- Eastport 385
- Cobscook Bay 389
- Passamaquoddy Bay and Letete Passage 395
- St. Andrews and the St. Croix River 398

FUNDY AND SAINT JOHN PASSAMAQUODDY BAY TO FREDERICTON 403
- Bay of Fundy to Saint John 405
- Saint John River and The Reversing Falls 412
- Grand Bay 418
- Kennebecasis Bay 419
- Belleisle Bay 426
- Washademoak Lake 430
- Grand Lake 433
- Fredericton 439

APPENDIX 441
- Emergency numbers 441
- Environmental advocacy groups 441
- Marine facilities 442
- Boat charters 448
- Public transportation 448
- Ferry service 449
- Saltwater fishing 450

INDEX 451
- Sidebars 465

PREFACE

ON a chart, the coast of Maine is a Jackson Pollock painting. Rivers dribble to the sea, the coast is flung far with abandon, and islands splatter the surface as if fallen from an overloaded brush. Nature seems to have gotten experimental on this part of our seaboard, as if by the time it ran up from Florida it was fed up with gentle bays, shallow sounds, and meek tides. Here, it says, take this, take that. You want a bay? *This* is a bay. You want to know what real islands look like? Have a gander over here. Tides? Rocks? Bigger. More. An open stretch of water? That's a perfect spot to spritz a few hundred lobster buoys. Like any good painting, it will surprise you each time you visit, each area you explore, each corner you examine. You are one of the lucky few if you have the chance to sail through this masterpiece.

Carol and I and our children have had that privilege for 18 years, and we are still stunned by the coast's staunch ruggedness, its independent island communities, its bold headlands, and snug coves and harbors. This guide, like all the editions before it, brings you as much specific detail as possible about everywhere we sail, all the gunkholes, the nooks, and the crannies. It helps you find safe harbors and tight tickles, beaches and exhilarating hikes, marinas and mountains. You can use it to plan or to dream. Or use it to stop dreaming and follow those plans.

This guide is the culmination of a collaborative effort, beginning with its original authors, Hank and Jan Taft, and aided by the indispensable comments, suggestions, and information provided by hundreds of cruisers. Thank you all. The unmatched accuracy of this guide would not be possible without you. Thanks, also, to Lee Bumsted, whose sharp eyes and keen editing have buffed this 20th-anniversary edition to a deep shine.

The rocks don't move, thankfully, but just about everything else is changing or could change. We publish a new edition about every five years, depending on how much change we find along the coast, and between editions, we post updates at *www.mainecoastguide.com*. Buoyage changes, harbors change, shoreside businesses come and go or expand or contract. An uncrowded anchorage one year may be packed the next. The owners of a private island who allowed visitors last year may forbid them the next. Many cruising guides shy away from including information that may change, so they will stay accurate. But they are also less informative and less fun. We take an opposite tack. If something is of interest to cruisers or to us personally, we include it, even if it may not be in existence for the five-year lifespan of the edition. The test is simple. If it adds to our adventure and joy and understanding of cruising in Maine, then it is here so it can add to yours.

Change hides in details. Since the last edition, Casco Bay has become a no-discharge zone for boats, so that even treated sewerage is forbidden to go overboard. But pump-out facilities have proliferated. Other regions of the coast are slated for similar designations in subsequent years. Cruisers can play an important part in keeping Maine waters clean.

Clean water comes at a price now. Charging for potable tank water is becoming more common, particularly in Penobscot Bay where many facilities do not have municipal water.

Several new invasive species have been found in Maine waters, most notably European green crabs and Asian shore crabs, both of which threaten native populations of commercially valuable shellfish. Other invasions include several attempts by business consortiums to locate and build a terminal for liquefied petroleum gas in Maine. Their efforts began in Harpswell and were rebuffed, so they hopscotched up the coast as far as Passamaquoddy Bay.

To prevent invasions of a different kind, U.S. Customs and Border Patrol continue to fine-tune their border crossing policy. They now require stricter documentation and collect data with James-Bond-like biometric devices. A network of sonar and robotic devices are being deployed in the Piscataqua River to protect the vital bridges linking Maine and New Hampshire. And the Department of Homeland Security has even established an anonymous hotline to report suspicious activity along the coast.

2006 was a shaky year for Frenchman Bay. Six minor earthquakes rocked the region in fall and early winter. The largest had a magnitude of 4.2 and caused rock slides that closed the Acadia Park Loop Road. The 2006 total of 11 quakes in Maine is nearly triple the average of the previous five years, but in that time they were the only ones centered directly on the coast. Earthquake officials speculate that a small fault line is forming in Frenchman Bay.

Chebeague Island rocked its world in 2007 by seceding from Cumberland and becoming their own town. Peaks Island mounted a similar, though less unified, campaign, hoping to secede from Portland, but the Maine legislature refused to allow it to come to a vote. These islands, like all of coastal Maine, are being crushed by real-estate taxes based on second-home market values that few of the local economies can support. Yet after years of struggling, the population of lonely Frenchboro is on the upswing.

One coastal economy continues to thrive: lobstering. Despite dire predictions of decline, the lobster population and catch continue to defy scientists' expectations. One obvious explanation is that lobstermen are feeding them with tons and tons of bait. It has been discovered that dominant lobsters guard a well-baited trap as their own personal larder. Lobsters may be as much farmed as fished. The fear now, though, is that the populations of bait fish may collapse.

Rockweed is making an economic comeback as a nutritional supplement and as a vegetable, but its harvesting has reopened the dilemma of who owns the intertidal zone. When Maine was part of the Massachusetts colony, the intertidal zone was granted to the shore owners in an effort to encourage private wharf-building at a time when the colony could not afford to build them itself. Today, Maine and Massachusetts are the only states where the intertidal zone is private.

Stripers continue to return to Maine in increasing numbers, as do eagles and ospreys and summer visitors. Native salmon haven't been as fortunate and have all but disappeared. Also disappearing, at an alarming rate, is access to working waterfronts for fishermen. As more and more waterfront falls to the second-home market, fishermen face the predicament of not being able to get to their boats. In Cundys Harbor, a land trust was formed to purchase an historic wharf to maintain access for fishermen to this working harbor.

Local land trusts and the larger Maine Coast Heritage Trust have made huge strides in protecting coastal land and islands from the temptation of development by creating conservation easements and by outright purchase. The Petit Manan National Wildlife Refuge has quietly expanded into the Maine Coastal Islands National Wildlife Refuge by purchasing bird-nesting islands along the entire coast.

Access for recreational boaters is in ever-increasing demand. In response, Portland has a new major marina and a new seasonal slip facility. Other large marinas are planned in Wiscasset and Belfast. Rockland's mooring field continues to expand eastward, and someday may actually fill the entire harbor behind the breakwater. And Bangor's waterfront is blossoming.

Technology continues to make cruising in Maine easier and safer. The amazing GPS units of five years ago have now morphed into multi-function chart-plotters and radar displays, providing vision and position and great peace of mind along a sometimes fogbound coast.

Ironically, this astounding technology brings us to a much simpler place, where the bright screens are dim in comparison to the faces of family and friends snug in the cockpit in a fresh breeze or huddled below around a kerosene lamp, where the air is tinged with spruce and rockweed, where a day's plan can be changed by a shift in the breeze. The essence of the Maine coast, its wild abandon, is still here, unchanged, in the cry of an osprey, in the lapping tide, in the mussels or lobster steaming in the pot. Or, if your fishing skills are better than mine, you might catch it at the end of a looping line, in the shape of a bucking pollock.

Enjoy,

Curtis Rindlaub

HOW TO USE THIS BOOK

GENERAL INFORMATION

The introduction covers general information about the coast of Maine: its weather, fog, tides, supplies, services, and communications, as well as approaches to the Maine coast. We suggest reading this section first.

REGIONS

This book divides the coast of Maine into seven regions, from southwest to northeast. An eighth region covers Canada's Fundy coast and the Saint John River. The regions are shown on the sketch charts on the endpapers and listed in the Table of Contents. Region headings are in the margins of this book.

OVERVIEW CHARTS

An overview chart for general orientation is located at the beginning of each region of the coast. Stars on the chart show the location of the harbors described within that region. Hurricane holes, the safest harbors of refuge in the most severe weather, are marked with ■, the symbol for maximum protection.

SKETCH CHARTS

Sketch charts are included where appropriate to illustrate piloting descriptions, harbor features, or shoreside amenities.

Sketch charts are for illustration and orientation only and not for navigation.

HARBORS and PASSAGES

Harbors, daystops, and important passages are listed in geographical order from west to east, starting at the Isles of Shoals. Passage headings are in italics. To find a particular harbor, use the Index or the Table of Contents, or follow the region headings to locate the geographical vicinity of that harbor.

HARBOR RATINGS

Harbors suitable for overnight anchorages are rated for protection, beauty and interest, and facilities and services.

PROTECTION. Overview charts at the beginning of each region show all hurricane holes with the symbol for maximum protection shown below.

■   Maximum protection available; hurricane hole.

4   Well protected anchorage under most conditions

3   Well protected from prevailing summer southwesterlies.

2   Reasonably protected from prevailing winds; some exposure.

1   Exposed in two or more directions; use as a temporary anchorage.

0   No protection; totally exposed.

BEAUTY AND INTEREST. These are subjective ratings for both natural and man-made features.

★★★★★   Both beautiful and interesting. Not to be missed.

★★★★   Very attractive or interesting. Worth going out of your way.

★★★   Attractive or interesting.

★★   Nothing special by Maine standards but still pleasant.

★   Not very attractive. Of little interest to cruisers.

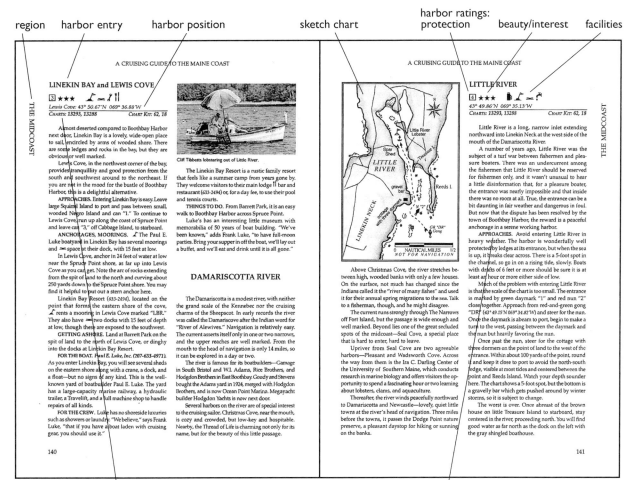

region — harbor entry — harbor position — sketch chart — harbor ratings: protection — beauty/interest — facilities

chart references (preferred charts in bold type) — navigation details

FACILITIES AND SERVICES.

 Gas or diesel

 Moorings

 Dockage

 Drinking water

 Holding-tank pump-out

 Repairs

 Groceries

 Laundry facilities

 Showers

 Restaurant or takeout

HARBOR POSITION

Latitude and longitude positions for each harbor or passage are provided for reference and location only. *Use of these positions as navigational waypoints is not recommended and is at the sole risk of the navigator.*

UPDATES and CONTRIBUTIONS

Many coastal changes can occur between editions of this guide. These changes or any errors we discover in the guide are posted on our website at *www.mainecoastguide.com.*

We also post any comments or cautions relayed to us by other sailors, as well as our list of contributors. We always welcome your feedback.

GUIDE CARE

As much as is practical, this guide has been constructed to withstand the abuse of onboard environments, both in the cockpit and down below. Yet any book in these extreme conditions has obvious construction limitations. The text paper is extra heavy, but it is not waterproof. The spine of the hardcover edition is sewn, but it can not withstand being sat (or stepped!) upon. Both the hardcover dustjacket and the paperback cover have been laminated with plastic film for maximum moisture, curl, scuff, and tear resistance. But prolonged exposure to direct sunlight will shrink and delaminate this film.

Please keep your guide in a safe, dry, shaded location. We recommend making a protective book cover and taping several marker ribbons near the spine.

LIMITATIONS and PRECAUTIONS

In preparing this book, every effort has been made to provide information that is accurate and up to date. It is, however, impossible to guarantee the accuracy of this ever-changing subject matter or how it is used or interpreted. The authors and publisher disclaim any liability for any loss or damage to persons or property that result directly or indirectly from the use of this book or the information contained within it.

> This guide should be used as an aid to prudent seamanship and only as a supplement to official U.S. and Canadian charts, light lists, and other publications.

Introduction

ENJOYING MAINE

WHEN TO COME

The typical cruising season in Maine lasts from the first of July through Labor Day. The days are warm, the nights are cool, and the prevailing southwest winds are light to moderate. Maine's sailing season is notoriously short, so there are more boats in Maine during July and August. Don't worry. It still won't be too crowded.

May and early June have some wonderful spring sailing days, but you are bound to encounter rain and cold as well. The last half of June is often delightful. It may still be chilly, and many facilities will not have opened for the season, but you will own the whole coast.

If life allows it, consider coming—or staying—in September. In the first half of the month, Maine truly comes into its own. Loons and lobstermen are out in full force, but you will be alone in many a cove. Toward the end of the month, it gets cold in the mornings, with occasional biting winds from the north, and the days are noticeably shorter.

October has wonderful bright sailing days as well, but by then you start to press your luck.

WHERE TO GO

Only 250 miles as the crow flies, Maine's convoluted coastline measures more than 3,500 miles. A newer computerized study using satellite images logged the length at a whopping 5,500 miles with 6,200 islands. By any count, there is enough coast here for a lifetime of cruising.

Arguably the best cruising grounds are Penobscot Bay, Blue Hill and Frenchman Bay, Mount Desert, and the offshore islands. These waters are dotted with islands and offer an enormous variety of fishing villages, resort towns, and remote coves framed against the spectacular backdrop of Mount Desert and the Camden Hills.

But that doesn't mean you must go way Down East to see the real Maine. There is wonderful sailing in the western part of the state as well, from Kittery to the Kennebec River. Casco Bay has many of the same natural features as Penobscot Bay, although being closer to population centers, it also has more boats. The midcoast and Muscongus Bay have a reputation for being difficult—a lot of rocks and a lot of rivers—but they also have fewer boats and a wealth of places to explore.

East of Mount Desert, Maine becomes more austere and challenging. To anchor off beautiful Roque Island beach is the goal of many sailors, and others aim beyond, way Down East to the Passamaquoddy Bay region, with its sparsely settled shores and awesome tides. And those who just can't get enough set their course for Saint John and the warm, freshwater pleasures of the Saint John River. Perhaps these are things to do, not on your first cruise to Maine, but after you have savored the more accessible cruising grounds of the western and central coast.

PREPARATIONS for CRUISING in MAINE

Because of the rocky coast, frequent fog, chilly water, and relative scarcity of marinas, cruising in Maine requires a greater degree of self-sufficiency than cruising in the more forgiving waters to the south. Particularly if you are a newcomer to Maine waters, consider the following checklist:

EQUIPMENT. *Anchors.* All the familiar types of anchors are in use along the coast of Maine, but the most prevalent are the plows, claws, and Danforths. The Tafts used a 45-pound CQR plow fitted with 18 feet of 3/8-inch chain and 250 feet of 5/8-inch nylon rode to hold their heavy displacement, 43-foot ketch. With our somewhat lighter displacement, we feel very secure on an 33-pound Bruce anchor on 30 feet of 5/16-inch chain and 5/8-inch rode. Both types are heavy enough to penetrate weeds and kelp and will hold well in sand and rock as well as mud. Although smaller boats need less heft, your ground tackle should be as heavy as you can reasonably handle. You should have the ability to rig a tripline when you suspect a rocky or foul bottom.

You can cruise most places in Maine without one, but sometimes a yachtsman or a fisherman anchor is sometimes the only type that will work, for example in a cobblestone bottom. We carry a second anchor and rode, a large Danforth, sized to easily hold the boat alone, which we use as a stern anchor or as a second anchor in a blow. A third anchor, a little lunch hook, saves the work of raising the primary anchor at temporary stops.

Fenders and Fenderboards. If you are going way Down East, be prepared for the boat to take a beating. Sometimes you will be coming alongside pilings instead of floats and often tying up to draggers and lobsterboats. Oversized fenders or commercial poly balls work well alone or rigged with a fenderboard made from a two-by-six.

Knife, Face Mask, Snorkel, Wetsuit. Lobster buoys are a fact of life on the Maine coast and their connecting lines—the potwarp—pose a constant threat to the keels, rudders, and propellers of cruising boats.

Your first line of defense is to stay alert and steer to avoid directly running over the buoys. If your prop is about to pass over a buoy, put your engine in neutral. If you do snag a trap, you may need to go over the side to clear it.

Maine's cold water makes a wetsuit for this purpose standard onboard equipment, along with a well-fitting mask and snorkel. Single-handed sailors should have plans on how they will accomplish this procedure alone. Ubiquitous buoys inshore will render an autopilot nearly useless.

Spurs or other cutting devices can be installed on your shaft. The downside of this option is the expense of installing them and their indiscriminate cutting.

Perhaps the ultimate way to prevent wrapping a lobster buoy around your prop is to install a prop cage fitted around the propeller. The prop cages on lobsterboats are what enable them to tear through the buoys without concern. This solution, however, doesn't keep the warp from snagging your rudder if it hangs free of the keel, and it will cost you some speed as well as some dollars.

For suggestions on how to avoid lobster buoys, see the sidebar on page 12.

Warm Clothing and Heater. Nights are cool in Maine, even in midsummer, and the weather is frequently damp and chilly. Bring plenty of warm clothes, especially wool hats, socks, and sweaters. A large oil lamp in the cabin will help keep things dry, and a stove or cabin heater stretches the sailing season.

Screens. Don't leave home without them. Some say that the mosquito is the Maine state bird. Other pests include black flies and no-see-ums. Your screens should have a fine mesh, and they should be in place well before dusk and past dawn. In certain marshy spots there are stories of yachtsmen being carried away....

Lobster Cooking Pot. The day will surely come when nothing will do but lobsters steamed over an open fire on some remote and rocky beach, and your big, black 19-quart lobster pot will be invaluable. We also carry two grates for cookouts, one a standard rectangle, the other a two-sided, fine-mesh arrangement that holds fish and small vegetables. Please review fire regulations and cautions on page 10.

Water Container. In some Maine ports and many Canadian ones, drinking water is not piped to the docks. A five-gallon container is useful to lug it aboard.

ENTERING CANADA. If you plan to enter Canada, you will want to have a Canadian courtesy flag and long mooring lines (two lines of 100 feet or more). Both Canadian and U.S. customs will ask for certain documents including crew identification and citizenship, boat papers, and rabies certificates for dogs. For more details, see *Entering the United States and Canada* on page 368.

FLOAT PLAN. The Coast Guard urges cruising sailors to file a detailed float plan with a friend or relative. This plan should include the intended route of the vessel, the anticipated ports of call, the number, ages, and relevant medical histories of the entire crew, and a complete description of the vessel, including all navigation and lifesaving equipment onboard.

ISLANDS

More islands lie off the coast of Maine than any one person can know in a lifetime. There are more islands in the Maine archipelago than in the Caribbean, more than in Polynesia or on the Dalmatian Coast.

About a third of these islands are 10 acres or more in size, and 15 are inhabited year round, mostly in Casco Bay, Penobscot Bay, and south of Mount Desert. There is scheduled ferry service to about 20 of them—some of it private, some of it public and run by the state. It's good to know about these ferries because you will need to keep an eye out for them and also because they come in handy for meeting friends or swapping crew. For complete ferry listings, see *Appendix F*, page 449.

Although it is difficult to imagine when you discover these remote, secluded little communities today, the outlying islands were, in fact, the first footholds of European settlement in the New World. The history of islands such as Damariscove and Monhegan date back to the earliest years of the seventeenth century when they were not only ideal fishing locations, but relatively safe from attack by mainland Indians. The mainland was not settled permanently until the end of the French and Indian Wars in 1763. Even then, as commerce grew along the coast, the islands remained the most convenient locations, easily serviced by coasting schooners. Fish, lumber, and ships were the early items of trade, followed later by granite, ice, and lime.

With the advent of railroads, coastal transportation was doomed. Steamships replaced sail, and highways changed the pattern of commerce even more. Now the islands were out of the mainstream, and their people started to drift toward the mainland. Island populations peaked around 1900 and have declined ever since. By the end of World War II, the islands were almost uninhabited.

Today islands are increasingly prized for recreational use and for second homes. In response, many conservation groups work diligently to protect coastal properties (see *Environmental Advocacy Groups*, page 11).

PUBLIC AND PRIVATE PROPERTY. Although more than half of Maine's islands are publicly owned, many of these are very small. Not counting Acadia National Park, federal and state agencies and conservation groups own over 1,500 islands totaling over 16,000 acres. With the recognition of the scarcity of bird-nesting habitat and the need to preserve the character of Maine's coast, this number is increasing annually. The rest of the islands—and most of the island acreage—are in private hands. In terms of acreage, 94 percent of Maine islands are private property.

Mainers have always felt free to land anywhere and use the beach below high water for the traditional purposes of fishing, fowling, and navigation, as provided in the Massachusetts Colonial Ordinance of 1641. But the world has changed. When it comes to the use of private beaches for recreational purposes such a sunbathing, picnicking, or exploring, recent court decisions have upheld the rights of coastal property owners, threatening the centuries-old perception that the public has access to the intertidal zone, between low water and the high-tide mark.

In the 1970s the State of Maine researched the titles to every Maine island, rock, and ledge to compile the Coastal Island Registry. Ownership of islands without clear title reverted back to the state.

Larger state-owned islands fall under the jurisdiction of the Bureau of Parks and Lands, and their use is managed jointly with the Maine Island Trail Association. MITA links about 150 public and private islands and other waterfront sites in a water-trail system that promotes both small-boat exploration and coastal stewardship.

Many of the small state-owned islets and ledges are managed by the Maine Bureau of Parks and Lands. In recent years, the U.S. Department of Fish and Wildlife has quietly but aggressively purchased easements or ownership of larger islands with significant bird-nesting or wildlife habitat, expanding what was the Petit Manan National Wildlife Refuge into the Maine Coastal Islands National Wildlife Refuge. Other islands are under federal control as part of Acadia National Park.

And finally, many islands and coastal tracts are protected by nonprofit organizations such as the Nature Conservancy, the Maine Coast Heritage Trust, and local land trusts.

This cruising guide details both where you can and cannot land. It identifies publicly accessible state and federally owned land, islands, and parks, as well as nature preserves and areas owned or managed by nonprofit organizations. We encourage cruisers to adopt the Maine Island Trail's sense of stewardship of those islands, that those who use them are responsible for taking care of them.

But what about privately owned islands? When is it permissible to go ashore? "Private" here is not an evil word. Maine is blessed with many generous landowners who welcome respectful visitors as a way of sharing the beauty and the bounty under their control. For this, the cruising community cannot be thankful enough, nor respectful enough. Their careful stewardship is what has preserved most of Maine's islands, keeping them ruggedly beautiful in their isolation and providing a home to much of our coast's wildlife.

The increasing popularity of small-boat exploring, however, and the increasing numbers of cruising boats are straining relationships between landowners and uninvited visitors. Once-remote islands are now more accessible to more people. Certain islands have been overused, and the curious are invading others.

To respect the efforts and the privacy of island landowners and to help maintain the delicate balance between island owners and uninvited visitors, here are some guidelines we follow ourselves:

• If an island is obviously inhabited, we stay aboard, whether or not the owner is there.

• We assume uninhabited islands are private unless we know otherwise, and we generally don't go ashore, particularly if there are nesting seabirds or any signs asking us to stay away. Occasionally, there are signs indicating how the owner would like you to treat the island. One of the most effective signs we have seen is on a small island near Stonington: "In our absence, the owners know they are dependent on your good judgment and sense of fitness as to what you do. If you value privacy and beauty, surely you will understand that the owners treasure the land on which you stand."

• If we land on an island that we thought was uninhabited but meet someone ashore, we introduce ourselves and ask permission to explore.

• We try to respect the privacy of people living on islands and their desire for peace and solitude.

• We limit our group ashore to six people or less. Even on public and private islands where visitors are welcomed, crowds are not. Club cruising groups should be particularly sensitive to this issue, since often their collective members can resemble small armies.

• We realize that our example ashore and in the harbor not only affects how we are welcomed onto the islands, but how everyone following us will be received.

A leave-no-trace sensitivity toward the islands is promoted by the Maine Island Trail Association. We urge adventurers in small boats who rely on island camping to join and support MITA. See *Appendix B,* page 441.

RESPECT for the ENVIRONMENT

Maine is unique. Its beautiful coast is relatively unsettled, untraveled, and unspoiled. Its waters sparkle. The air is clean. This environment, however, is fragile, and what is carelessly destroyed in an hour may never grow back. Here are some basic guidelines to help preserve this wonderful coast:

• *Leave only your footprints.* Adhere to the fundamental wilderness ethic—take out what you brought in, leave only footprints.

• *Trash.* Nothing should go over the side unless it has come from the ocean—especially not cans, oil, styrofoam, or plastic bags.

Don't leave your boat's garbage on islands, even if you find receptacles. It only adds to the problem of island trash disposal. Save it for the mainland.

Dispose of engine oil or oily bilge water properly. We strongly recommend oil-absorbent pads or sponges in the bilge. Beverage bottles in Maine are returnable, and while it is difficult to sort trash onboard, they don't take up any more space if they are in a separate bag to be returned. If you return enough of them, it almost feels like you are being paid to drink.

• *Human waste.* Federal law prohibits the discharge of untreated sewage within 3 miles of the coast. Yacht sewage pollutes the surrounding waters by decreasing the amount of dissolved oxygen available for marine organisms. Sewage also can carry pathogens that may affect the health of marine food sources and, ultimately, the health of humans.

Please be diligent in managing your onboard waste. Technically, all cruising boats with heads are required to have holding tanks or treatment systems, and they should be used anywhere within 3 miles of the coast. In the past, however, finding a pump-out station, or one that worked, was a major challenge. You no longer have that excuse. Recent legislation has mandated pump-out stations at marine facilities of a certain size, and the number of stations has more than quadrupled. All of Casco Bay became a no-discharge zone in 2008, and other coastal regions are due to follow. Pump-out stations are noted in harbor descriptions and listed in *Appendix C* on page 442.

Nobody likes being forced to provide services, however, particularly of the smelly kind. Please be patient and respectful of the marine facilities you use. Call ahead, and try to buy your fuel or make other purchases at the same time.

•*Fires. To report a fire in an emergency, call 911 and/or 888-900-FIRE.*

Fires are dangerous, particularly in July and August, and particularly on islands. This has prompted the requirement that you have a permit for any fire and the extreme reluctance of the Maine Forest Service to issue them for island locations. To request a permit, you must call or write the Forest Service and then physically pick it up at a MFS office—nearly impossibile for cruisers with no land transportation.

In the event that you clear these hurdles and plan on having a fire, use extreme care. If the weather has been dry for a long time, don't build fires at all. NOAA weather broadcasts often contain fire-danger information. Build fires only below the high-water mark, the farther down the beach the better. Use driftwood, not deadwood from ashore, and certainly not live limbs. Also, know that the wood from old lobster traps may be treated with preservative and shouldn't be used for cooking fires. When done, make sure your fire is thoroughly doused and scattered.

Permits may be obtained on the web no more than 48 hours in advance at *www.maine.gov/burnpermit* or by contacting the Maine State Forest Service at one of the following offices:

Southern Region Headquarters, Augusta
207-287-2275, 800-750-9777
Saco District Headquarters, Gray
207-657-3552
Damariscotta District Headquarters, Jefferson
207-549-7081, 207-549-3802
Central Region Headquarters, Old Town
207-827-6191
Machias River District Headquarters, Jonesboro
207-434-2621
Narraguagus River District Hdqtrs, Cherryfield
207-546-2346

•*Wildlife.* More than 400 coastal islands are used for nesting by colonies of seabirds and water birds in the spring and summer. During this time, from April 1 to mid to late August, nesting islands owned by the state or federal government are closed to visitors. Avoid disturbing any nesting birds, especially in sanctuaries and important nesting sites. If the parents are driven away, the eggs or young may be abandoned or exposed to predators.

Seals love to bask on remote halftide ledges. Their uneasy actions will indicate when you are getting too close. Pupping season is in May and June. Often pups are left alone on the shore while their mothers are off feeding. Please do not disturb them. If you think you have discovered a stranded seal or any other marine mammal in distress, including whales, please report it to the the marine mammal reporting hotline at *800-532-9551*.

•*Fishing.* Adhere to all fishing laws, particularly the size and bag limits on striped bass. See *Appendix G*, page 450, for regulations, contacts, and tips.

•*Foraging.* Wild foods are abundant on the coast. Islands are covered with blackberries, raspberries, and blueberries, as well as other more exotic foods such as mushrooms and edible plants. The shores are rich with clams, mussels, periwinkles, and even seaweed. Be sure to properly identify your food, particularly mushrooms, before you eat it. Do not take shellfish from areas affected by red tide. Closed areas are reported on the red tide hotline at *800-232-4733*.

•*Camping.* Most landowners discourage camping on their land. Do not camp without prior permission. The major fear is fire, which can ruin an island or a headland for generations.

• **Pets.** If you sail with a dog or cat, remember that the *need* to take your pet ashore is not *permission* to do so, particularly on private islands or those with sensitive nesting habitat. Once ashore, keep you pet on a tight leash.

The threat posed by pets is summarized well in the *Maine Island Trail Guidebook.* "On small islands, especially, wildlife is vulnerable to harassment by pets, and a dog or cat can wipe out a bird's production for an entire year in just a few joyous seconds."

ENVIRONMENTAL ADVOCACY GROUPS. The pristine beauty of the Maine coast is not just a lucky accident; it is the result of continuing vigilance on the part of many dedicated people who value Maine's unique quality of life. A generation ago, the greatest threat to the Maine environment came from industrial pollution. Today, it is posed by development pressure.

Many environmental and conservation groups work hard to preserve this beautiful area. Please, if possible, support their efforts. See *Appendix B,* page 441, for contact information.

FISHERMEN, LOBSTERBOATS, and WORKING HARBORS

Generally speaking, the native Maine fisherman is a person of honesty, character, and self-reliance as well as a shrewd observer and good judge of human worth. Fishermen are also friendly, in our experience, and will go out of their way to help you. More than once, a lobsterman has pointed out a ledge we were approaching, suggested the best place to anchor, or told us whose mooring to use for the night. On the other hand, fishermen have a job to do, and they have little time for demanding yachtsmen or their complicated needs.

Lobsterboats often run in circles as they haul their traps, and they may appear to be shooting off in unpredictable directions. Or they may be hauling or setting a string of traps that are joined together, usually set in a line parallel to the shore.

The lobsterman uses a hydraulic hauler to bring each trap aboard. He discards seaweed, crabs, female lobsters carrying eggs, and lobsters that are too short. The trap is rebaited and dropped over the side—then on to the next trap.

Each lobsterman has his own buoy colors, registered with the state and displayed by a buoy mounted on his boat. You can scan the waters for his colors and guess which way he'll head next. Even so, he may surprise you. Stay clear, give him plenty of room, and forget your right of way. These people are earning a living.

Never haul a lobster trap, even if you are merely curious or plan to leave money. If there are higher laws in this world, most lobstermen would put this one at the top of the list. Offenders might find their mooring lines mysteriously cut and little sympathy on the docks.

Lobstering is the primary fishery in Maine, but not the only one. Fishing boats may be towing nets or scallop drags astern and be unable to stop or alter course. Don't insist on your right of way. Be alert to the rare possibility of nets strung across the mouths of coves or in circles marked by little floats and dories. Fouling a net could be a real mess, and an expensive one, especially if you release a large catch. Farther Down East and in Canada, weirs and fish pens cram the sides of most coves, so unfamiliar shores should be approached with caution.

Many harbors in Maine shelter yachts and fishing boats side by side in peaceful coexistence, but there are also strictly working harbors, jammed full of lobsterboats and draggers, with no facilities for yachts. Cape Porpoise, Mackerel Cove, Friendship, South Bristol, and Carvers Harbor are good examples, as are most of the harbors east of Mount Desert.

BUOY, OH BUOY

Cruising in Maine is ruled by an unforgiving triumvirate: rocks, fog, and lobster buoys. The rocks are charted, the fog comes and goes, but the brightly painted lobster buoys are everywhere.

To cruising sailors, the buoys can be a minor inconvenience or a serious navigational or safety threat. Understanding the function and placement of buoys can help minimize the risks they pose and may, on occasion, steer you out of trouble.

Over three million lobster traps are fished in Maine waters, and, together or in groups, they are all marked by lobster buoys. Despite trap limits, the number of lobster buoys continues to increase as more and more lobstermen fish their maximum number of traps.

Each buoy is attached by a long length of line, the potwarp, to one or more traps (or "pots" as they are known farther south) weighted to the bottom. Most buoys are made of painted foam, sometimes with a stick of wood or plastic through the middle as a handle.

Hitting a buoy itself poses little danger. Pulling a buoy into your prop is less benign. The buoy may pop out from under your boat and bob to the surface in your wake. Or, the prop may cut it up and spit it out like chewed popcorn.

The worst case is when the prop catches the potwarp and wraps it in a knot on the shaft. The engine may stall, and the boat will either be anchored by the trap or adrift if the warp has broken. Night, fog, an exposed location, hard current, or heavy traffic all multiply the seriousness of the situation.

Before you cruise to Maine, you can install a cage around your propeller or Spurs or other cutting devices on your shaft. Both solutions are expensive. The Spurs cut indiscriminately, and cages cost you some speed. And neither keeps potwarp from snagging your spade rudder.

Your first—and best—line of defense is to stay alert and steer around the buoys. In many places, it's like weaving your way through a minefield. Turn off your autopilot and adjust your itinerary to avoid running inshore at night.

Look carefully at the buoys you pass. You may be in a region where the lobstermen fish singles, one buoy for each trap. Or, more likely, you will be in an area that fishes "strings"—many traps roped together and set in a line, with buoys at each end of the string.

To keep strings from crossing, strings are set parallel to the shore. Look for same-colored buoys spaced at regular intervals parallel to the shore. You can pass safely between them, since the potwarp between the traps is all on the bottom. The longer the distance between the buoys, the more traps there are on the string. Or, once you recognize the strings, you can run down the channels between the strings.

Next, note the potwarp in your area. Most potwarp sinks, but in rocky areas where extra warp is prone to getting caught, lobstermen may use floating potwarp. Most lobstermen will still use a length of sinking warp just below the buoy to keep the floating warp from lying on the surface where it is likely to foul a prop. In the rare places where you see floating warp, use extra caution.

Toggles also keep sinking warp from snagging the bottom. A toggle is a second float that supports the warp ahead of the buoy, usually smaller than the buoy, unpainted, and lying 20 to 30 feet up current or upwind of the buoy.

The toggle is used in areas of greater tidal range or in areas of strong current. The farther east you sail, the more toggles you will see. When possible, avoid passing between the toggle and its companion buoy, because the warp between them hangs in a shallow arc. More current or wind will stretch the toggle farther from the buoy and bring the connecting warp closer to the surface.

Now note the current. Lobster buoys point in the direction of current or wind, and the wake and trailing seaweed around them make good gauges of the current's strength. When narrowly avoiding lobster buoys, always pass down current or downwind of the buoy to avoid being swept onto it.

Finally, watch your depthsounder. Lobsters move throughout the season, and so do the lobster buoys. Early in the summer, before they molt, the lobsters hide among the shoreside rocks and in the warmer shallower waters. In June, the presence of lobster buoys may indicate that you are getting out of a channel or into rocks or shoals.

After molting, in early July, the lobsters move to deep-water holes and less rocky bottoms until their shells can harden. Most of the buoys move with them. In July and early August, most buoys will mark deep water, often with dramatic contour changes. At this time of the season, if you are searching for deep water through a tricky passage, you may find it by steering for the buoys.

By the height of the lobster season in August and September, the lobsters can be anywhere, still in the deep water or back in the rocks and shallows, so the buoys aren't as useful to steer by.

If you are about to pass over a buoy, immediately put your engine in neutral. If you do snag a trap, you may be able to remove it with a boathook or back it off with the engine. An experienced sailor, Stuart Gillespie, offers this advice: "Reach for the warp with your boathook, bring it aboard, and wrap it a couple of times around a winch. Nudge the clutch to turn the prop; you'll be able to sense which way it will come off by the pull on the winch. You'll feel the warp come off the prop one blade at a time—thump, thump, thump."

If this procedure fails, someone will have to go over the side and cut the warp. A knife, face mask, snorkel, and wetsuit should be standard onboard equipment. Single-handed sailors should plan on how they will safely accomplish this feat alone. Synthetic warp may fuse into a tough plastic blob. Try to save valuable gear for a lobsterman by tying the cut ends together.

When you reach your secluded anchorage, you may find that it, too, is peppered with lobster buoys. How do you anchor among them? Look for strings, and avoid dropping the anchor on a line between like-colored buoys. Adjust your rode to hang in as clear a patch of water as possible. Most importantly, understand if your underwater hull shape is prone to snagging warp if it swings across a trap. Fins and spade rudders are more vulnerable than full keels, but warp can even find the crack between a full keel and its rudder. Remember which buoys are nearby and check that they are still visible before you get under way in the morning. If you don't see them, check your keel and rudder.

For the yachtsman, working harbors can be difficult. They have no guest moorings, no yacht clubs, no fuel floats. A great bustle of lobsterboats come and go. For this very reason, the working harbors are fascinating places to visit. These communities have a culture and a language all their own, where self-sufficiency and reticence are a way of life. This is Maine the way it has been for 200 years.

When visiting, be a respectful guest. Don't feel you have the absolute right to be there. Don't demand to be serviced. Give way to the lobsterboats at the fuel dock, and wait for a lull in the activity to seek services and ask questions. Perhaps the less help you expect, the more you will get.

MOORINGS

Rental moorings for transients are common in the western and central part of the Maine coast. East of Mount Desert, they are rare. Be prepared to anchor.

Many cruising yachts seem driven to pick up a mooring, any mooring, as long as it's close to their desired destination, without any apparent regard for who owns it or whether it is adequate. Remember, all moorings are not created equal. Nor are they equally maintained. Nor are they unused just because they are empty. Here are some suggestions on the subject of mooring etiquette:

• Find out who is in charge of the moorings. In harbors with mostly pleasure boats, it may be the yacht club or the harbormaster, or one or more of the boatyards. Often transient moorings are marked "rental" or "guest" or identified by a special color or by a club or yard's initials. In working harbors, there is usually a harbormaster. If not, you can inquire at the lobster co-op or ask a nearby fisherman.

• If there is no one around to ask, either anchor or else pick up a mooring temporarily and go ashore to inquire. Leave aboard crew competent to move the boat in case the owner of the mooring returns.

• Never leave your boat unattended on an unknown mooring. Imagine a fisherman or yachtsman coming home late at night and finding your boat on his mooring, locked up and silent. What does he do? Cut you loose? Where does he moor his own boat? If there is room, we still prefer to anchor unless the holding ground is tenuous.

• There used to be guest moorings everywhere, maintained by yacht clubs or towns for the convenience of visiting yachts. Unfortunately, with a few notable exceptions, the day of the free guest mooring has passed. Expect to pay a mooring fee of $15 to $25 a night.

SUPPLIES and SERVICES

The supplies and services that you are used to elsewhere are available on the coast of Maine, only stretched a little thinner. Marinas become scarcer as you sail farther east and almost nonexistent beyond Mount Desert. There you will be dependent on the facilities that service fishermen—not as convenient, but a lot more interesting. In fact, this adds a great deal of flavor to cruising Down East.

Fuel. Gas and diesel can be found everywhere along the Maine coast.

Fuel docks serving recreational boaters are easy to find anywhere west of Mount Desert Island. East of Schoodic Peninsula, the majority of fuel docks are small operations serving their local fishing fleets. This guide indicates whether they will interrupt their regular business to serve cruising boats. Know your boat's fuel consumption and plan ahead.

Water. Good water in reasonable quantities is generally available all along the coast. In certain places dependent on wells, the supply is limited.

Ice. Most marinas and many markets have freezers with cubed or block ice. Way Down East you may have to go to a fish plant or a lobstermen's co-op to get chipped ice.

Propane and CNG. Propane is available in the larger harbors. In smaller harbors you may be able

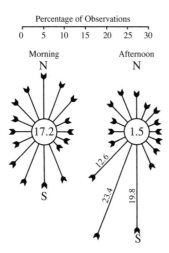

Average direction of morning and afternoon winds for 21 years in Brunswick, Maine, May–September. Center figures indicate the percentage of calm.

to get refills or exchange cylinders at shops that cater to trailers and mobile homes. Often hardware stores are also refill stations. CNG can be found in a few places, but in Maine, it is rare.

Repairs. Hull and engine repairs can be made in boatyards all along the coast. Most of them are described in this book. Additional help can be found inland. Farther east, as the number of marinas and boatyards dwindles, you will become more dependent on your own resources. Remember that most lobstermen and fishermen work on their own boats, so most of them are good mechanics.

In addition, hydraulic trailers transport boats almost anywhere along the coast where there are adequate ramps or other hauling facilities. Boatyards with hydraulic trailers are noted in this book, or the yards can often recommend someone reliable.

NAVIGATING

WEATHER

Weather in Maine is changeable—not only from day to day, but from place to place. It is not at all surprising to find quite different conditions in two coastal locations less than 50 miles apart. As Mark Twain put it, "I reverently believe that the Maker who made us all makes everything in New England but the weather. I don't know who makes that, but I think it must be some raw apprentices in the weather-clerk's factory..."

WEATHER PATTERNS. Summer weather in Maine is generally pleasant, especially on the water and in coastal areas. On a typical summer day, the breeze is light in the morning, gradually building up to 25 knots or so from the southwest, reaching its greatest strength in the afternoon, and diminishing before sunset.

The second half of June is often delightful for sailing, cool and sunny. Temperatures are mildest during July and August, but this is also when fog is most frequent. September, too, can bring wonderful sailing weather. In late September and early October, you will encounter an increasing number of strong winds with a chilling northerly component.

And then there are the exceptional years, both good and bad. After five or six summers of normal weather, with a preponderance of bright, sunny days and southwesterlies, a summer will come when nothing is normal—the sun is seldom seen, the fog lingers for a week or two at a stretch, and there are frequent winds from the north or northwest.

PREVAILING WINDS. The prevailing wind in the summer is from the southwest, usually no more than 15 or 20 knots, and often less. Morning calms are frequent in sheltered waters along the coast. Winds from anywhere in the east usually bring fog, drizzle, and generally gray, unpleasant weather.

Winds from the northwest are typically clearing winds following a cold front, sweeping moisture away and shining the visibility to a sparkle. But these winds tend to be gusty—strongest right after the front has passed, then tapering off and, after several hours, shifting to the southwest.

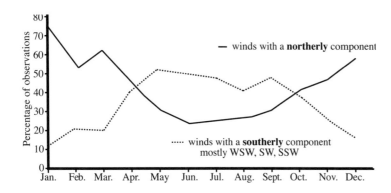

Wind direction by month, based on observations made over a five-year period in Rockland, Maine. (U.S. Coast Pilot)

DOWNDRAFTS. There are many places on the coast of Maine where hills and mountains are so close to the coast that you may encounter fluky winds and downdrafts unrelated to the general weather pattern. Downdrafts are common north of Camden, in Eggemoggin Reach, in Somes Sound, by Devils Head on Marshall Island, and along the St. Croix River.

FOG. July and August are the most common months for fog, but it can occur any time of year. There is a general belief that fog comes with calm or light winds, but this is often not the case in Maine. Don't be surprised to find fog and a brisk wind at the same time. Be prepared for fog to close in with little or no warning. Take bearings while you still can, and have in mind a destination that is easy to reach in low visibility. See page 17 for more details.

GALES AND HURRICANES. During spring and summer, the strongest winds are generated by fronts and thunderstorms. Although the prevailing summer wind is a moderate southwesterly, don't assume that everything from that direction is benign. When the forecasters predict southwest winds of 20 or 30 knots, expect periods of higher winds as well. In a given summer, there may be two or three gales (34 to 47 knots) from the south.

Hurricane season is June through October, but hurricanes are infrequent in Maine. Since hurricanes are tracked from their inception in the Caribbean or the Gulf of Mexico, there is always several-days warning. For more on hurricanes and hurricane holes, see next column.

WEATHER FORECASTS. NOAA Weather Radio broadcasts reasonably current weather forecasts on VHF weather channels, and there are similar broadcasts in Canada. Usually these forecasts are quite good, though the weather picture may be painted with a broad brush—from the Merrimack River to Stonington and then to Eastport and up to 25 miles offshore. Local conditions tend to vary considerably, so the offshore forecasts can be seriously misleading. NOAA's regional forecasts are done by county, and Maine counties are shown on the endpaper charts in this book.

Take all weather forecasts with a grain of salt. Consult the sky and consult your own barometer. A rapidly falling glass may mean much worse weather than the radio is predicting. Always have a fallback anchorage in mind to run for if the wind blows up too strong or from the wrong quarter. Use more scope than is needed for the conditions at the time you anchor.

HURRICANES

Since the 1938 hurricane which devastated New England, Maine has been seriously affected by only seven other hurricanes, most recently by Hurricane Bob in 1991. Here is the recent history:

September 21, 1938	Hurricane of '38
August 31, 1954	Carol
September 12, 1954	Edna
September 12, 1960	Donna
October 29, 1963	Ginny
September 27, 1985	Gloria
August 20, 1991	Bob

Hurricanes can strike at any time from June through October, but for the past 50 years, all except two have hit Maine in September or close to it. Of course, hurricanes and their tracks are highly erratic, and there is no assurance that the next one will arrive in the fall.

PREPARATIONS. Despite their infrequency, hurricanes do hit Maine. Early warnings are broadcast on commercial radio and TV as well as on NOAA Weather Radio. So if one is on its way, what should you do?

The first decision is whether to haul the boat or leave her in the water. There seems to be a general consensus that boats ashore fare better than boats in the water because their safety is not jeopardized by storm surges, dragging anchors, chafe, and other drifting boats. If a hurricane is approaching sometime near the end of the season, the argument for hauling the boat becomes even more compelling. Make these plans early, though. Boatyards all along the coast will be hauling boats around the clock, but they still might not get yours out in time.

If you decide to leave the boat in the water, then you must find the best possible harbor to protect your boat. Scattered along the coast of Maine are a number of coves, harbors, rivers, and hideaways that are so well protected by land, they are known as hurricane holes. Good ones are completely landlocked with good holding ground, depth deep enough but not so deep as to require excessive scope, and enough swinging room, but they are not so large that a serious fetch can build. Remember waves, more than wind, cause anchors to drag.

Some places that meet these criteria are already crowded with moored boats, and a transient yachts-

A sloop on the ledges of Rockport Harbor in the aftermath of Hurricane Gloria, September 1985. (Neal A. Parent)

man wouldn't have a chance of finding room (e.g. the western end of Boothbay Harbor). Others are difficult to enter even under the best conditions and impossible in bad visibility or heavy weather, particularly for strangers (e.g. Eastern Branch in Johns Bay). You might be able to use one of these if you arrived well ahead of time. Some of the best hurricane holes are not normally crowded, but they fill up quickly. Again, plan your strategy early and arrive before the last minute.

If you stay in the water, should you stay aboard? Is there anything you can do? Opinions vary, but your decision should be an opinion based on your experience, your knowledge of your vessel, the condition of your crew, and the state of your insurance policy. If you are at anchor or on a mooring, you may be able to keep chaffing under control, reset the anchors if you drag, or fend off other boats dragging into you, but you might put your safety and that of your crew in danger. On the other hand, the forces involved in hurricanes can be awesome, and often there is simply nothing you can do. Some boatyard owners will tell you bluntly to put out every anchor you have and "get the hell off the boat." In either case, make an informed decision and plan early.

For each of the eight regions of this cruising guide, an overview chart identifies good hurricane holes. We have anchored in each of these hurricane holes and talked with people who have used them under severe conditions. Nevertheless, each blow is different, and you must make your own choice.

FOG

Fog is a fact of life on the coast of Maine. It is most likely to occur in the months of July and August, but quite common from May through September. The amount of fog varies from area to area, and locally it can be totally unpredictable. Sometimes you will be shut in for a day or two, sometimes a week or more.

In the western part of the state, the foggiest areas are Halfway Rock, at the entrance to Casco Bay, and Cape Elizabeth. Fog is also frequent around the outlying islands of Matinicus and Monhegan. The farther east you go, the more likely you will encounter it, especially in the area between Petit Manan and West Quoddy Head. Fog is much less frequent as you go up rivers such as the Kennebec and the Penobscot.

The least likely time for fog is during or after a period of the northwest winds of a clearing cold front.

Fog is formed by warm, continental air flowing over colder waters, causing the moisture in the air to condense. In Maine this occurs when the prevailing southwest summer winds blow off the land, producing a "smoky sou'wester." Often you will see a distinct bank of fog lying offshore, or it may just thicken around you without warning. Morning fog may burn off by noon or early afternoon.

In his famous book *Summer Island,* photographer Eliot Porter described typical conditions for a foggy day. "A day that starts with a glassy bay and a clear sky, but with a white band of haze barely obscuring the southern horizon—the kind of day that promises to be warm—is a day to avoid. The faint white blending of sea and sky... usually indicates offshore fog that will come rolling in as the prevailing southerly afternoon breezes spring up...."

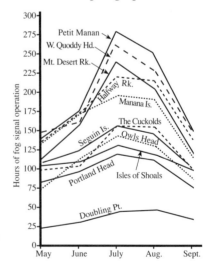

Monthly fog patterns from Isles of Shoals to West Quoddy Head based on observations at U.S. Coast Guard stations.

COPING WITH FOG. Although fog requires alert seamanship and careful navigation, it is not the end of the world. In light or patchy fog, it can be an exhilarating experience to find your way from buoy to buoy, occasionally catching a glimpse of an island or headland to confirm your track, all the while sharpening your piloting skills. But when fog shows signs of shutting down thick and heavy, then it is good to be safe in port. Break out the books and card games and settle in. This is a rocky and dangerous coast, and it is no fun to be out there cruising in zero visibility, GPS or no GPS.

If the fog closes down while you are at sea, get a good visual fix before everything disappears and head for that safe and easy-to-enter harbor you had in mind all along. Keep a careful dead reckoning, making allowance for current, which can be judged by observing lobster buoys. Take your time and use all your senses, especially hearing and smell. Set a bow watch. Use your depthsounder to confirm your dead reckoning.

If you have electronics onboard, make sure they are up and running before you are engulfed. It may take several minutes for your GPS to lock on to strong signals. Program your waypoints before you need them. And your radar will be easier to read in unfamiliar waters if it is on before all your visual references disappear.

Run for buoys with sound, either under sail or power. Sail has the advantage of being silent; power can help you run straight courses and get across areas of crosscurrent faster. Remember that fog often distorts the direction of the sound of a bell or horn.

When sound buoys are scarce, navigate by heading toward bold shores (Great Wass Island and Mount Desert are examples). You will hear the waves breaking and probably see a dim white line of surf through the fog in plenty of time to tack offshore again.

One of the greatest dangers in fog is being run down by another boat, and, even in Maine, there are plenty of them out there. Your best insurance is to move slowly, listening for the sound of engines. Have a good radar reflector aloft and sound your horn regularly.

If you are running a course that is a thoroughfare or if you are heading for a buoy that marks a major turning point along a well-traveled route, be especially aware that other boats are likely to be searching for the same mark and their courses will converge with yours as you draw near it.

Fishermen aren't likely to be running specific straight-line courses unless they are heading in or out. Lobstermen are still wheeling around their traps, and draggers are steering along bottom contours. Fishermen are not likely to be sounding their horns, and they are often too busy to see you as a blip on their radar.

Cruising boats, too, seem reluctant to blast their horns regularly. You may see a large yacht materialize close by and then vanish without so much as a sound. Even if they have seen you on one screen or another, not using a horn is a mistake. If you have radar, do not assume that every blip on your radar screen sees you as a blip on theirs. Likewise, those who don't have radar can't assume that everybody else does. Horn signals provide a warning and a sense of what is nearby, and even in this age of electronics, they can prevent a collision.

In areas of heavy traffic, commercial or otherwise, monitor VHF channel 16 for *securité* calls. Or in busy areas, make one yourself. The *securité* call (pronounced "say-curitay") is made to all concerned traffic and describes a vessel's position, course, and speed. Nearby vessels should respond, switch to another channel, and make arrangements on how to pass each other safely. For example: "*Securité, securité, securité*. This is the vessel *Intrepid* off Portland Head Light and inbound on course 340 magnetic at 6 knots. All concerned traffic please identify. This is *Intrepid* standing by on channel one six."

GPS and chartplotters have made a great difference in low-visibility navigation. Most boats that go out regularly in thick fog depend on satellite systems. Programming and operation, however, is still fraught with human error. Be familiar with your systems before you need to rely on them. See page 18 for more details.

Radar, of course, is of great value—some would say indispensable—in fog, both for navigating and avoiding collisions. It is used by most fishing boats in Maine and, increasingly, by pleasure craft. We cruised in Maine for over 10 years without radar, but we now appreciate the security it brings. In a pinch, radarless boats might meet somebody with radar they can follow.

CHARTS, GPS, and RADAR

Despite the new and wonderful instruments that have become widely available, the most important navigational tools for the coast of Maine remain the basic skills of piloting and dead reckoning. You will become alert to the color of rockweed under water (indicating a submerged rock) and grateful that swells break on submerged ledges. Most of the piloting along the coast in clear weather is done by eye, instinct, and chart. Fog, of course, requires more formal approaches.

CHARTS. The charts that cover Maine waters are extremely good. In exploring the whole coast, we have found very few errors, and we have had difficulty only when the scale of the chart is too small.

The Coast Guard periodically subjects regions of the Maine coast to waterway analyses, often resulting in dramatic changes to the buoyage. Use current charts and update them with the corrections published by the Coast Guard in *Local Notice to Mariners*. *Notice to Mariners* can be obtained by calling 800-848-3942, ext. 8335 or online at *www.navcen.uscg.gov*.

Government charts, in our opinion, are the best charts available. A Maine collection of reproduced government charts known as a Chart Kit is produced by Maptech (*www.maptech.com*) in a spiral-bound folio that may be more convenient to use on smaller boats.

For each harbor described in this guide, we have included the appropriate government chart number and Chart Kit page. Where the pilotage, anchorage, or shoreside amenities need illustration, we have included a sketch map to help you interpret the terrain.

Canadian charts can often be ordered through your usual U.S. chart agency or directly by visiting *www.chs-shc.gc.ca/pub/en/products*. The Canadian *Notices to Mariners* can be obtained at the same site.

ELECTRONIC CHARTS. Vector or raster-scanned digital charts of U.S. waters are available through most chart dealers. They give the appearance, particularly when interfaced with GPS, of absolute accuracy. They are, however, only as accurate and as current as the scanned originals.

Electronic versions of United States charts are available free by download at *www.navcen.uscg.gov*. The are also available commercially at *www.maptech.com* through Maptech, which includes them with their paper chart kits.

Canadian electronic charts can be obtained at *www.charts.gc.ca*.

GPS. As in most locations, GPS has become the standard electronic navigation tool in Maine. Differential transmitters are located in Portsmouth, New Hampshire and Brunswick, Maine.

Cruising sailors should be thoroughly comfortable with the use and limitations of GPS before they rely on it in the fog. The Coast Guard cautions against the common practice of relative navigation—using a previously observed position as a waypoint without first transferring it to a chart to check its accuracy—particularly for GPS. Likewise, waypoints obtained from sources other than charts should be plotted before being used. The Coast Guard recommends always calculating or cross-checking your waypoints directly from the chart.

This guide lists harbor positions only for locating harbors, not for use as navigational waypoints.

RADAR. If the incentive to develop radar hadn't been war, it probably would have been the Maine coast. Even when you can't see your own bow, the bold shores show on the screen as sharp images, and you can identify navigational aids and monitor the movements of nearby vessels. Most lobsterboats and commercial fishermen have radar, as do many yachts. In heavy fog it is a blessing (see *Fog*, page 17), but not essential. One thing no yacht should be without, however, is a good radar reflector, mounted high.

PUBLICATIONS. Below is a list of the U.S. and Canadian publications that we have found essential. Most can be found at the websites listed for charts.

United States
U.S. Coast Pilot (Eastport to Cape Cod)
U.S. Tide Tables (Atlantic Coast
   of North America)
Tidal Current Tables (Atlantic Coast
   of North America)

Canada
Sailing Directions (Nova Scotia and Bay of Fundy)
Sailing Directions (Saint John River)
Tidal Current Tables, Vol. 1 (Atlantic Coast
   and Bay of Fundy)

COMMUNICATIONS

VHF RADIO. The VHF radio is an essential piece of safety equipment. Not only can it save lives in an emergency, it is a convenience as well.

The Coast Guard and most boats along the coast monitor channel 16, the emergency channel. Ideally, boat-to-boat or boat-to-shore hailing should be on channel 09 to keep channel 16 clear for emergency or *securité* calls. To hail another boat or a marina, try to raise them first on channel 09. Then, if unsuccessful, try 16. If possible, set your radio to monitor both channels simultaneously. Commercial vessels such as large ships announce their intentions on channel 13.

Marine-related shoreside facilities such as marinas and harbormasters will usually monitor both channel 16 and 09, or, if they are on a special channel, we indicate it in this book. Most should be hailed on 09. Fishermen are usually on their own channels for chit-chat—lobstermen on one, draggers on another. Popular channels are 06, 08, 72, and 79.

The prevalance of cellphones and increasingly better coverage has made obsolete Maine's marine operator service, which once connected the land-based telephone system with your VHF radio. Maine no longer has such service.

CELLULAR PHONE SERVICE. In theory, all of the coast of Maine has cellular coverage via a collage of cells covered by various mobile service providers. In actuality, there are many areas that lack reception, and the size and control of the cells is in constant flux. Contact your carrier for advice.

A cellular phone aboard, while convenient when there is coverage, is by no means a substitute for a VHF radio. Phones are limited in ship-to-ship communications simply because you rarely would know the number of the vessel you would like to call. Consider the cellular phone a complement to the VHF and a wonderful convenience.

Cellular signals are in the 900 MHz range and are "line-of-sight." As such, they are affected by topography, so their range over water may be somewhat greater than over land. Still, reception extends only a few miles offshore. It can be vastly improved with a mast-mounted antenna.

The area code for all Maine locations is 207. The area code in New Hampshire is 603. The area code for coastal New Brunswick, Canada is 506. In an emergency, the relative privacy of a cellular call might encourage users to contact the Coast Guard without the added embarrassment of a VHF broadcast. For search-and-rescue operations, however, the Coast Guard prefers to be homing in on a VHF signal.

WI-FI. Connecting to the internet from your boat is typically accomplished with a wireless laptop card or via PDAs or "smart" phones with service through cellphone providers. As such, their coverage and range has the same limitations as the phones, as outlined above.

Alternatively, some harbors such as Camden are covered as fee-based hotspots which can be accessed via conventional wireless networking. Finally, if you bring your computer to town, libraries are often hotspots, even when closed.

We note these hotspots when possible, but this technology is constantly changing.

TIDES and TIDAL CURRENTS

For the cruising sailor accustomed to a tidal range of one or two feet in Chesapeake Bay, Florida, or the Caribbean—or even Long Island Sound's 7 feet—the rise and fall of 11 feet in Penobscot Bay or 20 or more feet in Passamaquoddy Bay and beyond presents a new and sometimes intimidating challenge to navigating and piloting. Remember, though, that sailors and fishermen have been dealing with these tides for hundreds of years. Once you know what to expect, it's just a matter of gaining a little experience.

SPRING TIDES AND NEAP TIDES. As every schoolchild learns, tides are caused by the gravitational influence of the sun, the earth, and, primarily, the moon. Tides of increased range, called spring tides, occur twice a month when the moon and sun are in conjunction or opposition (a day or two after a new moon or a full moon). Tides of decreased range, called neap tides, occur twice a month when the sun and moon partially counteract each other (first or third-quarter moon).

TIDAL RANGE. The tidal range increases as you sail from west to east along the coast of Maine and New Brunswick, Canada. Starting with a range of 9 feet or so in Kittery, it reaches about 19 feet in Eastport, and 25 feet in Saint John, New Brunswick.

To understand these great tidal ranges and why they increase as you move east, imagine the Atlantic

HARBOR SEALS

If you have the uncanny feeling that you are being watched as you ghost along a rocky shore, it is probably because you are. Turn around and look behind you. Liquid black eyes in a bobbing head might be staring back at you. Seals are naturally curious and friendly and irresistibly appealing, a cross between a Labrador retriever and a mermaid.

Of the four species of seals most likely to be seen in the Gulf of Maine, the harbor seal is by far the most common. The others are gray seals, harp seals, and hooded seals. They vary in color from brown to gray to reddish and may appear light or dark from a distance, depending on how heavily spotted they are. Their summertime population, sunning themselves on half-tide ledges all along the coast of Maine, amounts to 30,000 or more. This is a major increase from several decades ago, when a seal sighting was a rare and wonderful event.

Seals eat fish, and fishermen have long resented the competition. Seals are also accused of destroying nets and other valuable gear and suspected of spreading the cod worm. To reduce their numbers and presumably increase the fish catch by humans, Maine and Massachusetts in the late 19th century instituted a bounty on each seal destroyed. The seal population plummeted—without a noticeable effect on the fish catch.

When Maine repealed its bounty in 1905 and Massachusetts finally repealed theirs in 1962, seals began to increase in number. Further aid was given to all marine mammals in 1972 with the passage of the federal Marine Mammal Protection Act, and the seal population has rebounded dramatically. Only since 1972 have fishermen grudgingly abandoned the habit of taking a shot at any seal within range. The increase in the seal population is undoubtedly a direct result of federal protection.

Female seals have their single pups in the spring and rear them for three to four weeks without help from the males. The pups are sometimes in the water swimming within a minute or two of birth, and most are weaned by the end of June.

After the May-June pupping season, the cycle starts all over again with the breeding season, June through August. This is followed by a period of molting, July through September, when the seals retreat to the outer islands and ledges to replace their coats. In captivity, seals can live 30 to 35 years. In the wild, males live 15 to 20 years and females 25 to 30 years. A female may bear as many as 25 pups in her lifetime.

"Seal Bay," "Seal Cove," and "Seal Harbor" are common names along the coast of Maine, but there are concentrations of seals in southern Penobscot Bay, Jericho Bay, Machias Bay, and near Mount Desert Island and Swan's Island.

Seals are disturbed by nearby people and fast motorboats. Instead of heading straight toward a group of seals, sail a course that will take you past. If they appear nervous, you're close enough. If they are alarmed, you're too close. It is particularly important during the May and June pupping season to approach slowly and keep your distance, so as not to risk separating a pup from its mother.

Unless the seals are accustomed to your boat, they will wiggle and slither down to the water and disappear, often popping up suddenly to inspect you with great curiosity from a different direction. Seals can stay submerged for 20 minutes or more, though the normal dive is much shorter. They are acrobatic swimmers, with their streamlined shape and powerful flippers. They can cover great distances, and they are formidable in pursuit of fish.

Seals have been observed fishing cooperatively. One sailor reports watching four of them attack a school of fish in unison, from north, east, south, and west. The dark, sleek heads disappeared under water, and a few seconds later a great white patch appeared, accompanied by violent splashings, tiny fish catapulting through the air, and an occasional quick glimpse of a swiftly plunging seal. When the waters quieted, the seals reappeared, swimming away in different directions.

STRANDINGS. Please report any marine mammal in distress, including seals and whales, to the Maine Marine Mammal Reporting Hotline at 800-532-9551.

Ocean as a large basin of water being sloshed back and forth. The motion of the water in the middle of the basin is relatively small, but when it enters the narrow confines of the Bay of Fundy, the tidal range reaches an extraordinary 50 feet or more, culminating in a dramatic tidal bore at Moncton, New Brunswick.

ANCHORING WITH LARGE TIDES. The large tidal range requires caution when anchoring. Many a cruising sailor has anchored in a large open bay only to find in the night that the water has drained away, leaving him high and dry on an unsuspected ledge or wallowing in the mud. Study the chart carefully before anchoring. Try to visualize the anchorage at low tide and anticipate where you will lie if the wind or current shifts. In crowded harbors, be suspicious of any large areas free of moored boats—there is probably a reason for their absence, and it is probably lurking under water.

LOCATION	TIDAL RANGE (feet)	
	Mean	Spring
Kittery Point	8.7	10.0
Portland	9.1	10.4
Boothbay Harbor	8.8	10.1
Tenants Harbor	9.3	10.6
Pulpit Harbor (North Haven)	9.8	11.1
Bar Harbor (Mount Desert)	10.6	12.2
Roque Island Harbor	12.3	14.0
Cutler (Little River)	13.5	15.4
Eastport	18.4	20.9
Calais	20.0	22.8
North Head (Grand Manan)*	17.9	24.0 (large)
St. Andrews (Passamaquoddy Bay)*	19.6	27.2 (large)
Saint John*		0.0 (large)

*Canadian tide tables show "large" tides rather than spring tides. The two countries also use different data to calculate tidal ranges; for the same location, the tidal range is somewhat larger in Canadian tide tables than in U.S. tables.

When anchoring, scope needs to be calculated for the highest tide, especially when the tidal range is great or the depth is shallow. For example, at Mount Desert, where the tidal range is 12 feet, you could anchor in 16 feet at low and put out 96 feet of scope for a respectable scope-to-depth ratio of six to one. But at high tide, the 16 feet of water will become 28, and your ratio will shrink to an inadequate 3.4 to 1. If you had been in Passamaquoddy Bay, with a tidal range of 20 feet, your ratio would be even worse. The specifics of anchoring and tying up way Down East are discussed in detail on page 366. Meanwhile, use plenty of scope wherever you cruise in Maine.

Large tides have their advantages, too. As we have noted throughout this book, certain anchorages have dangers unseen at high tide which reveal themselves at low. An approach that can be a game of guesswork at high tide can be straightforward at midtide or less when the dangers are visible.

DRAIN TIDES. During the time of spring (large) tides each month, the extreme lows are called drain tides. Where the tidal range is small, these are of little importance. In Maine, however, you must pay attention to drain tides, shown in the tide tables as negative numbers. This means that the tides fall below the mean water level on which the chart soundings are based. For example, the depth in Kennebunkport at an average low tide is around 6 feet. That's fine if you draw 6 feet or less. But on days of drain tides, the water may be as much as a foot or two below that level, and your keel will be deep in the mud. If the charted depth in a harbor is close to your draft, be sure to consult your tide tables. Canadian charts use lower datum to indicate soundings, so negative tides are rarer and smaller.

When this book mentions the depth in an anchorage or alongside a float, we are referring to an average low tide. *There will be less water during drain tides.*

RIVERS. Tides in big rivers such as the Kennebec and the Penobscot vary considerably as you go upstream. At the mouth of the Kennebec, for example, the maximum range of tide is 9.7 feet. By the time you reach Bath, upstream, the maximum range is only 7.4 feet, and farther upstream at Richmond, it is only 6 feet. On the Penobscot River, the reverse is true. The range of tide increases as you go upstream, from about 11.8 feet at the mouth to a maximum of 15.5 feet at Bangor. The massive tides at Saint John, New Brunswick are hardly felt once you are above the Reversing Falls on the Saint John River.

TIDE TABLES. Have a current copy of the *U.S. Tide Tables (Atlantic Coast of North America)*. If there is a chance you will approach Canadian waters, you will also need the Canadian *Tide and Current Tables (Vol. 1, Atlantic Coast and Bay of Fundy)*.

Many commercial firms print complimentary tide tables for specific locations such as Boston, Portland, Rockland, and Bar Harbor. They are convenient and can save time looking up the tides in the tables. Errors are not uncommon, however, so they should be used with caution and double-checked periodically with official tables.

For estimating how much the sea level rises or falls at different stages of the tide, remember the "Rule of Twelfths." In the six hours between low and high tide, the tide will rise $1/12$ of its range during the first hour, $2/12$ the second hour, $3/12$ the third hour, $3/12$ the fourth hour, $2/12$ the fifth hour, and $1/12$ during the sixth hour.

For example, if the tidal range is 12 feet, the tide will rise about one foot the first hour after low slack, 2 more feet the second hour, 3 more feet the third hour, etc.

The state of the tide is so important to your planning when cruising Down East that most people find it handy to look up the times of high and low tides each day and post them somewhere aboard for easy reference.

TIDAL CURRENTS. Large tidal ranges create strong currents, and strong currents get stronger where they are constricted by coastal geography. The U.S. government does not publish tidal current charts for the coast of Maine, but general information can be obtained in the *Coast Pilot*, the *Tidal Current Tables for the United States* (a separate publication from the *Tide Tables*), or, for Canadian waters, the Canadian *Tide and Current Tables*.

Generally, all along the coast of Maine, the tide floods to the north and east and ebbs to the south and west. Along the coast and in the wider bays, the current seldom exceeds 2 knots. In narrower passages, though, such as Upper and Lower Hell Gate, Oven Mouth, and the constricted portions of the rivers, it will run much harder.

As you approach Canadian waters and the Bay of Fundy, currents get stronger as the tidal range gets larger. In Grand Manan Channel, the maximum tidal current runs 3 knots. In Head Harbour Passage, Letete Passage, and Lubec Narrows, the tidal current runs 8 knots or more, with boils, eddies and whirlpools. At Saint John, New Brunswick, the current literally cascades upriver at the Reversing Falls.

The direction of the ebb and flood varies in passages tending east and west. In the Fox Islands Thorofare, the flood comes in from both directions, meeting at Iron Point, and it also ebbs in both directions. In Eggemoggin Reach, the current floods

hours after slack	1	2	3	4	5	6
fraction of tide	1/12	2/12	3/12	3/12	2/12	1/12

The Rule of Twelfths is an easy way to estimate the tidal range for any period between high and low.

northwest and ebbs southeast. In Deer Island Thorofare and Merchant Row, the current floods east and ebbs west, but strong winds can reverse the flow. In Casco Passage and Moosabec Reach, the current floods east and ebbs west. At Bass Harbor Bar, the current floods west.

In making a passage east or west, especially Down East where currents are stronger, plan to use the current as much as possible. Fighting 2 or 3 knots of current from Mount Desert to Roque Island or Grand Manan is frustratingly slow. With the current pushing you along, the headlands slip by effortlessly, and you arrive with daylight to spare.

In limited visibility, an estimate of current is critical for good dead reckoning, especially crossing a big bay such as the Penobscot with the current on the beam. Watching the current on lobster buoys helps.

RIVER CURRENTS. In the rivers of Maine, the ebb is usually much stronger than the flood, since the strength of the river is added to the tidal current. You can expect stronger currents in narrow stretches of river. Consult the *Coast Pilot* and *Tidal Current Tables* for details.

When you are picking up a mooring or approaching a dock in a river, the current is usually a more important consideration than the wind except at slack water. In the Piscataqua or the Kennebec, for example, you would normally head into the current regardless of wind direction.

WIND AGAINST CURRENT. In certain places such as the mouth of the Kennebec River, Bass Harbor Bar, and Petit Manan Bar, the river or tidal current running against a strong wind will produce a vicious chop that is dangerous to small craft and highly uncomfortable even to 40-footers.

EMERGENCY HELP

You are in trouble. The engine has conked out. The dinghy has gone adrift. You are lost. Your propeller has wrapped a lobster buoy. Or you are hard aground or have been hit by another boat. Where do you turn for help? Here is a quick checklist. **See *Appendix A*, page 441, for emergency numbers.**

NEARBY BOATS. Don't hesitate to hail the nearest boat that can help you when you really need it, whether the boat is a cruising boat or fishing vessel. In the best tradition of the sea, mariners are glad to help. But fishermen are busy making a living. Don't ask for trivial assistance, but by all means, ask them for help in an emergency. Generally, lobstermen are not on channel 16, but on their favorite local channel with their friends. Try 06, 08, 72, or 79, or spin the dial until you hear local chit-chat.

BOATYARDS. Many yards have a working skiff or tow boat that can bring you a pump, haul you off a ledge, or tow you in. To most, it is part of the business. Most waterfront yards monitor channel 16 or 09 or the local fishermen's channel. Check this guide for specifics. Telephone numbers are included for those with cellphones aboard.

HURRICANE ISLAND OUTWARD BOUND SCHOOL (HIOBS). The Outward Bound School has bases on Hurricane Island off the southwest coast of Vinalhaven, on Burnt Island in Muscongus Bay, on Cross Island in Machias Bay, and at their headquar-

---

THE EARLY EXPLORERS

Norse sagas chronicle the discovery of the New World by Leif Ericson and by Biarni Heriulfson before him, and a Norse settlement has been found and excavated at L'Anse aux Meadows on the northern tip of Newfoundland. But it seems there was no follow-up to these voyages, and another half-millennium passed before explorers braved the western ocean again and discovered the bountiful coast to the south.

The history of Europeans along the coast of Maine starts in 1524 with the Florentine explorer Giovanni da Verrazano. Financed privately by French and Italian bankers, Verrazano sailed from North Carolina to New York and then to Maine, visiting Block Island and Narragansett Bay and rounding Cape Cod. His first landfall in Maine was probably at Cape Small. There he encountered Indians "exhibiting their bare behinds and laughing immoderately," as Samuel Elliot Morison put it, which led Verrazano to call the place "Terra Onde di Mala Gente." Estevan Gomez, a Portuguese mariner in the employ of Spain, followed in Verrazano's wake in 1524-25 and explored the coast of Maine from the other direction.

Meanwhile, the French were busy fighting wars. Though French fishermen had earlier found their way to the rich banks off Newfoundland, it was not until Carter's voyage to the Gulf of Saint Lawrence in 1534 that France started to develop a real interest in the Americas. Samuel de Champlain, known as the "Father of Canada," established a settlement on an island in the Saint Croix River and founded Quebec in

> There he encountered Indians "exhibiting their bare behinds and laughing immoderately."

1604. He also ventured southward, exploring the coast of Maine as far as Penobscot Bay, giving us such names a Mount Desert and Isle au Haut.

Many of the early explorers, expecting to drop anchor in a harbor hitherto unknown, were dismayed to find two or three dirty fishing vessels already there. The fishermen, of course, wrote no accounts of their voyages and kept their favorite fishing banks secret. Thirty-six English, French, Spanish, and Portugese ships were counted fishing off Newfoundland in 1583.

After the discovery of Newfoundland by John Cabot in 1497, the English discoverers lagged far behind the French. But by the 17th century, the English arrived in force, rapidly charting the gap between Florida and Nova Scotia. John Walker explored Penobscot Bay in 1580. Bartholomew Gosnold coasted southward from Cape Elizabeth to Martha's Vineyard and Cuttyhunk in 1602. And Martin Pring, who named the Fox Islands, roamed the Maine coast and Cape Cod Bay in 1603.

Two years later, in 1605, George Waymouth made his landfall on Monhegan, visited Allen Island, and explored a great river, probably the Saint George. He also captured a few Indians to take home as souvenirs, arousing great interest in London, but poisoning the relationship between the English and the Indians for two centuries.

In 1607, George Popham and Raleigh Gilbert founded the Popham Colony near the mouth of the Kennebec River, but this ill-fated attempt at settlement succumbed to one of the severest Maine winters in history, and Jamestown, Virginia, founded the same year, went on to claim the title of the first permanent British colony in the New World.

Captain John Smith, Governor of Virginia, sailed up the coast in 1614 to the Isles of Shoals and Monhegan. It was Smith's lyric descriptions of the beauty, lushness, and potential of New England that fired the English imagination and spurred the growth of the new colonies. And it was Smith who realized that the real treasure of the Maine coast was not gold or silver, but codfish.

ters on Wheeler Bay. Each location has search-and-rescue capabilities including well-equipped powerboats with licensed captains and emergency medical technicians on staff. They monitor channel 16 around the clock from mid-May to late September.

COMMERCIAL SERVICES. Commercial towing and salvage companies operate in most Maine waters. A call to the U.S. Coast Guard will usually be referred to a commercial concern if your situation is not life threatening. See *Appendix A*, page 441.

UNITED STATES COAST GUARD. There are currently six Coast Guard stations along the Maine coast plus one in Portsmouth, New Hampshire. They are charted in each region of this book. Each station has at least one boat that can be under way within 30 minutes. In the summer, an additional patrol cutter lies off the coast or can be under way within two hours.

For help, call the Coast Guard on channel 16. They will want to know the name of your vessel, your location, the number of persons aboard, and your situation. Remember, too, that if you call the Coast Guard, they will routinely inspect your boat for safety equipment.

The Coast Guard strongly emphasizes filing a detailed float plan with friends or relatives. In emergencies, they will always respond in some way and will always dispatch units in life-threatening situations.

CANADIAN COAST GUARD. The Canadian Coast Guard and all government coastal radio stations monitor channel 16. The closest Coast Guard ships are based in Saint John, New Brunswick and Dartmouth, Nova Scotia.

FUNDY TRAFFIC. The Canadian Coast Guard operates a vessel traffic service in the Bay of Fundy, call name "Fundy Traffic." Based in Saint John, it provides vessel traffic control and assistance to mariners. There are no requirements that yachts participate in this system, but doing so can help in many ways. Fundy Traffic can provide information on weather and navigational aids and detailed reports about nearby vessel traffic, and they will monitor you on radar. They monitor channel 16 and use channel 14 west of The Wolves and channel 12 east of The Wolves, or call 506-636-4696.

APPROACHES to the COAST of MAINE

Sailing Down East is always an adventure, whether it is the first or the twentieth time. At its best, the passage to Maine is a marvelous cruise itself.

Coming from Long Island Sound and points south, you will first savor the pleasures of Block Island and Narragansett Bay. You might explore the Elizabeth Islands or perhaps take a side trip to Martha's Vineyard or Nantucket. Then you will sail through Buzzards Bay and the Cape Cod Canal.

From here, you can either make an overnight passage to the Gulf of Maine or work your way up the coast in easy stages. If you prefer to avoid an overnight passage or are headed for cruising grounds in southern Maine, there are a number of approaches. For example, after leaving the Cape Cod Canal, you could spend a night at Provincetown or Scituate. The next stop might be Gloucester and the Annisquam Canal, or you might decide to go outside Thacher Island and Cape Ann.

If you want to get Down East in a hurry, there are a number of good landfalls on the coast of Maine. Most of those described below are on small, bold islands well off the coast with lights visible for 20 miles or more, foghorns, and some radiobeacons and racons. See the accompanying sketch map.

Depending on where you plan to cruise, set your course for Portland LNB, Seguin Island, Monhegan Island, or Matinicus Rock. The distance from the Cape Cod Canal to these various landfalls ranges from 108 to 142 miles. For example, the passage to Monhegan Island, which makes a fine landfall for Penobscot Bay or Muscongus Bay, is a comfortable 132 miles.

PORTLAND AND CASCO BAY. For a destination near Portland, one good landfall is Wood Island Light (*Al WG 10s 71ft 16M*), near Biddeford Pool. Another is Cape Elizabeth Light (*Fl (4) 15s 129ft 27M*).

In reduced visibility, however, the safest landfall is Portland Lighted Horn Buoy "P" (*43° 31.60'N 070° 05.49'W*), also referred to as "Portland LNB" for "large navigational buoy." This enormous buoy is well out to sea, 5 miles southeast of Cape Elizabeth, and free of all dangers except, perhaps, the shipping traffic which uses it as a landfall for the same reasons. The red-and-white buoy has a circular base 40 feet in diameter, a 42-foot tower, and the markings "P" and "Coast Guard." It also houses meteorological instruments and a flock of deaf seagulls.

A CRUISING GUIDE TO THE MAINE COAST

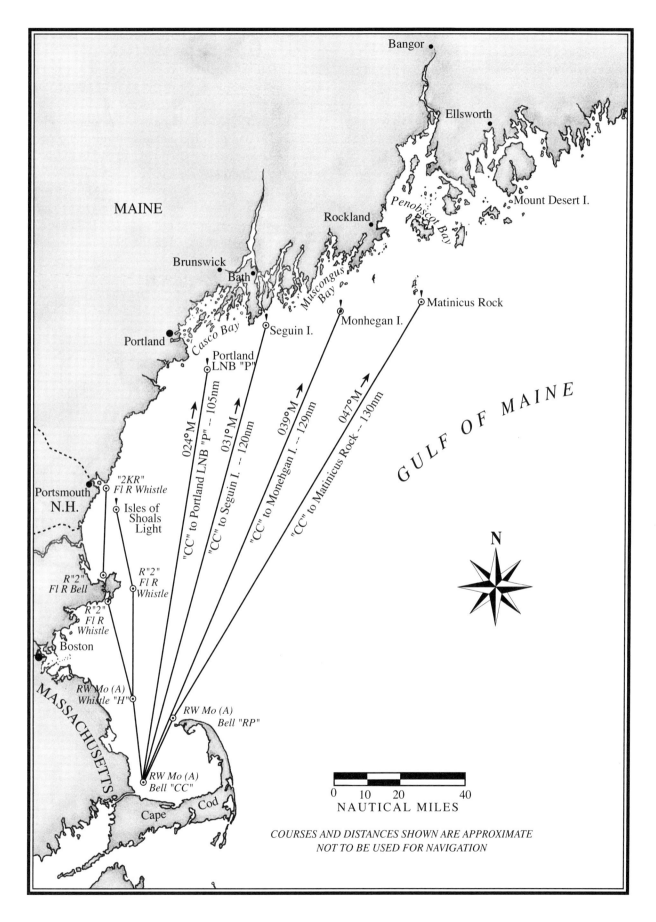

MIDCOAST RIVERS, BOOTHBAY HARBOR. Head for bold Seguin Island lying 2 miles south of the entrance to the Kennebec River. Seguin Light (*F 180ft 18M*), established in 1795, sits atop a 53-foot white, cylindrical granite tower connected to a dwelling which itself is 145 feet above the sea on the grassy summit. There is a foghorn at the light. Be aware of a magnetic disturbance near the Ellingwood Rock, as marked on the chart, and strong currents from the Kennebec.

Another possibility is the Cuckolds Light (*Fl (2) 6s 59ft 12M*) off Cape Newagen and Booth Bay. Although not offshore like Seguin, The Cuckolds has the advantage of a radiobeacon. The light flashes in a white group, and there is a foghorn.

MUSCONGUS BAY OR PENOBSCOT BAY. The best landfall is Monhegan Island, 9 miles off the mainland, used as a landmark since the earliest days of exploration along the coast. The island is 1.7 miles long and 165 feet high, with bold cliffs along its southeast side.

Monhegan Island Light (*Fl 15s 178ft 20M*) was established in 1824. The light flashes white from a 178-foot high gray conical tower connected to a white building in the middle of the island. Within three miles of the island, the light is obscured between the west and southwest. The foghorn and the radiobeacon are on 100-foot-high Manana Island, lying just to the westward of Monhegan.

Another possible landfall is Matinicus Rock, the outermost island in the approach to Penobscot Bay. The rock is 57 feet high, an enormous, barren granite rock swept by wind and sea. Matinicus Rock Light (*Fl 10s 90ft 20M*) was established in 1827 and flashes white from a 48-foot gray granite tower 90 feet above sea level on the southern part of the rock. The light also has a foghorn and radiobeacon. A second, abandoned light tower stands farther to the north.

MOUNT DESERT ISLAND AND POINTS EAST. Make your landfall at Matinicus Rock, as described above.

Alternatively, head for Mount Desert Rock, 17.5 miles south of Mount Desert Island, marked by a flashing white light (*Fl 15s 75ft 18M*). The rock itself is small and only 20 feet high, so it makes a more difficult landfall in heavy fog.

Mount Desert Island itself is the highest land feature on the coast of Maine. It can be seen in good conditions from as far away as 60 miles out to sea.

APPROACHES FROM CANADA. All pleasure boats entering the United States, whether of American or foreign registry, must check in with U.S. Customs and Border Patrol at a port of entry. See *Border Crossings* on page 368 for details.

Vessels headed for Maine from Passamaquoddy Bay or the Bay of Fundy area will normally pass through Grand Manan Channel and continue coasting toward their destination.

Boats approaching the coast of Maine from Newfoundland, the Gulf of Saint Lawrence, and Nova Scotia will round Cape Sable and then use the same landfalls discussed above for boats approaching from the south.

The Southern Coast
Isles of Shoals to Cape Elizabeth

A CRUISING GUIDE TO THE MAINE COAST

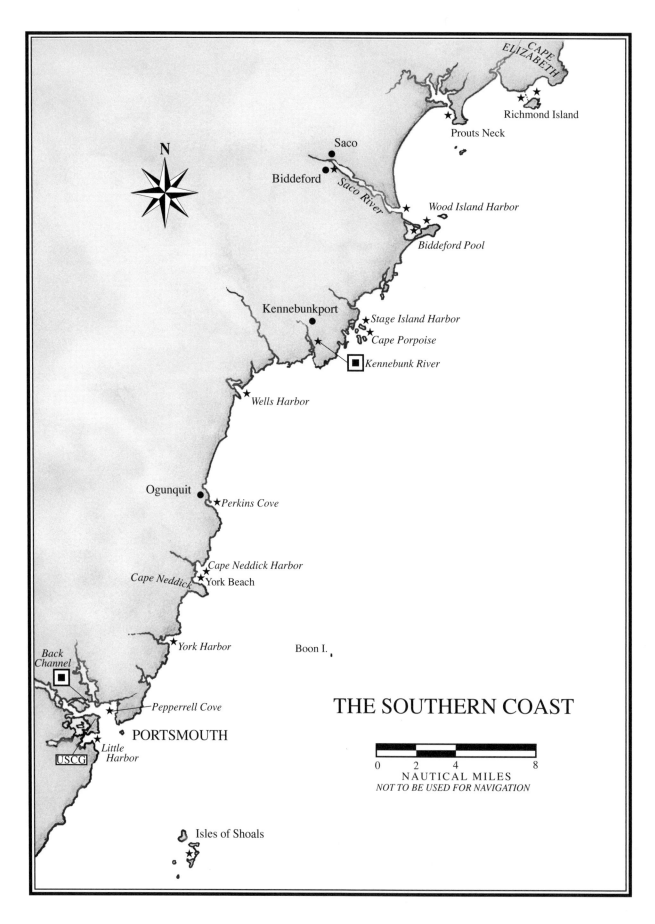

SOUTHERN COAST

> Sweet sounds on rocky shores the distant rote;
> O could we float forever, little boat,
> Under the blissful sky drifting alone!
> —Celia Thaxter

ON a large chart, the southern coast of Maine, from Kittery to Cape Elizabeth, appears almost featureless, sweeping northward from Cape Cod in a series of shallow bays, punctuated by an occasional rocky headland or tidal river, dotted with relatively few islands. Clearly this is not the fabled rockbound coast of Maine.

Instead, here are Maine's great beaches: Old Orchard Beach, favorite of French-speaking Canadians, Moody, Wells, Kennebunk, Scarborough, and others, some stretching for mile after mile. Here too are tidal rivers, from the mighty Piscataqua to the meanders of the York, the Kennebunk, and the Saco.

In comparison to the rest of Maine, this portion is short on harbors, and only a few are first-rate. But there are a number of interesting anchorages that are perfectly adequate in summer conditions and delightful to explore. The Isles of Shoals cluster offshore, and the thriving towns of Portsmouth and Kittery flank the surging Piscataqua at the very beginning of Maine.

Then come historic York and the contrasting harbors of Kennebunkport and Cape Porpoise—one swarming with "summer complaints," as tourists are known locally, and the other, only three miles away, a working harbor aswarm with lobster buoys. Farther north are the marshes and birds and beach of Biddeford Pool and the tranquil anchorages from which you can watch the sheep graze on Richmond Island.

Over the lower half of the coast broods old Mount Agamenticus, a landmark for Europeans from the early days of exploration. Several of the earliest settlements in America were attempted along this inviting coast. In the terrible winter of 1616-1617, ten years after the discouraging failure of the Popham Colony on the Kennebec River and three years before the Pilgrims would land at Plymouth, Captain Richard Vines wintered over at Biddeford Pool to prove that the Maine climate was not too severe for Europeans. He called it Winter Harbor. He returned to establish a permanent settlement on the Saco in 1623, the same year Kittery was founded. York was settled eight years later. During six Indian Wars, from 1675 to 1763, however, mainland farms were wiped out time and again by the Indians.

A slaughter of a different kind took place at the funeral ceremony of the revered Pawtucket Indian Aspinquid, who died in 1692 at the age of 94. He is said to be buried atop Mount Agamenticus, where more than 6,000 animals were sacrificed in his honor.

For those coming to Maine from points south, the southern coast is a wonderful alternative to the overnight passage across the Gulf of Maine and much more than a stopover en route to better cruising grounds. Here, along the stepping stones of history, sailors with the time and inclination will discover what the early explorers found—abundant beauty, safe harbors, expansive beaches, offshore islands, tidal rivers, and an ocean as bold and challenging as any.

ISLES OF SHOALS

3 ★★★★★

42° 58.75'N 070° 36.65'W

CHARTS: 13283, 13278   CHART KIT: 56 (D), 14

In 1614, when the venerable Captain John Smith dropped anchor among the Isles of Shoals, he was so captivated by them that he named them "Smith's Isles" for himself. "Of all foure parts of the world that I have seene not inhabited," he wrote, "could I have but the meanes to transport a Colonie, I would rather live here than any where."

Lying just six miles southeast of the entrance to Portsmouth Harbor, five of the nine islands are in Maine (Duck, Appledore, Smuttynose, Malaga, and Cedar) and four are in New Hampshire (Star, Lunging, White, and Seaveys). The islands are spectacularly scenic, and their long history is crammed with tales of buried treasure and bloody Indian attacks.

Smith's name didn't stick. More impressive than his praise or his unmatched contributions to the exploration of New England were the schools, or shoals, of cod near the islands, in an abundance never before seen by European fishermen. By the eighteenth century, the Isles of Shoals were considered one of England's most valuable colonies because of the astounding quantities of cod caught, dried, and shipped home.

The early cod-fishing communities were based primarily on Smuttynose and Appledore Island, but when the Massachusetts Bay Colony started to levy onerous taxes on the islanders in 1680, they crossed over to New Hampshire's Star Island, named their new town Gosport, and continued their thriving fishing industry for another century. During the Revolution, however, everyone was evacuated to the mainland. Those who returned to Star Island after the Revolution acquired a reputation for "laziness, drunkenness, lawlessness, and cohabitation," and the community never returned to its former prominence.

The islands had a modest revival in the nineteenth century, and a widely read poet, Celia Thaxter, grew up in the midst of it. Thaxter was the daughter of Thomas Laighton, the lighthouse keeper on White Island. Aware and sensitive at an early age, she referred to the area as "these precious isles set in a silver sea."

Enterprising Laighton built Appledore House around 1850 and publicized it as the first resort hotel between Nantucket and Eastport, Maine. At the peak of their combined popularity in the 1890s, both Appledore House and Celia Thaxter attracted such literary and artistic figures as Nathaniel Hawthorne, Henry Wadsworth Longfellow, Sarah Orne Jewett, Childe Hassam, and John Greenleaf Whittier.

The rival Oceanic House, built on Star Island in 1872, advertised itself as "an ideal summer resort of the highest class and full of historic associations. Preeminently the place for the tired worker. No noise, no dust, no trolleys."

With the advent of the automobile, the Isles of Shoals resort business declined, and in 1915 Star Island was sold to an association of Unitarians and Congregationalists as a conference center for religion, natural history, and the arts.

APPROACHES. From any direction the approach to the Isles of Shoals is easy. Coming from the south, aim for the 82-foot light on White Island, leaving it to port. Pass between nun "4" on Halfway Rocks and Lunging Island to find red-and-white bell "IS" (42° 58.87'N 070° 37.26'W) at the mouth of Gosport Harbor.

Coming from the north on a clear day, you may be able to see the Appledore tower and the cupola on the former lifesaving station from as far away as Boon Island. Stay well clear of low and featureless Duck Island, which tends to merge with Appledore. In particular, beware of long Southwest Ledge if you are running a course from the red groaner "24YL" to nun "2" off the northwest tip of Appledore. From nun "2" head for red-and-white bell "IS" at the harbor mouth.

From the east, you can enter Gosport Harbor through the "back door" between Appledore and Smuttynose. Favor Smuttynose to avoid the ledge making out southward from Appledore. Round Malaga Island into the harbor.

ANCHORAGES, MOORINGS. The only harbor in the Isles of Shoals is Gosport Harbor, between Star, Cedar, and Smuttynose Island, so it is likely to be crowded. On warm weekends you will be competing with day-trippers from Portsmouth and Kittery as well as cruising boats. Even on weekdays there may be 10 or 20 boats here, so arrive early to choose a spot.

Holding ground is very poor, in kelp and rock beds. In most harbors you can trust your anchor more than unknown moorings, but in Gosport Harbor it is better to pick up a mooring. There are a number of large, private moorings scattered around the harbor. Six of them are marked "PYC" and maintained by the Portsmouth Yacht Club for use by its members. The club generously welcomes visiting yachtsmen to use the moorings if they are unoccupied but asks that you vacate your mooring if a member needs it.

If you are forced to anchor, try to set your hook in the mud up in the cove between Star and Cedar Island without fouling lobster buoys and moorings while leaving enough swinging room should the wind swing around to the west. Failing that, find a 21-foot spot farther out, avoiding the 48-foot areas. Use a good heavy anchor with a tripline.

Gosport Harbor is totally exposed to a good northwest blow, in which case, get out of the harbor and head for Portsmouth or go to sea. With enough visibility, you can power around Smuttynose and reset the hook on the eastern side of the breakwater between Cedar and Smuttynose. Depth varies from 8 to 18 feet. The bottom has good holding in sand and boulders, but expect a swell.

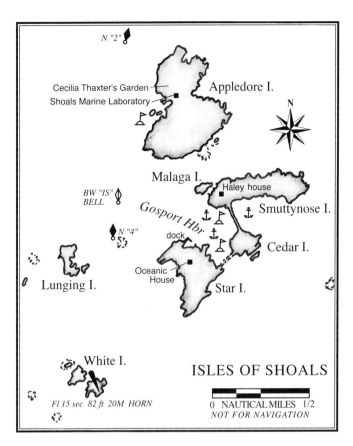

ISLES OF SHOALS

STAR ISLAND. The former Oceanic House is no longer a hotel, but the Star Island Corporation (*603-430-6272; www.starisland.org*) generously allows yachtsmen to visit the island from 10am to dusk on weekdays and from 10am to 3:30pm on Saturdays and Sundays. Please land only at the float, where there is room to tie a maximum of eight dinghies.

One of the young dockhands (known as "Pelicans") will orient you to the island, its buildings, its trails, and its rules, particularly the policy of no open flames in the buildings. The Oceanic House has a snack bar, a gift shop, and restrooms. The Vaughn Cottage, open daily from 1pm to 3pm, displays Celia Thaxter memorabilia. There is a small playground for children.

Near the eastern end of the island is Betty Moody's Cave, described in Robert Carter's 1858 *Summer Cruise on the Coast of New England*. "Early in the old Colony times," Carter relates, "the Indians from the mainland made a descent upon the islands, and killed or carried off all the inhabitants except a Mrs. Moody, who hid herself under the rocks with her two small children." As the Indians combed the island, "the unhappy mother, unable to keep her infants quiet, killed them with a knife to prevent them from crying."

From there, walk south along the coast to find Miss Underhill's Chair, a rocky perch from which a romantic young island schoolteacher was swept away by a great wave.

CEDAR ISLAND. On tiny Cedar Island, between Star and Smuttynose, are several houses, shacks, and a pier. These are the homes of the Hall and Foye families, who have fished these waters for six generations.

SMUTTYNOSE and MALAGA ISLAND. Named for the dark rocks at its eastern end, uninhabited Smuttynose boasts enough history for several volumes. In a little graveyard behind his 1750 cottage lies Captain Samuel Haley who lived to be 84 and was, according to his headstone, "a man of great ingenuity and industry." On this little island Captain Haley built a 270-foot ropewalk, a saltworks for curing fish, windmills to grind wheat and corn, blacksmith and cooper shops, a bakery, a brewery, and a distillery.

Pirates such as Captain Kidd and Quelch are reputed to have visited Smuttynose. In 1720 Edward Teach, known as Blackbeard, arrived on Smuttynose with his new young bride, who happened to be his fifteenth. The arrival of the British fleet, though, brought a quick end to the honeymoon. Being short on sentiment and long on self-preservation, Blackbeard fled, leaving his wife behind to wait for him in vain until her death fifteen years later. The ghost of Blackbeard's wife, they say, still roams the shores crying, "He will come back. He will come back."

For once, stories of buried pirate treasure have proved true. Captain Haley found four bars of silver under a flat stone and used the proceeds to build a breakwater connecting Smuttynose to little Malaga Island, creating Haley's Cove.

Smuttynose was also the site of a grisly double murder on the cold night of March 6, 1873. Six Norwegian immigrants—three couples—lived in what would become known as the "Hontvet House," which was owned and operated as a boarding house by the family of Celia Thaxter on Appledore Island. The men had sailed to the mainland on a bait-buying trip but had not returned by nightfall. The women turned in. Suddenly, in the depth of the night, they were awoken by a brutal attacker swinging an ax. Two of the women were killed, but the third, Maren, managed to escape and hide among the rocks with her dog Ringe.

Maren survived the night in nothing but her bloody nightdress. She later identified the murderer as Lewis Wagner, a Prussian fisherman who had helped the Norwegians bait trawls the previous summer. Lewis supposedly rowed the 12 miles to Smuttynose to rob the Norwegians of money he knew they were saving for a new boat. He was hastily tried and convicted and was hanged two years later despite his proclamations of innocence. The story has been intriguing the American imagination ever since, most recently in Anita Shreve's book *The Weight of Water*, and the ghoulish can find the supposed murder weapon on display in the Portsmouth Athenaeum.

Row ashore—if you dare—at Haley's Cove and land at the little beach behind the breakwater. This is a lovely spot for a picnic, peaceful and secluded, and you can swim in clear water off the beach. Nearby are two houses. The Haley cottage has been restored, and little Roz's cottage is used as a base for camping groups. Smuttynose is still privately owned by descendants of Celia Thaxter, but they have protected with a conservation easement to the U.S. Fish and Wildlife Service, and they generously welcome respectful visitors. It is managed by volunteer rangers and shared with 3,000 pairs of gulls who nest here. A guest log is outside Roz's cottage.

WHITE and SEAVEYS ISLAND. White and Seaveys Island, six acres in total, form a lonely outpost where two Coast Guardsmen once ran the light. The lighthouse is now automated, but the keeper's cottage is still there, and the state-owned islands are open to the public.

Choose calm weather to explore these islands. Pick up the white Coast Guard mooring at the north side and take your dinghy ashore near the boathouse. White Island is managed by a volunteer summer resident program, so check in with whomever you find. They will probably be glad to see you.

The history of White Island is filled with sagas of gales, shipwrecks, and dramatic rescues. In heavy gales, solid water has reached halfway up the light tower, and spray has cleared the top.

LUNGING ISLAND. Lunging Island is a small, privately owned island with a house at one end and a small cove on the east side. The name is a corruption of the island's original name of Londoners, reflecting its colonial use as a trading center for the London Company.

In 1974 the island was bought by oil-refinery interests as a site for an offshore terminal for unloading oil tankers and piping the oil to Portsmouth. Fortunately, the plan was defeated.

APPLEDORE ISLAND. The public is welcome on Appledore until sundown, and there is much to see. It is inhabited mostly by seagulls and summer students from Cornell University and the University of New Hampshire who work at the Shoals Marine Laboratory field station. The island is also a migratory way station for more than 125 species of birds, and harbor seals, whales, porpoises, and dolphins frequent its waters.

Shoals Marine Lab (*www.sml.cornell.edu, www.sml.unh.edu*) maintains several moorings on the west side of the island. Visitors are welcome to use them for limited periods, but this is not a good place to spend the night. Row ashore to the dock, sign in, and pick up a map of the island. Just to the left of the dock there is a delightful, well-protected tidal pool for swimming.

To the north, beyond the lab building, is the Laighton family cemetery, the foundation of the old Appledore House, and a recreation of Celia Thaxter's garden, as described in her book *An Island Garden*, illustrated by Childe Hassam. Lichen-covered whale bones lie near the Lab's dinner bell, up near Kiggins Commons. Stay on the trails to avoid nesting gulls and rampant poison ivy.

DUCK ISLAND. This 11-acre, low-lying, treeless outpost has, ironically, been devoted both to preservation and destruction. It once was a naval aircraft bombing target; now it is a wildlife refuge managed by the U.S. Fish and Wildlife Service as part of their Maine Coastal Islands National Wildlife Refuge. The island is home to large colonies of cormorants and gulls, and access is restricted. Leave it for the birds... and other sitting ducks.

PISCATAQUA RIVER

On the docks at Prescott Park.

The mighty Piscataqua River marks the boundary between Maine and New Hampshire. On the south side of the river lies the charming city of Portsmouth, New Hampshire, while the quiet towns of Kittery and Kittery Point are on the north side, in Maine. The Piscataqua rates as the second-fastest-flowing river in the continental United States (after the Columbia River, in Washington and Oregon). Narrow tributaries drain miles of New Hampshire and Maine countryside to Great Bay before the fresh water rushes to the sea.

Kittery was Maine's first town, organized in 1647. Early forts were built to protect against Indian attacks. Later they evolved into coastal defense batteries during the French and Indian War, the American Revolution, and the Spanish-American War.

Fort Constitution guards the mouth of the harbor. Originally called Fort William and Mary, it played a critical role in the Revolutionary War. As discontent with the British grew, Governor John Wentworth and his family moved into the fort in 1774 with a small force of British soldiers and a large supply of gunpowder. The ubiquitous Paul Revere rode to Portsmouth to persuade the citizens to storm Fort William and Mary. Two hundred ardent revolutionaries then marched to Market Square and stormed the fort with no loss of life.

They imprisoned the governor and the soldiers and sent the cannons and gunpowder upriver by gundalow to Durham. Eventually these munitions were used against the British at the Battle of Bunker Hill, so the War of Independence actually began in Portsmouth.

Shipbuilding started early on the banks of the Piscataqua. Dories, gundalows, frigates, and clipper ships were launched from shipyards on Seavey and Dennetts Island. Maine launched the U.S. Navy in 1777 when John Paul Jones' 18-gun sloop *Ranger* slid down the ways at Badgers Island, where remnants of these ways are still visible. The first U.S. naval shipyard began here too, in 1800. The Portsmouth Naval Shipyard is now is critical in maintaining the country's fleet of nuclear submarines. It dominates Seavey Island, in Kittery. When nuclear subs arrive for repairs, they tie up to cannons from old men-of-war now used as bollards. The shipyard's motto, most appropriately, is "Sails to Atoms."

During World War II, an antisubmarine net was stretched between Wood Island and New Castle. Ships proceeding into Portsmouth had to radio ahead to arrange for dropping the net. Sometime in 1943 a German U-boat crept through behind a merchant ship and lay on the bottom just off the Portsmouth Naval Shipyard for three days and three

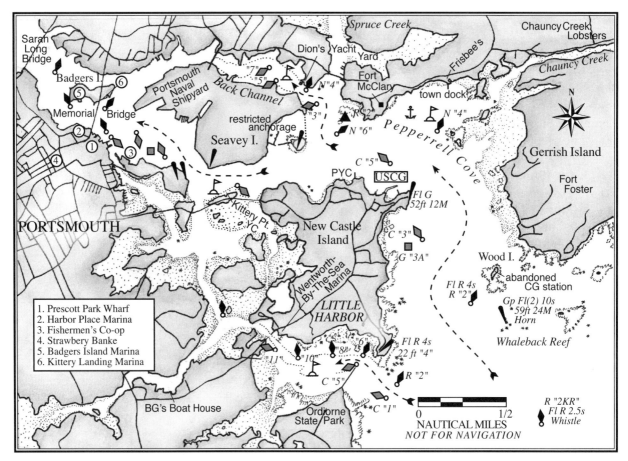

nights, rising after dark to make observations. The sub escaped undetected, but after the war, detailed drawings of the shipyard were discovered in files captured from the Germans.

In more recent years, Portsmouth Naval Shipyard has faced a potentially more devastating enemy. The U.S. cutback in its military spending has threatened to close the shipyard altogether.

APPROACHES. Tidal and river currents in this area are relentless and inescapable. Combined, they may run at 6 knots or more in places; opposing, they can form standing waves and rips. When approaching the Piscataqua, keep close track of which direction you are being set, and by how much. While the mouth is broad and well marked, it is flanked by the low dangers of Whaleback Reef and Jaffery Point, and it is shared by the heavy commercial traffic of tankers, tugs, cruise vessels, and Coast Guard ships, as well as a weekend flotilla of pleasure boats. Large ships often anchor off the mouth of the harbor awaiting berthing, pilotage, a favorable tide, or good visibility. In fog, Little Harbor or Pepperrell Cove are the easiest landfalls. Monitor channel 13 for Portsmouth vessel traffic information.

Coming up the coast from Cape Ann, leave bell "1" on Gunboat Shoal to port and head for flashing red whistle "2KR" (43° 02.96'N 070° 41.46'W) off Kitts Rocks. If the weather is thick, you'll hear the horn on White Island, one of the Isles of Shoals, to starboard. On a clear day you will see the white mass of Little Harbor's Wentworth-by-the-Sea Hotel, an obvious landmark whose cupola is marked on the chart.

From the north, run down the line of whistle buoys to whistle "24YL," southeast of York Ledge, leaving Boon Island, with its 133-foot lighthouse, to port. Then head in for whistle "2KR" off Kitts Rocks, leaving it to starboard. If approaching from closer to shore, from York or Cape Neddick, it is possible to run down the coast inside York Ledge and Murray Rock, but the swells will get a little lumpy as they pass over the shoaling, jagged bottom.

From Isles of Shoals, the approach is unobstructed except, perhaps, by a tanker or two anchored on the shelf between Gunboat Shoal and Kitts Rocks.

With whistle "2KR" to starboard, the channel is obvious and straightforward. A large dayglow range near Fort McClary designates the border between Maine and New Hampshire, which runs down the midline of the channel. Pass Whaleback Light to starboard. Although this light is visible for 24 miles, the 59-foot tower is dark brown and merges with the land behind it. (In 1943, the lighthouse keeper at Whaleback reported that three waves crested over the top of the light while he was in it!) Leave to starboard buoy "2" off treeless Wood Island, which has an abandoned lifesaving station. The town of Kittery has plans to restore it. Be sure to identify can "3" off Stielman Rocks on the port side of the channel. The white Fort Point lighthouse, with a fixed green light, is conspicuous.

ANCHORAGES, MOORINGS. Swift tidal current makes mooring preferable to anchoring. There are only a few spots to lie comfortably to an anchor, particularly if you plan to leave the boat for any length of time. Most visiting yachtsmen head for Little Harbor, Pepperrell Cove, or Back Channel behind Seavey Island, although there are alternatives. For anchorage specifics, see the entries that follow.

LITTLE HARBOR

43° 03.35'N 070° 43.00'W

CHARTS: 13283, 13278, 13286   CHART KIT: 57, 14, 15

Little Harbor lies on the New Hampshire side of the mouth of the Piscataqua River, tucked between the white sand beaches of the undeveloped shore of Odiorne State Park and the grounds of the venerable Wentworth-by-the-Sea Hotel on New Castle Island. In many ways, Little Harbor is the ideal anchorage in the area. For a busy harbor so near the bustle of Portsmouth, it is surprisingly beautiful. Yet it has all the amenities and convenience of a full-service marina. It is easy to enter in all weather with good protection in anything but a strong easterly.

But there are some drawbacks. First, the harbor is part of the mighty Piscataqua and can't escape its mighty currents. Perpendicular winds and currents, or wind and southeast swells, can combine for a rolly night. Also, because of its proximity to the city of Portsmouth, runabouts and day-trippers are out in force on summer weekends, rafting six or eight boats to a mooring with joyous partying on the beach at Odiorne. And finally, for those whose financial sensibilities are ruffled by paying for moorings, there is no room to anchor.

APPROACHES. Two overlapping breakwaters protect Little Harbor. Each is well marked, and there is plenty of room and water between them. From seaward, it is a straight run across the mouth of the Piscataqua from the red whistle "2KR" (*43° 02.96'N 070° 41.46'W*) to the mouth of the harbor, but remember that the river current will set you strongly. A strong southwest current blowing against an ebb tide can form a dangerous swell at the entrance.

From Kittery or Portsmouth, head out the river mouth, being sure to leave the can at Stielman Rocks to starboard. Aim outside Odiorne Point until you see the entrance nun "2" and can "1." The huge red-roofed Wentworth-by-the-Sea Hotel is not visible until you are opposite the first breakwater with its flashing red light. The inner breakwater is marked by can "5" with nun "6" opposite. Stay in midchannel until past the inner breakwater, then stay close to nun "8" to avoid the shoaling to the south of the channel.

ANCHORAGES, MOORINGS. Wentworth-by-the-Sea Marina has 170 slips, with 10 feet of water alongside, and can provide room for transients as big as 100 feet or more. Most of the moorings in the harbor are private, and there is no room to anchor. The Portsmouth Yacht Club (*Ch. 09; 603-436-9877*) maintains two moorings marked "PYC" on the south side of the harbor near the entrance. Visitors are welcome to use them until they are needed by club members.

GETTING ASHORE. The marina operates a launch service (*Ch. 71*), or you can use your own dinghy to land at their docks if you register. You can also dinghy to Odiorne Park.

FOR THE BOAT. *Wentworth-by-the-Sea Marina (Ch. 09, 68, 71; 603-433-5050; www.wentworthmarina.com).* High-speed gas and diesel pumps are available at the fuel dock on dock "A," along with water, ice, and pump-out facilities. Slips have electricity to 100 amps (three-phase available), cable TV and phone hookups, and high-speed Wi-Fi.

FOR THE CREW. At the marina, you will find showers and a laundromat. The "concierge" at the marina office can make reservations for the tennis courts or for one of the restaurants in the complex, or he can direct you to the hotel swimming pool. The Portsmouth Trolley stops at the marina on its hourly loop, or the marina can give you a one-way courtesy ride into town or to the nearby golf course, and you can take the trolley back.

Wood Island and Whaleback Light from Pepperrell Cove.

PEPPERRELL COVE (Kittery Point)

2 ★★★

43° 04.75′N 070° 42.35′W

CHARTS: *13283, 13287, 13286*   CHART KIT: *57, 15*

Pepperrell Cove is named after Kittery's most illustrious citizen, Sir William Pepperrell, Bt. Pepperrell was granted the hereditary honor of a Baronetcy (hence the "Bt.") by H.M. King George II for his service as the Commanding General and Chief Tactician for the New England forces when they captured the French Fortress of Louisbourg in 1745.

The cove is strategically positioned just inside the mouth of the Piscataqua but outside of the brunt of its currents. Moorings fill the cove, but it is broad enough to have room to anchor. On a calm summer's evening, the boats lie spread like an amphitheater from which to view the passing traffic on the river—the returning lobsterboats, an outbound ship, a sail ghosting in on the tide.

At other times, these virtues can be drawbacks. The very breadth of the cove leaves it exposed to the south, the moorings are tight, and Piscataqua eddies and tidal crosscurrents hang boats precariously askew. The lobsterboats rev up at dawn, the wash from large ships rolls the boats, and at dusk, the mosquitoes and no-see-ums descend.

APPROACHES. Pepperrell Cove is easy to enter. After passing Fort Point, head for nun "4" off Fishing Island, and you will see 100 or more boats almost filling the cove. A channel leading to the dock is marked by green and red spar buoys.

ANCHORAGES, MOORINGS. Both the town of Kittery and the Portsmouth Yacht Club (*Ch. 09, 78; 603-436-9877*) rent moorings for transients. The six yacht club moorings are located in a line below Fort McClary and marked "PYC." Their rental includes launch service to the clubhouse across the river on New Castle Island for showers or Wednesday-night dinners.

There is room to anchor west of the boats, off Fort McClary in 12 to 14 feet with good holding ground.

GETTING ASHORE. Row your dinghy to the crowded town floats. Take your oars and oarlocks with you.

Casual eating options include the floating Dinghy Dock Cafe (*603-766-0293*) at the marina, which will also deliver to your boat, or The Ice House, a classic ice cream and burger takeout .3 miles south on Route 1. A walk of about a mile in the same direction will bring you to BG's Boathouse restaurant (*603-431-1074*), which serves lunch and dinner and has a liquor license. It is also possible to take your dinghy to BG's small marina: head upriver, under the bridge, and bear left.

More formal meals are served in the dining room of the Marriott Grand Hotel and Spa (*603-422-7322*), once the venerable Wentworth-by-the-Sea Hotel, or at Latitudes, next to the marina.

THINGS TO DO. The Wentworth-by-the-Sea Hotel above the marina was built in 1874 during the Gilded Age. But perhaps that age has come again: not long ago developers sank $27 million into restoring and modernizing the hotel. Now the Marriott Grand Hotel and Spa, most of its amenities are available for a fee for use by non-guests.

Odiorne State Park on the south side of the harbor has lovely beaches, raspberries in profusion, a nature center, the remains of Fort Dearborne, and a boat ramp. Sagamore and Witches Creek beg for dinghy exploration for those with strong outboards or three-horsepower biceps.

FOR THE BOAT. *Kittery Point Town Dock (Harbormaster John McCollett: Ch. 09; 207-439-0912).* Gas, diesel, and water are available at the floats of the main dock with 10 feet alongside at low. Ice can be bought at Frisbee's Market. A dumpster for boat trash is in the parking lot by the harbormaster's shed.

Portsmouth Harbor Towing and Pumpout (Ch. 09, 16; 603-436-0915). This BoatUS tow boat also operates a pump-out boat for a moderate fee.

FOR THE CREW. Well-stocked Frisbee's Market, just up from the restaurant, claims to be "North America's oldest family store." It is also an agency liquor store. The Kittery Point post office is across the street, next to a long-term parking lot managed by Frisbee's.

Cap'n Simeon's Galley (*207-439-3655*) restaurant and takeout is just up from the wharf. If you yearn for a lobster dinner, Cap'n Simeon's sometimes has lobster bakes under a tent on the rocks. Or dinghy up Chauncey Creek, east of Pepperrell Cove, to the Chauncey Creek Lobster Pier (*207-439-1030*), a renowned eat-in-the-rough takeout. BYOB and BYOBJ (bring your own bug juice).

The small town of Kittery is 2 miles away. It has The Outback Café and several galleries, but no public transportation.

THINGS TO DO (Kittery Point and Kittery). Within walking or rowing distance, there are a number of pleasant excursions. Left from Frisbee's, a short walk down Route 103 takes you to the Fort McClary picnic area, a lily pond, and picnic tables in a grove of tall pines. The fort itself is farther along on the left. Its white, six-sided blockhouse dominates the west shore of Pepperrell Cove. You can also reach it by dinghy, but use caution as you approach–this is a popular site for scuba divers. The granite fortifications command outstanding views of the harbor, and kids love the climb. Farther down the road is the privately-owned, handsome 1760 Lady Pepperrell House.

Gerrish Island forms the east bank of Chauncey Creek and imparts its wild feel. Herons wade through the shallows, and sea creatures skitter among the mud flats. At the southwest corner of the island, a long pier extends from the Fort Foster recreation area. Good jogging and walking roads lead from the pier through the park to barbecues, picnic tables, grassy areas with benches, and to the concrete remains of World War II bunkers and a watch tower.

BACK CHANNEL (Kittery)

★★★

43° 05.00'N 070° 43.50'W

CHARTS: 13283, 13286   CHART KIT: 57

When the weather is brewing, head for Back Channel, almost completely landlocked behind Seavey Island. Back Channel is the major alternative to more exposed Pepperrell Cove, but it is a little farther up the Piscataqua River and less convenient for grocery shopping. The snug channel slows the current to an average of 1.5 knots and a maximum of 2.5 knots, considerably less than in the main part of the river. The presence of the Portsmouth Naval Shipyard on Seavey Island to the south is remarkably subdued, and there is a full-service boat yard to the north.

APPROACHES. Run up the Piscataqua River past the Pepperrell Cove area, heading for the huge, white former naval prison on Seavey Island, a half-Disney, half-Hitchcock structure designed to look like a medieval castle. The chart marks it as a square tower.

Keep nun "6" and daybeacon "2" to starboard while you are in the main river channel and then turn north toward Spruce Creek. Note that nun "6" is a mark for the main river channel and daybeacon "2" is on a ledge, so continue to keep them well to starboard after making your turn.

Stay clear of the restricted areas around Seavey Island. The pleasure boat moorings to port, between Clark Island and Jamaica Island, are owned by shipyard personnel, and the bigger yachts are nuclear subs, not noted for their hospitality at cocktail time. After a short distance, round can "3" to port and head into Back Channel. Nun "4" lies close inshore to starboard. Kittery Point Yacht Yard lies just beyond the point on the right. When maneuvering in Back Channel, be sure to locate the green can "7," which sits atop a 6-foot spot among the moored boats.

ANCHORAGES, MOORINGS. The Yacht Yard has ten reservable moorings and space alongside their floats in 22 feet of water. The slips on the Seavey Island side near the bridge are used by shipyard personnel. Back Channel is too crowded for anchoring, but you might find swinging room in the mouth of Spruce Creek just outside.

GETTING ASHORE. Row to the floats at Kittery Point Yacht Yard.

Portsmouth, New Hampshire from Badgers Island, Maine.

FOR THE BOAT. *Kittery Point Yacht Yard (Ch. 08; 207-439-9582; www.kpyy.net)* is a full-service boatyard with facilities to work in wood, fiberglass, aluminum, and steel for repairs of all kinds. The marine railway can handle craft up to 80 tons. They have water, pump-outs, showers, and ice, but no fuel.

FOR THE CREW. It is about a mile east to the post office and Frisbee's Market in Kittery Point (see Pepperrell Cove above) and a mile west to the post office and galleries in Kittery. To the west, Warren's Lobster House is hard by the bridge to Badgers Island. It is possible to go by dinghy by passing under the bridge to Seavey Island, but beware of strong currents.

THINGS TO DO. The Kittery Historical and Naval Museum (*207-439-3080, Tues.- Sat. 10am - 4pm*) is next to the town hall at the intersection of Route 1 and Rogers Road Extension. Among their many exhibits, they have the second-order Fresnel lens from the Boon Island Light. As part of its automation program, the Coast Guard removed the lens in 1993 and put it in storage. After a heated debate between the towns of Kittery and York, Kittery was finally awarded the lens.

Portsmouth Naval Shipyard runs a museum on Seavey Island, but due to increased security measures, it is difficult to visit. Its exhibits depict the history of the shipyard and display collections of nineteenth century naval artifacts, ship models, artifacts retrieved from surrendered German submarines, and large, six-foot-long models of modern subs. To make an appointment, call *207-438-1000*.

PORTSMOUTH

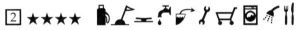

43° 04.75'N 070° 45.00'W

CHARTS: 13283, 13287, 13286   **CHART KIT: 57, 15**

Emergency: 911
Hospital: 603-436-5110
Coast Guard: 603-436-4414

The attractive city of Portsmouth can be a difficult place to visit by boat. There is a growing shortage of moorings for both residents and transients in the Portsmouth and Kittery area, so you may need to seek out alternatives to Little Harbor, Pepperrell Cove and Back Channel. Several anchorages or docks offer closer proximity to downtown Portsmouth, but these anchorages are exposed to the full river current and they may require passing under the Portsmouth bridges.

When the tide is running, moorings are tricky to pick up and may even be pulled under. For those not accustomed to tidal rivers, here is Rule Number One: approach the mooring into the current, regardless of wind direction. Rule Number Two: be exceptionally careful in your dinghy.

The Bridges. There are three bridges across the Piscataqua River between Kittery and Portsmouth. The first two are lift bridges and the third is fixed. Closed, the lift bridges have vertical clearances of 10 and 19 feet, so you will have to arrange for the spans to be lifted in order to go up the river.

The river runs very swiftly, and the bridges constitute a hazard for incautious skippers or boats with too little power to deal with the current. A small cruise ship once was swept into the Sarah Mildred Long Bridge. At full ebb, you will need at least 6 knots of power to fight your way upstream. If the current is flooding, you will need the same kind of power to avoid being shoved into the bridge while you are waiting for it to open. To minimize the problem, approach the bridge only near slack water.

Slack water on a tidal river usually occurs sometime other than high or low water due to the outflow of the river itself, which opposes or augments the effect of the tidal currents. At the Portsmouth bridges, slack occurs roughly three hours after the times of high and low water to seaward and roughly two hours after high and low water in Great Bay. The current is minimal an hour and a half either side of slack water, so aim to go through the bridges then.

In the summer the bridges only lift at specified times, but they don't lift automatically; you still need to request it. The first bridge, Memorial Bridge (*603-436-3830*), opens on the hour and on the half hour. Call on channel 13 or use your horn to give one long and one short blast. This bridge is scheduled for renovation in 2009, which may affect its schedule. The second bridge, called the Sarah Long (*603-436-2432*), is about half a mile farther upriver and opens 15 minutes before and 15 minutes after the hour. Call again on 13 or sound your horn.

The third bridge, some 900 yards beyond Sarah Long, is a fixed highway bridge with a vertical clearance of 135 feet.

SLIPS AND MOORINGS east of Memorial Bridge.
Portsmouth Yacht Club (New Castle) (Ch. 09; 603-436-9877; www.portsmouthyc.org). This attractive club is located just west of Salamander Point, on the south side of the Piscataqua. It has good facilities for visiting yachtsmen including two rental moorings with launch service and showers. Gas, diesel, water, electricity, and ice are available at the outer floats with plenty of depth. The club honors reciprocal privileges for members of other yacht clubs and allows them to use the handsome 1898 clubhouse. The Portsmouth Trolley swings by on its loop from town every hour.

Kittery Point Yacht Club (Ch. 09; 603-436-9303). The Kittery Point Yacht Club is not at Kittery Point, but at the eastern end of Goat Island on the south side of the Piscataqua. From the water, the initials "KPYC" can be seen along the porch of the gray clubhouse. This club is for members only.

Prescott Park Wharf (603-431-8748). The city of Portsmouth provides a series of slips just east of Memorial Bridge on the south side of the river, approached between nun "2" and can "3." The slips are exposed to very strong current and to wash from passing traffic, and on summer days they are very busy. Their advantage is their proximity to downtown Portsmouth, only steps away. Use extreme care in approaching. The reservable slips can be rented reasonably by the hour or for overnight berthing for boats up to 40 feet, with a 24-hour limit. Deep water, reported to be 8 feet, is limited to the outer floats. A dockmaster will help place you.

Kittery Landing Marina (Ch. 09; 207-439-1661). Kittery Landing Marina is on Badgers Island, hard by the bridge, with varying space for transients. There is plenty of depth, but beware of the strong currents when docking. Water and electricity are on the dock; showers and laundry are ashore.

SLIPS AND MOORINGS west of Memorial Bridge. Several new facilities have grown up in recent years west of Memorial Bridge.

Harbor Place Marina (603-436-0915). This small marina is located in Portsmouth on the south side of the river just past Memorial Bridge. They try to keep 200' of dock space open for transients. Depth ranges from 18 feet, and the docks have electricity, water, and stray Wi-Fi signals.

Badger's Island Marina (Ch. 09; 207-439-3820, 207-451-8174). This marina is on Badgers Island, on the north bank of the Piscataqua, opposite nun "16." They welcome transients when space is available. The outside of their large T dock can accommodate vessels to 70 feet. Water, electricity, cable, showers, and laundry are available. Their Island Marine Service can handle repairs (see below).

FOR THE BOAT. *Portsmouth Fisherman's Co-op* can be identified by the small fishing fleet alongside its wharf, just southeast from the docks at Prescott Park, at the northwestern point of Pierce Island. Pleasure boats are welcome to use the 24-hour credit-card pump for gas and diesel. Come into the floats east of the fishing fleet. Water may be available during business hours with your own hose from the co-op building.

Portsmouth Harbor Towing and Pumpout (603-436-0915) This BoatUS tow boat also operates a 🍴 pump-out boat for a moderate fee.

Island Marine Service (207-439-3810) on Badgers Island next to Badgers Island Marina specializes in mechanical and outboard service. They haul boats to 50′ with a hydraulic trailer and a 100-ton marine railway.

Back Channel Canvas Shop (207-439-9600) is a full-service marine canvas shop on Badgers Island, just over Memorial Bridge.

Jackson's Hardware and Marine (207-439-1133) is the closest chandlery, with a good selection of stainless fasteners, gear, and even dinghies. It's located in Kittery on the Route 1 bypass, the road that crosses Badgers Island, farther north.

West Marine (207-436-8300) is about 2.5 miles from Prescott Park, south of town at 775 Lafayette Road (Route 1).

FOR THE CREW. Portsmouth is a marvelous little city. Everything the crew may need is close at hand, with the exception, perhaps, of a good market. 🛒 State Street Provisions, just up from Prescott Park, is a small market with wine and beer, deli meats and sandwiches. A Hannaford supermarket is a long mile south of town out Islington Street. In the same plaza you'll find a laundromat and one of the New Hampshire State Liquor Stores, renowned for their dangerous discounts. A short walk or dinghy ride south from Prescott Park behind Pierce Island will bring you to Saunders Old Mill Fish Market (and lobster pound).

🍴 Pierce Island Channel also brings you alongside Geno's Chowder and Sandwich Shop, which serves breakfast, too. Another funky breakfast option is Friendly Toast, out Congress Street past Market Square.

Lobsters in the rough can be found at Morrison's Lobster across Memorial Bridge on Badger's Island or you can indulge in a full seafood menu at the Weathervane (*207-439-0920*) at the western end of the island. They have a dock, but patrons are not allowed to arrive by boat.

Portsmouth has many options for finer fare. The Dunaway Restaurant at Strawbery Banke (*603-373-6112*) is directly across from Prescott Park. The Rosa Restaurant (*603-436-9715*) is around the corner on State Street. They have been serving pasta specialties since before the Depression and drinks since Prohibition. Turn right on Penhallow Street to find Anthony's for more Italian, or try Lindbergh's Bistro on Ceres Street. Hit Ceres Street Bakery (which is actually on Penhallow St.) or La Brioche for dessert—or desserts.

The Portsmouth Public Library, a pharmacy, a 24-hour convenience store, and an auto parts place are out Islington Street on the way to the Hannaford Market. The HCA Portsmouth Regional Hospital (*603-436-5110*) is at the east end of town.

THINGS TO DO. A trolley bus makes a loop through Portsmouth and the neighboring communities about every hour, with stops at the Portsmouth Yacht Club and the Wentworth-by-the-Sea Marina. If you are at the Yacht Club, a pleasant walk leads out to Fort Constitution on New Castle Island, where you can take in the commanding views of the entrance to the Piscataqua.

In addition to the civilized pleasures of a small city, Portsmouth has some special attractions of its own. The old brick warehouses now house numerous wonderful little waterfront restaurants and bars. Check to see what's playing at the Seacoast Repertory Theatre (*603-433-4472; www.seacoastrep.org*) on Bow Street near Memorial Bridge or at Rings Theatre (*603-436-8123; www.playersring.org*) right in Prescott Park. Across the street is Strawbery Banke (*603-433-1100; www.strawberybanke.org*), a 10-acre historic waterfront neighborhood with fascinating old homes, gardens, and interesting exhibits. Both Strawbery Banke and Prescott Park host many special summer events, from open-air concerts and arts festivals to craft fairs, gardening workshops, and a farmers' market on Saturday mornings.

One of the best ways to see Portsmouth and appreciate its history is to take a harbor cruise. Portsmouth Harbor Cruises (*603-436-8084*) leave from a dock next to the tugboats on Ceres Street and the Isles of Shoals Steamship Company (*603-431-5500*) depart from a dock slightly to the west. One cruise takes you the back way from the Piscataqua to Little Harbor, under the bridge between Shapleigh and Goat Island, passing the Wentworth-Coolidge mansion, home of the oldest lilacs in the United States. Returning, you will get a good look at forts, lighthouses, the Coast Guard Rescue Station, and the Portsmouth Naval Shipyard. Other cruises take you into Great Bay with its quiet tributaries and abundant bird life.

When you want to tire out the kids, take them to the municipal pool and playground on Pierce Island, due east of Prescott Park and identifiable from the water by the chain-link fence around the pool.

About a mile to the west of Prescott Park, just beyond the second bridge, is the Port of Portsmouth Maritime Museum (603-436-3680), featuring the *U.S.S. Albacore*, a research submarine. You can walk through the submarine and marvel at how people live and work in such tight quarters. And you thought your boat felt small!

PISCATAQUA RIVER and its TRIBUTARIES: LITTLE BAY AND GREAT BAY

Patten: 43° 06.52'N 070° 47.67'W
Great Bay Marine: 43° 07.36'N 070° 51.42'W

CHARTS: *13285*   CHART KIT: NOT AVAILABLE

Anyone unfamiliar with the Piscataqua River should use extreme caution when heading above Portsmouth. As the *Coast Pilot* says, "General navigation throughout the entire length of the Piscataqua River is severely hampered by rapid tidal currents."

If this is your home territory or if you are comfortable with fast tidal currents, the Piscataqua River has a number of tributaries above Portsmouth worth exploring, with marshes, birds, and good gunkholing. Plan to run upriver with the flood; otherwise it takes forever.

In addition to the three bridges described above, there is a fourth, the General Sullivan, between Dover Point and Newington. Vertical clearance varies from 46 feet at high tide to 52 feet at low.

FOR THE BOAT. *George A. Patten Yacht Yard (Ch. 16, 68; 207-439-3967; www.pattensyachtyard.com).* This major repair facility is located about 1.5 miles above the I-95 bridge in Great Cove in the town of Elliot. They haul by hydraulic trailer and can perform repairs of all kinds. Or they can build you a new boat.

Great Bay Marine, Inc. (Ch. 68; 603-436-5299; www.greatbaymarine.com). Lying in Newington on the north side of Fox Point in Great Bay, Great Bay Marine is a large full-service marina. Facilities for transients include moorings, deep-water slips, water, electricity, gas, diesel, and ice as well as pump-out, a well-stocked marine store, showers, and laundry facilities. The yard hauls with a 35-ton boatlift and can handle a wide range of hull and engine repairs.

BOON ISLAND

due east of island: 43° 07.23'N 070° 28.05'W

CHARTS: *13286*   CHART KIT: *15*

Low and treacherous Boon Island lurks at the edge of the horizon northeast of Portsmouth Harbor. If it wasn't for the 133-foot lighthouse tower with a light visible for 18 miles, Boon Island might not be seen at all. And so it has been since boats have explored these shores. The number of its tragedies is proportional to the size of its tower.

Take, for instance, the story of the *Nottingham*, which inspired a novel by Kenneth Roberts. In a howling gale of snow on a winter night in 1710, the *Nottingham*, 135 days out of England, drove onto the rocks of Boon Island. Fourteen seamen made it ashore drenched by spray and frozen by bitter winds. When the storm abated, they could see the mainland, but with no means of signaling for help or of building a fire they were forced to survive on raw shellfish and gulls. Weeks passed. They built a small boat, but the sea destroyed it. They contrived a raft, but the two men on it were drowned. The discovery of the raft, which eventually washed ashore on Cape Neddick, led to their rescue, but not before the body of "the old carpenter" had provided grisly food for survival.

The granite lighthouse was built in 1855, with a tower and light visible from as far away as the Isles of Shoals to the south and Cape Porpoise to the north. There is no soil on these dismal rocks, and light-keepers used to carry a few sacks of earth in their boats to plant tiny gardens only to have it all disappear with the winter storms. Keepers were lost when waves crashed over earlier towers, but one, William Williams, liked the lonely post so much that he stayed for 27 years. The light is automated now.

During World War II a small Coast Guard observation tower was erected on the island, and ten men manned it 24 hours a day, armed only with rifles. Every two weeks they would be rewarded by two days of shore leave. Today, the lighthouse is leased to the American Lighthouse Foundation (*www.lighthousefoundation.org*).

THE YORKS to WELLS

YORK HARBOR

3 ★★★★

43° 07.82′N 070° 38.65′W

CHARTS: 13283, 13286   CHART KIT: 56 (E), 15

York Harbor is the most secure harbor between Portsmouth and Portland, a fact that did not escape explorer Christopher Levett in 1624. "This is a good place for a plantation," he wrote, "a good place for ships." And so it is today. Tucked behind Stage Neck, just up the York River, the harbor is a wonderful surprise, full of yachts and lobsterboats, fishermen and summer folk, tranquillity and... current. It is also a fashionable summer resort with antique stores and art galleries, beautiful homes and historic sites, beaches and walking paths.

The harbor's security, however, comes at a price. The entrance requires a sharp turn in formidable current, with ledges close by on both sides. Despite the lighted bell buoy at the harbor entrance, a stranger would be courting disaster to attempt entering in fog or at night. A safer option is to run for Portsmouth Harbor, a few miles south.

APPROACHES. York Harbor can be entered at all tides, but the sharp turn at nun "8" can be tricky during the maximum flood. If possible, plan to enter at slack water or against the ebb. Find the lighted red-and-white bell "YH" off York Harbor (*43° 07.75′N 070° 37.03′W*) and then head halfway between the buoys, leaving can "3" to port and nun "4" to starboard. Stage Neck is a peninsula lined with condominiums running from northeast to southwest. Aim for Stage Neck Light (*Fl 4s 10M*), a red, triangular shape on a small, white box with a light on top.

Observe nun "6" and can "7" with care and stay to port of the line between can "7" and nun "8." On a flood, the current will try to sweep you first toward Stage Neck and onto the ledge at nun "8" and then onto the opposite ledge at Harris Island. If you don't see nun "8," beware. On rare occasions it is pulled under by the current. Approach this turn with good steerage, keeping nun "8" a couple of boat lengths to starboard. Once nun "8" is abeam, make the sharp turn to starboard keeping your eye (not your boat!) on green daybeacon "11" marking the Harris Island Ledge to port. Beware also, particularly on weekends, that this part of the approach is often peppered with kayakers.

Once past Harris Island Ledge, the harbor is divided by Bragdon Island into North Basin and South Basin. Both basins are prone to silting. They were dredged to approximately 8 feet in 1996 but now have a controlling depth of about 6 feet. You can round out of the current in the mouth of the South Basin to catch your breath and get oriented to the harbor or to contact the harbormaster.

Leaving York Harbor is the reverse of entering, with the changing factors of tide and current. To avoid Harris Island Ledge, keep just to the left of midchannel and turn sharply to port around nun "8." But be sure to keep track of where you are. A flooding tide will set you to *port*, and it is easy to be aiming at can "7" and end up on Stage Neck. There is a lot of bottom paint on these ledges, and even the great and famous have come to grief here. *Ticonderoga* is said to have cut the nun close many years ago when leaving and ended up high and dry on Stage Neck. The embarrassed captain offered the then-astronomical sum of $50 to be towed off in secrecy during the night.

ANCHORAGES, MOORINGS. Anchoring is not allowed in York Harbor. Nor, thankfully, is jet skiing. The town has five transient moorings and also manages vacant moorings for visitors. Contact Harbormaster John Bridges (*Ch. 09, 16; 363-0433*). The moorings have controlling depths of about 9 feet and can accomodate boats to 50 feet. If you can't reach the harbormaster, York Harbor Marine on Harris Island or Donnell's Marine to the north may be able to help.

FOR THE BOAT. *Agamenticus Yacht Club (207-363-8510).* Located on the north side of the harbor, the AYC clubhouse and dinghy floats are the second farthest docks to the right, nearest Stage Neck (the farthest docks to the right belong to the Harborside Inn). Club facilities are limited to a two-hour dinghy tie-up and water at the floats, but people there are cordial and helpful.

Donnell's Marine (207-363-4308, 363-5324). Donnell's dock is to the left of the yacht club and has four transient slips. They can accomodate large boats, with 10 feet along the outside at low. Water and electricity are on the dock.

York Harbor Marine Service (Ch. 09, 16; 207-363-3602). Once past the narrows, York Harbor Marine

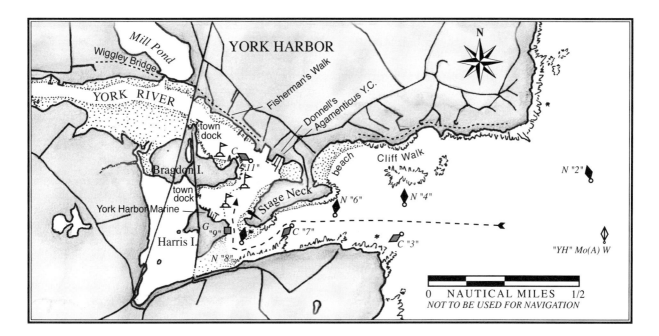

is on the left, on Harris Island. They can provide transient dock space for lengths to 40 feet, with 5 feet of depth at low. Gas, diesel, water, ice, and electricity to 50 amps are available on the docks. Their ship's store stocks parts for Johnson and Honda outboards. They can handle most boat and engine repairs, hauling with a 60-ton marine railway or a hydraulic trailer.

FOR THE CREW. York Harbor Marine has showers and laundry machines ashore.

The town of York Harbor is a three-block crossroads at the north side of the harbor. It offers pleasant strolling, but little commerce. Land your dinghy at the yacht club or Donnell's, or at either of the two town floats, one on the north side of Bragdon Island near the bridge and one on the south side of Bragdon.

The nearest provisions are at the Cumberland Farms convenience store in York Village, one mile away via Route 1A. On the way you'll pass the large Foster's Clambake Restaurant (*363-1424*). A laundromat is next to the Cumberland Farms and several banks are nearby.

The Dockside Restaurant and Guest Quarters (*363-2868*) on Harris Island is wonderfully situated at the turn of the entrance channel, overlooking the ocean and river. Lunch and dinner are served in a separate building with a screened porch. Walk out to the gazebo on the point and watch new arrivals negotiate (or fail to negotiate) the sharp turn.

The manicured lawns of the Stage Neck Inn lead to a lounge overlooking the adjacent town beach. The word "stage" comes not from stagecoaches but from early fishing days when cod was dried and salted on "stages," or racks. The inn serves lunch and dinner by reservation (*363-3850, 800-340-1130*).

THINGS TO DO. The town beach is a crescent of white sand where you can swim in gentle surf. At the north end of the beach, you'll find the beginning of the Cliff Walk, which winds around the northern side of the harbor entrance between the sea and beautiful old summer cottages, making it one of the loveliest coastal walks in Maine. Beware, though, that the base of the walk gets covered at high tides or during storm conditions.

In recent years a storm of a different kind has brewed over the Cliff Walk. Adjacent property owners threatened to close the trail, claiming ownership and associated liability and privacy concerns. The town fought to keep it open, and, as far as we know, prevailed, but please respect the privacy of the landowners.

Another good walk that offers a taste of York history starts at the Agamenticus Yacht Club or Donnell's Marine. Go left on a path along the riverbank called the Fisherman's Walk. Cross the road and walk over the mouth of Barrells Mill Pond on the pedestrian Wiggly Bridge, said to be the smallest suspension bridge in the world, and probably the most aptly named. The path leads through a wooded nature preserve until it reaches paved Lindsay Road. Take a right and stroll half a mile into York Village.

York Village is full of historic sites (*www.oldyork.org*). York originally was the Indian settlement of Agamenticus, wiped out by plague and resettled by colonists in 1630. Here Sir Ferdinando Gorges, friend of King Charles I, created the first city in North America in 1640. Anticipating becoming governor of New England, he called it "Gorgeana." But with the triumph of Oliver Cromwell, Gorges lost his power, and the city of Gorgeana was demoted to the Village of York.

At the corner of Lindsay Road and York Street is Jefferds Tavern, and nearby is the Old Gaol. Built in 1719 with timbers from its 1653 predecessor, the Gaol is the oldest remaining English public building in the United States. Here you can visit cells and dungeons and the restored quarters where jailer William Emerson and his family lived cheek-by-jowl with the prisoners. His bedroom window even opened onto one of the cells.

Near the Old Gaol are several other buildings open to the public, including the Old Schoolhouse and the Old Burying Ground, with some fascinating headstones dating back to the early 1700s. A plaque in the cemetery says, "Near this spot are interred the remains of the victims of one of the worst massacres of colonial days. On Candlemas Day, 1692, on the dawn of a January morning, Abenaki Indians attacked the settlement of York, burning the houses and capturing 300 of its inhabitants. About 40 were killed. The rest marched to Canada, many dying on the way." There the captured inhabitants of York were sold as servants to the French.

If you have a dinghy with an outboard, the upper reaches of the York River beg for exploration. Remember the strong current. You can land the dinghy at the John Hancock Warehouse and Wharf, now a museum, to shorten the walk into York Village.

Leaving York Harbor, if the weather suits, drop the lunch hook off the town beach outside Stage Neck in 10 to 12 feet of water at low, and have a swim. Head in for the middle of the beach, keeping halfway between nuns "6" and "4," and anchor a little way past Fort Point.

YORK BEACH

2 ★★ 🍴

43° 10.58′N 070° 36.15′W

CHARTS: **13283** (INSET), **13286**   CHART KIT: **56 (A), 15**

York Beach is a broad crescent beach tucked behind the north side of Cape Neddick but open from east to north. It is better protected than Cape Neddick Harbor from prevailing southerlies by the bulk of the cape and provides a fine anchorage in settled summer weather with small seas. Entry is easy. Ashore is a resort that time seems to have forgotten—large inns and hotels, boarding houses and cottages, gift shops and ice cream stores all centered on the beautiful—and very popular—beach. Here, you can still track sand into the restaurants, chew saltwater taffy, and bowl at the Fun-O-Rama.

APPROACHES. Coming from the south, pick up the light or horn on Cape Neddick Nubble and skirt the northern shore of Cape Neddick, which is steep-to. Head for the center of the beach. There are no obstructions.

Popularly known as The Nubble, Cape Neddick Light was built in 1879 to protect mariners from the "Savage Rock." It is one of the most photographed lighthouses on the coast of Maine. In fact, it is photographed at least once every three minutes (*web cam: http://216.71.193.135/WC/default.htm*). A narrow channel separates the Nubble from the mainland, and generations of lighthouse keepers' children have walked or been rowed across to school.

ANCHORAGES, MOORINGS. When the fathometer shows about 15 feet at low, it is time to drop the hook. Those tempted to go farther in will roll in the surf. The sand bottom holds well.

GETTING ASHORE. The calmest spot to land is usually in the southeast corner of the beach, but have your pants rolled high and don't wear your best shoes.

FOR THE CREW. 🍴 There are restaurants in the town of York Beach, within a block or two of the water. Fox's Lobster House overlooks the Nubble.

PERKINS COVE, OGUNQUIT

3 ★★★ 🍴

entrance: 43° 14.08′N 070° 35.22′W

CHARTS: 13286 (INSET)   CHART KIT: 58 (B), 15

Perkins Cove is a gem, but it is difficult to visit by boat and nearly impossible to stay the night. The cove is ringed by a tiny, working fishing village whose fish shacks have been transformed into shops and galleries, but the fishermen and charm remain despite a swarm of tourists. The tiny harbor was created in 1941 by dredging Josiah's River where it trickled into Oarweed Cove.

The cove is extremely well protected from every direction, but the entrance is always rather difficult. A unique, manually-operated wooden bascule bridge spans the narrow entrance gut, and once inside, room for visiting yachts is almost nonexistent among the resident lobsterboats. And during strong northerlies or easterlies, the water can pile in, making it dangerous to enter and completely untenable inside. Still, it is possible to make a short visit to Perkins Cove by boat by landing at the town dock, but don't count on finding room for the night.

In 1889, the surf-pounded rocks, lobster shacks, and brightly painted dories caught the eye of Charles Woodbury, a Massachusetts artist. He founded the Ogunquit Art Association on Adams Island, where most of the lobster and fish shacks stood. Fishermen, summer people, and artists have mingled in this idyllic spot ever since.

APPROACHES. It is a good idea to try and raise Harbormaster Harry Horning (*Ch.16, 72; 207-646-2136*) before beginning your approach into Perkins Cove to be sure there is room for you on the town float or elsewhere.

Begin at the red-and-white "PC" bell offshore (*43° 14.45′N 070° 34.17′W*) and head for the can and nun marking the entrance.

The ledges on both sides of the entrance break heavily and are easy to see, even at high water. From nun "2," head for the 13-foot spot until you can see right up the slot. Then turn to starboard and proceed up the middle toward the bridge.

ANCHORAGES, MOORINGS. Do not pass through the bridge unless directed to do so by the harbormaster. Depth and maneuvering room are equally scarce commodities. ⚓ Instead, land at the public floats before the bridge, between the

The Nubble, Cape Neddick.

CAPE NEDDICK HARBOR

2 ★★   🛒 🍴

43° 11.08′N 070° 36.00′W

CHARTS: 13283 (INSET), 13286   CHART KIT: 56 (A), 15

Cape Neddick Harbor is hardly a harbor at all and exposed from east to south. In settled summer weather, though, it can be comfortable, if rolly. The entrance is easy, but several unmarked dangers within the harbor make it difficult to go far inside.

There are no facilities except the 🍴 Cape Neddick Lobster Pound Harborside Restaurant (*363-5471*) across the street from the anchorage and 🛒 Shore Road Market and Deli down the road to the south. The view of the trailer park on the western shore is hardly inspiring. But the sand bar at the mouth of the Cape Neddick River is a nice place for wading and swimming.

APPROACHES. Head in between red bell "2" and can "1" toward a small gray house on the beach.

ANCHORAGES, MOORINGS. Once you are past the northern entrance ledge, the northeast shore to starboard is fairly steep-to, and you can anchor anywhere along it in 10 to 14 feet of water at low. Resist your instinct to seek better protection by anchoring in the southwest corner. Although you can pick your way among the rocks there and anchor on short scope, it is tricky, with several ledges awash and several ledges just below the surface.

lobstermen's fuel dock at the seaward end and the charterboats on the inner end. Depth is about 7 feet at low. Maximum allowable tie-up is 30 minutes unless you make arrangements with the harbormaster. His office is ashore by the float.

⚓ The sign on the wooden footbridge says, "Guest Moorings for Emergency Only—$20 per night." But the harbormaster may be able to put you on the one town mooring or the public dock for the night if you call well in advance.

To proceed into the cove, you may need to open the bridge. Walk up on it and press the button that raises the span. Steer carefully through the narrow opening and watch your upper spreaders. Once inside, stay to the right until you can turn toward your assigned mooring. All boats are moored fore and aft to chains laid sideways across the harbor.

How much water is there in the cove? As a local sailor put it, "If you draw more than five feet, you're in trouble." The chart shows 4, 5, and 6-foot spots, some soft and silty, others hard rock. On drain tides, the depth might be a foot or two less.

A web cam of the cove is at www.barnbilly.com.

FOR THE BOAT. *Town dock (Harbormaster Harry Horning: Ch. 72; 207-646-2136).* Water is available at the town dock, and in a pinch gas could be obtained at the lobster wharf.

FOR THE CREW. Perkins Cove has public restrooms and pay phones but no market, so you need to go into Ogunquit for provisions, and these tend to be of the high-end variety. A trolley-bus leaves from the harbor, and the fare includes the grand tour of the village, the beach, and the famous summer theater.

🍴 There are several waterfront restaurants in Perkins Cove. Notable is Barnacle Billy's (*646-5575*), Maine's most glorified takeout. Founder Billy Tower was a fisherman from the age of fourteen before he retired from the sea to become an energetic host. The Oarweed (*646-4022*) serves seafood in a slightly more formal atmosphere. Reservations are recommended.

THINGS TO DO. The hallmark of this artists' colony is its many small shops and galleries and the famous Ogunquit Playhouse summer theatre (*646-2402, tickets: 646-5511; www.ogunquitplayhouse.org*). Follow Shore Road south to find the wonderful Ogunquit Museum of American Art (*646-4909; www.ogunquitmuseum.org*).

For inspiration from nature, stroll Marginal Way, which parallels the open ocean for half a mile from Oarweed Cove to Ogunquit Beach. You will pass spectacular views, wildflowers, stunted pines, and surf shattering on the rocks. Marginal Way was once a local farmer's path for driving his cattle to summer pasture. He donated it to the town in 1923. The walk is not strenuous, and there are memorial benches for resting and dreaming. The white sand beaches of Ogunquit, Moody, and Wells stretch to the north.

The bascule bridge at Perkins Cove must be opened by hand.

WELLS HARBOR

2 ★★

RW "WH": 43° 18.83'N 070° 32.87'W

CHARTS: 13286 (INSET), 13286   CHART KIT: 58 (E), 15, 16

Wells is another difficult harbor to visit. The harbor lies in the shallow mouth of the Webhannet River, which is prone to silting. Not long ago it was dredged to 8 feet at low. It is now around four and slated for another dredging.

If your boat's draft is on the skimpy side, visit Wells in good weather on a high tide. The miles of unspoiled marshes behind the coastal beaches, including the Rachel Carson National Wildlife Refuge, have their own special beauty.

APPROACHES. Over time, the river tends to dump sand across the harbor entrance. Use extreme caution in entering. Swells sometimes break across the entrance, even in moderate seas, and especially at low tides.

The entrance is marked with red-and-white lighted bell buoy "WH" (*43° 18.83'N 070° 32.87'W*), and flashing green and red lights mark the ends of the jetties to the south and north of the channel. Once inside, the town dock is on the west side of the basin.

ANCHORAGES, MOORINGS. The town manages vacant moorings for visitors, but the mooring field is even shallower than what is left of the channel. Harbormaster Chick Falconer may be able to accommodate you overnight on the town dock.

FOR THE BOAT. *Town dock (Harbormaster Chick Falconer: Ch. 09, 16; 207-646-3236).* The town dock has water and pump-out facilities.

Webhannet River Boatyard (207-646-9649; www.webhannetriver.com) is a storage and repair facility. They haul boats to 40' with a 15-ton hydraulic trailer and can perform repairs of all kinds. They have ice, a ship's store, and a bait and tackle shop.

FOR THE CREW. The popular Lord's Harborside Restaurant (*646-2651*) is next to the town dock, and the Fisherman's Catch (*646-8780*), a pleasant little seafood restaurant and takeout, is clearly visible across the marshes.

A community park with a playground and restrooms is next to the boatyard, surrounded by salt marsh. A trolley stop nearby can take you to shops in town or to the long Wells Beach for a day on the shore.

THE KENNEBUNKS

KENNEBUNK RIVER

■ ★★★

entrance: 43° 20.74'N 070° 28.60'W

CHARTS: 13286 (INSET), 13286   CHART KIT: 58 (C), 15, 16

Kennebunkport is a madhouse, perhaps the epitome of tourist Maine. The sidewalks are jammed with people, the cars creep along the crowded streets, and the waterfront is more boat than river. The fact that it is the location of George H. W. Bush's summer compound adds to the fervor. Nevertheless, Kennebunkport has appeal for visiting yachtsmen, with places to explore and tranquil oases here and there. There are few, if any, moorings available for transients, and anchoring is not allowed. But if you can find—or afford—a place to tie up, this is a snug and extremely well-protected harbor. The river is prone to silting, but periodically it is dredged. Currently a 6-foot draft is about the most you can carry into this harbor at low water.

The Kennebunk River was probably discovered by Bartholomew Gosnold in 1602 or Martin Pring in 1603. In its natural state, it was a barred harbor with only two feet of water at the entrance at low tide and navigable for only about half a mile. Kennebunk and nearby towns were settled in 1643 but depopulated several times during the Indian Wars. During the nineteenth century, the banks of the Kennebunk River boasted six shipyards that launched 638 vessels.

APPROACHES. Entering the Kennebunk River is easy. From the lighted green entrance bell "1," keep Fishing Rock and its can "3" to port.

If you like to cut corners and are approaching from the south, you can stay close inshore and leave Fishing Rock to starboard, passing between its daybeacon "F" and daybeacon "O" to port, but don't cut either too close. Keep well off the rock between daybeacon "3" and can "5" near the mouth of the jetties.

Drop your sails outside and proceed between the jetties. Do not enter if heavy swells are running into the river. In 1995, the trough of a wave landed a fishing boat hard on the bottom of the river and it sank. The river was dredged most recently in 2005 to a controlling depth of about 8 feet at the entrance and 6 feet farther upriver, but expect continued silting.

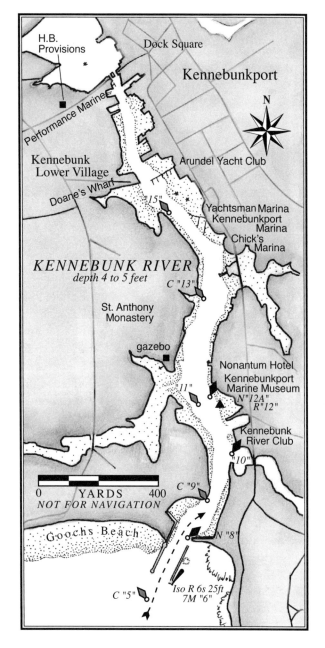

them, but the town maintains three for transients. Call Harbormaster Jim Nadeau (*Ch. 09, 207-985-7270*) to check availability.

Because space is at a premium, transient slip space should be reserved in advance. Chick's Marina, Kennebunkport Marina, and the Yachtsman Marina all welcome vistors. The Arundel Yacht Club and Performance Marine, farther upriver, may have space depending on availability. Combined with seasonal dockage, the Kennebunk River packs in a staggering number of boats and a number of staggering size.

FOR THE BOAT. Most of the yacht facilities lie on the east bank of the river in Kennebunkport. They are listed in order heading up river.

The Nonantum Hotel and Marina (207-967-4050; www.nonantumresort.com). This informal marina is primarily for guests of the hotel, but occasionally there is space available for transients. The docks have 6 feet alongside but no other facilities.

Chick's Marina (Ch. 09, 68; 207-967-2782; www.chicksmarina.com). Chick's is the largest marina on the river. The docks have 6 feet of water at low, and they can accommodate boats of well over a hundred feet if you should be so lucky. They pump gas and diesel and have water, electricity, pump-outs, Wi-Fi, and a marine store. Showers, laundry, and a pay phone are ashore. Chick's hauls boats by hydraulic trailer and can perform engine and hull repairs.

Kennebunkport Marina (Ch. 09, 16; 207-967-3411; www.kennebunkportmarina.com). The docks here are rented seasonally, but transients are welcome if dock space is available. The docks have 7 feet of depth, water, electricity, and pump-outs, but no fuel. Showers and ice are available, along with the Aft Deck lounge with a stereo and a television. A ship's store and mechanic are ashore.

Yachtsman Lodge and Marina (Ch. 09, 16; 207-967-2511; www.yachtsmanlodge.com). This upscale marina welcomes transient boaters when space is available. They can berth boats to 130 feet and provide water, 100-amp electricity, gas, diesel, water, pump-outs, showers, ice, and even a gas grill and outside tables overlooking the docks. Yachtsman also has dockage on the opposite side of the river at Doane's Wharf. You're in good company here—this is where George H. W. Bush keeps his Cigarette boat and where the Secret Service gases up theirs.

Moored boats can sometimes obscure the channel buoys, so it is wise to keep track of them as you go. Current is a modest 2 or 3 knots except upriver where it begins to be squeezed before passing beneath the bridge. Don't wait until it is too late to make your U-turn; you might discover the hard way that the bridge's 5-foot vertical clearance is somewhat less than the height of your mast and that what is shown as a swing bridge on the charts is welded shut.

ANCHORAGES, MOORINGS. Anchoring is not permitted in the river. Most moorings on the river are private, with a long list of residents waiting for

Arundel Yacht Club (Ch. 09; 207-967-3060; www.arundelyachtclub.org). The clubhouse is a wonderful wooden building that was once part of a rope walk where rope was laid and twisted. ⇌ A limited number of rental slips are available here, with up to 5 feet of depth, water, and electricity. Ice and club showers are ashore. There is no fuel, and the water shoals rapidly toward the inner slips. Note that large shoal areas lurk just before and after the club, the corners of which are marked by pilings.

Performance Marine (Ch. 09, 16; 207-967-5550) lies on the west bank of the river, past the seasonal slips of Riverside Landing and Harborview Docks and hard by the bridge between Kennebunkport and Kennebunk. After a long history of serving the working boats on the river, they are now balancing their repair business with seasonal slips for pleasure and excursion boats. ⇌ When available, they may have space for transients. ⛽ Gas and diesel are available, but use caution in the current around the fuel dock.

FOR THE CREW. The Kennebunk River separates two towns. The Lower Village of Kennebunk is to the west of the bridge, and Kennebunkport is to the east. The bridge is known locally as the Taintown bridge because the town on the far side, to Kennebunkporters, 'tain't Kennebunkport—it's Taintown.

🛒 H.B. Provisions (*967-5762*), located just beyond the bridge in the Lower Village, carries limited groceries, fresh baked goods, a full deli, and wine. Take a right at the next intersection to find Port Bakery and Cafe (*967-2263*) with soups and sandwiches, a hardware store, and farther out of town, Market Day (*967-5577*), a natural foods store and gourmet deli. A laundromat (*967-5066*) is tucked in the plaza by the bakery. A small playground is next door, so your kids and clothes can tumble at the same time.

🍴 The landmark Clam Shack takeout and fish market is hard by the bridge. Nearly next door, backed up against Performance Marine, is the excellent Pilot House (*967-9961*), a favorite hangout of local watermen. Around the corner, among the Shipyard Shops, is Federal Jack's Brew Pub (*967-4322*), the original home of Shipyard Ales. On the Kennebunkport side, east of the bridge, you'll find busy Dock Square. The Kennebunk Book Port is upstairs in an old rum warehouse. Numerous stores, ice cream shops, and restaurants, including The Hurricane (*967-9111*), are within walking distance.

More boats than water on the Kennebunk River.

THINGS TO DO. To appreciate the town's splendid architecture, start at the river and walk two blocks east to Maine Street, then along quiet, shaded streets lined with colonial and federal homes. Follow the street north along Grist Mill Pond. White Columns, built in 1853, was donated to the historical society in 1983 along with its completely original furnishings and decor, and it is now open to the public.

Take a moment to look up from the fray into the trees above. The largest collection of mature American elms casts its shade on the town of Kennebunkport, saved from Dutch Elm disease by the foresight of the town's tree warden in the early 1960s and through constant municipal and volunteer efforts.

For another walk, in the most peaceful part of town, land at the Nonantum Hotel and walk south along Parson's Way to the oceanfront. En route is "The Floats," used by novelist Booth Tarkington as a boathouse and summer afternoon retreat and now converted into the Kennebunkport Maritime Museum and Shop (*967-4195*), with a fine collection of nautical antiques.

Just south of the museum is the classic shingle-style boathouse of the Kennebunk River Club, built in 1889 to house canoes and social functions. It is still a private club.

At high tide you can land your dinghy on the west bank of the river near a brown, wooden gazebo. A boardwalk and path lead through trees and gardens to the chapel of St. Anthony Monastery (*967-2011*), the home of Lithuanian Franciscan monks. The handsome grounds and some areas of the monastery are open to the public. Or, if it

LOBSTERS AND LOBSTERING

Lobsterboats safe inside Cape Porpoise Harbor.

The history of Maine lobsters starts with James Rosier, who chronicled the Waymouth expedition of 1605. Near Burnt Island in Muscongus Bay, Rosier reported, "With a small net…very nigh the shore, we got about thirty very good and great Lobsters." In the old days, lobstering was child's play. Haul a net or lean over the side of the boat and gaff the lobsters in shallow water. Or wade along the shore and catch them with your hands, as the Indians did.

Four-foot lobsters were not uncommon, and they grew even larger. In 1858, Robert Carter dropped a hook over the side in Pulpit Harbor, got it tangled in the tail of a lobster, and pulled in a 12-pounder. Sometimes storms would drive ashore windrows of lobsters, which were used as fertilizer.

For a long time there was no lobster industry, because anyone could catch all he wanted for himself. After 1750, city folk in Boston and New York created a small market, but simple methods of catching them still sufficed. A century later, the French invention of canning changed the world of lobsters forever. The first lobster cannery was established in Eastport around 1843, and by 1880 there were 23 lobster canneries in the state of Maine shipping lobsters throughout the United States and even overseas.

As demand increased, so did the sophistication of the lobster fishery. The simple hoopnet trap was replaced by the wooden lobster pot, with a net funnel and a stone to hold it down. These were dumped over the side from rowing boats and hauled by hand. Sailing boats evolved for lobstering, such as the Muscongus Bay sloops, and fishermen could range farther afield to tend their traps. Then engines were developed, providing power to go long distances and haul traps in deep water.

The rounded wood-slat lobster pot has been replaced by rectangular wire traps. They stack better, are less susceptible to rot and worm, and last longer. Disintegrating panels that release their catch after a period of time have been invented for the wire trap, so lost traps, or "ghost traps," as they are called, won't continue to attract and kill lobsters.

Wooden lobsterboats are being replaced, in part, by fiberglass. Their bridges are loaded with GPS, bottom plotters, and radar to make it easier for the lobsterman to find his catch, his traps, and his way home.

The surge in demand, which began in 1850, had another result. The average size of lobsters started to drop immediately, and lobsters became scarcer.

In 1887, there were about 2,000 lobstermen hauling 109,000 traps by hand for a total catch of almost 22 million pounds. A century later, lobstermen fished *two million* traps and caught the same poundage.

But the value of lobsters has steadily increased, and the number of lobstermen and traps has kept pace. Maine issues about 7,000 lobster licenses a year, and of those perhaps 3,000 are full-time lobstermen. Altogether, nearly three million traps are fished in Maine waters. With such unparalleled pressure on the resource, scientists predicted an imminent population crash.

To the surprise of nearly everyone in the late 1980s and early 1990s, however, Maine lobstermen began to catch more and more lobsters. The fishery has now reached epic proportions, with 68.1 million pounds landed in 2005. With so much fishing, is the lobster resource healthy? Or is it on the verge of collapse?

continued, page 52

is sunny, land at the little sand beach on the west bank out past can "7" and walk over to the sandy expanse of Gooch's Beach.

Another marvelous exploration by dinghy is up the Kennebunkport marshes. At midtide, head under the Taintown Bridge and into the basin. Keep following the stream to the right, wending through the tranquil marsh grasses. Bring your binoculars for birdwatching.

Shipyards on these banks once built ships up to 200 feet long, and the remains of the locks needed to hold the water back for their launchings can still be seen. If your curiosity takes you as far as the first highway bridge, stroll up the hill on the east bank to visit the Landing School of Boatbuilding and Design (*985-7976, www.landingschool.org*).

If you still can't get enough of the water, the Kennebunk River claims to have more tour, sightseeing, whale-watching, deep-sea fishing, and ex-president-peeping boats than anywhere else on the coast.

CAPE PORPOISE HARBOR

43° 21.78'N   070° 25.87'W

CHARTS: 13286 (INSET), 13286   CHART KIT: 58 (A), 15

Cape Porpoise has, over the years, gained a reputation among cruising yachtsmen as having an entrance fouled with ledges and strewn with lobster buoys, and the taciturn fishermen who use this harbor might well like to keep it that way. The ledges haven't moved, but matters have improved. The Coast Guard has added navigational aids where they are most needed and removed and repositioned others, and now the entrance is straightforward in most weather even if the lobster buoys are still spread like confetti. The entrance is narrow, though, and would be difficult at night or in fog and foolhardy when the seas are up and breaking on the outer ledges.

The anchorage is broad but shoal, and the encircling islands and ledges provide reasonable protection. Forty or fifty lobsterboats make their home port here, with only several clusters of pleasure craft. For the first time, heading east, the buzz of tourism seems to have dropped astern. The Goat Island Light sweeps the evening sky, gulls cry, fishermen call to each other in the hole of sound left by their cut engines.

In his 1837 *History of Kennebunk Port*, Charles Bradbury wrote, "Cape Porpoise is a small but very convenient harbor. It lies at the extremity of the cape, and is the only safe harbor for coasting vessels between Portsmouth and Portland, being equidistant from them. Great numbers put in there during the dangerous seasons of the year. Nearly a hundred have harbored there in one day."

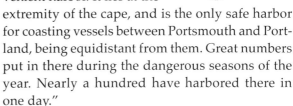

APPROACHES. The harbor approach is marked by a string of nuns leading to the entrance. Start at nun "2," which sits in deep water on all sides, and find nun "4" due south of the ledge called Old Prince. Keep it to starboard. Nun "6," also left to starboard, marks the 10-foot spot south of Goat Island (with the lighthouse) and due east of Folly Island (with its leaning, weather-beaten shack).

A daybeacon and can "7" mark Folly Island Ledge. To enter the harbor, steer midway between can "7" and the red daybeacon "8" off Goat Island, and then between can "9" and nun "10."

The presence of the ubiquitous lobster buoys in midchannel demands caution, but the careful skipper can wind his way through the maze, occasionally gliding in neutral, as in many Maine harbors. Entering at night multiplies the difficulty and is strongly discouraged except under ideal conditions.

ANCHORAGES, MOORINGS. The moorings are all private and most are used by lobsterboats. You might learn of an available mooring for the night by inquiring with Harbormaster Dave Billings at his shack on the town dock.

Anchor anywhere where the channel widens, preferably in the half closest to the entrance to keep a distance from the many lobsterboats moored farther in. Keep a close eye on your depthsounder and the state of the tide. There is a line of floats topped by sheds—known as lobster cars—moored along the western edge of the anchorage. In a strong southerly or southwesterly, the harbor may get a little lumpy, particularly when the ledges are covered at high tide. But holding ground is good mud.

GETTING ASHORE. There are dinghy floats beyond the town dock at the southern end of Bickford Island.

FOR THE BOAT. *Town dock (Harbormaster Dave Billings: Ch. 16; 207-967-5040).* This town dock is primarily a fishing wharf, not designed to accommodate pristine yachts alongside its grimy pilings. Fishermen play lines over the side, and lobsterboats circle while waiting to take on bait or fuel or unload their catch. Not surprisingly, yachts are not encouraged to land here, and bait has been known to fall unexpectedly on spotless decks. But gas, diesel, and water are available, and if you had an emergency or serious problems, you would undoubtedly find many helping hands.

FOR THE CREW. On the wharf, you will find a pay phone, the Cape Porpoise Lobster Co., and the Cape Pier Chowder House and takeout. For fine dining, Pier 77 (*967-8500*) is at the head of the wharf.

The village of Cape Porpoise lies beyond the head of the harbor, a fifteen-minute stroll from the wharf. The quiet and peaceful town is clustered around a stop sign, a small hardware store, and Bradbury Brothers Market (*967-3939*). Bradburys is well stocked and has newspapers and ice as well as a superb meat counter. The post office is next door, a pizza parlor is across the street. The proud Atlantic Hall houses the Cape Porpoise Library.

The Cape Porpoise Kitchen and Bakery (*967-1150*), right at the town's crossroads, is a specialty market, deli, bakery, and kitchen with food—and wine—to go. If you'd rather cook up your own, Sanborn's Fish Market is .25 miles up the hill toward Kennebunkport.

Cape Porpoise has several salty restaurants with names like The Captain's and Wayfarer. For good seafood and unusual atmosphere, don't miss Nunan's Lobster Hut (*967-4362*), one hundred yards down the road to the right, past the hardware store. Don't expect anything fancy. Nunan opens for dinner only when he brings in his catch. Guests attack their lobsters with gusto and wash up in sinks next to the tables. As the menu says, "Wine is served without undue attention to the complication of the wine ritual."

In the low-key atmosphere of present-day Cape Porpoise, it is hard to imagine what it was like a hundred years ago when a lively casino featured a dance floor and restaurant and patrons took romantic strolls by the sea.

(continued from page 50)

All this has led to regulation. Sometimes it is self-regulation. Lobstering is prohibited on Sundays in summertime, and is limited to daylight hours. The number of traps a lobsterman can set was first limited to 1,200 traps; now it's down to 800. Entry into the fishery is limited, too. Fishermen must serve two years as apprentices and must wait for a certain number of lobstermen to retire in their coastal zone before they are issued licenses.

Every lobsterman carries a brass gauge, called a "measure," to check the minimum carapace length for "keepers." Any lobster that measures less than $3\frac{5}{16}$ inches is a "short," and the maximum is 5 inches. Any female with eggs (berried) must be "V-notched" on the tail flipper and tossed back in. The fines for having either shorts or berried lobster on board are heavy.

Lobsters hatch from eggs and start life as tiny larvae floating near the surface of the ocean. After several days they molt, a process of shedding the old shell and developing a new and larger one. After another few weeks, they molt again, getting a little larger and more complex each time. A lobster may molt seven times in its first year. By the time it has reached a one-pound size, the lobster will molt only once a year. Rather astonishingly, when it molts, a lobster leaves its shell intact, claws and all.

Adult lobsters molt, or "shed," at times partly determined by seawater temperatures. Off Long Island, New York and Massachusetts, shedders are common in June; in Maine you'll find shedders in July along the southern coast and in August farther east. The price for shedders is less than for hard-shell lobsters because they have less meat per pound of lobster, but some say that shedder meat is sweeter.

At the end of the day, the lobsterman heads back in and unloads his catch at a lobster car moored in the harbor or at the floats of the co-op. The catch is weighed, and he is given credit at the going rate. Then the lobsters are sorted by size into the compartments of the float filled with seawater, waiting for retail customers or for shipment to wholesale customers such as restaurants. When prices are low, lobsters are often kept in lobster pounds (large enclosures), waiting for an upturn in demand.

There are several places along the coast of Maine where you can visit displays about different aspects of lobstering including the Maine Maritime Museum in Bath, the Department of Marine Resources Aquarium in Boothbay, the Ira C. Darling Center on the Damariscotta River, and the Mount Desert Oceanarium in Southwest Harbor. Rockland's lobster festival on the first weekend of August serves up ten tons of lobsters. And in many harbors there are lobstermen who will take you out to observe how traps are hauled.

STAGE ISLAND HARBOR

2 ★★★

43° 21.90′N 070° 25.13′W

CHARTS: 13286 (INSET)   CHART KIT: 58 (A), 15, 16

Stage Island Harbor, just east of Cape Porpoise Harbor, is unmarked. Strangers should only enter near low tide, when the ledges on both sides of the entrance are exposed. It is peaceful and private during settled summer weather but should be used only as a temporary anchorage at other times.

Stage Island, like Stage Neck in York and Stage Island at Biddeford Pool, is named for the stages, or drying racks, used by early cod fishermen.

Just south of Stage Island is Fort Island, named for the onetime colonial bastion there which was besieged by Indians in the early seventeenth century. As their supplies of food and ammunition dwindled, one of their number, lame and alone, stole off after dark in a decrepit canoe and headed for Portsmouth for help. Less than a day later an armed sloop arrived and rescued the settlers. Unnerved by their near-tragic experience, the colonists abandoned the area for the next decade.

APPROACHES. Identify Goat Island by its lighthouse. Cape Island is wooded at its north end and connected at low tide to larger Trott Island, which is mostly wooded. Do not enter the cove between Cape and Trott islands. Little Stage Island, with a small, white house, will appear as you round Cape Island.

Ledges make out a long way to the north of Cape Island and to the south of Little Stage Island. Low tide exposes most of them. Run into the harbor midway in the 100-yard gap between the ledges. Note the rock awash and the ledge at the western edge of the harbor.

ANCHORAGES, MOORINGS. Anchor in the southwest corner in 10 to 14 feet at low. At low water, the rock and the ledge at the western edge of the harbor are easily identifiable. Be sure your anchor is well set. Much of the bottom is covered with grass that holds poorly, but at lower tides you will be able to see patches of good mud. Exposure is from the east and south and also from strong northwesterlies.

GETTING ASHORE. The beach in the western corner of the harbor between the rocks is a good place to land your dinghy. Trott Island and many others are preserved by the Kennebunkport Conservation Trust (*www.thekennebunkportconservationtrust.org*).

Working wharves are busy places. Be patient and respectful when interrupting the serious business of lobstering.

THINGS TO DO. Vaughn's Island, southwest of the harbor, is owned by the Kennebunkport Conservation Trust (*www.thekennebunkportconservationtrust.org*), and visitors are allowed to explore. The island can be reached by dinghy or by wading across Turbats Creek from the mainland at low tide.

Goat Island, at the entrance to the harbor, was the last lighthouse to be automated on the coast of Maine due to its geographical proximity to Walkers Point. When George H. W. Bush was elected President, the island, less than a mile away from his home, was used as a security station. Another keeper was assigned, and military and security helicopters would come and go. The inevitable automation took place in 1990. Recently, though, the island was purchased from the government through the Maine Lights Program by the Kennebunkport Conservation Trust, and the island is occupied again. Volunteer lightkeepers welcome respectful visitors.

SACO BAY

WOOD ISLAND HARBOR

2 ★★★

43° 27.13'N 070° 21.07'W

BIDDEFORD POOL

4 ★★★★

43° 26.78'N 070° 21.41'W

CHARTS: 13287, 13286   CHART KIT: 58 (D), 15, 16

Wood Island Harbor and The Pool are right next to each other but quite different in character. Wood Island Harbor is formed behind Wood Island between Fletcher Neck and Stage Island. The harbor is shoal and fairly exposed, particularly from the northeast, but there is anchoring room, and moorings are available. A narrow channel with swift current called The Gut runs through the anchorage southwest into The Pool, a delightful little marshy bay separated from the ocean by Fletcher Neck. The Pool is snug in all aspects—good protection, small, and crowded. Anchoring is out of the question, and during the height of the season, a mooring is unlikely. Still, The Pool, The Biddeford Pool Yacht Club, and the village of Biddeford Pool can easily be reached by dinghy from Wood Island Harbor.

APPROACHES. From offshore, the Biddeford Pool area can be identified by Wood Island Light and the tall water tank shown on the chart behind The Pool. In thick weather, find off-lying red-and-white whistle "WI" (*43° 27.71'N 070° 20.39'W*), then run a compass course to Wood Island Light.

From the south, stay well clear of the dangers off Fletcher Neck by rounding Wood Island to the north. Pass offshore of nun "2" marking Dansbury Reef, then skirt the north shore of Wood Island. Keep the grassy hump of Negro Island to port and Stage Island, marked by an old, stone navigational monument, to starboard. If visibility is low, set a course for red-and-white bell "SA" (*43° 27.93'N 070° 20.30'W*), and then steer a straight course for the middle of Wood Island Harbor.

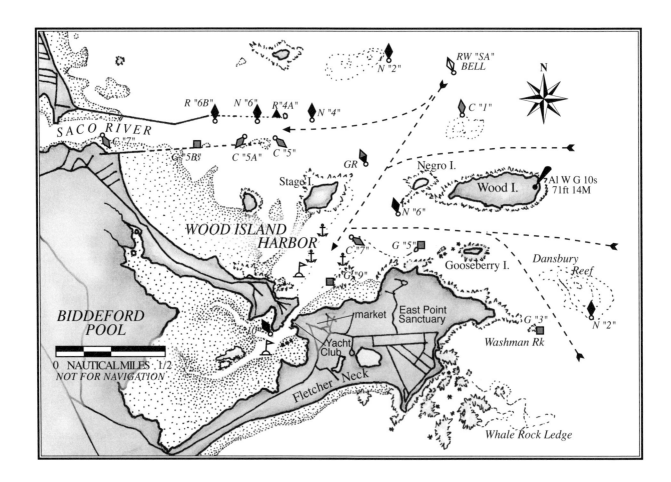

In good visibility, you can cut a corner; the channel between Wood Island and Gooseberry Island to the south is well marked and danger-free. Pass between Washman Rock, marked with can "3A," and Dansbury Reef, marked at its southern end by nun "2." Identify little Gooseberry Island and round between it and Wood Island.

From the north, head toward Wood Island Light and pick up red-and-white bell "SA" (*43° 27.93'N 070° 20.30'W*). Continue into Wood Island Harbor between Stage and Negro Island.

To get into The Pool, follow the channel to The Gut. The current runs in and out of The Gut with gusto. Stay in the middle or a bit to the right. A sloping beach making out from the left side claims several victims a year. It is also used by locals who beach their boats there to load and unload gear.

ANCHORAGES, MOORINGS. *Wood Island Harbor.* The Biddeford Pool Yacht Club maintains four moorings for transients, two for boats to 36 feet and two for boats to 50 feet. Their launch often meets arriving yachts and directs them to the moorings.

The mooring field lies on the west side of the channel with room to anchor farther out toward Stage Island. All of the water to the west of a line between Stage Island and the Gut is shoal, but the holding ground is good. For more depth, you can drop the hook east of the channel, between can "7" and green daybeacon "9" at Halftide Rock.

Wood Island Harbor is exposed from the east, northeast, and northwest, but it is fine in prevailing summer conditions. At low tide, you can watch the gulls wheeling and protesting as people wade out on the exposed sands to Basket Island and uninhabited Stage Island.

The Pool. At high tide, The Pool appears much bigger than it actually is. At low, it contracts into a puddle jammed with moorings. The Yacht Club has one mooring for boats less than 30 feet long. It can be reserved up to 14 days in advance.

In this tiny harbor, there is no time for indecision. Have your boathook ready. There probably will be just enough room and deep water to go in among the moored boats, make a tight turn, and head back into the current to pick up the mooring. Do not go west of nun "10" or east of the three granite mounds that break the winter ice, shown on the chart as little squares. The Pool is prone to silting, so if in doubt, seek local knowledge.

The Pool is not only beautiful, but also interesting. Experiencing the rushing of the waters at flood and ebb is rather like living in the neck of a bottle, but you will be well protected. The Pool is enormous at high tide, then shrinks to a sea of marsh grass a few hours later. It is popular among birdwatchers, who are likely to spot egrets, gulls, cormorants, loons, and night herons. Dusk brings out other marsh birds—mosquitoes and no-see-ums.

About a dozen lobsterboats and draggers have moorings in The Pool right among the pleasure craft. The pier of the Biddeford Pool Fishermen's Association is adjacent to the yacht club, and just down the street you may find fishermen mending their nets on their back lawns. Of course, a working harbor means waking at dawn as the lobstermen rev up, but that only adds to the enjoyment of the place.

GETTING ASHORE. The Biddeford Pool Yacht Club launch will pick you up (*Ch. 16 or 68*), or you can take your dinghy to the club floats or to the cobble beach at Vines Landing, at the mouth of The Gut just before the yacht club.

FOR THE BOAT. *Biddeford Pool Yacht Club (Ch. 16, 68; 207-282-0485; www.biddefordpool.org/bpyc/).* On the east side of The Gut, the club is small, informal, and friendly. Limited gas, diesel, and water are available at the float, which has plenty of depth beside it. Head into the current when docking.

FOR THE CREW. There is a shower in the BPYC clubhouse. Nearby, the Pool Lobster Co. (*284-5000*) sells lobsters, fish, and ice, as well as takeout lobster dinners, steamers, and chowdah. Provision at the F.O. Goldwaite Market, with groceries, beer and wine, and burgers, pizza, and ice cream. The post office is across the street. A short walk, bearing right from the club, then right again, leads past well-kept summer cottages to Hattie's Deli, with outside tables and great blueberry pie.

THINGS TO DO. The village of Biddeford Pool is peaceful, pretty, and very low-key. It is hard to believe that just 3 miles away from this tranquil place is the honkey-tonk of Old Orchard Beach. Old sepia prints in the Biddeford Pool Yacht Club show the 1885 herring fleet under sail in The Pool and the magnificent Fletcher Neck lifesaving station.

From Hattie's Deli, walk toward the ocean and the Biddeford Pool Boathouse and public beach, where there is swimming and even surfing. Bathhouses provide showers, dressing rooms, and phones.

THOSE WHO CAME BEFORE US

A handful of flint spears, crescent-shaped beads, knives, and arrowheads—these are the few remnants of a culture that flourished on the coast of Maine more than 5,000 years ago. Who produced these artifacts? We know almost nothing about these so-called "Red Paint People" except that the red iron oxide found in their burials has been traced to a single source on the slopes of Mount Katahdin, Maine's highest mountain.

More recently relatively speaking, when Nero was Emperor of Rome, "Oyster Shell Man" roamed up and down the coast of Maine. These people were creatures of habit and lovers of feasting. Year after year, they returned to the same locations for their outdoor feasts, and the evidence is still here—enormous heaps of oyster shells, sometimes rising 25 feet high, lining the banks of rivers such as the Damariscotta. Do you suppose our own civilization will leave traces as joyful and innocent?

The earliest European explorers of the coast encountered the "People of the Dawn," the Abenaki Indians. The exploits of the Abenakis, as well as those of the French and English who struggled to maintain a toehold here, are chronicled in myriad volumes of Maine's long history. Twentieth-century historical novelist Kenneth Roberts offered a paean to the Abenakis in his *Arundel*:

"Little enough is known of them now, God knows, and most of that is erroneous; and I fear that in another hundred years the only memory of them will be the names they gave to ten thousand hills and headlands and bays throughout our eastern country."

By way of background and explanation, Roberts provided the following:

"The Abenaki nation is a confederation of tribes living in the river valleys of our beautiful province of Maine, moving up the rivers in the autumn to hunt and gather furs, and down the rivers in the spring to fish and be cool. Between times they plant and harvest their crops on fertile spots along the rivers…

"In the Merrimac Valley were the Pennacooks, who went early to Canada to live on the St. Francis River because of the manner in which white men crowded them. In the valley of the Saco live the Sokokis, the Abenakis who came to Arundel for the summer fishing. In the Androscoggin Valley are the Assagunticooks, and in the Kennebec Valley dwell the Kennebecs, sometimes called the Norridgewocks, because the largest of their towns is at Norridgewock on the Kennebec. To my mind the Sokokis, the Assagunticooks, and the Kennebecs are the finest of all Abenakis, just as the Abenakis are the finest of all Indians.

"Farther to the eastward, in the Penobscot Valley and on the shores of Mt. Desert, which places have no equal for beauty in any of our provinces, live the Penobscots. Beyond them, along our wildest and foggiest shores, are the wigwams of the Passamaquoddies. All of these together, with the Micmacs of Acadia, which is also called Nova Scotia, form the Abenaki Confederation.

"It has been one of the peculiarities of our colonists that they have never kept faith with Indians. They have either stolen their lands outright, or made the Indians drink and persuaded them to sell vast stretches of territory for a few beads and a little rum and a musket or two, and they have made treaty after treaty with them—treaties which have always favored the white men; and never has there been a treaty that white men haven't broken.

"Everywhere throughout New England the colonists lied to them, cheated them, robbed them—an easy matter, since the Abenakis are brought up from childhood to think that all their possessions are safe, that no locks or bars are necessary to guard them. In trade they are fair and honest. Nothing causes them greater astonishment than crimes white men commit in order to accumulate property.

"For an Abenaki to tell an untruth to a friend except in jest or in the making of medicine, is accounted a crime. When an injury is done to one of them, all his friends make common cause against the guilty person. In friendship they are faithful and ardent, and grateful for favors, which never vanish from their memories; and if these be not returned in kind, then the Abenakis become contemptuous, revengeful, dangerous.

"Because of these traits the English might easily have gained and held their friendship and had their assistance against the French. Instead of that, by insults, cruelties and constant frauds they early aroused the enmity of many of them and drove them over to the French; and the French, by flattery and fair dealing, made them into faithful friends…

"It may be I shall be damned for saying so; but unless I have misread my Bible, I have found more Christianity and human kindness in…my Indian friends than in the venerated and violent Cotton Mather of Boston, who has declared in his writings that all red men are Scythians, and that the practicing of cruelties on them and the breaking of treaties with them are justified in God's sight."

Fishing is great in The Pool. Fishermen perch on every available pier around The Gut and putt about in little boats. Birdwatching is extraordinary here as well as at East Point Sanctuary (*www.maineaudubon.org/east*). To reach the sanctuary, walk east from the BPYC through an intersection, past The Pool Volunteer Fire Department, and go .7 miles to where the road meets the shore and turns south. The wire fence on the left leads past a golf course into East Point Sanctuary, maintained by the Maine Audubon Society. Bayberry and wildflowers abound, and there are lovely ocean views out to Wood Island. Wide, grassy paths and trails along the north shore make the walking easy and pleasant.

SACO RIVER

3 ★★★

entrance: 43° 27.70'N 070° 21.05'W

CHARTS: **13287, 13286**   CHART KIT: **16**

Emergency: 911
Hospital: 207-282-7000

The lower five miles of the Saco River are navigable to the head of navigation at the towns of Biddeford and Saco. At the mouth, the little summer community of Camp Ellis hunkers down among the dunes of Ferry Beach. Beyond that, the river is surprisingly pretty, though Biddeford is very much an industrial mill town. It also has the dubious distinction of being the site of a hydroelectric plant and a large trash-to-power incinerator.

APPROACHES. The Saco River is well marked and has only minor silting. Maximum current averages 2 to 3 knots, and controlling depths are about 7 or 8 feet at low tide.

Two stone breakwaters flank the mouth. From red-and-white bell "SA" (*43° 27.93'N 070° 20.30'W*), pass south of Ram Island Ledge, marked with nun "2" and south of flashing red bell "4." Head between can "5" and red daybeacon "4A" and follow the channel markers carefully between the breakwaters.

Once inside and past can "7," favor the south shore to avoid the ice breakers shown as small circles on the chart just downriver from the Camp Ellis town pier. Nun "8" takes you way over to the south shore, but keep close by the nun to avoid the ledges off Jordan Point. The tightest constriction lies upriver from nun "12," between two small islets. The river is relatively free of obstructions between the S-turn at Chase Point and the turn at Gordon Point known as Tarzan's Swing, but midway, favor Hills Point on the south shore to keep in the deep water.

Near the head, keep close to Cow Island. Several submerged pilings lie to the northwest, between Cow and a broad silt bank (covered at high) below Factory Island, and there are obstructions off the north shore, also shown on the chart but unmarked. Head carefully between the obstructions until you can pass closer to the yacht club on the north shore and the western tip of Factory Island. Rumery's Boat Yard can be reached by passing to the north and west of the silt bank (not to the south!), hugging the shores of Factory Island.

ANCHORAGES, MOORINGS. You can anchor almost anywhere along the sides of the river. The Camp Ellis harbormaster, Donald Abbott, may have a mooring just inside the breakwaters, and the Saco Yacht Club and Rumery's have some at the head of navigation.

FOR THE BOAT. *Town of Saco, Camp Ellis Wharf (Harbormaster Don Abbott: Ch. 16, 17, 18, 78; 207-284-6288, 284-6641).* Gas, diesel, and water are available, primarily for fishermen, at the Camp Ellis wharf. The town maintains one mooring for transients and a holding tank pump. Call *284-6641* for pump-out information.

Norwoods Marina (207-282-7411). This is primarily a small-boat marina, upriver from Chase Point on the north shore. Slips are rented seasonally, but they might sneak you in if they have space. Depths range to 10 feet. They also have a pump-out.

Bastille Boatworks (207-284-9535). Jim Bastille operates a small boatshop in the village of Camp Ellis and can help you with emergency repairs, particularly of the wooden kind.

Marston's Marina (Ch. 09; 207-283-3727; www.marstonsmarina.com). This small-boat marina is two miles upriver, on the Saco side. They have 120 slips and 10 moorings, which are mostly seasonal, and gas, water, and pump-outs.

Saco Yacht Club (Ch. 16, 69; 207-282-9893). The Saco Yacht Club has several moorings and might have dock space. They also have water and ice. The yacht club is home to USCG Auxiliary Flotilla 24.

Rumery's Boat Yard (Ch. 09, 16, 68; 207-282-0408). Rumery's is a real Maine boatyard in the heart of the hard-working town of Biddeford. They may have a mooring or be able to put you on the dock, with water and 16 feet of depth. A chandlery and a shower are ashore. Their crew can perform all types of hull, engine, rigging, or electrical repairs. They've developed a following among owners of Concordia yawls and a reputation for their gorgeous glass version of Nathaniel Herreshoff's Alerion sloop. They also build a line of custom trawlers based on a lobsterboat hull.

FOR THE CREW. The little town of Camp Ellis clings to a sandy spit that threatens to wash away. This classic, sandy summer community boasts several restaurants, the most notable of which is Wormwoods (*282-9679*). Huots (*282-1642*) offers more casual seafood. A small general store sells lobsters and ice and exchanges propane cylinders.

Saco has a laundromat, just up Front Street from the yacht club, and several restaurants. The supermarket is a couple of miles away, north on Route 1.

🍴 Biddeford must have more pizza and sub shops per capita than anywhere else in Maine. Main Street leads up from the bridge that crosses from Factory Island, and on it you will find the Chamber of Commerce and an abundance of secondhand shops and a gamut of Asian eateries: Chinese, Vietnamese, and Thai.

🛒 To get to a Hannaford supermarket, begin on Main where it crosses the river and take Hill Street south. When Granite Street forks to the left from Hill, follow it to the market.

The Southern Maine Medical Center (*282-7000*) is nearby.

THINGS TO DO. Camp Ellis, at the southern end of broad sandy Ferry Beach, is wonderful for walking, jogging, or swimming. Ferry Beach State Park is about a half-mile away, with trails through stands of tupelo trees, which are rare for this latitude. The rides and amusements of Old Orchard Beach are only a short cab ride away, down Route 9.

The City Theater of Biddeford (*282-0849*) is up Main Street on the left. It was designed by John Calvin Stevens, the famous Maine shingle-style architect. It staged its first performance in 1896. It survives as one of the finest examples of Victorian opera houses in the country, with almost perfect acoustics, and a listing on the National Register of Historic Places.

The elegant York Institute Museum on Main Street in Saco is only steps from the bridge crossing the river. It houses a wonderful collection of regional history, furnishings, and art.

STRATTON and BLUFF ISLAND

43° 30.53′N 070° 18.80′W

CHARTS: **13287, 13288**   CHART KIT: **16**

Just over a mile south of Prouts Neck are two little islands, low lying and grassy, with bushes and a few trees. They constitute the Phineas W. Sprague Memorial Sanctuary of the National Audubon Society. Stratton was colonized as early as 1630 by an Englishman, but now it is only home to gulls, cormorants, and certain exotic species like the glossy ibis.

To explore, anchor in the northeast bight between the two islands and land your dinghy on the west side of Stratton Island. Visiting hours are from sunrise to sunset. Audubon requests: no overnight camping; no fires; do not remove any plants or bushes; leave pets aboard; and stay away from bird nesting areas.

PROUTS NECK

0 ★★★ ⚓ 🚰 🍴

43° 31.78′N 070° 19.45′W

CHARTS: **13287**   CHART KIT: **16**

Prouts Neck is a beautiful, exclusive enclave jutting out into Saco Bay and flanked by two of Maine's most popular public beaches, Old Orchard and Scarborough. Winslow Homer spent the last 25 years of his life here and painted many of his most famous scenes, including *Eight Bells* and *Fog Warning*.

APPROACHES. Entering Saco Bay, you will see the headlands of Prouts Neck and then a concrete watchtower planted amid a number of large houses. Identify nun "2" at Bar Ledge (which instinctively seems like it should be a can) and nun "2SR" just off Prouts Neck and run between them toward the Prouts Neck Yacht Club, the low white building on the west shore of the Neck.

Note that because of sandbars, the channel for the Scarborough River runs all the way to the club, with the first can for the entrance, can "1," directly opposite the club's small breakwater, shown on the chart. Do not go north of the club or attempt the entrance to the Scarborough River unless you carry shallow draft.

ANCHORAGES, MOORINGS. Moorings are located in the little bight south of the club, north of the club and east of the river channel, and on the west side of the sandbar making out from Pine Point and marked by cans "1" and "3." Do not cut anywhere north of can "1" as you maneuver through the moorings!

At any of these locations, there is next to no protection. ⚓ Ask the launch operator or inquire at the club about the large guest mooring. Most of the other moorings are small, and there are tales of guests dragging through the fleet.

You can anchor west of the yacht club, south of can "1," where holding ground is adequate, but there isn't much between you and the great deep blue should the wind pick up.

GETTING ASHORE. Take the yacht club launch or row to the yacht club float.

FOR THE BOAT. *Prouts Neck Yacht Club (Ch. 09, 16; 207-883-9362).* There is daily launch service, and water is available on the dock.

FOR THE CREW. The club has showers and a pay phone. 🍴 The posh Black Point Inn (*883-2500*) is a short walk to the left and up the hill, and its elegant dining room is open to the public for three meals a day. Reservations (*510-1930*) are strongly recommended for dinner, and a jacket and tie are required.

THINGS TO DO. Walking north from the PNYC you will catch glimpses of a pristine white sand beach to your left—Ferry Beach. The path down to it is just opposite the entrance to the inn.

Homer's studio is a National Historic Landmark that is open to the public. Walk south from the yacht club and take Winslow Homer Road. The studio is a wing of the second house on the right past the gate. Homer's grave is on the lawn to seaward.

Near Winslow Homer Road is a path to the magnificent cliff walk around the perimeter of Prouts Neck and its "cottages" on the rim of Saco Bay. Vistas stretch past Stratton and Bluff islands to Wood Island Light and beyond. Here, life imitates art—the granite ledges, tidal pools, and coves still appear just as Homer painted them. You will end up back at the entrance to the Black Point Inn, where Scarborough Beach stretches northward.

SEAGULLS: SCAVENGERS OF THE COAST

The hook is set, the sails are furled, the dinner simmers, and the harbor lies still and silent. You sip your drink, nibble a piece of cheese, and toss a cracker to the gray-backed seagull floating patiently alongside. The cracker disappears into a yellow beak and suddenly three more gulls swoop down, then ten. They land gracefully on the water, close their wings neatly with a final twitch, and float in attentive formation, waiting.

Up high, another gull appears, answering some magic seagull summons, soaring gracefully on the air, wings never moving. Along the shore, a seagull turns and climbs and drops a clam to break open on the rocks below. Clouds of gulls wheel in the wake of fishing boats as they clean their fish.

Gulls are monogamous and can live for up to 30 years. Parents forage for their young and dutifully regurgitate it when the red spot on their beak is pecked. The gull can live on the sea for days. It eats fish, crabs, sea urchins, clams, and even kelp, and it drinks salt water and excretes the salt through its beak. On land, it drinks fresh water and becomes a connoisseur of garbage, frequenting open dumps and following fishing boats to feed on gurry. It has an expandable gullet, a cast-iron stomach, and the manners of a fishwife.

Jonathan Livingston Seagull was a herring gull, with silver-gray wings and black wing tips. Colored the same, but smaller, and with a black ring at the end of its beak, the ringbilled gull is seen in Maine late in the summer.

The black-backed gull is less common than the herring gull, but at 30-inches long it is bigger, tougher, and more aggressive. It is at the top of the pecking order and feeds on the young and eggs of other gulls, terns, eiders, and shorebirds.

Bonaparte's gull is a much smaller gull, easily mistaken for a tern. Its white body and black head are distinctive, and when you get closer, you can see its red feet. The wings are gray, with a thin white streak near the end.

Seagulls were not always common on the coast of Maine. Early settlers were able to lay out enormous stages of fish to dry in the open air wihtout fear of aerial theives. The few that were around were hunted—nearly to extinction by 1900. Their eggs were taken by coastal residents for food, and their feathers were collected for the millinery trade.

Protective laws were passed, and the gulls came back—with a vengeance. The voracious gulls eat tern eggs and gobble up tern chicks, decimating these populations. The gull resurgence has been so successful and so destructive to the other species that conservationists reluctantly approved the experimental poisoning of gulls on Green Island in 1984. The tern population still struggles to recover. What next? It is not a simple issue.

RICHMOND ISLAND HARBOR

1 ★★★★

43° 32.80'N   070° 14.58'W

SEAL COVE

3 ★★★★

43° 33.00'N   070° 14.17'W

CHARTS: 13287, 13288   CHART KIT: 16

Beautiful Richmond Island lies a half mile south of Cape Elizabeth and only seven miles from Portland. Its slopes are dark and wooded, with open meadows between, and its shoulders are rimmed with white sand beaches. There is a wonderful sense of remoteness and tranquillity here, a sense of being apart, even though the island is connected to the mainland by a breakwater.

The French explorer Samuel Champlain originally dubbed the island "Isle of Bacchus" because of the large number of wild grapes he found there. Later, in 1627, the island was settled by the English. In the seventeenth century a number of ships were built here, and a thriving fishery employed more than 60 people. More recently, various enterprises have been attempted including potato farming, herb growing, and sheep raising. A big barn on the island's north shore was once used as a root cellar to store potatoes. Several pairs of mink live here, as well as numerous woodchucks and deer, many with deep red coloration, supposedly from their diet of salt grass.

The breakwater divides the area behind Richmond Island into two anchorages, Seal Cove to the northeast and Richmond Island Harbor to the southwest.

APPROACHES. *Seal Cove.* From the south, pick up green bell "1" east of Richmond Island before heading into the middle of Seal Cove. This will take you clear of dangerous Watts Ledge. After passing East Point, coast along the north shore of Richmond Island, at least 200 yards offshore, toward the breakwater. There is plenty of room to clear the rocks and the shoals in the middle of Seal Cove.

APPROACHES. *Richmond Island Harbor.* Give West Ledge, off Richmond's western tip, a wide berth to starboard and head for the middle of the harbor, leaving green can "3" at Chimney Rock to port.

ANCHORAGES, MOORINGS. *Seal Cove.* In normal summer weather, Seal Cove, east of the breakwater, provides the best protection. Depending on the direction of the wind, anchor near either end of the breakwater in 10 to 15 feet at low. Watch for buoys that may mark either fish nets or pens. The holding ground is good, in sand, but ocean swells tend to round the island and roll you to sleep.

Although parts of the breakwater barely clear the surface at high tide, the line is easily visible. Drain tides expose the sands next to the breakwater, enabling the caretaker to drive his supplies out to the island. The lively park at Crescent Beach to the north is far enough away not to bother you but close enough to enjoy.

Richmond Island Harbor. If winds are from the east, choose Richmond Island Harbor. Again, anchor in either corner of the breakwater in 10 to 15 feet at low and expect a roll.

GETTING ASHORE. From any of the anchorages, it is an easy row ashore to gentle sand beaches.

THINGS TO DO. You are allowed to land on privately owned Richmond Island and picnic on the beach, but do not explore inland without checking first with the caretaker, who lives in the gray house at the crest of the meadow. We've had reports of black flies in the spring, so if you come then, come armed.

Beautiful, sandy Crescent Beach circles the northern shore of Seal Cove. Crescent Beach is a state park (*767-3625*). If you dinghy ashore for a day of basking, beware of Seal Rocks and The Sisters offshore and surf that may make landing difficult.

Casco Bay

Cape Elizabeth to Cape Small

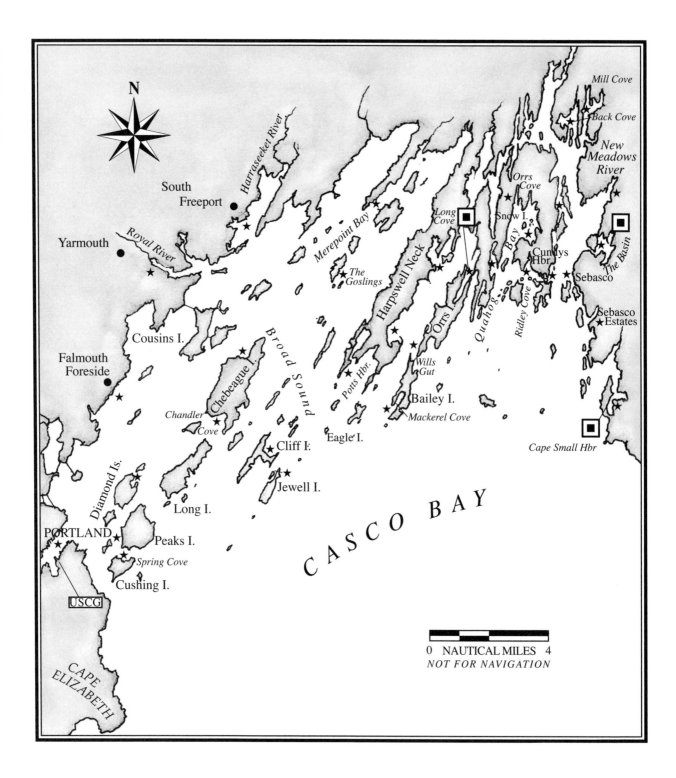

> "No element of beauty is wanting.
> Many of the islands are wildly picturesque in form,
> and from their woodland summits you behold on the one hand
> the surges of the Atlantic, breaking almost at your feet,
> and on the other the placid waters of the bay,
> spangled by multitudinous gems of emerald,
> while in the dim distance you discern on the horizon,
> the sublime peaks of the White Mountains."
>
> —Robert Carter,
> *Summer Cruise on the New England Coast*, 1858

CASCO BAY is where the Atlantic coast, tending north all the way from Florida, turns a sudden corner and breaks out in a flourish of islands. The great sand beaches of Maine's southern coast end abruptly and the rocky promontories and ledgy islands begin.

As the early explorers reported, there are hundreds of islands, large and small, in Casco Bay. Once known as the Calendar Islands, one for every day of the year, they are the peaks of three parallel ranges whose flanks are drowned in the bay, the valleys between gouged by glaciers that stood a mile thick on this land 13,000 years ago. Thus Great Chebeague Island, Long, Peaks, and Cushing are the extension of Merepoint Neck. Cliff Island and Jewell are remnants of Harpswell Neck. And Halfway Rock is the farthest tip of Orrs and Bailey Island.

The valleys between the ranges are now the great bays and sounds that make Casco Bay so interesting—Hussey, Luckse, and Broad Sound in the middle; Maquoit, Merepoint, and Middle Bay to the north; Merriconeag and Harpswell Sound and Quahog Bay and the New Meadows River to the east.

Casco Bay is an archipelago of islands busy yet unspoiled, rocky outcroppings barren and rugged, beaches fine and fair. Yet it is home to a small metropolis, a busy shipping port, an international airport, and a strategic harbor. While far busier with boats and people than Penobscot Bay and points east, Casco Bay is still surprisingly beautiful and unspoiled—a superb cruising ground.

The scale of the bay is small. From Cape Elizabeth to Cape Small is less than 20 miles. In clear weather, you can see the Portland skyline from Harpswell Neck. The broadest sound is only a mile wide, and the nearest island is usually only a short reach away. Sailing through the bay, one island slides against another, their dark silhouettes merging, dividing, blades of water cutting them apart.

One of the pleasures of sailing here is finding your way among these islands, chart in hand. Even in the fog, land is always close by. It is easy to become confused about whether red buoys should be left to port or starboard, so check the chart often. Winds fluctuate as you pass into the lee of one island or another, and the current often runs swiftly through the slots.

Unlike any place farther east, the presence of a substantial metropolitan area affects the look and feel of Casco Bay. Portland is small by most standards, but it overflows onto the islands in the shadow of the city. Most of the nearby islands are actually part of the city, both remote and urban. The briefcase is now more prevalent on these islands than the snowy egret. Little yellow-and-red ferries dart back and forth among the islands carrying summer folk and commuters alike. Fishing boats, daysailers, runabouts, excursion boats, and cruising yachts are ubiquitous. There are more marinas and moorings here than farther east. But do not bypass Portland just because it is a city. This is one city that is almost made to be visited by boat.

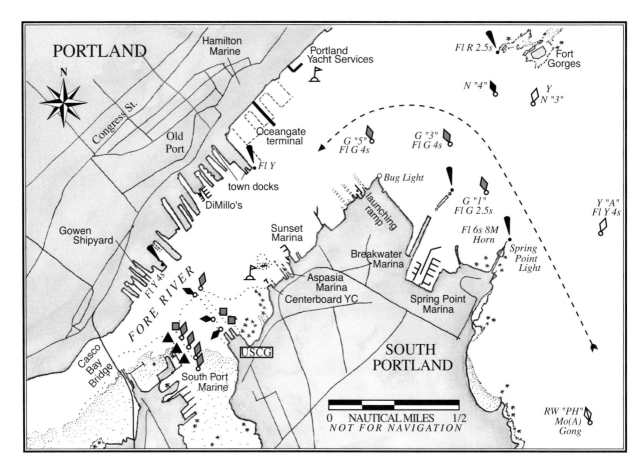

Another striking reminder of man's presence here is the layer of fortifications built over the centuries to guard Portland Harbor. Solid granite Fort Gorges has brooded over the harbor since 1864. World War II left less romantic ruins—towers for spotting German submarines, massive gun emplacements, concrete bunkers, decaying barracks, and old piers and military bases. The largest of the guns, the twin 16-inch guns on Peaks Island, could lob a 2,240-pound shell 26 miles!

There are four major yachting centers in Casco Bay. Portland Harbor holds the Centerboard Yacht Club fleet and is home to a handful of large marinas. Falmouth Foreside lies five miles to the north. Its harbor is a forest of masts on Portland Yacht Club and Handy Boat moorings. The banks of the winding Royal River are lined with marinas near the head of navigation in Yarmouth. And yachts and fishing boats cram the Harraseeket River off the Harraseeket Yacht Club and the pretty little town of South Freeport. Casco Bay is a newly designated "No Discharge Zone," which means even treated sewerage is not allowed to be pumped overboard.

There are plenty of good harbors in Casco Bay and lots of islands to explore. The choice and variety of daystops is infinite—little gunkholes between islands, picnic spots, and places of special interest like Eagle Island or the Chebeague Island Inn. Some narrow spots are worth sailing through for the sheer fun of it. Ghost along the cliffs of Whitehead Passage. Shoot through the tiny gut between Peaks Island and Pumpkin Knob. Or squeeze between Bustins Island and Little Bustins.

At the eastern end of Casco Bay are two delightful and less-visited areas: Quahog Bay and the New Meadows River. Snow Island, in Quahog Bay, may be one of the nicest anchorages anywhere on the coast. The New Meadows River has everything from working harbors to the resort at Sebasco Harbor to The Basin, a perfect hurricane hole so landlocked it feels like an inland lake. Farther inland, the water warms at the ends of the shallow fingers—perfect for swimming.

PORTLAND and the CITY ISLANDS

PORTLAND HARBOR

3 ★★★★★

43° 39.23'N   070° 14.75'W

CHARTS: 13292, 13288, 13290   CHART KIT: 59, 17

Emergency: 911
Coast Guard: Ch. 16; 207-799-1680
Sea Tow: Ch. 09, 16; 207-772-6724

L. DeMichele

Portland is a rare city. It is a major shipping port, the terminus of an oil pipeline to Montreal, a center of law and banking, and a cultural center. It is Maine's largest city. Yet, somehow, Portland retains the feeling of a small village with eclectic restaurants, concerts, theaters both mainstream and avant-garde, museums for all ages and interests, a local baseball team, and historic buildings. You will probably bump into friends on the street.

The city is bounded on all sides by water. To the west, below the elegant homes of the Western Promenade, flows the Fore River. Casco Bay, dotted with islands large and small, lies to the east. Back Cove fills to the north. And to the south, in the mouth of the Fore River, lies the harbor. The wharves, the very roots of the city, jam Commercial Street and comb the tides.

Bounded on the north by Portland and on the south by (what else?) South Portland, Portland Harbor bustles with an active fishing fleet, oil tankers, navy ships, tugs, island ferries, water taxis, excursion boats, and cruise ships. It is also a major yachting center.

The face of Portland, however, is changing. A new cruise ship terminal on the city's eastern shore is attracting ships in increasing numbers and size, and developers of high-end condominiums are following in lockstep. The Hilton has moved in; the Westin is on its heels. Blue-collar businesses are fleeing for cheaper pastures, and natural foods supermarkets are sprouting in the last industrial neighborhoods. This may be economic progress, but many who live and work here fear that the city could be well on its way to becoming someplace else.

APPROACHES. Portland Harbor is easy to enter under any conditions. The main approach is from the south. This is also the primary shipping lane. From offshore, pick up the red-and-white Portland buoy "P" (43° 31.60'N 070° 05.49'W) off Cape Elizabeth, which is identifiable by its lighthouse (Fl (4) 15s 27M).

Pass between flashing green bell "3" and flashing red whistle "4" and proceed to flashing gong "7" off Willard Rock and then to the brown and red "D" bell off Portland Head Light (Fl 4s 24M horn). Follow the channel northward past the lighthouse on Spring Point in South Portland and bear left into the harbor. All harbor amenities are east of the large Casco Bay Bridge, which spans the Fore River between Portland and South Portland.

When entering Portland, keep an especially good lookout ahead and astern for commercial traffic and frequent ferries. In fog, keep an ear glued to the radio, and don't hesitate to place a *securité* call if you are in the main channel.

The inner harbor, past the State Pier, is a no-wake zone.

ANCHORAGES, MOORINGS. Anchoring is not allowed in the inner harbor, west of the State Pier, so you should try to reserve a slip or mooring in advance. You can choose whether to stay on the Portland or the South Portland side.

On the north side of the harbor, DiMillo's Marina has a prime location at the foot of Portland's restored waterfront area, with city conveniences nearby. The disadvantages are the absence of moorings and the fact that you are amid the hurly-burly of downtown Portland. Portland Yacht Services has moorings at the eastern end of the harbor.

Two town floats allow limited docking for brief periods, one at the end of the State Pier, just east of the Casco Bay Lines ferry terminal and their yellow-and-red ferries, and another deep in the slip just to the west of the terminal. Neither is ideal. The east float is newer, longer, and has more depth than the other, but it is very exposed to wash from passing boats and so narrow that it is difficult to leave. Avoid the tarred, wooden planking on the State Pier or you might end up with a new, black deck job. The inner float is hard by the Portland fireboat with very little maneuvering room and shoal water at low.

On the South Portland side of the harbor several marinas and a yacht club squeeze among oil tanks and commercial docks. Here you can view the city in peace and quiet. The disadvantage here, though, is that you are somewhat removed from the city of Portland and will have to take a cab. Taking a dinghy across the harbor is not recommended. As noted sailor Dodge Morgan puts it, "A dinghy ride across the harbor channel is like making your way across an active airport runway on a tricycle."

Portland's newest marina offers a third option. Maine Yacht Center is located north of the Portland Penninsula in East Deering. Like the marinas in South Portland, these slips offer access to the city without being right downtown.

PORTLAND

FOR THE BOAT (waterfront from east to west). *Harbormaster (Ch. 09, 16; 207-772-8121).* In season, the harbormaster operates a patrol boat with large identifying lettering. His office is located in the Marine Trade Center on the Portland Fish Pier.

Portland Yacht Services (Ch. 09, 16; 207-774-1067; www.portlandyacht.com). The facilities of PYS lie at the eastern end of the harbor, directly below the Portland Observatory up on the hill. Owned by circumnavigators Phineas Sprague, Jr. and his wife Joanna, this yard is interested in cruising sailors. The moorings and docks at PYS have considerable exposure, but they are within walking distance of downtown Portland. PYS moorings are massive, 4000-pounders. The floats, with water and pump-outs, can accommodate vessels to 120 feet with depths to 20 feet. PYS handles major restoration and repair projects to wood and glass boats, including hull, engine, outboard, rigging, and electrical work, and they have an extensive engine parts department.

The yard is also home to the intriguing Maine Narrow Gauge Railway Museum and, in March, the astounding Maine Boatbuilders Show.

DiMillo's Marina (Ch. 09; 207-773-7632; www.dimillos.com). DiMillo's is easily recognized from the water by its large blue-and-white floating restaurant *(773-2216)* made out of an old ferry. The ferry was once the *Newport,* running between Newport and Jamestown and later the clubhouse and dry-sailing dock of the Pawtucket Yacht Club in Port Jefferson, New York.

The marina has gas, diesel, water, pump-outs, ice, cable television hookups, and electricity to 50 amps. Piling breakwaters protect 125 slips with room for transients in 8 to 21. They can accommodate several vessels to 200 feet on the faces of their outer floats. A marine store, showers, and laundry facilities are ashore. A rich variety of restaurants, shops, and entertainment is right across Commercial Street in the Old Port Exchange.

Gowen Inc. (Ch. 09, 16; 207-773-1761). Portland's Fish Pier is identifiable by a large beige metal building on the Portland side of the harbor and by all the fishing boats docked to the west of it. Gowen's, further to the west, was once a shipyard, but now caters to recreational boats as well as commercial. They haul everything from trawlers and Casco Bay Lines ferries to yachts of all sizes with their huge 150-ton boatlift. Hull, engine, and electrical repairs can be done by the yard or on a do-it-yourself basis. Their chandlery faces Commercial Street.

Baykeeper II (Ch. 09; 207-776-0136). This mobile pump-out boat based in South Portland and operated by Friends of Casco Bay will, for a nominal fee, come to your boat anywhere from Portland to Freeport and do your duty.

FOR THE BOAT (East Deering). *Maine Yacht Center (Ch. 09; 207-842-9000; www.maineyacht.com)* is Portland's newest marina and full-service yacht yard. They are located on the shore of Portland's East Deering neighborhood, due north of the Portland Peninsula, where the chart reads "dolphins." Boats of nearly any size can dock at their state-of-the-art floats imported from Sweden, with all amenities: water, electricity to 100 amps., phone, cable, and internet, laundry, kitchenette, and lounge. They have high-speed pumps for gas and diesel and pump-outs. They haul with hydraulic trailer and can perform all types of repairs.

FOR THE BOAT (ashore). *New England Fiberglass (207-773-3537; www.nefiberglass.com)* is located in the Portland Yacht Services complex. They can handle large and small glass repairs on your boat or at their shop. Their slogan: "You goon it, we glue it."

Hamilton Marine (207-774-1772; www.hamiltonmarine.com). This branch of the famous Searsport chandlery carries almost everything you might need

THE BEAUTIFUL TOWN BY THE SEA

Henry Wadsworth Longfellow immortalized and idealized Portland. "Often," he wrote in *My Lost Youth*, "I think of the beautiful town that is seated by the sea…" Now, more than a century later, he would scarcely recognize the metropolis that has come a long, long way in the last 350 years.

First settled as Casco in 1632, the town became Falmouth in 1658 and finally was incorporated as Portland in 1785. By 1775, most Falmouth families were sympathetic to the revolutionary cause and rebelled against the newly imposed Stamp, Sugar, and Navigation acts. They moved and concealed upriver ship masts that had been marked with the King's broad arrow and destined for the British Navy. Young men of Falmouth participated in the battles of Lexington and Bunker Hill. In October 1775, the British sent a Royal Navy fleet, under Captain Henry Mowatt, into Falmouth Harbor to retaliate. Giving the inhabitants two-hours' warning, Mowatt bombarded the city relentlessly for 12 hours, completely destroying it by fire.

It was during the ensuing rebuilding phase, in 1790, that Maine's first lighthouse was erected at Portland Head in what is now Cape Elizabeth. President George Washington appointed its first lightkeeper.

For the next hundred years or so, the city slowly grew and flourished. The economy was based on shipbuilding and the export of lumber from the deep, ice-free harbor with its easy access to the open sea. The Portland Observatory, a tall wooden structure, was built in 1807 as a signal tower, and a system of flags alerted townspeople, shipowners, and families to incoming ships hours before they docked. Nearly 200 years later, it still stands proudly on the skyline.

In 1866, Portland was once again destroyed by fire—this time allegedly started by a firecracker from the festivities celebrating the end of the Civil War. "It seemed as if the fires of Hell had erupted on earth," said one observer. More than 1,500 buildings burned, leaving hundreds of homeless who established a tent city.

This time, the rebuilding was planned carefully. Brick was used for shops and warehouses and granite for the public buildings. Streets were widened and lined with Victorian mansions and newly planted trees.

Portland's most recent revitalization began in 1966 with the restoration of a host of rundown 19th-century brick buildings along the waterfront. Today the Old Port Exchange, with its cobblestone streets lined with shops, restaurants, and galleries, is the core of an unprecedented commercial boom that extends to the Arts District on Congress Street, centered around a stunning museum and art school.

The city has become the state's banking, law, and insurance center, but the harbor remains its focus. The harbor ranks third in commercial activity among eastern seaboard ports and second in oil shipments. A large fish pier supports a dynamic fishing fleet which some years rivals the landings of New Bedford. For many years, Bath Iron Works repaired huge naval vessels at their floating dry dock. Now plans are afoot to develop a major cruise-ship terminal so others can come and enjoy the beautiful town by the sea.

---

for your boat, whether it is a polished yacht or a working boat. They are within walking distance from the Old Port at 100 Fore Street—the eastern end, next to Portland Yacht Services.

The Chart Room at Chase Leavitt (207-772-3751, 800-638-8906; www.chaseleavitt.com). Chase Leavitt is one of Portland's oldest marine businesses, located right in the Old Port where, as shipping agents, they serviced ships berthed along Commercial Street. They sell a worldwide array of charts and nautical publications and service and repair liferafts and inflatables. You can find them opposite the Casco Bay Lines ferries.

Sawyer and Whitten Marine Electronics (207-879-4500; www.sawyerwhitten.com). Located on Commercial Street just beyond Gowen, Sawyer and Whitten can handle all your electronics sales, service, and installation needs.

West Marine (207-761-7600; www.westmarine.com). This branch of the famous national marine supplier is a mile from Commercial Street. From Commercial Street, follow Franklin Arterial towards Route 295. Turn left on Marginal Way to number 127, on the left. Not only do they have a large selection of yacht hardware, cordage, and plumbing, they have plenty of sound advice.

FOR THE CREW. To provision in the Old Port, start at Harbor Fish Market on the tiny, cobbled Custom House Wharf, opposite the impressive granite customs house on Commercial Street. Harbor Fish is an authentic fish market, where the fish are gutted fresh off the boats and the floors are cleaned with a hose.

Follow your nose to the Standard Baking Company across from Casco Bay Lines for fresh breads, sticky buns, and pastries.

Walk east on Commercial Street and then a block north on India Street to Micucci's Italian Market. This is the place for prosciutto, cheeses, pasta, sauces, olives by the barrel, and a great selection of wines. Browne Trading Company, at the opposite end of Commercial Street, specializes in caviars, fish spreads, wines, and other delectables.

Every Wednesday, a farmers' market is set up right in the middle of town at Monument Square, featuring local produce and local farmers. Walk up Market Street or Exchange Street, then work your way to the left a couple of blocks.

For major provisioning, a Whole Foods natural foods supermarket is just up and over the hill on Franklin Arterial. A bus or cab can take you to the vast Hannaford supermarket on Forest Avenue or the Shaw's at the Mill Creek Shopping Center in South Portland.

Restaurant choices alone could keep you here several weeks. In the immediate vicinity of the waterfront, try Fore Street (*775-2717*) on Fore Street above Standard Baking, or Street and Company (*775-0887*), on Wharf Street, for the best of the best. The Old Port Sea Grill (*879-6100*), on Commercial, serves 12 types of oysters in a modern setting. The Porthole and Gilbert's Chowder House, tucked on gritty Customs House Wharf, and the Dry Dock and J's Oyster Bar, nearby, are local favorites. Other options include Benkay (*773-5555*) for maki rolls and sushi, and DiMillo's Floating Restaurant (*772-2216*). Becky's, at the west end of Commercial, is the place to be at the crack of dawn for heart-stopping eggs over easy.

Your laundry can soak at Lookin' Good Laundromat and Cleaners (*772-6676*) on Congress Street near India Street, where there is also a large Rite Aid pharmacy.

For crew logistics, the Portland Jetport (*774-7301*) is a short cab ride outside of town. Most car rental agencies are at the airport, but Enterprise (*772-0030, 800-736-8222*) is within walking distance of the waterfront on the corner of Forest Avenue and Marginal Way. The Greyhound (*800-231-2222*) bus station (*772-6587*) is at the west end of Congress Street. Concord Trailways (*800-639-3317*) is a short cab ride away on Sewell Street. Metro buses (*774-0351*) have routes over most of the city. For long-term parking, talk to Portland Yacht Services.

THINGS TO DO. Choices in Portland can be overwhelming. This little city packs more culture than most cities twice its size. Its shops are varied and interesting, its restaurants daring and reasonable. The main shopping districts of the Old Port and the Congress Street area are all within walking distance. The free Portland Phoenix newspaper or the Portland Press Herald might help you plan your time. You may need to stay longer than you planned.

Major fires destroyed the Portland peninsula on at least two occasions. The worst, in 1866, was set by a Fourth of July firecracker. The fires gave rise to the current architecture of the city—the brick warehouse area along the waterfront now known as the Old Port, the rebuilt, workingman's Munjoy Hill to the east, and the grand mansions of the West End, many of which were spared. Portland's motto, *Resurgam*, means "I will rise again."

Walking tour information is available at the Wadsworth-Longfellow House on Congress Street. The house where Longfellow grew up is now home to the Maine Historical Society. Their history museum is next door. Or you might explore the Victorian Mansion on Danforth Street.

On Munjoy Hill, climb all 102 steps to the top of the Portland Observatory (*www.portlandlandmarks.org/observatory.htm*), which was built in 1807 to alert merchants when their ships entered the harbor. Kids might like the Portland Fire Museum on Spring Street or the Narrow Gauge Railway Museum at Portland Yacht Services.

At the heart of Portland's art scene is the Portland Museum of Art on Congress Street, with a superb collection of Winslow Homer paintings, works by other Maine artists, Impressionist paintings, and traveling exhibits. Next door is the Children's Museum of Maine, which features a rooftop camera obscura which projects sights of the harbor onto a viewing table. The Maine College of Art, with several galleries, is nearby, and, for discounted gear, so is an L.L. Bean Outlet store.

For music or theater, investigate the Portland Performing Arts Center on Forest Avenue and the schedule of the Portland Symphony. In the summer, free lunchtime concerts are performed outside at Tommy's Park near Exchange Steet.

If you still yearn to be waterborne, numerous excursion boats leave from the Commercial Street area as well as the Casco Bay Lines ferries (*774-7871*), which have served the islands continuously since 1871, the oldest such ferry service in the country.

SOUTH PORTLAND

FOR THE BOAT (from east to west). *Spring Point Marina (Ch. 09; 207-767-3254, dock 767-3213; www.portharbormarine.com).* This is the first marina to port as you enter Portland Harbor, just before the large oil tankers' wharf. They have slip space for transients, but no moorings. The channel leading to the marina splits at the floats with the fuel dock to the right and slips to the left. To the left, at dead low, there is a less-than-5-foot spot just off dock "H." The fuel dock has 9 feet at low, gas, diesel, water and ice, and a sewage pump-out, and the slips have electricity and water. The marina has a 35-ton boatlift and can perform hull, engine, rig, and electrical repairs. Ashore there is also a well-stocked ships store, a canvas shop, laundry facilities, and showers. Joe's Boathouse (*741-2780*), at the head of the docks, serves excellent food.

Spring Point also manages the Breakwater Marina, just to the west, but the slips there are private with no room for transients.

Sunset Marina (Ch. 09; 207-767-4729). After passing the long oil pipeline terminal pier, can "5," and storybook little Bug Light, you will see the slips of Sunset Marina directly across the harbor from Portland's Old Port. If you should be so lucky, their face floats can accommodate transients to 250 feet with ample depth. Gas, diesel, water, pump-outs, ice, and 50-amp shore power are available, as well as showers and laundry facilities ashore. The Saltwater Grille (*799-5400*) serves excellent lunches and dinners overlooking the docks and the city.

Aspasia Marina (207-767-1914) occupies the site of the former South Portland Shipyard, due west of Sunset Marina. The shipyard was one of two in

The Portland Pipeline pumps crude to refineries in Montreal.

South Portland to build Liberty Ships in record time and numbers for the war effort in the 1940s. It is now a marina for seasonal customers only, but they do have a pump-out.

Centerboard Yacht Club (Ch. 68; 207-799-7084). This friendly and unassuming yacht club has a substantial fleet of cruising boats and moorings for transients. Centerboard is on the south side of the harbor, west of the extensive docks of Sunset and Aspasia marinas. The small clubhouse has big "CYC" letters on the roof.

The club has several guest moorings and keeps track of members who are away cruising. No fuel is available, but the long finger float has water. Depth is six feet at low at the outer end. A shower and ice are ashore.

It is quite pleasant to lie at a CYC mooring, gazing across at Portland's twinkling lights and listening to the low growl of the city in the distance, rocked occasionally by the wash of some large vessel passing. The harbor is beneath the path for jets from Portland's airport, but pilots tend to sleep later than lobstermen.

South Port Marine (Ch. 09; 207-799-8191; www.southportmarine.com) is the marina farthest into the harbor on the South Portland side, just before the new Casco Bay Bridge crossing into Portland. It has the advantage of being closer to Portland by land than any of the other South Portland marinas, and it is within walking distance of the supermarkets at the Mill Creek Shopping Center. If your boat draws more than 4½ feet, however, enter at half-tide or better, and follow the channel markers carefully.

⚓ The outside floats have 8 feet of depth at low and can accomodate boats to 135 feet, with hook-ups for water, 🚽 pump-outs, electricity to 50 amps, phone and cable. ⛽ The fuel dock pumps gas and diesel. Boats can be hauled with the 60-ton boatlift for repairs to hull, rig, or engine. Showers, laundry, a small chandlery, and ice are ashore.

FOR THE BOAT (ashore). *Reo Marine (207-767-5219; www.reomarine.com).* Located a couple of blocks inland from Sunset Marina, Reo Marine specializes in powerboats. Their shop services outboards, outdrives, and inboards of all kinds, and they have a small ship's store.

FOR THE CREW. A laundromat, a hardware store, and two major grocery stores are located in the Mill Creek Shopping Center in South Portland near the Casco Bay Bridge.

THINGS TO DO. The Spring Point walkway winds along the South Portland waterfront. From Spring Point Marina it runs east to the breakwater and the Spring Point Light, the grounds and buildings of Civil War-vintage Fort Preble, and the pre-Revolutionary "Old Settlers' Cemetery," and it continues to Willard Beach. On the way it passes the headquarters of Friends of Casco Bay (*799-8574, www.cascobay.org*), a group dedicated to preserving the health and the beauty of Casco Bay. You will also find the growing Portland Harbor Museum (*799-6337, www.portlandharbormuseum.org, Mon.- Sun., 10AM - 4:30PM*), which houses maritime exhibits on Portland Harbor and shipbuilding in the 19th century.

Heading west, the Shoreway heads to a second lighthouse, little Bug Light, which perches on a breakwater extending into the harbor from a large town park.

From South Portland, Portland can be reached by crossing the arching span of the Casco Bay Bridge. This new bridge replaced the Million Dollar Bridge, so nicknamed by outraged taxpayers because that was the appalling sum earmarked to build it in 1913. It actually cost a little less. Measure time in dollars: the replacement bridge cost $165 million. But what that money bought was a much-needed wider opening for tankers heading up the Fore River. In 1996, navigating the squeeze under the old bridge, the *Julie N* struck one of the piers and gashed a hole in its hull, spilling 180,000 gallons of oil.

L. DeMichele

FORT GORGES
43° 39.72'N 070° 13.40'W

CHARTS: 13292, 13290   CHART KIT: 59, 17

Fort Gorges, on Hog Island Ledge, was commissioned in 1858 as part of the Eastern Seaboard's Harbor Defense Plan during the Civil War. Built of solid granite blocks several feet thick, it was designed to mount 195 short-range guns. However, long-range guns were developed later in the war, so the fort was nearly obsolete by the time it was finished in 1864. The fort was armed but never garrisoned during the Civil and Spanish American wars. It never fired a gun or was shot upon. During World War II, antisubmarine mines were stored here.

The Fort is now owned by the City of Portland and kept in a marvelously rough state. Inside, two stories of arched gun bays form a dramatic amphitheater. A scramble to the top is rewarded by breathtaking views of the entire harbor and the long channel out past Portland Head Light. Watch your footing.

At low tide, the last remnant of a wreck reveals itself on Diamond Island Ledge to the north, out near nun "2." The six-masted schooner *Edward J. Lawrence* caught fire in 1925, possibly with a load of lime, while lying at anchor in Portland Harbor. The flaming boat was towed to the reef so she wouldn't obstruct the harbor when she went down.

APPROACHES. The fort is encircled by rocks except on the southwest. From nun "4," steer to the west of the Fort's granite wharf until you are almost abeam of the flashing daymark on the ledge farther to the west. The bottom shoals quickly, so check the tides, watch the depthsounder, and don't try to get too close before you drop the hook.

GETTING ASHORE. Beach the dinghy on the bar to the west of the fort, next to the granite wharf.

SPRING COVE

43° 38.74′N 070° 11.83′W

CHARTS: *13292, 13290*   CHART KIT: *59, 17*

Spring Cove is a bight on the north side of Cushing Island in Whitehead Passage between Cushing and Peaks. With just a glimpse of the Portland skyline, this is a pretty daystop, well protected from southerlies. Ocean swells, however, can curl around White Head to the east making for a rolly night.

Cushing Island is owned by a private association and kept as a small, exclusive community of summer homes, many of them shingle-style. The grounds were designed by the firm of Frederick Law Olmstead, famous for its design of New York's Central Park. The association strongly discourages landing on the island.

APPROACHES. From offshore, Cushing Island is clearly identified by two charted World War II observation towers atop the cliffs of White Head, the highest point in Casco Bay. A strong current moves through Whitehead Passage. Pass between green daymark "3" and Trotts Rock marked with daybeacon "4." Stay well off green daymark "3"; the ledge projects a considerable distance out from the mark. Once by nun "6" on the Peaks Island side, turn left into the cove.

If you are approaching from the south between Cushing and House Island, the buoyage is less intuitive. The red nuns stay to starboard (you are returning) until you turn between Cushing and House Island. Keep can "9" to starboard and red bell "8" close aboard to port (you are now leaving).

ANCHORAGES, MOORINGS. The cove is free of obstructions, but note that it is a pipeline area. Use caution when anchoring. There is plenty of room in about 20 feet at low off the association's pier, where you will be out of the strong current of Whitehead Passage. Don't be tempted to head to Spicers Cove, to the west, because that is a cable area.

PEAKS ISLAND

43° 39.40′N 070° 12.03′W

CHARTS: *13292, 13288, 13290*   CHART KIT: *59, 17*

In 1689, Indians gathered on Peaks for a successful attack on Portland. Three hundred years later, Peaks is under attack from the mainland as the growing population of Portland presses out into Casco Bay. Only 2.5 miles from the city, Peaks is a hybrid of summer community and suburb with paved roads, cars, city water, and cable TV. Yet Peaks is unmistakably an island, with an island's relaxed atmosphere and subtle blend of independence and interdependence, of isolation and community. The main street between the ferry landing and the market is full of kids on bikes and happy dogs. It is hard to believe this is part of a city. In fact, there is an ongoing debate here about seceding from Portland. The year-round population numbers about 800 and almost quadruples that in the summer. It includes fishermen, artists, and, more recently, professionals, who make the 20-minute commute to Portland by ferry.

Peaks' proximity to Portland has always shaped its history. Paddlewheel steamers provided service to the island as early as 1822. By 1830, it sported an open-air bowling alley and, in 1870, America's first summer theater. By the late 1880s, Peaks blossomed into a full-fledged amusement park with numerous rambling hotels, the huge Gem Theater, bowling lanes, balloon ascensions, and a boardwalk which ran the length of the western shore.

For the cruising sailor, Peaks is not exactly a remote, spruce-clad, rocky island. Instead, with its peaceful, panoramic view of the busy Portland harbor and twinkling skyline, it is isolated more in perspective than in fact.

Since moorings and slips may be hard to come by in Portland, an easy way to enjoy the city is to leave your boat at Peaks and take the ferry into town. The ferries run almost every hour, and the last one leaves town at 11:30 PM. Don't miss it, or you'll have to call a water taxi to get back.

APPROACHES. Peaks Island's facilities and landings are all on the island's west side, opposite the north end of House Island. You can approach from the south from either side of House Island. Beware of the Casco Bay Lines ferries.

ANCHORAGES, MOORINGS. Peaks has a large ferry wharf, identifiable by the tall green steel structure that raises and lowers the car-ferry ramp. A town float projects from the wharf to the north. For limited periods, you can dock on the town float in depths of about 5 feet or more at low. Jones' Landing Marina is just to the north, with temporary docking for patrons of the Inn on Peaks Island on the outer float.

Farther to the north, and most prominent, are the floats of Peaks Island Marina, which has moorings for transients and dock space. The floats have room for more than fifty boats with depths from 35 feet at the outside to 10 feet closer inshore.

This side of Peaks is exposed to the west but otherwise reasonably well protected except for wash from passing boats. If you want to anchor, chose a spot north of the marina floats in 16 to 22 feet at low in mud.

FOR THE BOAT. *Jones' Landing Marina (207-766-5652; www.joneslanding.net).* Jones' slips are rented seasonally, with no overnight space reserved for transients. Temporary docking, however, is allowed on the outer float for patrons of the Inn on Peaks Island. Depending on availability, you could, perhaps, arrange overnight dockage.

Peaks Island Marina (207-766-5783, 766-2508). The Peaks Island Marina is recognizable by the construction equipment ashore. Gas is available at the floats, but not water. Diesel is available at the top of the hill above the marina, but you will have to lug it. Propane tanks can be exchanged or filled, and the island laundromat is right by the pumps.

FOR THE CREW. Jones' Landing is next to the ferry wharf and town dock. If you land on a Sunday afternoon, you might think you've hit Jamaica. Jones' hosts a live reggae jam which has become somewhat of a Woodstock on the water. Depending on your temperament, this is either the place to avoid or the place to be. Otherwise Jones' Landing operates as a function hall for weddings and other occasions.

The Peaks Cafe, halfway up the hill, serves up steaming coffee and outstanding breakfast and lunch sandwiches. The Inn on Peaks Island (*766-5100*) is open for lunch and dinner, and the pub serves fresh beers and ales brewed by its parent company, Shipyard Brewing. The Peaks Island House (*766-4400*), up from the wharf and to the right on Island Avenue, has a dining room open for three meals a day, or you can head to the Asian-influenced cooking of the Cockeyed Gull (*766-2800*) by taking a left.

Walking left along Island Avenue, you will find an ice-cream shop, the post office, and Hannigan's Market, one of the most convenient places to provision in Casco Bay. It is well stocked with groceries and meats, wines and liquor, ice cream, ice cubes, and ice blocks. They make deli sandwiches and pizza and carry daily editions of local papers, the *Boston Globe*, *The Wall Street Journal*, and *The New York Times*. Bob Hannigan will arrange to get major orders to your boat.

Fresh lobsters can be bought at Forest City Seafood, right by the ferry wharf.

Beyond Hannigan's is the laundromat, Brad's Recycled Bike Shop, and the small library in the community building.

THINGS TO DO. The center and the east side of Peaks Island have been left largely undeveloped, and the views out over Casco Bay to sea are unending. The island is about a mile across and five miles around, so it is not too far to walk, or you can rent used bikes from Brad's (often self-service) Bike Shop (*766-5631*).

The Fifth Maine Memorial building (*www.fifthmainemuseum.org*) on the south side of the island was built by families of veterans to commemorate the Fifth Maine Regiment of the Civil War. The building is usually open to the public to display a moving collection of war memorabilia with the names of the astounding number of young men from Maine who gave their lives in the Civil War, etched in the stained-glass windows overlooking the sea. The Eighth Maine Regiment is commemorated in another memorial building farther to the east, but it is not open to the public.

On the Back Shore, as the ocean side of the island is called, you will find cobble beaches, nature trails, and lovely coastal walks among beach peas, *Rosa rugosa*, and marshes. Here, too, are the eerie remains of huge 16-inch gun emplacements at Battery Steele, where you can explore overgrown tunnels, bunkers, and towers left from a time when a camera wasn't allowed on the island. Bring yours.

THE DIAMOND ISLANDS

off Little Diamond wharf: 43° 39.77'N 070° 12.54'W
off Great Diamond wharf: 43° 40.30'N 070° 11.85'W

CHARTS: 13292, 13288, 13290   CHART KIT: 59, 17

The islands of Great and Little Diamond lie less than two miles from Portland. At low tide, a sandbar stretches between them and you can walk from one to the other. Casco Bay Lines ferries stop at the state piers at the south ends of each island, and they also service Diamond Cove, at Great Diamond's north end. In the eastern cove between the two islands lie the bones of a barge, a schooner, and the old Casco Bay Lines steamer *Maquoit*.

Great Diamond was originally named "Hoggiscand," meaning "Hog Island" for the pigs penned there by water at a safe distance from mainland wolves. By 1882, however, residents of the island thought the name Hog Island unbefitting their exclusive association, and they renamed it Great Diamond, presumably after the large quartz crystals found there.

Half of Great Diamond was once Fort McKinley, which served as an army base from the 1890s until it was abandoned after World War II. After nearly four decades as a forlorn ghost town, the three-story brick houses surrounding an oval parade ground have been developed as condominiums. The best anchorage and access to the island by boat is at Diamond Cove.

DIAMOND COVE

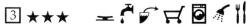

43° 41.10'N 070° 11.41'W

CHARTS: 13292, 13290   CHART KIT: 59, 17

Diamond Cove is a small, protected cove at the northeast tip of Great Diamond Island. For years it was known locally as "Cocktail Cove" where, in the shadow of abandoned Fort McKinley, it earned its reputation as a place to whoop it up. Now, with the development of the fort into luxury townhouses and house sites, the Diamond Cove exudes casual elegance and a carefully struck balance between the exclusivity of a private association and the welcoming of the public to the marina and restaurant ashore.

The stately brick houses of old Fort McKinley still ring the parade ground where soldiers once marched. Now the barracks and officers' quarters have been transformed into townhouses, and private homes are sprouting on surrounding land. The fort's tennis courts and the gymnasium have been refurbished for the residents, and others amenities have been added, including an art gallery and a heated swimming pool.

At the height of the summer, the ferry from Portland arrives eight or 10 times a day filled with residents and visitors alike. If you arrive by boat, the small marina can put you up for a visit to the restaurant or for the night. If you want to see Portland but keep your boat out of the fray, you could dock here and take the ferry to town.

APPROACHES. From the east, pass between small, overgrown Crow Island to starboard and can "1" to port. Favor the can and round into the harbor.

From the northwest, keep nun "2" to starboard and squeeze between Cow Island and Great Diamond paying close attention to the rocks at the southeastern tip of Cow. Then turn south inside tiny Crow Island into the cove.

ANCHORAGES, MOORINGS. The ferry dock and floats for visitors are on the west side of the cove with deep water.

The cove is full of private moorings, allowing little room for anchoring. Note the cable area and the ferry's approach path. Empty moorings seem to be used on a first-come, first-served basis unless the owner shows up. A swell from Hussey Sound to the east can sometimes find its way into the cove.

FOR THE BOAT. *Diamond Cove Marina (Ch. 09; 207-766-5694; www.diamondsedge.com).* Slips with water, electricity, and pump-outs should be reserved ahead of arrival. Patrons of the restaurant are also welcome to use the floats temporarily, though they may be asked to pay a small fee which is reimbursed in food or drink. The marina has showers, ice, and laundry ashore.

GETTING ASHORE. If you are on a mooring, row in to the dinghy float on the main pier.

FOR THE CREW. The former storehouse at the head of the main pier is now the four-star Diamond's Edge Restaurant, open for lunch and dinner (*766-5850*). Just beyond is the small General Store, open seven days a week during the height of the season, where you will find beer, wine, ice, grocery staples, coffee, and breakfast items. Often on weekend evenings, live bands play outside at the head of the cove.

THINGS TO DO. The Diamond Cove development is striving to maintain a balance between securing the privacy of the residents and welcoming the public. An information office is located next to the restaurant where you can check in for tours of the oval parade ground and the handsome brick houses of Fort McKinley.

The main, north-south road on the island is open to the public. It passes by some of the buildings of Fort McKinley, then through the woods and past some of the beautiful shingle-style summer cottages toward the southern end of the island.

LONG ISLAND

2 ★★

43° 41.40'N 070° 10.16'W

CHARTS: *13292, 13288, 13290*   CHART KIT: *59, 17*

During World War II, the Navy took the center of Long Island, including 90 cottages, by eminent domain and blasted the bedrock to build massive underground fuel tanks for refueling the convoys of the North Atlantic Fleet. After decommissioning, the Long Island Fuel Depot was sold to an oil company, whose plans to use the tanks to land crude oil from offshore wells never materialized. Other plans included Long Island as the final destination for crude shipped from Alaska or as a supertanker terminal for Middle Eastern crude. In the end, changing economics spared Long Island, and in the summer of 1994, the tanks were emptied and cleaned. The remains of the three huge refueling piers can still be seen on the west side of the island.

Long made history in 1993 by successfully seceding from the City of Portland and becoming its own town. Other Casco Bay islands that are part of Portland, including Peaks, Cushing, and Little and Great Diamond, have been carefully watching and weighing the successes and difficulties of the fledgling Town of Long Island.

ANCHORAGES, MOORINGS. Long should only be visited in settled weather. Its few small harbors are shallow, rocky, and exposed.

Most of the island access is from the northwest shore, where there is plenty of room to anchor. The southern end is divided by an obvious breakwater with a lobster wharf behind it. A new ferry wharf and town float are to the north of the breakwater. The old ferry wharf, Ponce's Landing, is to the south.

FOR THE CREW. The Spar, Long Island's oldest and only restaurant, is perched above Ponce's Landing and recognizable by its white awnings. The restaurant has moorings for patrons and its own dock for the dinghy to the south of Ponce's Landing. The small island market, Clarke's Store, is to the right when viewed from the water. They bake pizza, make sandwiches, and sell snacks.

THINGS TO DO. Long has a wonderful sand beach on its southeast shore which makes a nice lunch stop. Approach from the south and anchor between Long and Vaill Island. Don't go too far in because a bar and a ledge run between the two. In season, landing dinghies may not be allowed on portions of the beach.

THE FORESIDE

FALMOUTH FORESIDE

2 ★★★

43° 43.66'N 070° 12.38'W

CHARTS: *13292, 13288, 13290*   CHART KIT: *59, 17*

Falmouth Foreside is one of Maine's great boating centers and home to the Portland Yacht Club, the third-oldest yacht club in the United States. PYC's handsome old clubhouse overlooks an enormous fleet of handsome yachts, and members heartily welcome visiting yachtsmen. Handy Boat Service next door can meet every need.

APPROACHES. The forest of masts off Falmouth Foreside is visible from as far away as Portland. The various dangers are well marked, and the approach is easy from north or south. Look for seals on Clapboard Island Ledge to the south. In the harbor, watch for York Ledge hidden among the boats, marked by small green beacon "YL" and can "17."

ANCHORAGES, MOORINGS. The big complex of buildings and docks to the south belongs to Handy Boat Service. The dock farther to the north belongs to Portland Yacht Club. Check with the PYC launch to see if a mooring is available. If not, hail the Handy Boat launch. Handy's maintains a sizable number of very heavy moorings which can be reserved (*781-5110*).

Falmouth Foreside is on an exposed stretch of coast. The only protection comes from Clapboard Island to the east, so a strong chop can build up over the long fetch from the south. There is plenty of room to anchor out beyond the moored boats, and the bottom mud holds well. Be sure to use an adequate anchor light. ⇒ If there is not too much chop, it may be possible to lie along the floats at Handy's. Check with the dockmaster.

GETTING ASHORE. The Handy Boat launch monitors channel 09, and the PYC launch is on 68. Row into either dock or to the town landing, one dock farther to the north, for convenient access to the Town Landing Market, .2 miles up the hill.

FOR THE BOAT. *Handy Boat Service (Ch. 09; 207-781-5110; www.handyboat.com).* Founded more than fifty years ago as a boat yard, Handy Boat now offers almost every conceivable service including an elegant chandlery and the Hallett sail loft. Major repairs of all kinds can be performed, and facilities include ice, water, electricity, and ☞ holding tank pump-outs. There are two docks at Handy's, the southern one for ⛽ gas and diesel and the northern one for service, with 8½ feet of water at the ends and 6 feet along the sides.

Portland Yacht Club (Ch. 68; 207-781-9820; www.portlandyachtclub.com). The Portland Yacht Club welcomes members of recognized yacht clubs and asks that you sign the guest register. The "facility user fee" includes the mooring, launch service, and the use of club facilities.

The club was founded in Portland just after the Civil War when Portland Harbor was still crowded with great sailing ships and coasting schooners. Having survived fire and the Depression, the PYC almost foundered due to lost membership during World War II. It was born again with the acquisition of a Falmouth Foreside summer cottage on a bluff with grand views of Casco Bay. Today the club is a bustling, family-oriented sailing center, with a vigorous sailing program for juniors and popular races on weekends and Thursday evenings.

Launch service is provided from 8AM to 9PM. Water and ice are available at the dock, but no fuel.

Town dock (Harbormaster Gregg Fraser: 207-781-7317). The town dock and float are at the north end of the harbor. Limited docking is allowed on the float, with 3 feet of depth at low tide. Town Landing Market is directly up the hill.

Falmouth Pumpout Boat (Ch. 09; 207-781-2300). ☞ This mobile unit will suck you dry on the mooring.

Hallet Canvas and Sails (207-781-7070; www.hallettcanvasandsails.com). Located at Handy Boat, Hallet can repair your sails or build you a new set.

FOR THE CREW. PYC will hold mail for your arrival. There are showers, a washing machine, and a dryer. 🍴 The dining room is open to visiting yachtsmen for lunch and dinner. Dress is informal, but not too informal. BYOB. Credit cards are accepted.

🍴 The Falmouth Sea Grill (*781-5658*) is located right at Handy Boat, and in season they have an open-air bar overlooking the harbor.

🛒 The Town Landing Market, directly up from the town dock or half a mile north on Route 88 from Handy Boat, has the closest groceries and sandwiches. A huge supermarket, a hardware store, bookstores, a Radio Shack, a Wal-Mart and a Staples (if work must go on!) are located in the Falmouth shopping centers, 2.5 miles away.

BASKET ISLAND
43° 44.16′N 070° 10.10′W

CHARTS: 13292, 13288, 13290   CHART KIT: 59, 17

Basket Island lies north of Clapboard Island off Falmouth Foreside. Cumberland Mainland and Islands Trust protects all of its nine acres. This is a popular spot for picnics and exploration, but only in the daytime. No fires or camping are allowed (much of the island was burned in 1979).

Nun "4" marks a rock to the west of Basket. From the south, keep the nun to starboard and anchor off Basket's northwest or north shore. You can land your dinghy on the gravel beaches at the island's northern tip.

COUSINS ISLAND

off Doyle Point: 43° 45.10'N 070° 08.25'W
CHARTS: 13290, 13292, 13288   CHART KIT: 59, 17

Cousins Island is hard to miss. A huge oil-fired power plant dominates the southwest end with a green blockhouse, one enormous stack, and another merely large stack. Lights on these stacks blink at you day and night from almost any part of Casco Bay, but time heals all, and eventually you will come to ignore them. A small blue ferry plies the waters between the east side of Cousins and Great Chebeague Island.

Cousins Island is connected to the mainland by a bridge at Drinkwater Point in Yarmouth. The point's name comes from an early 19th century family that produced 14 sons. Each Drinkwater son in turn became a master mariner and captain of his own square-rigger.

Being Mainers of independent spirit, the Drinkwater captains found government regulations irksome and decided to strike back. One summer day—so the story goes—14 square-riggers entered New York Harbor and applied to Customs for clearance. Each vessel was commanded by a Captain Drinkwater, and each was registered from Cousins Island, Maine. It was a rough day for the fledgling federal bureaucracy.

The island's public wharf is on Doyle Point, on the southeast shore.

CHANDLER COVE

2 ★★

Chandler Cove: 43° 43.06'N 070° 07.68'W
Little Chebeague: 43° 42.69'N 070° 08.59'W
CHARTS: 13292, 13288, 13290   CHART KIT: 59, 17

Chandler Cove is formed by three islands—Great Chebeague, Little Chebeague, and Long. It may be a good anchorage for substantial vessels, as the *Coast Pilot* suggests, but not for cruising yachts except in the calmest weather. The cove is wide open to the southeast and southwest, deeper than comfortable for anchoring, and too large to provide a feeling of protection. The Portland ferry docks at the wharf on the east side, and the cove is also a main thoroughfare for Portland and Falmouth-based boats heading to popular Jewell Island and beyond.

But with Little Chebeague to explore, Chandler Cove makes an excellent daystop.

APPROACHES. Chandler Cove is well marked for entry from either the southwest or the southeast. The schooner wrecks between Long Island and tiny Crow Island were deliberately sunk here by the Navy during World War II as a breakwater and, along with antisubmarine nets, to block the harbor from enemy submarines and torpedoes.

ANCHORAGES, MOORINGS. The northeast corner of the cove provides good protection from the north. Or, in settled summer weather, you can anchor off the eastern shore of Little Chebeague. Note the cable areas. Both anchorages may be rolly.

THINGS TO DO. Little Chebeague was once a flourishing farm with open meadows and, later, several homes on the west side. In 1865, the island farmer decided to milk the new tourist trade that was developing in tandem with the advent of steamers. He built a steamer wharf and enlarged the farmhouse into a hotel. The hotel was expanded again and again in the next couple of decades to become the 70-room Waldo, which was reached from the wharf by a majestic "clamshell walk" lined with elms. A firework sparked the hotel in 1893, and it burned to the ground.

The Navy took over the island during World War II for a firefighting school. Practice blazes were set and extinguished in the rusty steel structure that is still on the east beach. The island is now owned by the State of Maine, so you can explore its sandy beaches and old buildings. Beware of poison ivy.

GREAT CHEBEAGUE ISLAND (west)

3 ★★★

43° 45.21'N 070° 06.53'W
CHARTS: 13290, 13292, 13288   CHART KIT: 59, 17

Chebeague's name is derived from the Indian word *Chibidisco* meaning "Island of Many Springs." Three and a half miles long and a mile or so wide, Great Chebeague is the largest of the Casco Bay islands. It was once a summer resort for the Indians, who retreated to the mainland in the winter, and much the same pattern is evident today. Two thousand summer folk retreat to the mainland on Labor Day and leave this beautiful island to about 350 year-round residents.

The first white owner of Great Chebeague was Sir Ferdinando Gorges, known as the "Father of Maine." In the 18th century, the island was settled by Scots, among them a real father of Maine, Ambros Hamilton. Hamilton had anywhere from 12 to 26 children—depending on who is doing the counting— who gave him as many as 194 grandchildren! The Hamiltons built a large fleet of "stone sloops," schooners that carried granite from quarries farther east to build libraries, post offices, and other public buildings in New York, Boston, and Philadelphia.

In 1997, Chebeague Islanders rescued a bit of Americana—and many a good bowl of chowdah—when they saved the Crown Pilot Cracker from being discontinued by megamanufacturer Nabisco. The flat, rectangular crackers were first made in 1792 by Massachusetts baker John Pearson as an unleavened bread for sailors to take to sea. A hundred and six years later, Pearson's Bakery became the National Biscuit Company, or Nabisco. Then, after more than two centuries of the tradition, Nabisco decided to discontinue the Pilot crackers.

From this offshore outpost, Donna Damon mounted a massive media campaign, bringing unbearable pressure upon the food giant to continue making this New England staple. Nabisco, reeling from the islanders' common sense and the prospects of a major PR coup, paraded cases of crackers by boat from Boston to Chebeague along with a hefty donation to the historical society. You can buy yours at Doughty's Market.

Chebeaguers showed their resourcefulness again when, in 2006, they seceded from the town of Cumberland to govern themselves as their own town. The catalysts for the move were skyrocketing taxes and the threat that the small island school might be closed.

APPROACHES. The island's primary anchorage is on its west side, opposite Littlejohn Island. Other anchorages and places to land include Chandler Cove or at the boatyard on the east coast.

When approaching from the south, find nun "18," well against the shore of Littlejohn Island, and leave it to starboard to avoid the constellation of rocks that make out from Great Chebeague. From there you will see the large, white Chebeague Island Inn.

ANCHORAGES, MOORINGS. The best anchorage is in the northwest bight off the stone pier and the Chebeague Island Inn. ⚓ The inn maintains several guest moorings marked "Chebeague Inn," and they operate a launch service (*Ch. 09*) for their customers.

There is plenty of room to anchor comfortably in 12 feet at low outside the moored boats. Try not to drop the hook in the path of the small blue ferry that makes its way back and forth between Great Chebeague and Cousins Island. The bottom is mud and grass, and the holding ground is good.

Although wide open to the north, this anchorage is well protected from prevailing winds and a long way from the ocean swells.

GETTING ASHORE. Bring your dinghy in to the float. Leave the west side free for the ferry.

FOR THE CREW. Sometimes lobsters can be bought fresh from cars tied off at the ferry wharf. As you come ashore, you'll pass a tiny golf tee on the right for the challenging par 3 waterhole—fun to watch.

Walk up the road or up a fairway of the public nine-hole course to the Chebeague Island Inn (*846-5155, 800-569-2288*). Many people come here to enjoy 🍴 picnics made to go or dinner in the dining room. Dress is informal. A large porch spans the front of the building, lined with rocking chairs. This is how summers used to be.

Chebeague's newest eatery, Calder's Clam Shack, is also in that direction, about 2 miles away on the North Road. From the Stone Wharf, take a right on South Road for about a mile, then bear right onto North Road at the fork by the Historical Society and go another mile. They are closed Mondays.

🛒 Doughty's Market (*846-9997*) and the post office are a mile from the inn, to the right.

THINGS TO DO. Great Chebeague has wonderful, flat walking. There are several nice beaches, the most convenient of which is Hamilton Beach (heard that name somewhere before?), on the northeast tip of the island, close to the inn. For a nice daystop, Hamilton Beach can be approached from the sea by anchoring just south of can "7."

CHEBEAGUE ISLAND (east)

3 ★★★

43° 43.90′N 070° 06.35′W

CHARTS: 13290, 13288   CHART KIT: 59, 17

Chebeague Island's east shore has a large bight behind tiny Crow Island. It is wide open to the northeast and, to a lesser extent, to the southeast, but in settled weather it makes a pleasant anchorage. It provides access to the quiet side of Great Chebeague and to several nearby islands for exploration.

APPROACHES. The northern end of tiny Crow Island is marked by green can "5." When approaching from Broad Sound to the north, be sure to pay particular attention to Goose Nest Ledge, between Stockman Island and Chebeague. Only its southern end is marked, and it lurks in water that seems wide and clear. To avoid it, stay close to Chebeague and work your way into the bight by Crow. Once can "5" is abeam to port, keep a close eye on your depthsounder as the bottom shoals.

From the south, enter the bight midway between Crow and the sandy flank of Chebeague, then favor Crow Island as you turn northward to avoid the 4-foot spot marked on the chart.

ANCHORAGES, MOORINGS. Most of the deep water in the bight is northeast of Crow, and much of that is occupied by the mooring field of the Chebeague Island Boatyard. There is still plenty of room to anchor, in good mud. If you expect any kind of weather, you might feel more secure on one of the boatyard's rental moorings.

FOR THE BOAT. *Chebeague Island Boatyard (207-846-4146).* Chebeague Island Boatyard is one of those places where time seems to have stood still. Boats are still hauled by tractor and cradled in a field that slopes gently to the beach. Kids plunge into the water off the float, and the smell of boat paint mingles with wafts of cut grass and salt air.

Time never stands still, however. In 2005, after 35 years of running the small yard, the owners decided to sell. Asking price and market value conspired to make the boatyard worth much more as house lots. Fortunately, a buyer was able to balance the price and payments so that the yard could continue to operate.

The fuel float has gas, diesel, and water but only 3 feet at low. Approach only at halftide or better. The boatyard can handle many repairs.

FOR THE CREW. Doughty's Island Market (*846-9997*), Calder's Clam Shack, and the Chebeague Island Inn (see previous entry) are within walking distance of the boatyard.

THINGS TO DO. Chebeague was famous for its littleneck clams. Through the 1920s, Fenderson's clam factory shucked, cleaned, cooked, and canned clams north of the steamer's Central Landing.

State-owned Crow Island is part of the Maine Island Trail network, and it often has campers ashore. Its small uninhabited cabin should be treated with respect. Stockman and Bangs Island to the north and east are also state-owned, but Stockman is mostly bare rock, and spruce-clad Bangs is nearly impenetrable.

Sand Island, to the southeast, is privately owned, but the owners generously allow careful day visitors. They ask that you please avoid the eroding southern end as well as the poison-ivy-infested middle and not to hang overnight on the exposed owner's moorings, to the west of the island.

THE OUTER ISLANDS

CLIFF ISLAND

3 ★★★

western anchorage: 43° 42.17′N 070° 05.69′W
eastern anchorage: 43° 42.03′N 070° 05.42′W

CHARTS: 13290, 13288   CHART KIT: 59, 17

On a chart, Cliff Island forms a striking H-shape with a sandy crossbar connecting a cliffy eastern end to the long narrow body of the island to the west. The shape gave Cliff Island its original name—Crotch. Islanders developed the Crotch Island pinky, one of Maine's most distinctive small craft.

Two fairly protected anchorages lie to the northeast of the crotch, far from civilization, with beaches and cliffs to explore. The little working harbor of Fisherman's Cove—Boat Cove to the locals—is tucked in the corner south of the crossbar.

The ferry dock and the public float are on the west shore, about .7 miles from the south end.

APPROACHES. The dangers in the anchorage north of the island are unmarked, so it's best to arrive near low water. You can approach easily from the northwest between Cliff and Stave Island.

The approach from the southeast is also easy, though less straightforward. Can "5" marks the end of the long ledges that run southwest from Bates Island. At high tides, the water seems more open on the northeast, Bates Island side of the can, but the deep water is to the southwest, much closer to the pointy end of Cliff. Stay close to can "5" and *pass south of it.* Depending on the tide, the current can be strong through these narrow passages.

ANCHORAGES, MOORINGS. Protection from southwesterlies is excellent north of the crossbar, but it is wide open to the northeast. You can anchor on either side of the large ledge that divides the anchorage. Note the 3-foot spot in the eastern section.

The chart shows a rock in the western half of the cove that uncovers 6 feet at low. Pass to the east of it and anchor south of it in 15 to 17 feet.

GETTING ASHORE. Row in to the sand beach.

FOR THE BOAT. *Fisherman's Wharf (Ch. 72; 207-766-2046),* in Fisherman's Cove, has gas, diesel, beer, wine, and limited groceries. Water is not available. The wharf has no float, so you will have to come alongside the pilings. Depth at low is 8 feet.

THINGS TO DO. From the anchorage north of the crossbar, you will see only one or two houses and a long beach. The cliffs and trails at the east end are protected by a conservation trust and are open to the public.

A trail also leads through berry bushes, across the crossbar, to the local ball field. A cottage on the cliffs on the south side of the crossbar was the site for the filming of *The Whales of August,* starring Bette Davis and Lillian Gish.

The road south leads past one of the last one-room schoolhouses in Maine. Turn left to Fisherman's Cove, where a small store perches on the wharf.

On a blustery weekday, you might think you've landed in some isolated outpost Down East, and in a way, you have. Even though Portland is only seven miles away, Cliff is struggling to keep its year-round community viable. In recent years islanders started a national advertising campaign to encourage new residents, particularly those who could employ themselves on an island and those with children. Suddenly the Cliff islanders discovered they had a shortage of a different kind—housing.

They also relearned what they probably already knew: that island life is not for everyone. One prospective family stepped off the ferry onto the wharf, took two steps, quickly glanced around, and hopped right back on the boat.

JEWELL ISLAND

4 ★★★★★

43° 41.30'N 070° 05.42'W

CHARTS: **13290, 13288** CHART KIT: **59, 17**

Tales abound of pirate treasure on Jewell Island, which, the stories go, was one of Captain Kidd's or Blackbeard's lairs. Disappointingly, Jewell is not named for the dreamed-of treasure, but for George Jewell of Saco who bought it from the Indians in 1637. Booty has never been found here, but natural treasure abounds.

Jewell lies on the outer fringes of Casco Bay, ideally situated for boats traveling east or west. This state-owned island is a mile long and no more than a quarter-mile wide, with a rich history and overgrown trails leading to craggy beaches and sub-sighting towers.

Its small, tight harbor lies on the northwest side of the island between Jewell and Little Jewell Island, which are connected at low tide. The anchorage is usually crowded on weekends, and the island is popular for camping, but even with the company, this is a special place.

APPROACHES. From seaward, make your landfall at Halfway Rock Light (*Fl R 5s 19M horn*). Be careful that a flooding tide does not push you onto the dangerous Drunkers Ledges to the north.

Run halfway between the northern tip of Jewell and West Brown Cow, staying well clear of the ledges that make out from Jewell. West Brown Cow is a little rocky plateau, 36 feet high, grassy on top, and normally quite visible, even in haze. In clear weather, you can use the higher face of Cliff Island as a reference.

After you round the north tip of Jewell and the ledge making out from it, head southwest into the harbor. Note that nun "4" and can "3" are *not* entrance buoys for the harbor at Jewell. Instead they mark the tight channel between Cliff Island and Little Jewell. To enter the harbor, keep nun "4" to starboard.

Jewell can also be approached from the south, to the inside. From a long way out you will see the tall observation tower on the southern tip of Jewell. Run between Cliff and Jewell, and pass between can "3" and nun "4" before rounding to starboard into the anchorage. Again, nun "4" should be well to starboard as you enter the harbor.

ANCHORAGES, MOORINGS. Jewell is a popular anchorage, particularly on weekends. Holding ground is good mud everywhere in the cove, but the narrowness of the harbor and the typical crowds tend to force anchoring on minimal scope. Anchor along the midline and not too close to the cliffs on Jewell to the east where the beach shelves gradually. There is good water past the house on Little Jewell Island, but be careful of two pilings just beyond, one of which is submerged at high tide. Beyond the pilings, the cove shoals rapidly.

This small harbor is protected from every direction but the north, and the wind can funnel down the slot to the south. You probably won't be able to put out an ideal amount of scope without swinging through the entire harbor, but try to put out enough that you won't drag through the fleet, which on a busy summer weekend may be anchored out as far as nun "4" or beyond. Rode weights might help.

In absolutely settled weather, if the harbor is full, some boats anchor off the old wharf north of Indian Rock on the west shore. Holding ground is good—perhaps too good. The chart still shows the area southwest of the wharf as a cable area, and you will be naked to the southwest. Another alternative is to head over to Cliff Island.

GETTING ASHORE. Clamber up the small cliffs at the obvious low spot. At anything but high tide, there is a gradual slate beach at its foot.

THINGS TO DO. For a quick and delightful way to stretch your legs, follow the well-trodden path straight across the island to the spectacular Punchbowl on the eastern side. At low tide, ledges completely encircle a broad crescent beach and a shallow body of water warmed by the sun. Pick berries on your way and bring your picnic. The beach has an abundance of driftwood, birds, beach roses, beach peas, and real sand. If it is near dusk or overcast, bring your bug dope too.

For a longer walk, follow the paths to the ruins of the World War II installation to the south. Carefully climb the makeshift ladders up the eight-story concrete observation tower and gasp at the view of

Casco Bay in its entirety. On clear days, you can make out Mount Washington in New Hampshire. The round pads poured into the floors mounted observation telescopes which took fixes on enemy ships or periscopes and, together with eleven other such towers, helped aim the long-range guns guarding Portland Harbor at Two Lights, Peaks Island, and Jewell Island. In 1942, a German submarine surfaced off Jewell Island and was pursued and attacked by destroyers.

Farther along the path, past collapsing tar-papered barracks, you will find an enormous concrete circle with a hole in the middle for Jewell's twin 6-inch gun emplacement. The 6-inch guns could fire a 100-pound shell as far as 15 miles. Dripping concrete tunnels lead underground into the bunker with massive, rusting doors, open floor pits, and branching tunnels with unknown destinations, all in a blackness thicker than ink. If spelunking runs in your blood, bring a flashlight. Again, be careful. This is not a convenient place for a serious accident.

Above all, please respect Jewell and its fragility. It has become a heavily used and sometimes abused island. It is kept in a marvelously natural state as an "unimproved state park," but that status can only last as long as it is well cared for by those who use it. Already, the Maine Island Trail Association has deemed it necessary to build several outhouses to control the human waste problem, and they have placed a summer ranger on the island.

Use exceptional care if you have a fire—an uncontrolled fire would be devastating. As on all islands, a fire permit is required. Fires should be made in fire rings only, and firewood should be collected, not cut (see *Introduction*, page 5, for guidelines on camping and fires).

BROAD SOUND

Broad Sound is an expanse of water centered in Western Casco Bay, stretching from the western rivers—the Royal and Harraseeket—to the northwest, out to Eagle Island to the southeast, and running north of Cousins and Great Chebeague Island.

Southwest winds make Broad Sound an easy reach in both directions, though you may encounter currents of a couple of knots that may kick up a chop. From offshore, enter the Sound at red-and-white bell "BS" (43° 41.72′N 070° 03.44′W).

BATES and MINISTERIAL ISLAND
43° 42.95′N 070° 04.34′W

CHARTS: **13290, 13288**   CHART KIT: **59, 17**

Bates and Ministerial are low and partly wooded islands northeast of Cliff Island, connected to each other by a bar. A small slot opens up between the two islands at their north ends where you can sneak in and drop the hook in 13 feet of water for a daystop. Swells bending in from Broad Sound can make it a rolly and uncomfortable anchorage for the night.

The islands are private, with only a few houses. Please do not land.

EAGLE ISLAND
43° 42.77′N 070° 03.28′W

CHARTS: **13290, 13288**   CHART KIT: **59, 60, 17**

Eagle Island is a lonely 17-acre outpost at the eastern end of Broad Sound—one of the outermost of the Casco Bay islands. As a boy, the solitary but ambitious Robert Peary fell in love with Eagle Island and eventually, in 1881, he bought it for $500. He built a house like a ship, with a deck all around, and summered here when he wasn't away exploring. His sled dogs summered on Upper Flag Island to the north.

Eagle Island and Peary's house are now a state park open to the public. The island is high with spectacular views of Casco Bay in every direction. A beautiful trail circles around the perimeter, and the ship-house is full of the admiral's furnishings and arctic artifacts. Visit on weekdays to avoid crowds.

APPROACHES. From seaward, an obvious landmark is Little Mark Island, to the east of Eagle. A tall monument, which now has a flashing light, was built in 1809 as a memorial to shipwrecked seamen. Its base once stored water and provisions for marooned sailors.

The dock at Eagle Island is near the northwest tip. Nun "2" marks the end of a long ledge from the west side of Eagle Island. It marks the mouth of Broad Sound, though, *not* the approach to the moorings off Eagle. Be sure to keep to the south or west of the nun, and don't cut around it to head directly toward the moorings and the dock. Instead, approach only from the northwest.

ANCHORAGES, MOORINGS. Several moorings are available, but you may have to anchor in 16 to 30 feet of water.

The anchorage is exposed, and the current in Broad Sound runs swiftly. If you have doubts, leave someone aboard while you explore the island, or visit Eagle via one of the tour boats that leave from Portland or South Freeport.

THINGS TO DO. The resident caretaker will greet you and collect a modest fee for your visit.

Admiral Peary announced his discovery of the North Pole in 1909. Another explorer, however, had made the same claim five days earlier. To determine who deserved the honor of being the first to reach the Pole, the documentation of Peary's expedition was subjected to intense scrutiny. The other claim turned out to be a hoax, but enough discrepancies and inconsistencies were unearthed in Peary's accounts that his achievement was mired in doubt.

Finally, in 1911, Congress officially named him the discoverer and awarded him a pension. Peary used the money to help complete the cottage on Eagle Island, where he spent the last nine years of his life well away from the critical public.

The question of whether Peary actually reached the North Pole continues to fascinate people, perhaps because the issues are as much those of character and obsession as they are of fact and scientific proof. For more arctic history, visit the Peary-MacMillan Arctic Museum located on the first floor of Hubbard Hall on the campus of Bowdoin College in Brunswick.

MASTS AND THE BROAD ARROW

Not more than one man in a thousand who looked at a ship of the line reflected that her great mainmast had been cut in the forests of Maine... Robert Albion, *Forests and Sea Power*, 1926

When English explorer George Waymouth brought the *Archangel* into Pentecost Harbor, near the St. George River in 1605, he found "notable high timber trees, masts for ships of four hundred tons." By 1609, the Jamestown Colony was sending home the first shipment of masts from the New World.

Softwoods such as firs and pines are superior to hardwoods for use as masts since they are far more flexible and carry less weight aloft. England's forests were cut for firewood in the Middle Ages, and by the 17th century, the closest supply for mast timbers was in the Baltic. England competed with the French, Spanish, and Dutch for the great Baltic firs, prized by the British Admiralty for their resilience and durability.

But the Baltic was 1,000 miles away and not under England's control, and the route could be blocked easily at the straits of Denmark. New sources were needed.

By about 1650, the colonists in America had established a flourishing trade in masts, lumber, and other naval stores to Europe and the Caribbean. They were not pleased when the British Admiralty awoke to the fact that its supply of mast trees in North America was in danger. In 1685, a Surveyor of Pines and Timber was appointed to survey the Maine woods "within 10 miles of any navigable waterway" and mark all suitable trees with "the king's broad arrow," the symbol used since early times to designate Royal Navy property. Any "trees of the diameter of twenty-four inches and upwards at twelve inches from the ground" with a yard of height for each inch of diameter at the butt was blazed with the broad arrow. Woe to anyone who damaged or stole the king's property. The fine was £100! The Broad Arrow Policy was observed with all the enthusiasm that greeted Prohibition more than two centuries later, and the same native ingenuity was applied to circumvent it.

Mast trees were partially burned in mysterious fires or splintered in unusual gales. Loopholes apparently excluded certain properties, whose great pines were promptly felled and sawn into profitable lumber. And never, under any circumstances, would the floorboards of any colonial home exceed 23 inches in width. The law was tightened again and again, but still the king's pines continued to disappear.

Mast trees usually were cut in the fall, when they were full of resin. The trees were carefully felled on prepared beds, limbed, and squared. During the winter, the rough timbers, or baulks, were dragged by brute strength onto sleds and hauled out of the woods by teams of oxen. Since one great tree could weigh as much as 18 tons, this was a difficult and dangerous process, requiring great skill and the efforts of as many as 100 oxen.

Where mast roads met, communities sprang up, and often the shape of the town square was determined by the clearance needed to turn the long timbers. An example of this remains in Freeport, opposite the L. L. Bean store, where the former Bliss-Holbrook Tavern (now stores) was sited at a strange angle for that very reason.

Every mast road led to the nearest navigable waters, often a tidal marsh. "Mast Landing" is a title that still appears on charts for many locations along the Maine coast. Here the baulks were delivered to a mast agent, slipped into the water, and towed to a mast depot.

At the mast depot, the baulks were graded and hewn to the specified 16 sides. Then they were loaded aboard special mast ships through large stern ports for transportation to England, where the final trimming and fitting of the mast was done. Some of these mast ships were of 400 to 600 tons burden and could carry 30 to 50 of the enormous baulks below decks.

Although the Broad Arrow Policy could never be effectively enforced and the colonists continued to cut mast trees for sale on the black market, the supply of great pines was sufficient to provide a crucial resource for the Royal Navy for 125 years, until the monopoly was finally ended by the American Revolution.

ROYAL RIVER

★★★

43° 47.79′N 070° 10.45′W

CHARTS: **13290, 13288**   CHART KIT: **59, 17**

The Royal River is a narrow, winding stream that is navigable from Casco Bay to a fixed bridge and falls below the town of Yarmouth. Its marshy banks are reminiscent of *Wind in the Willows*, or it conjures up images of Indians silently slipping downstream in their birchbark canoes.

Above the marshes, high, forested banks almost completely block winds from any direction, making it one of the best protected harbors in Casco Bay.

Following the river's meandering channel, however, can be a challenge for deep-draft boats. At worst, you'll ground out on a soft mudbank.

Although surrounded by busy boatyards and civilization, the basin at the head of the river attracts its share of wildlife. You might see herons at dusk or perhaps a flock of ducks flying upstream in the early morning.

APPROACHES. Royal River was last dredged to about 8 feet at low. It has since silted in somewhat, and the channel, despite excellent buoyage, can still be hard to find. In theory, a 6-foot draft should be able to make it upriver at low, but it is safest to head up the river on the flood at mid-tide or better.

Approach between Little Moshier and Cousins Island and keep flashing green "1" to port. From there, the channel is carefully marked by red and green buoys. The standard nuns and cans are supplemented with the town's smaller green and red cones. The channel sometimes is no wider than 50 feet, and it wanders in surprising places. Where the Cousins River joins from the north, the channel takes a 100-degree turn to port. Be particularly careful between Browns Point and Callen Point around nun "12." A ledge on the very edge of the channel to the southeast was marked for several years with a floating sign which read "LEDGE 1 FOOT UNDER." It now has a real buoy, can "11." Keep close to the red buoys and go slowly.

ANCHORAGES, MOORINGS. As you approach the basin at the head of the navigable portion of the river, you will pass three yards. The first, on the right, is Royal River Boat Yard, which maintains floats and a fuel dock parallel to the channel, with 7 feet at low, and 80 slips.

Yankee Marina is on the left, where the channel broadens into a turning basin. They have 110 slips and deep water alongside. Yarmouth Boat Yard is the last yard on the left, with 150 slips and 6 to 8 feet alongside. Go slowly. Boats jam-pack the tiny basin, and there is no room to anchor.

By reservation, Yankee rents transient slips with low-tide depths to 12 feet. Yarmouth Boat Yard has dock space with 6-foot depths and one mooring that will be rented if available.

FOR THE BOAT. *Royal River Boat Yard (207-846-9577).* This yard has been owned and managed by three generations of the Dugas family, whose grandfather started a small operation here many years ago. With a 50-ton boatlift and a 100-ton marine railway, the yard can make extensive repairs on both commercial and pleasure vessels in steel, fiberglass, or wood. Gas, diesel, water, ice, electricity, and holding-tank pumps are available at their fuel dock, the only one along the river.

Yankee Marina, Inc. (Ch. 09; 207-846-4326; www.yankeemarina.com). Water, electricity, and pump-out facilities are available at the slips, but no fuel. Yankee can haul boats of most sizes and perform repairs of all kinds. The yard dredges alongside to 12 feet, and if you get stuck in the river, Yankee can tow you off. "We do a lot of that," they say.

Yarmouth Boat Yard, Inc. (Ch. 09; 207-846-9050; www.yarmouthboatyard.com). This yard specializes in smaller powerboats and outboards: Honda, Suzuki, Mercury, and Yamaha. Water and electricity are on the floats, and they have a marine store with an extensive parts department. The yard has a 4-ton forklift for small boats and they can repair your engine or outboard or fix your wood or fiberglass hull.

Bayview Rigging and Sails, Inc. (207-846-8877; www.bayviewsails.com). Bayview's sail and rigging loft is located at the Royal River Boat Yard and headed by well-known sailmaker Jan Pedersen.

Landing Boat Supply (207-846-3777) is a complete chandlery located in the Lower Falls Landing building (see below).

Maine Sailing Partners (207-846-6400; www.mesailing.com). This full-service sail loft has moved inland, but they are still in Yarmouth. They can patch up your old sail or cut you a new one.

Reed's Machine Shop (207-846-5681). Reed's is located just past the highway bridge toward the town of Yarmouth. They can weld, repair, fabricate, or otherwise perform miracles with metal.

FOR THE CREW. The former fish factory between Yankee Marina and Yarmouth Boat Yard has found new life as Lower Falls Landing, a handsome complex of marine-related facilities. Landing Boat Supply is a well-stocked chandlery, and East Coast Yacht Sales' indoor showroom is the perfect place for dreaming on a rainy day. The Royal River Grillhouse (*846-1226*) serves lunch, dinner, and Sunday brunch on the waterfront.

You can obtain groceries and other supplies in the town of Yarmouth, about a mile away from the harbor. Head under the highway bridge and take Route 115 into the pretty town. A Hannaford supermarket, a Rite Aid pharmacy, and a hardware store are on Route 1.

THINGS TO DO. Every July, on the third weekend, Yarmouth holds its annual Clam Festival with three days of music, festivities, and feasting.

The Cousins River (see next entry) is wonderful to explore by dinghy or kayak.

COUSINS RIVER

43° 47.99'N   070° 09.16'W

CHARTS: 13290, 13288   CHART KIT: 59, 17

The Cousins River branches off the Royal River. It is unmarked and has next to no water at low tide. Don't be tempted to tuck in here for the night if you've got a keel. It would make a delightful exploration by fast dinghy from the basin at the head of the Royal River, with two intriguing boatbuilders and good eats along the marsh.

FOR THE BOAT. *Greene Marine (207-846-3184).* Greene Marine is located on the left bank, surrounded by gossamer trimarans. Walter Greene and his crew specialize in building and repairing world-class multihulls in composite-wood construction.

Even Keel Marine (207-846-4878; www.lowell-brothers.com) is next to Walter Greene's. Owners Jamie and Joe Lowell are direct descendants of Will Frost, the pioneer of the modern lobsterboat hull. They build custom boats or finish out new downeast hulls to your specs.

FOR THE CREW. The Muddy Rudder Restaurant (*846-3082*) is farther up the river on the left. They have a float and an extensive menu. To eat in the rough or to buy seafood, continue to Day's Crabmeat and Lobster, hard by the bridge where Route 1 crosses the river.

HARRASEEKET RIVER
(South Freeport)

4 ★★★★★

43° 49.18'N   070° 06.26'W

CHARTS: 13290, 13288   CHART KIT: 59, 17

The tidal Harraseeket River creates a splendid, well-protected harbor at the north corner of Casco Bay, in the small village of South Freeport. This major yachting center has almost every facility, yet it remains simple and attractive. The fabled L.L.Bean store and an abundance of outlet shopping is close by in Freeport.

Although a bit out of the way for the cruising boats heading east or west along the coast, the Harraseeket is a good place to change crew, find a diesel mechanic, or make some short excursions.

This area has a long shipbuilding history, one that parallels the history of the whole coast of Maine. Boats have been built on the Harraseeket River for two-and-a-half centuries, ever since the first settlers cut pine in the virgin forests to build England's Royal Navy. One of the most famous, the *Dash*, built at the Porter Yard in 1813 at the beginning of the war of 1812, was immortalized in Whittier's poem, *The Dead Ship of Harpswell*. Ships of 300 to 400 tons were built at what is now South Freeport Marine, and forty percent of the men in Freeport were employed by the yard. Since then, this location has seen boatyards come and go, but some of the pilings and stone abutments at South Freeport Marine still date from the mid-19th century.

The papers which separated the State of Maine from the Commonwealth of Massachusetts and admitted Maine to the Union were signed in 1820 at the Jameson Tavern, which still stands next to L.L.Bean.

APPROACHES. The entrance to the Harraseeket River is narrow. It bends sharply right, the current is usually strong, and the inadequate scale of the chart doesn't inspire much confidence. In thick weather, this can be a tricky entrance for a stranger.

The approach starts at Little Bustins Island. Pick up can "1" and steer between Crab Island and nun "2" off Googins Ledge. Note the line of mudflats on the left, from Moshier Island to the entrance. Crab Island is almost on the edge of these flats, so leave it well to port. There are two good landmarks for the entrance itself—tiny Pound of Tea Island with its single house and a few trees (yes, that was the

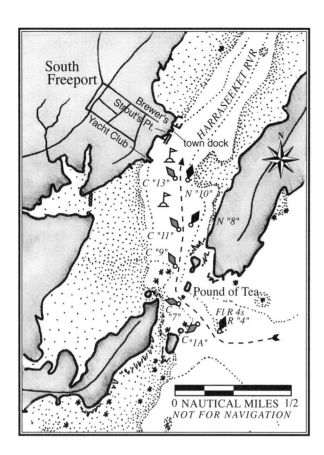

landing are next, recognizable by the red buildings of Harraseeket Lunch and Lobster nearby. The slips and fuel docks of Brewer's South Freeport Marine are the farthest to the north. Both Strouts Point and Brewer's South Freeport Marine have rental moorings in this crowded harbor, or they can put you on their docks. There is no room to anchor.

It doesn't really matter where you are in the harbor, and when it is busy, you probably won't have much choice. The protection is best and there is the least wake north of the docks, but on a mooring near the entrance you'll have a good view of The Castle, hear the cows mooing on Wolf Neck, and see a constant and ever-changing parade of interesting boats, splashing fish, and occasional seals.

GETTING ASHORE. Dinghy ashore to any of the floats, or hail the South Freeport Marine launch.

FOR THE BOAT. *Harraseeket Yacht Club (207-865-4949; www.hyc.cc).* This is an informal and friendly place whose burgee carries a drawing of The Castle. To assist visiting yachtsmen, members who are off cruising post notices on the bulletin board giving the locations of their moorings. There is no launch and no one visibly in charge, so ask a sailing instructor or a member for help or advice. Water, but not fuel, is available at the floats, with 10 feet alongside.

Strouts Point Wharf Company (Ch. 09, 16; 207-865-3899; www.stroutspoint.com). This marina has moorings and slips with 8 feet alongside at low. Gas, diesel, water, electricity, and holding tank pump-outs are available as well as ice and showers ashore. Their marine store is in the wonderfully-designed little building by the docks. Boats are hauled with a 25-ton boatlift, and repairs of all kinds can be performed.

Town Landing (Harbormaster Jay Pinkham: Ch. 09; 207-865-4546). Thirty-minute docking is allowed on the town floats, hard by the harbormaster's office.

Brewer's South Freeport Marine (Ch. 09, 16; 207-865-3181; www.byy.com/bsf). Transient slips, with depths to 12 feet, and moorings are available by reservation. Brewer's pumps gas, diesel, and water and provides launch and pump-out service, electricity, ice, showers, and a laundry machine and dryer. They haul with a 25-ton boatlift and can handle boat repairs of all kinds.

North Sails Maine (207-865-2100). This loft is located at Brewer's. They can fit you with a new suit or keep your old set pulling.

original purchase price from the Indians), and The Castle, which stands high on the shore in South Freeport and shows as a tower on the chart.

Leave flashing red buoy "4" to starboard and hug the left side of the channel, particularly at can "7." The rocks making out southward from Pound of Tea are covered with bottom paint. After leaving can "7" close aboard to port, curve gradually to starboard around Pound of Tea, leaving can "9" also to port. Local boats take great joy in passing inside Pound of Tea Island, especially at high tide. Resist the temptation.

A deep and fairly narrow channel leads through several more town-maintained nun-and-can pairs to the marinas. Do not stray from this channel, and do *not* go where there are no moored boats. At high tide, the Harraseeket River looks like an enormous quiet lake; at low, it looks like clam flats.

ANCHORAGES, MOORINGS. The first dock on the left belongs to the Harraseeket Yacht Club, which has one guest mooring for boats to 30 feet from reciprocal clubs. There is a gap after the yacht club, and then the remaining facilities and docks are all cheek-by-jowl. First are the floats and fuel dock of Strouts Point Wharf. The town docks and public

Cumberland Marine Electronics (207-865-3395; www.geocities.com/cmelectron) can help you with repairs, sales, and new installations.

Falls Point Marine (207-865-4567; www.fallspoint.com) is located upriver at Porters Landing. Carter Becker is an ace rigger with a boat-based mast crane that can come to your boat. He also offers dive services and custom fabrications.

Maine Compass Services (207-865-6645). Charlie Cook is southern Maine's premier compass adjuster. He can adjust, refill, or rejewel your compass or just replace those little red bulbs.

FOR THE CREW. Harraseeket Lunch and Lobster (*865-4888*) provides takeout, restaurant service, and live lobsters in the red building across the street from Brewer's, but you'll need to BYOB.

A ten-minute walk straight up the hill will bring you to the Village Store, which sells sandwiches, soda, beer, wine, and milk. For a more extensive grocery and liquor store, you might be able to get a ride to the Bow Street Market (*79 Bow St.; 865-6631*) or Shaw's (*200 Lower Main St.; 865-0094*) about three miles away in Freeport. The Freeport Taxi is at *865-9494*.

THINGS TO DO. If you have been repressing your consumer urges all this time on the boat, don't miss L.L.Bean, open 24 hours a day in Freeport. Attracted by the store's popularity, more than 80 other stores, factory outlets, and boutiques now cluster in the town.

The Castle is now private, with a new house built on the property, but wonderful photos of The Castle and early South Freeport can be seen at the small post office next to the Village Store. Casco Castle was a large Victorian hotel built around the turn of the century by Amos Gerald, the "Electric Railway King." Gerald had built an electric trolley which ran from Brunswick to Yarmouth and brought guests to the hotel. The trolley was completed in 1904, but the hotel burned in 1914, leaving only the stone tower. The trolley was discontinued in 1927.

For a much longer trip, take your dinghy up the estuary at the northeast end of the harbor to Mast Landing, named for the enormous pines felled and shipped to England for masts. Just beyond is Mast Landing Sanctuary, operated by Maine Audubon, with trails and nature exhibits.

Or explore the trails of Wolfe's Neck Woods State Park on the eastern side of the river. To get there, take your dinghy ashore at one of the two little indentations shown on the chart to the right of the word "Harraseeket."

MEREPOINT BAY

43° 49.78'N   070° 00.45'W

CHARTS: 13290, 13288   **CHART KIT: 59, 17**

Merepoint and Maquoit are shallow bays on either side of Merepoint Neck, only about three miles away from the town of Brunswick. This corner of Casco Bay feels remote and peaceful. As with many of Maine's long, shallow fingers that stretch inland, the water here is often comfortable for swimming.

APPROACHES. Beyond Bustins Island, a row of lovely little wooded islands—Sow and Pigs, Pettingill, Williams, and Sister—line the route to Merepoint Bay. Except for Bustins Ledge, which is visible, and a rock just south of Sister Island, there are no unmarked dangers along the way.

As you enter Merepoint Bay, the flagpole and clubhouse of the Mere Point Yacht Club will be to port near the tip of the peninsula. The club is a small one, however, and offers no facilities for the visiting yacht.

ANCHORAGES, MOORINGS. A substantial fleet of boats is moored up the bay, opposite the northern end of Birch Island and off Paul's Marina, on the west shore. The bay shoals rapidly at this point, so stay a good distance off Merepoint Neck. The outer moorings are in deep water. You can anchor, but the bottom may have some grass. Alternatively, Paul's has rental moorings.

The harbor is exposed to the prevailing southwesterlies but fine in settled summer weather. From any other direction, it is well protected.

FOR THE BOAT. *Paul's Marina (Ch. 09; 207-729-3067; www.paulsmarina.com)* has gas, water and pump-out facilities, but their floats have only a couple of feet of water at low. Their small general store carries basic marine supplies, ice, and ice cream.

THE GOSLINGS

43° 47.29'N   070° 02.43'W

CHARTS: **13290, 13288**   CHART KIT: **59, 17**

The Goslings are two small, privately owned islands south of Upper and Lower Goose Island. The anchorage just north of the Goslings is one of the finest in Casco Bay. With the wind anywhere in the south, you can lie here with a surprising degree of protection and choose from several islands to explore.

On a weekday, you might be alone here, but on summer weekends, the Goslings take on the air of a small, festive town with the constant arrival and departure of picnicking groups.

APPROACHES. Run halfway between Irony Island and Grassy Ledge.

ANCHORAGES, MOORINGS. Private moorings have proliferated here, to the point where anchoring boats are forced into less protected waters. When one yachtsman tried to anchor, he became so upset with the number of moorings he decided to do something about it—and he could. It turned out he was the head of the New England Division of the Army Corps of Engineers, which shares the permitting of moorings with local jurisdictions. Moorings without permits were removed, and new moorings are no longer allowed.

To anchor, find a spot in 12 to 16 feet of water, north of the eastern Gosling or a bit farther west. The heavy mud on the bottom holds well.

GETTING ASHORE. Land the dinghy on either of the Goslings to explore.

THINGS TO DO. There is a picnic table on the eastern Gosling and a lovely little shell beach. The water is clear, and the swimming is excellent. At low tide, the reason for the anchorage's protection becomes evident as you watch people wading across the sandbar between the two Goslings. The owners request that there be no fires except on the beach and that no wood be cut for the fires.

When the tide is right, you can watch the seals sunbathe and belch on the northern tip of tiny Irony Island. By July, seal pups have joined the entourage and are often basking on the rocks with their mothers. A bronze plaque on Irony memorializes a captain of the Portland Pilot Boat who was lost off his cutter near this island in 1962.

UPPER GOOSE ISLAND

north cove: 43° 48.67'N   070° 01.66'W
west beach: 43° 48.47'N   070° 02.04'W

CHARTS: **13290, 13288**   CHART KIT: **59, 17**

You can't have goslings without geese. Upper Goose Island is a 94-acre wooded island to the north of the Goslings on the western side of Middle Bay. The Nature Conservancy owns most of the island, which was once the largest great blue heron rookery in New England. At one time the Nature Conservancy counted more than 250 nests in the island's hemlock, beech, and yellow birch trees. Some trees held as many as ten nests. Since then, for unknown reasons, most of the herons have moved away to Little Whaleboat Island and others.

The Nature Conservancy has granted conservation easements to the U.S. Department of Fish and Wildlife, which asks you to refrain from visiting the interior of the island from early April to mid-August to avoid disturbing nesting sea birds.

To explore the shoreline's rocky ledges, gravel beaches, and salt marshes, land at the large gravel beach on the western shore or in the north cove. Osprey, eiders, and other sea ducks are common sights, as are harbor seals on the southeast ledges.

POTTS HARBOR

43° 44.32'N   070° 02.35'W

CHARTS: **13290, 13288**   CHART KIT: **59, 17**

Potts Harbor is a great, wide-open bay at the southern end of Harpswell Neck. At first glance, it appears exposed to prevailing winds. Inside, however, the surrounding islands and ledges provide reasonable protection all the way around. Easy to enter, with adequate facilities, Potts is a good stop for yachts traveling east or west or exploring Casco Bay.

APPROACHES. From outside Casco Bay, run from Halfway Rock, with its 76-foot lighthouse, to red-and-white bell "BS" (*43° 41.72'N   070° 03.44'W*) at the entrance to Broad Sound. In Broad Sound, expect a considerable current on the ebb or flood. Pass between green gong "1" and nun "2" southwest of Eagle Island.

The working harbor of Mackerel Cove.

Round to starboard between red nun "4," on the ledge west of Upper Flag Island, and red bell "6" off Little Birch Island. From there, the entrance between Horse Island and Thrumcap into Potts Harbor is wide and easy.

From the east, there is a well-marked, if circuitous, channel into Potts which should be attempted only in good visibility. Start with nun "2" north of Haskell Island, which should be left to starboard, and follow the channel buoys—reds to starboard and greens to port—around the hairpin to the south and then north and west again into Potts.

ANCHORAGES, MOORINGS. The charts show the anchorage inside the southwest tip of Harpswell Neck. This is the location of Dolphin Marine where you can obtain a ⚓ mooring for the night or lie alongside the ⛁ floats in depths of 6 feet at low.

If you prefer to anchor, the mud bottom is shallow, and there is plenty of swinging room near the moored boats.

Potts Harbor is almost a mile across, so it may not be a quiet anchorage in heavy weather. Quite a chop can build up. In a blow from the north or east, you might consider anchoring east of Ash Point.

If a hurricane threatens, consider Basin Cove which is virtually landlocked, but definitely don't attempt the entrance in anything but your dinghy without local knowledge.

GETTING ASHORE. Row your dinghy in to the floats at Dolphin Marine.

FOR THE BOAT. *Dolphin Marine Services, Inc. (207-833-5343; www.dolphinchowderhouse.com).* This laid-back, locally-oriented marina ⛽ pumps gas and diesel in and 🚽 holding tanks out. Water is unavailable, however, because it is drawn from a well. Ice and a pay phone are ashore. Dolphin hauls boats to 40 feet with their hydraulic trailer and ramp, and a mechanic is on hand.

Finestkind Boat Yard (207-833-6885; www.finestkindboatyard.com) is a full-service yard located inland, just east of Basin Cove. They haul with a 25-ton hydraulic trailer and can perform repairs of all kinds.

FOR THE CREW. Whether you are nestled in a booth or seated at the counter, 🍴 the Dolphin Chowderhouse (*833-6000*) at the marina has a wonderful warm, steamy atmosphere on a blustery gray day. They serve lunch and dinner. In August, the nearby field is loaded with succulent blueberries.

THINGS TO DO. At the end of the narrow slot leading northeast from Potts Harbor is a large lobster dock, dredged to 10 feet some years ago. Row west through the narrow spot across the remnants of a rock dam and into Basin Cove, a primordial and magical place. For a long way, the cove's western shore is privately owned and a sanctuary for birds. You may see great blue herons roosting high in the trees, or night herons, little green herons, or snowy egrets.

Make this trip at slack water. The tide pours hard in and out of Basin Cove. And be sure to row; an outboard will frighten the birds.

MERRICONEAG and HARPSWELL SOUND

Merriconeag Sound is a long finger of water tending northeast between Harpswell Neck and Bailey and Orrs Island. Mackerel Cove, at its mouth, is a quintessential fishing harbor. Halfway up the sound, Wills Gut cuts between Orrs and Bailey Island beneath a unique cribstone bridge. Harpswell Sound is to the north, and Long Cove, near the head of navigation, is tricky to enter but offers near-perfect protection.

Prevailing summer southwesterlies tend to funnel up the Sound, so you will have a glorious run in, dodging lobster buoys as you fly.

MACKEREL COVE

3 ★★

43° 43.67′N   069° 59.97′W

CHARTS: 13290, 13288   CHART KIT: 60, 17

Mackerel Cove is a very deep harbor on the south end of Bailey Island, jammed with lobsterboats and tuna boats. The shores are lined with pine-topped cliffs and docks piled high with lobster traps.

This harbor is difficult to visit by boat, and most of the fishermen like it that way. There is hardly any room, no services, and little concern for the needs of visiting yachtsmen. Still, for the intrepid, Mackerel Cove's very authenticity makes it worth a visit in settled weather.

The high shores offer excellent protection from all directions except from the southwest, which is wide open to winds and ocean swells which funnel into the cove.

APPROACHES. Mackerel Cove lies conveniently at the mouth of Merriconeag Sound. The entrance is fairly easy even at night or in poor visibility.

From offshore, the southern end of Bailey Island is clearly identifiable by two concrete observation towers on the high shore. Find the flashing red gong "8" at Turnip Ledge and keep it to starboard. Proceed northeast, straight into the cove, with flashing green "1" off Abner Point to port.

The reported rock shown on the chart at the head of the harbor "categorically does not exist," states sailor John Haug, who has kept his boat moored at that exact spot and has sailed out of Mackerel Cove for over 60 years. The rock never appeared on earlier charts, and he has tried for years to have the notation removed. "Apparently," he says, "it is much easier to put something on the chart than to take it off."

ANCHORAGES, MOORINGS. Mackerel Cove has no facilities for docking or mooring and very little room for anchoring. Pick settled weather and drop the hook out beyond the moorings. The bottom is good mud, but you'll be quite exposed, and you'll need a lot of scope to make up for 50 feet of depth.

If in doubt, there are several alternative anchorages nearby. Do not pick up a fisherman's mooring without permission, or you might discover just how authentic a working harbor can be.

GETTING ASHORE. A crescent of sand rings the head of the harbor. Land at the west end, by the town launching ramp.

FOR THE CREW. A fifteen-minute walk up the road will bring you to the post office and Bailey Island Market (*833-2800*), which has a good selection of groceries, beer, ice cream, and a small cafe. The local hangout for breakfast is next door at Jack Baker's Last Stand.

WILLS GUT (west)

43° 45.15′N   069° 59.30′W

CHARTS: 13290, 13288   CHART KIT: 60, 17

Wills Gut is spanned by a unique cribstone bridge. Several marine-related businesses lie in a small indent to the west of it. The Orrs-Bailey Yacht Club is a small, friendly, long-established place on the southwestern end of Orrs Island. A substantial fleet of boats is moored off its floats. Cook's Lobster House is a commercial lobster pound, a restaurant, and the farthest stop on the Casco Bay Lines ferry's "mailboat run." The small peninsula it occupies at the northwest tip of Bailey Island cradles a small cove full of working boats.

APPROACHES. See the sketch chart on page 90. You must go north of the red daybeacon on Cox Ledge and north of nun "12" at the entrance before you can swing south into the harbor.

In the inner harbor, be particularly careful not to hit the ledge shown on the chart to the west of the north end of the bridge. According to the locals, it's a bottom magnet.

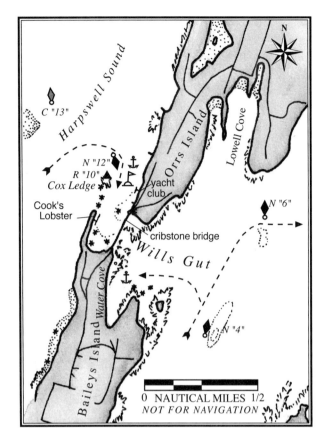

FOR THE BOAT. *Orrs-Bailey Yacht Club (www.obyc.org)* has 15 feet of water alongside their fuel dock, where they pump water and gas but no diesel. The fuel is for club members only, but they might be able to help you in an emergency.

Cook's Lobster House (207-833-2818; www.cookslobster.com) sells gas, diesel, ice, and lobster. They also rent dock space for the night.

FOR THE CREW. Pay phones are available either at the club or at Cook's.

Great seafood can be found at Cook's or at Jack's Bailey Island Grille (*833-0640*) in the gray building at the south end of the cribstone bridge. Jack Baker began his restaurant career by cooking lobsters in an old washing machine.

For more casual fare, the Salt Cod Café is in the red building at the north end of the bridge. In the morning they serve steaming coffee and muffins at the tables on the wharf and swap that for steaming chowder at lunch.

THINGS TO DO. Be sure to take a walk over the famous Orrs Island-Bailey Island cribstone bridge. Or take your dinghy under it. The cribstone bridge was built in 1927 with 10,000 tons of split Maine granite laid in an open cribwork held in place only by gravity. Its graceful span is engineered to allow the free flow of tidal currents and to withstand the corrosive effects of salt and the abrasion of ice. As the only bridge of this type in the world, it is listed in the National Register of Historic Places and is a National Historic Civil Engineering Landmark.

H_2Outfitters (*833-5257*) in the red building north of the bridge offers sea kayak tours and instruction. Nature tours of Bailey Island leave from Cook's wharf aboard the Casco Bay Lines ferry. Or, for a full day, you can take the mailboat run all the way to Portland and back.

For a nice walk, especially at sunset, head north from the bridge and take the first major left toward the peninsula marked "Beals Cove" on the chart. Here you'll find a small park with the Giant Steps, rock formations stepping down to the water.

Cook's hosts the annual Casco Bay Tuna Tournament during the last full week in July. The largest tuna ever caught by the Casco Bay Tuna Club weighed 1,009 pounds. Some tuna is eaten here, but most of it is put on ice and flown directly to Japan, where it is highly prized and dearly paid for as sushi.

ANCHORAGES, MOORINGS. The Orrs-Bailey Yacht Club rents four guest moorings with a two-night limit. Inquire at the club dock, or, if no one is around, take the mooring and drop $20 in the box on the dock. Three are marked "OBYC Guest" and can handle boats to 35 feet. A fourth mooring can handle boats to 48 feet. Its mooring ball is located in the outer row and has a protuding stick. You will need to rig your own pennant.

Be aware that the red-and-yellow ferry from Portland comes right past the club's moorings on its way to Cook's Wharf, so there is no anchoring room.

Cook's has a long line of floats running north and south with 10 to 12 feet of water at low. The floats are in constant use by lobstermen, who congregate toward the south. Space along the middle is intended for visitors and restaurant patrons and can be rented for the night. Keep the northern end free for the Casco Bay Lines ferry from Portland.

The harbor is well protected from the south, but according to the harbormaster, it "can be clobbered with a brisk wind across Harpswell Sound from the northwest."

GETTING ASHORE. Land at the club floats, at Cook's, or by the northern end of the bridge.

HARPSWELL HARBOR

3 ★★

43° 45.57′N   070° 00.24′W

CHARTS: 13290, 13288   CHART KIT: 60, 17

Harpswell Harbor is a big, wide-open bay on Harpswell Neck at the entrance to Harpswell Sound. It is easy to enter and well protected from prevailing winds, so even though this is not the area's most scenic anchorage, it may come in handy.

APPROACHES. Entering the harbor, leave can "13" to port and take a wide swing to the left and into the harbor to avoid the rock and the 1-foot spot off the elbow of Stover Point. One way to avoid the hazard is to leave most of the lobster buoys to port.

ANCHORAGES, MOORINGS. The Merriconeag Yachting Association is based here, and they graciously maintain a guest mooring, marked "MYA," for visiting yachtsmen. Or you can anchor in about 12 feet of water at low, just outside the lobster-boat moorings, where the holding ground is good. You will be well protected here from the southwest wind, without a hint of roll.

GETTING ASHORE. Land at the road where the chart says "West Harpswell." Don't make the mistake of going ashore (or returning) at low tide, or you will be squelching through clam flats.

THINGS TO DO. The road leading up the hill is pretty and bucolic, but there are no stores nearby.

MILL COVE

43° 48.01′N   069° 58.20′W

CHARTS: 13290, 13288   CHART KIT: 60, 17

This shallow cove near the head of Harpswell Sound stretches north on the west side of High Head peninsula. It is shoal and contains several houses, but you might want to explore it as a gunkholing adventure.

Head for the tip of land between Mill Cove and Widgeon Cove, then enter Mill Cove just east of this tip. Even at dead low, there is as much as 14 feet of water here. But if you go much farther in, your keel will be digging for clams.

HIGH HEAD YACHT CLUB

2 ★★

43° 48.03′N   069° 57.53′W

CHARTS: 13290, 13288   CHART KIT: 60, 17

High Head Yacht Club is a pleasant little club six miles up Harpswell Sound on the west side, just north of High Head.

APPROACHES. Coming up Harpswell Sound, aim for a modern white house on the southern tip of High Head to keep well off Dipper Cove Ledges, which make well out from the eastern shore. Opposite Wyer Island, coast around the high, wooded section of High Head. You will first see the moored boats, then the yacht club. The fixed bridge at Ewin Narrows is visible to the north.

ANCHORAGES, MOORINGS. The yacht club maintains a guest mooring, or a member might be able to steer you to a vacant one. There is also plenty of room to anchor among the moored boats in 8 to 10 feet of water at low, but there is plenty of exposure too, both from the northeast and south.

GETTING ASHORE. Land at the yacht club float.

FOR THE BOAT. *High Head Yacht Club.* This is a small club comprised primarily of members who live on High Head Road, but it welcomes visitors. Water and gas are available on an emergency basis only at the club's floats (10 feet alongside at low).

FOR THE CREW. There are showers in the clubhouse, but no supplies nearby.

LONG COVE

■ ★★★

43° 47.46′N   069° 56.98′W

CHARTS: 13290, 13288   CHART KIT: 60, 17

Long Cove is an intriguing slot that bisects the northern part of Orrs Island. It is out of the way and seldom used by yachts. The high, tree-lined shores of spruce, cedar, and oak hardly show a sign of habitation. The only drawback is the whir of cars on the Harpswell Islands Road.

Long Cove is a bit tricky to enter. The entrance is flanked by unmarked ledges, so don't try it in poor visibility. Study the chart carefully and eyeball your way in, preferably near low tide when the ledges are uncovered. Once in, you will be extremely snug.

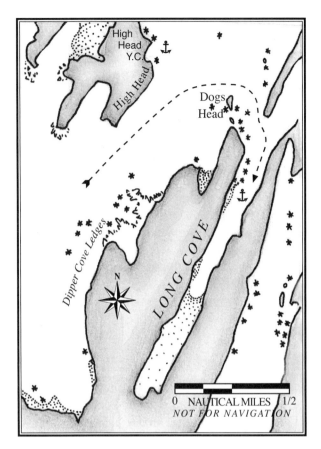

EASTERN CASCO BAY

East of Bailey Island, Casco Bay is as rugged and beautiful as any spot on the coast of Maine. It is defined by a series of ledge ridges, extensions of the southwest-northeast spines of the bay's islands and mainland. Between the ridges, deep fingers reach far inland, creating some of the most protected anchorages anywhere. Quahog Bay stretches so far to the north that it nearly splits Sebascodegan, Casco Bay's largest island.

Offshore, the ledge ridges form strings of rocks and islands—beautiful and treacherous pearls. Deepwater passages weave around them, but only a few of them are marked. Strangers beware; this is an easy place to become confused.

Up the fingers, spruces line the shore, snowy egrets prod the flats or lift into flight, bemused seals investigate passing boats, and osprey nests command almost every point.

In some years, feeding bluefish herd schools of menhaden, known here as pogies, into the upper reaches of Quahog Bay and the New Meadows River with such ruthlessness and in such vast numbers that in their panic the pogies use up all the oxygen in the water, and they suffocate and rot *en masse*. On a hot summer day, the resulting smell is nothing short of suffocating for humans as well. Mysteriously, some years there aren't any pogies at all.

PASSAGES through Eastern Casco Bay

Eastern Casco Bay is characterized by an abundance of ledges and rocky islands, matched by a scarcity of buoys. If you are unfamiliar with these waters, enter only in good visibility with the chart in hand. Pick a course that puts the least emphasis on identifying unmarked ledges (which may or may not break) and makes the best use of easily recognizable islands and the few prominent buoys.

Relatively straight slots of deep water—and the ledges between them—tend in a southwest to northeast direction. Prevailing southwest winds are usually astern or on the nose, and strong currents, particularly off the New Meadows River, add to the ride.

Only one route is marked across the grain of the ledges, from Orrs and Bailey Island across the mouth of Quahog Bay, and over to the New Meadows River.

APPROACHES. Make your way up Harpswell Sound, favoring the west side to avoid Dipper Cove Ledges and aiming for High Head.

At High Head, turn east, leaving to starboard tiny Dogs Head Island (with a house with a cupola and a sign which reads "Dogs Head"). Go well past Dogs Head to avoid the rocks, visible at low, to the east. Then start your turn southward, heading for a point south of the narrow peninsula bordering Lumbos Hole. Note the ledge that makes out from its tip.

Once inside Long Cove, favor the east side to avoid a long line of rocks that make out halfway across the cove about 400 yards inside the entrance, visible only at dead low. The channel is 8 feet deep at low and about 100 yards wide.

ANCHORAGES, MOORINGS. Anchor in good mud in 8 to 10 feet at low in the eastern half of the cove, either before the rocks or beyond them.

Long Cove would make a fine hurricane hole, although strong northwesterlies can funnel down its length and whip up a chop.

The following passages are the safest approaches to Quahog Bay:

APPROACHING FROM THE WEST. From green bell "1J" (*43° 42.32′N 069° 59.75′W*) off Jaquish Island, pass either side of can "1" on Charity Ledge, and follow the east shore of Bailey Island, leaving Middle Ground and nuns "4" and "6" to starboard.

You are now in the buoyed cross-channel between Orrs and Bailey Island and the New Meadows River. Head for Gun Point Cove, then turn east toward nun "8" and leave it close aboard to port. Leave Long Point Island (it has a house and telephone poles) to port and bend north into Quahog Bay.

APPROACHING FROM THE EAST. From the east, Quahog Bay can be reached directly or indirectly, with proportional anxiety levels as you guess your way between the unmarked ledges. The safest course is to approach from the entrance to the New Meadows River. Round Fuller Rock and follow the red marks north, all to starboard, first red bell "2BH" off Bald Head, then flashing red bell "4" on Wood Island Ledge, then past nun "6" with Jamison Ledge can "7" to port, and finally to flashing red "8" off Harbor Island. Once Harbor Island is abeam, cut across to can "9" off Rogues Island and leave it to starboard. Keep the red-and-white daybeacon on Goudy Ledge to port.

You are now in the buoyed cross-channel which cuts westward across the ledges from New Meadows to Orrs and Bailey Island. Keep nun "2," Jenny Island, and nuns "4" and "6" to starboard. Then turn north into Quahog Bay.

OTHER APPROACHES. For a more direct, white-knuckle approach, keep Round Rock's flashing green buoy "3" to port and Ragged Island to starboard and head directly north toward Quahog Bay. This requires positive identification of islands and ledges and should not be attempted in poor visibility.

It is also possible to enter Quahog Bay from Ridley Cove to the east, but it requires tricky navigation between a ledge off the southwestern end of Bethel Point and a 3-foot spot just offshore. The passage is complicated by the inadequate scale of the chart and the cartographer's interpretation of the Bethel Point ledge. As pointed out by local resident and avid small-craft aficionado Henry Bird, older charts marked it with a star, but more recent editions, including the current one, show it as a sketched ledge and a 5-foot spot. Our recommendation: if you need this kind of excitement, go at low and go slow (see *Ridley Cove*, p. 95, for more information).

WILLS GUT (east)

2 ★★

43° 44.82′N 069° 59.17′W

CHARTS: 13290, 13288   CHART KIT: 60, 17

Two anchorages lie on the east side of the crib bridge between Orrs and Bailey islands. They might prove useful if you don't want to take the time to explore farther up Quahog Bay. At the north end of Bailey Island is Water Cove, a small, tight crotch between cliffy sides, where coasting schooners used to take on water. The cove is still beautiful despite the houses on the cliffs which loom over the cove.

If you can find anchoring room among the handful of private moorings deep in the cove, you will be protected in all directions except from the north or northeast. If the wind swings to those directions, you can move to Lowell Cove, nearby to the north on Orrs Island, though anchoring room will be more scarce.

APPROACHES. See the sketch chart on page 90. The approach to both coves is easy. Find green "1J" (*43° 42.32′N 069° 59.75′W*) off Jaquish Island and pass either side of Charity Ledge, marked by green can "1." Follow the eastern shore of Bailey Island, keeping nun "4" on Middle Ground to starboard. When the cribstone bridge becomes visible, round into Wills Gut. Stay off the ledges that surround the small islet at the northeast tip of Bailey as you turn south into Water Cove.

To proceed to Lowell Cove, keep nun "6" on Littlejohn Rock well to starboard and head north into the cove. Favor the eastern side of the cove to avoid the scruff on the west.

If you are approaching Lowell Cove from Water Cove, be sure to note the unmarked 7-foot spot just northeast of Wills Gut.

ANCHORAGES, MOORINGS. Anchor anywhere in Water Cove in 10 to 15 feet. The bottom mud holds well. Lowell Cove is deeper with a mud bottom, but it is more crowded with moorings.

FOR THE CREW. The gray building at the south end of the Cribstone Bridge is Jack's Bailey Island Grille (*833-0640*), which serves fine dinners. Or dinghy under the bridge to Cook's Lobster House (*833-2818*) on the west side of the gut (see *Wills Gut west*, page 89).

QUAHOG BAY

On the chart, Quahog Bay is a fist punched deeply into the belly of Sebascodegan Island. It offers several beautiful anchorages, listed below, well off the usual east-west passages across Casco Bay.

APPROACHES. See *Passages through Eastern Casco Bay,* above, to arrive at Quahog's mouth. From a distance, you will see two dark, high, rounded islands in the mouth of Quahog Bay. The eastern and smaller of the two is Raspberry Island, just west of Yarmouth Island. Larger Pole Island lies to the north, identifiable by a conspicuous white boathouse at the southwest tip.

Dodge the fields of lobster buoys and leave Raspberry Island to starboard. Pole Island can be left to either side; just be sure to stay off South Ledges, marked with nun "4." The deeper channel passes to the west of the nun. Run well past the north tip of Pole Island to clear the North Ledges and bear to starboard. Keep tiny Center Island and its rocks well to port. Pass south of Snow Island to the 16 and 14-foot areas. You've arrived!

CARD COVE

43° 47.70′N   069° 55.89′W

CHARTS: 13290, 13288   CHART KIT: 60, 17

Card Cove is a well-protected little harbor on the west side of the entrance to Quahog Bay ringed by summer residences and fishermen's homes. Most of the lobstering activity is outside the harbor on Pinkham Point, at the Quahog Bay Co-op.

APPROACHES. Run up Quahog Bay west of nun "4" on South Ledges. The entrance to Card Cove is tricky and best attempted at midtide or higher. Overlapping ledges make out from each side of the entrance, the larger southern ledge outside and the northern one inside. Line up the white boathouse on the southwestern tip of Pole Island with the middle of the small island (Ham Loaf) in Card Cove and run in gently on that line. Once safely past the northern ledge, turn northward past Ham Loaf.

ANCHORAGES, MOORINGS. Anchor north of Ham Loaf Island in 29 feet at low. This is a quiet and comfortable spot for the night. Farther north in the cove, your peace may be broken by the cars along a road by the shore.

The Cribstone Bridge at Wills Gut.

SNOW ISLAND

43° 48.66′N   069° 54.38′W

CHARTS: 13290, 13288   CHART KIT: 60, 17

Snow Island, at the head of Quahog Bay, is one of those idyllic anchorages you dream about in the dead of winter. It is a very special place—beautiful, tranquil, and, except when a yacht club cruise or boat-owners' association has chosen the anchorage for its rendezvous, uncrowded.

APPROACHES. Ledges extend northward from Pole Island. Tiny Center Island, with a spray of trees, is very obvious ahead. Bear to the right to avoid the rock south of center, then turn east and run south of Snow Island, which has a red house on its southeast tip.

ANCHORAGES, MOORINGS. Anchor south and east of Snow Island in 16 feet of water at low. The bottom is mud and holds well.

You can also anchor between Snow Island and the several islets to the east or work your way along the mainland east of Snow Island to anchor near Ben Island in 9 to 11 feet. Beware, though, of a possible uncharted rock located southeast of Ben Island, between the 2 and the 9-foot soundings. Conceivably this is what makes the 2-foot sounding 2 feet. But sailors Kym and Paul Cournoyer say Ratherby Rock, as it is known, is awash at low and when covered "has claimed its share of prides."

THINGS TO DO. At high tide, explore the beautiful inlet beyond Ben Island. The southernmost of the islets is Little Snow Island, owned by the state.

In these upper reaches of Quahog Bay, the water temperature is delightful and swimming is a pleasure, rather than an ordeal—even for those from the three-quick-strokes-and-you're-out school.

ORRS COVE

4 ★★★

43° 49.65'N 069° 54.91'W

CHARTS: **13290, 13288**   *CHART KIT:* **60, 17**

Orrs Cove is the most western of four fingers extending north from Quahog Bay, deep and well protected. Occasionally strong southerlies funnel up Quahog Bay and into Orrs Cove, but you can easily escape to the north of Snow Island.

APPROACHES. The cove is identifiable by the masts of boats moored off Great Island Boat Yard on the west bank. Midchannel is 8 feet deep at low and extends as far as the boatyard's first dock to port.

ANCHORAGES, MOORINGS. Great Island Boat Yard has five moorings for transients, or you can anchor south of the boats. Holding is good, in mud.

FOR THE BOAT. *Great Island Boat Yard (Ch. 09, 207-729-1639; www.greatislandboatyard.com).* The fuel dock here has gas, diesel, water, pump-outs, and ice. Depth at mean low is 6 feet, but it drops to 4 1/2 on a drain tide. The yard can perform all repairs, hauling to 55 feet and 35 tons with their hydraulic trailer. They are a Yanmar dealer, and a chandlery is ashore.

Marine Computer Systems (207-871-1575; www.marinecomputer.com) specializes in integrated computer systems for boats. They are located at GIBY.

FOR THE CREW. The yard has showers and a pay phone. There is no grocery store in Orrs Cove, but you probably can arrange a ride to the shopping centers of Cook's Corner, only five miles away in Brunswick.

MILL COVE, BRICKYARD COVE, and RICH COVE

Mill Cove: 43° 49.50'N 069° 54.74'W

CHARTS: **13290, 13288**   *CHART KIT:* **60, 17**

From west to east, the four little coves that extend north like fingers from the hand of Quahog Bay are Orrs, Mill, Brickyard, and Rich. Orrs is discussed in the previous entry. Of the other three, Mill Cove is the easiest to enter and makes a fine daystop. Anchor in 9 feet at low in a mud bottom.

Brickyard and Rich Cove are guarded by ledges, and they are difficult to enter except at the bottom of the tide. The adventure is worthwhile for the pure pleasure of gunkholing.

RIDLEY COVE

3 ★★★

43° 47.40'N 069° 54.33'W

CHARTS: **13290, 13288**   *CHART KIT:* **60, 17**

The southern portion of Ridley Cove is beautiful and unspoiled, but it holds unmarked dangers and is broad and wide open to the south. George and Big Hen Island, however, define a cove to the north that offers fair protection and fun exploration. Small guts twist between the islands, and low tide reveals the remains of an old shipwreck.

With a lot of care, Ridley Cove can be used as an entry to Quahog Bay, although other approaches are safer (see *Passages through Eastern Casco Bay*, p. 92).

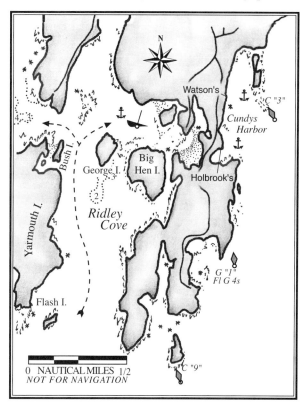

APPROACHES. To enter Ridley Cove from the south, leave Jenny Island to starboard and Flash Island to port. Proceed up the middle of the cove to avoid the ledges which make out from both sides. Just south of Big Hen Island there is a nice day anchorage.

Nearing the north end of Yarmouth Island, stay well to the west to avoid the shoals extending southwest from George Island, which are marked by a generous sprinkling of lobster buoys. Once clear of the shoals, head through "George Bush," the passage between George Island and Bush Island. Once abeam of George Island, look dead ahead for a large, charted ledge in the mouth of Hen Cove. To avoid the ledge, bear to starboard and favor the shore of George Island as you enter the harbor.

It is also possible to enter Ridley Cove from Quahog Bay, but it requires local knowledge, shallow draft, a rising tide, and nerves of steel. You need to feel your way between a 3-foot spot to the south and, to the north, a ledge off the southwest shore of Bethel Point, marked by ledge sketchings and a 5-foot sounding. Several yachts find bottom here every year.

The scale of the chart is small and confusing. Once past Bethel Point, a large ledge projects to the south. The ledge is exposed at low tide, but it extends farther to the south than it appears to on the chart. To avoid it, steer toward the center of George Island before rounding left into Ridley Cove.

ANCHORAGES, MOORINGS. The Ridley Cove approach will bring you to a lovely spot with only a couple of houses in sight. At low tide the muddy hulk of a sunken ship becomes visible north of the hole between George and Big Hen Island. At high, its southernmost extremity—the bow— is marked by a a buoy of styrofoam blocks. Drop anchor in 14 to 16 feet at low west of the hulk. The bottom is mud but may be foul. A tripline is recommended.

THINGS TO DO. Take your dinghy and poke around the eastern end of the harbor. If you use an outboard, be particularly careful of Mill Ledges extending south from Leavitt Island, which are marked by privately maintained buoys but only approximately. Do not land on Big Hen Island, which is private.

The sunken vessel in the middle of the cove is the *Philip E. Lake*, an old Grand Banks fishing schooner. Two Maine men bought her in 1986 at a Boston auction and towed her here for renovations. Before the work had begun, though, she sank at her rented mooring, and the two owners mysteriously vanished.

It is possible to walk across the peninsula to Cundys Harbor if you can find a place to land. There is no public dock. You might be able to squeeze in at the small marina of Oakhurst Island, Inc., or at the floats of Mill Ledge Seafood—a wholesale operation which might sell retail in slack moments.

NEW MEADOWS RIVER

The New Meadows River rises in lakes and salt marshes west of Bath and flows gently southward to form the eastern border of Casco Bay. It has neither the strength nor the majesty of the neighboring Kennebec River and seems more like another of the long bays and sounds of Casco Bay.

For the sailor with time to explore, the New Meadows River and its approaches offer a dozen attractive and remote anchorages that are little known to cruising boats—a mixture of working harbors and summer havens well worth the visit. Attractive Sebasco Harbor is only four miles north of Cape Small. Three miles farther north is The Basin, easy to enter and completely landlocked, a marvelous harbor of refuge. And Cundys Harbor is as authentic a Maine fishing town as any.

Proceeding in or out of the New Meadows River, it is fun to run inside Malaga Island and also inside Burnt Coat Island farther south. Both channels are well marked and easily navigated, despite the moored lobsterboats and draggers.

Where the river turns west above Sebascodegan Island, it feels like a long and peaceful lake. Only a tiny bit of horizon is visible, way to the south. On the east bank, near Houghton Pond, there are iron rings in the rock where old ice schooners used to tie up to take on the pond's winter harvest. Beyond the turn, the river winds north past Bombazine Island, approaches a highway and a small marina near coastal Route 1, and loses much of its charm.

CUNDYS HARBOR

2 ★★★

43° 47.38′N 069° 53.40′W

CHARTS: **13290, 13288, 13293**   CHART KIT: **60, 17**

Cundys Harbor lies tucked in a niche across from Bear Island on the west shore of the New Meadows River. This working harbor is simple and authentic. Fuel and lobster docks stand on ancient pilings, the little village clusters around the wharves, and the harbor is full of large draggers and lobsterboats. The harbor is broad, exposed, and notorious for its roll, but it makes for a good lobster stop before heading farther upriver to The Basin or beyond.

APPROACHES. The harbor is easy to enter. First, find flashing green "1" at Fort Point. Then follow the coast north until you find can "3" at the entrance, which marks the end of the ledges that form the eastern side of the harbor. Leave the can to starboard (the can marks the river, not the harbor entrance) and go straight in toward the docks.

ANCHORAGES, MOORINGS. Find an open spot anywhere in the harbor and anchor in 22 to 29 feet at low. If you are arriving in the middle of the day, you might be able to find a fisherman's mooring you can use temporarily, but be sure to inquire nearby.

If you are here to eat lobsters, dock at Holbrook's Wharf at the southern end of the harbor in 8 feet at low. If you need fuel, head to Watson's. For the night, you might be able to use the mooring of one of the draggers that leave for several days at a time. Again, inquire.

Cundys Harbor is not well protected, particularly from the prevailing southwesterlies, so there is an uncomfortable roll even on a calm night. As one local observer put it, "It ain't much of a harbor. Sea piles in when the wind blows southwest. But it's deep water, and the fishermen keep their boats here."

GETTING ASHORE. Tie your dinghy up to the floats at Holbrook's Lobster Wharf to the south or at Watson's dock, farther north.

FOR THE BOAT. *Watson's General Store (207-725-7794).* Watson's is worth a visit even if you don't think you need anything. Here are all the necessities of life right next to each other—shackles and candy bars, pot warp and newspapers, rubber boots and lobsters. At the float, they pump gas and diesel. Approach from the east to avoid a large rock awash at midtide, just south of the float.

Bottom paint or bubble gum can be found at Watson's.

Harbor Propeller Service (207-725-7528; www.prop-dr.com). Should you be unfortunate enough to need him, the "Prop Doctor" can miraculously undo what rocks do to propellers.

FOR THE CREW. Cundys Harbor is a case study for the pressures on Maine's small coastal towns and working waterfronts. For decades, the post office was located in the small general store above Holbrook's Wharf. In 1998, after the postmistress passed away, Holbrook's Store closed. Its working wharf continued to operate as a commercial seafood buyer and open-air seafood restaurant until it was put on the market in 2005. Facing the prospect of the wharf being bought by developers and by the loss of working waterfront, the community formed the Holbrook Community Foundation, which raised money to buy the property and continue to run it as a commercial enterprise, including the takeout. Now more than ever, your lobster dinner at Holbrook's will support the fishermen of Cundys Harbor.

Hawke's Lobster, next door to Holbrook's, sells live lobsters and lobster specialties. Unless the Community Foundation reopens Holbrook's Store, the only groceries are at Watson's General Store, where the limited selection runs the gamut from bread and soda to Petit paint and roller frames. But who needs groceries when you can buy live lobsters? Started by the Watson family in 1819, this claims to be Maine's oldest family-operated general store. It is currently run by the fourth and fifth generations of Watsons.

The Block and Tackle (*725-5690*) is about half a mile out of town. This classic, no-frills, small-town eatery serves up huge fishermen's breakfasts and their catch for lunch and dinner. The owner's brother sums it up nicely: "Remember, half a lobster in a cup of lobster stew and a whole lobster in a bowl."

THINGS TO DO. You can find food for thought at tiny Cundys Harbor Library, right on the harbor, halfway between Holbrook's and Watson's.

SEBASCO

4 ★★ 🔧🍴

43° 47.14′N 069° 52.57′W

CHARTS: 13290, 13288, 13293   CHART KIT: 60, 17

A sizable harbor used by large draggers and fishermen lies north of Malaga Island and east of Bear Island, near the fishing village of Sebasco. It is an attractive spot, an unspoiled working harbor well protected from the prevailing winds and open only to the north.

Malaga Island has a sad history of hope and desperation. In 1794, nearby Horse Island (now known as Harbor Island in Sebasco Harbor) was settled by Benjamin Darling, a black man, and his white wife, Sarah Proverbs. It is rumored that Darling was a slave who saved his master in a shipwreck. His master, in turn, gave him his freedom and some land.

The couple had two children. One moved to Bath. The other remained on Horse Island until he sold it and moved to Malaga with his own family. For more than a hundred years, Darling and his descendants squatted on the island, fishing and scratch farming in the rocky soil. Eventually they were joined by a mixed group of blacks, Indians, and mixed breeds who built a ramshackle community of shacks of driftwood and flotsam, clinging to the very shores of survival. They clammed and lobstered and heated with driftwood, and they occasionally worked for farmers on the mainland. It was rumored that social order was loose, and incest was rampant.

By 1903, the Malagoites were so desperate that they sought help from the town of Phippsburg. Phippsburg quickly argued that Malaga belonged to the town of Harpswell, and the dispute eventually had to be settled by state legislators.

The ensuing publicity focused charity on the poor islanders, both for good and bad. A school was built, and food and money were raised. But after visiting Malaga eight years later, Governor Frederick Plaisted suggested burning down the shacks. In 1912, Maine took the island for $400, evicted the 56 residents, dug up the remains of the dead, and torched the hovels.

Many of the Malagoites and the bones of their ancestors were installed in the Maine School for the Feeble-Minded in Pownel. Others were left to fend for themselves. Newspapers had dubbed Malaga as "Maine's Scandal Island." Now they praised the "cleaning up" of this "salt-water skid row" on a coast that was increasingly promoted as a vacation destination.

The island has since grown over in spruce and firs and the local fishermen use its ledges to store their traps. In 2001 it was purchased for preservation by the Maine Coast Heritage Trust.

APPROACHES. The harbor can be entered from either north or south. From the north, pass between the mainland and nun "4" at the north end of Bear Island.

From the south, clear the rocks and 2-foot spot off the west side of Harbor Island before turning northeast and leaving cans "1" and "3" to port. Can "3" seems a surprisingly long way out from Malaga Island, but be sure to observe it. Then head for midchannel to avoid the ledge jutting out from the southeastern flank of Malaga. The rock shown on the chart at the east side of the channel is marked with a pipe.

ANCHORAGES, MOORINGS. There is plenty of anchoring room among the moorings north of Malaga and east of Bear Island in 10 to 22 feet at low. Very little swell enters the harbor. The channel east of Malaga may be even more protected, but it is crammed with lobsterboats and draggers.

GETTING ASHORE. Sebasco Wharf has floats.

FOR THE BOAT. *Sebasco Wharf (207-389-2756).* Sebasco Wharf is a working lobster wharf by the Water's Edge restaurant. They have no fuel, but they could doubtless help you in an emergency. Depth at the outer face is 6 feet. Approach cautiously to avoid a ledge that lurks just north of the wharf.

Casco Bay Boat Works (207-389-1302; www.cascobayboatworks.com). This yard is tucked in the little indentation northeast of Harbor Island. They are a full service repair facility, custom boat-builder, and Mercruiser dealer. The yard hauls with a 35-ton hydraulic trailer and maintains some ⚓ moorings in nearby Sebasco Harbor which they will rent if available.

Approach from around Harbor Island to the north, but only at high tide. At low, the approach can be walked.

Years ago, when the yard was called Brewer's, when asked if they monitor a VHF channel, they replied, "Half the time, we don't even monitor the phone."

FOR THE CREW. 🍴 The Water's Edge Restaurant *(389-1803)* is at Sebasco Wharf, where you can dine in or take out.

DINGLEY COVE

 ★★

43° 48.22′N   069° 53.10′W

CHARTS: **13290, 13288, 13293**   CHART KIT: **60, 17**

Dingley Cove, on the west side of the New Meadows River just north of Cundys Harbor, is a small working harbor with no facilities for yachtsmen. Strangers should enter this harbor carefully, and not at night.

APPROACHES. Enter near the southern tip of Sheep Island and head towards Hopkins Island to avoid Green Ledges.

ANCHORAGES, MOORINGS. Anchor outside the moored lobsterboats in 16 to 20 feet at low in mud. The harbor is exposed to the south, but Cedar Ledges offer some protection.

THE BASIN

 ★★★★

43° 48.28′N   069° 51.35′W

CHARTS: **13290, 13288, 13293**   CHART KIT: **60, 17**

The Basin is what many yachtsmen dream about when they ponder the ultimate safe harbor—a small lake surrounded by rocky points and dark green trees, with barely a sign of human habitation, the water as still as a millpond, completely protected from every direction.

This extraordinary harbor on the New Meadows River lies about two miles north of Sebasco Harbor. It is worth going out of your way to make the passage through the deep, narrow entrance and into the broad sanctuary beyond. In this part of the world, this is the best hurricane hole around.

Unfortunately, the very protection and tranquillity produces glassy water which has, in recent years, attracted occasional swarms of jet skis.

It is also loved by quahogs. In the 1940s, the Basin was ground zero for gathering quahogs along the New Meadows River, America's largest quahog producer. The Basin still produces 52 bushels an acre, nearly twice that of other quahog bottoms.

APPROACHES. Coming up the New Meadows River, pass Cundys Harbor and leave can "5" on Sheep Island Ledge to port. Look along the east shore for a slight indentation which marks the entrance to The Basin. A house with a pointed chimney is just to the north of the entrance.

Turn right into the channel, perhaps 150 feet wide at low tide, and run straight down the middle. The channel narrows to about 75 feet at the left turn where slowed tidal current drops its suspended sediment. Controlling depth is 7 feet here, but turn wide around the bend to find it. Like magic, The Basin will open up before you.

ANCHORAGES, MOORINGS. Anchor in the middle of the western part of The Basin, in 14 to 20 feet of water at low. The bottom is mud. Even at the height of the summer season, The Basin usually has only a handful of boats, and there is room for a fleet.

Some boats work their way in close to the small island, Basin Island, to the east, or close to the rocky point of land that juts down from the north, but *use caution*. As benign as The Basin seems, the east side is shoal, and there is a ledge almost dead center, shown on the chart between two 11-foot spots, which claims its share of unsuspecting yachts. Pass either north or south of it but not over it.

The Basin on a calm night feels like a high mountain lake far removed from the sea. To dispel the illusion, taste the water.

THINGS TO DO. Beautiful Basin Island can be explored. However, this is a popular stop on the Maine Island Trail, and the steep banks of this tiny island are eroding. Walk softly.

For longer land-based exploration, you are free to explore the rest of The Nature Conservancy's Basin Preserve. In 2006, in an "act of staggering generosity," an anonymous donor gave the conservancy nearly 2,000 acres, encompassing all of the eastern and southern shores.

The Basin holds some of the warmest salt water on the Maine coast (67° F). The timid won't find a better place to take the plunge.

LONG ISLAND

 ★★★

43° 49.53′N   069° 52.97′W

CHARTS: 13290, 13288, 13293   CHART KIT: 60, 17

A slot between Long Island and Dingley Island provides shelter in winds from all quarters except from the south, and it is easy to enter.

APPROACHES. Staying off the ledge at the southwest tip of Long Island, head north up the slot.

ANCHORAGES, MOORINGS. Anchor in 17 to 20 feet between the northern part of Dingley Island and the southern part of Long Island. This section is lovely, with woods on both sides. You may be able to find better protection farther past the northern tip of Dingley, but the view is less appealing.

THREE ISLANDS

43° 50.62′N   069° 53.00′W

CHARTS: 13290, 13288, 13293   CHART KIT: 60, 17

Three Islands, just south of Bragdon Island at the turn of the New Meadows River, are small, rocky, and wooded, with a couple of small cottages. They are connected with each other and with Bragdon by sandbars at low tide.

The anchorage to their west is exposed in several directions, but it makes a pleasant lunch stop.

Run in halfway between the southernmost of the Three Islands and the tip of Long Island, in 60 feet of water. Then turn north around this first island until you are opposite the northern tip, where you can anchor in 18 to 25 feet. Beyond this spot the water shoals rapidly.

BRIGHAMS COVE

 ★★

43° 50.03′N   069° 50.93′W

CHARTS: 13290, 13288, 13293   CHART KIT: 60, 17

Brighams Cove feels like a small lake at the north end of Winnegance Bay. Perhaps 100 yards wide, it is surrounded by rocky shores, summer houses, and docks. The cove is protected from every direction except the southwest.

APPROACHES. Approach through pretty little Winnegance Bay. Leave red daybeacon "6" on Hen Island Ledge well to starboard and keep closer to the western side of the bay. Beware of the rock off the tip of the west side of the entrance and, once inside, of the rock off the west shore.

ANCHORAGES, MOORINGS. There are several private moorings, but you can usually find room to anchor outside of them. The bottom is mud and between 10 and 15 feet deep at low.

MILL COVE

 ★★★

43° 52.33′N   069° 51.98′W

CHARTS: 13290, 13288, 13293   CHART KIT: 60, 17

Where the New Meadows River turns west, two long, narrow fingers of water continue north on either side of Rich Hill. Mill Cove, to the east, is pretty but more inhabited than Back Cove to the west.

APPROACHES. Red daybeacon "10" on Bragdon Rock is visible against the trees from a long way off. Go to either side. There is a huge osprey nest on the beacon, with birds in residence.

The farther into Mill Cove you go, the prettier it gets, with rocky shores and handsome houses. The passage is easy and open. Favor the western shore to avoid the ledges to starboard, especially before the little islands near the upper end.

ANCHORAGES, MOORINGS. There is a beautiful spot to drop the hook opposite Dam Cove in 17 feet at low, or you can go in farther to anchor. The bottom is mud. Mill Cove provides good protection from any winds except a strong southerly.

BACK COVE

3 ★★★

43° 52.04′N 069° 52.47′W
CHARTS: 13290, 13288, 13293   CHART KIT: 60, 17

The finger of water west of Rich Hill in the New Meadows River is Back Cove, a tranquil and lovely spot.

APPROACHES. Leave red daybeacon "10" on Bragdon Rock to starboard and head straight up the middle of the slot between Merritt Island and Williams Island. It is an easy glide between the high, rocky, tree-lined shores with only a single house in sight.

ANCHORAGES, MOORINGS. You can anchor in 17 feet near the northern tip of Williams Island, but the current against the breeze may cause you to swing, so you might want to consider a stern anchor if you spend the night. Holding ground is good in mud. The tranquillity may be broken by the large airplanes lumbering overhead from the Brunswick Naval Air Station.

THINGS TO DO. The deepest indent on the western shore of Back Cove is part of the Hamilton Sanctuary, a 75-acre farm owned by the Maine Audubon Society. At high tide it is possible to land your dinghy and explore its nature trails, but low tide will leave you stranded in mud.

BOMBAZINE ISLAND
and Upper New Meadows River

★★

east of Bombazine: 43° 51.70′N 069° 53.50′W
New Meadows Marina: 43° 54.61′N 069° 52.19′W
CHARTS: 13290, 13288, 13293   CHART KIT: 60, 17

Just above the turn in the New Meadows River lies small, wooded Bombazine Island, home of the Indian Sagamore Bombazine, who was killed by the English in 1724.

"Granny" Young visited the island half a century later to pick berries, as related in *Beautiful Harpswell* by Margaret and Charles Todd.

"After filling her pail, she set out for home in her canoe. Hearing a noise behind her, she turned to see a huge black bear gaining on her fast.

"Granny had only a stave for a paddle. The bear quickly overtook the canoe and tried to overturn it; but Granny, with her pioneer resourcefulness, struck the bear a stunning blow on the head with her stave, and then held his head under water until he was dead. Then, after tying the carcass to the canoe, she continued to paddle homeward, towing the bear until she reached shore."

Beyond Bombazine, the New Meadows heads north through an ever-narrowing channel. This might be fun to explore in a dinghy, but unless you need to reach the New Meadows Marina at the head of navigation or use the public launching ramp nearby, there is little reason to take a deep-draft boat further up river.

FOR THE BOAT. *New Meadows Marina (207-443-6277)* is a full-service marina on the west side of the river, hard by the Bath Road, with gas, water, pump-outs and dockage. They have a chandlery and an outboard parts department, and they haul, store, and repair boats of nearly all sizes.

H & H Propeller (207-442-7595), located next door to New Meadows Marina, can service, repair, or replace propellers and shafts of all kinds.

THE EASTERN SHORE

SEBASCO HARBOR

3 ★★★★

43° 46.05′N 069° 52.02′W
CHARTS: 13290, 13288, 13293   CHART KIT: 60, 17

Sebasco Harbor is a welcoming spot, well protected under most conditions and easy to enter. It is probably the most convenient harbor at this end of Casco Bay, and it is not far out of the way for boats traveling east or west.

Before the days of refrigeration, this harbor sometimes was filled with trade ships taking on ice from the Cornelius Ice Pond for delivery to all parts of the world. Now it is the site of Sebasco Harbor Resort, a long-established and well-appointed resort whose resources are available to the visiting yachtsman. With its ample parking and facilities, the resort makes an elegant place for crew changes.

APPROACHES. Coming from the south, leave the Jamison Ledge can "7" to port and head for flashing red "8" at Harbor Island Point. Note unmarked

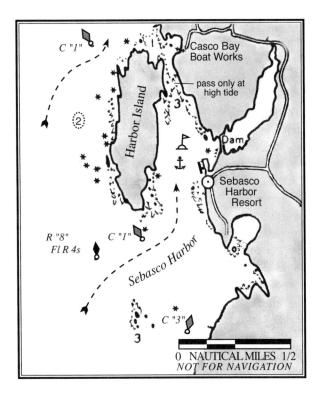

Dry Ledges on the chart and make sure you are well to the west of them. At flashing red "8," turn east and continue well beyond can "1" before heading north into the harbor.

The ledges making out a long way from the southeast tip of Harbor Island are exposed except at high tide. The less prominent ledges on the right side are marked by a privately maintained beacon. It is safe to turn north into the harbor when you can see the channel between the boats moored to the west side of the harbor and those on the east side.

ANCHORAGES, MOORINGS. During normal summer winds, the ledges at the entrance to the harbor provide good protection, and this is a comfortable anchorage. Strong wind from the southwest, though, blows in swells, and ⚓ you will sleep more peacefully on one of Sebasco Harbor Resort's dozen or so stout moorings, grouped on the west side of the harbor. They are marked "SHR." Expect to pay slightly more than the going rate for moorings here because your fee includes launch service, the use of the saltwater pool, and guest rates on golf and tennis.

If no moorings are available, there is usually room to anchor in midharbor, in 15 or 26 feet of water. Casco Bay Boat Works (*389-1302*) also maintains some rental moorings in the harbor (see *Sebasco*, page 98).

GETTING ASHORE. The usual three toots will summon the Sebasco's launch, or you can take your dinghy to the float.

FOR THE BOAT. *Sebasco Harbor Resort (Ch. 09; 207-389-1161, 800-225-3819; www.sebasco.com).* The cupola marked on the chart is a many-layered octagonal green-and-white wedding cake of a house, called the "Lighthouse" by the resort and used for guests. Sebasco's dock and float are just north of the Lighthouse, with 5 feet of depth at low tide. ⛽ Gas, water, 🚽 pump-outs, and ice are available, but no diesel. For repairs, head around Harbor Island to Sebasco.

FOR THE CREW. Sebasco Harbor Resort has showers and laundry facilities as well as 🍴 the Pilot House restaurant (*389-1161*) and The Ledges pub. For morning snacks or lunch, try the Patio Café next to the resort's Repairatorium.

The Rock Gardens Inn (*389-1339*), next door to the Resort, is another option for elegant lodging.

THINGS TO DO. Sebasco Harbor Resort has carefully maintained its essence of service and entertainment in a beautiful setting, served up in a wonderfully relaxed and welcoming way.

Even if you haven't rented a mooring, for a fee you can indulge in the resort's golf or tennis, dip in the saltwater pool, work out in the fitness room or on the dance floor, take in a movie, or even—if you can't get enough of the water—sign up for boat tours or sea kayaking. The kids can discover candlepin bowling, shuffleboard, a small playground, and plenty of open space to fly kites.

CARRYING PLACE COVE
northern side: 43° 44.98'N 069° 52.00'W
CHARTS: 13290, 13288, 13293   CHART KIT: 60, 17

One local fisherman describes Carrying Place Cove as "just a little hole in the wall." Fishing shacks balance on spindly pilings in the lee of Carrying Place Head. A tiny island with a fish wharf sits in the center, dividing Carrying Place into two coves, one to the north and one to the south. Either makes a pleasant daystop.

Don't try to take your boat through Carrying Place Cove to get to the northern cove. If the power cables overhead don't get you at 30 feet, the ledges beneath you will. Instead anchor in 21 feet at low at the northeast tip of the island, just outside the moored lobsterboats, and explore by dinghy. The bottom holds well in mud—or in old potwarp. Have your tripline rigged.

The southern part of Carrying Place Cove is known locally as West Point Harbor. It is totally exposed, so only stop here in the most settled weather. Not long ago, the village of West Point had two general stores. Both are now gone.

In a real pinch, you might be able to get fuel at the commercial Sea Horse Wharf on the mainland shore of the northern cove.

SMALL POINT HARBOR
(Cape Small Harbor)

43° 43.85'N 069° 50.79'W

CHARTS: 13290, 13288, 13293   CHART KIT: 60, 17

Lying to the west and north of Cape Small between Hermit and Wood Island, the Small Point Harbor shown on the chart is exposed to the southwest and wide open to the ocean. This would be a very uncomfortable place to spend the night. As one local put it, the chart might as well label it "Open Ocean." A "Harbor" it may be, but certainly not a recommended one.

The real harbor at Small Point is Cape Small Harbor. In fact, *this* is the harbor the locals call Small Point Harbor. The harbor is an interesting little hideaway tucked between Hermit Island and the mainland. It is extremely well protected but difficult to enter—not a place to visit casually or to try to enter on a dark and stormy night. In difficult conditions, it would be far safer to make for easily entered Sebasco Harbor or The Basin, a short way north.

Cape Small Harbor is relatively undeveloped, though packed tightly with a mixture of commercial craft and pleasure boats. Blue herons pose silently along the tree-lined shores at low tide, side by side snowy egrets and other shore birds.

APPROACHES. Cape Small Harbor is guarded by two overlapping bars, one of sand and the other of mussels and ledge. The approach depends on the state of the tide, and it is complicated by the inadequate scale of the chart.

Run toward the harbor along Hermit Island. Keep nun "4," on Pitchpine Ledges, to starboard and head for Goose Rock, which at a distance seems to merge with the mainland. As you draw near, you will begin to see a narrow channel east of it. The chart shows the channel being marked by nun "2." It isn't.

A long, sandy beach curves south from Flat Point ending in an outcropping of white rocks. This is followed by a shorter sand beach and a second set of white rocks, which mark the beginning of the channel obstructions.

First, a sandbar extends westward from the rocks on the mainland. Then, a short distance farther south, a large rock ledge and mussel bed protrudes eastward from the southeastern end of Goose Rock. At mean low water, the channel around these bars is about 30 feet wide and about 4½ feet deep.

John Gardner, captain of the charter boat *Yankee*, navigates this passage daily. He describes the approach as follows: Begin by turning southward, favoring Goose Rock to starboard. A fleet of small boats is moored in a line off the eastern shore. The center of the channel is about 30 feet to the west of those moorings. On the western shore, off Goose Rock, a float is moored for dry sailing a fleet of Optimists. The channel is about 40 feet to the left of this float.

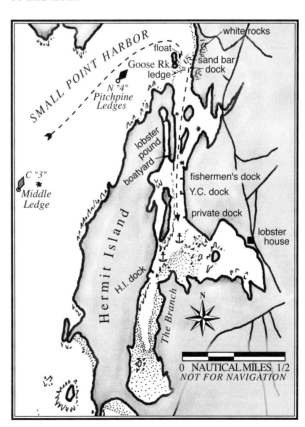

Once abreast of the Optimist float, begin to steer eastward, to the left, toward the center of a dock projecting from the mainland shore. Stay on the line between the Opti float and the mainland dock. When you are about 10 feet off the dock, turn to the right, parrallel to the dock and run straight to clear the second ledge-and-mussel bar off Goose Island.

Once past the bars, stay to the eastern side. Pass Mill Point to starboard and the two islets to port. Opposite the first islet, a ledge extends halfway across the channel from the west side—shown on the chart as an asterisk—so keep to the east, tight against the islands.

After this ledge, you have an almost clear shot to the anchorage. There is deep water near the lobster docks on the west side of the channel. Go slowly right up the middle among the moored boats.

ANCHORAGES, MOORINGS. Watch your depthsounder carefully and know the state of the tide before you drop the hook. Anchor among the moored boats if there is room or just beyond them opposite the tip of Tennants Island, in the mud bottom in 6 to 8 feet of water at low. Note that the channel curves west after Tennants Island into The Branch, and then it shoals.

The anchorage appears to be a wide open pond at high tide, but at low it becomes a narrow slot lined with clam flats and mussel beds. The prevailing southwest wind funnels down The Branch and usually keeps you lined up with the channel. But any shift in wind or surge of incoming tide can put you into the mud. You might sleep better with a stern anchor out.

One local resident says that there can be as many as eight or nine transients here on a given night, but that's about the limit.

Hermit Island Campground, privately owned, sometimes has a mooring available, or space alongside the floats extending south from the lobster pound. It is best to call ahead (*443-2101*).

GETTING ASHORE. Depending on your goal, take your dinghy either to the Small Point Yacht Club float or to the Hermit Island dock.

FOR THE BOAT. Gas and diesel are available at the floats of the lobster dock.

Robert Stevens Boatbuilders (207-389-1794). If you need any woodwork repaired or replaced, Robert Stevens can take care of you in his idyllic little yard. This is where he built a replica Viking ship, which voyaged from Greenland to Newfoundland.

His next historical project is a replica of the first ship built in Maine, the *Virginia* of Popham (see *Fort Popham*, page 110.)

The Small Point Yacht Club has a float just south of the obvious fishermen's dock on the east side of the channel. You will search in vain for a clubhouse. No boat larger than 12 feet is allowed to stay on the float. The next float to the south is private.

FOR THE CREW. Land at the yacht club float, scramble up the hill to the paved road, and walk right about half a mile to the main road. Turn right, and it's a quarter mile to the Lobster House, featuring fish and lobsters without the frills. About an hour either side of high tide you can reach the Lobster House in the dinghy by heading east from the southern tip of Tennants Island. At other times, you might be stranded.

The campground store has a snack counter and is well stocked with fresh food. Their restaurant, the Kelp Shed, is across the sandy road on Head Beach.

THINGS TO DO. From the Lobster House, it is half a mile east to the spectacular Small Point Beach, with views of the mouth of the Kennebec River and Seguin offshore. The beach is an Audubon sanctuary for the least tern.

The lobster pound next to the Stevens boatyard claims to be the southernmost lobster pound in Maine. You can walk across a wonderful wooden footbridge that spans the mouth of the pound, where lobsters are stocked in the fall to even out market fluctuations in late winter and early spring. The pound keeper gambles that the "hoped for" price will be well above the "going in" price. This pound will hold 40,000 to 80,000 pounds depending, as a small sign puts it, "on the state of the owner's ulcer."

The Hermit Island Campground dock lies up The Branch. From there, walk to the entrance of the campground, right at Head Beach. Hermit Island Campground (*443-2101*) has been owned and operated by Nick and Dave Sewall since 1953. Pick up an island map. Numerous trails lead all over the island, from one pocket beach to another and along stunning cliffs where the sunset spills out of the sky.

THE MIDCOAST
Cape Small to Marshall Point

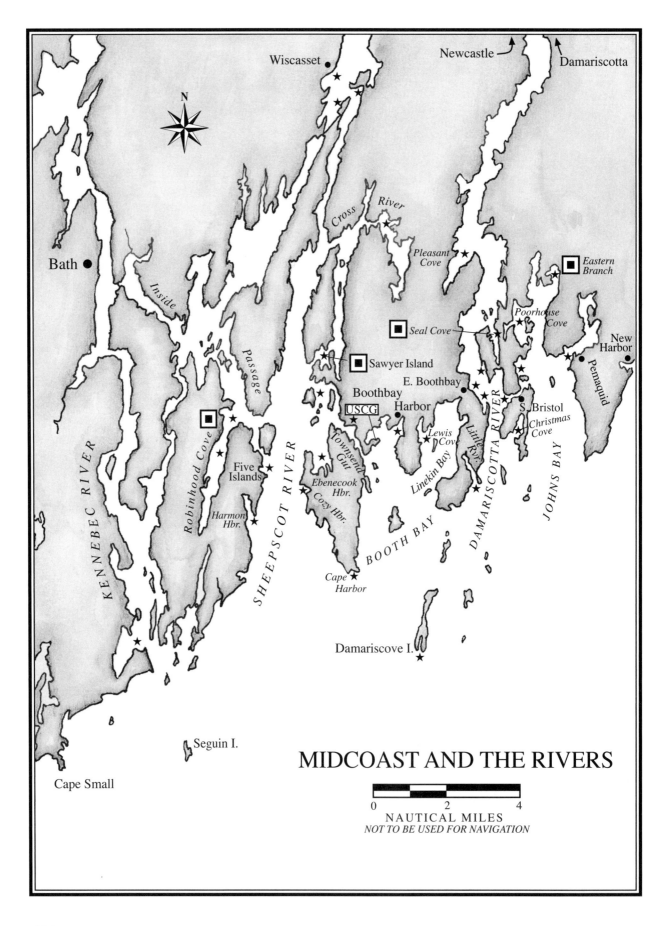

THE magic of the midcoast, from Cape Small to Marshall Point, is less familiar to cruising sailors than almost any other part of the coast, except, perhaps, for way Down East. Study the chart, and the reason becomes obvious. For here, the land and sea hold equal sway. There is no straight line course, no rhumb line, no as-the-crow-flies. Here, geography forces wandering, meandering, poking around.

This is the land of the geographical cul-de-sac. Long, narrow bays and salty fingers cut between chains of ledges and mountains running north and south. Rivers great and small run down to the ocean in a complex pattern of coves, estuaries, marshes, and back channels and swirl around the large islands offshore. Small towns isolated at the ends of the long peninsulas are still communities of the sea.

The mighty Kennebec, rising far to the north, surges past the cities of Augusta and Bath, past the site of the ill-fated 17th-century Popham Colony, and past Seguin Island as it spills into the Gulf of Maine. The Inside Passage cuts across faults of granite from Bath to Boothbay Harbor and winds through narrow guts where current rushes with impetuous force. The Sheepscot and Damariscotta rivers, each with its own history and special character, glide past small towns and quiet anchorages.

Beyond Pemaquid Point, the land gives way to the sea, and the character of the coast changes again. Muscongus Bay floods between dozens of islands and countless rocks and ledges, and a maze of passages twists and winds among them. Ashore, spruce-clad hills and gentle farmland roll down to the banks of the Saint George River.

Offshore, Seguin guards the coast with its towering light. Damariscove, once a busy fishing station long before Plymouth was settled, now sprouts wildflowers among its ghosts and ruins. And Monhegan, known by Verrazano and Champlain, inspires artists and hermits alike.

On the midcoast, the cruising sailor will discover unexpected adventures and unheralded charms among the hidden islands, the teeming wildlife, the picturesque villages, and the outposts of history.

THE KENNEBEC RIVER

The mighty Kennebec River, linked inextricably to Maine's history, is the longest river in the state, flowing from Moosehead Lake some 150 miles to the sea. From the head of navigation at Augusta, the Kennebec runs 45 miles to its guardian offshore island of Seguin. The tide pushing up the river salts the water as far as Bath. A few miles farther north, the water is fresh. The majestic southern reaches flow through wooded islands and past bold promontories. The northern reaches mingle with the Androscoggin River and lesser streams and marshland at Merrymeeting Bay. But everywhere, the Kennebec is big and strong and worthy of the mariner's attention and respect.

For two centuries, until not so long ago, the Kennebec was white with the sails of ships newly launched from a dozen yards along its banks, heading downstream to every port in the world. Today, a sailboat on the lower reaches of the river is rare enough. Farther north, they are seldom seen.

For much of its past, the waters of the Kennebec were choked with logs floating down to build the ships and cities of America. There are still logs in the depths of the river, and every spring the deadheads emerge to threaten the unwary sailor.

More recently, the waters of the Kennebec have recovered from a scourge of toxic waste. The bluefish are back, and so are the salmon and striped bass. In 1999 the Edwards Dam in Augusta was dismantled, and less than a year later salmon, bass, sturgeon and shad were sighted upriver.

Cruising the Kennebec is an extraordinary experience, and often a difficult one. Most of the time, you will be under power, and often the current will be against you, even when you have patiently waited for the tide. Much of the uniqueness comes not from the places you visit (though some are special indeed), but from the river itself.

Visit the historic city of Bath, where a centuries-old tradition of shipbuilding persists to this day. Or continue on to Swan Island, a fascinating state-owned wildlife preserve opposite Richmond.

To make a trip upriver to Bath, enjoy the town, and return, allow two days. To go all the way to the fixed bridge at Gardiner, visiting both Bath and Swan Island, allow three or four days. Running down under power from Gardiner with the ebb, you should be able to reach Bath in five hours. Allow at least another two and a half hours to reach Pond Island, at the mouth of the river.

Another cruising itinerary, both challenging and fascinating, is to follow the meandering Inside Passage from Bath to Boothbay, via the Sasanoa, through the Hell Gates, out Goose Rocks Passage, across the Sheepscot River, and through Townsend Gut.

A trip up the Kennebec has its drawbacks. There are few safe anchorages, the current is strong, and the river can be very rough when the wind is against the current, particularly on the ebb. Yet with alertness and familiarity with basic river seamanship, it can be a world of your own.

SOME THOUGHTS ON RIVER SEAMANSHIP

The dominant fact of life in the Kennebec is the current. You go upstream with the flood and downstream with the ebb. Unless you have a powerboat, any other schedule means slow going indeed.

With a summer southerly you can sail upriver as the schooners used to do, but be prepared to lose your wind in certain stretches like Fiddler Reach. Sometimes you may be lucky enough to get a wind that will help you downstream. Because of the funneling effect of the high banks and escarpments, the wind usually blows either straight upriver or straight downriver.

As in most rivers, the natural flow of the river is added to the tidal effect, so the ebb is much stronger than the flood. The flood normally runs one to two knots, and the ebb runs three and a half to four knots. Ebb velocities can reach six knots or higher during spring freshets. Heading upriver, the mean range of tide is 8.4 feet at Fort Popham, 6.4 feet at Bath, 5 feet at Gardiner, and 4.1 feet at Augusta.

According to the *Coast Pilot*, the channel above Bath is subject to considerable annual changes caused by the spring freshets. Have an anchor on deck ready to let go on short notice. Be prepared to kedge yourself off.

A huge fixed bridge carries Route 1 across the Kennebec at Bath. There is no need to go under the bridge in order to visit Bath, but if you want to go farther upstream, your height will be limited to 70 feet at high tide.

A swing bridge at Richmond opens on demand. Then, just above the towns of Gardiner and Randolph, a fixed bridge with a vertical clearance of 35 feet prevents most cruising boats from going farther upriver to the state capital, Augusta.

The river can be extremely rough with a southerly wind against the ebb current. The mouth of the Kennebec is notorious for an ugly chop, as is the stretch between there and Bath. Disquieting tide rips occur in certain places, with strips of foam or brown discoloration, but no change in depth. The chart will reassure you.

It is usually easier in a river to pick up a mooring or come alongside into the current, rather than into the wind. Also, a swiftly running river can deceive the eye. It can be very hard to judge whether your anchor is dragging or whether it's an optical illusion. Watch the land rather than the surface of the water.

Safe anchorages on the Kennebec are few and far between, so allow plenty of time to reach your destination and have a fallback plan. Do not travel on the river at night; there are few lights and many dangers. If necessary, you can always anchor in the river, but it is hard to get out of the current, and you are likely to spend a sleepless night.

SEGUIN

43° 42.68'N 069° 45.39'W

CHARTS: *13295, 13293, 13288*   CHART KIT: *61, 17, 18*

Two miles off the mouth of the Kennebec, high atop lonely Seguin Island, a light burns 180 feet above the sea. Seguin Island Light is Maine's most imposing lighthouse, and one of its most important. It is the second oldest lighthouse on the coast of Maine, built after the Portland Head Light, and President George Washington himself signed the deed for the island. In clear weather, the high, almost treeless island can be seen from a great distance. The powerful foghorn, however, is in frequent use. A visit to Seguin is an unforgettable experience.

Seguin stands like a rock in the slipstream of the great Kennebec River current. From the island's summit you can clearly see the demarcation of the ebb tide. If the wind is running against the ebb, the waters north of Seguin form a nasty chop.

After almost 200 years of manned service, Seguin Light was automated in 1985. Unfortunately, the new automatic devices lack the human qualities of watchfulness, devotion, and judgment. When a sailboarder was blown helplessly offshore, he was sighted and saved by the Coast Guardsmen then stationed at Seguin. In 1991, a sea kayaker who was carried offshore by the Kennebec's current was not so fortunate. One of the great comforts and safeguards of the coast is gone.

Rather than let the splendid Coast Guard buildings fall into disrepair, the Friends of Seguin (*207-443-4808, www.seguinisland.org*) negotiated a lease with the Coast Guard and, in 1997, acquired the light through the Maine Lights Program. Volunteer caretakers live here during the summer months and run a small but growing museum in one of the rooms of the keeper's house, which displays memorabilia and photographs of Seguin. If you are lucky, they will take you to climb the beautiful spiral staircases of the light tower and watch the rainbows refracted on the walls by the huge, first-order Fresnel lens.

APPROACHES. From the east, head for bare Ellingwood Rock, then head into the cove on a southerly course. Note the ledge making out northward from the northeast tip of Seguin.

Coming from the west, leave Ellingwood Rock close aboard to starboard, passing south of Seguin Ledges. Then head for the cove.

Be aware that within a mile of Ellingwood Rock there may be a local magnetic disturbance that could throw your compass off by as much as 8 degrees.

ANCHORAGES, MOORINGS. All of the north cove of Seguin Island is a cable area and *anchoring is not allowed*. The Coast Guard will board, inspect, and possibly fine anyone they find anchored here. Or worse, you could damage the cable and put out the light. Instead, pick up one of the two Coast Guard moorings or the moorings placed by the Friends of Seguin. These are used on a first-come, first-served basis, but allowing others to raft to you is standard policy when it is busy. Row ashore to the tiny sand beach.

THINGS TO DO. Walk past the beach peas, the seagulls roosting in the trees, and the boathouse. Then follow the path by the ramp and railway up to the lighthouse on the crest. From there you have magnificent vistas westward to Cape Small and the dim islands of Casco Bay and eastward to The Cuckolds and Monhegan on the horizon. The small museum is in the keeper's quarters.

Recently, as part of its program to convert most of its lighthouses to cheaper solar power, the United States Coast Guard planned to remove Seguin's rare, Fresnel lens and place it in a museum. The lens was first installed in 1857, replacing the series of mirrors that were used previously, and now Seguin's is the only first-order lens north of Rhode Island that is still operating. Not wanting to diminish the light's historical significance—or its beam—the Friends of Seguin collected more than 7,000 signatures for a petition and successfully lobbied the Coast Guard to leave the lens where it belongs. As the Maine saying goes, "Don't fix it if it ain't broke."

FORT POPHAM

1 ★★★

43° 45.34′N 069° 47.19′W

CHARTS: 13295, 13293, 13288   CHART KIT: 61, 17, 18

The shoreline at the mouth of the Kennebec River came very close to being the site of the first permanent English settlement in the New World. The Plymouth Company, assigned the rights to colonize "Northern Virginia," from 38 to 45 degrees north latitude, sent out colonists in early 1607 led by George Popham and Raleigh Gilbert. During that summer, the group crossed the Atlantic in two small ships, the *Mary and John* and the *Gifte of God* and established the Popham Colony.

According to the journal entry of Captain Robert Davies on August 17th, "Captain Popham in his Pynnance with 30. persons, & Captain Gilbert in his long boat, with 18. persons more, went early in the morning fromm their shippe into the River of Sachadehoc, to view the River, and to search where they might fynd a fitt place for their plantationn." Fort St. George and a number of houses were constructed that fall on Sabino Head, along with the 30-ton pinnace *Virginia*, the first English vessel to be built in the New World.

But winter struck with extraordinary severity that year, and many of the colonists died, including Popham. Although resupplied the next summer, the survivors refused to face another winter, and they were evacuated in the fall of 1608. And so it chanced that Jamestown, founded by the London Company in May 1607, became instead the first permanent English colony in the New World.

Fort Popham was built near the site of the original colony during the Civil War to defend the important Kennebec shipbuilding industry against the Confederates. It was manned again during the Spanish-American War. The massive fort is now part of a state park and fun to explore.

The anchorage contains numerous moored boats, but on an anchor, it is not very secure and exposed both to the ceaseless river current and to wind from all directions but the south. Fort Popham is best explored as a daystop in all but the most settled weather.

ANCHORAGES, MOORINGS. To stop here for a visit, round Fort Popham on Hunnewell Point and anchor outside or among the moored boats in about 20 feet of water. Note that all of the cove formed by Hunnewell Point dries out at low tide and that the narrow slot left in Atkins Bay to the west is full of private moorings, with no room to anchor. Row in to the dinghy float at the dock on the northwest tip of Hunnewell Point.

THINGS TO DO. Bud Warren of Phippsburg is leading a crusade of modern pilgrims to raise funds to build a replica of the *Virginia*. If you would like to become involved, call or visit their website (*207-389-2990, www.mainesfirstship.org*).

Just south of Fort Popham, you can walk along beautiful Popham Beach or go for a swim.

Spinney's Restaurant (*389-1122*) is near the fort, and groceries, takeout, and ice can be found at Percy's, almost next door. The small Popham Library is down the street. Fort Baldwin, built on the top of Sabino Head during World War II, is now a park and open to the public. Hike up the short trail from the parking lot to the lookout tower, shown on the chart, and climb its four stories. It is hard to know which makes you more breathless, the climb or the view.

PERKINS ISLAND

43° 47.22′N 069° 46.95′W

CHARTS: 13295, 13293, 13288   CHART KIT: 61, 17, 18

Small, rotund Perkins Island lies three miles up from the mouth of the Kennebec, just east of can "13." Perkins is owned by the Bureau of Parks and Lands and is open to visitors and campers. It makes an interesting daystop with its lighthouse, bell tower, abandoned keeper's buildings, and its lengthwise views of the Kennebec.

Note the cable area on the northeast side of the island. Anchor instead to the east of the island and dinghy to the old Coast Guard landing about halfway down the east shore.

If approaching from the south, keep nun "12" to port and favor the Mark Island side to stay in the deepest part of the channel. The east side of the island is steep-to, the water is deep, and the current can be strong. The bottom, however, is mud and holds well.

If you prefer shallower water, you can anchor closer to Mark Island and carefully dinghy across. A trail leads from the old landing to the Coast Guard buildings and lighthouse on the west side.

BATH

3 ★★★★

south of bridge: 43° 53.71′N 069° 48.80′W
north of bridge: 43° 54.83′N 069° 48.70′W

CHARTS: 13296, 13298, 13293, 13288   CHART KIT: 61

Emergency: 911
Hospital: 207-443-5524

Bath is an interesting small city of 10,000 inhabitants on the west bank of the Kennebec, about 12 miles above the river's entrance. It has a long history of shipbuilding, elegant old houses, many marine-related activities, and a flourishing arts scene. This is the home of the Maine Maritime Museum and Bath Iron Works, one of the country's leading naval shipyards. It is easy to spend an enjoyable day or two here.

Opposite Bath, on the east shore of the Kennebec, is the entrance to the Sasanoa River, the beginning of the Inside Passage that leads east and south to Boothbay Harbor.

Ever since the launching of the pinnace *Virginia* at the nearby Popham Colony, the history of the Kennebec River and of Bath has been linked to ships and shipbuilding. Century-old Bath Iron Works (*www.gdbiw.com*) maintains that tradition, adhering to its motto "Ahead of schedule and under budget."

Founded in 1884 by young Brigadier General Thomas W. Hyde, BIW has had a roller-coaster history, soaring to periods of greatness and surviving a devastating fire and two bankruptcies. When it went up for public auction in 1925, the yard was sold to a scrap metal dealer and stripped, and for a time, in spite of its proud heritage, BIW launched nothing but pie plates.

But new entrepreneurs stepped forward, the yard was rebuilt, and the next decades saw the launching of a series of famous yachts. In 1930, J.P. Morgan's 343-foot black-and-gold *Corsair* slid down the ways. "What does it cost to maintain her?" This was the yacht about which Morgan replied, "If you have to ask, you can't afford it." Then followed trawlers for F.J. O'Hara's fishing fleet, destroyers for the U.S. Navy, and Harold Vanderbilt's famous J-boat *Ranger*, which defeated *Endeavour* in 1937 for the America's Cup and won every race she ever entered.

Guided missile frigates at Bath Iron Works under "Number 11," the tallest crane in the western Hemisphere. (BIW)

During World War II, BIW built 82 destroyers, more than the number produced by the entire Japanese Empire. More recently, it has been building guided missile frigates, cruisers, and the Arleigh Burke class of destroyers.

Bath Iron Work's behemoth cranes brood over Bath like prehistoric monsters. With capacities to 300 tons, the cranes can handle enormous modular sections of ships, a building method that has contributed to BIW's success. The yard also takes pride in its tradition of hard work, high quality, and low turnover in the experienced workforce.

APPROACHES. Bath's waterfront is split by the green Carlton Bridge and the new Sagadahoc Bridge. The Carlton's lift span will open on demand if you call Jenny or one of her assistants on Ch. 13 or at *443-2482*. She reminds powerboaters, though, that there is plenty of depth and clearance under the fixed, east end of the bridge.

The Sagadahoc, which carries busy Route 1 across the river, is fixed with a minimum vertical clearance of 70 feet.

Be particularly careful of the strong currents when passing under the bridges. Currents at

SHIPBUILDING IN MAINE

The mammoth cranes that dominate the sprawling shipyard of Bath Iron Works, on the banks of the Kennebec, represent the contemporary incarnation of an industry that has thrived on the Maine coast since 1607. Sleek gray destroyers and missile frigates are today's versions of the 30-ton pinnace *Virginia*, built at the turn of the 17th century by the ill-fated settlers of the Popham Colony. Much has changed in the ensuing 400 years, but the shipbuilding tradition remains strong in communities all along the Maine coast.

For more than three centuries after the construction of *Virginia*, ships were built up and down the coast, along riverbanks, and on islands, wherever there were settlers. They were built from the materials at hand, supplemented by special items cannibalized from older boats or brought from Europe. They were built mostly by farmers and by fishermen-turned-boatbuilders, often for their own use.

Shipbuilding as a business flourished from the middle of the 18th century, with a series of booms and busts, as entrepreneurs from Kittery and Kennebunkport, Wiscasset and Bath, Waldoboro and Camden, Belfast and Roque Island built ships for the Caribbean trade and the trade with Europe. All this halted abruptly with the Napoleonic Wars and President Jefferson's Embargo Act of 1807.

The great era of building wooden sailing ships in the United States lasted about a century, from 1800 to 1900. The boom years were even shorter, from 1843 to an all-time national peak in 1855.

During most of the era of sailing ships, Maine built the largest number of them in the United States. The center of shipbuilding was Bath, where shipyards lined the banks of the Kennebec. In the boom year of 1854, only Boston and New York City built more. The Waldoboro region was second to Bath in that peak year, and their cumulative total placed Maine well out ahead as the nation's leading shipbuilding state. Ships and barks, barkentines and brigantines, brigs and schooners slid down the ways to meet an endless demand for Maine-built vessels to carry the world's cargo.

The efficient carrying of freight was the watchword until the 1840s, when speed became an important factor for the China trade. When gold was discovered in California in 1848, speed became everything.

The beautiful, fast clipper ships were developed to meet this demand, and shipbuilding surged during the 1850s. It is 15,000 miles from New York to San Francisco around Cape Horn. The clippers were able to make the passage in an average of 130 days, and some did better. The Bath-built *Flying Dragon* once made it in 97 days! But the Gold Rush was short-lived, and so was the clipper-ship era.

Then it was California again, with wheat as the new gold—wheat for the markets of Europe. Both speed and cargo capacity were important, and the "Downeasters" built for this trade were the pride of Maine.

But the profitability of voyages halfway round the world declined with foreign competition, and new markets developed for coastwise trade—mundane traffic in coal, lumber, lime, ice, and granite. To transport these bulk cargoes, Maine developed the economical "coasting" schooner, with its fore-and-aft rig and much-reduced crew ("two men and a boy," they used to say).

The two-masted schooner soon evolved into the three-master, correspondingly larger and more profitable. Four and five-masters were built after the 1880s. The first six-master was built in Camden in 1900, and the second at Percy and Small Shipyard in Bath that same year.

Shipbuilding ultimately declined due to its own pace of innovation. One Bath builder saw the possibility that oceangoing tugs, towing strings of barges, could carry cargo at a fraction of the cost of the great schooners. In 1890, 36 wooden sailing vessels were built in Bath; in 1910, none were built.

It's interesting to speculate on what has happened to all these hundreds and hundreds of great sailing ships. Many were lost, of course, on rocks and ledges the world over, from Bishop Rock, off Land's End, to the Straits of Magellan, on lonely Pacific reefs, and in the roaring 40s and the Tasman Sea. Many more were wrecked within sight of home, on Portland Head and Matinicus and the Libby Isles. Fire claimed many of the lime ships, and storms caught the unfortunate granite schooners.

Those that survived 25 or more years of profitable voyages often were converted to ignominious uses such as storage hulks or barges. The famous clipper *Red Jacket* ended her days in obscurity as a coal hulk in the Cape Verde Islands. For years, the four-masters *Luther Little* and *Hesper* lay rotting in Wiscasset Harbor for tourists to photograph.

Many ships were simply junked when their useful days were over—run up in some convenient cove and left to fall apart. Their ribs emerge from the bottom at times when the tide is low or the sands are disturbed by winter storms.

A few have fared better, like the schooner *Lewis R. French*, built in 1871 and now the oldest documented pure sailing vessel in the U.S. merchant fleet. She and the *Stephen Taber*, also built in 1871, in New York, still sail as windjammers out of Rockland. Admiral Donald MacMillan's famous arctic research schooner *Bowdoin* was refurbished at Percy and Small in Bath and launched anew in 1985.

In modern times there has been something of a renaissance for traditional wooden sailing ships, and they are still being built in small numbers on the coast of Maine, most destined for the windjammer trade.

Large nontraditional wooden sailing ships have also been built in Maine in recent decades, designed by Bruce King of East Boothbay and built by Phil Long. The magnificent cold-molded ketch *Whitehawk*, reminiscent of *Ticonderoga*, was launched in Rockland in 1979, followed by the 90-foot sloop *Whitefin*, constructed on a Rockport tennis court in 1983. In 1990 the luxurious 100-foot ketch *Signe* went down the ways at the Renaissance boatyard in Thomaston.

maximum ebb can reach 6 knots or more, and much weaker currents can still sweep you onto the bridge abutments. Also remember that you will usually want to come alongside or pick up a mooring heading into the current rather than into the wind.

ANCHORAGES, MOORINGS—Below the Bridge. *Percy and Small Shipyard, Maine Maritime Museum (207-443-1316; www.bathmaine.com).* Below the bridge, the only place for a berth or mooring is the Percy and Small Shipyard, located about a mile south of the bridge and part of the Maine Maritime Museum. Both dockage and moorings are available for visiting cruisers, and the rental of either includes admission to the museum and the use of their yachtsmen's building, with laundry, showers, and heads.

The museum has 170 feet of floating dock space, with 8 feet of depth at low and water and electricity. Large vessels can lie along their fixed pier, where depth is 17 feet at low. Calling ahead is recommended.

Ten moorings are set in two lines. The outer, heavier moorings can accomodate sailboats to 50 feet and powerboats to 45 feet. The lighter, inner moorings can handle boats to 32 feet. They are marked by numbered white balls and lobster-buoy pickups. Don't make the mistake of trying to pick up the white cylinder with red markings near the *Sherman Zwicker* dock—it marks a ledge with 4½ feet at low.

If no mooring is available, anchor in some convenient spot on the river in 15 to 25 feet at low. Use a generous amount of scope.

GETTING ASHORE. The museum does not offer a launch service, so you will need to dinghy in to the floats at Percy and Small. Remember that the current runs hard. If you don't have an outboard, be sure you can row against it. Avoid overloading your dinghy, and take all precautions.

ANCHORAGES, MOORINGS—Above the Bridge. There are three possible places to stop just north of the bridge, all on the west bank. Though convenient to town, they cater mostly to small powerboats and are exposed to strong current.

BFC Marine (207-443-3022). The first small-boat slips just above the bridge belong to BFC Marine. The slips are rented seasonally. Gas and water are available, but not diesel. BFC is an Evinrude dealer and has a small chandlery, which sells charts.

Town Float Landing (harbormaster, police 207-443-8339). Transients may stay at the main float (between the large dolphins) for limited periods or overnight with reservations or permission. Depths are 15 feet at low. Pump-outs and fresh water are available, and restrooms are in the adjacent park.

Kennebec Tavern and Marina (207-442-9636) lies north of the town landing. They rent heavy moorings to transients and keep space open on their floats for restaurant patrons. If there is a vacancy, they would be happy to put you on their docks, most of which are in range of their Wi-Fi signal. The fuel dock pumps gas and water and sells ice.

FOR THE CREW. Just about anything you need can be found in Bath close to the waterfront, including a drugstore, a hardware store, banks, beauty and gift shops, and restaurants.

To provision, walk several blocks north on Front Street to Brackett's Market (*443-2012*). A Shaw's supermarket (*443-9179*) is about a mile farther west on Route 1. A pharmacy and laundromat are in the same shopping center as Shaw's. A natural foods store is on Centre Street, one block from Front Street, or you may be lucky enough to catch the farmers' market in the town park by the river.

Mae's Café and Bakery serves breakfast, lunch, and dinner (*442-8577; Tues-Sun*). It is worth the ten-minute walk up Centre Street for good food, warm bread, wonderful pies, and delectable baked goodies to bring back to the boat. Or tuck into some ribs, gumbo, and shrimp at Beale Street Barbeque (*442-9514*) on Water Street where it parallels Front.

Midcoast Hospital (*443-5524*) is nearby at 1356 Washington Street.

THINGS TO DO. There is plenty to do here for both grown-ups and children, much of it related to small boats, shipbuilding, the river, and the great maritime tradition of the "City of Ships." For security reasons, however, the public is not allowed to visit Bath Iron Works except on rare occasions such as launchings. The Chamber of Commerce or BIW (*443-3311*) will give you the launching schedule.

The Maine Maritime Museum offers a wide variety of sights and activities for the entire family. The core of the museum is the Percy and Small Shipyard, the only surviving shipyard in the country where large wooden ships were built. In the summer, the yard is home to the Grand Banks fishing schooner *Sherman Zwicker,* and it is visited by sailing vessels such as *Westward* and the restored arctic schooner *Bowdoin.* Construction of a replica of the Popham colony's ship *Virginia* is slated to begin here as soon as funding is secured (see page 110).

There is a collection of classic wooden small craft at the yard, impressive exhibits on boatbuilding and lobstering, and a section of the bow of the last of the clipper ships, the *Snow Squall*, built in South Portland in 1851 and recovered from the Falkland Islands. Live exhibits on sailing, shipbuilding, navigation, and other nautical subjects are presented during the summer, and the Maritime History Building holds a trove of marine art, ship models, and displays of life at sea, as well as a museum store. If you still don't feel salty enough, the museum runs frequent narrated boat cruises on the river.

The Center for the Arts is the formal name for what is locally known as the "Chocolate Church," an old dark brown church on Washington Street, near Centre Street, used for a wide variety of exhibits, concerts, and plays. There is an annex with an art gallery and a small theater downstairs. Call *442-8455* for schedule information.

WOODS, CRAWFORD, and RAM ISLAND

3 ★★★

43° 57.12'N 069° 49.66'W

CHARTS: 13293, 13288   CHART KIT: NOT AVAILABLE

Three miles above Bath, the river turns sharply, broadens, and forms three channels leading to the Chops and on into Merrymeeting Bay. Several small and beautiful islands, all trees and granite, nestle among these channels.

APPROACHES. Of the three channels leading to The Chops, by far the most difficult is the one to the right, east of Thorne and Lines Island. The section above Lines Island is full of rocks—don't try it.

The channel to the left, south of Woods, Crawford, and Ram Island, also contains some unmarked rocks, but is far easier.

The easiest channel of the three is straight through, north of Woods, Crawford, and Ram. On an ebb tide, leave nun "10" well to starboard to avoid being set down on Thorne Island Ledge.

ANCHORAGES, MOORINGS. Anchor between Crawford and Ram in 7 to 20 feet at low. Do not anchor in the area west of Ram Island, where you will be in the full strength of the current.

MERRYMEETING BAY

43° 58.97'N 069° 50.12'W

CHARTS: 13293, 13288   CHART KIT: NOT AVAILABLE

Above Lines Island the Kennebec passes into Merrymeeting Bay through the Chops, a deep passage between high Chops Point and West Chops Point. The overhead power cables across the Chops have a vertical clearance of 145 feet. Sailors Peter and Jean Willard run Chop Point Camp (*737-2725*), directly on the point. They have a heavy guest mooring and welcome visitors to explore or to join them for a meal in the mess hall. Sturgeon Island and its small companion just to the west are state owned and have lovely beaches. Anchor in the 13-foot spot north of Sturgeon Island to explore.

Merrymeeting Bay is formed by the confluence of six rivers—the Androscoggin, Muddy, Cathance, Abagadasset, Eastern, and Kennebec. The extensive shoals and marshes of the bay make it a haven for waterfowl but no place for deep-draft boats.

EASTERN RIVER

4 ★★★

43° 03.18'N 069° 46.64'W

CHARTS: 13293, 13298, 13288   CHART KIT: NOT AVAILABLE

This is a short river for true gunkholers. Starting at the southern tip of Swan Island, Eastern River meanders for about two miles through expanses of marsh until it reaches a fixed bridge with a vertical clearance of 16 feet. Feel your way up the river very carefully at dead low so that the banks and all the hazards are clearly visible.

The south end of the "training wall" shown on the chart is marked by a granite block, and there are some small and indecipherable private marks in the area. Using chart and fathometer, work carefully past the rocks at Carney Point. At the left turn, the channel shoals. Stay well over to the right. From there on, it is relatively easy. Just run up about halfway between the marshy shores.

This would be an excellent place to hide during a storm or to find refuge from the Kennebec current if no mooring is available farther up in Richmond.

RICHMOND

4 ★★

44° 05.28'N 069° 47.85'W

CHARTS: 13298   CHART KIT: NOT AVAILABLE

Richmond is a small and remote town tucked away on the west bank of the Kennebec, 11 miles north of Bath and 23 miles above the river's mouth. Several handsome old red brick buildings form the downtown. One of the buildings has been recycled to manufacture electronic parts. Fascinating Swan Island is off Richmond.

APPROACHES. The approach to Richmond is up the main, eastern channel of the river, on the eastern side of Swan Island. Just before you reach the swing bridge, curve left around the north end of Swan Island, leaving to port can "33," which marks the end of a submerged jetty.

You will see a pier and float on the north tip of Swan Island to port before you get to the red brick buildings of Richmond.

ANCHORAGES, MOORINGS. Do not anchor here. The water main for the town was not placed with the mariner in mind; it runs straight through the anchorage.

As you come around the northern end of Swan Island, the first landing on the Richmond side belongs to the state and is used by the Department of Inland Fisheries and Wildlife to ferry visitors to Swan.

Next comes the Richmond town landing and float at a waterfront park, with 16 feet of water alongside at low. You can tie up here for two hours, but not overnight without permission from the harbormaster. There are no facilities.

The Swan Island Yacht Club may have a few guest moorings near the northwest bank, just before you reach the first floats. Or you may find a vacant private mooring if you inquire at the town dock.

GETTING ASHORE. Row in to the town landing.

FOR THE BOAT. *Ideal Auto and Marine (207-737-8401)* sits on the hill just north of town. They haul with a small hydraulic trailer and can help you with repairs.

FOR THE CREW. The Front Street Market is just across the street from the town dock, but it stocks more convenience items than groceries—beer, ice, and snacks. For provisioning, head up the hill on Route 197 to Pierce's Country Store, passing a bank and a hardware store on your way. The Railroad Cafe across the street, the only act in town, is said to serve great breakfasts. For dessert, try the Old Goat, a tavern on lower Main Street.

THINGS TO DO. Time seems to have passed by Richmond. Settled in large part by Russian immigrants, the town's architecture still combines an onion-dome church with a few blocks of some of Maine's most unspoiled Greek Revival houses. Take the time for a stroll.

A new innovative business has chosen Richmond as its home. Shucks Maine Lobster (*www.shucksmaine.com*), at 150 Main Street, does just that: shucking raw lobster with high water pressure and packing the uncooked meat for markets. Sorry, no retail.

During the last week in July, Richmond hosts "Richmond Days," which culminates in fireworks exploding above the moorings. "It's small-town America at its best," writes sailor Stephen Mayotte.

SWAN ISLAND

4 ★★★★★

44° 05.28'N 069° 47.75'W

CHARTS: 13298   CHART KIT: NOT AVAILABLE

Swan, the Kennebec's largest island, is four miles long and a half-mile wide, ringed by a bright green fringe of wild rice. The Indians named it *Sowango*, the "Island of Eagles," and today several still nest there. If you are lucky, you may experience the thrill of seeing one of these magnificent birds flying up the river.

The island is now the Steve Powell Wildlife Management Area (*www.maine.gov/ifw/education/swan-island*), owned and managed by the Department of Inland Fisheries and Wildlife as a sanctuary for migrating waterfowl, particularly Canada geese. Some 300 acres are kept as open field, and in the spring as many as 5,000 geese may be grazing in the meadows.

Swan Island is home to a large herd of white-tailed deer that come and go as they please by swimming the river or crossing the ice. The woods on Swan Island are bare of underbrush or low branches to the height of a deer's reach when it stands on hind legs. This vegetation contrasts vividly with the areas inside various fenced "enclosures" that are used for protecting gardens and for experimental purposes.

European visitors inhabited Swan Island as early as the beginning of the 18th century. In 1750 the Widden family lived there when Indians attacked, captured 11 family members, marched then to Canada, and sold them as slaves to the French. The price for four of the adults was reported as $29 apiece.

During the fall and winter of 1775, when Benedict Arnold's bold and futile expedition to Quebec struggled up the Kennebec, Aaron Burr and Benedict Arnold spent the night on Swan Island in the Dumaresq House, which still stands today.

The pre-revolutionary Pownalborough Courthouse sits high on Courthouse Point on the eastern bank.

Swan prospered for centuries as a farming and fishing community and at one time as an exporter of lumber and ice. But the Depression brought hard times, and eventually Swan was abandoned. The state bought it for back taxes during the 1950s.

APPROACHES. Follow the approach to Richmond in the preceding entry.

ANCHORAGES, MOORINGS. Leave your boat in Richmond.

GETTING ASHORE. To maintain control over this important and delicate waterfowl staging area, Inland Fisheries and Wildlife staff request that visitors come through Richmond.

THINGS TO DO. Swan Island is open to the public on a limited basis—only 60 people a day. No pets are allowed. Make reservations ahead of time, whether for a brief visit or for an overnight stay at the island campground by calling *547-5322*. People who arrive without a reservation sometimes can get on the island, but it depends on how many already have reservations for that day.

Visitors are ferried across from the state landing at Richmond, loaded on a rack truck, and driven the length of the island with a guide. A mile from the Swan Island landing, a beautiful meadow looks south to the river and across to Little Swan. Three-sided lean-tos with fireplaces in front are available by reservation for overnight campers, who are free to wander around the northern part of the island. Visitors are permitted south of the campground only with the resident wardens.

THE BRIDGE above RICHMOND
44° 05.42′N 069° 47.00′W

CHARTS: 13298   **CHART KIT:** NOT AVAILABLE

The road running east from Richmond carries relatively little traffic, but so does the river. Operation of the Richmond-Dresden Bridge is, therefore, understandably casual. Bridgetender Elin Page can be reached on VHF 09 or at *737-4751*.

GARDINER and RANDOLPH
[3] ★★
44° 13.76′N 069° 46.10′W

CHARTS: 13297   **CHART KIT:** NOT AVAILABLE

Gardiner lies on the west bank of the Kennebec, 33 miles from the sea, opposite its sister town of Randolph on the east bank. This marks the limit of navigation for most cruising sailboats. Just above the towns, an old swing bridge has been replaced by a fixed bridge with clearance of only 35 feet.

APPROACHES. From Richmond, continue north about 10 miles through the buoyed channel at Tarbox Flats to the twin towns of Gardiner and Randolph.

This is historic territory. One mile north of the Richmond Bridge, on the east bank, you may catch a glimpse through the pines of a large, white, foursquare house on a bluff labeled on the chart as Courthouse Point. This was the pre-revolutionary Pownalborough Courthouse, the first seat of government in the area. Beyond that, though not visible from the water, is the Colburn House (*582-7080*), where, in 1775, Benedict Arnold loaded his 1,100 men and their supplies into 220 bateaux, built by Major Reuben

Colburn, for his expedition to Quebec. Though not easily accessible by boat, both landmarks are open to the public. Farther upriver you will pass Oakland, an unmistakable castle-like estate set back on the west bank. This is still owned by the family of Sylvester Gardiner, the founder of the town that bears his name.

ANCHORAGES, MOORINGS. The small Smithtown Marina is on the east side of the river opposite Oakland. They may have space for transients. Farther upriver, the Gardiner town landing is on the west bank next to a small park. You can tie up to the town float in 12 feet of depth for a reasonable daily rate, although you will have to cope with the glare of arc lights and the rumble of an occasional truck. Be sure to stay clear of can "47A" upstream from the landing; it marks the end of submerged pilings left over from the old swing bridge.

FOR THE BOAT. *Smithtown Marina (207-582-3153)*, on the east bank below Gardiner, has gas, ice, and pump-outs. Slips are seasonal, but dockage may sometimes be available for transients. In the winter you can rent a smelt camp here.

Gardiner Town Landing can arrange water and electricity hookups on their float.

FOR THE CREW. On the waterfront in Gardiner, the town has posted a useful map showing all the nearby stores and facilities, including a market, laundromat, bakery, hardware and liquor stores, and restaurants. From the town landing head north several steps to a shopping plaza with a Hannaford supermarket, a bank, a hardware store, and a Radio Shack.

Randolph is less convenient and smaller than Gardiner, but it has an IGA, a pharmacy, and one of the better named cleaneries: the Polyclean Laundromat.

THINGS TO DO. If you are in need of a nostalgic throwback, bowl a few strings at Lucky Strike Lanes, right on the river across from the supermarket, and then hit the classic A1 Diner for a meal or snack. For an evening out, check the schedule of live theater and concerts at the Johnson Hall Performing Arts Center (*582-7144*) on Main Street.

Over the Fourth of July weekend, Gardiner is the finish point for the annual Great Kennebec River Whatever Week Race—an eight-mile regatta run downriver from Augusta that features homemade craft of great imagination and doubtful ancestry.

INSIDE PASSAGE—*Bath to Boothbay*
CHARTS: *13296, 13293, 13288*   CHART KIT: *61, 62*

The Inside Passage from Bath to Boothbay Harbor is one of the great adventures of the Maine coast. Like marriage, however, it is not to be entered into lightly, but advisedly… and soberly. The current in two of the stretches, Upper and Lower Hell Gates, can be awesome for those of us not accustomed to running whitewater rivers in deep-keeled boats.

The 11-mile passage cuts southeast and northwest across the grain of glacial valleys, broadening into placid bays and narrowing into sluices through which the current rushes with great force. You must pass under a fixed bridge with a 51-foot vertical clearance at the western end near Bath and through a swing bridge at Townsend Gut that connects Southport Island to the mainland.

If you contemplate this passage, be sure you know the height of your masthead above the water, including antennas. Study the chart carefully ahead of time, and study the *Tidal Current Tables.*

The western part of the Inside Passage, including Upper and Lower Hell Gates, is by far the most difficult. Theoretically, the current floods northwest from Boothbay Harbor to Bath and ebbs southeast. But currents here are also driven by a difference in the heights of the Kennebec and Sheepscot rivers, and a number of bays and rivers empty into the passage, making the timing of the currents complex and variable.

Powerboats that know the territory come through at any tide, but cruising sailboats should come through near slack water or with a minimum of current. Much of the time, the current in the narrows will be running at 3 or 4 knots and peaking at 8 knots or more in Upper and Lower Hell Gates. What makes these spots so dangerous is not so much the force of the current as the turbulence, which may cause you to yaw or lose control.

Most cruising sailboats power all the way through the passage. It is possible to sail in certain areas, but you will often be blanketed and be at the mercy of the current. There is no room to tack in the narrows, and there may be floating debris like logs and branches. During summer there is often a lot of traffic, especially powerboats, and at certain tides, the narrows are full of sport fishermen.

There are four narrow places in the Inside Passage where the current is strongest: Upper Hell Gate, Lower Hell Gate, Goose Rock Passage, and

Townsend Gut. The amount of current and difficulty of the passage are greatest at the western end and least as you approach Boothbay Harbor.

The description below assumes that you are entering the Inside Passage at Bath and heading southeast. Pointers on making the passage the other way, from Boothbay Harbor to Bath, are at the end of this section.

ARROWSIC BRIDGE ENTRANCE (BATH). Your ability to make the passage will be determined unequivocally by whether your mast will fit beneath the fixed Arrowsic Bridge, spanning the Sasanoa River. The bridge has a vertical clearance of 51 feet at mean high water. Remember that every tide is different, and remember your flagstaff and antenna. The mean tidal range at Bath is 6.4 feet, so on a normal low tide, you should be able to carry 57 feet under the bridge… just.

Assuming that your mast will clear the bridge near a high tide, the best time to enter the Sasanoa River and begin making the passage from west to east is just before high slack water at Upper Hell Gate. You will have a minimum of current at Upper Hell Gate, and the ebb will help you through the rest of the passage.

Use the *Tidal Current Tables* to determine the time of high slack water, which is not the same as high tide. If you are too early or need to come under the bridge at lower tides, you can anchor on the eastern side of Hanson Bay in the area between can "35" and nun "34" in a mud bottom. A former captain of the cruise boat *Argo* reported that sometimes slack high water at Upper Hell Gate lasts only a few minutes. "Then it starts to ebb like hell."

It is possible, but inadvisable, to make the passage starting at low slack water instead. Your margin of error in the narrow and shoal places will be reduced by the low water levels, and the current will turn against you with gusto.

THE SASANOA RIVER AND UPPER HELL GATE. This is the narrowest and most difficult part of the Inside Passage. Follow the channel buoys around the slow S-turn though Hanson Bay and into the narrows, leaving green to starboard, and red to port.

Watch carefully now for turbulence and cross-channel currents. As you approach Money Point, watch your chart carefully. Leave cans "31" and "29" fairly close to starboard to keep away from Carlton Ledges, which are covered at high and have snagged

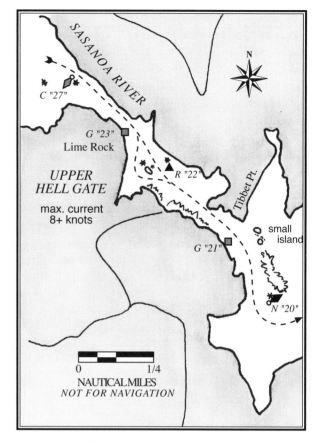

many boats. (John Hilpman, whose dock is 200 feet east of Carlton Ledges, invites stranded mariners to come ashore, use the phone, have a beer, and wait for the next high tide to refloat their vessels. Now there's an invitation to hit a rock!)

Shortly after can "27" you will come to the narrowest part of the passage at Upper Hell Gate, only about 60 yards wide. The chart shows several tiny islands. At high tide they are very hard to find, consisting of flat rocks and grass only barely above water. Stay close to the left bank as you approach green daybeacon "23" at Lime Rock, which you leave to starboard. You can almost reach out and touch the rocks on the bank with your left hand. Then pass between red daybeacon "22," left close aboard to port, and the little island (barely visible above the water at high), which you leave to starboard.

It is also possible to go to the other side of this little island, leaving it to port, and at low tide that would be preferable. There is deeper water here, but it requires a sharp turn to starboard after green daybeacon "23."

Once past red daybeacon "22" the worst is over, but note the small island off Tibbet Point and the ledge beyond marked by nun "20."

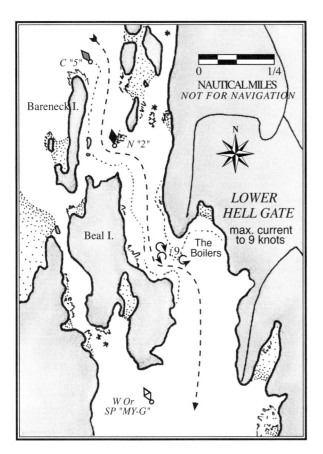

At its strongest, there may be 8 knots of current or more in Upper Hell Gate, churning with boils, upwellings, and standing waves in the narrowest spots. Bill and Barbara Hadlock once tried to fight their way against the full current in their Friendship sloop, which could do 6 knots under power. They didn't make it. It was too narrow to turn around, so they were actually flushed back.

HOCKOMOCK BAY AND LOWER HELL GATE. At Mill Point you enter Hockomock Bay, a broad calm area filled with flats and shallows, fringed by marsh grass and featureless shores, and named for the Hockomock Indians. Montsweag Bay and the Back River trend off to the northeast. Follow the buoys through Hockomock Bay, north of Castle Island, checking them off carefully. Can "5," north of Bareneck Island, is often dragged under water.

The maximum current in Lower Hell Gate occurs near The Boilers, where a velocity of 9 knots has been observed and where dangerous boils and eddies can spin you around.

The main danger is the ebb current that sets you cross-channel to the southeast, toward The Boilers. After rounding Bareneck Island, steer well to the right of nun "2," making a sharp turn to port when you are sure you can clear The Boilers.

Fortunately the passage is clear and reasonably wide. But stay alert because you never know what kind of traffic you might meet. With the current flushing fish out of Hockomock Bay, Lower Hell Gate is a popular spot for sport fishing, and anglers' boats may be in every conceivable eddy with lines everywhere. Beal Island belongs to the Appalachian Mountain Club, which has built portaging trails so canoeists can avoid Lower Hell Gate, but some canoeists and kayakers crave the current.

KNUBBLE BAY AND GOOSE ROCK PASSAGE. Breathe a sigh of relief. Now you pass through Knubble Bay, a broad and beautiful expanse of calm water bordered by handsome saltwater farms. Give Little Knubble a comfortable berth and enter Riggs Cove, where you can pick up a Robinhood Marine mooring, or continue on through Goose Rock Passage (*Robinhood* and *Riggs* are discussed in detail on pages 124 and 125).

The current flows hard through Goose Rock Passage, but the channel is quite wide and straightforward. Entering from the west, run past Lowe Point and southward along the shore in deep water toward the flashing green light "5" on a skeleton tower ashore. Find nun "4" at Boiler Rock as early as possible and keep track of its location. Until it was enlarged, the buoy was often dragged under until only its tip showed—or nothing at all. Leave it well to port. If you can't see it, run along a line from light "5" on shore to a point north of can "3," favoring the right shore and keeping the can to starboard.

The remainder of the passage is wide and clear. Leave green daybeacon "1" north of MacMahan Island to starboard as you enter the Sheepscot River (see *Goose Rock Passage*, page 124).

TOWNSEND GUT. Work your way across the Sheepscot River and south, and approach Townsend Gut north of Dogfish Head and Ebenecook Harbor. Dogfish Head can be recognized by the skeleton tower shown on the chart ¼ mile south of the Head and by the round, white-roofed house on the bare rocks of the point.

A second approach to Townsend Gut is north of the Isle of Springs. Head up the Sheepscot until you have located green daybeacon "1" north of the Isle of Springs, and keep it to starboard. Proceed beyond the daybeacon before you turn to starboard and head south between Sawyer Island and Isle of

Springs. Favor the Isle side to avoid the ledge making out southward from Sawyer.

The passage through Townsend Gut, the anchorages, and the swing bridge are discussed later (see *Townsend Gut*, page 127).

The Inside Passage—East to West. In order to encounter a minimum of current, Glen Baldwin of Robinhood Marine recommends that you leave Robinhood about one hour after low slack water at Upper Hell Gate. The alternative, leaving Robinhood about one hour *before* high slack water at Upper Hell Gate, seems to work equally well.

MONTSWEAG BAY and BACK RIVER
CHARTS: 13296, 13293, 13288   CHART KIT: 61, 62

As you shoot through the Inside Passage and enter Hockomock Bay, your on-board Geiger counter might start ticking. Bailey Point, on the west side of the Back River at the head of Montsweag Bay, was the site of the Maine Yankee nuclear power plant. The plant went on line in 1972, but ran into increasing trouble in the 1990s when it needed the most expensive repairs in the history of nuclear power.

Soon after the plant was up and running again, new problems surfaced. But this time the cost of the needed repairs outpaced the profitability of the plant. Suddenly, it was shut down for good, and almost overnight, most of Wiscasset's tax base vanished. The long dismantling process continues, with various plans and timetables for the storage and removal of its radioactive components.

A narrow channel runs through Montsweag Bay and the Back River north to Wiscasset, but it is recommended only for smaller, shallow-draft boats with good power. Vertical heights are limited to 48 feet by a fixed bridge at Cowseagan Narrows, and the area south of the bridge is filled with ledges and very turbulent.

SHEEPSCOT RIVER

Of all the rivers on the coast of Maine, the Sheepscot offers the best cruising, perhaps because it is more like a bay than a river. The entrance opens invitingly between Southport and Georgetown Island, two miles wide and almost free of dangers. For an hour's sail, the river remains a broad sound, with islands and passages of great beauty and interest beckoning on either side. Flags flutter on the headlands, ospreys nest on every beacon, and white granite glistens on the shores.

The Sheepscot has many good harbors and anchorages—enough so that you could discover a different place every night for two weeks. You can find good places to hole up in heavy weather and one or two spots that even qualify as hurricane holes. Two major harbors, Robinhood and Ebenecook, provide supplies and repairs.

Unlike many of the other midcoast rivers, there are few narrow or shoal spots here and usually only a moderate current. Deepest of all Maine rivers, the Sheepscot is also cold, and lobsters hauled from here are reputed to be the tastiest.

Harbors such as Cape Harbor, Cozy, Five Islands, and Love Cove are a delight. One short but lovely sail runs north of Ebenecook Harbor, tucks in behind the Green Islands, passes Cameron Point, and merges back into the river again north of Isle of Springs. Or you can circumnavigate MacMahan Island through Goose Rock Passage and the Little Sheepscot River.

The Sheepscot River bisects the Inside Passage, an interesting, meandering, and sometimes hair-raising route between Bath and Boothbay Harbor (see previous entry). The portions of this passage adjoining the Sheepscot are relatively benign and enjoyable to explore.

Wiscasset lies 14 miles from the sea. A low causeway prevents any further exploration upriver. This historic town is packed with antique stores.

Until her death in 1964, writer Rachel Carson spent summers along the shores of the Sheepscot, north of Hendricks Head on Southport Island. The river was both her home and her laboratory. Her first great and influential book, *The Sea Around Us*, introduced us to the beauties and complexities of the oceans. And then her *Silent Spring* opened our eyes to the fragility of our surroundings and helped launch the environmental movement.

CAPE HARBOR

2 ★★★

43° 47.14'N   069° 39.42'W

CHARTS: *13296, 13293, 13288*   CHART KIT: *61, 62, 18*

This beautiful little harbor lies at the southern end of Southport Island at Cape Newagen and provides reasonable protection except in a strong southwesterly. The entrance is narrow, deep, and exhilarating. It will test your convictions about "red right returning." If the harbor is new to you, do not enter it at night or in poor visibility.

Both Hunting Island and Cape Island (locally known as Witches Island) to the south of the harbor are privately owned. Cape Harbor is the site of the handsome old Newagen Seaside Inn, whose meals and facilities are available to visiting yachtsmen.

APPROACHES. Enter Cape Harbor from the west. Leave nun "2" to starboard, but do not run a straight line to the entrance. Instead, to avoid the 3-foot spot outside, run north from the nun before turning in toward the entrance beacon.

The entrance is only 50 to 75 feet wide, but it is deep, and Southport Island is steep-to. Pass between Southport Island to port and red daybeacon "4" to starboard. Daybeacon "4" is not on the edge of the channel—you must be 10 to 15 feet north of it.

ANCHORAGES, MOORINGS. The Newagen Seaside Inn maintains a couple of guest moorings for visitors. To reserve ahead, call *633-5242*. There is little room to anchor among the lobster buoys and boats, and the northern cove is foul.

GETTING ASHORE. Row your dinghy to the inn's float at the western end of the harbor or to the town floats at the eastern end.

FOR THE CREW. The inn's Cape Harbor Grill serves dinner (*633-5242*), and more casual eats can be found at their pub. Ice and showers are available. There are no nearby stores or markets, but the post office is just up the road from the town landing.

THINGS TO DO. The Inn has a heated freshwater pool, a large saltwater pool, croquet, horseshoes, badminton, and volleyball. It is set in a forest preserve with nature trails and rocky beaches to explore.

Cape Island, forming the south side of the harbor, was actress Margaret Hamilton's retreat. She was the Wicked Witch in *The Wizard of Oz*. The island is proof positive that there is no place like home.

HARMON HARBOR

3 ★★★

43° 48.63'N   069° 43.11'W

CHARTS: *13295, 13293, 13288*   CHART KIT: *61, 62, 18*

As you enter the Sheepscot River from the sea, this long narrow cove is the first on the west side. The harbor is well protected except from southerly gales, but it is likely to be rolly when the entrance ledges are covered at high tide. The shores are rocky and wooded with an occasional house and a dozen or so moored lobsterboats.

Few cruising boats use this harbor, so it is far more peaceful than neighboring Five Islands. Harmon Harbor is home to two impressive screech owls and swarms of other birds of prey—the mosquitoes. Have your netting ready by dusk.

APPROACHES. The harbor is easy to enter only in good visibility. After you round Griffith Head, coming in from the sea, the sun will gleam on the white granite of Five Islands. The green indentation of Harmon Harbor is to the left of Five Islands. Head for Grey Havens Inn, the massive, gray, turreted building high on a hill.

At the entrance leave nun "2" about 20 feet off to starboard, staying in water that is about 16 feet deep at low. The ledges to port and starboard are both clearly visible at low.

ANCHORAGES, MOORINGS. Come in past the first two or three docks until the depth shoals to 16 or 20 feet at low, and anchor in the middle of the harbor. There is plenty of swinging room, and the mud bottom is good holding ground.

The Grey Havens Inn has a mooring available, or it may be possible to lie along their float in 17 feet of water at low. Reserve ahead by calling *371-2616*.

GETTING ASHORE. Land at the Grey Havens Inn float.

THINGS TO DO. The Inn does not serve any meals other than breakfast, and that is for guests only. It is still worth a visit to this authentic Victorian resort. Its lounge is dominated by a huge stone fireplace, and the stunning view is framed by a picture window that, when it was built in 1904, was Maine's largest piece of glass.

Little Wood Island, off Dry Point, is a nature preserve, and an Audubon sanctuary is near the inn.

COZY HARBOR

4 ★★★

43° 49.07'N   069° 40.95'W

CHARTS: 13295, 13293, 13288   CHART KIT: 61, 62, 18

This is truly a cozy harbor, a charming surprise tucked into the west flank of Southport Island. Home of the family-oriented Southport Yacht Club, Cozy Harbor is full of kids splashing in and out of the water and tearing around in the club dinghies, while the lobstermen get on with their work.

The harbor is exceptionally well protected, but "cozy" can mean small too. Boats of all sizes are squeezed on their moorings like sardines in a can, and maximum depths are only 8 feet. The lack of room discourages even moderate-sized yachts from visiting. Occasionally a windjammer might spend a night in the mouth of the harbor.

APPROACHES. Coming up Sheepscot Bay, you will have no trouble spotting the square, white lighthouse on Hendricks Head. Look a quarter of a mile south of the lighthouse for the nun and the red daybeacon which mark the entrance to Cozy Harbor.

The entrance approach is an east-south-east S-curve. Nun "6" is a river mark, so you leave it to *port* and head east. Keeping red daybeacon "2" on Nick Ledge to starboard, turn south and pass between the daybeacon and can "3," to port. Then round to the east again. Leave green daybeacon "5" close aboard to port, but no closer than 10 feet, and enter the harbor.

Although the entrance is very narrow, it is easy when there's good visibility. Don't try this harbor for the first time in the fog.

Once in the harbor, go slow, and think two steps ahead. There is very little maneuvering room. As sailor Steve Mayotte puts it, "you get the feeling of taking a battleship into a bathtub."

ANCHORAGES, MOORINGS. There are two possibilities. Pick up the single mooring identified by a club burgee. Depth at the mooring is about 6 feet at low tide. Deeper-draft boats might rest gently in the mud.

The second possibility—and this is what the windjammers do—is to anchor just inside the mouth of the harbor after you pass the green daybeacon. Wherever you lie, swinging room is limited, so you may need to put out fenders.

GETTING ASHORE. Dinghy in to the yacht club float, which is used by everybody in the harbor, even the lobstermen, who are also members.

FOR THE BOAT. *Southport Yacht Club (633-5767).* There is water on the club float with 4 feet alongside at low.

FOR THE CREW. A white building with a sign saying "E.W. Pratt, General Merchandise" stands next to the club. This is Gus's place, a turn-of-the-century ice cream parlor, sandwich counter, and candlepin bowling alley. Gus, however, has passed away, and the future of this throwback is uncertain.

Walk half a mile up the road to the left to the Southport General Store, well stocked with plain and fancy items and big-city newspapers. The post office is another block down the road. Sweet Dreams bakery, next door, serves everything you want but shouldn't have.

THINGS TO DO. Take the dinghy under the bridge to Pratts Island or row north through the rocky guzzle inside David Island to the beach in the corner of Hendricks Harbor. The first road to the right from the yacht club makes a nice walk over to Pratts Island. Or, on a rainy day, bowl a few strings of candlepins at Gus's.

The classic lanes at Gus's.

FIVE ISLANDS

3 ★★★★

43° 49.44′N 069° 42.50′W

CHARTS: 13295, 13296, 13293, 13288 CHART KIT: 61, 62

On the west side of the Sheepscot River, five bold, granite-fringed islands ring a beautiful and well-protected harbor. The entrance is easy, and the atmosphere is wonderful.

The harbor is surrounded by a working waterfront, modest homes on the mainland, and the handsome summer cottages on Malden Island, which have been in the same families for six generations. This is both a working harbor and a haven for pleasure craft.

APPROACHES. The usual entrance is just north of Malden Island, the largest of the group, high and wooded with several substantial summer cottages on the north side. Find green can "1" inside the entrance and head for it on a westerly course about halfway between Malden Island and the ledge to the north. Keep the can close by to port.

There is also a northern entrance. Identify wooded Crow Island and find red daybeacon "2" in the north channel. Approaching the beacon, favor the Crow Island side to avoid the unmarked ledge making out from Georgetown Island. Pass westward of the beacon, leaving it close to port.

There are other unmarked entrances from the south, between the islands, but they are difficult without local knowledge.

ANCHORAGES, MOORINGS. The Five Islands Yacht Club of Malden Island maintains four guest moorings, free for the first two nights. They have white balls with blue stripes and are marked "FIYC." Sheepscot Bay Boat Company rents several heavy moorings to transients, and they have dock space with depths to 8 feet.

GETTING ASHORE. To get to the village of Five Islands, use the town floats south of the red buildings of Five Islands Lobster. The floats to the north are commercial only, and the southernmost float is private, for Malden Island residents.

FOR THE BOAT. *Five Islands Town Dock.* The ramp to the town float projects off the wharf just south of the red buildings of Five Islands Lobster. It provides a good place to tie up briefly or to land the dinghy.

Five Islands Yacht Club. The club maintains four free guest moorings for visiting yachtsmen—an almost extinct courtesy. Thank you. There is no clubhouse or other facilities, so please don't land on Malden Island to look for one.

Sheepscot Bay Boat Company (Ch. 09, 207-371-2442). The yard's docks and blue crane are in the northwest corner of the harbor. In additon to moorings and dock space, SBBC pumps gas and diesel, and has a small chandlery. The yard can repair engines, outboards, and hulls on boats to 30 feet.

Five Islands Boat Yard (207-371-2837). Bill Plummer, a fouth-generation boatbuilder, can handle major fiberglass repairs or build you a replacement.

FOR THE CREW. Five Islands Lobster steams clams and lobsters to be enjoyed at the picnic tables on the dock. The Love Nest Snack Bar next door has milk, eggs, papers, ice cream, and food from the grill. Try the Jenny Special.

A short walk up the hill out of town will bring you to Five Islands Farm (*371-9383*), which sells exquisitely fresh produce, cheese, smoked fish, natural beef, pastas, sauces, and fine wines.

Five Islands is also a stop on the weekly route of a floating vegetable boat. Check the notice board on the wharf for the schedule.

THINGS TO DO. Row under the footbridge between Malden Island and its neighbor to the west or between the northern islands. The residents of Malden Island, who are kind enough to maintain the guest moorings in the harbor, request that visiting yachtsmen do not land on Malden Island.

If your timing is right, and if there are any fish left in the oceans, you can watch tuna being landed here. The size and magnificence of these fish are matched by the high stakes and high risks of catching them. The tuna fishery hinges on the spotter's good eyes, the harpooner's steady arm, lady luck, and the state of the economy in Japan, where the tuna is immediately flown to be eaten raw as sushi. Buyers meet the incoming boats, core-sample the fish for temperature and fat content, ice the fish, and cut checks on the spot in amounts that could buy a small car.

For a pleasant walk, take Ledgemere Road, inland from the harbor and the first left. This beautiful little road runs a mile through piney woods. Dry Point, at its end, is private.

A walk of about .25 miles out of town on the main Five Islands Road will bring you to a local swimming hole, known as Charles's Pond or, before plumbing, as the Georgetown Bathtub.

LITTLE SHEEPSCOT RIVER

43° 50.55'N 069° 42.90'W

CHARTS: 13296, 13293, 13288   CHART KIT: 61, 62, 18

The Little Sheepscot River is a pleasant passage of a mile or so, west of MacMahan Island. It makes a fun sail with the current, though you are likely to lose your wind partway along.

Current floods north and ebbs south and runs hard in the narrow places. If it is flooding, the Little Sheepscot is the best route to the Inside Passage. If it is ebbing, you will do better to go outside MacMahan Island and through Goose Rock Passage.

Entering the Little Sheepscot from the south, leave Turnip Island to starboard. Turnip is rocky, with a granite and cedar shore. Run through the narrows, leaving nun "2" to starboard.

Just past the narrow section is the landing for MacMahan Island, with its pleasant summer community. Anchor off the dock and row in to the float to explore the island. Walk inland from the landing, then take a left up the hill to the Episcopal church, an unusual landmark on such a small island.

GOOSE ROCK PASSAGE

43° 50.99'N 069° 43.05'W

CHARTS: 13296, 13293, 13288   CHART KIT: 61, 62, 18

Goose Rock Passage can be entered either from the Sheepscot River north of MacMahan Island, leaving green daybeacon "1" to port, or from Little Sheepscot River. Strong current runs through the passage. Leave can "3" to port and locate nun "4" at Boiler Rock as soon as possible. Nun "4" is sometimes pulled under by the current. If you can't see it, run a straight line course from north of can "3" toward flashing green light "5" on the skeleton tower on the south shore, then follow the shoreline around Lowe Point.

Two little coves in Goose Rock Passage can provide anchorage or shelter from the current if you need it. On the north side of MacMahan Island, the chart shows a break in the ledges. Head straight in for the docks, and you can anchor outside the moored boats in 13 to 15 feet at low. This is easiest to do at halftide or lower, when the ledges are visible.

Brooks Cove is on the north side of Goose Rock Passage. Head straight up the middle of the cove and anchor near the moorings in 15 to 20 feet at low.

RIGGS COVE (Robinhood)

43° 51.20'N 069° 43.90'W

CHARTS: 13296, 13293, 13288   CHART KIT: 61, 62, 18

Riggs Cove, just south of Knubble Bay, is full of the moorings of Robinhood Marine. This is a good place to get anything you need for the boat (except water, perhaps) and to wait for the correct tides for passing through Lower and Upper Hell Gates. You'll be in the good company of many large cruising boats.

APPROACHES. Approach from the east through Goose Rock Passage. The boats moored at Robinhood are visible as you come around Lowe Point. The only danger is Blacksmithshop Ledge, just south of Robinhood Marine, marked by green daybeacon "1."

ANCHORAGES, MOORINGS. Robinhood Marine has 135 slips and at least 70 moorings, of which a dozen are usually available for transients. No one objects if you pick up a mooring temporarily while waiting for the tide to turn. Anchoring among the moorings would be a tight squeeze, but you can anchor outside the mooring field toward Robinhood Cove where you will have plenty of swinging room even though the water is deep.

GETTING ASHORE. Take your dinghy in to the Robinhood Marine floats.

FOR THE BOAT. *Robinhood Marine Center (Ch. 09; 207-371-2525; www.robinhoodmarinecenter.com).* This large, full-service yard has every facility for the boat and can perform all types of repairs or even build you a new yacht. The fuel dock is midway along the floats with gas and diesel, ice, water, electricity, and sewage pump-out facilities. The water comes from a well, however, and may have a salty or minerally tinge. Taste before you tank.

FOR THE CREW. In addition to their exceptionally pretty yard, Robinhood Marine has showers, laundry facilities, and a great library with a book exchange. A free nautical lecture series is held there on most Wednesday nights, and a live band plays at the gazebo on Fridays, weather permitting. If you need it, they have a free loaner car.

The marina's Osprey Restaurant *(371-2530)* serves an extraordinary lunch and dinner. Or hit their Tavern at Riggs Cove for lighter food and heavier beer. The five-star Robinhood Free Meeting House restaurant *(371-2188)* is a short walk away. Reservations for dinner are recommended at both.

ROBINHOOD COVE

◼ ★★

43° 50.29′N 069° 44.24′W

CHARTS: 13296, 13293, 13288   CHART KIT: 61, 62, 18

South of Knubble Bay and south of the moored boats in Riggs Cove, a long, narrow slot stretches for more than two miles into Georgetown Island. A seldom-used, unbuoyed tongue of deep water extends southward for about a mile from the entrance. The shores are low and wooded, with few houses.

Robinhood Cove is landlocked and offers good protection in every direction except the southwest, where the length of fetch detracts somewhat from its value as a hurricane hole.

APPROACHES. Head straight down the middle of Robinhood Cove. As shown on the chart, just past the narrows at the entrance, the remains of the *Mary Barrett*, an old five-masted schooner, lie near the east bank. Feel your way down the deepwater tongue as far as you wish.

ANCHORAGES, MOORINGS. Anchor at the edge of the channel in 8 to 20 feet in the mud bottom. You won't have much company of the human kind, but the mosquitoes might throw you a party.

EBENECOOK HARBOR

43° 50.39′N 069° 40.52′W

CHARTS: 13296, 13295, 13293, 13288   CHART KIT: 61, 62

The first major harbor on the east side of the Sheepscot is Ebenecook, at the northwest corner of Southport Island. The harbor is easy to enter and consists of three separate arms of water, each quite different in character. From west to east, these are Maddock Cove, Pierce Cove, and Love Cove.

MADDOCK COVE (Ebenecook)

④ ★★

43° 49.92′N 069° 40.73′W

CHARTS: 13296, 13295, 13293, 13288   CHART KIT: 61, 62

Maddock Cove, the westernmost cove in Ebenecook Harbor, is the well-protected home of Boothbay Region Boatyard, whose docks and moorings fill most of the available space. On a typical summer day there might be 70 or 80 boats here.

APPROACHES. Head into Ebenecook Harbor between Green Islands and Dogfish Head. Dogfish Head is easily identified by the white-roofed, circular house on the point ¼ mile north of the skeleton light-tower on Southport Island's west shore.

Before nun "2," turn south into Maddock Cove and run down the channel between the two lines of moored boats. Note the ledges to your left, which separate Maddock Cove from Pierce Cove to the east. There is a pole on the highest part of the long ledge.

ANCHORAGES, MOORINGS. Check with Boothbay Region Boatyard for moorings or slips. It may also be possible to anchor in midchannel, northwest of the marina in 8 to 16 feet at low, but it will be tight.

GETTING ASHORE. Row your dinghy in to the floats at Boothbay Region Boatyard.

FOR THE BOAT. *Boothbay Region Boatyard (Ch. 09; 207-633-2970; www.brby.com).* This large, full-service boatyard pumps gas and diesel. Water, ice, pump-out facilities, and 50-amp electrical service are available at the floats with 6 feet alongside at low. The yard hauls with a 50-ton boatlift and crane, and they can perform all types repairs. The chandlery carries charts and other marine supplies.

Maloney Marine Rigging (207-633-6788) is located at Boothbay Region Boatyard. Jay can repair, replace, or redesign any standing or running rigging, from advanced hydraulics, rod rigging, furlers, and carbon fiber spars to traditional wire splices hand-parcelled, tarred, served, and leathered.

FOR THE CREW. The yard has a separate little house with showers and laundry facilities.

Southport General Store, the post office, and the Sweet Dreams bakery and cafe are a half-mile walk away. Turn south from the boatyard.

Townsend Gut Swing Bridge opens on every hour and half-hour.

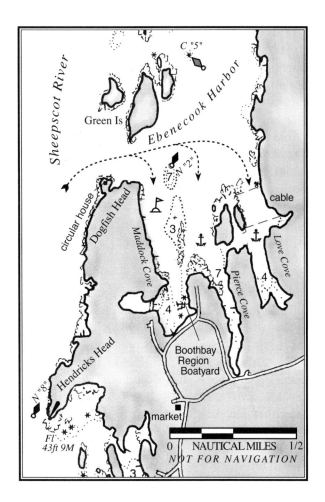

LOVE COVE (Ebenecook)

4 ★★★★

43° 50.00'N 069° 40.26'W

CHARTS: 13296, 13295, 13293, 13288   CHART KIT: 61, 62

The farthest east of Ebenecook's arms and the best protected of all is serene Love Cove. The entrance is a bit tricky, and much of the anchorage is a cable area, but once inside it is a delightful place to watch seals, herons, and ospreys and to explore by dinghy and savor the tranquillity.

APPROACHES. After you pass Dogfish Head, leave nun "2" to starboard and continue east until you can see down into Love Cove. Unmarked ledges guard both sides of the entrance, so coming in at high tide requires some guesswork. Better still, enter at midtide or lower when the ledges are visible.

ANCHORAGES, MOORINGS. Note the cable area shown on the chart. There are signs on both shores showing the location of a pipeline. Anchor south of this area in 8 to 10 feet at low. Eliot Winslow was a resident harbor pilot and tugboat tycoon. You are welcome to use his mooring here. Appreciative cruising folk leave messages—or poems—in a pickle jar attached to the float.

PIERCE COVE (Ebenecook)

3 ★★

43° 49.95'N 069° 40.49'W

CHARTS: 13296, 13295, 13293, 13288   CHART KIT: 61, 62

The middle arm of Ebenecook Harbor, Pierce Cove, is less protected than Maddock Cove or Love Cove when the winds are from the northwest. The cove harbors a dozen moored boats but has no facilities.

APPROACHES. After passing Dogfish Head, leave nun "2" to starboard before heading down into the slot of Pierce Cove. Favor the east side to avoid the ledges on the west.

ANCHORAGES, MOORINGS. Anchor in 10 to 15 feet of water at low either before you reach the moored boats or among them. Do not go past the tip of land on the west side of the cove.

TOWNSEND GUT

NW of bridge: 43° 50.59'N 069° 39.27'W

CHARTS: 13296, 13293, 13288   CHART KIT: 61, 62, 18

In his 1858 book *A Summer Cruise on the Coast of New England*, Robert Carter described Townsend Gut: "We…had a delightful sail through a most singular strait, narrow like a river of moderate size, and bordered on both sides by meadows green to the water's edge, with occasional groves ringing the banks. We should have had no suspicion that this passage was not a river had it not been for seaweed growing on its rocky edges."

Townsend Gut is toward the eastern end of the Inside Passage leading from Bath to Boothbay Harbor. Narrow in several places and about a mile long, the banks are still lined with meadows and trees and handsome summer homes. Along the way, there are secure anchorages protected in every direction. The passage is clear and very well defined.

From the Sheepscot River, the entrance to Townsend Gut is easy to find. Look for flashing light

Duane Lewis commands the Townsend Gut Swing Bridge.

"7" off Cameron Point, rounding it wide and leaving it to starboard. The current runs from zero to moderately strong at the entrance near Cameron Point and in the other constricted areas, but it is no problem for sailboats with power. Pass through the narrow space between flashing light "7" and nun "6," which marks the long ledge to port. Then turn left into the channel.

Townsend Gut Swing Bridge. This bridge connects Southport Island to the mainland, spanning busy Townsend Gut. It is manned 24 hours a day and will open on the hour and half hour and on demand in the off season. Call the bridgetender on channel 09 to let him know you will need the opening. If you don't have a radio, sound one long and one short blast. As you pass through, keep to the right. The bridgetender will call for the name of your vessel and its owner for his log.

For half a century or more, the bridge has been tended by members of the Lewis family. Currently the bridgetender's booth is occupied by twin brothers Dwight and Duane Lewis. Their brother Don, their father, their uncles, and their sister Ruth all swung the bridge before them. Together the family has logged more than 136 years of service. When Ruth passed away in the summer of 1998, the bridge was opened for nearly an hour while a parade of hundreds of boats of all shapes and sizes passed through the bridge in her memory.

The current can be as much as 5 knots here, and there is about 20 feet of depth in the channel at low. After passing through the bridge, note nun "4" marking a midchannel rock just beyond. South of the bridge, the channel narrows. Favor the deeper Southport Island side.

OAK POINT

4 ★★★ ❙❙

43° 51.04'N 069° 39.70'W

CHARTS: *13296, 13293, 13288*   CHART KIT: *61, 62, 18*

On the north side of Townsend Gut, just west of Oak Point, is a lovely cove, unnamed on the chart (Oak Point is not named on the Chart Kit charts—it defines the western side of Hodgdon Cove). Several classic yachts lie to private moorings before handsome homes with lawns rolling down to the water. The cove is well protected from every direction but the northwest, and the sunsets are splendid.

APPROACHES. Head for the center of the cove. There are no dangers.

ANCHORAGES, MOORINGS. Anchor just outside the moored boats in mud 17 feet deep at low.

FOR THE CREW. If you feel like eating out, there is a dock at the Ocean Gate Motor Inn in Southport, across the bay, or you can go under the swing bridge to ❙❙ Robinson's Wharf, an excellent informal seafood place (see *Hodgdon Cove*, below).

HODGDON COVE

4 ★★ ❙❙

43° 50.88'N 069° 39.30'W

CHARTS: *13296, 13293, 13288*   CHART KIT: *61, 62, 18*

Hodgdon Cove lies halfway through Townsend Gut, east of Oak Point and well out of the current. There are just a few other boats in the pretty, broad cove, and usually there is plenty of room to anchor. The protection is excellent.

APPROACHES. The approach is straightforward. Favor the eastern side of the harbor to avoid the ledges in the north corner, but do not go east of the moorings.

ANCHORAGES, MOORINGS. Anchor near the moorings on the east side of the cove in 13 to 20 feet at low in the muddy bottom.

GETTING ASHORE. While waiting for the tide or if anchored for the night, run the dinghy under the swing bridge to ❙❙ Robinson's Wharf (*633-3830*) in Deckers Cove. Tie up at the docks and enjoy one of those atmospheric, clam-and-lobster, eat-in-the-rough restaurants, while watching the parade of passing boats in Townsend Gut.

INDIANTOWN ISLAND

43° 51.90'N 069° 40.22'W

CHARTS: 13296, 13293, 13288   CHART KIT: 61, 62, 18

Indiantown Island lies just north of Cameron Point, at the western entrance to Townsend Gut. The island is high and wooded with a rock bluff on the southeast tip. Huge shell mounds on the southern side remain as mute testimony to the large Indian summer village for which the island was named. Developers once proposed building a bridge to the island, but it was spared that fate. Instead, it has one residence at the southern end. The rest is owned by the Boothbay Region Land Trust with public trails on the northern two thirds of the island.

Indiantown Island is a pretty daystop. ⚓ The Trust maintains a guest mooring off the dock and float at the northwest corner of the island. A courtesy dinghy is usually tied to the float. They request that you please stay on the trails.

ISLE OF SPRINGS

3 ★★★

43° 51.86'N 069° 40.75'W

CHARTS: 13296, 13293, 13288   CHART KIT: 62, 18

Large, wooded Isle of Springs lies two miles north of Ebenecook Harbor. For more than a hundred years it has been a rustic summer community. Some of the several dozen summer cottages peek through the trees. Residents arrive at the landing on the northwest side via a ferry from Sawyer Island.

APPROACHES. Approach from the east around the northern tip of the two Spectacle Islands. Then coast along the east side of the Isle of Springs to avoid the 4 and 5-foot shoals in the middle of the bay.

From the west, leave green daybeacon "1" at the northern end of Isle of Springs to starboard and make a wide right turn to avoid the ledge extending from the northern end of the island. Pass halfway between Isle of Springs and Sawyer Island.

ANCHORAGES, MOORINGS. There may be a guest mooring, or you can anchor outside the moored boats in 10 to 15 feet of water at low, in mud bottom. Protection here, in the lee of the island, is good.

GETTING ASHORE. Row in to the back of the float, leaving the sides open for the ferry.

THINGS TO DO. The island is crisscrossed by a network of grassy paths and boardwalks among classic summer cottages. Bring your letters just so you can stop at the world's most idyllic post office.

MILL COVE

◨ ★★   🛒 🍴

43° 52.77'N 069° 40.49'W

CHARTS: 13296, 13293, 13288   CHART KIT: 62, 18

Sawyer, Hodgdon, and Barters Island nearly landlock a small bay behind them. The chart leaves it unnamed, but it is known locally as Mill Cove, for the mill pond along its southern shore. It is not particularly beautiful, but the islands protect it from every direction, making it an excellent hurricane hole. The entrance is straightforward, between Sawyer and Barter Island.

Mill Cove forms part of the approach into Back River via the Trevett swing bridge, which is the last hand-operated swing bridge in the country. It is still manned and opens on demand in season, but overhead power cables limit heights to 50 feet and strong currents run through the narrows. Even if you could pass safely under the cables, there is little in the Back River to attract you.

APPROACHES. Approach the entrance from the west between Sawyer and Barters Island, starting near nun "18" through a reasonably wide channel. At the end of this channel, leave can "1," which marks a ledge, to port.

If you feel compelled to squeeze your boat into the Back River, Jon Hodgdon suggests the following: go at high water unless you draw less than 4.5 feet; keep red nuns tight on your right and go straight from nun to nun without drifting to the west.

ANCHORAGES, MOORINGS. Anchor in the middle of the Mill Cove in 8 to 20 feet of water at low. The mud and sand bottom holds well. You will have plenty of swinging room and not much company.

FOR THE CREW. Hodgdon Island is the town center of Trevett, with a post office and a small general store. The store has a float just past the swing bridge on the right side of the channel where you can land your dinghy. ⚓They sell milk, wine, beer and essential groceries and serve 🍽pizza and sandwiches.

THINGS TO DO. From the store's dock, walk east over the fixed Knickerbocker Bridge and continue for about .5 miles to the Coastal Maine Botanical Gardens (*633-4333; www.mainegardens.org*). This 250-acre preserve has trails through gardens and woods and includes about a mile of the shoreline south of the bridge. Beware of taking your dinghy directly there, however. The current runs hard around the southern end of Hodgdon Island and there are numerous ledges waiting for wayward propellers.

If you walk south on Hodgdon Island and cross to Sawyer Island, you will find the Gregory Hiking Trail on land preserved by the Boothbay Region Land Trust. This .8-mile loop leads along the shores of the Back River. Another BRLT Preserve, the Porter Preserve, is at the southern tip of Barters Island. You may be able to land there by dinghy because the preserve includes a beach and the small island north of nun "18." Donations toward their conservation efforts are welcome (*633-4818; www.bbrlt.org*).

CORMORANTS

On every ledge in Maine it seems, cormorants stand in solemn rows drying their wings. With long, black, snaky necks, green eyes, and hooked beaks, they look malevolent— throwbacks to an age of dinosaurs and sharks and pterodactyls. And so they are.

While later-model ducks have oil glands with which to preen their feathers, cormorants do not. The water soaks their feathers, and they stand sopping wet, like someone who has fallen in wearing all his clothes, elbows up, hung out to dry. Cormorants have no sweat glands either. They dissipate body heat via their bright-orange throat pouch. The most common species in Maine during the summer is the doublecrested cormorant, although its crests are hard to see.

On land the cormorant is awkward and ungainly, with short legs and a fan-shaped tail, but in the water it is a champion. Like the loon, it uses its legs to propel itself underwater with powerful thrusts of its webbed feet, and it can stay submerged a minute or more.

Cormorants are neither loved nor much admired. In Maine they're called "shags" when fishermen are feeling kindly toward them, which is not often. At other times they are called "lawyer birds," or "crow ducks." No sooner has a fisherman made a catch of herring in a weir or net than shags arrive. They gorge themselves on young fish and panic the fish in the nets, causing many of them to be wounded or destroyed. They descend ravenously on the young salmon that the state releases in the Machias, Penobscot, and other rivers, thus earning the enmity of sportfishermen.

Cormorants often roost in large dead trees, which have been killed by the acidity of their guano. Entire small islands have been denuded in this way. Shag droppings whiten the tops of ledges, where they do no harm, and the tops of open boats, where they are less popular.

With such annoying habits, it is not surprising that cormorants were shot by fishermen or that their eggs were sprayed with kerosene to reduce the population. By the early 1900s, they were close to extinction. In 1972, however, Maine passed a law that protected many birds including the cormorants, and the shags have made a phenomenal comeback. A recent count indicated there were more than 28,000 pairs nesting in Maine. Ironically, some scientists attribute the increase in population to the overfishing of the groundfish which used to prey heavily on the small bottom-dwellers that make up most of the cormorants' diet.

The cormorants build their nests of twigs and dry seaweed, three to five eggs to a nest, and the parents take turns incubating the eggs. The urgent cries of the cormorant chicks induce the adults returning with their catch to open their beaks and regurgitate the food into their throats, from which the babies feed.

Duck Island in the Isles of Shoals is one of the important nesting sites for shags. Another is Ross Island in Muscongus Bay, where the National Audubon Society has banded young cormorants for years to track their movements and breeding success. One early result of the banding program was the discovery that cormorants are particularly fond of Tampa Bay, Florida, for the winter months.

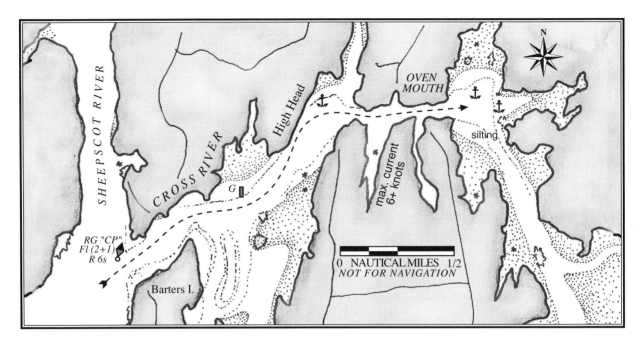

CROSS RIVER and OVEN MOUTH

4 ★★★★

inner basin: 43° 56.12'N 069° 38.22'W
off High Head: 43° 56.18'N 069° 39.07'W

CHARTS: 13293, 13288   CHART KIT: 62, 18

At the north end of Barters Island, the Sheepscot divides into two branches. The smaller, eastern branch forms the pleasant diversion of the Cross River. Both American and British vessels hid here during the Revolution, and it was the site of one of the area's earliest shipyards.

Cross River is lovely and open, with just a few houses in sight. But at Oven Mouth, the river narrows down to 50 yards and the current shoots you between bold cliffs until you pop out into the basin beyond. This is a dramatic and delightful passage.

Unfortunately, it's also a passage that has been discovered by Boothbay tour boats, which sometimes seem propelled more by the force of their PA systems than by their engines.

APPROACHES. Chart 13293 makes Oven Mouth look narrower and more difficult than it really is, and Chart Kit 62 is even smaller.

The fork on the Sheepscot is marked by a red-and-green lighted buoy, left to port to enter the Cross River. Give the northern tip of Barters Island a wide berth. The privately maintained daymarks shown on the chart on the mudbank to the south of the Cross River may or may not be there. Sometimes another privately maintained daymark is set on the 2-foot extremity of the mudbank to the north, just below the "v" in the word "River" on chart 13293. Keep this mark to port. Proceed toward High Head and keep to its bold shores before turning into Oven Mouth.

In passing through Oven Mouth, stay in midchannel until you emerge into the basin beyond. The current is strongest at the eastern end, where it might reach 6 knots or more. Once in the basin, do not stray to the south, where it shoals rapidly.

ANCHORAGES, MOORINGS. The basin to the east of Oven Mouth is a little too large with a little too much fetch from the south to be a perfect hurricane hole, but it comes close. You can anchor anywhere in the basin where there is enough water. Perhaps you can locate the 13-foot spot in the middle.

Another anchoring possibility is west of Oven Mouth by High Head, in about 25 feet at low, just north of the 45-foot spot, near some private moorings. This will put you north of the ebb that pours through Oven Mouth and hits the middle of High Head. Watch your depths and do not go too far north, where it shoals rapidly.

THINGS TO DO. The two points to the south of Oven Mouth, due south of the word "Mouth" on the chart, are part of a preserve of the Boothbay Region Land Trust (*www.bbrlt.org*). A footbridge connects the two peninsulas and trails lead through beautiful woods to a series of nature stations. Hiking on the western peninsula is more strenuous than on the east.

WISCASSET

3 ★★★★

Eddy Marina: 43° 59.45'N 069° 39.08'W
Town dock: 43° 59.92'N 069° 39.88'W

CHARTS: **13293, 13288**   CHART KIT: **62, 18**

Wiscasset claims itself to be "the prettiest village in Maine," and it might be. It certainly is worth the 14-mile trip up the Sheepscot River. This is a delightful and peaceful spot for strolling. There are wonderful old homes open to the public, antique shops, museums, a small coastal railway, specialty stores, and good restaurants.

Once alive with shipyards and great schooners, the waterfront today is strangely silent. Change, however, is in the wind. At the time of this writing, developers are planning to refurbish the old Mason Station, an obsolete generating plant on Birch Point, south of town, and the projects includes a large marina and service yard.

APPROACHES. The approach to Wiscasset could not be simpler. The Sheepscot River makes a dramatic left turn around the northern tip of Westport Island. Leave to port one of the world's largest osprey nests on can "25" at the turn. The wooden blockhouse at Fort Edgecomb will be to your right. On your starboard bow is the village of Wiscasset, through which Route 1 passes before it crosses the Sheepscot on a graceful fixed bridge (vertical clearance of 25 feet).

ANCHORAGES, MOORINGS. You can anchor anywhere south of the docks off Wiscasset in 25 to 30 feet, in good mud holding ground. Protection is good unless a heavy southerly is blowing. The boats usually ride to the strong current.

Wiscasset has a large town pier and a smaller town landing just to the south. The pier has two floats. Visitors are allowed to dock free for three hours on its northern float or overnight for a fee. Depth is 15 feet at low. The southern pier float is reserved for commercial and small craft. A third float, further south, is part of the town landing and is used for launching boats at the town ramp. The harbormaster (*882-7230*) is usually somewhere nearby.

The Wiscasset Yacht Club, just south of the town pier, may have a member's mooring available. Another less convenient possibility is a mooring or dock space at the Eddy Marina (see below).

The wrecks of *Hesper* and *Luther Little*, the last four-masted schooners in the world. The masts fell in an autumn blow in 1995 and the hulks were removed in 1998.

GETTING ASHORE. If you are on a mooring, take the dinghy to the town floats or to the yacht club. As in all swift-moving tidal rivers, be careful, and do not overload the dinghy.

FOR THE BOAT. *Town Pier.* A tall flagpole marks the big town pier, which has 15 feet alongside at low. With permission, the harbormaster allows dockage for up to two days. Diesel or gas can be arranged by truck.

Town Landing. Next southward is the smaller town landing and float.

Wiscasset Yacht Club (Ch. 09; 207-882-4058). The docks and floats of the Wiscasset Yacht Club are farther south, alongside the low, white clubhouse with a green roof and a sign on top. This is a nice, friendly little club which offers water and showers. There is 15 feet along the floats at low.

The current off the yard is strong—5 knots or more. It is also peculiar, almost always running in the same northerly direction, tapering off in strength near high tide, hence the name, Back River. If your dinghy is inadequate for the passage, you could ask the yard to run you over to town with their outboard.

The Eddy Marina (Ch. 16; 207-882-7776). The Eddy is located on the east side of the Sheepscot just at the hard left turn southeast of Davis Island in North Edgecomb. Sailors Willis and Merry Clifford have moorings and dock space with 17 feet of water at low. Gas, diesel, water, electricity, ice, and a shower are available, and there is a small chandlery that carries charts, has a book exchange, and sells live lobsters. They are also happy to serve as a mail drop for cruisers.

Fort Edgecomb and ⁕ Bintliffs Ocean Grill (*882-9401*), serving brunch and dinner, are within walking distance. It is about a mile across the bridge to Wiscasset. The house prepared for Marie Antoinette (see sidebar below) overlooks the marina.

FOR THE CREW. Restrooms are located at the town pier. Within a few minutes' walk of the Wiscasset waterfront, you will find just about anything you need, except, as in the case of so many small towns these days, a large market within walking distance.

🛒 Treats, on Main Street, sells fine cheese, wine, imported beer, and gourmet coffees. ⁕ Reds Eats at the other end of the street—and spectrum—serves up classic takeout. Sarah's Café, across the street, and the Schooner Pub can take care of sandwiches, seafood, chowder, and beer. For fine dining, try Le Garage overlooking the town docks.

The post office and a bank and ATM are on Main Street. A hardware store is downstairs in the Old General Store. The nearest liquor store is out Route 27, a quarter of a mile out of town, and an inconvenient convenience store is another quarter of a mile beyond that.

THINGS TO DO. Except for the waterfront, most of Wiscasset looks today as it did a century ago. Much of it is included in a National Historic District. A free Wiscasset area guide identifies the historic buildings and suggests a walking tour. Many are open to the public.

The elegant 1807 Castle Tucker is a replica of a manor in Dunbar Scotland. It sits on a hill overlooking the harbor. The Nickels-Sortwell house, also built in 1807, is in town. The Musical Wonder House, circa 1852, is on High Street. It has a fantastic collection of antique music boxes and Steinway player pianos that are shown and played in rooms furnished with antiques of the period.

The Carlton house, built in 1804, is also on High Street. In the days before banks, Moses Carlton bought it for 100 puncheons of rum. He had his cash wheelbarrowed up the hill in kegs and dumped into chests in the cellar.

WISCASSET: DAYS OF GLORY, DAYS OF GLOOM

Wiscasset, with its large, protected and ice-free harbor, was once the busiest port north of Boston. It has known both great prosperity and economic oblivion.

For the first settlers, the Sheepscot River yielded fish in great quantities, and the surrounding forest provided almost limitless lumber. The early fortunes of Wiscasset men were made in the Caribbean trade—lumber and fish to build the plantations and feed the slaves, barrel staves to ship their rum.

Shipbuilding began in earnest in the late 1760s, and more fortunes were made in trade with Europe. Francs and pieces-of-eight were accepted as good currency in Wiscasset, and the local schools taught French, Italian, and Spanish. Citizens of Wiscasset were citizens of the world.

Perhaps this explains why such a pragmatic sea captain as Samuel Clough risked himself and his ship in an attempt to save Queen Marie Antoinette from the guillotine. His ship, the *Sally*, was loaded with furnishings, objets d'art, and the queen's personal belongings in preparation for her escape, while back home in North Edgecomb, Captain Clough's simple farmhouse was being prepared to receive a queen.

As Maine author Louise Dickinson Rich suggests, Mrs. Clough may have resented her husband's news that "he was refurnishing the house for this woman." Mrs. Clough undoubtedly was nervous about entertaining a royal guest and wondering if "she'd be satisfied with a good clam chowder and a slab of johnnycake."

As it turned out, before she could be brought aboard the *Sally*, Marie Antoinette was seized by the mob. All of Captain Clough's generous and romantic plans went for naught, and he was lucky to escape with his own head.

This golden age came to an abrupt end with the Napoleonic Wars. Both France and Britain seized neutral ships trading with the enemy. President Jefferson's resulting Embargo Act of 1807 and the War of 1812 left New England's shipping tied up at the docks. Topmasts were sent down, rigging was stowed, and the mainmast trucks were covered with tar barrels, known as "Jefferson's nightcaps."

Fortunes dwindled, and a period of economic decline set in from which Wiscasset never recovered. In 1866 and 1870, great fires consumed the waterfront. They turned to ashes all the piers, docks, and marine enterprises that had been the heart of the town and many of the old houses. These setbacks destroyed the will to dream and risk that had marked Wiscasset men. In 1913, the Customs House was closed, and for the first time in more than a century, Wiscasset no longer invited ships from abroad.

A couple of blocks up Main Street and then north out Federal Street (Route 218), you will find the Ancient Cemetery and beyond that the Old Gaol and Lincoln County Museum (.7 miles from the waterfront). The jail is built of solid granite blocks and windows are barred with strap iron. Several deaths and a birth have been recorded here, but no escapes. The original graffiti is still on the walls.

For another good outing, walk across the bridge over the Sheepscot, take the first right and go right again at the sign for Fort Edgecomb (about 1.25 miles each way). Here, surrounding the wooden blockhouse built to defend Wiscasset during the War of 1812, you will find a lovely state park with picnic tables and a wonderful view down the river.

Another pleasant picnic spot is Whites Island. Follow the railroad tracks past the yacht club and cross to the island on the footbridge. Whites was once the site of a busy shipyard; today it is deserted.

Service has been restored to the coastal railroad from Brunswick to Rockland (*866-637-2457*). The train stops directly in front of the town dock.

The waterfront once teemed with shipbuilding and the shipping trade. For years, the only reminder of what once was were the two hulks of the schooners *Hesper* and *Luther Little* that sat in the mud between the town landing and the bridge. In 1932, near the end of their useful life, the boats were bought for $600 each by visionary Frank Winter of Auburn. He had them towed to Wiscasset with the intention of cutting lumber on his Auburn property, hauling it to Wiscasset by narrow-gauge rail, and shipping it to Boston aboard the schooners. They would take on coal for the return trip.

Winter died before he could carry out his plans. His legacy lay rotting along the waterfront for years, largely unappreciated. Over the years, however, the boats came to symbolize Wiscasset's and Maine's maritime heritage. When the suggestion was made to get rid of the wrecks, one of the town selectmen responded "If they touch them, they'll be dead."

Before any formal preservation efforts could be mounted, however, a winter storm in 1995 toppled the last mast and battered to pieces the awe-inspiring hulls that once linked our seaports to the world and us to our past.

They are gone, but not forgotten. Photos of the wrecks are sold in many Wiscasset stores, and if you are unlucky enough to get a speeding ticket, look carefully at the officer's badge. The schooners sail on...

BOOTHBAY HARBOR REGION

DAMARISCOVE

3 ★★★★★

43° 45.25'N 069° 36.90'W

CHARTS: 13293, 13288   CHART KIT: 61, 18

Long before the settlements of Plymouth and Jamestown, Damariscove was a bustling harbor and a major trade center. "Here was the chief maritime port of New England," according to historian Charles Bolton.

The earliest years are unrecorded, but it's clear that fishermen have known Damariscove for at least 400 years. Damarill's Cove was named for Humphrey Damarill, the resident entrepreneur in 1608. Ships arrived from European ports to trade for fish and furs, and fishermen brought their catch ashore to split and dry on stages. By the 1620s, there was a year-round fishing community here under the ownership of Sir Ferdinando Gorges.

According to her log, the *Mayflower* put into Damariscove in 1620 with 102 pilgrims aboard "to take some coddes before sailing on to Maffachufeets Bay." Two years later, Edward Winslow of Plymouth Plantation returned to Damariscove to plead for rations of food for the starving pilgrims. Thirty ships crammed the harbor, food was given generously, and no payment was accepted.

In 1629, a second plea from Plymouth for charity was met with less enthusiasm by Damariscove's fishermen, yet the bounty here was never far from the colonists' minds. In 1671, the Puritans of Massachusetts laid claim to the island, and three years later they installed a military officer and a constable. Inevitably, taxes soon followed.

When the first Indian War broke out in 1676, coastal settlers fled to the safety of the outlying islands. Some 300 went first to Damariscove and then to Monhegan, narrowly escaping the first of the Indian attacks which would last, on and off, for almost fifty years.

As every local child has heard by the fire on a winter's night, 1689 was the year that the Indians beheaded Captain Richard Pattishall, then the owner of Damariscove, aboard his sloop at Pemaquid and threw him into the sea. His body washed ashore on Damariscove, and to this day the lonely traveler

walking the foggy headlands may encounter the headless Captain Pattishall or hear the howling of his faithful dog.

For another two-and-a-half centuries, until modern times, the people of Damariscove fished and farmed. Alberta Poole grew up on Damariscove between 1910 and 1922 and related her experiences in *Coming of Age on Damariscove Island*. At the time there were 25 or 30 residents, including the fishing community, the seven members of the lifesaving station, and her father Isaac and her Uncle Chester, who raised garden produce, poultry, eggs, and milk for sale to the summer resort on Squirrel Island. When the Depression hit, Isaac and Chester Poole and their combined 13 children moved to the mainland. The Coast Guard station was manned until it was abandoned in 1959, when the only residents remaining were the summer fishermen.

In 1966 Damariscove was given to the Nature Conservancy, and it has reverted to the remote and uninhabited spot it was before the Europeans came. It is now owned and maintained as a preserve by the Boothbay Region Land Trust, which welcomes respectful vistors to what may well have been the first permanent settlement in the New World.

If you are making a passage along the coast and have no need for services, Damariscove is a convenient and reasonably protected harbor under normal conditions. It is also a place of wild and desolate beauty worth going a long way to visit. The proximity to Boothbay Harbor, however, attracts harbor cruises and daytrippers, particularly on weekends or holidays.

Damariscove's 1.7-mile length is almost divided in two by a narrow neck. The northern portion, called Wood End, was once forested before a series of disastrous fires. It is now a nesting island for several species of birds, and the major nesting site for eiders in this country.

The southern section is almost treeless as well. There are a few lobstermen's shacks, a small house for the BRLT caretakers, one cottage dating from the years when there was a farming and dairy community on the island, and the old Coast Guard station at the mouth of the harbor, now restored as a private home.

APPROACHES. The harbor itself is a deep, narrow slot in the southern end, no more than 100 feet wide at low tide. In settled weather it is easy to enter. From red-and-white gong "TM" (*43° 44.76'N 069° 37.09'W*), steer a course toward the center of the harbor, leaving The Motions to port. Swells breaking on these ledges show them clearly.

As you come abreast of the western tip of the island, favor the left side of the channel to avoid a ledge making out from the eastern shore. From there on, the harbor is free of dangers, and you can run down the middle.

Heavy swells from the south roll well into the harbor and sometimes even break across the entrance. In that case, find another anchorage.

ANCHORAGES, MOORINGS. This narrow harbor is divided into an inner and outer pool. The outer pool has 11 feet of water over a sand bottom, with room for three or four boats to anchor comfortably far enough north of the lifesaving station to be reasonably protected from ocean swells and southwesterlies. On a busy night, however, you might find six or seven boats here, bow-to-stern and side-by-side, with bow and stern anchors out on short scope. Prepare your stern anchor, and consider weighting your rodes.

Anchoring is further complicated by a smattering of lobster buoys and several private moorings just off the lifesaving station. In settled weather, boats might be anchored as far out as the mouth of the cove.

The inner pool has 6 feet of depth and better protection, and at high tide it looks open enough to tempt many sailors to give into their head-of-the-harbor urges. The bottom, however, has a lot of kelp, it shoals rapidly beyond the halfway point of the inner pool, and several moorings occupy most of the space. The tide settles yachts on the bottom

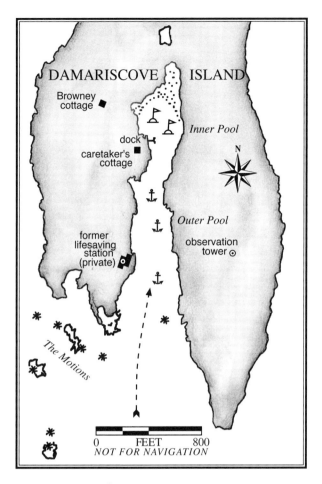

THINGS TO DO. Damariscove is open to the public for "careful day use." That means no camping, no open fires, no trash, and no dogs. Subject to those regulations, you are free to hike around the island. A little natural history museum is perched at the head of the harbor. In season there are abundant blueberries, blackberries, strawberries, raspberries, wildflowers, and—beware—poison ivy.

The island trails lead past the fieldstone foundations of early dwellings and over open hummocks with expansive views. Some trails lead through the meadows to the spring at the head of the harbor, which used to supply the fishing ships. The smaller trails are made by muskrats, which can be ferocious when cornered. From there, you can work your way north to the large freshwater pond shown on the chart, but stay high along the eastern ridge to avoid deep puckerbrush and poison ivy. Bird and muskrat watchers will love this pond; swimmers won't.

Another walk leads to Milk Slip. Isaac Poole rowed his dory from here to Squirrel Island to sell his milk to summer residents. From the first of June until October, he never missed a day, no matter what the weather.

The southern tip of the island is dominated by the handsome shingled Coast Guard station, originally built in 1897 as a U.S. Lifesaving Station. The building has been restored as a private home, and visitors are not allowed. The far northern end is also off limits until mid-August since it is a nesting area for more than a thousand common eiders.

Stewardship and access to this island comes at a price. BRLT greatly appreciates donations toward the preservation of this special place (*633-4818; www.bbrlt.org*).

here regularly. The good news is that two of those moorings are maintained by the Boothbay Region Land Trust and are available to daytime visitors for an overnight stay.

If the harbor is crowded or a strong southwesterly is piling in a surge, go somewhere else for the night. It is easy to get in trouble here.

GETTING ASHORE. Land your dinghy at the BRLT float just north of the stone wharf. Please do not land your big boat here or on the adjacent stone wharf, which is used by fishermen.

SQUIRREL ISLAND
43° 48.42'N 069° 38.14'W

CHARTS: 13293, 13288   **CHART KIT: 61, 62, 18**

Right in the middle of Booth Bay stands high, wooded Squirrel Island, an exclusive summer retreat for generations. Lawn tennis championships were held here in the 1920s and 1930s, and there are wonderful turn-of-the-century cottages with an air of calm assurance. Members have 99-year leases on their homes, renewable for a dollar.

Despite this apparent exclusivity, visiting yachtsmen are welcome, and you can land to explore. The island's exposed location makes it a poor choice for

an overnight anchorage, but it is a delightful day-stop, with inviting paths around the island.

On the western side are two coves. The smaller, more northerly one is used by the boat from Boothbay Harbor to unload supplies. Squirrel Cove, to the south, is the home of the Squirrel Island Boating Association, ⚓ which usually maintains a guest mooring on the outer fringe of the cove.

Pick up the mooring and row ashore to the SIBA floats. There is deep water alongside the floats if you want to drop off passengers, but don't tie up the wharf for long. Behind the boathouse are two pay phones, but there are no other facilities.

When you land, you will immediately see a concrete path with railings. This takes you around the island, perhaps an hour's walk. By all means, take it. The path leads to every house on the island. Between them, there are wonderful views of the surrounding islands and peninsulas.

Only service vehicles are allowed on Squirrel. No bikes, roller skates, or skateboards are permitted on the path. Summer residents cart their groceries in little red wagons. The northern section of the path turns into a beautiful woodland trail along high rocky cliffs, and there is a private sand beach on the north side.

If you walk left from the boathouse, past another beach, the chapel, and the ball field, you will reach "downtown" Squirrel Island, which boasts a post office, pay phones, town hall, and a tearoom (open to members only).

Squirrel Island is a wonderfully relaxed and friendly place of astounding beauty and old-world hospitality. The cruising community is blessed by the generosity of its residents in sharing it with us. Thank you.

BOOTHBAY HARBOR

4 ★★★

Inner Harbor: 43° 50.90'N 069° 37.61'W
W of McKown Pt.: 43° 50.72'N 069° 38.54'W

Charts: 13293, 13296, 13288   Chart Kit: 61, 62, 18

Emergency: 911
Hospital: 207-633-2121
Coast Guard: 207-633-2643

Boothbay Harbor is one of the best harbors on the Maine coast. Not only is it large and well protected, it is easy to enter under all conditions. Here are all the services a yachtsman might need and then some. If you are coasting west or east, it is very convenient, less than 5 miles off the direct route.

The harbor bustles with boats of every description and has the air of a city of the sea. Like Kennebunkport, the town is crowded and touristy, but festive. Shops and restaurants jam the streets, and the calendar is filled with summer activities.

The town's annual Windjammer Days festival is a wonderful spectacle, but it is popular. Reserve your berth or mooring well ahead of time.

APPROACHES. Coming from the west, round green bell "1C" off Southport Island, leaving the Cuckholds (two little bare islets) to port. The Cuckholds Light, a good landfall, is a distinctive white octagonal tower 48 feet high on a dwelling with a foghorn and radio beacon.

Pass either side of high Squirrel Island, observing red lighted buoy "4" if you are to the west. Head for red lighted buoy "8" off Tumbler Island, leaving Burnt Island, with a lighthouse and a horn, to port. Don't be tempted to go inside Tumbler Island. Keep it to starboard. Leave green lighted buoy "9" to port at McFarland Island and enter the inner harbor.

Coming from the east, find red-and-white bell "HL" (*43° 48.39'N 069° 34.79'W*) at the beginning of Fisherman Island Passage, sailing north of Hypocrites, Fisherman Island, and small, grassy Ram Island with its lighthouse and foghorn. Keep the three nuns marking the ledges to the south of Linekin Neck to starboard and turn northward. Leave high Squirrel Island to port and enter as described above.

In fog, keep your radio on to listen for *securité* calls—some whale-watching boats waste little time getting to the feeding grounds, even when the visibility is zero.

ANCHORAGES, MOORINGS. Because Boothbay Harbor is one of the busiest boating centers on the coast, anchoring is prohibited everywhere in the harbor except in exposed Mill Cove, on the backside of the town, north of Wotten's Wharf. However, there are a dozen or more places to find a mooring or berth (see sketch map). The most convenient locations are in the inner harbor, at the northeast corner of Boothbay Harbor. Tugboat Inn Marina and Boothbay Harbor Marina are right in town.

Carousel Marina and Brown's Wharf Marina are well down Spruce Point along the eastern side of the inner harbor, about half a mile from town. Wotton's Wharf is just around McFarland Point. The short distance from town makes these locations less noisy.

The Boothbay Harbor Yacht Club has a good number of moorings in the less busy western part of Boothbay Harbor, inside McKown Point. Facilities are good, although it is not close to town. Other possibilities are listed below.

FOR THE BOAT: Inner Harbor, Western Side. *Boothbay Harbor Marina (Ch. 09; 207-633-6003).* Right next to the footbridge in town, this is a full-service marina. ➤ There are several finger floats and transient dock space that can accommodate boats to 150 feet, with 16 feet of water along the outer floats. Electricity to 50 amps, cable, and water are available, and showers and a laundromat are in the building on the dock.

Public Landing (Harbormaster Earl Brown: Ch. 09, 16; office: 207-633-3671; cell: 207-380-5635). Just south of Fisherman's Wharf Inn and Pier 6 are the floats of the public landing. ➤ Their 120-foot length can accommodate boats of up to 50 feet in depths to around 10 feet, with a generous three-hour maximum tie-up.

Boothbay Harbor Pumpout Boat (Ch. 09; 207-633-4220). ☞ This mobile pump-out is operated by the town and can come to your boat anywhere in the harbor.

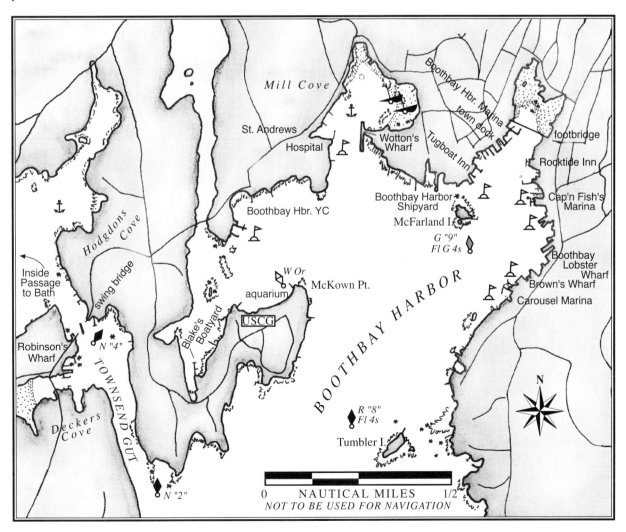

Tugboat Inn Marina (Ch. 09; 207-633-4434; www.tugboatinn.com/marina.htm). The Tugboat Inn is the first facility on your left as you enter the inner harbor. Its many moorings lie just south of Pier 8, with bright polyball floats marked "TBI." The marina is wonderful about letting you use them temporarily, even if they are reserved by somebody else for the night. Or you can reserve them yourself throughout the year (cancellations are allowed up to 9AM the day you are due to arrive).

The marina also provides a large amount of dockage, which is also reservable. The docks can accommodate boats to 110 feet, with 12 feet of water at the outer floats. Electricity, water, ice, pump-outs, and phones are available, but no fuel. Ashore, they have very pleasant coin-operated showers, a laundromat, a seafood restaurant, and lounge and deck overlooking the marina.

Down East Yacht Club. DEYC operates from the northern end of the Tugboat Inn Marina, and they have moorings just off it. The slips and moorings, however, are for club members only.

Boothbay Harbor Shipyard (Ch. 09, 68; 207-633-3171). Around to the west from the inner harbor, BHS is recognizable by its yellow buildings and two enormous marine railways. One of these is a 700-ton giant which regularly hauls huge windjammers, Coast Guard vessels, and recently, the replica ship *Bounty*. As Sample's Shipyard, it has been around "since almost forever." During World War II it built wooden minesweepers, and for a while it was leased by the builders of Steve Dashew's Deerfoot yachts. It now specializes in large wooden-vessel repairs and new construction. A mechanic and diver are on site. The yard maintains 23 moorings. Ten are reserved for transients. They are quite exposed, but four of them are big enough to hold yachts to 100 feet.

Signal Point Marina (Ch. 09; 207-633-6920). This marina is near McFarland Point, at the southern end of Wotton's Wharf. Its floats are tucked behind a small breakwater, but they are all rented seasonally.

Wotton's Wharf (Ch. 09; 207-633-7440). Wotton's is run by Boothbay Region Boatyard (see *Maddock Cove*, page 125) as a full-service marina and repair center. They are located on the east shore of the mouth to Mill Cove and can be identifed by their large blue-and-white fuel sign. Boats to 80 feet can be accommodated on their floats, with 7 to 16 feet of water at low, power to 100 amps, water, and

Harbormaster Earl Brown can often be found at the Public Landing.

Wi-Fi. Eight reservable moorings are lined up in the entrance to Mill Cove. The fuel dock has high-speed pumps for gas and diesel and there is ice and a small chandlery ashore.

The East Boothbay General Store (207-633-7800). After careers as crew and chef aboard yachts, owners Dominic and Elizabeth Pochee offer full provisioning services to yachts visiting the Boothbay region.

FOR THE CREW: Inner Harbor, Western Side. In this part of town, you are within a hundred yards of every kind of shop imaginable, including the small but well-stocked Village Market, right in the center of town. A huge Hannaford market and liquor store is in the shopping center north of town, .8 miles from the waterfront. Note that free trolley-buses leave from the Rocktide Inn (see below).

One of the unique facilities serving Boothbay Harbor is St. Andrews Hospital (633-2121), which overlooks Mill Cove opposite Wotton's. They have a dock and a float, so you can practically dinghy to the emergency room.

FOR THE BOAT: Inner Harbor, Eastern Side. *Rocktide Motor Inn (207-633-4455).* The northernmost floats on the east side of the inner harbor, next to the footbridge, are for the residents of Squirrel Island. The next floats to the south belong to the Boothbay Harbor Inn. The Rocktide Inn has the next floats further south, and they are available to anyone coming to eat or stay at the inn. The outer floats have 12 feet of water at low. Reservations are recommended.

Cap'n Fish's Motel and Marina (Ch. 09; 207-633-6605). Cap'n Fish's has several moorings and some dockage with water and electricity for transients. Reserve ahead. Breakfast is served in the restaurant.

Brown's Wharf Marina (Ch. 09; 207-633-5440; www.brownswharfinn.com). Brown's has 40 slips, with 8 to 20 feet alongside, and three transient moorings available on a first-come, first-served basis. Water, pump-outs, electricity, cable, Wi-Fi, ice, and showers are available, but no fuel. Their restaurant is right at the water's edge.

Carousel Marina (Ch. 09; 207-633-2922). Carousel is a large, full-service marina at the southern end of the harbor, identifiable by a yellow-and-red fuel sign and gray buildings. The fuel dock has gas, diesel, pump-outs, ice, and water. Transient dockage and rental moorings are available. The mooring floats have yellow tops and green bottoms. A small ship's store, laundry facilities, Wi-Fi, showers, and a sitting room with television are ashore, along with the Whale Tail restaurant.

Marine Supply, Inc. (207-633-0709). This small chandlery is located between the Rocktide Inn and Cap'n Fish's, at 55 Atlantic Avenue.

FOR THE CREW: Inner Harbor, Eastern Side. In addition to the restaurants at the marinas, seafood is served at the Lobster Dock (*633-7120*) and takeout at the Boothbay Lobster Wharf (*633-4900*), formerly the Lobstermen's Co-op, where a sign reads: "We are not responsible if seagulls steal your food." For supplies and shops and additional dining of all kinds, walk across the footbridge into town, or take the dinghy to the public landing.

FOR THE BOAT: West of McKown Point. *Boothbay Harbor Yacht Club (Ch. 09; 207-633-5750).* At the northwest side of the bay, opposite McKown Point, the Boothbay Harbor Yacht Club has a low, white building over the water on pilings, with a BHYC sign, a dock, and floats. The club maintains a large number of moorings, but it is a good idea to reserve ahead for weekends during big events. Launch service is provided. The floats have 20 feet of water alongside and water and electricity, but no fuel. Overnight dockage is not permitted. Bring your dinghy in to the inner float.

The club has showers, telephones, ice, and laundry machines. Visiting yachtsmen are welcome to use the bar and restaurant if they wear jackets for the evening meals. There are also two tennis courts that visitors may use. The West Boothbay Harbor post office is next to the tennis courts.

Blake's Boatyard (Ch. 09; 207-633-5040). Blake's is a small yard tucked down in the western corner of the bay, where there is the best protection in Boothbay Harbor. This is the place to be if the weather is threatening. Unfortunately, Blake's is not a transient facility. They have water and electricity at their floats, and the yard has three marine railways and can perform most repairs, but they give priority to their seasonal customers. Their well-stocked chandlery carries charts and hardware.

If you enter this cove, beware of a deceptive, submerged rock off the eastern shore. It is shown on the chart inset as a 3-foot spot, but it lies further from shore than the chart seems to indicate, and it is surrounded by moorings. At lower tides, it is best to get at least two moorings west of the shore before heading south.

FOR THE CREW: West of McKown Point. To get into town, call a cab or hitch a ride.

THINGS TO DO. Wandering around town, eating ice cream cones, and watching the tourists—these are the favorite occupations in Boothbay Harbor. Despite being a tourist mecca, Boothbay Harbor has some authentic Maine holdouts. Our favorites are Grover's Hardware (*633-2694*) on Townsend Avenue, which opens at 7 AM, and the Romar Bowling Lanes, whose eight candlepin lanes are in an old log cabin right in the middle of town.

A delightful evening outing is to row across the harbor to the Boothbay Lobster Wharf, tie up at the float, and have a lobster or clam dinner outdoors overlooking the lobsterboats—good food at reasonable prices. For finer fare, make reservations at Ports of Italy (*633-1011*) on Commercial Street. Cruiser Sterling Blake votes this "hands down the best restaurant in town."

From downtown, all sorts of excursions are available. Dayboats run trips up the Inside Passage through the Hell Gates and the Sasanoa River to Bath. If the current is running with any force, you may be glad you are in somebody else's boat. Other trips circumnavigate Southport Island, or run out to historic Damariscove Island. Day-long trips depart for Monhegan, and short jaunts will take you to Squirrel Island.

Maine's Department of Marine Resources operates a little aquarium at the end of McKown Point near the western part of the harbor. Their displays show various fisheries, and you can reach into a "touch tank" full of live marine animals—if you dare.

LINEKIN BAY and LEWIS COVE

3 ★★★

Lewis Cove: 43° 50.67'N 069° 36.88'W

CHARTS: 13293, 13288   CHART KIT: 62, 18

Almost deserted compared to Boothbay Harbor next door, Linekin Bay is a lovely, wide-open place to sail, encircled by arms of wooded shore. There are some ledges and rocks in the bay, but they are obvious or well marked.

Lewis Cove, in the northwest corner of the bay, provides tranquillity and good protection from the south and southwest around to the northeast. If you are not in the mood for the bustle of Boothbay Harbor, this is a delightful alternative.

APPROACHES. Entering Linekin Bay is easy. Leave large Squirrel Island to port and pass between small, wooded Negro Island and can "1." To continue to Lewis Cove, run up along the coast of Spruce Point and leave can "3," off Cabbage Island, to starboard.

ANCHORAGES, MOORINGS. The Paul E. Luke boatyard in Linekin Bay has several moorings and space at their dock, with 15 feet at low.

In Lewis Cove, anchor in 24 feet of water at low near the Spruce Point shore, as far up into Lewis Cove as you can get. Note the arc of rocks extending from the spit of land to the north and curving about 250 yards down to the Spruce Point shore. You may find it helpful to put out a stern anchor here.

Linekin Bay Resort (633-2494), located on the point that forms the eastern shore of the cove, rents a mooring in Lewis Cove marked "LBR." They also have two docks with 15 feet of depth at low, though these are exposed to the southwest.

GETTING ASHORE. Land at Barrett Park on the spit of land to the north of Lewis Cove, or dinghy into the docks at Linekin Bay Resort.

FOR THE BOAT. *Paul E. Luke, Inc. (207-633-4971).* As you enter Linekin Bay, you will see several sheds on the eastern shore along with a crane, a dock, and a float—but no signs of any kind. This is the well-known yard of boatbuilder Paul E. Luke. The yard has a large-capacity marine railway, a hydraulic trailer, a Travelift, and a full machine shop to handle repairs of all kinds.

FOR THE CREW. Luke has no shoreside luxuries such as showers or laundry. "We believe," says Frank Luke, "that if you have a boat laden with cruising gear, you should use it."

Cliff Tibbetts lobstering out of Little River.

The Linekin Bay Resort is a rustic family resort that feels like a summer camp from years gone by. They welcome visitors to their main lodge bar and restaurant (*633-2494*) or, for a day fee, to use their pool and tennis courts.

THINGS TO DO. From Barrett Park, it is an easy walk to Boothbay Harbor across Spruce Point.

Luke's has an interesting little museum with memorabilia of 50 years of boat building. "We've been known," adds Frank Luke, "to have full-moon parties. Bring your supper in off the boat, we'll lay out a buffet, and we'll eat and drink until it is all gone."

DAMARISCOTTA RIVER

The Damariscotta is a modest river, with neither the grand scale of the Kennebec nor the cruising charms of the Sheepscot. In early records the river was called the Damariscove after the Indian word for "River of Alewives." Navigation is relatively easy. The current asserts itself only in one or two narrows, and the upper reaches are well marked. From the mouth to the head of navigation is only 14 miles, so it can be explored in a day or two.

The river is famous for its boatbuilders—Gamage in South Bristol and W.I. Adams, Rice Brothers, and Hodgdon Brothers in East Boothbay. Goudy and Stevens bought the Adams yard in 1924, merged with Hodgdon Brothers, and is now Ocean Point Marina. Megayacht builder Hodgdon Yachts is now next door.

Several harbors on the river are of special interest to the cruising sailor. Christmas Cove, near the mouth, is cozy and crowded, but low-key and hospitable. Nearby, the Thread of Life is charming not only for its name, but for the beauty of this little passage.

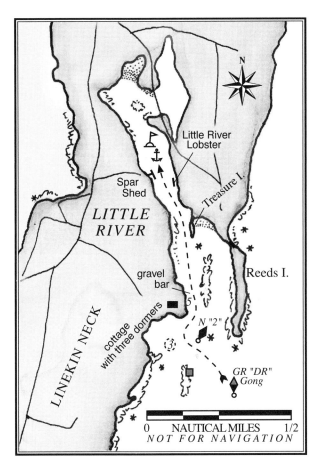

LITTLE RIVER

43° 49.86′N 069° 35.13′W

CHARTS: 13293, 13288   CHART KIT: 62, 18

Little River is a long, narrow inlet extending northward into Linekin Neck at the west side of the mouth of the Damariscotta River.

A number of years ago, Little River was the subject of a turf war between fishermen and pleasure boaters. There was an undercurrent among the fishermen that Little River should be reserved for fishermen only, and it wasn't unusual to hear a little disinformation that, for a pleasure boater, the entrance was nearly impossible and that inside there was no room at all. True, the entrance can be a bit daunting in fair weather and dangerous in foul. But now that the dispute has been resolved by the town of Boothbay Harbor, the reward is a peaceful anchorage in a serene working harbor.

APPROACHES. Avoid entering Little River in heavy weather. The harbor is wonderfully well protected by ledges at its entrance, but when the sea is up, it breaks clear across. There is a 5-foot spot in the channel, so go in on a rising tide, slowly. Boats with drafts of 6 feet or more should be sure it is at least an hour or more either side of low.

Much of the problem with entering Little River is that the scale of the chart is too small. The entrance is marked by green daymark "1" and red nun "2" close together. Approach from red-and-green gong "DR" (*43° 49.15′N 069° 34.82′W*) and steer for the nun. Once the daymark is abeam to port, begin to make a turn to the west, passing between the daymark and the nun but heavily favoring the nun.

Once past the nun, steer for the cottage with three dormers on the point of land to the west of the entrance. Within about 100 yards of the point, round it and keep it close to port to avoid the north-south ledge, visible at most tides and centered between the point and Reeds Island. Watch your depth sounder here. The chart shows a 5-foot spot, but the bottom is a gravelly bar which gets pushed around by winter storms, so it is subject to change.

The worst is over. Once abreast of the brown house on little Treasure Island to starboard, stay centered in the river, proceeding north. You will find good water as far north as the dock on the left with the gray shingled boathouse.

Above Christmas Cove, the river stretches between high, wooded banks with only a few houses. On the surface, not much has changed since the Indians called it the "river of many fishes" and used it for their annual spring migrations to the sea. Talk to a fisherman, though, and he might disagree.

The current runs strongly through The Narrows off Fort Island, but the passage is wide enough and well marked. Beyond lies one of the great secluded spots of the midcoast—Seal Cove, a special place that is hard to enter, hard to leave.

Upriver from Seal Cove are two agreeable harbors—Pleasant and Wadsworth Cove. Across the way from them is the Ira C. Darling Center of the University of Southern Maine, which conducts research in marine biology and offers visitors the opportunity to spend a fascinating hour or two learning about lobsters, clams, and aquaculture.

Thereafter, the river winds peacefully northward to Damariscotta and Newcastle—lovely, quiet little towns at the river's head of navigation. Three miles before the towns, it passes the Dodge Point nature preserve, a pleasant daystop for hiking or sunning on the banks.

ANCHORAGES, MOORINGS. After the inner harbor opens up north of Treasure Island, you may find room to anchor in midchannel in 13 to 18 feet of water.

⚓ Spar Shed Marina has seven rental moorings, which have conical white floats with blue stripes. Pete Yesmentes is kind enough to offer a free guest mooring maintained by the Boothbay Shores Association in the outer harbor. In moderate weather it can handle boats to 35 feet and drafts to 7. The rules? Use at your own risk; first-come, first-served; three-night limit. Thank you.

GETTING ASHORE. Row in to the dinghy float at the Spar Shed Marina or Little River Lobster.

FOR THE BOAT. *Little River Lobster (207-633-2648).* Look for a two-story shingled building on the eastern shore with a series of floats in front. Little River pumps ⛽ gas and diesel at the float, which has 14 feet alongside at low. They will also sell you lobsters live or cooked.

Spar Shed Marina (207-633-4389). The obvious dock and gray boathouse of Spar Shed Marina lie just beyond Little River Lobster, on the opposite shore. It's often identifiable by the fresh flowers on the docks. In addition to moorings, ⚓ Spar Shed has six slips with 7 feet of depth at low. Water, ice, and electricity are on the floats, but no fuel.

FOR THE CREW. 🍴 The Ocean Point Inn (633-4200), about a mile south out Linekin Neck, serves breakfast and dinner.

THINGS TO DO. Row your dinghy north to the far end of the cove. You can land at either end of the dam to the east and then take a pleasant walk north or south on a dirt road.

CHRISTMAS COVE

3 ★★★

43° 50.80′N 069° 33.32′W

CHARTS: **13293** (INSET)**, 13288**   CHART KIT: **62 (A), 18**

Ever since Christmas 1614, when Captain John Smith dropped anchor in this picturesque spot, Christmas Cove has been a favorite of sailors. The protection is excellent, and it is not far off the east-west path. Christmas Cove is a busy little place, always full of movement, but it retains its wonderful, low-key atmosphere. Seals, mallards, and ospreys share the cove, along with a multitude of cruisers.

APPROACHES. The usual approach is from the west of Inner Heron Island, at the entrance to the Damariscotta River. Pick up nun "4" north of Inner Heron, then aim for the square, shingled tower near the head of Christmas Cove.

The entrance to the cove is tight, about 35 yards wide, and you are likely to find kids racing small sailboats right in the narrowest part of the channel. As you can see on the chart inset, there are two tiny dots of land on each side of the entrance. Red daybeacon "2" is on your right coming in, but do not cut it close, since it is 50 feet or so from the end of the ledge. Green daybeacon "3" stands at the edge of the shoal water to port. Pass about halfway between the two beacons and expect to find boats moored just inside. In fog, finding these silent and unforgiving daybeacons could be treacherous.

ANCHORAGES, MOORINGS. Christmas Cove is packed with moorings, so it is doubtful that you will find anchoring room. ⚓ Most of the moorings belong to Coveside Marina, which keeps 14 available for transients. They have white, numbered floats marked "CS." Reservations are highly recommended. The moorings are often taken by early afternoon.

GETTING ASHORE. Row your dinghy to the Coveside dinghy float.

FOR THE BOAT. *Coveside Marina (Ch. 09; 207-644-8282).* Coveside Marina is the complex of red buildings on the north side of the harbor. ⚓ They have limited dockage but can handle large boats, with 12 feet alongside at low. The docks have electricity, Wi-Fi, ice, and 🚰 pump-outs. Water is for your tanks only, since it is drawn from a well.

FOR THE CREW. Because of the well water, there is no laundromat, but there are showers and a pay phone at Coveside Marina. 🍴 The marina restaurant serves lunch and dinner (*644-8282*). In good weather you can eat on the deck overlooking the harbor.

This is not an easy place to provision. For a limited food selection you will need to walk 1.2 miles along the Damariscotta River to South Bristol. From Coveside, turn left, and take the next left on the Westside Road. The small Island Grocery stocks wines, cheeses, breads, and pre-cooked delicacies (see *The Gut*, next).

THINGS TO DO. The walk to South Bristol is pretty and shaded. When you arrive, it is fun to observe one of Maine's most active swing bridges at The Gut and the busy scene in quaint South Bristol Harbor as lobsterboats unload their catch.

For a shorter walk, turn right from Coveside Marina and then walk south on Route 129. You will find a spectacular view of the Thread of Life and maybe see a sailboat or two tacking through.

THE GUT (South Bristol)— West of the Bridge

3 ★★

43° 51.68'N 069° 33.64'W

CHARTS: 13292 (INSET), 13288   CHART KIT: 67 (F), 62, 18

The Gut is a narrow passage that snakes between the Damariscotta River and Johns Bay, separating Rutherford Island from the mainland at South Bristol. The west side of The Gut is home to the boatyard made famous by shipbuilder Harvey Gamage. The east side of The Gut is an active lobstermen's harbor with several commercial docks. The narrow gut between east and west is flankd by the authentic village of South Bristol and spanned by a small swing bridge that will open on short notice.

Check your mast height and the fine print on the chart carefully before trying this passage. Overhead power cables with a clearance of 55 feet at mean high water span the gut, and the bridgetender warns that the cables may have sagged lower. Depth at the bridge has silted to about 5 feet at low, and the current runs up to 2 or 3 knots. Also give the ledge on the north side of the channel, just west of the bridge and marked by daybeacon "3," plenty of room.

APPROACHES. Coming from the Damariscotta River, leave red daybeacon "2" well to starboard to avoid the projecting ledge and rock and run for the middle of the opening. The bridge is manned 24 hours a day and will open for the conventional long and short blasts, one or two toots, or if the bridgetender sees you coming. Or you can be sure by hailing him on channel 09. This may be the busiest swing bridge in New England, but it is very small-scale and friendly.

ANCHORAGES, MOORINGS. There is very little room to anchor in The Gut. Your best bet is to get one of Gamage's four moorings or lie alongside their floats by the long wharf, with 10 to 11 feet of water at low.

GETTING ASHORE. Row in to the float at the Gamage yard.

FOR THE BOAT. *Gamage Shipyard* (**207-644-8181**). Gamage is west of the bridge on the north side of The Gut, identifiable by their new, red-roofed dock house. The fuel dock, with 11 feet of depth at low, pumps gas and diesel and the docks have electricity and water. The dock house carries some boat supplies, ice, snacks, and "Items of Intrest, and Miscalanious (*sic*)." Showers are ashore. The yard no longer builds boats. Their huge marine railways are now blocked by the shallow-draft floats. But they have a 25-ton boatlift and can perform most hull and engine repairs.

Many famous boats were designed and built at the Gamage yard, including the 95-foot gaff-rigged sloop *Clearwater*, used to spur the cleanup of the Hudson River.

FOR THE CREW. On the north side of the bridge and slightly up Route 129 you will find Harborside Grocery and Grill (*644-8751*), a very convenient

Passing westward through The Gut.

convenience store with an extensive beer inventory, weighted towards quantity rather than quality. They also serve pizza or burgers from their grill until 8:30 at night.

Osier's Wharf, just south of the bridge, caters primarily to fishermen, serving a crack-of-dawn breakfast in the morning and selling seafood, lobsters, and beer for lunch.

A seafood store is south of the bridge, up the hill and on the right. The first right turn will lead to the small Island Grocery, which sells wines, cheeses, coffees, breads, and pre-cooked delicacies.

THINGS TO DO. A delightful, shady walk of just over a mile along the Damariscotta River will take you to Christmas Cove and Coveside Restaurant. From Gamage, cross the swing bridge and take the first right.

The Thompson Ice Harvesting Museum is on the right side of the road north of town. Icehouses have been on this site for decades, and fearful that this bit of history would slip into the past, the Thompsons donated the land, and a new, working icehouse was built there in 1990. Each winter volunteers rediscover how much sweat there is in ice as they scribe, cut, and store ice blocks from the ice pond. The museum is open in July and August from 1 to 4PM or by appointment by calling *644-8551*.

FARNHAM COVE

43° 51.56'N   069° 34.57'W

CHARTS: 13293, 13288   CHART KIT: 62, 18

Farnham Point is directly opposite the west side of The Gut and within sight of East Boothbay. A little cove to the west of the point is out of the current and well protected from all directions but north. The sides are wooded and steep-to, and at the head is a lobster pound. This can be a good alternative if there is no room at Christmas Cove.

APPROACHES. Avoid the 6-foot spot north of Farnham Point and run down the middle of the slot.

ANCHORAGES, MOORINGS. Anchor outside the moored boats, in mud, in about 16 feet at low.

EAST BOOTHBAY

43° 52.01'N   069° 34.94'W

CHARTS: 13293, 13288   CHART KIT: 62, 18

East Boothbay is a small community on the west bank of the Damariscotta River, about 4 miles from its mouth. More of an open roadstead than a harbor, it is exposed to wind and current up and down the river. There are good moorings, almost any repairs can be accomplished, and, if you're lucky enough, you can come here to build your new tug or megayacht.

East Boothbay is famous for its boatbuilding. This was the home of Goudy and Stevens for more than half a century as well as the Hodgdon Brothers and Rice Brothers. Goudy and Stevens launched many famous vessels, including a replica of the 104-foot *America,* for which the coveted trophy was named.

Hodgdon Brothers built *Bowdoin,* the famous schooner which logged 300,000 miles in her 26 trips to the Arctic under Admiral Donald MacMillan. Today, Hodgdon Yachts constructs some of the largest and most elegant boats afloat. At the time of this writing, they had just launched the 98-foot *Windcrest*. Before that, *Scheherazade* reached nearly 155 feet.

APPROACHES. The moored boats and boat sheds become visible as you pass Farnham Point on the west bank of the Damariscotta River.

ANCHORAGES, MOORINGS. It is too deep to anchor off East Boothbay, so pick up a mooring off Ocean Point Marina (in 50 or 60 feet of water) and inquire at the office. If none is available, there may be space at the floats. Or you can run back down to Farnham Cove for the night (see previous entry).

GETTING ASHORE. Take your dinghy to a convenient spot at the floats at Ocean Point Marina, or tie up at the Lobsterman's Wharf and Inn next door.

FOR THE BOAT. *Ocean Point Marina (Ch. 09; 207-633-0773)* is at the north end of the waterfront, with obvious slips, several large sheds, and a boatlift. They pump gas and diesel along the north side of the floats, with 12 feet alongside. They have water, electricity, ice, transient dock space, and rental moorings. Showers, laundry facilities, a chandlery, and pump-out facilities are ashore. A 35-ton boatlift and two cranes can pull you or your masts out for all types of repairs.

Nathaniel S. Wilson (207-633-5071). Well-known sailmaker Nat Wilson is based on Lincoln Street in East Boothbay. Turn right when you leave Ocean Point, right again on Lincoln, and go up the hill. He can handle sail and rigging repairs or cut you new sails. Wilson's specialty is making new "traditional" sails. He helped develop a sailcloth of modern materials with the look and soft hand of cotton. He even landed the unenviable contract to build a new suit of sails for the USS *Constitution*. Be thankful that you weren't paying for them. Well, actually, you were.

Hodgdon Yachts (207-633-4194; www.hodgdonyachts.com). The large boat sheds of Hodgdon Yachts are south of the Lobsterman's Wharf and Inn, which is recognizable by the umbrellas on their deck. Hodgdon has no facilities for the visiting cruiser, unless you happen to be in the market for a jaw-dropping

ICE

Soft music floated on the warm evening breeze. On the verandah, ladies in long white gowns whispered to their escorts. Ice tinkled in the mint juleps…

Ice—from the midcoast rivers. All up and down the coast of Maine, ice was harvested in every available pond and river to satisfy the growing markets down south. The center of the industry was the Kennebec River, with 60 ice houses between Bath and Augusta harvesting 1,185,000 tons of tidewater ice in 1886-87. Other important areas were the Penobscot River, Boothbay, Wiscasset, and Bristol.

An acre of ice 12 inches thick produces 1,200 tons of ice, or 1,000 tons after shrinkage. In 1870 Kennebec ice was sold for $10 a ton delivered onboard.

During its heyday in the latter half of the 19th century, the ice business was very substantial, employing considerable numbers of men and considerable capital. Enormous icehouses were built to hold the harvest. In 1868 the "Ice King of the Kennebec," James Cheeseman, built a dozen icehouses, each 700 feet long by 40 feet wide by 30 feet high. Together they could hold 70,000 tons of ice.

Ice harvesting usually began as soon as possible after the river froze in mid-December. Once the ice was thick enough to bear the weight of men and horses, the area to be harvested was scraped after every snowfall. When the ice thickened to a foot or more, the field was laid out and marked by hand, then grooved by plow-like implements drawn by spans of horses. A hole was chiseled, and the cutting began. As the saw cuts gradually opened up the river, blocks of ice were prodded along with picks to chainbelt conveyors, which lifted them into the icehouse. Inside, they were stacked in tiers and packed with insulating hay or sawdust.

In the spring, when the river was again open for navigation, the process was reversed. The ice was run down from the icehouses, loaded aboard the hundreds of waiting schooners, and shipped south. In the later years, when the schooners were replaced by tugs and barges, the strings of barges were often a mile long.

New York was the biggest market, taking 500,000 tons a year by 1874. The main source of supply for New York was the Hudson, so a poor ice year on the Hudson meant "ice fever" on the Kennebec. Philadelphia took 400,000 tons, Baltimore 100,000 tons, and so on down the Atlantic coast to Washington, Norfolk, and Savannah. Shipments were made to Cuba, Panama, the coastal cities of South America, and even halfway around the world to India and China.

During the season's storage, about five to ten percent of the ice was lost to melting. Further shrinkage took place during shipment, despite huge quantities of dunnage. As an example, the schooner *John F. Randall* in 1898 carried a cargo of 2,169 tons of ice to Cuba and 200 cords of shavings as dunnage. Ice was an unpopular cargo with captains, awkward to load and turning into freshwater in the bilges.

And indeed, the handwriting was soon on the wall for Maine's flourishing ice industry. Refrigeration was growing in use by the beginning of the 20th century. No ice was cut at all on the Kennebec in 1907-08, and by 1921, only one icehouse was left.

The Thompson Ice Harvesting Museum is located north of The Gut at South Bristol.

Teams of horses scrape snow off an ice field on the Kennebec prior to cutting and storing the blocks in the huge sheds of the American Ice Company. (From *Tidewater Ice of the Kennebec River*, by Jennie G. Everson, courtesy Eleanor L. Everson.)

vessel, but if you are lucky, you may see one of these beauties pulled out of one of the sheds.

FOR THE CREW. The East Boothbay post office is next to the driveway of Ocean Point Marina.

🛒 The East Boothbay General Store (*633-7800*) is a short walk south on Route 96, up the hill and on the right. After having crewed on yachts, the store's new owners know what boaters need—essential groceries, local produce, fresh-baked breads, hot and cold sandwiches, pizza, beer and wine, and provisioning and delivery services. They also rent bikes.

🍴 Lunch and dinner are served at Lobsterman's Wharf, just south of the docks of Ocean Point Marina. Their sun umbrellas make it easy to spot from the water. Customers with shallow draft (or dinghies) can tie up at the restaurant float. Often on weekends, live music plays at the water's edge.

JONES COVE

43° 52.19'N 069° 34.11'W

CHARTS: **13293, 13288**   CHART KIT: **62, 18**

Jones Cove is a pretty and seldom-used anchorage on the east side of the Damariscotta River opposite East Boothbay. Horses sometimes graze in the meadows, and a small sand beach lines the head of the cove. It offers no protection from the prevailing summer winds, but it would be useful if the wind came around to the north.

APPROACHES. The cove is easy to enter. Stay on the west side to avoid the rocks shown on the chart. They are visible at all tides.

ANCHORAGES, MOORINGS. Anchor past the southernmost rock, in 12 to 14 feet at low, in a good mud bottom.

MEADOW COVE

43° 52.43'N 069° 35.43'W

CHARTS: **13293, 13288**   CHART KIT: **62, 18**

Meadow Cove is just above East Boothbay, on the west side of the Damariscotta River. It is protected from the southwest, but it is too deep for comfortable anchoring until you get close in to shore.

For a temporary stop, anchor in the little bight just west of Montgomery Point, in 13 feet of water at low.

SEAL COVE

43° 53.17'N 069° 34.07'W

CHARTS: **13293, 13288**   CHART KIT: **62, 18**

Seal Cove is one of the most protected and peaceful harbors in the midcoast. Still water is ringed by deep woods, with only a few houses set well back in the trees. A family of harbor seals is usually in residence, lounging lazily on their favorite haul-out rocks.

The cove is tucked into the eastern bank of the Damariscotta River. Entering takes some careful eyeball navigation, but once in, you will be perfectly protected and set for a tranquil night.

APPROACHES. Seal Cove is bisected into an outer harbor and an inner harbor by a ledge and a rock in the middle of its long length. It is difficult to enter the inner harbor at high tide when the ledges are covered. If you plan to go all the way in, make your entry on a rising tide, from an hour after low to midtide.

The timing of your approach is further complicated by the current at The Narrows off Fort Island. Here the current can run 4 or 5 knots on a strong ebb, with some eddies and boils. The passage, however, is 100 yards wide and well-marked, so you should have no real trouble. Nun "12," marking Eastern Ledge, sometimes tows under.

After coming through The Narrows, identify the long ledge to your right and Hodgsons Island beyond. Continue upriver until you can clear the north tip of Hodgsons Island, leaving it to starboard. Then turn south down the slot between Hodgsons and the unnamed island farther east and continue more or less down the middle.

Focus now on the long ledge extending south from the head of land on the east bank and the 5-foot spot just west of the ledge. Identify the head of land by the small, round island covered with trees which is separated from it. The long ledge begins about 100 feet off the little island and runs south. Only the northern bit of the ledge shows above the water at high tide—a piece of whitish rock covered with grass. At low, the whole ledge is visible.

Abeam of the ledge, you will pass over the 5-foot spot. Continue south, paralleling the ledge, but not too closely. On the opposite side, about 50 feet west of the long ledge, lurks another little ledge, identifiable by rockweed until midtide. A lump on its south end

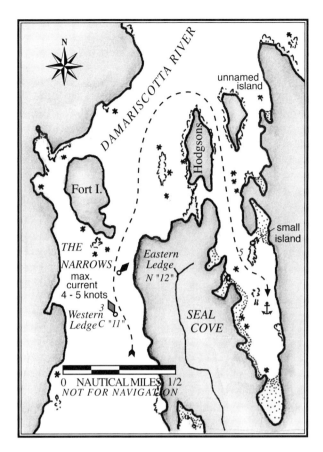

stays exposed a while longer, but at the top of the tide it also disappears. Keep it to starboard.

Now look for a break in the ledge to port. After the break, the geological continuation of the same ledge continues farther south, but you want to pass between the end of the long ledge and the chunk of ledge farther south. Turn to port to pass between them and enter the anchorage.

It would be wise to sight a compass course or a range of some kind—or turn on your chart plotter—so you can clear the southern tip of the ledge if you need to leave when it is covered.

ANCHORAGES, MOORINGS. If you prefer not to enter the inner harbor or if the timing is wrong, anchor almost anywhere along the slot between Hodgsons Island and the unnamed island to the east in about 20 feet at low. This is a well-protected and attractive anchorage, with more breeze and fewer mosquitoes than the inner one.

If you have made it into the inner harbor, anchor east and south of the long ledge in 12 to 14 feet of water at low. The mud bottom holds well. You can work your way south quite a distance along the east bank, but there will be less swinging room. Watch out for the chunk of ledge south of the long one.

PLEASANT COVE

4 ★★★

43° 55.39′N 069° 35.28′W

CHARTS: **13293, 13288**   CHART KIT: **62, 18**

This tranquil harbor lies off the west bank of the Damariscotta River upriver of The Narrows and Carlisle Point. It is well named. Here, surrounded by saltwater farms, you can spend a night out of the river current and away from the rest of the world.

APPROACHES. The entrance is wide and easy. As it narrows, the deeper water is near the south bank.

ANCHORAGES, MOORINGS. You can anchor in 10 to 20 feet at low water in the mouth of the cove, west of Carlisle Point and north of Pleasant Cove Island. For better protection, go half a mile farther in, anchoring east of the point that sticks down from the north, in mud 13 to 14-feet deep at low. Opposite this point are several moored boats and a dock.

THINGS TO DO. Explore the rocky points and piney nooks at the head of the cove, part of the 22-acre Marshall E. Saunders Memorial Park. Due to overuse, the owners request that you do not land on tiny Pleasant Cove Island or the southern shore.

WADSWORTH COVE

3 ★★★

43° 55.79′N 069° 35.34′W

CHARTS: **13293, 13288**   CHART KIT: **62, 18**

On the west bank of the Damariscotta River just north of Pleasant Cove, an encircling arm of ledges defines Wadsworth Cove, which is surprisingly protected and big enough to harbor a number of boats. A handsome hayfield sweeps down to the water from the north, and a private dock and float are at the rocky point in the southwest corner.

APPROACHES. The ledges that make up the eastern and southern parts of this cove are visible from low to halftide. From the south, leave can "13" to port and make a wide left turn into the harbor. The northernmost ledge is considerably west of the can, but don't be tempted to cut the corner.

ANCHORAGES, MOORINGS. Anchor east of the rocky point with the dock and float in 14 to 20 feet of water with a mud bottom. The surrounding ledges and Carlisle Point to the south will protect you.

Riverside Boat hauls with a model T-powered railway before skidding the cradled classics into their yard.

IRA C. DARLING CENTER (Wentworth Point)
43° 56.10'N 069° 34.92'W

CHARTS: **13293, 13288**   CHART KIT: **62, 18**

As part of the University of Southern Maine, the Ira C. Darling Center (*563-3146; www.dmc.maine.edu*) conducts fisheries research at Wentworth Point, on the east bank of the Damariscotta. Visitors are welcome. Projects range from the restoration of alewives in the Medomak River, smelt in Portland, and Atlantic salmon in the Kennebec, to developing a strain of blue lobsters to help track lobster migrations.

The laboratory is partly funded by the federal Sea Grant Program, and it actively supports Maine's growing aquaculture industry. For example, studies are made of water temperature, flow, and salinity to optimize the growth of shellfish, and genetics experiments are conducted to improve the breed of oysters, clams, mussels, and scallops. You may never have tasted a triploid oyster, but they grow 40 percent faster than their backward diploid cousins.

Stop here for a visit, particularly if you appreciate lobsters and other shellfish. You will receive an interesting briefing on the work in progress.

⚓ There is a guest mooring off the concrete dock and float at Wentworth Point, about three miles north of The Narrows. Call the center at *563-3146* to reserve the mooring.

SALT MARSH COVE
43° 57.06'N 069° 35.09'W

CHARTS: **13293, 13288**   CHART KIT: **62**

On the west bank of the Damariscotta River, two miles above Pleasant Cove, Salt Marsh Cove mostly dries out at low, and you are likely to see clammers here working the mudflats. It is possible to sneak in just far enough past the headlands to get protection and anchor in 13 to 20 feet at low in the mud bottom. Nearby Wadsworth or Pleasant Cove would be a better choice.

DODGE POINT
43° 59.02'N 069° 33.91'W

CHARTS: **13293, 13288**   CHART KIT: **62**

Dodge Point is a bold headland on the west bank of the Damariscotta River, three miles south of Damariscotta and Newcastle, opposite the town dock and floats of South Bristol.

During the development boom of the 1980s, Dodge Point was slated to become house lots, but it was rescued with a staggering $2.35 million in funds from the Lands for Maine's Future program. Dodge Point is now managed by the Bureau of Parks and Lands with assistance from the Damariscotta River Association.

There is no anchorage here at all, but it is possible to anchor temporarily in settled weather near the dock off the point in 13 feet of water at low in a mud bottom. Note that Dodge Upper Cove, north of the point, dries out at low. Row in to the dock.

From the dock, hiking trails follow an old farm road through a stand of towering red pines planted in the 1930s. A freshwater pond attracts waterfowl, and the remnants of a 19th-century brickyard are located nearby.

Near the river, you can discover an Indian midden, or shell heap—evidence of the river's abundance and the long history of this site. At low tide, you can watch clammers digging on the flats or just succumb to sunbathing on the small beaches.

DAMARISCOTTA and NEWCASTLE

3 ★★★

44° 01.91'N   069° 32.14'W

CHARTS: 13293, 13288   CHART KIT: 62

Damariscotta and Newcastle are twin towns separated by a short bridge at the head of navigation of the Damariscotta River, some 13 miles from the open ocean. There are beautiful old homes here, and the pace is slow.

APPROACHES. The channel in the upper two miles of the Damariscotta is narrow and winding, but reasonably well buoyed. After nun "20" off Hog Island, follow the chart carefully and do not run straight courses between the buoys. Favor the western side of the river until abeam of Hall Point, then keep to the eastern shore to avoid the shoal marked by can "21." After nun "24," the deeper water is again to the west, on the Newcastle side.

ANCHORAGES, MOORINGS. There are several choices here. Schooner Landing is right at the head of the harbor, next to the public landing, and they rent slips with depths of 9 feet. Riverside Boat Co., farther downriver on the west side, has moorings for rent. Or you can anchor outside the moored boats and dinghy in to the public landing. The harbormaster says, "Anchor anywhere and nobody'll bother you. We're pretty easygoing here."

GETTING ASHORE. If you are up near the bridge, come ashore in your dinghy at the town landing floats, and you will be right in town. If you are on a Riverside Boat mooring, either dinghy to town or land at the boatyard and walk.

FOR THE BOAT. *Riverside Boat Company (207-563-3398).* Riverside Boat is about .75 miles downriver from the bridge on the west side of the river. You will see a small dock and float (which dries out at low), a marine railway, and a flagpole. Riverside Boat Company has to rate as one of the most authentic boatyards in all of Maine. Paul Bryant has taken over where his father left off, hauling classic boats with an old Model T, scrubbing them with a brush (what's a pressure washer?) and skidding their cradles across the yard on greased rails. Forms of past boatbuilding projects hang on the sides of the sheds, and sawdust and carefully stacked rough-sawn oak give away the saw that mills the lumber to make them. The railway can handle boats to about 40 feet for repairs.

Schooner Landing (207-563-7447). Schooner Landing is the only marina in town, hard by the bridge on the Damariscotta side. Transients are welcome on their docks, where there water and electricity and 7 to 9 feet of depth at low. Their restaurant, however, is also right on the docks, so don't expect too much privacy on a weekend night. But if you can't beat 'em, join 'em.

Public Landing. The public landing in Damariscotta can be identified by the flagpole and launching ramp east of the bridge and just east of the Schooner Landing docks. A line of floats juts out into the river next to the ramp, with limited dockage.

Newcastle Marine (207-563-5550). Newcastle Marine is located about half a mile from the river. They specialize in mechanical work on smaller boats, particularly outboards, but they are also well-versed in auxiliaries and electrical work.

FOR THE CREW. Damariscotta is a remarkable small town. The historic buildings along the main street house thriving business, and any other conveniences you might need are neatly tucked in behind them. You'll find groceries, a hardware store, a laundromat, a barber, gift shops, an excellent bookstore, and even a Radio Shack. The classic Rexall Pharmacy is also a Trailways bus stop.

A fish market is by the public landing. A large supermarket and a health food store are a half mile further east.

Restaurants run the gamut from heart-stopping specialties at The Breakfast Place to tavern fare at Schooner Landing to several fine-dining choices including the Mediterranean Kitchen at the east end of town.

Miles Health Care Center is nearby (563-1234).

THINGS TO DO. The little Lincoln Theatre doubles as a movie theater and a community playhouse with performances throughout the summer. The Skidompha Public Library is at the east end of town.

For an archeological expedition, take your dinghy upriver a mile to the Indian oyster-shell heaps, called middens, on the west bank, just below the next fixed bridge. Though the heaps are of modest proportions compared to, say, the pyramids, it is fun to imagine the hundreds of years of picnics and oyster feasts required to create 12-foot high mounds of shells, now covered with trees. Some striking columbines thrive in the alkaline soil. There is a lot of current, so plan your trip to go up with the last of the flood and return with the ebb. Or bring your motor.

JOHNS BAY

Johns Bay, just west of the Pemaquid Peninsula and east of the Damariscotta River, is short but pithy. Named for Captain John Smith who explored the area in 1614, it is a very pretty bay and packs a lot of beauty and interest into a five-mile stretch.

The fort and the state park at Pemaquid Harbor can give you a glimpse of our tumultuous early history or spectacular views of the Bay. The eastern side of The Gut at South Bristol is a picturesque working harbor crowded with fishing boats and active wharves, sparing little room for yachts.

Of the three extremely well-protected anchorages in the bay only one, west of Witch Island, is easy to enter. Poorhouse Cove provides excellent shelter but should be entered at halftide or below. Eastern Branch, a hurricane hole, is very difficult to enter.

Perhaps because there is no major town at the head of Johns Bay, it is considered an unlikely cruising area. You will see few boats and discover unspoiled harbors. And that is what makes it so special.

THREAD OF LIFE
43° 50.21'N   069° 33.00'W
CHARTS: 13293, 13288   CHART KIT: 62, 18

This little passage at the western entrance to Johns Bay is as beautiful as its name. Running through it before a sunny southwesterly, the heart quickens, the spirits soar, and all seems right with the world. Coming the other way, when leaving Johns Bay heading west, there is just enough width to tack against the prevailing wind, and, depending on how big your genoa is, your heart might be pounding for different reasons. Other than an abundance of lobster buoys, there are no dangers in the channel.

From Christmas Cove or the Damariscotta River, run east of Inner Heron Island. Aim for the wooded portion of Thrumcap Island and identify can "1" off Turnip Island. Make a gradual left turn around the can and Turnip, which has a little brown house and a number of trees. Then run up the middle of Thread of Life, which has bold ledges on either side. From Turnip Island, you will be able to see nun "2" and can "3" that mark the exit to Johns Bay north of Crow Island.

Draggers at The Gut, South Bristol.

WITCH ISLAND

43° 52.28'N   069° 33.19'W
CHARTS: 13293, 13288   CHART KIT: 62, 18

Witch Island lies on the west side of Johns Bay, just north of the eastern entrance to The Gut. It is densely wooded and a Maine Audubon sanctuary.

To the west of the island, there is an excellent anchorage with protection from every quarter except the northeast. You'll be sleepily aware of lobstermen leaving South Bristol in the early morning.

APPROACHES. From Johns Bay pass between can "1" on Corvette Ledge and nun "2" off McFarlands Ledges, taking a wide curve around the north end of Witch Island.

ANCHORAGES, MOORINGS. Moorings flank both sides of the channel west of Witch Island, but they leave plenty of room to anchor in 21 feet of water at low, with excellent mud holding ground. We've had reports, however, of a bottom fouled with gear or a mysterious cable between the 12 and 21-foot soundings shown on the chart. Use a tripline on your anchor to be safe.

THINGS TO DO. Respectful day visitors are allowed to explore Witch Island.

McFARLANDS COVE
43° 52.65'N   069° 33.24'W
CHARTS: 13293, 13288   CHART KIT: 62, 18

McFarlands Cove is a small indentation north of Witch Island, on the west side of Johns Bay. It is full of moored small craft, with a rocky beach at the north end. There is heavy kelp on the bottom, so use it only as a daystop. Anchor outside the moored boats, making sure the hook is well set.

Pemaquid Point Light.

THE GUT (South Bristol)—
East of the Bridge
43° 51.77'N 069° 33.33'W
CHARTS: 13292 (INSET), 13288   CHART KIT: 67 (F), 62, 18

The eastern part of The Gut is a picturesque working harbor, full of lobsterboats going about their business, dories loaded with nets, seagulls quarreling over fish, and a busy little swing bridge connecting the town of South Bristol to the mainland. This side of The Gut is marvelous to see and pass through, but it is a difficult place to stop.

APPROACHES. Coming from Johns Bay to the east, pass between can "1" and nun "2," then curve wide around densely wooded Witch Island, leaving it to port. At the narrow entrance to The Gut, favor the right side of the channel to avoid the isolated ledge on the left. Inside, pick your way carefully among the moored and moving boats.

If you plan to pass through the Gut via the swing bridge, please read our cautions on page 143.

ANCHORAGES, MOORINGS. Don't even think of anchoring here. It is too busy, and you'll just be in the way. Moorings and dockage are available at the Gamage Shipyard, just west of the bridge (see *The Gut—West of the Bridge*, page 143).

FOR THE BOAT. The eastern part of The Gut is a working harbor, not geared at all toward visiting yachtsmen. Gas, diesel, moorings and dockage are available on the west side of The Gut.

Bittersweet Landing Boatyard (207-644-8731). This small full-service yard is located on the sheltered south shore of The Gut. They have a hydraulic trailer and a private launching ramp. They haul and store boats to 40 feet, and they will tackle hull and engine repairs.

FOR THE CREW. See *The Gut (South Bristol)—West of the Bridge*, page 143.

PEMAQUID HARBOR
3 ★★★★
43° 52.62'N 069° 31.56'W
CHARTS: 13293, 13288   CHART KIT: 62, 18

Four flags fly from the round stone tower overlooking Pemaquid Harbor—American, British, French, and the Maine state flag. All have been deeply involved in this historic place on the west side of Pemaquid Neck, three miles north of the entrance to Johns Bay.

The fort at Pemaquid was the eastern bastion of English territory, counterbalancing the French fort at Castine, farther up the coast. All the territory in between was in dispute, and there was open warfare between the British and French between 1613 and 1759.

Much of our early history can be read in the succession of forts at Pemaquid. The first Fort Pemaquid was erected in 1613, and a year later it was attacked and plundered by pirate Dixie Bull. He got only 500 pounds from the impoverished settlement, but it earned Pemaquid the dubious distinction of being the site of the first act of piracy in New England.

The fort's successor, Fort Charles, was built in 1677 but destroyed by the Penobscot Indians in 1689. Fort William Henry was erected in 1692. Although it mounted 28 guns, an attack in 1696 from both land and sea by a force of French and Indians led by Baron de Castin destroyed the fort once more. Governor David Dunbar restored it and renamed it Fort Frederick in 1729. This last fort was razed by the town of Bristol in 1775 to keep it from falling into the hands of the British. Finally, in 1908, the replica William Henry Memorial Fort was built here to defend history itself.

APPROACHES. There are two approaches to Pemaquid Harbor, both unmarked but easy. Both the round stone tower shown as "FORT" on the chart and Pemaquid Beach to the south are visible from a long distance. Coming from the south, pass halfway between high, wooded Johns Island and Knowles Rocks. The rocks are low and blend with the background, but they are visible at all tides.

From the west, pass halfway between the grassy mound of Beaver Island and the unmarked ledge to the north off Thurston Point. Some of this ledge is usually visible even at high water, but give it a wide berth, because the part that extends farthest is visible only near halftide or below.

In 1635, when the 240-ton *Angel Gabriel* was lost on the ledges of Pemaquid Point, it became the first recorded shipwreck on the Atlantic coast.

ANCHORAGES, MOORINGS. There are two anchorages with reasonable protection, either north or south of the fort. If you choose to anchor in Pemaquid Harbor north of the fort, pass midway between the point where the fort sits and the tiny island (usually a peninsula) on the left side of the channel.

Having entered the harbor north of the fort, anchor on either side of the channel. You'll find the most swinging room on the north side, outside the moored boats.

This little harbor is actually just inside the entrance to the Pemaquid River, and there is a fair amount of current. Your boat may swing to the current or wind, depending on which is stronger. The anchorage is exposed to the southwest.

It is possible to go farther up the Pemaquid River, near the Pemaquid Fishermen's Co-op on the north shore with all the workboats, but it is unmarked and tricky. Be careful of the ledge extending from the northern shore near the 8-foot spot. Check with the harbormaster, usually at the Co-op, for a possible mooring.

The second anchorage is where a number of boats are moored, outside the Pemaquid River and its currents, between the fort and Fish Point. Anchor near the private moorings in 17 feet of water at low with a sand bottom. This anchorage is protected from the south but exposed to the southwest.

GETTING ASHORE. The dock next to the fort is private. Take the dinghy farther east to the float at the end of the long Pemaquid Pier, which extends from the Contented Sole restaurant.

FOR THE BOAT. *Pemaquid Beach Boat Works (207-677-3726; www.pbbw.com).* PBBW's primary business is building beautiful new boats, but we include them in case you get into trouble and need repair help. Their yard is tucked in the southeastern pocket of the inner harbor, not visible from either anchorage but only steps away by land.

Pemaquid Marine (207-677-2024; lincoln.midcoast.com/~pemmar) is located inland, off Beach Road, halfway between Pemaquid Harbor and New Harbor. They can help you with repairs, hauling, or storage.

FOR THE CREW. The Contented Sole *(677-3000)* is located at the end of the Pemaquid Pier in an old clam factory. They serve lunch and dinner, or you can have takeout food on the dock. Later in the evening, their bar comes to life with locals and tourists alike. They have ice and pay phones.

A mile walk south on the road marked "Pemaquid Beach Rd. and New Harbor" brings you to the center of New Harbor and the excellent C.E. Reilly and Son market and liquor store *(677-2321)*. If you load yourself down at the market, they'll give you a ride back to the harbor. A post office, bank, gift shop, art gallery, and dairy bar are nearby.

THINGS TO DO. The Bureau of Parks and Recreation runs the wonderful little Colonial Pemaquid Museum right next to the restaurant. Excavations in the grassy park have unearthed the fieldstone foundations of early buildings, including the Customs House, where clearance was required of all ships between the Kennebec and St. Croix River. The Old Burial Ground across the street dates from 1695.

The round stone tower of William Henry Memorial Fort was built in 1908 as a replica of the last fort to have stood at this strategic spot. It is open to the public. Inside there are interesting exhibits and stairs that curve to the roof. The view from the battlements is outstanding.

A short walk down Snowball Hill Road and a small charge will admit you into beautiful, sandy Pemaquid Beach.

POORHOUSE COVE

43° 53.62'N   069° 33.36'W

CHARTS: 13293, 13288   **CHART KIT: 62, 18**

Poorhouse is a well-protected cove on the west side of Johns River at the head of Johns Bay. Although much easier to enter than neighboring Eastern Branch, it is best to come in at halftide or below so you can see the ledges.

Wooded High Island forms the eastern side of the cove. There are a number of houses and docks lining the west side, but only a few boats.

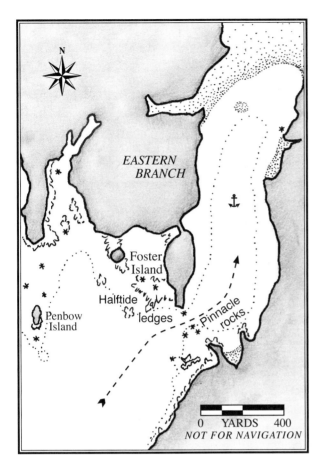

EASTERN BRANCH

◼ ★★★

43° 54.48'N 069° 31.95'W

CHARTS: **13293, 13288**   CHART KIT: **62, 18**

Eastern Branch is a beautiful and extremely well-protected arm of the Johns River at the head of Johns Bay. A few homes sit at the edge of the woods and tranquil meadows.

It is common to find 30 or more seals slouching and snorting on the ledges west of Foster Island as you enter. You are also likely to see terns, ospreys, gulls, herons, egrets, cormorants, and other sea birds.

The entrance is dangerous, and many yachts with experienced skippers have left bottom paint on the rocks. As you can see on the chart, the mouth is cluttered with unmarked rocks. For the uninitiated, safe passage through the 35-foot opening requires a combination of low tide, a keen eye, and good luck.

APPROACHES. Come in within an hour or so of dead low and proceed as slowly as you can. As you approach Foster Island, running roughly northeast, identify the halftide ledges on the northern side of the entrance but do not get too close—they make out quite far under water. On the right side of the entrance, you will see the tips of several pinnacle rocks. Others are submerged, even at low.

Make a sharp turn to starboard, passing about halfway between the pinnacle rocks on your right and the ledges on your left. Local resident Stuart Gillespie says, "Stay among the lobster buoys; they are most likely in deep water." Once past the ledges on your left, make a slow turn (about 90 degrees) to port and head up the branch, favoring the western shore.

ANCHORAGES, MOORINGS. There are a number of private moorings in the harbor, but you can find plenty of room to anchor outside of them in 13 to 15 feet of water at low. Holding ground is good, in mud, and there is little current.

Protection is excellent in all directions. This would be an excellent place to ride out a hurricane.

THINGS TO DO. The water here is about as warm as it gets in Maine, so it is a good place to plan your annual dip.

Land your dinghy way up the eastern shore, just where the green appears on the chart, and walk through the old cemetery. The Harrington Meeting House (1773) and Museum is across the road.

APPROACHES. Enter within an hour or two of low tide, so you can locate the ledge outside the northern end of High Island. Inside Sproul Point a larger ledge projects down from the north. Leave it to starboard, aiming for the north corner of the entrance to Bradstreet Cove. As you approach the boats moored near Bradstreet, turn to port and run down the west side of Poorhouse Cove. The east side is shoal.

ANCHORAGES, MOORINGS. A number of private moorings are set in the cove, but there is still room to anchor nearby in 8 to 11 feet at low, in good mud holding ground. You can go as far south as the fourth dock and still find good water.

The protection is good all around in Poorhouse Cove. You should be safe and comfortable here under almost any conditions.

There is an agreeable daystop, with 16 feet of water, just outside the entrance in the little bight south of Sproul Point.

MUSCONGUS BAY

Muscongus Bay has a bad reputation for rocks. As one sailor put it, "You have to navigate all the time, and I'd rather be enjoying myself." There *are* a lot of rocks and unmarked ledges, but also lovely islands and passages. In 1524 Verrazano glowingly described "...islands lying all near the land, being small and pleasant to view, high, and having many turnings and windings between them, making many fair harbors and channels, as they do in the gulf of Venice in Illyria and Dalmatia...." Verrazano had it right. Muscongus Bay is a beautiful place to sail and comparatively uncrowded. It has a number of interesting harbors, plenty of gunkholes, and yes, lots of navigation.

The scale of Muscongus Bay is fairly small, and in a week or so you can comfortably see most of it. Two narrow, winding rivers at the head of the bay, the Medomak and the St. George, lead respectively to Waldoboro and Thomaston. Both have strong tidal currents, so cruising is better in the bay, but Maple Juice and Otis Cove, up the St. George, make it worth the effort.

There are the large working harbors of Port Clyde and Friendship and the tiny ones of New Harbor and Pleasant Point Gut. The unspoiled cove at Harbor Island is lovely, as are Greenland and Davis Cove. The National Audubon Society maintains a fascinating preserve on Hog Island, in the northwest corner of the bay.

Among the many interesting passages in the bay are Georges Harbor, Port Clyde Harbor, the slot between Allen Island and Burnt, Davis Strait, the passages between Louds and Marsh Island, and north of McGee.

Best of all is magnificent Monhegan 10 miles offshore, guarding Muscongus Bay—a good landfall and possibly the most spectacular island on the coast of Maine.

PASSAGES through MUSCONGUS BAY

CHARTS: 13301, 13288   CHART KIT: 63, 64, 18, 19

PASSAGES NORTH AND SOUTH. The islands and ledges of Muscongus Bay are oriented in long, north-south lines, and the passages between them are fairly simple and well marked. Two major routes northward are from Burnt and Allen Island northward to the St. George River and from Pemaquid Point up Muscongus Sound.

PASSAGES EAST AND WEST. *The Outside Route.* Passages east and west through Muscongus Bay run against the north-south grain of the islands and are much trickier. For boats headed east or west with no particular interest in Muscongus Bay itself, the normal route would be outside everything—outside Burnt and Allen Island, outside Old Woman Ledge and seaward of red whistle "2 OM."

In poor visibility, unless you are intimately acquainted with the area, avoid the dangers in Muscongus Bay and take this outside route.

NORTHERN AND MIDDLE ROUTES. From the west, there are three main passages through the chain of islands running south from Friendship and many variations. The most direct is northeastward through Muscongus Bay, south of Western Egg Rock and Harbor Island, and north of Franklin Island with its lighthouse. Or, if you've explored farther north, you can cut eastward north of Harbor Island and south of Black Island and Black Island Ledge with a jog around a 5-foot spot. Or you can wend your way eastward through the lobster buoys, past the busy port of Friendship, and around Gay Island.

To pass through the next chain of islands, from Caldwell Island south to Allen Island, there are even more options. Coming past Friendship, you can hook around the top of the whole group by passing north of Caldwell Island. Or, for a more direct route to Port Clyde and beyond, you can wend your way between McGee and Seavey Island and south of Hupper Island. Or, if you are farther south, pass through Davis Strait between Davis and Thompson Island.

These routes are not necessarily the most direct routes, but they take advantage of easily recognizable islands, deep channels, and navigation marks. Discovering the many variations in and among the islands is one of the unique joys of cruising in Muscongus Bay.

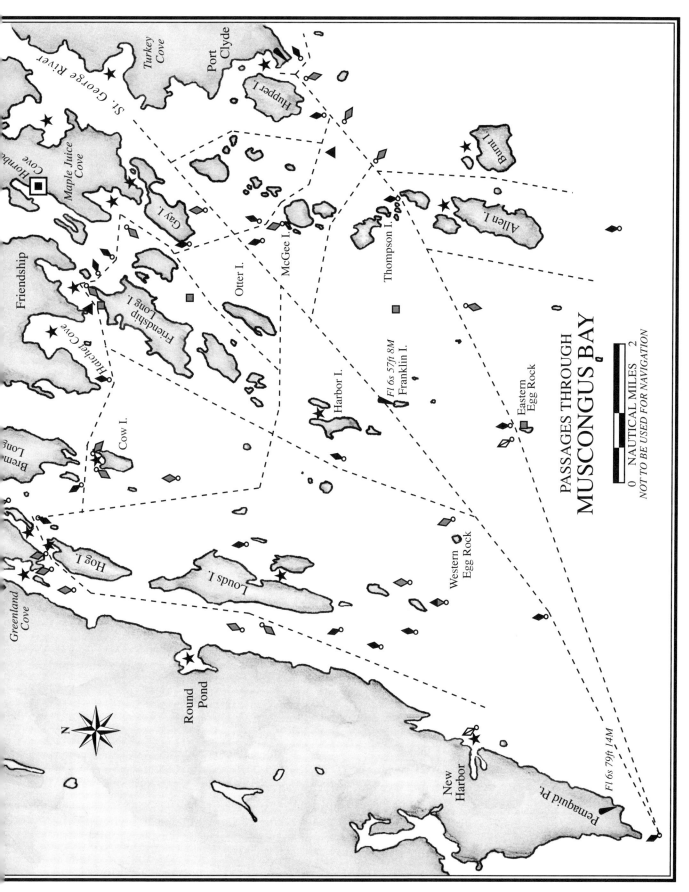

NEW HARBOR

4 ★★★

43° 52.48′N 069° 29.36′W

CHARTS: 13301 (INSET), 13288   CHART KIT: 63, 18

New Harbor is a tightly packed working harbor tucked on the eastern flank of Pemaquid Point. It is small and busy, filled with lobsterboats and draggers with hardly any room for visiting boats. Still, if you have the patience to negotiate a mooring and the humility to stay out of the way, New Harbor is an interesting and well-protected adventure.

New Harbor has the distinction of being the site where Indian Chief Samoset gave a deed to John Brown of New Harbor on July 15, 1625 and acknowledged it before Abraham Shurte. It was the first deed properly executed in New England.

APPROACHES. Approaching the harbor from the east, run from nun "16" to red-and-white bell "NH" (43° 52.43′N 069° 28.72′W) marking the entrance. The harbor is too small for much maneuvering, so drop your sails near the bell.

Do not run a straight line between the entrance bell and nun "4." Stay north of this line to avoid the ledge at the southern point of the entrance. Leave nun "4" to starboard and green daybeacon "5" to port.

There is a narrow path clear of lobster buoys in the winding channel which runs along the northern side of the harbor.

ANCHORAGES, MOORINGS. This is a tight harbor. There is no room to anchor and no harbormaster to help you. Shaw's Wharf or the New Harbor Co-op (second large dock on the right) might be able to steer you toward a vacant fisherman's mooring, but these are busy commercial operations with little time for "visitahs."

The Gosnold Arms (*677-3727*) has several 500-pound concrete moorings off their dock at the harbor entrance. Preference is given to hotel guests, but they rent their moorings and dockage, with 10 feet of depth, to transients when available.

If there is no room in New Harbor, your best bet is to go north to Round Pond or Greenland Cove.

GETTING ASHORE. Land the dinghy at Shaw's.

FOR THE BOAT. *Shaw's Wharf (207-677-2200)* is commercial, but they will pump gas, diesel, and water when things aren't too busy. Depth is 10 feet at low.

New Harbor Co-op (207-677-2791). Gas, diesel, and ice are also sold at the Co-op, with 8 feet alongside at low, but this is also a working wharf.

Pemaquid Marine (207-677-2024, lincoln.midcoast.com/~pemmar) is located inland, off Beach Road, halfway between New Harbor and Pemaquid Harbor. They can handle repairs, hauling, or storage.

FOR THE CREW. Shaw's serves lobsters that you can enjoy outside overlooking the harbor. The Co-op also serves lobsters and clams on their outside deck, or they will steam them for you to enjoy back on your boat.

Across the street, on the north side of the harbor, the Gosnold Arms serves breakfast, Sunday brunch, and suppers. The inn is named after Bartholomew Gosnold, who stopped here in 1602.

For supplies, downtown New Harbor is a pleasant 15-minute walk away. Make a left as you leave Shaw's parking lot and continue on the paved road to Route 130. Turn left up the hill and continue to the center, which includes a bank, service station, art gallery, dairy bar, gift shop, and C.E. Reilly and Son market and liquor store (*677-2321*), which has everything from *The New York Times* to fresh crabmeat, even on Sundays. To find the post office, take a right on Route 130 instead of taking the left.

THINGS TO DO. Row the dinghy to Back Cove, which branches off the southern side of the harbor. It is full of fishing boats and lobstermen's shacks, with a small reversing falls at the head. It is reputedly the most-photographed cove in Maine.

The *Hardy* (*882-7909*) runs daily excursions to Eastern Egg Rock and Monhegan from Shaw's Wharf.

Purse-seining for herring in Greenland Cove. This traditional fishery is threatened by the offshore herring fleet. After the fishermen in the dory close the purse and haul the net, the herring will be pumped into the sardine carrier. (Boutillier)

LOUDS and MARSH ISLAND

43° 55.29'N   069° 25.77'W

CHARTS: **13301, 13288**   CHART KIT: **63, 18**

There is a delightful passage between Marsh and Louds Island, near the western side of Muscongus Bay. Loudville, on the eastern shore of Louds, is a small fishing community.

Approaching from the west, leave nun "2" on Webber Sunken Ledge to port and Haddock and Ross Island to starboard. Haddock Island is wooded. Ross Island, beyond, is grassy and treeless and a nesting site for cormorants. Keep Ross to starboard and run north between Louds and Marsh Island.

From the east, the cleanest course is north of state-owned Thief Island, which is high and wooded, and north and then west of Killick Stone Island, which is low and rocky with grass. Be aware of the unmarked 7-foot rock north ot Killick Stone. It is also possible to pass between Killick Stone and Marsh, but note the 4-foot spot east of Marsh. Find the ledge that is partially visible at most tides making out to the east of Loudville. Leave the ledge to starboard and run down the middle of the passage between Louds and Marsh Island.

The chart labels this whole passage as Marsh Harbor, but it won't provide much protection. In fair weather the small bight on the western shore of Marsh Island, near its northern tip, makes a fine daystop. The chart shows it as 9-feet deep with barely enough room for one boat. A tiny sand beach is tucked among the rocks, facing south.

The cove at Loudville dries out at low and is studded with rocks. Stay clear. Pass south of Bar Island to enter Muscongus Sound.

ROUND POND

43° 56.60'N   069° 27.35'W

CHARTS: **13301, 13288**   CHART KIT: **63, 18**

In *Sailing Alone Around the World,* Joshua Slocum recorded his visit to this snug harbor on May 8, 1895. "The wind being free, I ran into Round Pond harbour, which is a little port east from Pemaquid. Here I rested a day while the wind rattled among the pine-trees on shore."

Today, Slocum might have trouble finding a place to anchor in Round Pond. The harbor is crammed with more than 100 boats, and there is not much room for transients. Round Pond is partly a working harbor for lobstermen, partly a port for pleasure craft.

APPROACHES. A bar makes out from the southern point of the entrance, so be sure to leave can "1" to port. Thereafter you will be meandering among moored boats and lobster cars. Avoid the shoals at the southwest and northeast ends of the harbor.

ANCHORAGES, MOORINGS. This perfect harbor is packed. If you arrive early, you might find room to anchor in 15 to 20 feet of water at the southeastern corner outside the mooring field. Padebco maintains a number of moorings ranging from 500 to 1,800-pound granite blocks with floats marked "Padebco" and the weight. On the average night, two or three moorings are available for transients.

Protection is excellent. Round Pond is landlocked except for due east, and even from there, Louds Island provides cover.

GETTING ASHORE. The most convenient spot to take your dinghy is the town landing. The floats at Padebco ground out at low.

FOR THE BOAT. *Muscongus Bay Lobster Company (207-529-5528).* Just below the Anchor Inn on the north shore are the busy lobster docks and floats of the Muscongus Bay Lobster Company. The outer floats have 10 feet of depth at low. Gas and diesel are available, but no water.

Padebco (207-529-5106; www.padebco.com). The family enterprise of Bruce and Paul Cunningham occupies the large gray building and shingled house farthest east along the waterfront. The yard builds the handsome line of Padebco power boats and also provides hauling, storage, repairs,

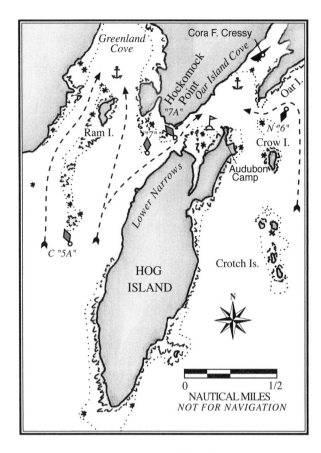

and overland transport with their hydraulic trailer. Their water is available but highly mineralized and not recommended.

Town Landing. The town landing's launching ramp and float are obvious, just west of Muscongus Bay Lobster Company.

Round Pond Lobster Co-op. The Co-op dock is to the left of the town landing. Fuel is sold only to members.

FOR THE CREW. There is no shortage of outdoor dining at the town wharf. Both Muscongus Bay Lobster and Round Pond Lobster Co-op have lobsters and steamers in-the-rough or sell them live. The Co-op has block ice. For more civilized fare, you can eat at the Anchor Inn (*529-5584*), right by the waterfront.

Just up the road is the Granite Hall store, which sells gifts, antiques, and ice cream. Curving left up the hill past the Little Brown Church, the road will take you to King Ro Market, which has limited groceries and beer, wine, newspapers, and propane exchange.

THINGS TO DO. A very pleasant walk starts at the waterfront. Turn right at the Granite Hall store, then head down to the old boathouse on the northern point of the harbor entrance. This is a peaceful road with little summer homes tucked among the trees.

Concerts are often performed on Wednesday evenings at the Little Brown Church. Head north on Route 32 to find tennis and basketball courts. You did stow your basketball, didn't you?

GREENLAND COVE

3 ★★★

43° 59.01'N 069° 25.73'W

CHARTS: **13301, 13288**   CHART KIT: **63, 18**

Large and attractive Greenland Cove lies in the northwest corner of Muscongus Bay. It is a favorite of cruising yachtsmen, well protected from all directions except the southwest and endowed with plenty of swinging room. There are only a few docks and summer homes on the shores, and the swimming is good.

For generations, Greenland Cove was a perfect place for purse-seining herring. Though the inshore herring fishery has all but disappeared, you may still see several dories moored on the west side of the cove.

APPROACHES. From Muscongus Sound, Greenland Cove can be entered on either side of Ram Island. Ram is wooded, with a house and a flagpole on its east side. Ledges make out a long way to the south of the island, so choose your entrance early.

The easier entrance is to the east, leaving can "11" at Halftide Ledge to port and running up the middle of the channel between Ram Island and Hockomock Point. Favor Ram Island to stay clear of the 3-foot spot marked by can "13" and the ledge to the north of it, which is partly visible at high tide.

Coming from the east and passing north of Hog Island, leave cans "15" and "13" in the Lower Narrows to starboard before turning north into Greenland Cove.

ANCHORAGES, MOORINGS. There is good water for anchoring over a large part of Greenland Cove, but holding ground varies from good mud in some areas to grass in others. Be sure your anchor is well set. With normal southwest winds, the best place to be is in the lee of Ram Island, but be careful. There is a ledge making out some distance to the north of Ram, as shown on the chart. In a strong southwesterly, Greenland Cove can get quite choppy.

HOG ISLAND

4 ★★★★

43° 58.76′N 069° 25.11′W

CHARTS: 13301, 13288   CHART KIT: 63, 18

Hog Island is one of the larger islands in the upper reaches of Muscongus Bay. Its 330 acres encompass a variety of wilderness habitats—rocky intertidal zones, mudflats, salt marshes, spruce-fir forests, and freshwater ponds. A small slot in its northern end forms a well-protected anchorage.

Hog Island was purchased by sailors Mabel and David Todd after they first noticed it on a cruise in 1908. Their daughter, Millicent, inherited it in 1932, during the depths of the Depression. Desperate for money, the owners of the one lot on the island that Millicent did not inherit made plans to log their island timber. Millicent held them off for several years until she could partner with Audubon to aquire the last lot.

In 1936, Audubon established their first Ecology Camp on Hog Island, and its workshops still attract people from all over the country to study the interrelationships among plants, animals, and their physical environments. Over the years, many famous naturalists have taught here, including Roger Tory Peterson, the bird-book author, and, more recently, Stephen Kress, founder of the Eastern Egg Rock Puffin Project.

Audubon graciously welcomes respectful visitors to explore this beautiful island and learn about their courses.

APPROACHES. From the west, coming up Muscongus Sound west of Louds and Hog Island, leave cans "11," "13," and "15" to port. Just past can "15" in the Lower Narrows, run close to the mainland at Hockomock Point, leaving to starboard the rocks making out from Hog Island.

From the east, run up outside the Crotch Islands and higher Crow Island, which are state-owned. Then leave nun "16" to port, turning westward between Hog and Oar Island and aiming just to the right of the dock at the northern tip of Hog Island. Sail past the dock to the mooring area.

ANCHORAGES, MOORINGS. Do not anchor here. There is a cable crossing that you are likely to foul. Audubon maintains four moorings in about 7 feet at low, three of which they rent for the night on a first-come, first-served basis. The moorings may be slightly above the going rate, but they include island access and are extremely well protected.

GETTING ASHORE. Take the dinghy to Audubon's float. Leave the outside open for the launches, which make frequent trips to the mainland and outlying islands.

THINGS TO DO. Introduce yourself and pay for your mooring at the Audubon office (*529-5148*). The office has maps of the trails circling this beautiful island, and restrooms are nearby. Please do not bring pets ashore, and be respectful of courses in progress. If you are lucky, you may be invited, for a modest fee, to join Audubon in a family-style dinner served up by the first-class camp chef. "It's not camp grub," says director Seth Benz. "We eat well out here."

OAR ISLAND COVE

3 ★★★

43° 58.97′N 069° 24.91′W

CHARTS: 13301, 13288   CHART KIT: 63, 18

Between the south end of Oar Island and Keene Neck is a little working harbor that yachts seldom visit. It is the final resting place of the five-masted schooner *Cora F. Cressy*, arguably the largest surviving wooden hull of a sailing schooner in the world. If the moorings at Hog Island are occupied, you might choose to spend the night here instead.

APPROACHES. The approaches are the same as for Hog Island, above.

ANCHORAGES, MOORINGS. Anchor outside the lobsterboats, in about 14 feet of water at low. The lobster buoys in the cove might be moorings for lobsterboats, so don't anchor next to one.

The anchorage is well protected except from the south, and the bottom is mud. This would be a good place to be in a northerly.

GETTING ASHORE. Row your dinghy to the floats of Keene Narrows Lobster on the west shore, hidden behind the schooner's great hull, or to the obvious landing farther south used by the Hog Island Ecology Camp.

FOR THE BOAT. *Keene Narrows Lobster.* In an emergency, gas and diesel are available at this working wharf, with about 4 feet at the floats at low.

THINGS TO DO. Take the dinghy around the wreck of the *Cora F. Cressy* just to feel the hulk looming over you. But don't literally go under any part of it that is overhanging. The fishermen who watch huge hunks of the old lady drop into the water are constantly amazed to see cruisers blithely dinghy beneath such dangers.

The *Cressy* (2,499 tons) was built in Bath by Percy and Small (now part of the Maine Maritime Museum) and launched in 1902. Her length is 273 feet, her beam is 45.4 feet, and her draft is 27.9 feet. When her sailing days were over, she lay alongside a dock in Boston as wharfage charges accumulated. Then she was transformed into a casino, and her last crew were ladies of the night. Finally, she was brought here and sunk to be used as a lobster pound.

Audubon runs a visitors' center and gift shop at their mainland landing. From there, two short nature trails lead out Hockomock Point. For more hiking, take the dinghy over to the Audubon camp on Hog Island to explore (see *Hog Island,* above).

MEDOMAK RIVER
CHARTS: **13301, 13288**   CHART KIT: **63, 18**

The Medomak River empties into the northwest corner of Muscongus Bay. Waldoboro, at the head of the river's navigation, was once one of Maine's major 19th-century shipbuilding centers. However, this is not an easy place to visit in deep-draft cruising vessels. The channels wind, the currents run hard, and unmarked rocks and ledges stud the river.

If you want to explore the river, do so at low slack, when most of the dangers are visible and the tide will be rising in case your keel finds the bottom. There are several approaches on either side of Bremen Long Island, but the widest and easiest to navigate is Hockomock Channel, to the west.

Entering Hockomock Channel at low slack, leave can "9," north of Oar Island, to port. The small working harbor of Bremen, jammed with lobsterboats, is tucked in around the ledge north of Oar Island.

Proceeding up Hockomock Channel, look north of Clam Island for a ledge that is low and hard to see. After passing this ledge, favor the right side of the channel to avoid the necklace of rocks and ledges on the west side. The farthest outlying rock on the west side, opposite the letter "C" in the word "Channel" on the chart, is visible for an hour of so on either side of low tide.

Having negotiated Hockomock Channel, leave nun "10" to starboard and wooded, state-owned Hardy Island to port. Head for the small island on Havener Ledge, which has some trees. Turn sharply to starboard and pass through the narrows. A strong current starts to run through the narrows within half an hour of low. Pass between nun "12" and can "13" in the channel between Locust Island and Havener Ledge.

This is where the Chart Kit chart ends, and it may be just as well so you aren't tempted to go farther. Here is what the *Coast Pilot* says about the channel leading farther north:

"For the next 2.5 miles to within 1.6 miles of Waldoboro, the channel leads between flats nearly bare at low water, and shoals gradually to 5 feet. The controlling depth to Waldoboro is about 3½ feet… The channel can best be followed at low water when the flats are visible or on a rising tide."

BREMEN
3 ★★★
43° 59.49′N   069° 24.40′W
CHARTS: **13301, 13288**   CHART KIT: **63, 18**

The small working harbor of Bremen is west of Bremen Long Island in the bight north of Oar Island. The well-protected harbor is full of lobsterboats.

APPROACHES. From the south, proceed up the Hockomock Channel of the Medomak River, passing between Oar Island and Bremen Long Island and keeping can "9" close by to port. Continue on a northwest course until you are abeam of the building on the western shore. Turn westward into the bight and pass between the large ledge off the north tip of Oar Island and the shoal area with a 3-foot spot off the town of Medomak.

ANCHORAGES, MOORINGS. Broad Cove Marine has several moorings for rent or dock space with 5 or 6 feet alongside. Or you can anchor outside the moored boats in about 15 feet at low in good mud.

FOR THE BOAT. *Broad Cove Marine Services (Ch. 09; 207-529-5186).* Blair Pyne runs Broad Cove as a wonderful blend of working wharf and casual marina. Lobsters are landed here, but they welcome recreational boaters too. In addition to the moorings and dockage, they pump gas and diesel and have water, ice, and a tank pump-out. Their small store sells snacks and convenience groceries, and their takeout serves up steaming lobsters and clams.

COW ISLAND

2 ★★★

43° 58.00'N 069° 23.52'W

CHARTS: *13301, 13288*
CHART KIT: *63, 18*

Large and wooded Cow Island, south of Bremen Long Island, offers an attractive little anchorage at its northern tip.

APPROACHES. Approaching from the west, pass south of nun "4" next to Palmer Island, then enter the anchorage south of can "3," giving it some space to avoid the 3-foot spot that it marks.

ANCHORAGES, MOORINGS. Work your way east to get protection from southwesterlies, then anchor in 13 feet at low.

WHARTON ISLAND

2 ★★

44° 00.05'N 069° 21.54'W

CHARTS: *13301, 13288*   CHART KIT: *63, 18*

East of Hungry Island, at the entrance to the Medomak River, the cove between Wharton Island and Delano Hill on the mainland provides a seldom-used anchorage that is easy to enter but exposed to prevailing winds.

APPROACHES. Run in to the Medomak River east of Cow and Bremen Long Island. As you approach the southern tip of Hungry Island, stay well to the east to avoid the ledges south of Wharton. A small house is perched on one of the ledges. Also note the 4-foot lump off the southeast flank of Wharton.

ANCHORAGES, MOORINGS. Anchor halfway along Wharton Island, midway between Wharton and the mainland, in 7 to 12 feet at low.

The looming hulk of the five-masted schooner *Cora F. Cressy* protects Keene Narrows Lobster.

HATCHET COVE

3 ★★

43° 58.60'N 069° 21.02'W

CHARTS: *13301, 13288*   CHART KIT: *63, 18*

Just west of Friendship Harbor, large Hatchet Cove is a precarious place to enter and has little to offer once you get in.

Pretty little Ram Island, in the mouth of Hatchet Cove, has been used by generations of Friendship people for Saturday-night parties.

APPROACHES. The cove is divided by a line of attractive islands and not-so-attractive ledges, starting at the mouth with Ram and Sand Island. Do not attempt the eastern entrance. Instead, use the entrance to the west of the island. Come in within an hour of dead low so you can see most of the rocks. The "Chocolate Chips," as they are known locally, start opposite the remains of a granite pier on the western shore and reach almost to the middle. Hug the shore about 100 feet off, where the channel is about 100 yards wide.

As the cove opens out, you will see a small, isolated ledge opposite the 15-foot sounding on the chart. Leave this ledge to starboard.

ANCHORAGES, MOORINGS. Anchor in 11 to 17 feet at low, just past the isolated ledge.

FRIENDSHIP LONG ISLAND

43° 57.63'N 069° 20.86'W

CHARTS: **13301, 13288**   CHART KIT: **63, 18**

Friendship Long Island forms the southern flank of Friendship Harbor, ground zero for lobstering in Muscongus Bay. It is not surprising, then, that this is also the epicenter for lobster research. A large lobster pound can be seen on the chart on the island's northwest shore. This is the Lobster Life Studies Center of the Lobster Conservancy, a group dedicated to learning as much as possible about lobsters' murky biology.

The cove off the pound offers very little protection from the prevailing southwesterlies, but this could be a pleasant and educational daystop.

FRIENDSHIP

3 ★★★

43° 58.21'N 069° 20.33'W

CHARTS: **13301, 13288**   CHART KIT: **63, 18**

If you want to see a real Maine coastal harbor with few yachts and tourist facilities, go to Friendship. Located on the mainland halfway across Muscongus Bay, Friendship shelters one of the largest fleets of lobsterboats in Maine. They work year-round, breaking harbor ice up with an old LCM landing craft, the *Sea Smoke*. In summer, the harbor is full of colorful comings and goings as boats load traps and gear and unload their catch at the various commercial docks.

Friendship resolutely remains a working harbor and can sometimes be a difficult—some might even say unfriendly—place for yachts. The harbor is busy with its own affairs and has little time to cater to yachtsmen's needs or problems. Be prepared to accept the place on its own terms.

This is the birthplace of the famous Friendship sloops. Their beautiful design, with its long bowsprit, proud topmasts, and graceful sheer, evolved a century ago as a workboat, before the advent of engines. These lovely craft have now become pleasure boats, and their devout owners still return to Friendship on special occasions.

APPROACHES. At certain times of the summer, depending on where the lobsters are, you could walk to Friendship across the lobster buoys. Otherwise, the approaches are fairly straightforward.

The easiest approach is from the west. Note that nun "12," at Martin Point, marks the Medomak River and not the entrance to Friendship Harbor. Leave Ram Island to port and sail between red daybeacon "10" to port and green daybeacon "9" to starboard. There is plenty of room between them, but from a distance, the green daybeacon might be difficult to see against the background of trees.

Coming from the east, you need to pass between Morse and Gay Island, heading north. Keep nun "2" to starboard and favor the Morse Island side. Can "3" lies surprisingly far off the northeast tip of Morse Island, but it must be kept to port to avoid the long Northeast Point Reef. Round the can, turning to port, and follow nuns "4" and "6" into the channel between Friendship Long Island and small Garrison Island. As the harbor opens up, don't blithely follow a lobsterboat inside of can "7," which lies a long way off Friendship Long Island. Keep the can to port.

ANCHORAGES, MOORINGS. Almost all of the moorings in Friendship Harbor belong to lobstermen. You might be lucky enough to find somebody who can direct you to one that is vacant for the night, but don't count on it. If you pick up a mooring temporarily, be prepared to move your boat on short notice.

You can anchor outside the moored boats, but the water is rather deep—21 to 28 feet at low.

Traditional Friendship sloops evolved as work boats before the advent of engines. They are now built or lovingly restored as pleasure boats. (Boutilier)

GETTING ASHORE. Row your dinghy in to the float at the town landing.

FOR THE BOAT. The commercial docks in Friendship cluster in two groups—one at each end of the harbor—with a beach in between. The town landing is in the eastern group.

Friendship Town Landing. The town floats can be identified by the road leading directly down to them on the east side of the harbor. There are no facilities and only about 1½ feet of water alongside at low. You can tie up briefly, but do not leave the boat for any length of time, or you will interfere with lobstermen who may need to use the dock.

Lash Brothers Boat Yard (207-832-7807). This yard is just north of town on Route 97. They specialize in all things lobster, particularly in finishing off lobsterboat hulls, but they could help you with emergency repairs.

FOR THE CREW. To find pay phones, the post office, ice, and limited provisions, walk about .7 miles north, to Friendship Village, passing the post office on your way. Basic provisions can be found at the Archie Wallace grocery store. You will also find a hardware store, an antique store, and the small Friendship Library. What you won't find is anything to stock the liquor cabinet or beer cooler. Friendship is a dry town.

THINGS TO DO. The Friendship Museum in the former schoolhouse has local memorabilia and an interesting exhibit on Friendship sloops. The school has a small playground.

HORNBARN COVE

1 ★★★

43° 58.63′N 069° 18.34′W

CHARTS: 13301, 13288 **CHART KIT: 63, 18**

Just east of Friendship Harbor, the three-mile-long Meduncook River might easily be overlooked. Hornbarn Cove, a short distance upriver on the east side, is easy to enter and has excellent protection.

APPROACHES. There are two entrances to the Meduncook River, on either side of Crotch Island. The wider, western approach is foul and not recommended.

Although the entry to Hornbarn Cove is a straight shot, you will be happier if you plan it between low and halftide when the rocks on the right side of the entrance and in the cove are visible.

Coming from Friendship, round nun "4," leaving it to port. Aim for the crotch east of Crotch Island, favoring the western bank. The river here is about 100 yards wide, deep, and clear of obstructions except for the ledge making out from the eastern shore just as you enter. There are several docks along the eastern shore of the river and usually some moored sailboats. Pass the end of Crotch Island and the ledge and the rocks just beyond. The midharbor ledge and 4-foot spot south of Bradford Point may be marked by an informal beacon.

ANCHORAGES, MOORINGS. Anchor in 14 feet at low, beyond the midharbor ledge and to the east of it, opposite a meadow on the right bank. There are a couple of private moorings. The holding ground is good mud, but there is some grass.

DAVIS COVE

2 ★★★

43° 57.78′N 069° 18.28′W

CHARTS: 13301, 13288 **CHART KIT: 63, 18**

Beautiful little Davis Cove lies a mile or so east of Friendship Harbor with spectacular panoramic views, including a glimpse of Franklin Island Light to seaward. There are a few summer cottages well hidden among the trees and two islets set in the northern end of the cove. Protection is good from all directions but the southwest. During settled summer weather, Davis Cove is likely to be quiet at night, but pretty lively during the day.

APPROACHES. Coming from the south, pass between Gay and Morse Island, leaving nun "2" to starboard and favoring the Morse Island shore. The rocks northeast of nun "2" uncover about 4 feet at low. As you reach the end of Morse Island, leave can "3," which marks the tip of North East Point Reef, well to port and steer for the center of Davis Cove.

ANCHORAGES, MOORINGS. Most of the cottages around the cove have docks and private moorings. There is still ample room to anchor in 8 to 15 feet of water in a good mud bottom.

CRANBERRY ISLAND
43° 56.46'N 069° 21.03'W

CHARTS: **13301, 13288**   CHART KIT: **63, 18**

Between the north end of Cranberry Island and Friendship Long Island there is a pretty little working harbor. Although it is well protected from the southwest, the holding ground is poor, so it is recommended only as a daystop. The bottom is soft mud and eel grass.

In approaching from the east, note that Morse Ledge consists of two separate ledges. Green daybeacon "1" stands on the south end of the more easterly ledge. The second ledge lies west of the beacon and is submerged at halftide.

OTTER ISLAND
43° 55.47'N 069° 21.06'W

CHARTS: **13301, 13288**   CHART KIT: **63, 18**

There is an intriguing little slot in the southwest corner of Otter Island, no more than 50 yards wide and 200 yards long. Reportedly, one or two schooners would snuggle up in here in the old days, but it is hard to imagine. The slot is wide open to the southwest, so it would be an uncertain place to spend the night unless winds were out of the north. And if there is another boat here, swinging room will be scarce.

The entrance is straightforward but full of lobster buoys. The centerline of the cove has 7 to 8 feet of water at low as far back as the large rock that sticks out from the eastern shore.

This is a delightful place for a lunch stop. The island is private, but you can row ashore at the head of the cove and walk through the woods either east of west to a nice little beach. No fires, please.

HARBOR ISLAND
4 ★★★★

43° 54.69'N 069° 22.52'W

CHARTS: **13301, 13288**   CHART KIT: **63, 18**

What a special spot. Harbor Island is privately owned, but the owners have put up a welcoming sign: "Welcome to Harbor Island, which for Us is Home. Please Respect our Privacy as We Would Yours—The Lev Davises."

From this beautiful and well-protected anchorage between Harbor and Hall Island, you can gaze south to the lighthouse on Franklin Island, to the lobsterman's house on Hall Island, and to the owners' old stone cape on Harbor Island. There may be guillemots in the cove, or eiders, or blue herons.

APPROACHES. The approach is less difficult than it appears on the chart. Hall Island to the east is wooded, while the northern tip of Harbor is mostly grassy. You must pass between two isolated rocks, one just north of Harbor Island and the other just north of Hall, but there is at least 250 yards between them. The rock north of Harbor Island sticks farther out from the shore, and rockweed is visible floating

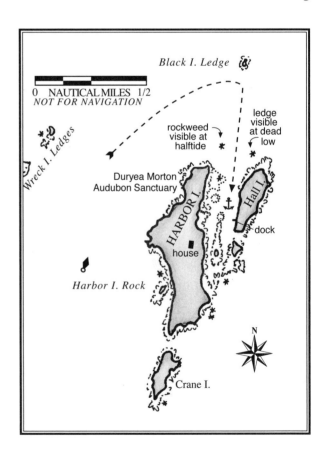

above it at halftide. The rock north of Hall Island is visible only at dead low, but it is in line with the western edge of the island and therefore easy to avoid. Favor the eastern side of the entrance.

It is helpful to start at small, bare Black Island Ledge to the north and then run a compass course for the entrance to the harbor.

ANCHORAGES, MOORINGS. At high tide, the harbor appears to be exposed to the south, but the whole passage between Hall and Harbor Island is blocked by ledges, which provide good protection at all tides. Most cruising boats anchor in the lee of the ledges, near the Hall Island side of the harbor, in 27 feet at low.

The Harbor Island side of the harbor is shoal for quite a distance from the shore, and, according to Mr. Davis, "for the most part foul for anchoring—kelp ledges and lots of lobster trap buoys." But it is possible to find an anchorage there in 12 feet or so.

Further advice from Mr. Davis: "The middle ground ledge makes up from the south of the harbor as far as a line at right angles to the lobsterman's dock on Hall. Stay north of this line when crossing the harbor. The five-foot spot charted has only about two feet of water on a drain tide."

GETTING ASHORE. Land on the beach southeast of the house and wharf on Harbor Island.

THINGS TO DO. The Davises have expressed their desire to be generous with the island, so please use it with respect, care, and discretion. Stick to the shorelines, no fires above the high-water mark, and no camping. The northern end of the island is the Duryea Morton Audubon Sanctuary, which should be avoided entirely up until mid-July during the eider nesting season.

Walk around the southern end of the island. Take the path north of the white house over to the western shore, then walk south along the beach. It takes an hour or more, past a cave, nesting ospreys, and interesting rock formations. The beaches are lined with beach peas, beach roses, and wild irises.

It is also fun to explore the tiny southern tip of Hall, which becomes a separate island when the tide comes in.

EASTERN EGG ROCK
43° 51.82'N 069° 22.91'W

Charts: **13301, 13288**   *Chart Kit:* **63, 18**

Eastern Egg Rock is a seven-acre patch of boulders and grass amid a spine of ledges in the middle of the entrance to Muscongus Bay. Plan to be nowhere near it when the fog closes in.

In sunlight, however, this is a fascinating island to sail around. Here Cornell University ornithologist Stephen Kress and the National Audubon Society have reestablished a colony of puffins, the most southerly nesting ground for these strange and wonderful birds, with their colorful beaks and comical postures.

Puffins and terns were plentiful here in the 1800s. But the terns were hunted for the millinery trade and puffins were eaten. Farmers collected their eggs from outlying islands as a cash crop for additional income. By 1900, both species were gone, replaced by aggressive herring gulls.

To nest, puffins need tumbled rock to provide burrows and drainage. They need good fishing grounds nearby, and they must be free of mammalian predators. Eastern Egg Rock, far out in the bay, has no mammals, not even a mouse, so it is ideal. In the 1970s, researchers laboriously dug burrows by hand for puffin chicks transported from Newfoundland. Their efforts have been highly successful. In the spring of 2001, 37 pairs of puffins returned to Eastern Egg Rock. They were joined by thousands of pairs of common terns and wholesome numbers of the rarer arctic and roseate terns.

You may be lucky enough to see puffins near Eastern Egg Rock during June and July, but the population is much larger on Matinicus Rock farther east. For the best view of puffins, plan a journey to lonely Machias Seal Island, way Down East, where, hidden in a blind, you can observe puffins, terns, and razorbills at close range. Please do not go ashore.

MONHEGAN ISLAND

1 ★★★★★ 🛒 🍴

43° 45.87'N 069° 19.35'W

CHARTS: 13301, 13288   CHART KIT: 18 (INSET), 18, 19

Places that are hard to get to are often the best. Certainly visit Monhegan if you can. Unspoiled and beautiful, 1.4 miles long and .7 miles wide, it stands majestically alone, 10 miles out to sea, with its own personality and a wonderful sense of remoteness.

Monhegan remains as independent in spirit and fact as it is possible to be in these United States. As one islander put it, "What makes Monhegan different is that it's hard to get to and hard to live on, and anything that makes it easier is a step in the wrong direction."

Giovanni da Verrazano sailed by Monhegan in 1524, while searching for a passage to China. Other early explorers associated with Monhegan include Estevan Gomez in 1525, who was also searching for Cathay on behalf of Spain. Then came Martin Pring, sailing for the English in 1603, soon followed by Samuel de Champlain for the French.

George Waymouth was probably the first to document the island, in 1605. James Rosier's contemporaneous narrative of this voyage says, "It appeared a meane high land, as we after found it, being but an Iland of some six miles in compass, but I hope the most fortunate yet discovered."

When Captain John Smith sailed north from the Virginia colony with two ships in 1614, he was looking for gold and copper. At Monhegan, however, he soon discovered the island's true wealth—"the strangest fish-pond I ever saw."

Smith was a wonderful publicist, and his enthusiastic accounts led to early attempts at colonization of these northern regions. "During the 16th and 17th centuries," as a placard at Monhegan's little museum explains, "Monhegan steadily gained a reputation among European traders and explorers as the richest fishing area in the New World. It became the landfall for European mariners westward-bound."

Today there is still a marvelously old-world feeling in the small town, where dusty footpaths cut between rambling inns and cottages. The remaining two-thirds of Monhegan is protected woodlands, and the hiking is unsurpassed. So is the birding. Monhegan is on the flyway for spring and fall migrations of hundreds of species of land and sea birds, and the woods are alive with their sounds. The cliffs tower 160 feet above the sea like an enormous beacon, the highest on America's east coast, making it a perfect landfall.

About 14 active lobstermen fish out of Monhegan. In the 1940s, Monhegan's lobstermen voluntarily elected to limit their lobstering to a unique, six-month fishing season to conserve the fishery. Lobstering starts on "Trap Day" on January 1 and ends on June 25th, just when the season is beginning for most Maine lobstermen. Ending the season in June might conserve something else, too. "After that," says Sherm Stanley, "we begin to catch shedders with soft shells, and the prices go down anyway."

By tradition, no one will begin lobstering until everyone is ready. If one lobsterman has engine or health trouble, all the others will wait—and help—until he is good to go.

Little Manana Island forms the western part of Monhegan Harbor, and tiny Smuttynose protects the north end. The harbor, though, is wide open to the southwest, and the bottom is foul. It is never easy to spend the night with your boat here in this difficult and exposed anchorage. If the weather is bad or threatening, you should not be here at all.

GETTING THERE. The most direct way to get to Monhegan, of course, is to sail there in your own boat, either on your way Down East or from a mainland base such as Tenants Harbor or Port Clyde. But since the anchorage is difficult, you may want

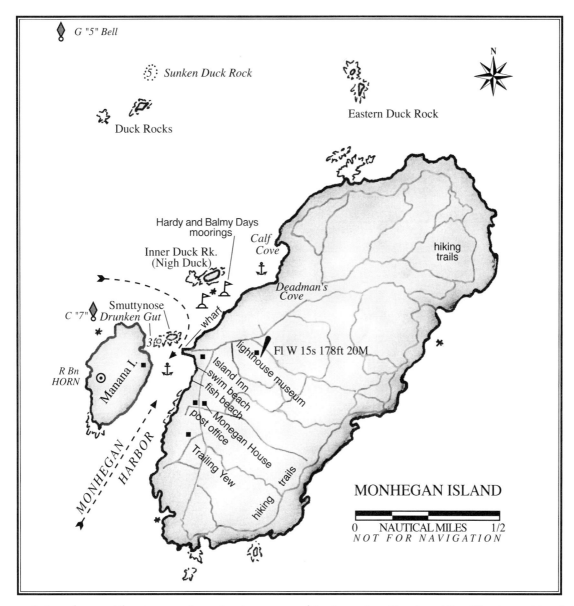

to spend time here without worrying about your boat. There is no airport on Monhegan, so the only alternatives are by water.

The easiest way is to take the mailboats *Laura B.* or *Elizabeth Ann.* The trip departs from Port Clyde two or three times daily and takes about an hour in good weather. Reservations are required (*372-8848*).

The *Balmy Days* sails out of Boothbay Harbor to Monhegan during the summer (*633-2284*) and the *Hardy III* leaves from New Harbor (*882-7909*).

APPROACHES. From any direction, the high island of Monhegan is easy to find. Its light is visible for 20 miles, flashing white every 15 seconds. The foghorn and radio beacon are located on neighboring Manana Island. Two miles west is Manana Island red, lighted whistle buoy "14M," where pilots board ships bound up Penobscot Bay. Plan to avoid arriving at Monhegan late in the day, when there might not be time to retreat to the mainland before dark if the few available moorings are taken.

The harbor can be entered from the south or from the north. The southern approach is obstruction free. From the west, keep can "7" and Manana Island to starboard and turn south into the harbor through Herring Gut, between the rock called Smuttynose and the town wharf.

From the north or east, plot a course to gong "3" north of Eastern Duck Rock and then coast down the western side of Monhegan. Or find green bell "5" and pass to the west and south of Duck Rocks. Head into Monhegan Harbor through Herring Gut, between Smuttynose and the town wharf.

There is 12 feet of water in Herring Gut, and it is wide enough to easily pass through even when a boat is lying alongside the wharf. Drunken Gut is the aptly named little passage between Smuttynose and Manana Island. It is very narrow and only has about 3 feet of water at low. Of the few who use it, even fewer are sober.

ANCHORAGES, MOORINGS. Do not anchor in Monhegan Harbor. It is crowded and exposed, and the bottom is fouled by 300 years of heavy use.

⚓ There are no guest or rental moorings here, but a vacant fisherman's may be available. Try to find Sherm Stanley, the harbormaster, either at the town dock or aboard his black-hulled lobsterboat *Legacy*, and ask. It is especially important in such a limited harbor not to leave your boat unattended on one of these moorings without permission.

The harbor is exposed to the southwest and, to a lesser degree, to the northeast. Even in calm weather, the ocean swells funnel into the harbor and make it rolly. During a blow from the south, the swells are awesome, and it would be an extremely difficult and dangerous place to be.

The lobsterboat moorings are great granite blocks, half the size of a house, and the boats have ridden out winds of 120 m.p.h.. Notice their chain pennants. In a blow, you might be safe enough on one of these moorings if one was available, but you wouldn't be comfortable.

For comfort and safety, the best solution is to be on the mooring of either the *Balmy Days* or the *Hardy III*, just inside Inner Duck Rock (locally called Nigh Duck). These excursion boats arrive about 10AM and leave about 3PM, and they are extremely generous in accommodating yachtsmen who wish to hang on their moorings for the night. But you will need to be gone by 10AM the next morning.

A final possibility is to anchor inside Nigh Duck or toward Deadman Cove, where there is fair protection from southerly winds. Note, however, that the entire area from Manana to Nigh Duck to Monhegan's western shore, including Deadman Cove, is a cable area. The bottom is deep, and there is a lot of tidal current.

It seems impossible to combine safety, comfort, and convenience on Monhegan. If you are uneasy, do not stay. The nearest alternatives are Burnt or Allen Island, Port Clyde, or Tenants Harbor.

GETTING ASHORE. You can bring your boat up to the town wharf temporarily, but do not stay for

The bulletin board is ground zero for Monhegan Island news.

long, especially if one of the passenger boats is due to arrive. Dinghy passengers can land at the ladders on the north side of the town wharf anytime, but do not leave the dinghy there.

If you intend to spend some time ashore, land at one of the two tiny beaches south of the town wharf. The one closest to the wharf is Swim Beach; the one farther south is Fish Beach. Fish Beach has some broken glass and rusty iron, so Swim Beach is preferable.

FOR THE BOAT. Come to Monhegan well prepared. In an emergency, diesel fuel is available at the town wharf by truck through the Monhegan Store, but not drinking water. There are no repair facilities. If you have engine troubles, ask for the island mechanic, and you will probably have more experts than you know what to do with.

FOR THE CREW. 🛒 Limited fresh produce and groceries, wine, and beer are available at Carina, behind the Island Inn or at North End Market, farther to the south. The Barnacle Cafe by the wharf has baked goods, wine, and an ATM. The inns have pay phones.

🍴 The Island Inn (*596-0371*) serves three meals a day overlooking the harbor. Monhegan House Cafe (*594-7983*) serves breakfast. Lunch options also include the Novelty, behind Monhegan House, for pizzas and sandwiches, the Scruffy Dog, a casual grill, and takeout lobsters, crab rolls, and chowder at Shermie's Fish Market by Fish Beach. Dinners are covered more formally by the Island Inn, casually by Scruffy's, and family-style at The Trailing Yew (*596-0440*). Make reservations early and expect to pay a premium to be eating food that has been brought out to the island for you.

Years ago things Monhegan was pretty Spartan, but nowadays most of the comforts of home are available, thanks not only to generators, but also to solar energy. Monhegan has an impressive amount of photovoltaic power, including the world's first solar-powered post office.

THINGS TO DO. Watching the daily boats arrive at the wharf is probably the biggest entertainment on Monhegan. Don't underestimate the show. Everyone crowds down to the dock to meet guests or send them on their way, or just to see the arriving passengers. Pickups are waiting to carry bags to the inns, and a torrent of propane bottles, lumber, and all the other fascinating miscellany needed for island life pours from the hold of the *Laura B.*

Be sure to allow time to hike the beautiful trails that run across and around the island. The protected woodlands are controlled by Monhegan Associates, and this group of citizens has preserved an extraordinary private park for the public to enjoy. The family that established the present colony in 1784, Henry Trefethern and his two brothers-in-law, believed that "the rocks belong to everyone." By the mid-20th century, however, summer people had arrived in force, and cottages were encroaching on the wild back side of the island. The formation of Monhegan Associates in 1954 halted this trend and even reversed it with the removal of some existing buildings.

The result today is a great expanse of beautiful land more or less in its virgin state, with 17 miles of rustic trails and breathtaking walks along the cliffs. There are forests and swamps, rocky headlands, and spectacular views to the distant mainland and islands. Trail maps are available at any of the inns or stores. Be aware that deer ticks carrying Lyme disease have been found on Monhegan, so consider wearing long pants, socks, and a good dose of insect repellant.

A trek around half the island is more than enough for most people unless you are in great shape and want to be thoroughly tired. The maps include several warnings: "The trails are rough and wild, especially cliff trails. Walk with care." No smoking is allowed outside of the village, nor are any outdoor fires.

One further warning: on the headland side of the island, the eastern shore, stay well above the tide line. If you were to fall in, the currents and undertows would make rescue difficult, if not impossible. The town library was founded in memory of two children who were swept out to sea by a huge wave.

The great natural beauty of Monhegan has attracted artists to the island for more than a century. Today, there are 20 or more studios of working artists scattered around the town, some of which are open to the public at specified hours. A listing of these can be found on the maps posted on various bulletin boards all over the island. Among the familiar artists who have been linked to Monhegan are George Bellows, Rockwell Kent, and Jamie Wyeth.

The lighthouse, high on the hill above the town, is now automated and run from the Coast Guard station on Manana Island. The lightkeeper's cottage is a wonderful little museum—one article described it as "a well-organized communal attic" —with an interesting collection of old photos, books, artifacts of the lobstering trade, and mementos of this extraordinary island and the hardships of living here.

Recent donations have included works from some of Monhegan's more illustrious summer residents, including a self-portrait by George Bellows and pencil sketches by Rockwell Kent. The late philanthropist Betty Noyce donated 20 more paintings and half the money to build a new gallery to display them. The new gallery is a replica of the assistant lightkeeper's house, which was torn down long ago. Life, however, imitates art. The plans were drawn from Edward Hopper's painting *Monhegan Lighthouse,* painted between 1916 and 1919.

Manana Island. Row your dinghy over to the Coast Guard ramp at Manana Island, at the northwest corner of the inner harbor. There is no float, so tie up to the railings and walk to the top of the hill for a magnificent view. The Coast Guard fog signal station is on the west side of the island.

As you approach the Coast Guard station, look for a yellow "X" on the rocks across the gully to your right. This marks the "Norse runes" reportedly inscribed by Norse explorers.

For years, Manana was home to "The Hermit." Ray Phillips was a food inspector in New York City during the 1920s. During the Depression, he came to Monhegan, bought one-sixth of Manana Island, and built a 12'-by-15' driftwood shack. He lived there, unmarried, with his goats, his sheep, and his geese until he died in 1975 at age 83. His ashes are buried on the island. The ramshackle hut on the east side of the island is not his (his was razed because it was dangerous), but you can imagine him living there.

LIGHTHOUSES

The storm howls and the waves pound endlessly on shore, breaking into driven white spray. A small boat wallows in the surf and is flung, helpless, toward the rocks. In the beam of the lighthouse, frightened faces are lit for a flash, yellow-slickered arms flung outward, then they're lost in the black, chaotic night. The vessel strikes the rocks. Unperturbed, the great beam of the lighthouse sweeps its appointed arc, briefly lighting the splintered planks, the gear sucked out in the backwash of the waves, the yellow bodies tossed ashore. No one sees.

In 280 B.C., the Pharos of Alexandria was one of the seven wonders of the world, its fires blazing 440 feet above the sea. More than 2,000 years later, the 63 lighthouses in Maine are direct descendants, gems of functional architecture that gladden the heart of the mariner. The first was the Portland Head Light on Cape Elizabeth, built during George Washington's first term, in 1791. Next came the tower on Seguin Island, off the mouth of the Kennebec, which much later became the only lighthouse in Maine with a first-order Fresnel lens.

The lights were fueled first by whale oil and later by lard oil and, eventually, by kerosene. One of the duties of the keeper was to trim the wicks, and every few hours he wound the weights that turned the panels to make the light flash. Electricity made things easier, and the invention of large refracting prism lenses in 1922 by Augustin Fresnel greatly improved the range and brightness of the lighthouse beam. Powerful airway beacons cast the lights today.

Fog bells were struck by hand in the early days, then by a wound-clock mechanism. Bells were replaced by steam whistles and later by air horns to produce the familiar and sonorous notes that reverberate through the fog. Whooo-aaah. Whooo-aaah.

The lights were first managed by a Lighthouse Board and then by the U.S. Lighthouse Service. In 1939, they came under the control of the U.S. Coast Guard. During the austerity programs of the late 1980s, the Coast Guard decided to deactivate many lighthouses and to automate the remaining ones over a period of several years. This process is now complete.

Some lighthouses, like Hendricks Head at the mouth of the Sheepscot River and Indian Island off Rockport, are now private residences. The lighthouse on Marshall Point at the mouth of the St. George River is now an historical museum. The facilities on Matinicus Rock are leased to the Audubon Society, and the keeper's house on Seguin Island is occupied during summer months by the Friends of Seguin.

These lighthouses remind us of Captain Joshua Strout, who served more than half a century at Portland Head Light; of William Williams, who spent 27 years on lonely Boon Island, off York; and of 17-year-old Abbie Burgess, who was tending the light on Matinicus Rock when a great wave swept over the rock in 1856. For two hundred years, there was a great tradition of lighthouse keepers—families willing to live alone on a rock in the sea, their only communications and resupply by small boat, often isolated for weeks on end during winter storms. They minded the lights, they rescued sailors in distress, and they raised their families. That way of life has passed, and it will take a while to understand what we have lost.

You can visit these lighthouses: Portland Head Light (museum); Spring Point Light (part of museum); Seguin Island Light (caretakers and a small museum); Monhegan Island Light (keeper's cottage is a museum); Marshall Point Light (museum); Owls Head Light (museum); Matinicus Rock (used by the National Audubon Society); Robinson Point Lighthouse, Isle au Haut (a bed and breakfast); Grindal Point Light, Islesboro (museum); Mount Desert Rock (used by Allied Whale Watch); West Quoddy Head Light; Partridge Island, New Brunswick (museum). Also, visit the Lighthouse Museum in Rockland.

ALLEN ISLAND and GEORGES HARBOR

4 ★★★

43° 52.55'N   069° 18.87'W

CHARTS: *13301, 13288*   CHART KIT: *63, 64, 18, 19*

On Sunday, May 19, 1605, English explorer George Waymouth anchored his vessel *Archangel* in "Pentecost Harbor," probably the bight formed by the Georges Islands—Allen, Benner, and Davis. They fished, filled their water tanks, and gave thanks for their safe arrival and fortunate discovery.

Waymouth explored the St. George River from this base. He mounted a cross on one of these islands to guide future explorers. Although the original has long since disappeared, a granite cross has been erected on the northwest tip of Allen Island to commemorate his expedition.

After meeting a group of Pemaquid Indians and trading with them in cautious but friendly fashion, Waymouth and his men captured five Indians and took them back to England. This turned out to be one of the most important events in the early history of America. The appearance and dignified demeanor of the Indians greatly impressed the people of London and the court, creating a surge of interest in New World colonization. The Indians, of course, were not quite as pleased.

The capture of the Indians by Waymouth was reported by Chief Arrossac shortly afterward to Samuel de Champlain, sailing in the same area. "He told us," wrote Champlain, "that there was a vessel ten leagues from the post that was fishing, and that those on the vessel had killed five savages from this river, under the guise of friendship; and according to the way they described the people of this vessel, we believe them to be English…"

The enmity and distrust of the British that this engendered among the Indians was encouraged and exploited by the French, who allied themselves with the Indians in a series of bloody wars that destroyed English settlements along the coast for 100 years.

Allen Island lies due west of Burnt Island and four miles south of Port Clyde. Its 400 acres are high and mostly wooded. Over the years, Allen's owner has experimented with island self-sufficiency to see whether there are practical and economic ways to use Maine's islands today instead of abandoning them to nature or the onslaught of vacation homes. The northeast point has been cleared, and the lumber has been used to build houses and a wharf. Sheep graze in the clearings, and fish pens are sometimes in the harbor. These experiments led to the formation of the Island Institute, so their lessons could be applied to other islands.

Signs on the beach at the northern tip of the island say, "No fires, no dogs, beware dangerous ram." All of these are serious warnings. To protect the sheep, the shepherd will shoot dogs found on the island, and the ram has his own set of rules.

Georges Harbor, between Allen and Benner Island, is a cozy anchorage about 100 yards wide. The ledge on Benner breaks the seas, and the harbor is often calm when it's roaring outside.

APPROACHES. Enter Georges Harbor from northeast or southwest. From the northeast, run down the middle of the slot, leaving the dock on Allen to port. From the southwest, favor Allen to avoid the ledgy southern tip of Benner.

ANCHORAGES, MOORINGS. Most of the harbor is shown on the chart as cable area. A submerged mooring cable is also shown crossing Georges Harbor at its narrowest point, where you wouldn't have much swing room anyway. Space is further limited by several private moorings.

You can anchor comfortably in midchannel northeast of the large dock on Benner Island in 19 feet of water at low. Or run past the cable area and anchor where the channel widens on both sides. Holding ground is good.

GETTING ASHORE. Allen is a private island. While visiting is allowed, it is not encouraged. Scramble ashore near the Waymouth cross, or take the dinghy around to the little sand beach at the northern end of the island. Do not use the island's only wharf. Keep an eye out for the ram, who cares more about sheep than you do and can run much faster.

BURNT ISLAND

3 ★★★

43° 52.36′N 069° 17.44′W

CHARTS: *13301, 13288*   CHART KIT: *64, 18, 19*

High, wooded Burnt Island marks the southeast corner of Muscongus Bay, due east of Allen Island. Up until just after World War II, Burnt was the site of a Coast Guard lifesaving station. In an emergency, coastguardsmen could launch their rescue boats from the exposed outer shores. Today, some of the Coast Guard buildings still stand on the point toward Little Burnt Island.

Little Burnt Island and Burnt form an anchorage with adequate protection from the southwest, but it is exposed to the blasts from the northwest or north.

Burnt Island is private, but the owner shares the island with the Hurricane Island Outward Bound School (HIOBS). In early or late summer—*not in July or August*—he also generously allows limited day use by respectful yachtsmen (see below).

APPROACHES. Coming from Monhegan, run between Dry Ledges and Burnt Island. From Port Clyde, observe cans "7," "9," and bell "11," all of which should be left to port.

ANCHORAGES, MOORINGS. The best anchorage in settled summer weather is north of Burnt Island and east of Little Burnt. Feel your way in until you find a comfortable anchoring depth. The bottom is a thin layer of mud and sand over ledge, so rig a tripline for retrieving your anchor in case it gets lodged in the rocks.

Please *do not* pick up any vacant moorings. HIOBS may have set moorings off the former Coast Guard dock, but they are light—intended only to hold the school's open, 30-foot pulling boats. Others belong to the owner.

GETTING ASHORE. Dinghy ashore, but please *do not* use the owner's dock.

THINGS TO DO. The old Coast Guard station next to the dock is occupied during the summer months. Please stay in the general vicinity of the dock and beach area and do not interfere with ongoing island lumber operations or with the activities of HIOBS. No camping or fires are allowed.

You may see HIOBS students heaving at the oars of their pulling boats or jumping off the dock. Or, if it's 5:30 AM, they might be running around the island before an icy plunge. Care to join?

McGEE ISLAND PASSAGE

43° 55.16′N 069° 18.41′W

CHARTS: *13301, 13288*   CHART KIT: *64, 18, 19*

McGee Island is part of the chain of islands running north and south from Caldwell Island to Allen Island. The passage north of McGee is a good one to know about—easy to find and well marked. The island itself is private, and visiting is not encouraged.

Coming from the south or west, pass well to the south and east of Jenks Ledge, marked by nun "2JL." Pass between can "1" and nun "2" and favor McGee Island as you pass Seavey Island to port. Or approach by squeezing between McGee and Twobush Island.

Exit the passage south of Bar Island and north of Inner and Outer Shag Ledge. Beware of the unmarked 5-foot spot northeast of Outer Shag and of Old Horse Ledge, marked by red daybeacon "10."

From the east, use Old Horse Ledge and Bar Island to steer into the passage.

CALDWELL ISLAND PASSAGE

43° 55.85′N 069° 17.54′W

CHARTS: *13301, 13288*   CHART KIT: *64, 18, 19*

The Caldwell Islands, along with neighboring Teel Island, are some of the most attractive in Muscongus Bay. They lie at the mouth of the St. George River, west of Port Clyde.

A beautiful passage wends its way south of the Caldwells and north of Teel. It is a tempting shortcut along the route between Friendship and Port Clyde, but it can be tricky. A ledge lurks halfway between Stone Island and Caldwell Island, and a rock lies just south of Caldwell. Strangers should try the passage near low tide so they can find the ledge with their eyes and not their keels.

From the west, locate Goose Rock and approach the passage by keeping close to the southwestern flank of Caldwell Island. When off the southwest tip of Caldwell, look for a small, privately maintained radar reflector daybeacon just east of the ledge. This beacon also marks the north-south passage between Stone and Teel Island. Work your way toward the ledge and keep it close aboard to starboard, thereby avoiding the rock due south of Caldwell.

Once past the ledge, head between Teel and the southernmost of the Little Caldwell Islands, identifiable as a high rock with grass on top, crowned by a cairn of piled rocks.

From the east, identify the southernmost of the Little Caldwell Islands, a high rock with a grass cap and a rock cairn. Pass south of it and north of Teel. Then head for the ledge marker and leave it close aboard to port.

There is a large home and dock on Teel Island, and often a yacht is moored off it. Anchoring, however, is not recommend. The harbor is not well protected, it is prone to current, and the mud and grass bottom is poor holding ground.

ST. GEORGE RIVER

CHARTS: *13301, 13288*   CHART KIT: *64, 19*

In 1605, English explorer George Waymouth and his crew encountered an impressive river along the Maine coast. The expedition's chronicler, James Rosier, wrote, "The River it selfe as it runneth up into the main very nigh forty miles toward the great mountaines beareth in bredth a mile, sometimes three quarters, and halfe a mile is the narrowest."

Rosier's description, intended to impress and perhaps to mislead certain folks at home, is a trifle exaggerated. After comparing the river favorably to the Orinoco, Rio Grande, Loire, and Seine, he modestly adds: "I will not prefer it before our river of Thames, because it is England's richest treasure."

The first reaction of most people reading Rosier's account is that he must have been talking about the Kennebec or Penobscot or some other great river, and a hot historical debate has raged on the subject for centuries. But despite some discrepancies, it seems evident that it was the St. George River that Waymouth explored.

At the eastern entrance to the river is the busy working harbor of Port Clyde, which is not well protected, but fun to visit. At the western entrance is the much smaller and better-protected harbor of Pleasant Point Gut, which you can hardly squeeze into.

The river is easy to enter, and, for the first 5 miles or so, it is broad and deep. The banks are wooded or cleared for saltwater farms and meadows. On either shore are the wide coves described by Rosier—Turkey Cove, Maple Juice, Otis, and Watts. Most of these are pleasant and secure anchorages for the cruising boat, although relatively shoal.

The whole river from Marshall Point to the head of navigation at Thomaston is less than 10 miles long. The upper section is narrow, bordered by mudflats, with strong tidal currents. The channel near Thomaston shifts constantly, and the buoys need to be repositioned frequently. This can be a tricky place for a deep-draft boat, so it is wise to go up on a rising tide.

PLEASANT POINT GUT

■ ★★★

43° 57.40′N 069° 17.91′W

CHARTS: *13301, 13288*   CHART KIT: *64, 19*

Pleasant Point Gut is a working harbor on the west side of the entrance to the St. George River, inside Gay Island. It offers extremely good protection under almost any conditions, and the current through the Gut is not strong. People who were born here know how to get through the western end of the Gut safely at high water, but strangers shouldn't attempt it. A few hours before low tide, the western entrance will be completely dry.

APPROACHES. The best time to enter is at dead low tide, from the east, so that the ledge making out from tiny Flea Island (unnamed on the chart), midway through the Gut, is fully visible. At other tides, the ledge can be avoided by heavily favoring the mainland side.

Note that can "5" is a river marker, not a harbor buoy, and you leave it to *starboard* on entering the Gut. After rounding the northern tip of Gay Island, favor the northwestern, mainland shore of the Gut. The major danger is the ledge making out north and northeast of Flea Island, the ledge being about as long as Flea Island itself.

ANCHORAGES, MOORINGS. You may be lucky enough to find an available mooring by asking around, but be prepared to anchor in tight quarters. If you can find space among the moorings, drop the hook in 10 to 13 feet of water at low. Note the large shoal area on the Gay Island side, beyond Flea Island. The bottom mud holds well.

TURKEY COVE

3 ★★

43° 57.55′N 069° 15.70′W

CHARTS: **13301, 13288**   CHART KIT: **64, 19**

The first of the large, open coves on the east side of the St. George River is Turkey Cove. This well-protected anchorage is large enough to hold a small fleet. Forty cruising boats have been known to rendezvous here at one time. The south shore is wooded, but there are a number of houses on the north shore, and there's a road at the head of the harbor.

APPROACHES. The entrance is wide and easy.

ANCHORAGES, MOORINGS. Anchor in the center or southern part of the cove in 8 to 15 feet of water at low. For the best protection, get east of the rounded point of the south shore. The water starts to shoal rapidly just before the dock on the south side. Holding ground is good, in mud.

MAPLE JUICE COVE

3 ★★★

43° 58.60′N 069° 16.64′W

CHARTS: **13301, 13288**   CHART KIT: **64, 19**

Maple Juice Cove is a large, round bay half a mile wide on the west side of the St. George River about 2 miles above the entrance. Access is easy, and the cove provides a calm and secure anchorage. This must have been one of the "very gallant Coves" described by James Rosier, chronicler of George Waymouth's expedition in 1605.

The arm of the cove extending to the north is the home of a small fishing community, and some handsome old houses are set on the meadows of Stones Point. It is not unusual to see ten or more cruising boats in Maple Juice Cove on a summer day, nor would it be crowded. Seals like it here too, and the water is warm enough to join them for a dip.

The Olson House was immortalized by Andrew Wyeth in *Christina's World*.

APPROACHES. Coming upriver, leave can "7" to port and turn to the west into the cove.

ANCHORAGES, MOORINGS. Anchor anywhere in the cove in 9 to 15 feet of low water, in mud.

GETTING ASHORE. Row your dinghy in to the far end of the floats at Blue Water Seafood, but don't get in their way. Do not be tempted to land at Stones Point, or you are likely to encounter an irate property owner.

FOR THE BOAT AND CREW. *Blue Water Seafood (207-354-6798).* Blue Water is a lobster company occupying the red buildings at the northern entrance to the cove. They generously welcome cruising boats or their tenders to temporarily tie up to their wharf while crews take a look at the farmhouse made famous by Andrew Wyeth's painting *Christina's World*. Hint for a truly artistic experience: they also sell lobsters retail.

THINGS TO DO. A quarter of a mile up the road, on the right, is the distinctive, weathered farmhouse which was the backdrop for Andrew Wyeth's most famous painting, which made its debut in 1948. Christina was a crippled woman who lived in the farmhouse with her brother Alvaro. "I kept thinking about the day I would paint Christina in her pink dress," Wyeth told his biographer, "like a faded lobster shell I might find on the beach."

When the tide is down, row over and examine what is left of the hulk near Blue Water Seafood. Referred to as "the old tug," her superstructure now forms the restaurant for the Tugboat Inn in Boothbay Harbor.

OTIS COVE

3 ★★★

43° 59.19′N 069° 14.34′W

CHARTS: 13301, 13288   CHART KIT: 64, 19

Otis Cove, on the east side of the river, is one of the loveliest spots on the St. George. It has pretty, wooded, rocky shores with only a few houses in sight and broad views all around, including fields and farmhouses across the river and the Camden Hills to the north. In the middle of the cove is small, uninhabited Ten Pound Island. Protection is good.

APPROACHES. To stay clear of the rock and 6-foot spot at the southern end of the cove, head in for the island on a southeasterly course.

ANCHORAGES, MOORINGS. There is a good anchorage either southwest or northeast of the island. The northeast one provides a little better protection from prevailing winds and somehow seems cozier. Holding is good in mud and penetrable grass.

WATTS COVE

44° 00.00′N 069° 13.36′W

CHARTS: 13301, 13288   CHART KIT: 64, 19

Watts Cove, lying on the east bank of the St. George River opposite Broad Cove, is shallower than some of the other anchorages on the river and therefore less useful to cruising sailboats. If you draw less than 5 feet, however, this would be a secure anchorage. It is mostly wooded, with only a few houses and a lovely encircling cove.

APPROACHES. The entrance is wide open. Go in as far as your draft allows to get as much protection as possible from the southwest. Note the ledge extending from the south shore.

ANCHORAGES, MOORINGS. With shoal draft, you will be comfortable here. Boats drawing 5 or 6 feet might ground gently in the mud, which holds anchors well.

THOMASTON

4 ★★★

44° 04.29′N 069° 10.97′W

CHARTS: 13301 (INSET) 13288   CHART KIT: 64 (A, B), 19

In 1605, Captain George Waymouth landed at the head of navigation of the St. George River, less than 10 miles upriver from Muscongus Bay. Today, just above the Thomaston waterfront, a wooden cross commemorates the event: "That We, The Menne of Englande, have marked this spot for home." Thomaston was settled twenty-five years later as a trading post between Plymouth Plantation and the Indians.

The town's first industry—burning lime to make plaster—sprang up in the 19th century and spawned its second industry—making ships to transport the lime. For half a century, eight major yards launched huge schooners and other vessels which carried the name of this small town to ports around the world. Whoever was not a sailor or a captain was a rigger, cooper, sailmaker, or merchant. It was indeed, as its motto proclaims, "the town that went to sea."

Thomaston remains a working town linked with past and present. Dragon Concrete Company still works the lime beds outside of the town, and although the great shipyards are long gone, there has been a boatbuilding renaissance.

APPROACHES. The channel at Fort Point is still wide and deep enough that you should have no trouble working your way from buoy to buoy. Above nun "16," opposite Hospital Point, the situation changes. There are frequent shifts in the channel near the sharp turn in the river toward Thomaston, and many boats have grounded in the mud here.

As noted in the inset to chart 13301, "Buoys '17,' '18,' '19,' and '23' are not charted because they are frequently shifted in position." The Coast Guard tries to keep up with the shifting channel, but there is no guarantee that these buoys are in the right places. Flashing green light "21" is fixed in position on a pile of granite blocks, but don't cut it close. The channel is probably some distance to the east.

Come in on a rising tide, either at dead low, when the mudflats appear and the channel becomes obvious or after midtide, when there is plenty of water. The current may run 2 or 3 knots above Fort Point and 4 or 5 knots off the Thomaston waterfront.

OSPREYS

One of the great experiences of sailing the Maine coast is watching an osprey hover 50 feet in the air, plunge into the sea, seize a fish in its talons, and carry it away to its nest in triumph. The osprey, or fish hawk, is a magnificent creature, with a wingspan of five or six feet and a curved beak that lends it an air of ferocity. It is white underneath, dark brown on top, and has a beautiful brown-and-white pattern under its wings and tail.

Like storks, ospreys build large, messy nests of sticks, and the same nest is occupied year after year, presumably with a few renovations. Dead spruce trees are favorite spots, and so are aids to navigation. One of the biggest osprey nests along the coast is on a can buoy at the turn of the Sheepscot River near Wiscasset. Another nest familiar to yachtsmen has been at the mouth of Pulpit Harbor as long as anyone can remember. Scientists have dated a nest on Sutton Island, and in 1995 it became 100 years old.

Ospreys fish tirelessly all day and carry their catch back to the nest full of noisy, insatiable young. The osprey defends its nest vociferously, screaming "Kee, kee, kee" and diving at intruders.

It is crucial to avoid disturbing an osprey during nesting season (late April until early July), because the chicks may be abandoned as a result.

Not so long ago, fish hawks had disappeared almost entirely from the coast of Maine. The pesticide DDT accumulated in the food chain reaching heavy concentrations in fish, the birds' total diet. Subsequently, osprey chicks were deformed or failed to hatch. Thanks to the alarm raised by the 1962 publication of Rachel Carson's *Silent Spring*, the life-threatening pesticides and other chemicals are no longer legal, and the osprey has been saved.

Large as it is, the osprey has enemies other than man. Crows are fond of osprey eggs, and bald eagles have the strength and flying ability to rob ospreys of their catch in midair—a spectacular duel of giants that is infrequent but fascinating to witness.

ANCHORAGES, MOORINGS. The head of navigation is a fixed bridge just beyond the docks of Thomaston. The protection is excellent, but with the strong current in the river and that fixed bridge nearby, you may be uncomfortable on an anchor.

For limited periods, dock at the Thomaston town dock. The Harbor View Restaurant and Marina, just to the west, may have overnight dock space. They also have three rental moorings. In a pinch, Lyman-Morse has two small rental moorings for boats less than 25 feet, but they might be able to find larger boats space on their moored service docks.

GETTING ASHORE. Row in to the town dock.

FOR THE BOAT. *Lyman-Morse Boatbuilding Company (Ch. 09; 207-354-6904; www.lymanmorse.com).* The St. George makes a sharp bend to the west in Thomaston. Most of the buildings on the north shore before the town docks belong to Lyman-Morse. This is the yard that first became famous building John Alden's Malabar schooners and the Friendship sloops. Since then their reputation and repertoire has grown to include custom sail and power craft of the highest caliber.

The yard can haul all sizes of boats and make repairs of all kinds. Their service floats near the eastern end of the waterfront have 10 feet alongside at low. Pump-outs and showers are available.

Thomaston Town Dock (harbormaster, police 207-354-2511). The town dock is obvious to the west of Lyman-Morse, past a sloping wall of granite blocks. Docking is limited to 30 minutes on weekends and to three hours on weekdays. The town landing has public restrooms.

Harbor View Restaurant and Marina (207-354-8173). This long dock is west of the town floats.

Jeff's Marine Service (Ch. 09; 207-354-8777; www.jeffsmarine.com). Jeff's is hidden on the left side of the harbor just before the fixed bridge. They have floats with water and electricity and 8 feet of water alongside at low. The yard specializes in the sale and repair of outboards and stern-drives and has a related chandlery.

Midcoast Marine Electronics (207-354-0012) is at 70 Water Street, at the head of the harbor.

FOR THE CREW. The Harbor View Restaurant sits just west of the town landing at the end of a long pier. Patrons can lie alongside their floats while they gam at the bar or eat dinner. This is a casual, boaty kind of place, with a rudder on the wall and a sign that reads "If you don't stand for something, you'll fall for anything."

A half-mile walk up the road leads to Main Street where you will find a post office, drugstore, laundromat, bank, and a small market.

THINGS TO DO. Just up Knox Street from the waterfront is the museum of the Thomaston Historical Society. It is housed in the last of nine brick buildings that formed a semicircle at the rear of Montpelier, Major General Henry Knox's mansion built on this spot in 1795.

Knox, a military historian and tactician and the Commander of Artillery under George Washington in the American Revolution, was Thomaston's most renowned citizen. He served as the fledgling nation's first Secretary of War. Through his marriage to Lucy Flucker, Knox acquired large landholdings in Maine and built his splendid mansion here. Knox's demise was less noble—he choked to death on a chicken bone in 1806.

Montpelier fell on hard times and was torn down, but a replica was built east of the town in the 1930s. The magnificent building is open to the public and contains original furniture and other Knox memorabilia. The 1.2-mile walk is well worth it.

The route to Montpelier takes you up Knox Street, lined with beautiful old houses, then to Route 1 where you will pass the town library. Farther along Route 1 is an imposing cemetery, a good place to make some unusual grave rubbings. One of the stones is for the "Captain of the unfortunate *Pacific*," and one is for the first officer of a ship lost a sea between Le Havre and New York. Edward O'Brien stands there full size in stone, wearing his frock coat and bow tie, leaning on an anchor, and looking out over a garden of lesser relatives.

The walk eastward along Main Street also passes a wonderful row of handsome homes built by merchants and sea captains. Now designated a National Historic District, this area's homes reflect the sophistication of world travelers. Their European-inspired architecture ranges from Federal and Greek Revival to Gothic, Italianate, and French Second Empire—splendid reminders of those days of glory.

Certainly explore the waterfront. Each of the many boat sheds of Lyman-Morse contains fascinating high-end work in all phases of yacht construction. The modern classics *White Hawk* and *White Fin* were also born in this understated harbor when Renaissance Yacht Company was located in the gray building at the western end of the river. If you are inspired to do what you know you should do, the headquarters for Epifanes varnish is in the metal building next door.

PORT CLYDE

43° 55.52'N 069° 15.61'W

CHARTS: 13301, 13288   **CHART KIT:** 64, 18, 19

Port Clyde is easy to enter and a reasonably good harbor. It can be miserable, however, in strong winds from the southwest or northwest. It is primarily a working harbor for draggers and fishermen, whose picturesque houses and wharves line the northern shore.

This is also the starting point for the *Laura B.* and the *Elizabeth Ann*, which service Monhegan Island. Although the ferry traffic generates good business for local residents, the parking lots for Monhegan passengers threaten to overrun the small town. The old-fashioned and delightful Port Clyde General Store is the center of activity.

Before the Revolution, the channel inside Hupper Island bore the euphonious name of Lobsterfare. The pragmatic British changed it to Herring Gut at about the time the town was first settled. In 1891, the town's name became Port Clyde.

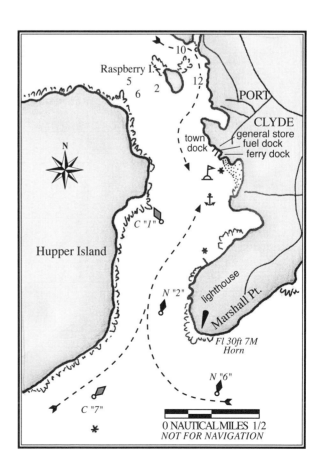

Shipbuilding was the major industry here during the 1800s. When that declined, the town turned to canning lobsters, clams, and fish, and finally back to lobstering and fishing.

APPROACHES. Approaching from the west or south, there are two usual routes. The first is through Davis Strait between Thompson and Davis Island. This passage is narrow but straightforward. Or, if you are outside Allen Island, you can approach Port Clyde by running northward between Allen and Burnt Island, leaving Dry Ledges to port and more or less following the cable area shown on the chart to green bell "11."

From both routes, leave green bell "11," can "9," and can "7" to *starboard* (they mark the east-west route). From can "7," head north into the harbor, where the marks follow the usual red-right-returning: nun "2" to starboard and can "3" to port.

There is a northern entrance to Port Clyde between Hupper Island and Hupper Point, which can be used by most boats except at dead low because of the 5 and 6-foot spots between Hupper Island and Raspberry Island, the small, wooded island to the northeast, unnamed on the chart. Come through slowly on a rising tide.

There is also an exciting back-door entrance with deep water between Raspberry Island and the mainland, but it is difficult and should be avoided without local knowledge. A ledge extends northeastward from Raspberry 150 feet or more beyond the portion of ledge visible at low tide. The trick is to stick to within 15 or 20 feet of the docks on the mainland and to creep along with a lookout on the bow.

Coming from Tenants Harbor or points farther east, the passage is easy and well marked. Turn at red bell "2" south of Mosquito Island and pass north of The Brothers, keeping can "5" to port and nun "4" to starboard. Turn northward into the harbor around nun "6" off Marshall Point and the Marshall Point Light. This lighthouse has a white tower, and the distinctive bridge leading to it looks like a small Roman aqueduct.

ANCHORAGES, MOORINGS. The Port Clyde General Store puts out 25 heavy rental moorings (inquire at the store). There is plenty of room if you prefer to anchor, but the depths in midharbor are substantial at 26 to 35 feet at low.

GETTING ASHORE. Land at the Port Clyde General Store floats to the left of the large wooden wharf or land at the town float just to the left of that.

FOR THE BOAT. *Port Clyde General Store (Ch. 09, 28; 207-372-6543).* The docks of the Port Clyde General Store lie between the Monhegan boat landing to the southeast and the town float to the northwest. The floats have 10 feet of water at low and gas, diesel, water, and electricity. Tie-up time is limited to 20 minutes. Approach with caution. The docks are generally busy with fishing boats, the Outward Bound launches from Burnt Island, and pleasure boats of all kinds.

If you need fuel, beware that the channel between the float and the stone fill to the left is narrow and that the outer end of the float is prone to tidal currents that can veer your boat as you approach.

Town Landing. The town landing, to the west of the floats of the General Store, has short-term dockage on the front of the float and dinghy space on the back. There are no other facilities.

FOR THE CREW. The Port Clyde General Store stocks almost anything you need, from groceries and ice to lobsters, newspapers, and wine. Their lunch counter makes sandwiches, pizzas, chowders, and baked goods. The Dip Net takeout can set you up with pizza, crab rolls, or lobsters on the wharf.

The General Store has a shower that is intermittently available to boaters, depending on the state of the septic system.

There is a pay phone at the Monhegan Boat Landing. Your kids can get jazzed at the Village Ice Cream and Bakery, up the hill from the General Store. Then they can burn it off at a welcome little playground just beyond the small post office. Left down the dirt road by the playground is the Fishermen's Co-op.

THINGS TO DO. In any town with a ferry, one of the principal occupations and greatest pleasures is to go down to the dock and watch the boats loading or unloading. This is particularly true in Port Clyde, where the *Laura B.* and *Elizabeth Ann* load up for the Monhegan trip. Or, if you have the time, make the trip yourself. See *Monhegan*, p. 166.

For a pleasant walk of a mile each way, take the first right along the shore to the Marshall Point Light. The view is spectacular. The lighthouse dates from 1833, and the light keeper's house has been restored as a historical museum.

Penobscot Bay

Marshall Point to Isle au Haut

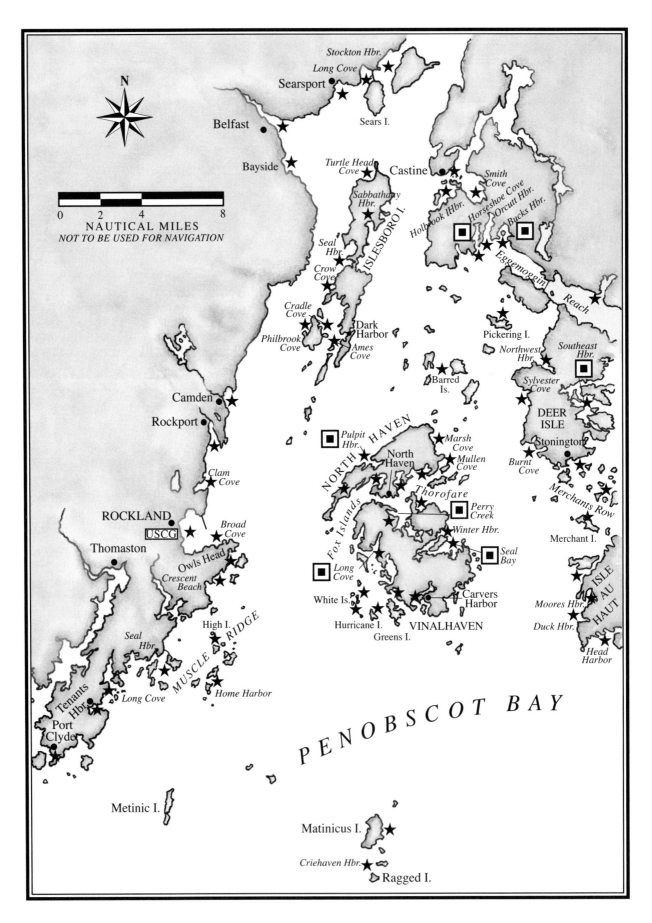

> "It's a beautiful, beautiful bay!
> Spruce-covered headlands jut boldly from its shores,
> jewel-like islands float on its surface,
> and the gentle, glacier-rounded contours of the Camden Hills look down on its broad reaches."
>
> —Louise Dickinson Rich,
> *Coast of Maine.*

DREAM of perfect cruising grounds, of islands large and small, grand and modest, of intriguing harbors and alluring towns, of broad reaches and narrow tickles, of gritty fishing villages and sophisticated summer resorts, of lonely outposts lost in time. There is such a place, and the place is Penobscot Bay.

Against the mountainous backdrop of the Camden Hills to the west, Penobscot Bay spreads 40 miles long and 15 miles wide, graced by more than 200 islands. Between them lie great stretches of open water and small, winding thoroughfares. Along their shores are bustling fishing communities, quaint villages, isolated outposts, and uninhabited beauty. This is the heart of cruising in Maine, and some of the best cruising in the world.

The winds are generally moderate and predictable, and the dangers are well marked. Here there are gentle sails with sunsets behind the Camden Hills, or exhilarating passages, rail down, surrounded by an ever-shifting scene of dark islands and distant headlands. There are winding thoroughfares to thread, endless gunkholes to explore, and a hundred harbors tucked away.

At the entrance to the bay are the outlying islands, remote and hard to visit: lonely, sea-swept Matinicus Rock, where puffins fly between your masts at sunrise; Ragged Island and Matinicus, the most seaward communities on the coast of Maine; and Metinic, Green, Seal, and Wooden Ball Island.

The shores of Penobscot Bay form a variety of interesting harbors, from safe and welcoming Tenants Harbor to the very different towns along the western shore—the vast harbor of Rockland, small and charming Rockport, and beautiful Camden nestled at the base of the Camden Hills.

Camden Harbor holds the largest fleet of windjammers on the Maine coast. Seeing these majestic vessels under the press of their sails is another joy in cruising the bay.

Tankers, tugs and tows, cargo ships, and container vessels steam northward up Penobscot Bay to the shipping ports of Searsport and Sears Island. A century ago, Belfast and Searsport were homeports to ships that sailed for every corner of the world and home to hundreds of sea captains.

Bangor lies up the Penobscot River, 24 miles past Fort Knox. Originally, it was a brawling frontier town and the lumber capital of the world. Castine sits strategically near the eastern head of the bay along the mouth the Bagaduce River. It was once a stronghold of the French, then later the British, and it was where America suffered its first great naval defeat during the Penobscot Expedition. It is now the home of the Maine Maritime Academy.

Bucks Harbor, Little Deer Isle, Deer Isle, and the broad passageway of Eggemoggin Reach lie south of Cape Rosier. On the eastern side of the bay, the fishing town of Stonington looks out on Deer Island Thorofare and the magnificent islands of Merchants Row. And at the eastern entrance of Penobscot Bay stands rugged Isle au Haut, much of which is part of Acadia National Park.

The islands of Penobscot Bay are as interesting as its shores. At the western entrance lies the archipelago of Muscle Ridge, once important for its granite quarries, now sparsely settled by fishermen. Among these islands and ledges are two or three anchorages that have hardly changed since Indian times.

In the center of the bay are the Fox Islands—the twin islands of North Haven and Vinalhaven. Separated by the narrow, winding, and altogether delightful Fox Islands Thorofare, they are separated also by a wider gulf. North Haven is a fashionable community settled long ago by Boston yachtsmen. Vinalhaven is a working island, where Carvers Harbor is the base for lobstermen, seiners, and draggers. Around the convoluted shores of the Fox Islands are some of the best harbors in the bay—Pulpit Harbor, Perry Creek, Winter Harbor, and many more.

South of Vinalhaven, volcanic Brimstone Island stands all alone with twin crescent beaches of polished black stones. And at the southwest edge of Vinalhaven is Hurricane Sound, with five separate entrances and a multitude of islands. Hurricane Island Outward Bound School has their island base here, and their students roam the coast of Maine in open, ketch-rigged pulling boats.

Another scattering of islands lie beyond North Haven—Pond and Pickering, Butter, Eagle, Great Spruce Head, and the Barred Islands, to name a few. Dividing the northern part of Penobscot Bay is Islesboro, a long and lovely island also settled by Boston Brahmins, who established the fashionable summer and sailing community of Dark Harbor a century ago.

Lime was quarried on the western shores of Penobscot Bay for plaster and for mortar. Granite was cut in dozens of quarries, from Dix Island to Stonington. It was shipped in the holds of stone sloops and steamers to build the banks and post offices, museums and monuments, breakwaters, bridges, and cobblestone streets of Boston, Washington, Philadelphia, and New York.

Penobscot Bay has always been inextricably linked to the sea—with building ships, with shipping, with fishing, with lobstering, and with those who come here to enjoy it by boat.

THE OUTLYING ISLANDS

Metinic and the Green Islands, Matinicus and Ragged, Wooden Ball and Seal Island are scattered for 20 miles across the mouth of Penobscot Bay. Beyond them all, standing as the first rampart to the sea, is lonely Matinicus Rock, whose lighthouse has guided mariners into Penobscot Bay since 1827. These rocks and islands are the ends of mountain ranges depressed by the glaciers long ago, then drowned by the returning sea.

Ocean islands are different from bay islands. Each stands isolated, exposed to unending waves that sweep across 3,000 miles of ocean to crash upon their shores. Even on the calmest days, swells break in spray and spume on every ledge, and in the worst gales, great waves have broken entirely over Matinicus Rock.

The people, too, are different here. Like the ocean islands themselves, they stand alone—tougher perhaps, and independent to a fault. Most were born here and stay because they love the islands, or perhaps because they have no choice.

Of all these outlying islands, only one—Matinicus—has a permanent year-round settlement, serviced in the winter by one monthly ferry from Rockland. Ragged Island was inhabited for generations, but now it supports only a small seasonal community of fishermen and summer people. Lonely Metinic is home to a few staunch fishing families.

Peregrine falcons, soaring over these barren islands, glide down to land on Wooden Ball and Matinicus Rock during spring and fall migrations. The National Audubon Society is trying to reestablish a puffin colony on Seal Island. The other islands are visited occasionally by fishermen, naturalists, Outward Bound students and sometimes by intrepid yachtsmen.

METINIC ISLAND

2 ★★★

43° 52.44′N 069° 07.48′W

CHARTS: 13303, 13302   CHART KIT: 64, 19

Metinic is an Indian word meaning "far-out island." In 1750, enterprising Ebenezer Thorndike took a 99-year lease on Metinic Island from the Abenaki Indians. For 250 years, the island and its waters were farmed and fished by his descendants. Metinic is long and low. It has woods in the middle and meadows at both ends. A group of homes huddles toward the southern end, and two unoccupied houses stand gray and lonely to the north.

Now it is largely owned by the Post family of fishermen, although in 1994 and 1995 the U.S. Fish and Wildlife Service acquired nearly half the island to preserve seabird-nesting habitat on the northern end. This remote outpost is the home of eiders, gulls, guillemots, arctic terns, 150 white-faced Metinic sheep, and fishermen as self-reliant as any in the world.

Walter Wotton, who grew up on Metinic, remembers his father rowing him to church in Tenants Harbor every Sunday.

The flock of wild, white-faced sheep has been here since the Civil War, roaming the island at will and living off the land. The rams are kept on a neighboring island until February so that lambs are born in the late spring. Then the sheep are driven the length of the island into a corral for shearing.

This is a hard place for a cruising yachtsman to visit. Attempt it only during settled weather and not when the wind is from the east or the north.

APPROACHES. From the north, the island appears as a smooth, rounded meadow with a fringe of dark trees. Head for green bell "1W" off the north end of Metinic Island, then run down the east side until you are opposite the houses at the narrow waist of the island. Stay far enough out to be well clear of Cat Ledge. Identify The Nubble, which is small and rocky, and Hog Island, which is larger and grassy. Head north of The Nubble in toward the little cove on Metinic's eastern shore.

From the south, be aware of unmarked Southeast Breaker, well to the southwest of Metinic. Identify The Nubble and round north of it.

ANCHORAGES, MOORINGS. Do not enter the little cove without local advice. There are ledges at each corner of the entrance and not much water inside. Hail a fisherman and ask if a mooring is available. Sometimes there are one or two heavy stake moorings north of Hog Island, in 13 feet of water at low. The one "to nor'ard" has a 4-ton block, and the other two have 2-ton blocks. You probably will need to rig a pennant.

Hog Island, The Nubble, and Metinic together provide pretty good protection in normal summer weather, but if the wind comes in from the north or east, *leave*.

Anchoring temporarily is fine, but you probably will not be happy spending the night here except on a big mooring.

GETTING ASHORE. Take your dinghy in to the float, which dries out at low.

THINGS TO DO. This is a private island, but you are welcome to go ashore (ask permission from one of the fishermen). There are no public facilities of any kind.

Walk through the red spruce woods to the north end and contemplate the lonely lives of the families who lived in these old houses, exposed to the fury of the northern gales. Metinic is a major nesting area for many seabirds, especially eiders, until July 15, so step lightly and stay on the trails.

MATINICUS ISLAND

3 ★★★★★

43° 51.81′N 068° 52.88′W

CHARTS: 13303, 13302   CHART KIT: 19, 21

Matinicus is the outermost Maine island inhabited year round. With its true remoteness, air of independence, and long occupation by the same families, Matinicus is reminiscent of other extraordinary islands like Tangier in the Chesapeake and lonely Pitcairn. Here, even on the brightest and calmest days and in the best of company, everything is muted by a sense of isolation, and everywhere there are forlorn reminders that existence here is a daily struggle. Now imagine it during a bleak midwinter storm.

Islander Donna Rogers writes, "I guess what Matinicus offers most is her stubborn refusal to keep up completely with the rest of the world. There are no factories, no traffic, no noise, except for the sound of waves breaking on shore or the bell buoy on a blowy day."

Even the government of Matinicus is remote. Classified as part of Maine's wildlands, Matinicus is considered a Plantation of the State and administered by the Land Use Regulatory Commission in Augusta. According to one islander, "that's about as far away as you can get." Most, no doubt, would like to keep it that way.

This is a beautiful and fascinating place to visit, but also difficult. Choose your time carefully. It can be uncomfortable and dangerous in heavy weather, and frequent fogs arrive with little warning.

The local fishing fleet numbers 20 or 25 boats. On a given summer day, there may be half a dozen visiting yachtsmen, or you may be the only one. Don't expect to be greeted warmly, but Matinicus men, in the best tradition of the sea, keep a watchful eye on all passing or visiting vessels. If you are in real trouble, you can always count on them for help.

In the winter of 1992 on a subfreezing night in near-zero visibility, the ocean tug *Harkness* mysteriously began taking on water offshore. Hearing the distress calls over the VHF, several intrepid fishermen put out from Matinicus and miraculously located and pulled from the water the near-frozen crewmembers, guided by a flashlight that had frozen to the hand of one of the crew.

The central occupations of the island are lobstering and fishing, and the islanders like it that way. "We don't want to become like Monhegan," they say. And they will defend that livelihood at all costs. In 2006, a "lobster war" over turf included playing chicken with lobsterboats and a shotgun blast across the bow.

Matinicus is almost two miles long and a mile wide, with one good harbor on the east side. Its 750 acres are covered by meadows, woods, and white sand beaches of great beauty. The fields are alive with wildflowers that last into October.

A state-run ferry services Matinicus from Rockland, but it runs only four times a month in summer and once a month in the winter, and then it can't unload when the tide is down. "It's always low tide on Matinicus," says an islander waiting for the ferry, which is waiting for the tide. Instead, contact with the mainland is mostly by plane or private boat. Air-taxi service, provided by Penobscot Island Air (*596-7500*) from Owls Head, is an important lifeline, but its natural limitation rolls in in the form of fog. Captain George Tarkleson (*691-9030*) operates a private ferry service to the mainland and back.

Winter population varies from 35 to 50 and grows to 150 or more during the summer. Most important to the survival of Matinicus as a year-round community is the small one-room school, which sometimes has seven children, sometimes nine, and occasionally only three.

APPROACHES from the west. Once clear of Bantam Ledge, marked by black-and-red "DBL," head through Matinicus Roads, between Matinicus and Ragged Island. At the eastern end of the Roads, favor Tenpound Island (grassy on top) to avoid The Hogshead rock to the south. Identify Wheaton Island and West Black Ledge to the north and pass between them to the red-and-green bell outside Matinicus Harbor (*43° 51.92'N 068° 52.74'W*).

APPROACHES from the south. If you are approaching from Matinicus Rock, give the east coast of Ragged Island a wide berth as you head north for Matinicus. Pass Matinicus Roads and Tenpound Island and then head between Wheaton Island and West Black Ledge. Both East and West Black ledges are high and easy to identify by the surf breaking on them. Note the shoal spots at the southeast end of Wheaton Island. From West Black Ledge, steer for the red-and-green bell outside Matinicus Harbor.

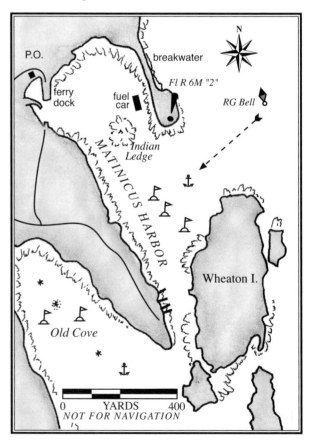

APPROACHES from the north. Set your course for red nun "2" (43°55.83'N 068°53.13'W) north of Matinicus and then to flashing green buoy "5" off Zephyr Rock, kept to starboard. Coast past barren, rocky No Mans Land—a forbidding place even in sunny weather and loud with herring gulls—to can "5" north of Mackerel Ledge. Leave this can to port.

From can "5," a straight course to the red-and-green bell at the mouth of Matinicus Harbor will clear The Barrel, a small rock, 10 feet awash at low, on which the swells break with enthusiasm. Note Harbor Ledge northeast of the bell.

Keep to starboard the red-and-green bell just outside the harbor and run directly through the middle of the entrance between the rocky tip of Wheaton Island to the south and the small, lighted beacon near (but not on) the end of the breakwater to the north.

ANCHORAGES, MOORINGS. ⚓ Just inside the harbor on your line of approach, local fishermen have put down heavy nylon cables to which are attached a number of rental moorings managed by Joshua and Ronnie Ames. Most have lobster buoys as pennant floats, but their color schemes vary depending on which lobsterman is fishing for yachtsmen. At this writing, Josh's pickup buoys are white with a blue stripe; Ronnie's are orange. If you are in doubt, ask any fisherman in the harbor. Expect to pay a premium for the convenience of a mooring in one of the least convenient places on the Maine coast ($25 and $20, respectively). You can reach Josh Ames on Ch. 19, aboard *Enterprise*, at 366-3937, or at www.islandmooringrentals.com.

There may be room to anchor just west of the moorings or in the entrance of the harbor (see sketch chart), but the bottom is hard sand and not good holding ground. Do not enter the area between Wheaton Island and Matinicus, where mooring cables stretched across could snag the keel of a deep-draft sailboat.

Protection is good here in settled summer weather, with the only real exposure to the east. The ocean swells, however, may treat you to a rolly night. If an easterly is predicted, be prepared to leave.

Most of the fishing fleet is moored at the north end of the harbor behind Indian Ledge, near the state ferry dock. It is possible to pass to either side of Indian Ledge, but the water inside is shoal, so this is no place for yachts.

Lying at anchor here is not the same as in the protected islands of the bay. During the night you will see the sweep of Matinicus Rock Light through the slot inside Wheaton Island, and the sounds of the sea are loud and relentless.

Old Cove. If the harbor at Matinicus is full, it's possible to anchor in Old Cove, just around the corner to the south. This seldom-used anchorage offers reasonable protection from southwesterlies but is open to the southeast. Ledges line the western shore, so stay well over to the east and do not go much beyond the first house on the eastern shore. ⚓ Josh has several rental moorings here too, but don't confuse them with an orange polyball that marks a rock. Stay to the east of the rock.

GETTING ASHORE. Landing your dinghy is difficult here. There are no floats in the harbor. The best solution is to run the dinghy up the beach just west of the ferry dock to land passengers and cargo. Then tie up to a ladder near the end of the ferry dock.

Do not bring trash ashore; there is no place to leave it. There are no public toilets on the island.

FOR THE BOAT. If you are desperate, ⛽ fuel may be obtained at the lobster car near the breakwater on the eastern side of the harbor. If you need engine repairs, ask around. Matinicus lobstermen are experts.

FOR THE CREW. Come to Matinicus with everything you need. Businesses that start up on Matinicus cling tenuously to economic viability. A post office and pay phone are located near the ferry wharf, but there is no market. Residents call in large orders to a Rockland supermarket, and they are shipped to the island by air. Bill Hoadley runs the Tuckanuck Lodge bed and breakfast. "I tell my guests that breakfast is included, but they sometimes get a little taken aback when I tell them they might want to bring their own lunches and dinners."

🛒 Eva Murray runs a small bakery (*366-3695*), near the center of the island that sells her homemade goodies. No problem with confusion here: it's called The Bakery. MKM Island Lobster (*Ch. 19; 366-3985*) can set you up with some local critters.

🍴 Matinicus Island women may serve lobster suppers in their homes Friday or Saturday nights—you will find notices on the small bulletin board by the post office or on various phone poles.

If you need transportation, call Mermaid Taxi (*Ann Mitchell, 366-3161*), or the Tuckanuck Lodge (*366-3830*) will rent you wheels with pedals.

THINGS TO DO. As Bill Hoadley puts it, "As soon as they ask what there is to do, I know they aren't going to be happy here." Walking is a delightful activity on Matinicus, though locals note with irony that "on Matinicus, everybody drives." And fast. It is almost as if driving were a way to keep pace with the rest of the world. Some say that Matinicus is the Indian word for "Land of the Rusted Automobile." Even after a major wreck removal effort, the island still boasts a fleet of cars that seem to defy the physics of oxidation and gravity. You won't be able to walk far before you are offered a ride.

The main road runs down the center of the island. Paths lead along the ledges and through the woods, and everyone is free to walk them. You are welcome to pick strawberries, blueberries, and raspberries along the roadside, but stay off private property.

Makries Beach is at the northeastern end of the island, and the magnificent crescent of South Sandy Beach, at the south end, looks out to Criehaven. Birding is good on Matinicus, especially during spring and fall migrations. Captain George Tarkleson (*691-9030*) runs puffin-watching trips to Matinicus Rock aboard his Albin *Robin*.

MATINICUS: "A DIFFERENT BREED"

Welcomes from Tiffany and Sheba.

In *Matinicus Isle: Its Story and Its People* (1926), Charles Long tells the story of Ebenezer Hall, the first permanent settler, who landed on Matinicus in 1751. Hall was not well received by the Penobscot tribe of the Tarratine Indians, who resented his presence in their traditional fishing and birding lands. They complained to the governor when Hall burned over Green Island to improve the hay, thus disturbing the habitat for eiders, which the Indians relied on for food and feathers. But Hall continued the practice.

In June of 1757, a band of Indians laid siege to Hall's house on Matinicus for several days. It ended with the scalping of Hall and the capture of his wife and daughters. Mrs. Hall was finally ransomed in Quebec, but her four daughters were never heard from again. A son, Ebenezer, who escaped only because he had been out fishing, returned to live on the island in 1763 and married Susannah Young. Many of the islanders today are descendents of this couple.

In an interview in the annual *Island Journal*, Dorothy Simpson of Criehaven reminisced, "Matinicus ancestors go way back to the Revolution and before that, even to the Indian Wars. There's something grown into them I think. They are a different breed… The people fish there and work there and their kids keep coming along in the steps of their fathers… They say, 'My grandfather fished over there and my father and my uncle and I got a right.'"

Much of the land on Matinicus has been owned by the same families for 200 years. Ames, Young, Philbrook, Bunker, and Ripley are some of the familiar names. There are more recent settlers, but fishermen still outnumber rusticators.

"Matinicus, eh? They have their own sense of justice out there. They're either the best people you ever met or the worst, depending on how they accept you," says one observer. "Don't cross 'em. Somebody was cutting lobster traps and one of the boys flew over his lobsterboat in his airplane and dropped this great big field stone right through the boat. Sank him.

"And don't fool around with the things they love, like churches and graveyards. A man threw a rock through the window of *Sunbeam* once, and before you knew it, he was off the island."

Sunbeam is the 65-foot vessel of the Maine Seacoast Missionary Society in Bar Harbor. It is important to this community, and its monthly winter visits are much anticipated. In addition to sponsoring religious and medical services and a scallop dinner each spring, *Sunbeam* is relied on for marriages and burials. Many are born, live all their lives, and die on the island. Then *Sunbeam* takes them to a funeral home in Rockland and returns them to Matinicus for burial. One elderly lady requested in her will that she be spared the rigorous trip to the mainland. Her wish was granted.

On a ledge near the post office, there is a plaque dedicated to Ebenezer Hall, the first white settler on Matinicus, killed by Indians June 6, 1757. Curiously, though, the Penobscot tribe wrote Spencer Phipps, Governor of the Massachusetts Bay Colony, four years earlier, politely requesting that he remove the "Englishman on the Island," who was encroaching on their livelihood. "If you don't remove him in two months we shall be obliged to do it ourselves," it read. Needless to say, the letter went unanswered.

RAGGED ISLAND (Criehaven)

3 ★★★★★

43° 50.08′N 068° 53.48′W

Charts: 13303, 13302	Chart Kit: 19, 21

Even more remote than Matinicus, beautiful Ragged Island is the farthest outpost on the coast of Maine. There's no store, no school, no post office. There's hardly any way to get there and no place to stay. It is possible, however, to visit by boat.

Lovely walking trails wind through woods and meadows, wildflowers bloom late into the fall, and the birding is excellent. This was "Bennett's Island" of Elizabeth Ogilvie's famous trilogy, *Storm Tide, The Ebbing Tide,* and *High Tide at Noon.*

The Indians called it *Racketash*, which the British later corrupted to "Ragged Arse." Prudish American cartographers later shortened the name. After the Revolution, a British soldier named John Crie retired to live on Matinicus. His grandson Robert set up housekeeping on neighboring Ragged Island in 1848, and for half a century he was the island's leading citizen and entrepreneur. Like many Maine islands, Ragged produced hay and lumber, sheep and wool, and fish for the city folk in Boston—all under Robert Crie's jurisdiction. Eventually, he bought up the whole island and was known locally as "King Crie." At the head of Criehaven Harbor, the path to the right leads to the dark and overgrown cemetery where Crie is buried.

The fishermen of Ragged Island tried for decades to get federal funding for a breakwater for their harbor. Matinicus got its breakwater in 1912, but Criehaven had to wait until 1938. It was breached during the 1978 Groundhog Day storm, and considerable damage was done to the boats and docks.

Ragged Island was abandoned by its year-round community in 1941. As George Putz wrote in the annual *Island Journal*, "Though island communities are elegant in their social orders, they are for the same reason more prone to reverses brought about by only one or two events. The death of an influential man, the loss of a mailboat, a teacher, a store, or a breakwater, may have influence all out of proportion to comparable events in mainland communities where more options are available."

Today the families of about 10 lobstermen live and fish here in the summer. Another 10 families summer at Criehaven in relative harmony with the fishermen. The rest of the island is well cared for by generous landowners and their caretakers.

The lobstermen prize their independence and the good lobstering grounds. By unwritten law, one of the lobstermen has to relinquish his rights before another lobsterman can start fishing here. They work out of Criehaven 10 or 11 months a year, usually returning to their mainland homes and families on weekends.

To row ashore in Criehaven on a fall day is to enter a ghost town. There are dinghies at every mooring and fishing gear piled on the surrounding docks. Little clapboard houses surround the harbor, but there is no visible movement—no dogs, no women and children, no sound except the gulls. In summertime, there is activity, of course, but, as a Mainer would say, "It's real peaceful."

APPROACHES. A 6 or 7-foot draft can be carried into Criehaven Harbor at low.

Coming from the west, clear Bantam Ledge, which is marked by red-and-black "DBL," and run into Matinicus Roads between Ragged Island and Matinicus. Nun "6" marks Harbor Ledges from the west and north. Harbor Ledges are covered two hours after low, but swells will continue to break on them. Give them plenty of room before you turn south toward the harbor—a straight course from nun "6" to the harbor entrance passes right over the ledges. Leave the breakwater, marked by a beacon, to starboard as you enter the harbor.

Coming from Matinicus Harbor, run south past West Black Ledge and turn westward between Ten-pound Island (long, rocky, grass on top) and The Hogshead, a tall and dark rock, with swells breaking heavily. Leave Pudding Island to port (35 feet high, rounded and cliffy, with a few windblown trees). Criehaven Harbor and the granite breakwater are hidden at first (except for the beacon), but they will open up as you coast westward through Matinicus Roads. Pass the 7-foot sounding west of Pudding Island before turning south into Criehaven Harbor. Leave the breakwater to starboard.

ANCHORAGES, MOORINGS. *Criehaven.* The harbor is very small and full of lobstermen's moorings. These are secured to cables laid on the bottom—four cables with three moorings on each. There is little or no room to anchor outside the moorings, the bottom is reported to be solid granite ledge, and the whole harbor is an *electric* cable area. Somewhere near the mouth of the harbor there is a pile of rusting metal junk, dumped by the lobstermen to snag the gear of draggers. This is not a healthy place to anchor.

If you want to spend the night, pick up a mooring near the entrance, wait until a fisherman appears, and ask him if there's an empty mooring with enough water and swinging room. If you go ashore, leave someone aboard who can move the boat. The harbor is well protected except from the north and northwest.

Wilson Cove. Wilson Cove, on the north side of Ragged Island, is unnamed on the chart. In summer weather it can provide a possible anchorage south of Pudding Island, protected from ocean swells and the prevailing southerlies. There is a big, grassy field above the cobble beach, and trails lead to Criehaven.

Pass inside either Pudding Island or Shag Ledge to reach the cove. A ledge makes out 50 yards from the eastern shore, so favor the deeper water in the western part of the cove. Stay a good distance off the beach, which slopes gradually. The bottom is cobblestone; use a kedge anchor and not a Danforth.

GETTING ASHORE. The wharves in Criehaven Harbor are high and private, and few of them have floats. To get ashore, row your dinghy in to the little beach at the head of the harbor and tie up to the heavy mooring cable that comes ashore at the right end of the beach.

THINGS TO DO. Criehaven is a welcoming place, but please be respectful. Avoid private lawns and infringing on the privacy of the islanders. Please do not light fires or smoke. As you walk the narrow paths from Criehaven toward Seal Cove, you can see Matinicus Rock in the distance. A path runs down the eastern shore, then cuts across to the western shore and down to the southern tip of the island, passing through a grove of yellow birch. Wildflowers and birds are everywhere.

MATINICUS ROCK
43° 47.10'N 068° 51.40'W

CHARTS: 13303, 13302   CHART KIT: 19, 21

Matinicus Rock is an impressive place—"the outermost, loneliest rampart," as photographer Eliot Porter put it in *Summer Island.* Here great swells surge and break on the cliffs as they did before the memory of man. Coming from the sea, the rock is the first outpost of land at the entrance to Penobscot Bay, an ideal landfall for early explorers and modern mariners alike.

A lighthouse was first built here in 1827. It was replaced again and again. Today's stone tower shows a powerful light 90 feet above the sea, visible 23 miles, with a foghorn and radio beacon. The last pair of native puffins chose Matinicus Rock for nesting in 1908, but they have since been reintroduced. Scientists and birdwatchers come here to see puffins, terns, petrels, auks, and other rare species.

The Coast Guard facilities on Matinicus Rock were manned until automation in 1983. In 1999 the island was transferred to the U.S. Fish and Wildlife Service, which manages it jointly with the National Audubon Society to preserve nesting habitat for colonial seabirds. You may encounter a group of Island Institute volunteers camped on the rock, counting migrating peregrine falcons, or Audubon members "puffin-grubbing"—crawling beneath the island boulders to find puffin chicks for weighing, measuring, and banding as part of their Puffin Project (*www.puffinproject.org*). They are also attempting to reintroduce the common murre—which is not common at all—by attracting it with decoys. The only other nesting site for this penguin-like bird is on Murre Ledge, 75 miles away.

In bad weather access to the island by sea is cut off altogether. Landing is extremely difficult even in normal weather, so don't try it except in a dead-flat calm. Go ashore on the western side, near the Coast Guard station. The instructions given in the 1891 report of the Lighthouse Board still apply: "The lightkeeper effects a landing by steering his boat through the breakers on top of a wave so that it will land on the boat ways, where his assistants receive him and draw his boat up so far on the ways that a receding wave cannot carry it back to the sea."

To avoid disturbing the nesting birds, do not go ashore at all before July 15. You can sense the power and presence of this lonely outpost from your boat without having to land.

The twin light towers on lonely Matinicus Rock, 25 miles offshore, in 1904. Today, only one tower is in use, automated in 1983. Landing here is always difficult, and impossible in heavy weather. (Frank Claes collection)

Samuel Burgess and his family are perhaps the best known of the lightkeepers who lived on Matinicus Rock. Burgess spent much of his time away fishing, and he depended on his daughter, Abbie, to maintain the light while he was gone.

Abbie described the great gale of January 19, 1856 in a letter to a friend when she was 17 years old. Her mother was sick, and her father had rowed (yes, *rowed*) to the mainland for supplies. "Early in the day, as the tide rose, the sea made a complete breach over the rock, washing every movable thing away, and of the old dwelling not one stone was left upon another. The new dwelling was flooded, and the windows had to be secured to prevent the violence of the spray from breaking them in. As the tide came, the sea rose higher and higher, till the only endurable places were the light towers. If they stood we were saved, otherwise our fate was only too certain. But for some reason, I know not why, I had no misgivings, and went on with my work as usual ….

"You know the hens were our only companions. Becoming convinced as the gale increased, that unless they were brought into the house they would be lost, I said to mother: 'I must try to save them.' She advised me not to attempt it. The thought, however, of parting with them without an effort was not to be endured, so seizing a basket, I ran out a few yards after the rollers had passed and the sea fell off a little, with the water knee deep, to the coop, and rescued all but one.

"It was the work of a moment, and I was back in the house with the door fastened, but I was none too quick, for at that instant my little sister, standing at the window, exclaimed, 'Oh look! Look there! The worst sea is coming.' That wave destroyed the old dwelling and swept the rock. I cannot think you would enjoy remaining here any great length of time, for the sea is never still and when agitated, its roar shuts out every other sound, even drowning our voices."

Abbie's story is now a wonderful children's book, *Keep the Lights Burning, Abbie,* ideal if you have young crew aboard.

During a storm in 1975, a great wave once again swept across the entire rock while the keepers watched in awe from the light tower.

WOODEN BALL ISLAND
Northeast Cove: 43° 51.67′N 068° 48.75′W
CHARTS: **13303, 13302**   CHART KIT: **21**

Three miles east of Matinicus, high, rocky Wooden Ball Island is narrow and a mile long. It has enough pasturage on the island to graze sheep and do some farming, but it is hard to believe that two families lived here year round during the 1840s and 1850s.

Today the only inhabitants are birds and occasional birdwatchers. This is a favorite nesting island for the usual gulls, cormorants, and eiders. Among the nests are two rarer species—laughing gulls and arctic terns. And peregrine falcons land here during their migrations, much to the distress of the gulls. Birds are nesting until mid-July.

The two most likely anchorages are Northeast Cove and Frenchman Cove, but neither offers much protection or an easy landing. Even in calm weather, the ocean swells break hard on the eastern ledges. Holding ground is rocky and poor.

SOUTHWEST PENOBSCOT BAY
Tenants Harbor to Camden

TENANTS HARBOR

3 ★★★

43° 57.85'N 069° 12.24'W

CHARTS: 13302, 13301   CHART KIT: 64, 19

This is an attractive working harbor and a convenient stop on your way east or west. Sarah Orne Jewett, who called it "this quietest of seaside villages," in *The Country of the Pointed Firs,* wrote much of that famous book here.

Tenants Harbor is very easy to enter under most conditions. It is deep and reasonably protected except from the east or southeast. The village is on the north side, and private homes are scattered around the harbor. The handsome bell tower at the seaward end of privately owned Southern Island has been immortalized by Andrew Wyeth.

On the south shore of the harbor is the boathouse that houses *Dyon,* a 52-foot gaff-rigged sloop designed and built by Luders in 1924 for Philip L. Smith of New York and Tenants Harbor. The old man had a reputation for pushing her hard and being reluctant to shorten sail (1,350 square feet in the huge main). The harbor looks right when this graceful, classic yacht is at her mooring, waiting for the third and fourth generations of Smiths to sail her east.

APPROACHES. The entrance to Tenants Harbor is clearly marked by the white buildings of the ex-lighthouse on Southern Island. Pick up green lighted bell "1" just to the east of Southern Island and leave it to port. Head up the middle of the channel, staying clear of the bar at the western end of Southern Island and the countless lobster buoys.

ANCHORAGES, MOORINGS. Room to anchor in the more protected portions of the harbor is scarce. The little space that isn't occupied by moorings is likely to be peppered with lobster buoys. Beware of occasional patches of kelp, particularly toward the head of the harbor. If you want to hang on the hook but can't find room, try Long Cove.

Moorings have proliferated in Tenants Harbor, but the good news is that many of them are rentals. They are available from several sources. Cod End moorings are a yellow-green chartreuse and are marked "Cod End Rental." Lyman-Morse's moorings are marked "THBY" and available on a first-come, first-served basis. Art's Lobsters' moorings have green lobster buoys marked "Rental," and Witham's Lobsters' moorings are white balls marked with lobsters.

Dockage is available at Lyman-Morse, with ten feet of depth at low.

GETTING ASHORE. Take your dinghy to the Cod End float or the town landing, and you will be right in the heart of Tenants Harbor.

FOR THE BOAT (from east to west). *Witham's Lobsters.* Witham's owns the first dock on the north side. Gas and diesel are sold here off the end of the dock (not the floats). You can also buy lobsters.

Art's Lobsters (207-372-6265). The second set of docks on the north shore belongs to Art's Lobsters. Except for mooring rentals, this is strictly a commercial lobstering operation. They also sell lobsters retail.

Lyman-Morse at Tenants Harbor (Ch. 09, 68; 207-372-8063). World-class boatbuilder Lyman-Morse now manages what used to be Tenants Harbor Boatyard on the steep north shore. They have 180 feet of dock space, with 10 feet of depth at low and limited power and water. Showers and a customer lounge with a phone and computer facilities are ashore. Mechanics and craftsmen are available from their Thomaston base.

Cod End (Ch. 09, 16, 69; 207-372-6782). Cod End occupies a shingled shack with red trim, with a dock and float with 4 feet alongside at low. Gas, diesel, ice, and water are available at the float. There is a small charge for water even if you are a fuel customer. They also sell charts.

Town Landing (Harbormaster David Schmanska: Ch. 09, 16; 207-372-6597). Just beyond Cod End are the launching ramp and town dock and float, with about 4 feet alongside at low.

FOR THE CREW. The "cod end" is the bag at the end of a funnel-shaped trawl where the fish accumulate. When the cod end is lifted from the sea, you find out whether you have a good catch. Cod End is owned by the Miller family, who offer fresh fish, lobsters, and other seafood both at their fish market and at their takeout restaurant on the deck. Cod End's delivery boat makes the rounds of the harbor, providing hot muffins, groceries, ice, cooked lobsters, clams, and trash pickup. In the mornings, you can come in for muffins and coffee. When asked if there have been any changes in recent years, one loyal customer answered, "Yep. The chowder just keeps getting better and better."

🛒 In additon to the Cod End fish market, Hall's Market has a good selection of fresh meat and produce, beer and wine, and the town's only pay phone. Walk up the hill to the right from Cod End and turn right to Hall's and the post office. The Schoolhouse Bakery and Coffee Shop is slightly farther up Route 131. Turn left to the library and the Farmer's Restaurant and takeout.

🍴 The East Wind Inn (*372-6366*), a large white building on the north shore, serves breakfast and dinner in pleasant surroundings. The inn runs The Chandlery on the wharf east of Cod End, but instead of stainless and bronze, it sells sandwiches and burgers. Picnic tables are on the wharf. ⚓ The inn also has a few rental moorings, but they say "they're not worth advertising."

LONG COVE

 ★★★

43° 58.24'N   069° 11.47'W

CHARTS: *13302, 13301*   CHART KIT: *64, 19*

If you prefer an anchorage that is quieter and more isolated than Tenants Harbor, tranquil Long Cove is right next door. The anchorage is broad and shallow but well protected in most weather. When the wind is blowing hard from the east or southeast, it is far preferable to Tenants Harbor.

Long Cove was a huge quarrying center. So many English stonecutters came to work in the granite industry here at the end of the 19th century that a section of Long Cove was called Englishtown. They were followed by Finns and Swedes and other nationalities, so that Clark Island and Long Cove developed a cosmopolitan air. In those days, the village boasted stores, a post office, boardinghouses, and even a bandstand. Only the silent quarries remain to remind us of that fascinating era.

APPROACHES. The approach to Long Cove is easy. Run between nun "2" at Northern Island and the ledge making out from the mainland, staying close to the nun.

ANCHORAGES, MOORINGS. With the exception of the 3-foot spot north of the entrance ledge and the shoal area southwest of the Spectacles and the reported rock just to the north of the word "Long" on the chart, you can anchor almost anywhere in Long Cove in 9 to 14 feet of water at low. There are no rental moorings, but plenty of lobster buoys.

Protection is good in normal summer winds, and the ledge between High and Northern Island blocks most of the swells, except at high tide or if the seas are up. The cove is long enough for northerlies to kick up some fetch, too, unless you tuck up beyond the Spectacles. Two yachtsmen who rode out a 50-knot northerly here report that the water barely rippled.

The protection is excellent deeper in the cove up by Clark Island, but the channel is ledgy and narrow and unmarked. It is swept by strong currents, full of lobster buoys, and not a good place for strangers.

GETTING ASHORE. Row to the floats of Atwood Brothers by their gray building on the west side, or scramble up on the rocks nearby. The Spectacles, Northern, and High Island are private and inhabited.

THINGS TO DO. A visit to the plant of Great Eastern Mussel Farms makes an interesting outing. It is tucked in a cove on the west shore opposite the north end of Clark Island. Arrange a visit by calling *372-6317*. Great Eastern has a gray building on a granite wharf, with a float for your dinghy. Be aware, though, that this upper part of Long Cove bristles with ledges, and the tidal currents are strong. Without caution, you might be needing those extra shear pins.

If you prefer to walk, it is about 1.5 miles. Land at Atwood Brothers and walk past the quarry, turning right on Route 131. Continue a half mile or more to the next right turn, Long Cove Road. There is a sign for the mussel farm. Walk another quarter mile and take the first right turn.

Wild mussels from low-tide ledges have always been a local Maine delicacy. Early attempts to culture oysters and mussels relied on rafts moored in bays and rivers, with the seeds arranged in trays or clinging to long vertical strings. Better methods have been developed since the passage of a 1973 law making it possible to lease parcels of sea bottom for aquaculture. In recent years, Great Eastern has learned how to improve the quality of wild mussels and reduce the time they take to grow.

Great Eastern sows mussel seed on acreage leased from the state. After two years, when they have grown to sufficient size, the mussels are harvested at the astounding yield of 15,000 pounds per acre and brought to the Great Eastern plant. There they spend a day in tanks of clear seawater to siphon themselves clean before workers declump, wash, and inspect them to remove broken shells and grit. Finally, the shiny black mussels are shipped all over the country in special refrigerated trucks. You can buy mussels right here, and Great Eastern provides recipes in English and French. They also sell a T-shirt printed with the comparative nutritional values of a serving of mussels and a T-bone steak (95 calories versus 395 calories, just for a start).

Despite the popularity of these succulent mollusks, the mussel industry remains highly controversial. Fishermen have objected strenuously to the harvesting of seed mussels and to the leasing of mussel beds. "One man's seed mussel is another man's catch," they say, and lobstermen complain of damage done to good lobstering ground.

MUSCLE RIDGE CHANNEL
CHARTS: 13302, 13303, 13305   CHART KIT: 64, 65, 19, 21

Muscle Ridge Channel runs northeast-southwest inside the islands of the Muscle Ridge archipelago, at the western entrance to Penobscot Bay. The name is an English variation of "mussel." Muscle Ridge Channel is the usual approach to Penobscot Bay for small vessels and cruising yachts. Deep-draft ships must use Two Bush Channel, which is farther out, much wider, and well marked with light and sound buoys.

Muscle Ridge is a good passage in daylight and clear weather. It has the advantages of being somewhat shorter than Two Bush, and it offers protection from the ocean swells and a degree of shelter from the wind. It is also an interesting passage, with a variety of unspoiled islands and headlands stretching eastward about 9 miles from the lighthouse on Whitehead Island to dramatic Owls Head Light.

The current through Muscle Ridge Channel is strong, so time your passage accordingly. The tide floods northeast and ebbs southwest. The channel is relatively narrow and bordered with a number

---

MUSSELS—
THE BEST THINGS IN LIFE ARE FREE

Lobsters you can buy from a lobsterman. Clams you can dig on a clam flat, if you have the equipment and the license. Oysters you can find in the fish market. But mussels, delicious mussels, are free for the taking and easy to collect from the cold, clear waters of Maine.

As tiny seed, mussels attach themselves to rocks and pilings in the intertidal zone or clump together with other mussels. They lead sedentary lives, straining the seawater for microscopic organisms brought to them by the rising and falling of the tides.

To collect them, just wait for low tide, survey the scene, and pry the clumps of mussels from the most convenient rocks. They come off easily, and you need no special tools. You can also collect them from mussel beds on the bottom, but that may be a bit messy.

As the Maine Department of Marine Resources points out, be sure that the location you select is not closed to the taking of mussels because of red tide, known technically as paralytic shellfish poisoning or PSP. Red tide is a bloom of diatoms, microorganisms that cause the shellfish to produce a toxin. When eaten by humans, the toxin affects the central nervous system.

Restricted areas are announced on the red tide hotline at 800-232-4733. In Canada, call 800-665-8045 or 506-456-2424. *Never eat shellfish from a closed area.* And, of course, do not take mussels in populated harbors or in areas that obviously are polluted. Also, some people have allergic reactions to shellfish that are unrelated to red tide.

The best mussels are those just exposed or below water at low tide. They will last several days if refrigerated, or indefinitely if hung over the side in moving water. However, it is best to eat them fresh right away.

Cleaning mussels is simple. With your fingers or a knife, pull out their "beards," which they use to attach themselves to rocks. Rinse them in cold water (salt water is fine) to get rid of sand. Throw away any mussels that remain open or are heavy with sand.

You can substitute mussels in almost any recipe calling for clams or oysters, but there is no need to be fancy. Steam them in white wine alone or with onion or shallots, garlic, and other herbs until the shells open (5 or 10 minutes). They're ready to eat. Serve with French bread and a nice white wine.

of ledges and rocks. Good piloting is essential. The dangers are well marked, but only one buoy has sound. The lighthouse on Whitehead Island was the site for several experimental fog-signaling devices, including one in 1838 powered by the tides which wound a spring that struck a bell. Going through in a moderate fog, you can catch enough glimpses of the land and the marks are just close enough together to make the passage without much difficulty. In a thick fog, it is very sticky.

There are no really good harbors on Muscle Ridge Channel, but there are several beautiful anchorages tucked among the islands. The closest refuge is in Tenants Harbor to the south or Rockland to the north. If you were fogged in or wanted to stop for some other reason, you might find a mooring in Seal Harbor or Owls Head Harbor. Both are easy to enter but crowded with lobsterboats.

The chart breaks right at the western entrance to the channel. If you are leaving the channel heading south or southwest past Whitehead Island, be sure to use chart 13301 or 13303 (Chart Kit 64), which shows red bell "2SB" marking South Breaker.

Consider these "Directions from Tennant Harbour to the Muscle Ridges," from *The American Coast Pilot* of 1806: "In sailing from this harbour you may steer E. by N. 1 league to Whitehead, but be careful not to haul in for it till it bears N.E.; . . . a pistol shot from the shore, is safe navigation. There is a good harbour called Seal Harbour, on your larboard hand as you pass this head (bound to the eastward), where you may lie safe from all winds."

WHITEHEAD ISLAND

43° 58.70'N 069° 07.25'W

Charts: *13305, 13302, 13303*   Chart Kit: *64, 65, 19*

The chart shows 17 feet in a little bight on the northeast side of Whitehead Island. It is fun to poke into, but it would be uncomfortably exposed for the night, being open from southeast to northeast and to the ocean swells.

From Muscle Ridge Channel, work your way in south of the large ledge with two little rocky islands (Seal), favoring the bold shore of Whitehead Island. There is a granite pier on Whitehead, after which the water shoals. The ways formerly used by the Coast Guard lightkeepers are just beyond.

SEAL HARBOR

43° 59.70'N 069° 07.40'W

Charts: *13305, 13302, 13303*   Chart Kit: *64, 65, 19*

Seal Harbor is a big open bay on the west side of Muscle Ridge Channel, between Whitehead Island and Sprucehead Island. In the days of sail, coasting schooners waited here for favorable tides in the channel. Today it is a working harbor with a fleet of some 50 lobsterboats, most of them concentrated along the south shore of Sprucehead Island. The rest of Seal Harbor is hardly used.

APPROACHES. Enter between green can "3" and the red-and-green can south of Burnt Island, which has a large gray house on the southeast end. Head toward nun "4" near Sprucehead Island, which marks rocks and ledges to the northwest. The fishing fleet is moored in the cove to the northeast of the nun.

ANCHORAGES, MOORINGS. This is a friendly place, though it is quite exposed to prevailing winds. Ask a lobsterman if there is an available mooring. You can spend a night here, but it may be somewhat rolly because of exposure to the ocean swells.

If no mooring is available, there are several alternatives, but none is particularly satisfactory. The first is to anchor southwest of nun "4" in the main part of Seal Harbor. This is a mile across and does not offer very good protection, and the bottom is reported to be poor holding in kelp. Parts of Long Ledge are visible until two hours before high.

The second alternative is to skirt Long Ledge, keeping to the north and then just to the west of it, to anchor north of the northern point of Whitehead Island. From nun "4," a range to the slot between Norton Island and the unnamed island to the west of Norton should take you safely past the ledge. Note the 2-foot spot and the rock to the northeast of Norton.

A third alternative is to run up north of Slins Island and anchor in the strip of deep water running westward above Rackliff Island. Give the ledges northwest of nun "4" a wide berth. This is narrow, crowded, and unmarked except by a line of moored boats. It would be easy to find yourself aground on the mudflats. Go on a rising tide.

GETTING ASHORE. Take your dinghy in behind the float at the Fishermen's Co-op.

FOR THE BOAT. The main part of the harbor is in the small cove northeast of nun "4." The larger gray-and-white building with a wharf is Atwood Lobster. The smaller gray-and-white building with a float is the Spruce Head Fishermen's Co-op. The Co-op float has gas and diesel and a 7-foot depth, but this is a busy working wharf. Another buying station is in a red-and-white building in the little cove to the east of nun "4," formed by Burnt Island.

FOR THE CREW. The Co-op only sells lobsters wholesale, but W.M. Atwood, across the street, will pull some for you from their enormous indoor lobster pound. Atwood also runs the small takeout next door. It's about 1.2 miles to the Off Island Store (594-7475), which sells groceries, newspapers, and wine.

THINGS TO DO. Turn right as you leave the dock and walk toward the private bridge to Burnt Island. There is granite everywhere, rough and partly finished blocks. It protects the shoreline, lines the road, and forms unusual fences between neighbors.

FALSE WHITEHEAD HARBOR

3 ★★

44° 00.12′N   069° 07.06′W

CHARTS: 13305, 13302, 13303   CHART KIT: 64, 65, 19

False Whitehead is the indentation on the north side of Sprucehead Island, west of Muscle Ridge Channel. Locally, the entire bight from Sprucehead Island to Ash Island is known as Lobster Cove, and False Whitehead Harbor is often called that too. Although it provides a lee in southwest winds and gets you out of the current in Muscle Ridge, the harbor is full of moorings, and you can't get into it as deeply as the chart might suggest.

APPROACHES. The approach from the channel is clear between Sprucehead Island and can "7" on Sunken Ledge, but note that a ledge makes out a good way from the northeast tip of Sprucehead.

ANCHORAGES, MOORINGS. The mud bottom holds well, but kelp and eelgrass have been reported.

Moorings are available for boats up to 50 feet from Merchant's Landing Moorings, located on the left side of the harbor as you enter.

FOR THE BOAT. *Merchant's Landing Moorings (207-594-7459).* Sally and Don Merchant primarily rent moorings for the season, but they often have some available for transients. Their float has about 7 feet of depth at low. Water, ice, electricity, pump-outs, and a telephone are available as well as a toilet and garbage and recycling facilities.

Spruce Head Marine (Ch. 16; 207-594-7545) is tucked in a shoal cove just north of the beginning of the causeway to Spruce Head Island. This small yard has no landings or floats, but it hauls boats for repairs of all kinds with a 30-ton boatlift. Approach only two hours either side of high tide.

FOR THE CREW. A walk of about 1.7 miles north takes you to the Off Island Store (594-7475), which has a lunch counter, groceries, newspapers, and wine. The Spruce Head post office is nearby. With strong arms or an outboard, you can take the dinghy to the Sprucehead Island bridge and cut your walk in half, or, on a high tide, you can take it all the way around to Spruce Head Marine, which is adjacent to the store. Be careful, though, of the rocks that make out north of Elwell Point.

HOME HARBOR

3 ★★★

43° 58.92′N   069° 04.48′W

CHARTS: 13305, 13302, 13303   CHART KIT: 65, 19

Home Harbor is tucked among the Muscle Ridge islands off Two Bush Channel. Under normal conditions it is a lovely harbor, seldom used, nestled in the encircling arms of several islands and ledges, with a distant view of Heron Neck and Hurricane Island to the east.

APPROACHES. Start your approach south of nun "2B," off Halibut Rock, or well north of it. The easiest course is north of Yellow Ledge, then southwest between Hewett Island Rocks and Hewett Island, favoring the shore.

Yellow Ledge, its top whitened with bird droppings, is a smooth rock 8 feet above high water. At first it merges with the shoreline, and it can be easy to confuse it with the small islet further north, between Hewitt and Andrews Island. But as you draw closer, you will see that there is plenty of room between Yellow Ledge and Hewett Island. Stay well north of the ledge and the shoal spot near it.

Hewett Island Rocks are visible at halftide but covered at high. The entrance between the rocks and Hewett Island is about 150 yards wide. Once past the southeast tip of Hewett Island, you are in Home Harbor. Aim for the little cobble beach at the head of the cove on Pleasant Island.

Lobsterboats often cut through the narrow, rock-infested channels from the west. A deep-keeled boat can do so too, but it requires great caution.

ANCHORAGES, MOORINGS. Work your way in to depths of 16 or 20 feet at low, noting the ledges that make out from the west side of the harbor. Holding ground is good in mud. You will be well protected from prevailing southerlies but exposed to the east and somewhat to the north-northwest. At high tide, a roll can work its way over Hewett Island Rocks and into the harbor.

DIX ISLAND HARBOR
44° 00.34'N   069° 03.91'W

CHARTS: *13305, 13302, 13303*   CHART KIT: *65, 19*

Despite its name, Dix Island Harbor is exposed to prevailing southwesterlies and more of a rocky gunkhole than a harbor.

It is tucked south of Dix Island and west of The Neck. The *Coast Pilot* notes that it can be entered "from the southwestward through a narrow and crooked channel leading between the ledges north of Hewett Island," but it advises that the channel and harbor are unsafe for strangers. If you dare, try it on a rising tide near low.

HIGH ISLAND HARBOR and DIX ISLAND

3 ★★★★

44° 00.75'N   069° 04.00'W

CHARTS: *13305, 13302, 13303*   CHART KIT: *65, 19, 21*

High Island Harbor is our name for the anchorage surrounded by Dix, Birch, and High Island at the north end of the Muscle Ridge archipelago. In summer weather, it is a delightful place for a short visit. The two larger islands, High and Dix, were important quarries during the 19th century.

APPROACHES. See chart on next page. The approach to the anchorage is direct, with few dangers. From Muscle Ridge Channel, run halfway between nun "10" and red daybeacon "12" on Otter Island. Otter is partly wooded with clearings at the west end and along the south shore. Leave Little Green Island (with a few low trees and a small house) to port and Oak Island (small, rocky, and treeless) to starboard.

It is also possible to approach the harbor south of Oak Island, favoring the island to avoid the long ledges to the west of Dix.

High Island is high and densely wooded, with two large granite wharves at the west end. Note the 3-foot spot reported west of it. Dix Island is lower and partially wooded, with a clearing and house at the northwest end. Granite rubble lines the northeast shore. Birch Island is low-lying, treeless, and grassy, with a number of small beaches.

ANCHORAGES, MOORINGS. You can choose a number of different spots to anchor, depending on the wind. For access to the islands, drop the hook in

"....many skeletons, much decayed, seem to have been buried in a circle, with their feet pointing towards the centre."

the middle of a triangle formed by Dix, Birch, and High, in about 15 feet of water at low. The bottom varies in spots from rocky to sand and mud. Exposed only to the northwest and north, the anchorage provides good protection from prevailing summer winds.

High Island. Row over to the granite wharf and scramble up. Then work your way inland, through the alders and apple trees, along the path of granite tailings. Be careful—you may suddenly come upon the water-filled quarry at the west end of the island, with its high ledges and clean granite. One visitor's exuberant Labrador was charging through the woods when the terra firma changed places with thin air. The dog hurtled downward to a ledge below, unhurt, but definitely bewildered. Usually the water is clear, but sometimes it is green with algae.

Quarrying by the High Island Granite Company started here in 1894—about 16 years after the last stone was cut on neighboring Dix—and lasted until World War I. Life was not easy in the boardinghouses where 200 quarrymen lived. As Charles McLane reports in *Islands of the Mid-Maine Coast*, "Local liquor laws led to periodic raids by the county sheriff and confiscation of the red wine necessary to the morale of the luckless Italians."

Dix Island. Dix Island is owned by members of the Dix Island Association, who aspire to self-sufficiency. Please respect their privacy. They live here simply, without electricity, and pump water by hand. The land is owned in common. The few houses, individually owned, are like summer cottages, built to blend into the landscape. The exception is the one at the northwest end that was the maid's quarters and laundry for Horace Beals' mansion.

You are welcome to come ashore, though recently residents have felt somewhat overrun by visitors in July and August. Land at the beach well north of the dock on the east side and walk south on the path marked by small white arrows. *Please stay on the designated path.* No camping, fires, or swimming in the quarries is allowed.

There are at least three layers of history on Dix Island starting with this mysterious account by a local historian (reported in the August 1982 issue of *Down East* magazine): "....many skeletons, much decayed, seem to have been buried in a circle, with their feet pointing towards the centre." One leg bone was said to have been several inches longer than those of today's tallest people.

Today there is no trace of these giant early inhabitants, but the later residents have left their mark everywhere. In its heyday, there were more than 2,000 quarrymen on Dix and more than 150 buildings, including the Shamrock Boardinghouse for the Irish, the Aberdeen for the Scottish, and even an opera house.

All this was started by Horace Beals of New York and carried on by his wife, Jennie, with cost-plus-fifteen-percent contracts for granite to build post offices in New York City, Charleston, and Philadelphia. Legends accumulated around beautiful Jennie Beals—sophisticated, blessed with business acumen, but not satisfied to spend her life here. After her husband's death in 1863, she moved to New York City, eventually becoming the Duchess of Castellucia by a second marriage.

CRESCENT BEACH

44° 03.90′N   069° 03.95′W

CHARTS: 13305, 13302, 13303   CHART KIT: 65, 68A, 19, 21

South of Owls Head Harbor and west of Sheep Island is Crescent Beach. Although protected from the southwest, it is open to the south and southeast and is likely to be rolly.

APPROACHES. From the north, pass through Owls Head Bay and then between can "3" and nun "4," west of Sheep Island. Crescent Beach is the more northerly of the two visible beaches. Favor green daymark "1" as you turn westward toward shore to avoid the rocks off the little island to the north.

ANCHORAGES, MOORINGS. Anchor in mud north and east of Emery Island in 15 to 20 feet at low. Note the piling shown on the chart.

GETTING ASHORE. Land on the beach.

OWLS HEAD HARBOR

44° 05.00′N   069° 03.00′W

CHARTS: 13307, 13302, 13305   CHART KIT: 65, 68A, 19, 21

Owls Head is more of a roadstead than a harbor, but it is home to a fleet of 35 or more lobsterboats and draggers and served by a large dock and lobster pound. The harbor is located just south of Owls Head Light and protected by Monroe and Sheep Island to seaward. It is very convenient as a tranquil layover before proceeding to busy Rockland.

In his 1858 book, *A Summer Cruise on the Coast of New England,* Robert Carter noted: "At 6PM, we reached Owl's Head, an exceedingly picturesque promontory where a large white lighthouse crowned a high rock rising abruptly from the water. Here we anchored in a broad channel between the mainland and two islands, amid a fleet of vessels. This channel is much frequented by coasters and fishermen, and five hundred sail have been seen passing Owl's Head in one day."

APPROACHES. Coming from the southwest, pass between can "3" and nun "4" west of Sheep Island. From the north, round Owls Head with its lighthouse, then leave to starboard can "7" and green daybeacon "5" on Dodge Point Ledge.

ANCHORAGES, MOORINGS. The harbor is full of moorings, so the only room to anchor is at the outside edge, almost into the channel. The holding ground can be poor, so be sure your anchor is well set. Windjammers often stop here before offloading their passengers the next day in Rockland.

Your best alternative is to pick up the guest mooring with the cordial inscription, "Strangers Welcome," located outside the fleet. Otherwise, inquire at P.K. Reed and Sons, who will know of any vacant moorings.

GETTING ASHORE. Row in to the dinghy float.

FOR THE BOAT. *P.K. Reed & Sons (207-594-4606).* Gas, diesel, and ice are available at this commercial dock, with 7 feet alongside at low. Traps and drums are piled high on this working wharf, and the air is perfumed with brine and baitfish.

FOR THE CREW. P.K. Reed has live clams and lobsters. A short walk up the hill will bring you to the post office and an upscale general store.

THINGS TO DO. The "picturesque promontory" on which the lighthouse stands is now Owls Head State Park, about a 1-mile walk from the harbor and a delightful picnic spot.

DEEP COVE

44° 05.65′N   069° 03.22′W

CHARTS: 13307, 13302, 13305   CHART KIT: 65, 68A, 19, 21

Deep Cove lies just west of Owls Head. Despite being adjacent to Rockland's huge harbor, it is woodsy and tranquil and well protected from southwesterlies.

APPROACHES. Deep Cove is easily identified by green daybeacon "9" on Shag Rock, just east of the entrance. Avoid the ledge making out from the west side of the entrance and work your way in slowly. It shoals very rapidly. Do not go beyond the first point to port after the entrance.

ANCHORAGES, MOORINGS. Anchor here in 15 feet at low. Your anchor is likely to come up clean, suggesting a rocky bottom.

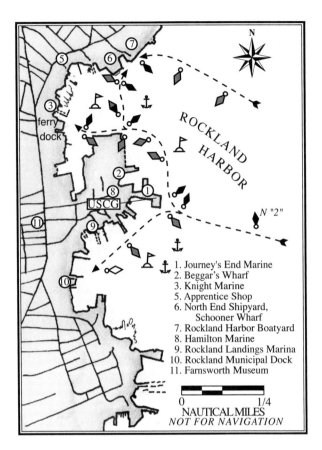

1. Journey's End Marine
2. Beggar's Wharf
3. Knight Marine
5. Apprentice Shop
6. North End Shipyard, Schooner Wharf
7. Rockland Harbor Boatyard
8. Hamilton Marine
9. Rockland Landings Marina
10. Rockland Municipal Dock
11. Farnsworth Museum

BROAD COVE

3 ★★★

44° 05.42'N 069° 03.76'W

CHARTS: 13307, 13302, 13305   CHART KIT: 65, 68A, 19, 21

Opposite the Rockland breakwater and slightly east of it, Broad Cove provides an attractive setting and good protection from prevailing summer winds. There are spectacular views of the Camden Hills and the city lights of Rockland. The New Jersey-based Bancroft School has a summer camp here, and there are several private homes.

APPROACHES. Coming from Rockland, between the breakwater and can "1," you will see Broad Cove open up to starboard. The complex of buildings at the eastern end of the cove is the Bancroft camp.

ANCHORAGES, MOORINGS. Note the ledge making out from the west side of the cove. Anchor in 10 to 15 feet at low near the private moorings in the southern end, or anchor in 20 feet at low in the eastern bight of the cove, outside a line between the end of the Bancroft dock and a point to the southwest. The bottom is both mud and kelp, so be sure your anchor is well set.

ROCKLAND

3 ★★★

44° 06.23'N 069° 05.95'W

CHARTS: 13307, 13302, 13305   CHART KIT: 65, 68A, 19

Emergency: 911
Hospital: 207-596-8000
Coast Guard: 207-596-6666

Rockland is the area's largest city and Penobscot Bay's largest harbor, full of commercial activity and, more recently, yachts. It lies behind a long breakwater on the west side of the bay. Fishing boats come and go, windjammers ply the waters, and ferries depart and arrive from North Haven, Vinalhaven, and Matinicus. Rockland also has a major Coast Guard base with search-and-rescue capabilities.

In 1765, Isaiah Tolman of Massachusetts moved with his 21 children to form an instant settlement at "the shore," part of Thomaston, which became Shore Village and eventually Rockland. Not long afterwards, John Lermond moved from nearby Warren and established a logging camp at Lermond Cove, called Catawamteak or "great landing place," by the Indians. Today it is the Rockland waterfront.

The story of Rockland is one of lime, granite, ships, and fish. Lime was shipped to market in 50 to 150-ton schooners, and it was natural that a shipbuilding industry should flourish in Rockland. For more than a century, half a dozen well-known yards grew up, and a large number of ships were launched until lack of business closed them down in the 1930s. Fishing has been a central part of Rockland from the earliest days. In the recent past, Rockland was known—and avoided—for the smell of its fish processing plants, but many of the plants have closed or moved. A seaweed processing plant still remains, rendering seaweed from as far away as South America into food fillers and stabilizers.

Tourists are the latest boon to Rockland, particularly the kind who arrive in boats. Rockland has the valuable asset that the nearby harbors of Rockport and Camden lack—space, and enough of it to hold the entire 150-boat New York Yacht Club cruising fleet. But Rockland is attractive in its own right. It is convenient and interesting, and it has undergone a cultural and aesthetic revival that has transformed it from an odoriferous fish-processing town to a hub of culture. Rockland operates an elaborate town dock and has encouraged the construction of many new marine facilities.

Once off the boat, visiting yachtsmen are discovering the gems in this small city where almost every amenity or service can found within a few blocks of the water. The famed Farnsworth Art Museum is only steps from the waterfront, the Maine Lighthouse Museum is hard by the harbor park, and the Owls Head Transportation Museum is nearby. The Island Institute has its headquarters here, along with a gallery of island art. The Samoset Resort, with a health club and golf course, is located at the north end of the breakwater.

The logistics of crew changes or rendezvous are simplified by Rockland's regional airport and the ferry service to the islands. Long-term parking can be arranged at the ferry terminal (*596-2202*), Knight Marine, Beggar's Wharf, or the Rockland Landings Marina.

APPROACHES. Rockland is easy to find. Even from a distance it looks like a city, with conspicuous high structures isolated in a comparatively low-lying area well south of the Camden Hills. Owls Head Light, on its dramatic wooded bluff, is visible a long way. Beware of ferries coming and going, not to mention barges, fishing boats, and lots of commercial traffic. In fog, keep an ear on your radio.

Approaching the harbor, you will pick up the lighthouse at the end of the mile-long breakwater, which forms the eastern part of the harbor. Leave the end of the breakwater to starboard and head in toward the radio towers on the hill. A clearly marked channel starts with nun "2" and runs in toward the town.

The Coast Guard station is on the large granite pier that projects out farthest to the east. Here the channel divides. Bear left and to the southwest for the fuel dock at Journey's End Marina, Rockland Landings Marina, and the public landing. Bear right and to the north for the rest of Journey's End Marina, the ferry terminal, Knight Marine, the North End Shipyard, and Rockland Harbor Boatyard.

If you are headed for the Samoset Resort, run north along the west side of the breakwater.

ANCHORAGES, MOORINGS. The harbor has a complete spectrum of anchoring and docking options from anchoring out and using the dinghy to moorings, to dockage with cable TV, Wi-Fi, and phones. Rockland schedules a summer full of concerts, festivals, and shows at its harbor park, so sometimes the facilities are chock full. At other times you will have no problem finding a slip or a mooring. It is wise to reserve ahead.

⇌ Dockage can be found north of the Coast Guard station at Journey's End Marina or Knight Marine. South of the Coast Guard station, you can rent dock space at Rockland Landings Marina or the town dock.

⚓ Moorings are rented north of the Coast Guard by Journey's End, Beggar's Wharf, and Knight's, and also by Rockland Harbor Boatyard, farther north. Moorings south of the Coast Guard station are rented by the town dock or Rockland Landings. On the north shore of the harbor, just inside the breakwater, you can anchor or get a mooring at the Samoset Resort to enjoy its amenities.

There is room to anchor outside the mooring areas, but you'll be quite exposed and have a long dinghy ride.

FOR THE BOAT—north of the Coast Guard. *Journey's End Marina (Ch. 09, 18; 207-594-4444; www.oharabait.com/marina.php).* Journey's End has floats both south and north of the Coast Guard pier. ⛽Gas, diesel, water, and electricity are available at the southern docks. Depths range from 6 to 12 feet, with deeper water at the southern dock. The northern dock has 🚽 pump-out facilities and is next to their office, chandlery, and showers. ⇌ Their docks can handle transient vessels to 225 feet, but space, particularly that much space, should be reserved. Journey's End are mechanical specialists, but they can haul boats and perform repairs of most kinds. They are owned by the O'Hara Group, who in the early 1900s ran a fleet of fishing schooners out of

RED JACKET

During all her long life, *Red Jacket* was considered the most handsome of the great clipper ships and the swiftest. Designed by young genius Samuel Harte Pook of Boston, she was launched November 2, 1853, at the Thomas yard in Rockland. *Red Jacket* was 251 feet long, had a 44-foot beam, drew 44 feet, and weighed 2,305 tons.

Red Jacket was built for speed, and she proved her worth on her first Atlantic crossing. She was towed to New York for rigging and then dispatched to Liverpool in January 1854, before her bottom was copper-sheathed. Manned by an indifferent crew but a determined captain, *Red Jacket* logged more than 343 miles a day for six days running, and on January 19 made a run of 413 miles, a record surpassed only two or three times in the history of sailing ships. In spite of constant snow, rain, and hail during the midwinter crossing, Captain Asa Eldridge stayed on deck day and night and drove her hard, arriving in Liverpool in 13 days, one hour, and 25 minutes—still a record for single-hulled sailing ships.

Shrugging off lines from waiting tugs, *Red Jacket* drove up the River Mersey, all sails flying in a brisk northwester, toward a great crowd awaiting her arrival. Then, with enormous daring, Captain Eldridge performed a feat of seamanship that still sends shivers down a sailor's back a century later. The maneuver was described by one of Eldridge's descendants. Still under full sail and abreast of the pier, "He took in his kites, his skysails, royals and top gallants, hung his course—or lower sails—in their gear, ignored the tugs which by that time had caught up, and throwing *Red Jacket* into the wind, helm down hard, he backed her alongside the berth without aid."

Sailing next from London to Melbourne, she made the passage in 67 days and 13 hours. There she was bought by a British firm that carried freight and passengers to Australia and India, never again to sail under the American flag. In her later years, she carried lumber from Canada. Sadly, she ended her career of more than 30 years as a coal barge, fueling steamers in the Cape Verde Islands. A bronze plaque at Jordan's Market in Rockland marks the location of the Thomas yard where the *Red Jacket* was built.

Boston. They are still in the fishing business, with an Alaskan fleet and an ice-making plant in Rockland that supplies most of the Maine coast.

Knight Marine Service (Ch. 09; 207-594-4068, dock: 596-7216; www.midcoast.com/~knightma). Knight's is a full-service boatyard. To get there, branch right at the Coast Guard pier and follow the channel around to the right of the ferry terminal. They have moorings and a limited amount of dockage with 8 feet of water at the floats. Gas, diesel, water, and ice may be obtained here, and ashore there are showers and laundry facilities. The yard has a tugboat, 20 and 35-ton boatlifts, a hydraulic trailer, and a crane. They can handle most repairs. The Captain Hornblower takeout is on the premises and it is next door to the Maine State Ferry Terminal, serving North Haven and Vinalhaven, and the Concord Trailways bus stop.

Beggar's Wharf (Ch. 09; 207-594-8500; www.beggarswharf.com) is tucked in the niche just west of Journey's End. They manage 40 rental moorings for boats as large as 150 feet, and they also offer a dive service. They unabashedly charge a "storm rate" when customers rent a mooring specifically to weather a storm. As owner Charlie Weidman explains, "That's when our gear takes the most beating and when we assume the highest risk." You'll get your money's worth, though; a concierge service is included, which, among its services offered, lists "Mutinies quelled" and "Barratry plots foiled."

Rockland Harbor Boatyard, Inc. (207-594-1766; www.rhby.com). Rockland Harbor Boatyard, located next to North End Shipyard on Front Street, is a full-service facility. They offer transients mooring rentals and water and pump-outs at their float. The yard can handle any type of repair on all vessels. Owner and marine surveyor Sam Slaymaker started International Classic Yachts here to buy, restore, and resell classic wooden yachts.

Ocean Pursuits (207-596-7357), also on Front Street, is a full-service boatyard run by Bill and Judy Cowan. They offer all types of boat repair, specializing in marine systems and electronic sales, service, and installation.

Samoset Resort (207-594-2511, 800-341-1650; www.samoset.com). The Samoset has three moorings for visitors in the inner corner by the breakwater. They are free and available on a first-come, first-served basis. While you are far from town here, you are close to their resort, which is open to the public, with a health club and pool, a public golf course, and elegant dining. This part of the harbor is exposed to prevailing southwesterlies because of the size of Rockland Harbor, although it is fine for settled summer weather or a short stop. There is kelp on the bottom, so if you anchor here, be sure your anchor is well set.

FOR THE BOAT—south of the Coast Guard. *Rockland Landings Marina (Ch. 09, 16; 207-596-6573).* The dock complex of the Landings Marina sprawls between the fish pier and the town landing to the south, right at the foot of Rockland's business district. In addition to several mooring rentals, they offer transient dockage for vessels to 150 feet, with 12 feet of water at low and all the electrical amenities. They pump gas and diesel in, and holding tanks out. Showers, laundry, and a restaurant are ashore, where long and short-term parking can be arranged.

Rockland Municipal Dock (Harbormaster Ed Glaser: Ch. 09; 207-594-0312). To reach the public landing, branch left at the Coast Guard pier. The floats and dock are opposite a yellow nun at the end of the channel. The dock itself has a green trestle bridge and lands you near a little park with a white building that houses the harbormaster's office, and boaters' showers and laundry.

You can tie up at the floats for a maximum of two hours (longer with permission), with 8 feet alongside at low, or you can rent dock space for the night. For additional fees, water, electricity, ice, and pump-outs are available. This is a Wi-Fi area. Obtain a rental mooring by checking with the harbormaster based here, or anchor outside the moorings. Moorings and dock space are not reservable. Rockland's business district is just steps away.

Harbormaster Ed Glaser can help arrange short and long-term parking and will hold mail or packages for cruisers. Address mail care of him at 270 Pleasant St., Rockland, ME 04841 or send FedEx or UPS to Harbor Park, Main St., Rockland, ME 04841.

FOR THE BOAT—ashore. *Lew Grant and Associates Marine Electronics (207-594-7073),* next to Journey's End, sells, installs, repairs, and performs warranty work for things that go beep on your boat.

Hamilton Marine (207-594-8181; www.hamiltonmarine.com). This is Maine's answer to serious boating needs. Where else can you walk down nearly a whole aisle of cutlass bearings? It's chock full of things you need at low prices—seaboots, paint, potwarp, sailmaker's palms, marine hardware, and whatchamacallits. It is tucked in among the fish plants on Wharf Street, just inland from the Coast Guard station.

Gemini Canvas and Sails (207-596-7705; www.geminicanvas.com), across the street from Rockland Boat, specializes in canvas and upholstery work.

Hallet Canvas and Sails (207-594-9810; www.hallettcanvasandsails.com). This branch of the Falmouth loft is located at the southern end of the harbor at 34 Atlantic Avenue.

Pope Sailmakers (207-596-7293; www.popesails.com) is a full-service sail and rigging loft located on Park Street on the outskirts of town.

FOR THE CREW. Rockland is following the usual pattern of growth—the more interesting the downtown becomes, the farther away the groceries get. The nearest supermarket is Hannaford, a mile north on Route 1 with a quick left at the golden arches. It's also a liquor store, and a laundromat is across the street—a dangerous rainy-day combination. Shaw's (*594-8615*), farther up Route 1, offers a Summer Express Service. If you phone in a grocery order by 11 AM they will box it for you for same-day pickup, or they'll make arrangements to get it to your boat—even if you are on North Haven or Vinalhaven.

For less extensive provisioning, combine the Brown Bag Bakery at the north end of Main Street with the specialties at Sage Market on Main Street. You can simply load up on indescribable donuts at the Willow Bake Shoppe (*596-0564*), on Willow Street, up from Knight's. It's open only on weekdays, excluding Wednesdays, from 5 AM to 10 AM. Basics, i.e. chips, wine, beer, and ice, can be bought at Rite Aid, directly opposite Harbor Park.

For dinner ashore, try Amalfi (*596-0012*) on Main Street for Mediterranean fare, Conte's on the waterfront for vast bowls of Italian seafood, or Cafe Miranda (*15 Oak St., 594-2034*) for a more eclectic menu. Primo (*596-0770*) serves incredible French and Italian-inspired meals in a restored Victorian house a mile south of town on Main Street (Route 73).

The Knox County Regional Airport in Owls Head serves Portland, Boston, and some of the Penobscot Bay islands. US Airways Express (*800-428-4322*) flies roundtrip to Boston. Rental cars are available. Concord Trailways (*800-639-3317*) buses stop at the Maine State Ferry. Penobscot Bay Medical Center is a few miles north of the city (*207-596-8000*).

THINGS TO DO. Highlights of the summer include windjammer races and other events ashore during Schooner Days in early July, the North Atlantic Blues Festival in July, and the Maine Lobster Festival in early August.

The Maine Lighthouse Museum (*594-3301, ww.mainelighthousemuseum.com*) is located next to Harbor Park. They display the country's largest collection of lighthouse lenses and artifacts, including Fresnel lenses removed from Petit Manan, Whitehead Island, and Matinicus Rock when their lighthouses were automated.

To see or sail some of the windjammers, visit the North End Shipyard (*594-8007*). This unusual yard was started on a shoestring by Doug and Linda Lee and John Foss to rebuild and maintain some of the area's windjammers. It is located north of the ferry terminal beyond Lermond Cove. There they launched the new 93-foot schooner *Heritage* in 1983 and the restored 90-foot fishing schooner *American Eagle* in 1986. They were joined by Ed Glaser and the *Issac H. Evans*. North End can handle major repairs, or you can do them yourself, as many other windjammer captains do, by leasing their space and heavy woodworking equipment. A dredged channel leads to their substantial marine railway.

True wharf rats will want to sniff the sawdust at the Apprenticeshop (*594-1800, www.apprenticeshop.com*) on Main Street at the north end of town. Here, apprentice boatbuilders build beautiful, three-masted, 38-foot open gigs that are replicas of a French admiral's barge captured off Ireland in the 1700s. The Atlantic Challenge program uses the boats to promote seamanship and self-confidence in youngsters, culminating in an international competition. If they aren't out sailing, the Bantry Bay gigs might be moored proudly out in the harbor. Their long pier extends from the north end of Lermond Cove.

Walk out to the end of the Rockland breakwater. Begun in 1881, it took 18 years to lay up its 700,000 tons of granite, most of which was brought from Vinalhaven by stone sloop. The 17-foot wide causeway starts near the Samoset Resort and extends a mile to the lighthouse.

The spectacular Samoset looks out to sea over an 18-hole golf course. Their facilities include tennis, swimming, a health club, and a good restaurant, and they are available for use by visitors as well as guests.

The Farnsworth Art Museum (*596-6457, www.farnsworthmuseum.org*) in Rockland is on Main Street, only a few blocks from the waterfront. Their nationally recognized collection of Amercan art includes works by Gilbert Stuart, Thomas Sully, Thomas Eakins, Eastman Johnson, Fitz Hugh Lane, Frank Bensen, Childe Hassam, and Maurice Prendergast. A $10-million expansion includes coastal landscapes and ship paintings, and an extensive collection of works by three generations of Wyeths.

The family homestead of the Farnsworth lime barons is next door to the museum and part of it. Its original decor and elegant furnishings make it one of the finest Victorian homes in the country.

The offices of the Island Institute (*594-9209, www.islandinstitute.org*) are on Main Street, just north of the Farnsworth. This organization studies and helps sustain the physical, social, ecological, and economic factors affecting island communities. Visit Archipelago, their shop that features crafts and goods from Maine's islands.

The National Audubon Society operates a visitor center (*596-5566, www.projectpuffin.org*) on Main Street to highlight their Project Puffin, which reintroduces puffins to coastal Maine islands (see *Eastern Egg Rock*, *Matinicus Rock*, and *Machias Seal Island*).

The famous Owls Head Transportation Museum (*594-4418, www.ohtm.org*) is 2 miles south of Rockland, near Knox County Regional Airport. On weekends, their unique collection of antique airplanes, cars, and engines hiss, puff, creak, and fly. There are horse-drawn carriages, Stanley Steamers, Model-T Fords, and Rolls-Royces, as well as World War I fighter planes and a Ford Trimotor. The museum has special events all summer long. The best way to get there from Rockland is by taxi. The taxi stand is on Rockland's Main Street, two blocks from the public landing.

CLAM COVE

44° 08.18′N   069° 05.00′W

CHARTS: *13307, 13302, 13305*   CHART KIT: *65, 68A, 19*

This little-used harbor between Rockland and Rockport is broad, but it offers good protection from prevailing winds, and it is easy to enter. Although the shores are attractive, busy Route 1 passes along the head of the cove.

APPROACHES. Enter north of can "3" off Ram Island and proceed up the middle of the harbor. There are no obstructions.

ANCHORAGES, MOORINGS. Most of Clam Cove is a cable area. It is possible, however, to find a 10 to 15-foot spot south of the cable area off the south shore. The bottom is mud with some kelp, so be sure your anchor is well set.

GETTING ASHORE. When the mud flats are covered, it's possible to land at the little park at the head of the harbor, but this would be an ugly slog at low.

FOR THE CREW. It's 1.5 miles south to the outskirts of Rockland. Turn right at McDonald's for a nearby Hannaford market. Penobscot Bay Medical Center is a short distance north of Clam Cove on Route 1 (*596-8000*).

ROCKPORT

44° 11.00′N   069° 04.35′W

CHARTS: *13307, 13302, 13305*   CHART KIT: *20(A), 65, 68A, 20*

Small and beautiful Rockport is on the west side of Penobscot Bay, between Rockland and Camden. Fifty years ago, Rockport was primarily used by lobstermen and fishermen who built their weirs in the harbor, but today the harbor is mainly filled with pleasure boats.

With all of the boats moored here, it is a wonder that the 70-foot windjammer *Timberwind* can make her way among them without an engine. It's a lovely sight to see her glide past Indian Island under sail, drop all canvas, and come boldly in, pushed by her little yawlboat, turning in the last possible open space at the head of the harbor to nestle alongside her float.

Although Rockport is popular as a harbor, it often is not a comfortable one. Exposed to the south, the harbor gathers up the ocean swells. The "Rockport roll" is well known to those who live aboard.

In the 19th century, Rockport was at the heart of the midcoast lime industry. The town was once the third largest lime producer in the country and supplied the lime for the Capitol in Washington. You can still see remnants of several old kilns on the waterfront. The kilns inspire images of the night sky aglow with fires, the harbor crowded with kilnwooder schooners unloading cords of four-foot spruce logs and limers loading casks for Boston and New York. Great piles of white lime tailings are still visible in Walker Park and on the banks of the Goose River. One story has it that a locomotive is buried in one of these piles, and small boys dig and dream.

APPROACHES. Rockport is an easy harbor to enter, the only danger being Porterfield Ledge, topped with a granite marker. In poor visibility, find red-and-white bell "RO" (*44° 09.50′N 069° 03.22′W*) off Indian Island, and then the light on Lowell Rock.

Do not cut the abandoned lighthouse on Indian Island close—a ledge makes out well south of the lighthouse to the present light structure on Lowell Rock, which you leave to starboard.

Continue in among the moored boats to the head of the harbor, noting red beacon "4" on Seal Ledge to starboard.

ANCHORAGES, MOORINGS. Rockport Marine, at the northeast end of the harbor, maintains moorings for rent and also offers space alongside, with 10 feet of depth at low.

At the northwest end of the harbor, Rockport Marine Park has several floats with 10 feet of depth at mean low water. An overnight tie-up can be arranged, but space is limited and this location is uncomfortable with any sea.

Across the Goose River from the park is the small red building of the Rockport Boat Club, with floats for temporary tie-up, dredged to 6 feet of water at low. The club also maintains a guest mooring.

Rockport Harbor is now full of moorings, with little or no space to anchor except well outside, where the depths run 50 feet or more.

GETTING ASHORE. Take your dinghy in to the floats at the public landing at the head of the harbor, to Rockport Marine on the east side, or to the Marine Park on the west side.

Work and pleasure mix on the docks at Rockport Marine.

FOR THE BOAT. *Rockport Marine Park (Harbormaster Ken Kooyenga: Ch. 16; 207-236-0676).* Rockport is a shining example of a marine park done right. They have 200 feet of dock space on the south side of the harbor for temporary, 1-hour docking, with 10 feet of depth at mean low water, and free water. Longer term dockage may be able to be arranged. Their building has immaculate showers and laundry, and the park has a small beach for the kids.

Camden-Rockport Pumpout Boat (Ch. 16; 207-236-7969). This federally funded pump-out service is free, but plan well in advance since the pump-out boat may have to come from Camden.

Rockport Marine, Inc. (207-236-9651; www.rockportmarine.com). Taylor Allen's Rockport Marine specializes in building, restoration, and repair of wooden boats. They can, however, provide hull and engine repairs for boats of all kinds, and they have a 35-ton boatlift. In addition to dockage and moorings, the yard offers gas, diesel, water, pump-outs, and electricity at the floats, with 9 feet or more alongside. Ice and marine hardware are available.

Rockport Boat Club. Water and electricity are available at the floats, but no other facilities. Visiting yachtsmen are welcome to use the guest mooring and the clubhouse, which has a pay phone.

E.S. Bohndell & Co. (207-236-3549). Half a mile from Rockport Harbor, on Route 1, the Bohndell sailmaking and rigging firm has been in business for more than a century.

FOR THE CREW. A footpath and small bridge connect the Rockport Marine Park with Rockport Marine. Just up the hill from Rockport Marine is

The Corner Shop, a wonderfully un-fancy eatery with pine bench booths, where carpenters and fishermen rub elbows with students from the Maine Photographic Workshop. Breakfast and lunch are available at reasonable prices, cardiologist's visit not included.

It's worth the .75-mile walk to The Market Basket (*236-4371*), a gourmet deli serving delicious panini sandwiches and stocking a good variety of wine, cheese, French bread, and fresh produce. Cross the bridge to Pascal Avenue and walk south along the western side of the harbor, then right on West Street to the intersection of Route 1 and Route 90. Stop at the Sweet Sensations Pastry Shop on your way. OK, stop on the way back, too.

Camden, with a wide selection of restaurants and shops, is only a few miles away. For a cab, call Don's Taxi (*236-4762*). For local use, you can rent a car at Smith's Garage (*236-2320*).

THINGS TO DO. Rockport's Marine Park contains the last of the lime kilns and a tiny Vulcan steam locomotive that used to haul limestone from the quarries on a 3-mile narrow-gauge railway.

It also has a statue commemorating Andre, an abandoned seal pup who was raised by Rockport resident Harry Goodridge. *Andre the Seal* is a wonderful children's book to have aboard.

Up Main Street, to the right, is the Rockport Opera House, where Bay Chamber Concerts are held Thursday nights (and some Friday nights) with well-known piano soloists, string quartets, and other performers. The Center for Maine Contemporary Art is on Russell Avenue, just beyond the Opera House, in a converted livery stable and firehouse. The public library is across the street.

For a beautiful walk, continue on Russell Avenue past the golf course and the Lily Pond. At Aldermere Farm you may see a herd of Belted Galloways, a distinctive breed of cattle, black with broad white belts. If you continue past a cemetery with an ocean view, you will eventually arrive in Camden.

For good views of the harbor, walk up Russell Avenue to Mechanic Street and then along the eastern side of the harbor toward Beauchamp Point.

Take a left on Chapel Road to the lovely open-air Children's Chapel and gardens.

The Maine Photographic Workshop is opposite the Corner Shop. People from all over the country come here for its acclaimed courses. During the summer, special lectures and shows are open to the public. The workshop also operates a well-stocked camera store. You probably will encounter shutterbugs from the school here or elsewhere along the coast, so if you do nothing else, smile.

LIME

"Kilnwooders" unloading wood in Rockport Harbor, 1894. It took 30 cords of wood to fire one lime kiln. Most of the houses shown on Amsbury Hill still exist. (Frank Claes collection)

From Thomaston to Warren, from Rockland to Rockport to Lincolnville, there are deposits of limestone all along the shore, perhaps laid down originally as seashells in a shallow sea. When limestone is burned to eliminate carbon dioxide, what's left is lime. From prehistoric times, lime has been useful as a building material and as fertilizer. Lime from Rockland was used mostly for plastering walls and ceilings and to make mortar for laying brick.

The first lime kiln in the area was built about 1733, in Thomaston to burn limestone for shipment to Boston. By 1828, there were 60 lime kilns in the midcoast area of Maine, and by the Civil War, they were burning more than a million casks of lime a year. Rockland split off from its parent town of Thomaston and became the lime-producing capital of the world.

It took 30 cords of wood to fire one lime kiln and still more wood to build the casks for shipment. Cordwood came from islands in the bay, from Down East, and from Canada. Farmers became coopers for winter income, cutting birch and alder trees and splitting and bending them for hoops and staves.

To fetch the cordwood and carry the lime casks to market, mostly for construction in New York City, ships were needed. More than 500 ships were engaged in this trade by mid-century. The countless "kilnwooders" were rough-built vessels, their decks piled high with cords of wood. The ships used to carry lime to market, on the other hand, were carefully constructed, because this cargo had one enormous drawback: if the slightest amount of water reached the lime in the casks, it caught on fire, and the fire often was unquenchable. As W.H. Rowe explains in his *Maritime History of Maine*, "The master needed a keen sense of smell. The odor of lime being slaked by water was an ominous danger signal... Every crack and crevice through which air might get into the hold and the doors, ports, and smokestack were quickly sealed with plaster made from the lime.

"Then the craft was headed for the nearest harbor and anchored some distance from the shore and away from other vessels. For at any time she might burst into flames. The schooner was stripped of all movables and the captain and crew sat down to await developments. Sometimes three months would go by before their patience was rewarded and the vessel saved. If, however, the fire could not be smothered, the vessel was towed to some secluded place and scuttled." The fire problem wasn't solved until the middle of the 20th century when steel barges replaced the sloops and schooners.

Technology came to the industry in the last half of the 19th century, with oil or coal-fired, continuous-process "patent kilns." In 1900, the three largest lime manufacturing firms merged to form the Rockland-Rockport Lime Company. By this time, railroads transported 100,000 tons of limerock a year from the quarries to Rockland's kilns for burning.

But within 30 years, the quality of Rockland lime declined and there was price-cutting and new competition. New materials and building methods cut the market for lime: wallboard instead of plaster, concrete instead of brick. The kilns went cold, and the railroads were abandoned. Today the only traces of the whole flourishing industry are a few kilns preserved on the waterfronts of Thomaston and Rockport and white limestone tailings on the shores.

CAMDEN

3 ★★★★★

44° 12.40'N 069° 03.40'W

CHARTS: *13307, 13302, 13305* CHART KIT: *20(A), 68A, 20*

Camden is one of the jewels of Penobscot Bay. This beautiful harbor is home to a large fleet of picturesque windjammers and cruising boats of every kind. Curtis Island, with its lighthouse, guards the entrance. The steeples of the small town are white against the hills. A stream cascades down into the head of the harbor. There are excellent marine facilities, good restaurants and stores, entertainment, mountains, and lakes.

All of these attractions have not escaped notice. The town is on the itinerary of nearly every visitor to this part of the coast, by land or sea.

The inner harbor is extremely well protected but also crowded. You are likely to end up on a mooring in the outer harbor. Exposure there is to the south and southeast, and some nights are rolly.

Shipbuilding was a major industry in Camden well before the Civil War. Holly M. Bean is the best known of the Camden builders. The Bean yard stood where Wayfarer Marine is today, and they built the world's largest five-masted schooner and the first six-master in 1900.

Among Camden's many noted citizens, Captain Hanson Crockett Gregory claims an unusual spot in history. In 1847, with a sudden stroke of genius, he poked a hole in his biscuit to place it over one of the spokes of his ship's wheel while he navigated, thereby inventing the donut hole. At the danger of diluting this myth with truth, another version of the story has a fifteen-year-old Hanson poking a hole in his mother's fried dough at the family homestead. Regardless, proud citizens of Camden have a tradition of celebrating the Hole in the Donut Festival in June or whenever they are hungry.

APPROACHES. Nestled at the base of the Camden Hills, Camden is easy to recognize from a distance. Two well-marked entrances lead into Camden Harbor—from the southeast and from the northeast—skirting either side of a group of ledges in the mouth. Most boats use the larger southeast entrance, though both are easy and well marked.

Coming from the south, pass The Graves to either side. You will see the small lighthouse (*Oc G 4s 6M*) on Curtis Island. The sea buoy is red bell "2," (*44° 12.01'N 069° 02.41'W*) off Curtis, and from there it is a straight shot into the outer harbor. Do not use the unmarked entrance west of Curtis unless you carry shoal draft.

The entrance from the northeast is narrower, but well marked and perfectly safe. The sea buoy is red-and-white bell "CH" (*44° 12.68'N 069° 02.27'W*) to the northeast. Keep Northeast Point and its light to starboard and can "1" and green daybeacon "3" on Inner Ledges to port. If it is blowing hard, be alert for strong puffs coming down off the mountains in the area north of Camden.

The geography of Camden Harbor divides it into an inner and an outer harbor. Two hundred or more boats may be moored in the outer harbor. A marked channel leads through them to the inner harbor. Do not cut the corner at Eaton Point—rocks make out from the end of the seawall just before the turn into the inner harbor.

The inner harbor has channels running down both sides and a place to cross over at the head. On weekends, when many of the windjammers are in port, things get pretty tight.

ANCHORAGES, MOORINGS. Most of the outer harbor is filled with moorings. Often the only place to anchor is near the entrance, north of can "7," or in the outer part of Sherman Cove, and both locations are exposed to the east. Don't anchor inside the moorings along Eaton Point, where it is shoal. Anchoring is not allowed in the inner harbor.

⚓ Most of the moorings for transients are managed by Wayfarer Marine and are located in the outer harbor. To request a mooring call Wayfarer on channel 71. During working hours its launch will come out to show you the way. The white mooring buoys for transients are marked with a Wayfarer burgee and the maximum length boat allowed. Their rental includes the use of showers and laundry facilities ashore. Willey Wharf also maintains a number of transient moorings.

⇒ Boats up to 42 feet may lie along the outer floats at the public landing for a maximum of two hours or overnight with permission from Harbormaster Steve Pixley (*Ch. 09; 236-7969*). His office is in the Chamber of Commerce building at the public landing, near the head of the inner harbor, and his launch, appropriately, is the *Welcome*. Wayfarer's docks or Willey Wharf can handle boats of nearly any size.

GETTING ASHORE. For a fee, Wayfarer provides launch service, hailed on channel 71, as does Camden Yacht Club, reached on channel 68. On busy summer weekends you will need to be patient.

The yacht club has a dinghy float, but the public landing, near the head of the harbor on the left, is closer to town. If you can squeeze in, you will be right next to the Chamber of Commerce, pay phones, and the heart of Camden.

FOR THE BOAT. *Camden Yacht Club (Ch. 68; 207-236-3014).* The low red building of the Camden Yacht Club is at the left-hand entrance to the inner harbor. The club has no guest moorings, but they may steer you to one that belongs to a member who is away. You can tie up at its floats for short periods in 7 to 9 feet at low. The inner harbor is extremely busy and crowded, and the easiest spot to fill your water tanks is at the yacht club floats. The steward of this friendly club will help you in any way he can. No fuel is available.

Wayfarer Marine (Ch. 09, 71; 207-236-4378; wayfarermarine.com). Wayfarer's docks and floats line the east side of the inner harbor and have more than 12 feet of depth at low. In midseason boats from all over the world are rafted three or four deep in every available space.

The Wayfarer fuel dock is the first float on your right as you enter the inner harbor. For international arrivals using Camden as their port of entry, they can assist with customs and immigration. They pump gas and diesel and have ice, water, electricity, and a holding-tank pump. Showers and laundry facilities are ashore, and there is a good marine store, with charts.

All services are available. This is a major yard with a large staff of experienced craftsmen capable of repairs on wood, fiberglass, or metal hulls and on engines or electronics. On a site where ships have been built for a century or more, the yard has a 110-ton travelhoist as well as 80 and 35-ton hydraulic trailers, a 35-ton rigging crane, and the beautiful yard tug, the *Barbie D.*

Willey Wharf (207-236-3256). Willey Wharf lies just across the harbor from Wayfarer and caters to very large power and sailboats. Reservations are often made far in advance, but anyone is welcome if there is space.

Willey does not have the boatyard facilities of Wayfarer, but it has the advantage of being right in town. Gas or diesel in large quantities may be obtained at the floats, where depth is 10 feet at low and 30, 50, and 100-amp service is available. Willey can refill propane tanks, and provide water, electricity, ice, and showers.

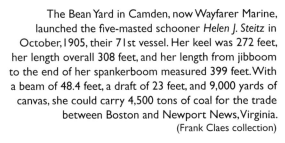

The Bean Yard in Camden, now Wayfarer Marine, launched the five-masted schooner *Helen J. Steitz* in October, 1905, their 71st vessel. Her keel was 272 feet, her length overall 308 feet, and her length from jibboom to the end of her spankerboom measured 399 feet. With a beam of 48.4 feet, a draft of 23 feet, and 9,000 yards of canvas, she could carry 4,500 tons of coal for the trade between Boston and Newport News, Virginia. (Frank Claes collection)

Camden-Rockport Pumpout Boat (Ch. 16; 207-236-7969). This federally funded pump-out service is free, but plan well in advance since the boat may have to come from Rockport.

Gambell & Hunter Sailmakers (207-236-3561; www.grambellandhunter.com). This traditional sailmaker and rigger is located on Limerock Street.

E.S. Bohndell & Co. (207-236-3549). In business for more than a century, sailmaker and rigger Bohndell is on Route 1 in adjacent Rockport.

FOR THE CREW. Visiting yachtsmen are welcome to use the handsome old clubhouse of the Camden Yacht Club, which serves an excellent lunch.

Pay phones are located by the restrooms at the public landing. The post office is a block away on Chestnut Street, and there are several banks. If the office can't survive without you, you can fax or overnight-mail them instructions from the Kax Office Center above Maine Coast Photo on Elm Street: "Sorry. I can't be reached—gone sailing. Press on regardless." Internet connections are available at Wayfarer and the Camden Public Library at the head of the harbor. The inner harbor is a fee-based Wi-Fi zone.

French and Brawn (*236-3361*) is a classic little grocery store jam-packed with all the basics, custom-cut meats, cheeses, and wines. It is right at the apex of Camden's downtown, steps from the town dock. It also provides provisioning service for larger yachts. If bigger is better, a Hannaford supermarket and liquor store is out of town, south on Route 1. A Rite Aid just up from French and Brawn also sells liquor (and aspirin), but you will want to be up early enough for the farmers' market, held a short walk out Colcord Avenue, on Saturday mornings from 9 to noon.

Wayfarer's is the most convenient place to do laundry. Another laundromat is located at Bishop's Store (*236-3339*), a fair trek out Washington Street (Route 105); they are open 7 days a week and do laundry by the pound. Rainy-day reads can be found at The Owl and Turtle, in the renovated Knox Mill; Sherman's, on Bayview Street; and Bookland on Camden Street.

In summertime, there is a wide choice of places to eat within a few minutes of the public landing, ranging from pizzas to nouvelle cuisine, delis, and takeouts. The Camden Deli (*236-8343*), on Main Street, serves knock-out gourmet sandwiches and also sells cheese, deli meats, and other specialties. Boynton and McKay (*236-2465*), across the street, serves breakfasts and lunches and industrial-strength capuccino. Or try the classic Cappy's Chowder House (*236-2254*) by the waterfront. Options for finer fare include the Waterfront Restaurant (*236-3747*), Peter Ott's (*236-4032*), or, highly recommended, Natalie's at the Mill (*236-7008*).

You can reach Don's Taxi at *236-4762*, or you can rent a car for local use at Smith's Garage (*236-2320*) in Rockport. Penobscot Bay Medical Center (*207-596-8000*), on Route 1 about 6 miles south of Camden, is a large, modern acute-care hospital with 24-hour emergency-room service and a helicopter pad.

THINGS TO DO. Walking the streets of Camden, window-shopping, and observing the activity in the busy harbor can easily fill your day. If you have a little extra time, however, there are many other interesting things to see and do. Start at the Chamber of Commerce, in the little house at the public landing, in the heart of town.

Check to see what plays or concerts are scheduled in the beautiful outdoor amphitheater at the head of the harbor. For movies, there is the old-fashioned Bay View Street Cinema.

The Camden YMCA (*236-3375*) is on Union Street just over the Camden-Rockport town line. It has a swimming pool, squash and racquetball courts, exercise rooms, showers, and a gym. All are open to the public. It is about .1 miles from the intersection of Union and Limerock Street.

If you want to cruise Maine in traditional style, or if you just want to dream on a drizzly day, visit Cannell, Payne, and Page (*236-2383*), one of the best-known brokers of wooden boats in the country. But bring either your self-control or your checkbook—they broker some gorgeous boats. Their office is in the shingled American Boathouse at the head of the harbor.

Camden Hills State Park is a mile or so north of town on Route 1. The park covers some 8,000 acres which include Mount Megunticook and Mount Battie, hiking trails, modest climbs, and spectacular panoramic views.

With the exception of Cadillac Mountain on Mount Desert Island, 1,385-foot Mount Megunticook is the highest point on the Atlantic seaboard. The road to the top of Mount Battie was originally built for cannon and later used by sightseers. From the summit, the view is breathtaking, with the peaks of Mount Desert in the distance, Isle au Haut, Deer Isle, Blue Hill, and dozens of spruce and granite islands dotting the bay. There is a modest entrance fee.

You can easily climb Mount Battie on foot. Walk north up Main Street and turn left on Mountain Street (Route 52). Take the first right on Megunticook Street and follow it up to the beginning of the trail, perhaps 15 minutes from the center of town. It is about a 25-minute hike to the top, with only a few spots where you have to scramble.

Bikes can be rented at Brown Dog Bikes on Elm Street or at Maine Sport, on Route 1 in Rockport. Camden Hills State Park is one good destination, or go out Mountain Street (Route 52) to Lake Megunticook, where you can swim and picnic at Barrett Cove. To work up more of a sweat, you can hike the Maiden's Cliff trail, which starts nearby and connects to the state park trail system.

The Camden-Rockport Historical Society has a pamphlet outlining a 2.5-mile tour on foot, and a more extensive bicycle or car tour. Get a copy at the Chamber of Commerce. Noteworthy on the tour is the Whitehall Inn, north of town on Route 1, where in 1912 Edna St. Vincent Millay, then a schoolgirl, first read her poem "Renascence," beginning the literary career that won her a Pulitzer Prize. Her statue looks out over the harbor.

The Camden History Center, in a former sea captain's home at 4 Union Street, has fascinating collections about the town and harbor. Be sure to check out the collections of yacht historian Fred Crockett, a distant relative of Captain Crockett of donut-hole fame.

Half a mile south of Camden on Route 1 is the Old Conway House (c. 1780), Blacksmith Shop, and Museum. At the end of Conway Road is Merryspring Nature Park, with 66 acres of fields and woods, an arboretum, and marked trails.

A lovely walk leads out Bay View or Chestnut Street through one of the town's quieter residential districts and then past Aldermere Farm, home of the intriguing Belted Galloway cattle. If you go beyond the farm, all the way to Rockport, it is about 3 miles each way.

Boats from around the world put in at Wayfarer Marine. Curtis Island (below, left) is a town park.

En route take Calderwood Lane on the left and follow it for about half a mile to a right turn. A sign for "Vesper Hill" on the far side of the turn marks the entrance to the beautiful open-air chapel and gardens of the Children's Chapel.

With your own dinghy you can reach Camden's waterside parks. Live concerts are often performed on Wednesday afternoons at small Laite Beach, on the south shore of the harbor east of the yacht club. All of Curtis Island, in the harbor entrance, is a town park. Land at the end toward town. There is a grassy area for a picnic, plus a working lighthouse and the whole little island to explore. In September of 1993 an extremely rare True's beaked whale beached itself on the shores of Curtis. None of these whales has ever been sighted alive in the wild, and only sixteen have been observed in North America, and those only by similar misfortune. What brought this whale to its sad fate on Curtis remains a mystery.

Windjammers have operated out of Camden for more than half a century. Not only do they offer a wonderful experience for their guests, but their tall sails provide a glorious sight on the horizon. Some are a century old and converted from the coastal trade, some are brand new and built for the purpose. On Mondays, a procession of these magnificent vessels depart Camden Harbor, their decks colorful with passengers. Some proceed under their own power, but many are pushed by yawlboats until they reach open waters and can hoist sail. After leaving Camden, the schooners spread to all points of the compass to explore the remote harbors, small villages, and deserted islands of Penobscot Bay and beyond.

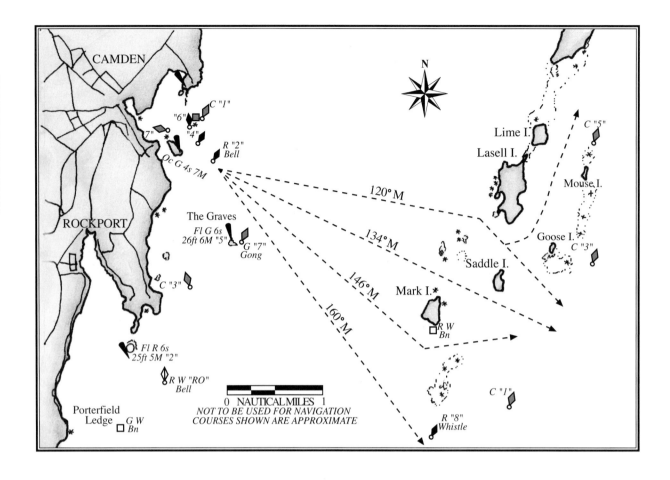

PASSAGES EAST FROM CAMDEN
Charts: 13305, 13302 *Chart Kit: 68, 20*

Going east from Camden, you will have to pass through the chain of islands that splits Penobscot Bay—Mark, Saddle, Lasell, Lime, and Islesboro. If you are heading for the Fox Islands Thorofare, the easiest route is south of the whole chain, picking up red bell "12" (44° 09.00′N 068° 59.08′W) at the end of Robinson Rock and leaving it to port.

If you are bound for Pulpit Harbor, on North Haven, or other points north, there are several slots through the islands. The cleanest and safest route is between Mark Island and Robinson Rock, leaving the daybeacon south of Mark Island close to port.

The next choice, a bit more direct but not quite as easy, is to pass between Mark and Saddle Island. Be sure to identify East Goose Rock, which is always visible, and leave it to port. To avoid the ledge east of East Goose Rock, stay south of a line between East Goose and the southern tip of Saddle Island. Give Mark Island a fair berth to starboard, noting the outlying rocks on the chart.

The next passage north runs between Saddle and Lasell Island—less comfortable because of the ledge (usually invisible) northeast of East Goose Rock. At least one schooner has fetched up on this ledge.

The safest route is to stay within 100 or 200 yards of Lasell Island. Rocks run out like a breakwater from the southwestern tip of Lasell and form a small cove, but they are visible, and they are usually privately marked. Leave the mark and the rocks to port.

Once past the tip of Lasell, continue southeast, leaving Goose Island to port. Or run northeast between Lasell and Mouse Island, emerging north of the ledges marked by can "5," where seals love to bask.

The daring can hold their breath and pass north of Lime Island. A treeless islet known locally as "Little Bermuda" lies just north of Lime Island, and just north of "Little Bermuda," the chart shows a 4-foot spot, which you can cross with caution on a rising tide. *Do not attempt going between Lasell and Lime Island.*

LASELL ISLAND

44° 12.15'N 068° 57.65'W

CHARTS: 13305, 13302   CHART KIT: 68, 68A, 20

There is an attractive beach at the north end of Lasell Island, east of Camden. Approaching Lasell from the northwest, find the long ledge extending north from Lasell, visible except near high tide.

After rounding this ledge, anchor in a comfortable depth off the pebble beach. This anchorage is exposed, but it makes a pleasant lunch stop with views of the Camden Hills and the surrounding islands. Please do not land on the private beach.

THE FOX ISLANDS

Known collectively as the Fox Islands, North Haven and Vinalhaven are a paradise for cruising sailors. Fox Islands Thorofare provides a convenient and fascinating passage from west to east, and there are enough harbors and coves, passages and bays in the deeply indented shores of these large islands to keep you well occupied for a week. Pulpit Harbor on North Haven is one of the finest harbors in Maine, and one of the most beautiful.

The Fox Islands were named by Martin Pring, in command of *Discoverer* and *Speedwell*, who saw foxes on the islands during his voyage of exploration in 1603. Separated by only a few hundred feet of water, North Haven and Vinalhaven are worlds apart.

North Haven and the Thorofare are home to summer people whose relatives first came here by yacht or steamer. Splendid summer cottages and docks line both sides of Fox Islands Thorofare, and a working lobsterboat here seems like a quaint anachronism.

Vinalhaven is entirely different. Although its population swells in summer, this island never has the gritty feeling of a working community where the year-round residents, the fishermen and lobstermen from generations back, are still in charge. Here the names are Dyer and Carver, Arey and Vinal, Calderwood and Coombs. The Reach, the entry to Carvers Harbor, is a working man's thoroughfare bordered with modest homes.

As if to emphasize the difference, each island has its own separate ferry departing from the same terminal in Rockland.

PULPIT HARBOR

■ ★★★★★   ⚓ 🛒 🍴

entrance: 44° 09.63'N 068° 53.38'W
Pulpit anchorage: 44° 09.30'N 068° 53.18'W
Cabot Cove: 44° 09.32'N 068° 53.49'W
CHARTS: 13308, 13302, 13305
CHART KIT: 68B, 68, 18, 20, 21

Pulpit is a true harbor, nearly landlocked, on the northwest coast of North Haven Island. It is easy to get to, easy to enter, and stunningly beautiful. As you approach, a hidden entrance reveals itself, guarded by an osprey nest on Pulpit Rock. Once inside, the protection is excellent for a hundred boats or more. In the evening, the sun sets through the harbor entrance and over the Camden Hills. And Pulpit is only a two-hour sail from Camden or less than an hour from the Fox Islands Thorofare.

Pulpit's beauty, accessibility, protection, and its size make this ground zero for Penobscot Bay cruising. It's on the harbor list of nearly every boat that is cruising the bay for the first time, and yacht club cruises rarely pass it by. If you want an anchorage to yourself, this is not it. But if you want one of the best sunsets on the coast of Maine in a setting you will never forget, here's a contender.

It is customary for several windjammers to anchor here on Friday nights to give them a short sail back to Camden on Saturday morning. It's a glorious sight to watch these stately schooners glide into Pulpit and drop the hook.

APPROACHES. Pulpit Harbor is hard to see, especially from the south or west. At a distance, look for a clearing in the woods and a meadow sloping to a barn-red house on the left side of the entrance.

The resident ospreys of Pulpit Rock.

From the north, find the large yellow house on the crest of the hill north of the entrance and look for a cluster of small gray cottages inside the harbor.

Pulpit Rock, a bold pinnacle that gives the harbor its name, is notoriously hard to see against the shore. Sometimes you can distinguish it at a distance by its speckling of white guano. A huge osprey nest balances on its top.

The entrance is northeast of Pulpit Rock. Leave Pulpit to starboard and turn south, passing halfway between it and the small cliffs on shore to port. Stay in the middle as you turn to the southeast, away from the ledges that make out from the north side.

As you enter, have your binoculars ready. The enormous osprey nest has crowned Pulpit Rock for more than 150 years. In the early summer, you can watch the parents returning to the nest to feed their squawking chicks, grasping fat fish in their talons and landing with loud chirps of triumph.

ANCHORAGES, MOORINGS. Just after you enter the harbor, a cove extends to your right. The land surrounding this cove has been owned for generations by the Cabot family, and it is known locally as Cabot Cove. The several moorings here are private. You can anchor a long way in, in 15 to 20 feet at low, with excellent holding and protection.

For an unobstructed view of the sunset on the Camden Hills, head for the southeast end of Pulpit Harbor, anchoring in 16 to 20 feet anywhere along the perimeter. Latecomers and the big schooners anchor more in the middle of the harbor, where depths are 20 to 30 feet. Private markers may designate a channel through the anchorage.

Another cove stretching out to the east has a number of moorings maintained by Thayer's Y-Knot Boatyard. Five are available to transients by calling *867-4701*.

Pulpit Harbor's mud bottom is generally good holding ground. There are also patches of kelp, however, so be sure your hook is set. The encircling hills provide good protection from winds from any direction, and it is a superb harbor under all conditions.

GETTING ASHORE. Row to the floats at the public landing, at the head of the cove stretching east.

FOR THE BOAT. *Thayer's Y-Knot Boatyard (Ch. 09; 207-867-4701).* Thayer's is based a short distance away in Southern Harbor (see entry, page 214). In addition to the Pulpit Harbor moorings, the yard sometimes keeps a work boat in Pulpit for tows and repairs.

FOR THE CREW. Pulpit Harbor was once North Haven's center, with three general stores, a post office, and a customs house. All this has disappeared.

North Haven Grocery (*867-2233*) is a short walk from the town landing. Walk south over the little bridge, following the harbor around past the graveyard and up the hill. They are open from 6AM to 8PM daily, and they will give a ride to big-order customers if you call. In addition to being well stocked with all the basics, they carry ice, *The New York Times,* and their cafe serves breakfast, lunch, and dinner. The Islander Grocery Store (*867-4771*) is just beyond the summit. They will also deliver substantial orders—and perhaps their orderers—back to the public landing. They have a good line of groceries, beer, wine, and ice.

THINGS TO DO. Please respect the tranquillity of this special place. The harbor holds many boats that won't want to share in your revelry—or hear the hum of your generator.

Rural walks lead from Pulpit in all directions. Allow an hour to walk to the town of North Haven, plus time for collecting raspberries. Aside from taking walks, it's fun to poke around with the dinghy. Row up under the little bridge to the east. This provides a lovely, peaceful passage, and you can go for a mile, if you wish, at high tide.

BARTLETT HARBOR

2 ★★★

44° 08.22′N 068° 55.30′W

CHARTS: 13308, 13302, 13305
CHART KIT: 68B, 68, 18, 20, 21

This quiet little harbor lies on the west side of North Haven Island, a couple of miles west of Pulpit Harbor. Although exposed to the west and southwest, it can be a pleasant place to anchor in settled summer weather, with a beautiful view of the Camden Hills.

APPROACHES. The only danger in the approach is the long ledge that makes out from the north side of the entrance. It is visible almost until high tide.

ANCHORAGES, MOORINGS. A number of boathouses and sheds cluster at the northeast end of the harbor. There is room to anchor outside several private moorings in 12 to 20 feet. The holding is good.

Pulpit Harbor and its gatekeeper, Pulpit Rock. Once inside, Cabot Cove is to the right. The town landing (not visible) is at the head of the harbor to the left. (Hylander)

FOX ISLANDS THOROFARE

CHARTS: *13308, 13302, 13305*
CHART KIT: *68B, 68, 18, 20, 21*

Fox Islands Thorofare is one of the great east-west passages of Maine. Winding seven or eight miles between the islands of North Haven and Vinalhaven, it is sometimes 500 yards wide and sometimes as narrow as 100 yards. Because of the many turns and shifting winds and places where you will be blanketed by the land, it is a challenge to sail all the way through. Like all passages in Maine, the Thorofare is buoyed from east to west.

In the town of North Haven, on the north side of the Thorofare, you can satisfy most of the needs of boat or crew. The ferry from Rockland docks here several times a day, making it a convenient place to meet people without sailing back to the mainland.

The Thorofare is lined with great summer cottages, each with its own dock and float. It provides its own form of entertainment, with a constant stream of every kind of boat—classic and modern, wood and glass, large and small. It is hard to come here without meeting someone you know.

Currents in the Thorofare are not usually very strong, unless it has been blowing hard from west or east. The critical area is in the narrows at Iron Point, where you may find strong currents and eddies. The ebb flows in both directions from Iron Point, and the flood meets there.

APPROACHES from the west. Approaching from the west, the entrance to Fox Islands Thorofare is hard to see because of the way North Haven curves down around Vinalhaven. From Camden, steer just north of the outermost islands on the horizon (Hurricane and the White Islands), passing north of The Graves and south of Robinson Rock.

The first mark you will pick up probably will be the large granite monument on Fiddler Ledge. Before reaching Fiddler, keep a careful watch for dangerous Drunkard Ledge, half a mile to the west. The currents are especially strong in this area, and the flood will set you toward this ledge, which is submerged at high tide, uncovering 7 feet at low. The red daybeacon marking Drunkard Ledge is near its eastern edge. Do not cut it close. Leave the granite monument on Fiddler Ledge to port. The ledge extends a short distance south of the monument, so give it a little room.

In thick weather, find red-and-white bell "FT" *(44° 05.29'N 068° 57.28'W)* or red gong "26" *(44° 05.81'N 068° 56.81'W)* in the entrance.

Near Fiddler Ledge, you will sight inconspicuous Browns Head Light in a clearing on the northwest shore of Vinalhaven. The white sector marks the safest approach to the Thorofare. This lighthouse is more than 150 years old, built during

Andrew Jackson's presidency. It is now the home of the Vinalhaven Town Manager.

The Sugar Loaves are high, barren rocks, grassy on top and usually dotted with cormorants drying their wings. Beyond them are the low, wooded, and rocky Dumpling Islands.

From there on, the Thorofare is obvious and well marked. Can "17" sticks out in the channel farther than you might expect, and it always seems to be at one of these narrow points that you look astern and find the ferry bearing down with a bone in its teeth.

There is a substantial fleet of boats moored off the town of North Haven that often obscure nun "14." It marks Lobster Ledge, covered 2 feet at low.

Iron Point can be a challenging passage under sail. The wind may head you just enough to force you to tack between Dobbin Rock and Iron Point Ledge, right where the current is being squeezed through the Thorofare's narrowest section.

The eastern end of the Thorofare is characterized by fluky winds, but there are few dangers. Huge meadows and saltwater farms run to the water. The lighthouse on Goose Rocks (known as "The Sparkplug") has been fitted with a solar panel that is probably highly efficient but looks incongruous.

There is plenty of room on either side of Widow Island, which has a beach on the western side and a house near its peak on the southeast. In 1888 the U.S. Navy built a hospital here for victims of yellow fever, thought to be caused by the damp and the rot so prevalent in wooden sailing ships. But the few actual cases of yellow fever were quarantined in New York, and the hospital on Widow never saw a single patient. Six years later, the Navy gave the hospital to the state, which sent mental patients here for "sanitary retreats" up until World War I. Then the buildings were used as a school for lighthouse keepers' children. But the cost of maintenance or the Depression closed the school, and the buildings were torn down.

APPROACHES from the east. Approaching from the east in thick weather, run for red bell "4" (*44° 07.75'N 068° 48.48'W*) off Channel Rock and for the foghorn on Goose Rocks behind it. If you are fogbound or caught by darkness, Carver Cove in Calderwood Neck is extremely easy to enter, and it provides good anchorage and protection in winds from all directions except the northeast. Kent Cove is another easy-to-enter alternative.

In clear weather, you can identify the white tower of Goose Rocks Light at a distance.

SOUTHERN HARBOR

2 ★★★

44° 08.34'N 068° 53.37'W

CHARTS: *13308, 13302, 13305* CHART KIT: *68, 68B, 19, 21*

Southern Harbor is a large body of water on the southwest side of North Haven Island extending northward from Fox Islands Thorofare. The entrance is easy. Because of the exposure to prevailing southwesterlies, there may be an uncomfortable chop here at times. When the wind is out of the north or east, however, it offers very good protection.

APPROACHES. Coming from the west, leave the Sugar Loaves (high, barren, grassy on top) to starboard and continue up into the harbor.

From North Haven, leave the Dumpling Islands and nun "20" at Calderwood Rock to starboard, then double back toward Southern Harbor.

Avoid the 2-foot spot west of Turnip Island by favoring the left side of the channel at this point. Near low tide, Seal Ledge will probably be occupied by a large and lethargic group of these creatures. Seal Ledge, Lobster Island, and the subsequent little bits of land lie more or less in a straight line, so you should have no trouble steering a safe course past them, even if the tide is high.

ANCHORAGES, MOORINGS. Anchor on the centerline of the harbor in good mud, 10 to 16 feet deep at low, or pick up one of the moorings set by Thayer's Y-Knot Boatyard. The outer moorings are suitable for boats drawing 7 feet.

GETTING ASHORE. Row in to Y-Knot Boatyard.

FOR THE BOAT. *Thayer's Y-Knot Boatyard (Ch. 09; 207-867-4701)* is on the eastern side of the harbor, past the cove east of Pigeon Hill. Red-and-green stakes mark a channel to the stone dock and floats. The channel is dredged to 3.5 feet at low.

The yard has water, electricity, and basic boat supplies, but no fuel. The yard can haul and handle most repairs, including diesel engines and electronics, and a diver is available, should you be unlucky enough to need one.

FOR THE CREW. The Y-Knot has showers. The Islander Grocery Store (*867-4771*) is about a mile north of the boatyard, and they will deliver substantial orders. North Haven Grocery (*867-2233*) is a little farther in the same direction. The small village of North Haven is about .7 miles south of the harbor.

NORTH HAVEN

3 ★★★★

44° 07.50'N 068° 52.33'W

CHARTS: *13308, 13302, 13305* CHART KIT: *68, 19, 20, 21*

North Haven is a pleasant village on the north side of the Fox Islands Thorofare. This was one of the first summer communities in America, founded by Boston yachtsmen who saw it as a wonderfully protected but challenging area to race small boats.

The 14½-foot, gaff-rigged North Haven dinghies started racing in 1887 and are still racing today—probably the oldest formal one-design boat in the country. Another popular class in North Haven is the beautiful little Herreshoff 12½.

Although the Thorofare is lined with private docks and floats, most of those of interest to visiting yachtsmen are clustered around the ferry landing. The car ferry from Rockland makes frequent trips, so keep your eyes open.

APPROACHES. North Haven is midway through the Thorofare, centered on the ferry landing. While maneuvering among the moored boats, do not ignore nun "14," marking 2-foot Lobster Ledge to the east.

ANCHORAGES, MOORINGS. The yacht club here is officially known as the North Haven Casino. It has several guest moorings, marked "NHC," and extensive floats west of the ferry landing. Tie up to the yacht club floats, with 5 feet alongside at low, for short periods only. The J.O. Brown & Son boatyard also has moorings (marked "JOB") opposite the yard, at very reasonable rates.

If no mooring is available, anchor outside the moored boats in 16 to 25 feet of water at low. The holding ground is good, in mud. Exposure is to the west and southeast.

The ship house along the Thorofare.
Holding forth at J.O. Brown (above).

GETTING ASHORE. Row in to Brown's float, the town landing, or the yacht club floats. The small building at the seaward end of the yacht club dock is the Casino. The buildings on shore are private homes and not part of the Casino, so don't wander in looking for showers.

FOR THE BOAT. *J.O. Brown & Son, Inc. (Ch. 16; 207-867-4621).* Located east of the ferry landing, this yard was started in 1899 at the old clam and lobster processing factory. Fourth-generation Foy Brown is still building wooden boats here, ranging from traditional lobsterboats to a radical Phil Bolger design.

Gas, diesel, water, ice, and electricity are on the floats, and there is a chandlery. The water comes from an island well, and may be a bit skunky. Taste before you tank. The yard has a 15-ton boatlift and can handle hull and engine repairs. Or you can do them yourself. Brown's is not open on Sundays.

North Haven Casino (207-867-4696). Water is available at the floats.

Town Landing. There are no facilities at the town landing, which is tucked inside the ferry slip on the west side.

FOR THE CREW. North Haven is a little town—too little, it seems, to support a market, though not for lack of trying over the years. If you are here on Saturday morning, you can get local produce at the farmers' market, which is held at the ballfield, a block up from the waterfront.

Or you can get cleaned up and go out to eat. Brown's boatyard has showers and laundry, but they are unavailable when the boatyard is closed on Sunday. You can dine in or take out at H.J. Blake's restaurant in the middle of town, or you can buy some delectables to take back to the boat. Brown's Coal Wharf Restaurant (*867-4739*), east of the boat-

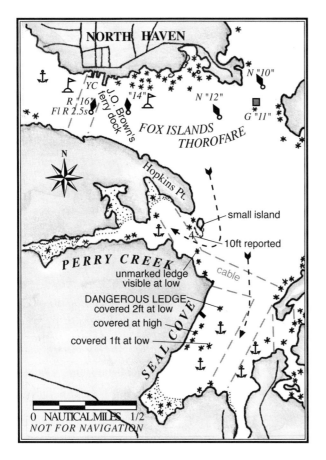

PERRY CREEK

 ★★★

44° 06.90′N 068° 52.05′W

CHARTS: *13308, 13302, 13305*   CHART KIT: *68, 68B, 19, 21*

Perry Creek is a beautiful well-protected anchorage, tucked away around the corner from North Haven village. It is small and very peaceful, with only one farmhouse visible, fields leading down to the water, and a picturesque little island guarding the entrance.

Perry Creek suffers the dubious geographic fate of being both beautiful and easy to get to, which makes it a magnet for boats passing through the Fox Islands Thorofare. It has become very popular—some would even say overused. Arrive early and treat this special place with extra care.

APPROACHES. Coming from Fox Islands Thorofare in either direction, bear south toward Hopkins Point and the entrance to Seal Cove and round little Mouse Island (unnamed on the chart) off Hopkins Point.

A substantial unmarked ledge extends southward from the island for a distance about equal to the length of the island itself.

Another unmarked ledge protrudes from the south shore of Perry Creek, and this one catches a number of unwary sailors each summer. The stake reported on the chart no longer exists. At low tide, the ledges on both sides of the entrance are visible. At high tide, come in halfway between the southern tip of the island and the shore to your left.

Local sailors report they have been unable to find the 4-foot spot shown on the chart just past the entrance. Curve northward past the rocky point to port until you are beyond the charted cable area.

ANCHORAGES, MOORINGS. Anchorage in Perry Creek is limited by the ever-increasing number of private moorings toward the head of the creek and by a cable area at its mouth. There is still room to anchor just west of the cable area, near the green boathouse on the north shore, in 8 to 11 feet of water at low. Holding ground is good, in mud.

You can also work your way farther up the creek, although the deep water narrows rapidly. The farther in you go, the prettier it becomes. Also the buggier. Prepare for mosquitoes early.

THINGS TO DO. On a high tide, you can row about a mile up to the head of the creek—a most

yard, serves lunches and dinners most weekdays. Dinner service continues through the weekend, and bar fare is served in their lounge until 11 PM.

THINGS TO DO. Waterman's Community Center (*867-2237*) dominates the center of town. Check their schedule of local performances. A pay phone and restrooms are at the ferry landing next door. Walking east on the main street from the casino, you will find the small library and the post office.

The little town has several craft shops and galleries. Eric Hopkins, noted for his abstract aerial paintings of Maine islands, has a gallery just west of the ferry landing. The Calderwood Hall Gallery, at the east end of town, occupies what once was the community center.

If you have your own bikes, there are 27 miles of paved roads and much to enjoy, from sparkling coves to archaeological sites.

This is a great area to explore by small boat. Take your dinghy and poke around in the Thorofare, or up Perry Creek. An interesting run, well seeded with rocks and ledges, goes through the Mill River (dry at low tide), under the bridge, and into Winter Harbor. Both sail and powerboats can be rented at Brown's yard.

delightful experience. The shores are lined with sloping rocks and bits of marsh where sanderlings and kingfishers play and ospreys lumber overhead.

Both shores are conservation land. Trails along both shores lead through stands of pine and spruce and birch to outcrops and clear vistas.

SEAL COVE

 ★★

44° 06.27′N 068° 51.70′W

CHARTS: 13308, 13302, 13305   CHART KIT: 68, 21

Seal Cove cuts into the northern end of Vinalhaven Island halfway through the Fox Islands Thorofare. It is well protected and easy to enter, but the charts are dangerously misleading.

Mr. Emmett Holt reports that the 6-foot spot shown on the west side of the entrance is actually a large ledge covered by 2 feet of water or less at low. An informal marker may be placed on the ledge during summer months.

APPROACHES. Coming through Fox Islands Thorofare from either direction, head southward into the middle of Seal Cove, staying well east of the ledge mentioned above.

ANCHORAGES, MOORINGS. Like Perry Creek, much of Seal Cove is charted as cable area, though the cable itself is reported to be nonexistent. To be safe, anchor in the southeastern part of the cove, favoring the eastern shore, just outside the cable area. Or try the other areas shown on the sketch chart. Mr. Holt reports that it is also possible to tuck inside, between the "6-foot" ledge and the western shore, with plenty of swinging room. The holding ground is good mud.

WATERMAN COVE

1 ★★

43° 08.28′N 068° 51.48′W

CHARTS: 13308, 13302, 13305   CHART KIT: 68, 68B, 21

Due north of the narrows at Iron Point, halfway through Fox Islands Thorofare, Waterman Cove is of little interest under normal cruising conditions. If the wind is out of the north, however, it provides a good refuge that is easy to enter.

APPROACHES. Leave nun "8" to port and head up the middle of the cove toward the huge meadow.

ANCHORAGES, MOORINGS. Anchor in the middle of the cove, west of Fish Point, in 7 to 10 feet of water at low. The holding ground is good, thick mud.

KENT COVE

1 ★★

44° 08.87′N 068° 49.74′W

CHARTS: 13308, 13302, 13305   CHART KIT: 68, 68B, 21

This large cove, with two bights, is at the eastern end of Fox Islands Thorofare, north of Goose Rocks Light. Like adjoining Waterman Cove to the west, it is of little interest during the summer months. However, the eastern bight here would provide good protection from winds from the north or northeast.

APPROACHES. Coming through Fox Islands Thorofare from the east, leave Goose Rocks Light to starboard and head up the middle of Kent Cove. Identify the high little wooded island and proceed into the northeast bight, leaving the island to port.

ANCHORAGES, MOORINGS. Anchor in 8 to 10 feet of water at low, just past the island. Holding ground is good, in mud, everywhere but in the western bight of Kent Cove where there is grass.

CARVER COVE

3 ★★★

44° 07.11′N   068° 50.17′W

CHARTS: 13308, 13302, 13305   CHART KIT: 68, 21

Carver Cove is at the eastern end of Fox Islands Thorofare, making into Calderwood Neck on Vinalhaven Island. It is extremely easy to enter and provides good protection under normal summer conditions. It is the prettiest of the several large coves in the Thorofare, and a convenient stop if you are coming in late at night from the east. Even if the wind is blowing a gale from the southwest, you can spend a comfortable night here.

There are only a couple of houses on the cove, with several little gravel beaches. Herons and ospreys live nearby, and a seal will probably come near to inspect his guests.

APPROACHES. Pass to either side of Widow Island and head southward for the center of the cove.

ANCHORAGES, MOORINGS. Go well up to the head of the cove and anchor in 9 to 17 feet of water at low. Holding ground is good, in mud.

LITTLE THOROFARE and EASTERN PASSAGES

CHARTS: 13308, 13302, 13305   CHART KIT: 68, 68B, 21

At the eastern end of Fox Islands Thorofare, Stimpsons, Calderwood, and Babbidge Island, together with Burnt Island, provide an interesting series of passages that require considerable attention. Both Stimpsons and Babbidge are heavily wooded, while Calderwood between them has several meadows and clearings.

Between Stimpsons and Calderwood. The easiest passage is between Stimpsons and Calderwood. Stimpsons is privately owned, but Calderwood is a good daystop with beaches and trails to explore.

Run along the south coast of Stimpsons until you see the southern tip of Calderwood, rocky with just a few trees. The only danger is the rock near the southeast corner of Stimpsons. It uncovers 9 feet at low and normally is easy to find. Run up halfway between this rock to port and Calderwood Island to starboard. The best place to anchor is on the bar (44° 08.52′N 068° 48.26′W), where the chart shows 7, 8, and 10-foot soundings.

Between Calderwood and Babbidge. The passage between Calderwood and Babbidge has more obstructions. One person reported hitting an uncharted ledge, covered only 2 or 3 feet at low, while sailing through this passage at 6 knots!

It is best to go through here at halftide or better, preferably on a rising tide. Coming from the west, stay close to either side of the 7-foot rock in midchannel to avoid the ledge making out from Babbidge. Then stay to the right of midchannel, to avoid the ledge off Calderwood.

Little Thorofare. Little Thorofare, between Stimpsons and Burnt Island, is clean at the eastern end (except for the three little islands off Burnt), but the western end is a rock garden.

Coming from the west, run close to the northern shore of the Thorofare heading toward the dock east of Indian Point. From the dock, turn sharply to starboard and strike across the Thorofare toward another dock west of the little island off the coast of Stimpsons. As you approach the islet and dock, turn to port and continue down the midline of Little Thorofare.

MULLEN COVE

2 ★★

44° 09.25′N   068° 48.55′W

CHARTS: 13308, 13305   CHART KIT: 68, 68B, 21

Mullen Cove, on the east side of North Haven Island, north of Little Thorofare, looks good on the chart, but it is not a secure anchorage. The cove makes a fine daystop, though, for exploring the town-run park and trails on Mullen Head.

APPROACHES. As you leave Fox Islands Thorofare and run north past Babbidge and Calderwood Island, high, wooded Mullen Head is easy to find. Note the unmarked rocks off Burnt Island and off Mullen Head, and stay well out until the cove opens up. Then run down the centerline. There is a red house at the head of Mullen Cove.

ANCHORAGES, MOORINGS. The cove is easy to enter, but it holds numerous unmarked rocks near its head. Don't go in too far. Anchor in 12 to 17 feet at low, and be sure your anchor is well set. The holding ground is likely to be poor, with an ample supply of kelp. Southwest winds may funnel into the cove through the slot west of Burnt Island.

MARSH COVE

44° 10.60'N 068° 49.23'W

CHARTS: 13305, 13302   CHART KIT: 68, 68B, 21

At the northeastern end of North Haven Island, just south of Oak Hill, Marsh Cove provides good protection from prevailing southwest summer winds. There is a pleasant beach and one of the few totem poles to be found in the Penobscot Bay region. A substantial stone dock marks the north side of the cove, and there are views of Isle au Haut and Deer Isle on the eastern horizon.

The family compound of the late Thomas J. Watson, Jr., former chairman of IBM and ambassador to the USSR, occupies 330 acres of Oak Hill. The estate is complete with half a dozen houses, a private airfield, a flock of deer, and a fleet of Model-Ts—designed, as Watson said, as "a grandchildren trap."

APPROACHES. The cove is easy to enter, south of Hog Island. Hog is actually two little islands connected at low. Head for the middle of the beach.

ANCHORAGES, MOORINGS. There are several private moorings, but plenty of room to anchor in 15 feet or so at low, west of Hog Island, off the center of the beach. Holding ground is fair, in mud and grass.

THINGS TO DO. If you arrive here after the nesting season ends in mid-July, row over to state-owned Dagger Island to the east.

WINTER HARBOR

4 ★★★★

44° 05.67'N 068° 49.22'W

CHARTS: 13308, 13302, 13305   CHART KIT: 68, 21

Winter Harbor is a long, narrow slot on the east side of Vinalhaven Island, south of Calderwood Neck. It is remote and beautiful, and it provides excellent protection from all winds except northeasterlies. The harbor is home to several fishing and lobsterboats, and the back reaches of the harbor are used for seeding and harvesting mussels.

APPROACHES. The approach to Winter Harbor is easy. As you come from Fox Islands Thorofare, coast around Calderwood Neck. The highest, rounded, wooded bump to the south is Bluff Head. Just west of Bluff Head is Big Hen Island, with

Goose Rocks Light, Fox Islands Thorofare.

conspicuous white granite shores. Steer for Big Hen until you are well past the southeast corner of Calderwood Neck, from which the rocks make out a long way. Then make your turn into Winter Harbor, going right up the middle. Note the 2-foot spot just past the island on the right side of the entrance.

ANCHORAGES, MOORINGS. Anchor anywhere you wish in 17 to 30 feet. The depth varies considerably, so find a reasonably shallow spot, such as near the first little island to port. Another good spot is midchannel between the second little island to port and the imposing 100-foot cliff of Starboard Rock.

It is also possible to go in farther, beyond Starboard Rock, squeezing right to avoid a visible rock to port and then another 2-foot spot—this one a solid rock with a wig of kelp that's visible at low. Once past the obstructions, depth drops to over 50 feet at low. Find the edge of the hole and anchor in 17 to 26 feet before the next set of little islands. If you are particularly adventurous, you can even work your way farther into the pool off the entrance to Mill River.

Winter Harbor connects with Mill River, so there is current in the anchorage. You may find yourself lying to either the current or the wind, whichever is stronger. Under normal conditions, Winter Harbor is exposed only to the northeast. Hard winds from the southwest, however, can funnel down the length of the harbor and build up an uncomfortable fetch.

THINGS TO DO. If the tide is up, the estuaries of Winter Harbor are fun to explore by dinghy. Please don't land on private Starboard Rock.

Hurricane Island Outward Bound students may be camping in various spots around Winter Harbor during their course's "solo" experience. Wave, but respect their solitude.

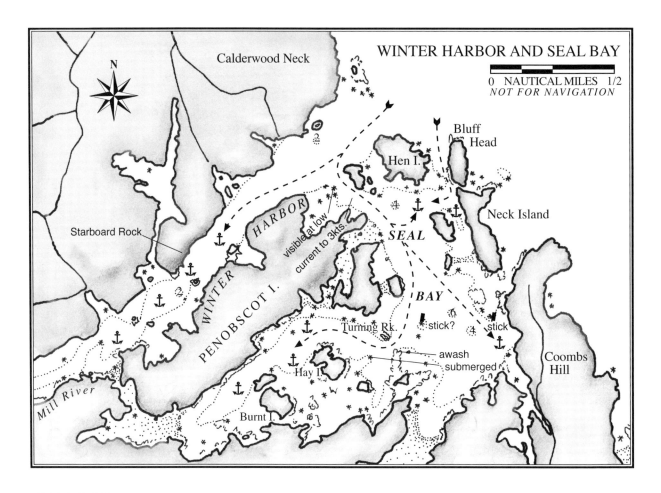

SEAL BAY

 ★★★★

S of Hen Island: 44° 05.84′N 068° 48.02′W
NNW of Hay Island: 44° 05.35′N 068° 48.60′W

CHARTS: **13308, 13302, 13305**   CHART KIT: **68, 21**

Seal Bay wraps around Penobscot Island to the east and south of Winter Harbor, at the northeastern end of Vinalhaven. Its corners provide a variety of beautiful and well-protected anchorages, but the bay is unmarked and requires careful navigation.

The wooded, rocky islands near the entrance to the bay are particularly beautiful, and yes, there still are seals here.

APPROACHES. Approach Seal Bay as you would Winter Harbor. Rounding Calderwood Neck from the north, pick up the high, rounded, wooded bump of Bluff Head. Turn in toward Winter Harbor when you can safely pass the rocks making out a long way from the eastern end of Calderwood Neck. The Hen Islands are easy to identify, with their long, sloping, white granite ledges leading up into the trees.

The usual entrance to Seal Bay is just west of Little Hen Island. Leave it close aboard to port to avoid the rocks off Penobscot Island (visible at low). The current running in and out of Seal Bay past Hen Island can be several knots, so be sure that it doesn't set you. There are a number of islands, rocks, and peninsulas in the bay, and it is helpful to check them off carefully as you enter.

It is also possible to enter Seal Bay through the narrow passage between Bluff Head and Big Hen Island, favoring the eastern side. Again, be aware that the current may set you.

ANCHORAGES, MOORINGS. There are several good anchorages in the bay. One of the most beautiful is south of Big Hen, in 10 to 14 feet of water at low, east of the 4-foot spot. The bottom is rocky, and there is some exposure to the northeast, but otherwise the protection is good all around.

You can also anchor nearby between Ram Island, the large, unnamed island south of Big Hen, and Neck Island, which at low is connected to Bluff Head. The deep water is quite narrow, and a rock lurks off the east shore of Ram.

Another possibility is to enter the deeper recesses of Seal Bay. Coast south and then southwest along Penobscot Island on the right, keeping close to the last little island to starboard. This should take you clear of the rock to the southeast, which sometimes is privately marked with a stick topped with a lobster buoy or a flag.

Locate Turning Rock at the end of a long ledge projecting southward from the island to northeast of Hay Island. The entrance to the inner bay is between Turning Rock and a rock awash 2 feet at low about 200 yards to the southeast. Follow the deep water around to your right, leaving Turning Rock about 50 yards to starboard. Note the submerged rock 150 yards southwest of Turning Rock and due east of Hay Island.

Pass north of Hay Island, curving left and avoiding the rocks on the northwest side of the channel. Anchor north or west of Hay Island, in 7 to 16 feet, with good holding ground in mud. The bay here is fairly wide, but you are surrounded by land.

To reach a more protected anchorage in a smaller bay, continue west of Burnt Island and anchor in 7 to 10 feet. Most of the western end of Seal Bay is still natural, and only one or two houses are visible.

It is also possible to anchor in the southeastern corner of Seal Bay, off Coombs Hill, but the approach is a bit daunting the first time. After entering Seal Bay west of Little Hen, run down the narrow tongue of deep water toward Coombs Hill. Leave to port the smaller of the two unnamed islands (rocky, with a bunch of trees). Also leave to port the ledge beyond, which is awash 10 feet at low and identifiable at high by a patch of rockweed. Beyond the ledge is a rock sometimes marked by a green stick and green flag, also left to port. Note the shoal spots to the west of the deep water.

Anchor near the first moored boats beyond the stake in 10 to 18 feet at low. Other boats are moored in shoal water farther in.

GETTING ASHORE. Many of the islands and points in Seal Bay are protected as conservation land through the Maine Coast Heritage Trust, the Maine Bureau of Public Lands, or other groups. Camping is not permitted. Please don't land on Penobscot Island, which is private and has a house on it. Please don't light fires ashore.

THINGS TO DO. Explore the nooks and crannies of this lovely bay in your dinghy, including state-owned Hay and Little Hen.

VINALHAVEN ISLAND
West Coast

CROCKETT COVE

2 ★★

44° 06.05′N   068° 53.95′W

CHARTS: *13308, 13302, 13305*   CHART KIT: *68, 21, 19*

Crockett Cove is a deep indentation at the northwestern corner of Vinalhaven Island, south of the entrance to the Fox Islands Thorofare, opposite Dogfish Island. It is exposed to the prevailing southwesterlies but otherwise well protected. Although the cove is easy to enter, nearly all of the anchorage is a cable area, so you take your chances.

APPROACHES. Note the long, unmarked ledge between Dogfish Island and the entrance to the cove. The more visible portions of the ledge are at the southern end, showing 2 feet at high. Pass south of the ledge and then head up the middle of Crockett Cove.

ANCHORAGES, MOORINGS. Rig a tripline on your anchor and drop it just east of the cable area where chart 13308 shows an 8-foot sounding. There is good holding ground here in mud, though there may be some rocks.

HURRICANE SOUND

CHARTS: *13305, 13308, 13302, 13303*
CHART KIT: *68, 21, 19*

On the southwest side of Vinalhaven Island, a series of beautiful islands form Hurricane Sound—Leadbetter, Lawrys, Cedar, Crane, Hurricane, and the Whites. There are five separate entrances to Hurricane Sound, several of which are exhilarating.

Hurricane Island Outward Bound School is based here, with a welcome for yachtsmen. Long Cove provides extremely good protection, and an unusual tidal pool forms The Basin.

LEADBETTER NARROWS. From the west, or from Fox Islands Thorofare, the usual entrance is at the northern end of the sound, inside Leadbetter Island. Use chart 13308. A large granite dock on the eastern end of Dogfish Island is conspicuous from either the monument on Fiddler Ledge or green daybeacon "25" on Dogfish Ledges. The slot of Leadbetter Narrows will also be visible, with land behind it.

Green beacon "25" is not at the easternmost point of Dogfish Ledges, so give it a wide berth. Note the unmarked ledge at the mouth of Crockett Cove, some part of which is visible at every tide. The narrow spot in Leadbetter Narrows is marked by can "1," which you leave to starboard.

LAWRYS NARROWS. South of Leadbetter Island is the interesting passage called Lawrys Narrows. Some of the passage is shown on chart 13308, but you will need 13305 as well.

Lawrys Narrows is hard to find from a distance and difficult in poor visibility. Coming from the west, work your way in from buoy to buoy, starting at red-and-white bell "FT" (44°05.29'N 068°57.28'W), leaving to starboard the cans at Inner Bay Ledges and Seal Ledge. The Narrows begins between flashing can "3" and nun "2." From there, aim for red lighted buoy "2A," keeping it close aboard to port.

As you slide through the passage north of Lawrys Island, you will probably lose your wind for a minute or two and may even need a boost from the engine.

The end of Lawrys Narrows is marked by green lighted buoy "1," which you leave to starboard. Turn south into Hurricane Sound immediately past this buoy, running west of the little unnamed island. The local name for it is Tobacco Juice, since it looks as though it has been chewed and spat out.

As you pass Lawrys Island on your way through the narrows, look for the bell mounted on the shore with a long lanyard reaching up to the porch of a beautiful log house. The owners like to sit on the porch and salute passing vessels. On the tip of Cedar Island, the bow of a wrecked steamship has been turned upside down and made into a little house.

The ferry from Rockland to Vinalhaven passes back and forth six times a day through this narrow passage, so keep your eyes open. In the summer of 1993, the *Governor Curtis* earned the dubious distinction of finding a rock here not once, but twice. Lest they forget, some helpful soul once nailed a huge piece of plywood to a tree on the north end of Lawrys Island, reading, with an arrow, "KEEP LEFT."

Once you find the outlying buoys, Lawrys Narrows is an easy passage in good visibility, but always an exciting one.

NORTH OF HURRICANE ISLAND. The passage between Hurricane Island and the White Islands to the north is wide and easy, well marked by two green cans.

Coming from the west, the standpipe at Carvers Harbor makes a good landmark that is visible from a good distance. The southwestern-most of the White Islands, Big White, is wooded and identifiable by high granite ledges at the southwestern end. State-owned Little Hurricane Island is low and has grassy granite ledges and only a few trees. Hurricane Island is high and wooded. As you approach Big White Island, look for cans "15" and "13" inside the passage to be sure you are in the right place, and leave both to starboard.

HERON NECK. The southern entrance to Hurricane Sound is also wide and easy. Use the conspicuous lighthouse on Heron Neck as a landmark. It sits high on a granite bluff with a spine of trees. Deadman Ledge, to port, is a 20-foot-high granite island with a little grass.

THE REACH. The fifth entrance to Hurricane Sound is The Reach (see the following entry on page 226).

LONG COVE

■ ★★★

44° 05.36'N 068° 53.03'W

CHARTS: **13308, 13302, 13303**   CHART KIT: **68, 21, 19**

At the northeastern end of Hurricane Sound, opposite Leadbetter Island, is the beautiful little harbor of Long Cove. It is easy to enter and the protection is excellent. Rock and spruce ring the shores. The owners request that you not go ashore.

APPROACHES. When approaching from the south, be careful of the rocks and the ledge midway along the eastern shore of Leadbetter Island. The ledge extends farther than it seems, and more than one cruising boat has spent a tide here. From the north, pass through Leadbetter Narrows, described above.

Enter Long Cove north of Fiddlehead Island, which is quite large, high, and wooded, with a small dock and house at the north end. The unnamed island to the north looks like a smaller replica. Pass halfway between Fiddlehead and the unnamed island. Near the top of the hill on your right, you will see a great pile of granite tailings and a granite loading dock.

Follow the curve of the channel around to the left and into the inner harbor, staying in midchannel.

ANCHORAGES, MOORINGS. A number of private moorings are in the cove, particularly along the northwest shore. These are often picked up by cruising boats who enjoy them until their owners come to collect their fee, which is so steep that the skippers are too shocked to move.

There is some room to anchor outside the moorings in midharbor in 8 to 13 feet at low. Beware of the southeast shore, which shoals. The holding ground is good mud, but there is some kelp, particularly along the west shore off Conway Point. Be sure your anchor is well set.

On summer weekends, there may be a lot of boats here. If there is no room, head back to the outer harbor, anchoring north of Fiddlehead in 13 to 20 feet. This is not quite as landlocked, but it is still a lovely and well-protected anchorage.

As shown on the chart, there is an inner sanctum, providing almost perfect protection. At midtide or higher, it is possible to work your way up beyond the rocks into the head of Long Cove, with 14 feet or more of water. Be aware of a ledge on your right before heading in, after which favor the right side.

THE BASIN

entrance: 44° 04.38'N 068° 52.71'W

CHARTS: **13308, 13302, 13305**   CHART KIT: **68, 21, 19**

The Basin is a magnificent tidal lake cut in the west side of Vinalhaven Island. Here you will find a scattering of little islands, depths as great as 111 feet, and good swimming. The only problem is a narrow, crooked, and obstructed entrance through which the current usually rushes like a millrace.

Entering The Basin is not recommended except for the most daring of gunkholers. Rarely does a year go by without a grounding here. But others have done it. According to local lore, schooners used to winter over inside The Basin.

The best time to enter or leave is within an hour of high slack tide (see sketch chart, page 222). As if the current weren't bad enough, a rock sits directly in the middle of the entrance, as shown on the chart. The channel runs south of this rock and immediately branches at a small island.

Once safely through the channel, you can enter The Basin either to the left or right of the small island, but the right branch is wider. Another rock lies to the southeast of the island. Head southeast past the island, keeping the rock to port, then turn slightly to port, rounding the rock and heading east into The Basin. Anchoring options are plentiful. So are the rocks. Be careful.

LAWRYS and CEDAR ISLAND

3 ★★★

44° 03.77'N 068° 53.86'W

CHARTS: 13305, 13302   CHART KIT: 68, 21, 19

Tucked between Lawrys and Cedar Island is a snug little anchorage that is not as secluded as it seems at first. There are houses and docks overlooking the anchorage, so you may feel as though you are intruding on someone's privacy.

APPROACHES. Approach from Hurricane Sound, either along the east coast of Crane Island or south of Cedar Island. Note the several tiny islets (one with trees) off the southern tip of Cedar. Enter halfway between Lawrys and Cedar.

ANCHORAGES, MOORINGS. Several private moorings are at the head of the cove, but you can anchor just south of these in 14 to 20 feet. Your anchor may hold well here, but the bottom appears to be rocky. The cove is exposed to the southeast and slightly to the northeast. Do not be startled when you see the Rockland-Vinalhaven ferry pass through Lawrys Narrows, a stone's throw away.

WHITE ISLANDS

2 ★★★★

44° 02.82'N 068° 54.64'W

CHARTS: 13305, 13303, 13302   CHART KIT: 68, 21, 19

Off the southwest shore of Vinalhaven lies a tiny archipelago that is among the most beautiful in Maine. Shown as the White Islands on the chart, the group includes Big and Little White and Big and Little Garden. Charles Lindbergh and his wife, Anne Morrow Lindbergh (author of *Gift From the Sea*), donated Big Garden to The Nature Conservancy, which also owns Little White and part of Big White. Little Garden is privately owned.

The islands are wooded, with tiny beaches and steep, sloping shores of white granite. Anchored in the lagoon, with vistas of the Camden Hills and not a house in sight, any yachtsman can feel fortunate indeed.

APPROACHES. The southernmost islands of the group are Big White to the west and Little White to the east. Enter from the south between these two islands, favoring the eastern side.

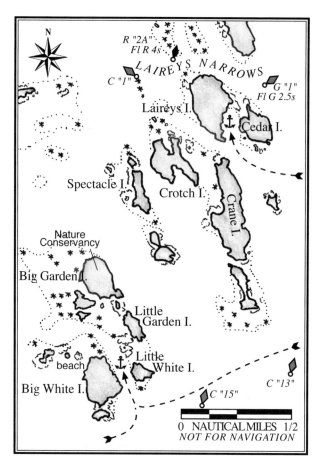

Although the entrance is clear and straight, it also is narrow, so it is more comfortable to come in near low tide, when you can see the ledges bordering the channel.

ANCHORAGES, MOORINGS. Anchor between Big and Little White in approximately 24 feet at low. There is not much swinging room, and tidal currents of as much as a couple of knots will affect the way you hang. A stern anchor might be appropriate. One of the windjammers that sometimes lies here puts a line ashore.

You can also continue past the end of Big White and anchor near the 20-foot mark on the chart. Here it opens up a bit, with more swinging room and better views, but there is also greater exposure to the west. Again, you might want to use a stern anchor in case of a middle-of-the-night wind shift. There is good holding ground here, in mud.

GETTING ASHORE. Land on Big White with your dinghy on the little beach at the northeastern tip. The landing on Big Garden (northernmost of the group) is in the little cove formed by South Big Garden, a separate little island at the southwest end. If your dinghy has an outboard, feel your way in carefully—there are ledges lurking just beneath the surface at higher tides.

THINGS TO DO. Visitors are welcome on Big Garden and Big White for day use only. There are no trails on the islands, but there is a salt marsh and a wide variety of plants, including abundant blueberries and beach peas. An old field on Big Garden has wild irises and primroses in early June and July. The birding is good, especially during spring and fall migrations. As Anne Morrow Lindbergh wrote, "I must remember to see with island eyes. The shells will remind me."

HURRICANE ISLAND

2 ★★★★

44° 02.08'N 068° 53.14'W

CHARTS: *13305, 13303, 13302*   CHART KIT: *68, 21, 19*

Hurricane is one of the outlying islands at the southwest corner of Vinalhaven. It is the home of the Hurricane Island Outward Bound School's summer sailing program, which was founded in 1964 to inspire students to take risks and grow through challenging outdoor expeditions.

The Hurricane Island pulling boats, with their distinctive spritsail ketch rigs, roam the coast of Maine from Muscongus Bay almost to the Canadian border. If you see a lone figure sitting in contemplation on a rocky point of some deserted island, it is probably a Hurricane student on "solo," an essential part of the Outward Bound experience.

HIOBS is part of a worldwide network of 35 Outward Bound schools throughout Maine and Florida and in Baltimore, Maryland, which offer a variety of courses for youths and adults. In the wilderness and on the sea, students cope with the challenges of the natural world and in the process develop self-confidence and self-esteem. They learn to work as a team, giving help and accepting it. They learn wilderness skills, and respect for and care of the environment.

Visitors are welcome on Hurricane Island, and there is much to see—not only student activities like rock climbing in the quarry and a hair-raising ropes course, but also the remains of an earlier period when Hurricane was quarried for granite.

If you need help, HIOBS has several well-equipped power vessels with search-and-rescue capability and highly experienced staff members. Fire-fighting equipment, diving gear, and first-aid supplies are at hand. Channels 09, 16, and 18 are monitored 24 hours a day.

Likewise, cruising sailors can help HIOBS in emergencies. Soloing students in trouble will alert other nearby soloists by blowing a whistle, and they in turn will blow theirs. If you should hear such a "whistle chain," please radio HIOBS immediately.

Hurricane's history is an intriguing one. In 1870, a Civil War general named Davis Tilson bought the island for the preposterous sum of $50, and within four years, the island had a population of 1,200 Italian and Irish immigrants. By 1880, the island had seceded from Vinalhaven and become its own town. It had a post office, six boardinghouses, 40 cottages, a pool hall, bowling green, bandstand, company store, and two major quarries and several smaller ones scattered throughout the island.

The island's polishing mill was so renowned that many other quarries shipped their stone here for final finishing. Hurricane Island became one of the nation's largest, most productive, and famous granite sources. The island's granite helped build such imposing structures as the New York City Public Library and the Metropolitan Museum of Art.

But, like much of the history of quarrying, the Hurricane era was short. The end of the Hurricane Island Granite Company came in 1915 with the death of the superintendent of works, who had been on the island for 20 years. A mild panic ensued when it was announced that the goods in the company store were being moved to the mainland, and workers and their families rushed to catch the last scheduled boat, in many cases leaving behind most of their worldly possessions. Hurricane was transformed overnight into a ghost town. It remained that way until Outward Bound established its base here.

APPROACHES. Hurricane Island's landing is on the east side, in Hurricane Sound. If you have come through any of the northern entrances (see pages 221-222), head down the east coast of Hurricane Island toward the moored boats. Be careful of the ledge shown on the chart east of the 41-foot spot, marked with a triangular beacon. Run through the moored fleet, between the ledge and Hurricane Island.

ANCHORAGES, MOORINGS. The anchorage here is only moderately well protected and may be rolly. It is wise not to anchor here, since the water is deep and the bottom rocky.

⚓ The school maintains several heavy guest moorings at the south end of the anchorage available on a first-come, first-served basis. Guest moorings have red ball floats and lie opposite the big gray mess hall.

GETTING ASHORE. Row to the float just below the rescue station. Sign the guest register and pick up information for island visitors, including a trail map.

THINGS TO DO. When you are on the island, the school asks that you keep your distance when observing student activities. It's fascinating to watch students rock climbing and rappelling on the granite cliffs of the quarry or negotiating the ropes, beams, and zip wire of the ropes course. There may be a capsize drill in a pulling boat or sea kayaking instruction. You might want to join the daily "rise, run, and dip" at 5:30AM for a run around the island followed by a plunge in the icy waters of Hurricane Sound. But then again, you might not.

A 2.5-mile trail leads around the island through woods and raspberries, past quarries, old machinery, and piles of granite blocks still waiting to be shipped. It takes you out on great granite slabs into the ocean with vistas all around Penobscot Bay. HIOBS asks that you stay on the paths, be unobtrusive, and resist the temptation to try the ropes course, climb the cliffs, or swim in the quarry. No fires or smoking are allowed, and all pets must be left on board your boat; none are allowed on the island.

SANDS COVE

2 ★★★

44° 02.21′N 068° 52.28′W

CHARTS: 13305 (INSET), 13302, 13303
CHART KIT: 65 (B), 68, 21, 19

Sands Cove is the northernmost cove on the west side of Greens Island, due east of Hurricane Island. It is unnamed on the chart, but it can be identified by its 11-foot sounding.

As a boy, Joseph Green was the sole survivor of an Indian massacre on Matinicus. In his adult years, he moved to the island that would take his name.

Like the surrounding islands, Greens was quarried for its granite. Today, a handful of fishermen and artists and their families live on the island.

APPROACHES. Be sure you have the right cove. Find the narrow, rocky point forming the southern entrance and the house at the head of the cove. Enter on the midline, leaving the ledge at the southern entrance well to starboard.

ANCHORAGES, MOORINGS. The two moorings in the cove will force you to anchor precariously close to the cove's mouth. The southern point may give you some protection from southwesterlies, but not much. Despite the cove's name, the bottom appears to be rocky, so be sure your anchor is well set. If swells are running, pick another anchorage.

THE REACH

The Narrows: 44° 02.45′N 068° 51.11′W

CHARTS: 13305 (INSET), 13302, 13303
CHART KIT: 65 (B), 68, 21

The Reach winds between Hurricane Sound and Carvers Harbor. The Coast Pilot describes it as a "narrow, much obstructed channel." It is also part of the ferry route from Rockland. It would be a more enjoyable passage if you weren't so busy piloting.

Strong current ebbs southeast and floods northwest, but the channel is well marked. Red marks are left to port going eastward, green marks to starboard.

The narrowest part of The Reach is at a sharp S-turn near the little island at green daybeacon "3." Here it is only 100 feet wide, so be especially careful. In defiance of all laws of probability, it is a near certainty that this is the very spot that the ferry will bear down on you.

OLD HARBOR

44° 03.09′N 068° 52.01′W

CHARTS: 13305 (INSET), 13302, 13303
CHART KIT: 65 (B), 68, 21

This attractive old working harbor is at the western end of The Reach, surrounded by small houses and lobster shacks. A number of lobsterboats and lobster cars are moored here, but there are no yachts whatsoever. Among them there is plenty of room to anchor.

APPROACHES. Enter The Reach from the west between can "11" and nun "10," then curve northward into the western portion of Old Harbor, leaving

Hanging out on Main Street, Carvers Harbor.

to starboard the wooded island that splits the harbor. From points east, turn northward into Old Harbor after passing the island and its extensive ledges at the east side of the entrance.

ANCHORAGES, MOORINGS. Anchor in 10 to 15 feet of water at low in the western portion of the harbor, but do not venture much north of the small Whaleback Rock to starboard. In an old working harbor like this, the bottom may be foul, and the Coast Pilot warns of "many old fish stakes," so use a tripline on your anchor. You will be exposed to prevailing southwesterlies but protected if the wind is from the north or east.

REINDEER COVE

3 ★★★

44° 02.78′N 068° 52.05′W

CHARTS: *13305 (INSET), 13302, 13303*
CHART KIT: *65 (B), 68, 21*

The little cove at the northern tip of Greens Island, just opposite Old Harbor, is unnamed on the chart, but locally it is called Reindeer Cove. This is the base of the Atlantic Challenge, an international experiment in learning and cooperation. The Challenge's reproduction French gigs may be in the cove, and a yurt or a teepee may be ashore.

APPROACHES. Coming from Hurricane Sound and the west, enter The Reach between can "11" and nun "10" and continue until you see the cove and the granite wharf. Head straight into the middle of the cove.

ANCHORAGES, MOORINGS. Anchor in a comfortable depth just inside the north end of the wharf. Holding ground is good in mud, but there is some kelp. There is good protection here from prevailing summer winds.

GETTING ASHORE. Row to the granite wharf and clamber up (quite a process at low tide). The clearing next to the wharf is a delightful spot for a picnic.

CARVERS HARBOR

3 ★★★★

44° 02.65′N 068° 50.18′W

CHARTS: *13305 (INSET), 13302, 13303*
CHART KIT: *65 (B), 68, 21*

Carvers is a busy fishermen's harbor, full of activity and interest. But it is a difficult place for a yachtsman to visit. There are few moorings and facilities, and little time to spare when there are lobsters to be caught. Yet it is well worth making the effort to stop here. Carvers Harbor is Vinalhaven's main port, its authentic front door, and it is a fascinating place to visit.

Curiously, while some large commercial harbors such as Rockland are blossoming into major yachting ports, the few services once available to visiting yachtsmen in Carver's Harbor have reverted to commercial operations.

The Carver family settled Carvers Harbor in 1766. Fishing, farming, and shipbuilding were the early occupations, and at one time the Carvers were the world's largest manufacturers of horsenets, which were used to protect horses from flies. Granite quarrying became the dominant industry of Vinalhaven from its beginning in 1826 until the last of the paving quarries closed in 1939.

Signs of the granite industry are everywhere here—the magnificent wharves that surround the harbor, the quarry tailings on the hillsides, the elaborate memorial horse troughs decorating many Vinalhaven intersections, the unusual granite curbing around family plots in the Carver cemetery, and the quarries themselves.

A rough granite column lies on rollers in a quarry on Vinalhaven, 1892. Some of the largest pieces of granite were quarried here for the Cathedral of St. John the Divine in New York City. (Frank Claes)

Sands Quarry opened in 1860. The Bodwell Granite Company employed more than 600, including workers from England, Scotland, and Ireland, and stone carvers from Italy, Finland, and Norway. Stone was cut here for the Brooklyn Bridge, but the most famous contract was for four enormous columns to surround the high altar of the Cathedral of St. John the Divine in New York. The rough granite monoliths were 64 feet high, 8 feet in diameter, and weighed 360 tons—the largest single pieces of stone ever cut and moved.

APPROACHES. Carvers Harbor is at the southern end of Vinalhaven Island, approached either from the west through The Reach, or from the east between Folly and Green Ledge. Red lighted buoy "4" marks the right side of the harbor entrance, and can "5" marks the left.

ANCHORAGES, MOORINGS. Carvers has no room for anchoring, and moorings are not easy to come by. This is the major challenge of a Carvers Harbor visit. Arrive early enough so that if you don't find a mooring, you still have time to go somewhere else for the night.

Fishermen sometimes rent spare moorings. They might be marked as a rental, sometimes with a plastic jar attached for your payment on the honor system. It's more likely that you will need to find somebody who can direct you. Hopkins Boatyard may be able to steer you toward a vacant mooring or to the harbormaster.

For a while, the town tried setting a couple of guest moorings, but they kept mysteriously "disappearing." Definitely do not leave your boat on an unknown mooring without leaving someone aboard who can move it when the owner returns.

Once settled on a mooring, be aware that this is one of the most tightly packed mooring fields in Maine, and in light winds it is not uncommon for boats to be nudging one another.

GETTING ASHORE. Row either in to the floats at the boatyard on the west side of the harbor, or, to get closer to town, to the public floats at the head of the harbor near the millrace.

FOR THE BOAT. *Hopkins Boatyard (207-863-2551).* Hopkins has no signs but it is recognizable by the 30-ton boatlift and the cradled boats on the west side of the harbor. The yard has some marine supplies and can do repairs. There is a small dock and a float with 8 feet alongside, but no fuel.

Fisherman's Friend (207-863-2140). Located in the Harbor Wharf building, Fisherman's Friend carries a limited selection of hardware and marine items and a full inventory of beer and ice.

In a pinch, fuel or water can probably be obtained at the Fishermen's Co-op, though they won't be happy about it.

FOR THE CREW. Carvers Harbor Market (*863-4319*) is known locally as the "Mother Ship." This IGA has groceries, a bakery, ice, wine, beer, and liquor. It is right across the street from the float at the public landing and is open until 7 PM. Wines and cheeses can be found at Island Spirits, by the Tidewater Motel. Fisherman's Friend, on the west side of the harbor by the boatyard, stays open later than the market. They carry the basics—bread, beer, and boots.

If you want to see the real Vinalhaven, start your day at the Pizza Pit for breakfast at 4:30 AM. By lunchtime you will be ready for the Harbor Gawker (*863-9365*), which serves lunch and dinner overlooking the millrace between Carvers Pond and the harbor. They begin their season by giving out free ice cream cones to every schoolchild on opening day. Their menu ranges from the Big Mort double burger plate to fried clams and overflowing haddock sandwiches.

The Haven Restaurant (*863-4969*) serves more formal dinners either in their front room or out back on the deck. They are located at the east end of the parking lot at the public landing. Because of their limited seating, reservations are vital.

The post office is on Main Street, as is The Paper Store, for books and newspapers. Pick up a copy of The Wind, an island weekly, to find out what's going on. Pay phones are located at the public landing and the ferry terminal. Surprisingly, there is no laundry in town.

THINGS TO DO. If you happen to be here on the Fourth of July, don't miss the entourage that parades up Main Street, turns around, and parades right back down Main Street.

The few blocks of Main Street are well worth a short stroll, past the millrace and the historic Star of Hope Lodge built in 1885. This extraordinary mansard-roofed building has been restored by pop artist Robert Indiana. Built during the heyday of Vinalhaven's granite industry, it is one of the few buildings remaining from those prosperous times.

For information on history (or anything else), you may be referred to one of the colorful institutions of Vinalhaven, the "Historians"—three old men who hold court in the well-stocked Port O'Call Hardware Store.

From Main Street, go left up over the hill on High Street past the Historical Society (worth a visit) to the John Carver cemetery, with a most interesting array of granite memorials overlooking Carvers Pond. Past the cemetery, turn left at the round, granite memorial horse trough. Walk a short distance along Old Harbor Road until you reach a dirt road on the left that leads to the cliffs of Sands Quarry.

Returning to Carvers Harbor along the Sands Road, you'll pass a great 60-foot granite monolith abandoned in the grass by the roadside. It had been cut for the General Troy monument in Troy, NY but discarded when a flaw was discovered. The Sands Road will take you back to the ferry terminal, next to which is pretty little Grimes Park on the waterfront.

For a longer walk and better swimming, take a right at the horse trough and trudge a couple of miles to the top of a long hill. On a hot summer day, half of Vinalhaven will be at Lawson's Quarry on the right. The other half will be at Booth Quarry, 2.5 miles east of town on the Pequot Road.

If you walk out of town to the east, you will reach a bandstand and a little park displaying a blue galamander. These awesome machines, with rear wheels 9 feet in diameter, were drawn by teams of oxen to carry granite blocks from the quarries to the cutting sheds at Sands Cove.

Follow the next right onto Atlantic Avenue and walk south until you reach the Island Community Medical Center (*863-4341*), on your left. Grassy trails start in the back of the parking lot and lead through the Armbrust Hill Wildlife Reservation, passing several of the oldest quarries on Vinalhaven.

A short walk will bring you back to Atlantic Avenue. Then continue south across the bridge to Lane's Island. Forty-five acres of Lane's Island were bought by residents of Vinalhaven and given to The Nature Conservancy as a preserve. Follow the signs to the preserve and walk around the perimeter of Lane's, enjoying the views of Indian Creek (site of Susquehanna Indian and Red Paint villages from 4,000 B.C. to colonial times), Brimstone Island in the distance, and Carvers Harbor. The terrain is heath, marsh, and bayberry, with many different kinds of grasses, wildflowers, and berries. At the tip is a simple memorial plaque to Vinalhaven men lost at sea. On your return, you will pass the old Lane family cemetery, dating from the early 19th century.

For explorations of the wheeled kind, bicycles can be rented from the Tidewater Motel (*863-4618*). The bikes are for adults only (and when we inquired, they had somehow been temporarily "borrowed" the night before). You can buy a local map at The Paper Store.

VINALHAVEN ISLAND
East Coast

The east coast of Vinalhaven is a lee shore exposed to open ocean. Even in the calmest weather, the swells burst on the rocky headlands and shores. Between Seal Bay and Carvers Harbor, there is no safe refuge for a cruising sailor.

The islands along this coast are mostly barren, battered rocks with grass and a few stunted trees. The only one where you can land easily is beautiful Brimstone.

The dangers lying off the coast are marked by a long line of buoys, from can "5" on Triangle Ledge, past nuns "2," "4," "6," and "8," to red beacon "10" on Point Ledge, and finally to the light on Green Ledge at the entrance to The Reach. Leave them all to starboard. It is an impressive coast to run, easy enough in clear weather but no place to be in the fog.

BRIMSTONE ISLAND
44° 00.88′N 068° 46.33′W

CHARTS: *13303, 13302, 13305*   CHART KIT: *68, 21*

High, beautiful Brimstone Island stands well out to sea, southwest of Vinalhaven Island. Its volcanic rock is covered with grasses and wildflowers and nesting gulls. The relentless pounding of the Atlantic has rolled and polished the island rocks into wonderfully smooth, black stones and pushed them into a high peninsula off the island's northwest shore, with crescent beaches on both sides.

From the 112-foot summit of the island, there are unsurpassed views in every direction—to Blue Hill and Isle au Haut, to Matinicus and the other seaward islands, and to the Camden Hills—not to mention a spectacular view of your own boat anchored far below. Brimstone and Little Brimstone belong to The Nature Conservancy.

Be aware of strong tidal currents running between Brimstone Island and the islands to the west.

For a daystop in prevailing summer winds, anchor off the north side of the pebble peninsula. Come in to the center of the crescent beach until you reach a comfortable depth, which will be quite close to shore. The beach of greenstone pebbles continues down to larger rocks, so consider rigging the anchor with a tripline.

The beach is steep. Depending on the size of the waves, it might be easier to get your dinghy ashore at the rocks on the west end of the beach rather than to land on the beach itself. From there, it's a good scramble up grassy trails to the summit.

Foragers will appreciate the island, not only for the usual selection of edible land plants (including a profusion of wild strawberries in season) and tasty intertidal varieties, but especially for the clumps of spearmint near the northwest end.

Brimstone Island is one of the few nesting areas for the increasingly rare Leach's storm-petrel, a small nocturnal bird that comes ashore only to nest. As Peter Blanchard, a volunteer for The Nature Conservancy, says, "They make a riotous laughter when they return from the sea, combined with a purring noise. It's like being in a room with thousands of children, all giggling at once. It's a magical experience." The female digs her burrow in the tussocky ground on the east side of the island, lays one egg, and incubates it in the nest for 90 days. Avoid the nesting area entirely—the petrels nest all summer long.

Although it is possible to anchor for the night between Brimstone and the little islands to the south, do it only when assured of fine weather. Brimstone is a lonely outpost in the ocean and much exposed to waves and wind. The best protection is north and slightly east of Little Brimstone, the largest and westernmost of the little islands.

ISLESBORO

Islesboro stretches north and south for 10 miles and divides the upper reaches of Penobscot Bay into east and west. At the end of the 19th century, the beauty of the island's bays and headlands caught the eye of a young Bostonian, Jeffrey Brackett, who, in 1899, began the exclusive community of Dark Harbor. It was much like those in Bar Harbor, Winter Harbor, North Haven, and Prouts Neck, and the great "cottages," the yacht club, tennis club, and golf club remain today for the fifth and sixth generations of families who have loved it ever since.

Gilkey Harbor is formed to the southwest of Islesboro by Seven Hundred Acre Island to the west. Gilkey itself is too broad to be a useful anchorage for cruising boats, but its shores are divided into several beautiful spots—Ames Cove with the Tarratine Yacht Club, Cradle Cove, the state park on Warren Island, and Grindel Point where the ferry lands.

There are several interesting anchorages on the west coast of Islesboro, particularly Crow Cove and Turtle Head. On the east coast of the island, several coves offer protection from prevailing winds, the nicest of which is Sabbathday Harbor. Dark Harbor, for which the summer community is named, was long ago dammed to form a pool.

The island is served by a small airport and by a state ferry that runs from Lincolnville Beach on the mainland. The frequency of the short ferry ride has made Islesboro less insular than one might expect, and some islanders commute to the mainland for work and school.

Islesboro is a lovely island to explore by boat. Facilities are few, but the islanders are friendly. It is also a biker's paradise, because it is so flat. You can't go anywhere on the north-south road without returning waves from every passing car.

Indians used to summer on Islesboro, fishing and trapping. Early white settlers earned a living with subsistence farming and fishing. In the 19th century, the island-based F.S. Pendleton fleet sailed to all corners of the globe.

The influx of wealthy summer people in the 1890s changed the character of the Dark Harbor area, and the islanders became increasingly dependent on the summer residents for their livelihoods. The summer people, in turn, depended on the fishermen-turned-carpenters-and-gardeners for many of life's necessities. However, relationships were not always entirely cordial, as Michael Kinnicutt describes:

"A salmon fisherman was setting out his nets, when a private steam yacht dropped anchor a ways out. Someone waved the fisherman over. He rowed across. A fancy fellow climbed into the boat and told the fisherman to take him ashore.

"As they approached the shore, the passenger said, 'My name is George Washington Childe Drexel and I just bought that land and intend to build a large house with stables.' The fisherman squarely regarded the passenger and replied, 'My name is George Robeson, and this is my punt.'"

The summer community continued almost unchanged for half a century, with horse-drawn carriages and a slow-paced, elegant way of life, keeping at bay the annoyances of the modern world. Finally, in 1932, the islanders waited until the summer people went home, then voted to allow cars on the island for the first time.

In 1984 the threat of further radical change hit Islesboro when two subdivisions were proposed. With the enthusiastic help of the summer community, a land trust was formed in conjunction with Maine Coast Heritage Trust. Hutchins Island and much of Spruce Island came under the trust's guardianship.

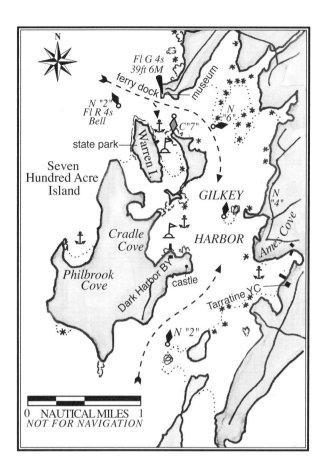

GILKEY HARBOR
44° 15.45'N   068° 55.87'W

CHARTS: **13309, 13302, 13305**   CHART KIT: **20 (B), 66, 20**

The handsome body of water between Islesboro and neighboring Seven Hundred Acre Island is Gilkey Harbor. More of a thoroughfare than a harbor, Gilkey is wide, deep, and exposed to the southwest. Most boats spend the night in Cradle Cove, at the Tarratine Yacht Club in Ames Cove, or at various private docks and moorings along the shore.

Gilkey Harbor and its approaches are wide enough so you can sail in under most conditions, enjoying the summer cottages, each larger than the last, and the changing vistas of the Camden Hills.

Dangers are well marked, except for Long Ledge, in the northeast corner of Gilkey Harbor. The red buoy and flag in its vicinity does not mark the ledge—it's a racing mark. Stay well over to the Spruce Island shore in this area.

APPROACHES. From the south, sail through the wide and well-marked entrance between large Job Island and the smaller Ensign Islands. Another entrance east of Job Island, called Bracketts Channel, is narrow in spots and tricky. In recent years, this has been privately marked.

Seven Hundred Acre Island can be identified by the complex of white buildings on its eastern tip and the miniature Norman castle (see *Cradle Cove*, below).

From the north or west approach between Warren Island and Grindel Point. This passage is also easy and well marked. Look for the former lighthouse and the ferry slip on Grindel Point. Stay alert for the ferry, which makes frequent runs between here and Lincolnville Beach.

AMES COVE

2 ★★★

44° 15.30'N 068° 55.30'W

CHARTS: *13305, 13309, 13302* CHART KIT: *20(B), 66, 20*

Ames Cove lies just opposite the eastern tip of Seven Hundred Acre Island, obvious by the fleet of handsome yachts moored off the Tarratine Yacht Club. Other magnificent vessels lie docked on the surrounding piers, and a fleet of juniors is usually tearing around the buoys in the harbor.

APPROACHES. To get your bearings, identify the rock shown on the chart at the north end of Ames Cove, and approach south of this rock. The yacht club is an understated single-story building on the east side of the cove with a flagpole, a green roof, a dock, and several floats.

ANCHORAGES, MOORINGS. Neither the yacht club nor Pendleton Yacht Yard have rental moorings. You can anchor outside the yacht-club fleet in 10 to 15 feet at low. You probably will be exposed to the prevailing southwesterlies, but in little danger of dragging. The holding ground is excellent, in mud. If you need better protection, hop over to Cradle Cove on Seven Hundred Acre Island.

GETTING ASHORE. Tie up behind the dinghy float at the yacht club. Or, if the tide is up and you are in a hurry for your ice cream, take the dinghy in to the float at Pendleton Yacht Yard to the north.

FOR THE BOAT. *Pendleton Yacht Yard (Ch. 09, 16; 207-734-6728; www. pendletonyachtyard.com).* At the northern end of Ames Cove, Pendleton has a dock and float that dries out at low and has 5 or 6 feet alongside at high. Keep your eyes peeled for rocks when entering this part of Ames Cove. Gas, diesel, and premixed fuel is available. The yard has a boatlift and a chandlery, and they can handle all types of repairs.

Tarratine Yacht Club (207-734-6994). Yachtsmen may visit the pleasant old clubhouse to eat lunch on the porch overlooking the cove or use the showers. Established in 1896, the yacht club runs a racing series for the beautiful old Dark Harbor 20s. You might be lucky enough to see half a dozen of these sleek little sloops some Saturday at the starting line.

FOR THE CREW. The yacht club occasionally has pancake breakfasts. The Dark Harbor Shop (*734-8878*) is within walking distance of the yacht club and yacht yard. Ice cream and sandwiches are served up from their magnificent marble soda fountain. The newsstand includes *The New York Times* and a good array of sailing magazines. Dark Harbor House (*734-6669*), near the yacht club, also serves dinner by reservation seven nights a week from 5:30 to 8:30 PM.

The post office and the well-stocked Island Market (*734-6672*) are a couple of miles north. Try to arrange a ride or hitchhike.

CRADLE COVE
(Seven Hundred Acre Island)

3 ★★★

44° 15.75'N 068° 56.40'W

CHARTS: *13305, 13309, 13302* CHART KIT: *20(B), 66, 20*

Seven Hundred Acre Island forms the west side of Gilkey Harbor. It is best known as the summer residence of Charles Dana Gibson, creator of the turn-of-the-century "Gibson Girl" (modeled after his Virginia-belle wife, Irene). A full-service yacht yard is in Cradle Cove, on the east side of the island.

A miniature Norman castle guards the eastern tip of Seven Hundred Acre Island, complete with dungeon, a great hall, and a maiden's chamber. This pleasant folly was built by Gibson in 1926 for his grandchildren. It is not open to the public.

APPROACHES. Entering Gilkey Harbor from the south, you will see a large white house on several levels at the eastern tip of Seven Hundred Acre Island. Just around the corner, in Cradle Cove, are the moored boats, fuel tanks, and stone dock of the Dark Harbor Boat Yard. Do not cut the point too close—a small yellow floating barrel marks the end of the ledge.

Coming from the north, you will see the sheds and moored boats ahead as soon as you round Spruce Island.

ANCHORAGES, MOORINGS. The Dark Harbor Boat Yard has about 20 moorings. There may also be room to anchor north of the moorings, in a mud bottom 14 feet deep at low. Note, however, that the southern half of the cove is a cable area.

GETTING ASHORE. Row your dinghy in to the boatyard floats.

FOR THE BOAT. *Dark Harbor Boat Yard (Ch. 09; 207-734-2246).* The yard has gas, diesel, electricity, water, and ice at the floats, with about 6 feet alongside at low. There is a bit less depth in the approach to the floats, however, so choose a time other than dead low. The end of the ledge to the west is marked with a pipe. The yard has a chandlery, a 20-ton marine railway, a hydraulic trailer, and the ability to handle most repairs.

FOR THE CREW. The Dark Harbor Boat Yard has showers and laundry facilities and, in season, wild strawberry jam made by an employee. That's wild strawberries, as in teeny, tiny.

THINGS TO DO. Although Seven Hundred Acre Island is private, there is a town-maintained dirt road leading off to the right from the boatyard. It traverses the island and provides a pleasant walk.

WARREN ISLAND

44° 16.54′N   068° 56.53′W

CHARTS: 13309, 13302   CHART KIT: 20 (B), 66, 20

Spruce and Warren Island lie due north of Seven Hundred Acre Island and are almost connected to it. An unspoiled cove lies between them, well protected from the prevailing southwesterlies but open to the north.

Seventy-six-acre Warren Island is a state park, "Given for the benefit and enjoyment of visitors to coastal Maine" by the Town of Islesboro in 1959. The state maintains nine or ten moorings in the cove and a landing float on the east side of the island. Spruce-needle trails meander from there through the woods past various rustic campsites, west to a view of the Camden Hills, and south to a shale beach overlooking Cradle Cove. For information about the park, call *236-3109* in season or *236-0849* off season.

APPROACHES. Approaching from east or west, head southward down the middle of the slot between Warren and Spruce Islands.

ANCHORAGES, MOORINGS. The state-maintained moorings have white pick-up buoys and white polyball floats marked "Warren Island State Park." They are available on a first-come, first-served basis. Shallow-draft boats should use the moorings closest to the head of the harbor, where the bottom begins to shoal, and leave the outer moorings for deeper draft boats. There is also room to anchor outside of the moorings, in about 15 feet of water at low. The bottom is mud that holds well.

GETTING ASHORE. Row to the state pier.

FOR THE CREW. There are a number of campsites, fireplaces, picnic tables, privies, and shelters. Well water is available inland, but it's for the crew, not the boat. You will need to pump it by hand.

THINGS TO DO. Woodsy trails lead around and through the island. If it is sunny and the tide is up, the beach at the south end is a tempting place for a dip.

As you explore, you can still see some of the stone piers that supported the 100 by 100-foot floor plan of one of the most expensive rustic cottages ever built in New England. The dream cabin of William Folwell from Philadelphia was named "Mon Reve" or "My Dream" and contained 22 bedrooms and a living room 60 feet by 30 feet. But a dream it was. Folwell died before its completion, and his descendants used it only occasionally in the years that followed. Seldom occupied, it became a magnet for local picnics and parties, and the grand structure burned in 1919.

From your mooring on a busy weekend, you can watch the parade of boats passing through the northern entrance to Gilkey Harbor—magnificent yachts, windjammers, modest cruisers, and ferries. If your tender is seaworthy and your arms or outboard strong, you can venture across to explore Grindel Point (see next entry).

GRINDEL POINT

🍴

44° 16.80'N   068° 56.49'W

CHARTS: 13309, 13302   CHART KIT: 20 (B), 66, 20

To visit the Islesboro Marine Museum on Grindel Point, anchor among the private moorings east of the ferry landing and row in to the floats (6 feet at low). Stay out of the way of the ferry, which makes frequent runs from Lincolnville Beach.

If your dinghy is substantial, you can reach Grindel Point from the moorings at Warren Island.

The museum is located in the old lighthouse on Grindel Point. It is open daily except Sundays and Mondays, with charming displays of local history and memorabilia. The lighthouse was built in 1850, but it was decommissioned in the 1930s. Over fifty years later, in 1987, the Coast Guard reactivated the light, and it is now automated.

When you visit, you may not be alone. The Grindel Point Lighthouse may be haunted. On July 17, 2004, a departing renter who was waiting for the ferry reported hearing a man's voice coming from the tower. Two hours later, a daytripper made an identical report.

The ferry landing has a pay phone, Wi-Fi, and restrooms, with picnic tables nearby. 🍴 If it's not Monday, try Jeanne's Place takeout (*734-8222*) for breakfast, lunch, or dinner. Bring your manners. Her sign reads, "Please no rudeness. If you must be rude, this is not the place."

PHILBROOK COVE

44° 15.65'N   068° 57.53'W

CHARTS: 13305, 13309, 13302   CHART KIT: 20(B), 66, 20

This pleasant little cove offers good summertime shelter on the west side of Seven Hundred Acre Island. It is easy to enter and affords a view of the sun setting over the Camden Hills and the lights twinkling at Lincolnville Beach.

APPROACHES. Entering is easy. Head in to the center of the cove from the north. The large ledge along the western shore is visible at halftide.

ANCHORAGES, MOORINGS. Anchor in 10 to 15 feet of water at low. A tripline might be a good precaution since the bottom seems to be rocky.

CROW COVE

44° 18.84'N   068° 54.46'W

CHARTS: 13309, 13302   CHART KIT: 66, 20

On the west coast of Islesboro, just south of Seal Harbor, is a little gunkhole called Crow Cove. True to form, the crows are still there to greet you in the morning and so are the seagulls and ospreys. In the evening the sun sets over the Camden Hills while you are lulled by the distant cries of a thousand herring gulls on Flat Island to the west.

The cove is extremely well protected except from the west, and you should have a restful night under almost any conditions. The Islesboro airport, serving small private planes, is a bit close, as is the main island road, but neither presents major problems.

APPROACHES. Flat Island, off the west side of Islesboro, is low and grassy, with only a few trees. Crow Cove is almost due east of Flat Island.

Both shores are reasonably bold except for some rocks making out from the southern shore. Stay centered and run past the moored boats.

ANCHORAGES, MOORINGS. Note the indentation to the north, which dries out at low tide. You can find good water for a short distance past the headland on the western side of this indentation. Anchor in 13 to 15 feet of water at low, in mud.

THINGS TO DO. The small Islesboro Historical Society is housed at the head of the southern branch.

SEAL HARBOR

🍴🛏

44° 19.30'N   068° 54.32'W

CHARTS: 13309, 13302   CHART KIT: 66, 20

Seal Harbor is a big, wide-open bight on the west coast of Islesboro, north of Crow Cove. The harbor has little to recommend it as an anchorage. It is exposed to prevailing winds from the southwest and too deep for easy anchoring. It could, however, be a useful daystop. A public landing and Big Tree Beach are where the main road comes close to the water's edge. 🛏 A short walk northward on the road leads to Durkee's General Store (*734-2201*), which stocks groceries, beer, wine, and liquor. 🍴 Their deli and takeout serve sandwiches, burgers, and pizza.

TURTLE HEAD COVE

2 ★★★

day: 44° 23.27'N   068° 52.94'W
night: 44° 22.77'N   068° 53.19'W

CHARTS: 13309, 13302   CHART KIT: 66, 20

Turtle Head forms the northern tip of Islesboro, pointing toward Sears Island, across the bay. The appropriately named head is high, wooded, bold, and delightful to explore. Turtle Head Cove is a broad bight to the west of the head, in the lee of Islesboro. In settled weather, it offers good protection from prevailing southwesterlies.

APPROACHES. To stay clear of the shoals and rocks at the western edge of the cove, run a course from red-and-white bell "11" northwest of Marshall Point to a point just north of Turtle Head. When you are well past Marshall Point and close to the peninsula of Turtle Head, turn south into the cove.

ANCHORAGES, MOORINGS. To explore Turtle Head, anchor opposite the narrowest part of the neck where the water shoals to 10 to 15 feet at low.

If you plan to spend the night, the best protection is in the southeast corner of the cove, off Spragues Beach, in 12 to 20 feet at low.

GETTING ASHORE. Row in to the beach.

THINGS TO DO. This is a conservation area. No fires or camping are permitted. To explore Turtle Head, land at a gravel beach near the narrowest part of the neck. Climb up and strike inland to a ferny, wooded path leading to the northern tip. You'll know you have arrived when the path skirts the very edge. (There are one or two cliff-hangers.) There is a beautiful little grassy plateau with interesting uptilted rocks, where a natural ramp leads down to the water. The whole walk should take less than an hour.

COOMBS COVE

3 ★★★

44° 20.88'N   068° 52.29'W

CHARTS: 13309, 13302   CHART KIT: 66, 20

Coombs Cove, at the northeastern end of Islesboro, is easy to enter and provides sweeping views from Castine down to the islands of upper Penobscot Bay. Hutchins Island is owned and protected by a local land trust.

APPROACHES. From can "9" at Islesboro Ledge, enter the middle of the cove, halfway between Hutchins Island and the point to the south. You can go in about three-fourths of the way toward the west end of Hutchins.

ANCHORAGES, MOORINGS. Anchor in 12 to 15 feet at low. The mud bottom holds well. You will be protected from the prevailing southwesterlies, but there is exposure from the south around to the east.

SABBATHDAY HARBOR

4 ★★★

44° 20.44'N   068° 53.21'W

CHARTS: 13309, 13302   CHART KIT: 66, 20

A dramatic wooded and rocky point guards the entrance to this pretty cove and the beach at its head. From within there are lovely views of the upper Penobscot Bay islands. Protection is good except from due south.

The cottages in Ryder Cove, in the northern part of the harbor, are the remnants of Islesboro's first development, in the late 19th century, which included a large pavilion, a dance hall, and a dock. This was a summer colony for Bangor residents who arrived by steamboat.

APPROACHES. Look for the prominent point that forms the east side of Sabbathday Harbor. It is crowned with spruce and has a house nestled in the woods. A small "cliff house" clings to its ledges.

The low, rocky points along the west side of the entrance are lined with summer cottages.

ANCHORAGES, MOORINGS. There are several private moorings in the harbor but plenty of room to anchor outside the moorings in 11 to 15 feet at low. The holding ground is good, in mud.

FOR THE CREW. Follow Ryder Road from the west point at the harbor entrance to the Main Road and turn south to Durkee's General Store (734-2201), which stocks 🛒 groceries, beer, wine, and liquor. 🍴 Their deli and takeout serve sandwiches, burgers, and pizza.

BILLY'S SHORE

44° 19.81′N 068° 53.51′W

CHARTS: *13309, 13302* CHART KIT: *66, 20*

The shoreline south of Sabbathday Harbor is known as Billy's Shore. It leads to Russell Point, a knob of land standing out to sea, high and wooded, crowned by one of Maine's classic 19th-century summer cottages. Anchor to the north and west of the cottage, in 12 to 15 feet at low. You can eat lunch here with dramatic views of Castine, Blue Hill, and the islands of Penobscot Bay.

BOUNTY COVE

 ★★

44° 19.03′N 068° 53.92′W

CHARTS: *13309, 13302* CHART KIT: *66, 20*

North of Hewes Point in the northwestern corner of Islesboro Harbor, two rocky headlands lead to small Bounty Cove, which is protected except from south to east. The houses on each headland and the road at the head of the cove are among the many signs of civilization.

The cove is of interest for its brief moment in history. In 1780, when Harvard College was mounting the first formal expedition to observe a total eclipse of the sun, the astronomers calculated that tiny Bounty Cove on Islesboro was the best viewing location. But Islesboro was British territory at the time, and the solar phenomenon would pay no heed to the affairs of men.

But what the natural workings of the world would ignore, civilized gentlemen could put right. Always devoted to the interests of science—and always very sporting—the British gave permission for the scientists to go behind enemy lines and witness the event.

The team was disappointed, however, because the eclipse was not total. A better place to observe it would have been Machias. Mathematician Rev. Samuel Williams calculated the center of the eclipse, but his maps had placed Islesboro 30 miles farther north than it actually was.

Don't make the same mistake.

APPROACHES. Pass north of can "1" at Hewes Ledge and head for the northwestern corner of Islesboro Harbor. Enter the middle of the cove, favoring the right side a little to avoid the ledges on the left, which are submerged even at halftide.

ANCHORAGES, MOORINGS. Anchor in the middle, in 11 to 15 feet at low. The holding ground is good mud.

GETTING ASHORE. Land at the small beach.

ISLESBORO HARBOR

 ★★

44° 18.43′N 068° 53.58′W

CHARTS: *13309, 13302* CHART KIT: *66, 20*

Islesboro Harbor is a wide-open bay, but it's in the lee of Islesboro in prevailing southwesterlies.

APPROACHES. Enter Islesboro Harbor north of can "1" on Hewes Ledge. There is a big house surrounded by birches on Hewes Point.

ANCHORAGES, MOORINGS. Several private moorings nestle along the south shore. Anchor near the moorings, in 10 to 20 feet of water at low. Go in cautiously, because it shoals rapidly here.

WEST PENOBSCOT BAY

LINCOLNVILLE BEACH

🍴

44° 16.90′N 069° 00.26′W

CHARTS: *13309, 13302* CHART KIT: *20*

Lincolnville Beach (Lincolnville on the chart) can be identified near the northern end of the Camden Hills by the cluster of houses and the terminal for the Islesboro ferry. This is not a protected spot for the night, but you might enjoy making a lunch stop at one of the restaurants or shops.

Staying south of nun "2" on Haddock Ledge, run in toward the boats moored north of the ferry terminal. Do not go much past the outside moorings because the beach is steep and shoals very rapidly. For the same reason, it is difficult to anchor here.

⚓ The Lobster Pound Restaurant (*789-5550*) maintains a couple of convenient guest moorings.

There is a float at the public landing north of the ferry wharf, with 3 feet alongside at low, where you can leave your dinghy and go ashore. Public phones are at the ferry wharf. The beach is flanked by the Lobster Pound Restaurant, at its north end, with a kids' menu, takeout, and a picnic area, and by McLaughlin's takeout to the south. Route 1 runs between the beach and a stage-set row of shops, giving the strip the feeling of a post-war auto destination. The line-up includes The Beach Store, the post office, and several shops and restaurants. Chez Michel (*789-5600*) is one of the best restaurants in the Camden/Belfast area, but it's closed on Mondays.

SATURDAY COVE
44° 20.28'N 068° 57.11'W

CHARTS: **13309, 13302** CHART KIT: **66, 20**

This is the fictional home of a cast of wonderfully wacky characters who appeared in Cap'n Perc Sane's columns in *National Fisherman* magazine. Cap'n Sane—in reality, journalist and author Allen D. ("Mike") Brown—peoples his columns with such figures as Shorty Gage, the lobsterman, Bubba Beal, the clam cop, and the Dodge girls, who sell "homemade dandelion wine n'chokeberry brandy, which is somewheres around 190 proof. They giggle a lot."

The cove is visited occasionally by yachtsmen, and in one of his columns Cap'n Sane describes the big cove event of the year. "The Marblehead Yacht Club pulled inta the Cove for n'overnighter. We were ready for 'em....There was 35 yachts, a self-propelled fuel barge n'a seaplane tender. They wanted just one mooring. Said they'd raft up. Short rented 'em just the ticket, his guest mooring, which is a 1932 Buick Roadster with a chain run right through her, stem to stern."

The little harbor is full of boats. There is not much room to anchor, and no place to go ashore. The dock near the head of the cove is for the use of resident boatowners only.

Saturday Cove was the scene of a skirmish in the War of 1812 in which fisherman Zachariah Lawrence single-handedly repelled a landing force of two barges of British Marines. Returning in greater strength, the Marines fired one-pound shot into several homes, pillaged and plundered, and took away Captain Amos Pendleton's gold watch. Wars were different in those days. Captain Pendleton went to Fort George in Castine to demand his valuables, and the watch was returned.

Saturday Cove, Cap'n Perc Sane's homeport.

BAYSIDE

44° 22.89'N 068° 58.00'W

CHARTS: **13309, 13302** CHART KIT: **20**

This is a charming summer community three miles south of Belfast where small Victorian gingerbread cottages surround a common. Although located on an open stretch of coast, Bayside has a bit of protection from prevailing winds.

Bayside was founded in 1859 as a Methodist campground—tents on platforms at first, later replaced by the small Victorian cottages that remain today. By the 1930s it was no longer a religious community and became part of the village of Northport.

APPROACHES. Bayside can be identified by the fleet of pleasure boats moored off a long dock and by the water tank above the trees (shown south of the town on the chart).

ANCHORAGES, MOORINGS. The Northport Yacht Club has its little clubhouse here and a vigorous sailing program, but there are no facilities for visiting yachts. The harbormaster may be able to help you find a mooring.

GETTING ASHORE. Row in to the dinghy float at the public landing or to the beach to the north.

FOR THE BOAT. *Town wharf (wharfmaster, 338-1312; harbormaster, 338-3419).* Water may be obtained at the public float, with 15 feet of depth at low.

FOR THE CREW. For the small crew there are swings and a small playground at the park by the waterfront. The nearest provisions are a mile away at a place known locally as Bell's Corner. Walk north 0.8 miles until the road intersects Route 1. Here you will find the brightly-painted Dos Amigos Mexican restaurant (*338-5775*) and a small market. The Bayside Country Store (*338-6118*) has groceries, beer and wine, ice, a deli, takeout, and pay phone.

BELFAST

3 ★★★

44° 25.58'N 068° 59.70'W

CHARTS: 13309 (INSET), 13302   CHART KIT: 20

Belfast, at the northwestern corner of Penobscot Bay, is an authentic Maine town with excellent facilities for visiting by boat. It is much less of a tourist destination than Camden, to the south, yet among its grand old buildings it has shops, galleries, and restaurants to explore. Route 3 runs due east from the Maine Turnpike in Augusta, so it is a convenient place to swap crew, and a large supermarket and a natural foods store fairly close to the waterfront make it a good place to reprovision.

Belfast was once an important shipbuilding town. Hundreds of schooners slid down the ways here into the Passagassawakeag River. Forest products were shipped all along the coast, and one historian reports that 10,000 cords of wood were stacked on Belfast's wharves in 1844. After the discovery of gold in 1848, the first direct voyage from Maine to California was made by the bark *Suliote* of Belfast.

It took two great fires, in 1865 and 1873, before Belfast was rebuilt in brick. Splendid blocks of Victorian Gothic and Greek Revival buildings mark the downtown section today, and most are on the National Register of Historic Places.

In more recent—and more ignominious—history, Belfast was a center for raising and processing chickens. But progress works in mysterious ways. Belfast moved out of the chicken business and into processed potato skins and then into the processing of credit cards. When the giant credit-card issuer MBNA moved to the area, it brought with it hundreds of employees and jobs and an economic boost to the entire western Penobscot Bay region. Their philanthropy included purchasing derelict property along the Belfast waterfront, razing the buildings, and building a waterfront park. MBNA has since become part of Bank of America.

Belfast is the homeport to the Penobscot Bay Area Pilots and their tugs, who guide large tankers and freighters up the bay to Searsport and up the Penobscot River to Bangor.

APPROACHES. Technically, the harbor of Belfast is in the Passagassawakeag River, which is more difficult to pronounce than to navigate. Leave red bell "2" at Steets Ledge to starboard, then run up the middle of the channel toward the bridge and nun "6." The nun is among the moored boats, which extend a long way past the town.

ANCHORAGES, MOORINGS. Belfast's waterfront has seen tremendous growth in recent years, with extensive new docks and floats, a boat ramp, and an ever-expanding mooring field. The city has five guest moorings, and others when available, or you can tie up at the city landing. The harbormaster has a booth on the wharf. Belfast Boatyard also rents moorings and dockage. It is also possible to anchor near the moorings in 7 to 15 feet at low.

GETTING ASHORE. Land at the dinghy floats at the city landing.

FOR THE BOAT. *Belfast Boatyard (207-338-5098).* Past the city landing and the tugboats on the south shore, Belfast Boatyard has 8 feet alongside the floats at low, with water, pump-outs, and electricity but no fuel. The yard offers storage, repairs, and marine supplies.

City Landing (Harbormaster Kathy Messier: Ch. 09; 207-338-1142). The Belfast city landing is on the south shore, opposite nun "6." Twenty-five transient slips are available. Small craft can dock on finger floats, and large boats can come alongside the outer floats and dolphins, with 13 feet of depth at low. Belfast also rents moorings. The harbormaster's attendants are usually in a booth at the city landing, and they can arrange a tie-up or direct you to a mooring. Gas, diesel, water, pump-outs, and electricity to 50 amps are available.

Hamilton Marine (207-548-2985; www.hamiltonmarine.com). This huge discount chandlery is 8 miles north of town on Route 1 in Searsport.

FOR THE CREW. The town provides restrooms, showers, ice, and pay phones at the city landing, and the Chamber of Commerce has a helpful information booth next door. The town landing can hold mail and packages and provides internet connections.

A Hannaford supermarket is .75 miles up Main Street. A natural foods store, the Belfast Co-op, is on High Street, just off Main. In addition to whole grains and fresh produce, they carry a full selection of wine and beer. A laundromat is next door. Twice a week a small farmers' market operates in a parking lot just up from the waterfront on the right.

To satisfy a hankering for homefries and eggs over easy, stop in at Dudley's Diner on Main Street. The Weathervane Restaurant (*338-3540*) serves seafood lunches and dinners overlooking the town landing. The Dockside Family Restaurant (*338-6889*) is halfway up Main Street with a casual menu. Darby's Restaurant (*338-2339*) serves fish and vegetarian dishes, or the Oriental Plaza (*338-4444*) has Philippine food and delivery service.

Young's Lobster Pound and Restaurant (*338-1160*) serves takeout in the large red building across the harbor, easily reached by dinghy. Or for the best in the area, take a cab (*338-2500, 338-2943*) to Chez Michel (*789-5600*) in Lincolnville and eat out on their second-floor deck.

Waldo County General Hospital (*338-2500*) is about a mile out Northport Avenue.

THINGS TO DO. Belfast's staunch brick buildings make it feel like a miniature city in a train-set layout, a feeling confirmed by its two-steps-and-you're-there scale. The Belfast City Park on High Street has a playground, tennis courts, a pool, and showers, all for free. The excellent public library has an active concert and lecture series. Be sure your kids say hello to Hawthorne, the Giant Elephant, on the roof of the restored Colonial Theatre, on High Street, to the right. Baby Hawthorne is in the lobby. A small Radio Shack is beyond the Colonial.

For a great harbor view of the waterfront, take a walk across the former Route 1 bridge, which has been converted to a footbridge at the head of the harbor. On your way, check out what the Belfast Maskers are performing in the summer theater in the old railway station of the B&M Railroad.

Wooden boat lovers should stop in to French and Webb, some of the finest yacht joiners on the coast, and see what is taking place in their shop above the town landing.

SEARSPORT HARBOR

1 ★★★★

44° 27.07′N 068° 55.38′W

CHARTS: 13309, 13302   **CHART KIT:** 67, 20

Because the harbor is wide open, exposed from southwest to southeast, and the view is dominated by elevators and oil tanks, Searsport would seem to have little appeal to visiting yachtsmen. But the town has a fascinating history of bustling shipyards and as the home of more than 200 ship captains who roamed the globe. This heritage has inspired the splendid Penobscot Marine Museum, which is well worth going out of your way to visit.

In the late 1800s, one of every 10 American shipmasters hailed from this small town. Several shipyards flourished in Searsport, supplying the demand and providing steady employment.

As Gretchen Ebbesson reported in the *Maine Times,* "Square-rigged Down-easters brought raw sugar from Hawaii and the West Indies; hemp from Manila; rice from India; Japan's silks and sulfur; China tea and coolies for railway labor; flax and tallow from Australia; copper ore, wool and hides from Chile; and guano from the Peruvian islands. To the homes of the seafarers came ostrich eggs, coral beads, clothes, pottery and carvings from China, teakwood chests, tropical shells, curry powders and Canton ginger.

"Some wives waited at home; others went to sea with their families. Many sea births were listed by latitude and longitude on the birth certificate …The Nichols family of Searsport had 35 members born at sea."

Searsport is also known for its antique shops, vying with Hallowell as the antiques capital of Maine.

APPROACHES. The approach is straightforward and without dangers. Sears Island to the east is high and wooded, with a tall radio tower near the southern tip. Mack Point, to the northeast, is readily identified by the elevators and oil tanks. Head for the group of boats and the pier at the northwestern corner of Searsport Harbor.

ANCHORAGES, MOORINGS. The town maintains a guest mooring. Or, if you plan to visit the Penobscot Marine Museum, call ahead (*548-2529*) to reserve their guest mooring, a big granite block often used by the windjammers. Otherwise, you can anchor anywhere near the moorings, in 15 to 23 feet at low, in good mud holding ground.

GETTING ASHORE. Land your dinghy at the town landing floats.

FOR THE BOAT. *Town Public Landing (Harbormaster Wayne Hamilton, at Hamilton Marine: Ch. 09, 10, or 16; 207-548-2985).* One float on the north side of the pier is just for loading and unloading. If you need to dock for longer periods, use the other floats as indicated. The floats have water, and the wharf is wired for power.

Hamilton Marine (207-548-2985; www. hamiltonmarine.com). Hamilton Marine is probably the largest chandlery in Maine. You'll find just about any marine item there, plain or fancy, including books and charts.

Hamilton is located north of town, 1.3 miles from the town landing. It is worth the walk or the taxi ride if you need something for the boat.

FOR THE CREW. If you are like us and get hungry the minute you step ashore, Waynk's takeout, in a little trailer in the landing parking lot, is the stop for you. Waynk's franks are a specialty. "PIZZA" is printed across the trailer. "I don't have pizza," says Waynk. "That's just false advertising."

Walk up Steamboat Avenue to Route 1 and turn right in to town. There you will find a market, drugstore, bank and ATM, post office, and several restaurants. Coastal Coffee (*548-2400*) serves breakfast and lunch and has internet-connected computers.

A laundromat is located farther out Route 1, just past Hamilton Marine.

THINGS TO DO. The Penobscot Marine Museum (*www.penobscotmarinemuseum.org*) is located in a group of seven 19th-century buildings right in downtown Searsport. The museum is a gem. For anyone interested in the history of Maine and of the great sailing ships, this is a wonderful place to visit.

One of the houses is furnished with the belongings of a sea captain and his family: beds, bureaus, commodes, paintings, and personal articles. Another exhibit traces the development of sailing vessels from the early days to the square-rigged Downeasters built in the 1870s and 1880s. And the museum's watercraft collection includes a salmon wherry, a dory, a peapod, a yawlboat, a smelt scow, a river punt, and a lumberman's bateau.

The museum has a wonderful collection of fine ship paintings, including 24 by noted artists Thomas and James Buttersworth, and many examples of ship portraits commissioned in Chinese and European ports.

LONG COVE

44° 27.50'N 068° 53.39'W

CHARTS: 13309, 13302   CHART KIT: 67, 20

Long Cove is one of the most industrialized areas in Maine, with cargo piers, conveyors, warehouses, and oil tanks. There is nothing here to attract yachtsmen except adequate protection.

Nearby Sears Island may also be developed as a cargo port. Like a bull's-eye in the center of Penobscot Bay, the island has been in developers' sights ever since Major General Henry Knox, George Washington's first Secretary of War, acquired ownership of it as part of his wife's dowry.

If Knox had not choked to death on a chicken bone in 1806, the history of the island might have been different. After his death, it passed through other hands to the Bangor and Aroostook Railroad, which envisioned a profitable resort with vacationers arriving, of course, by train. "Penobscot Park" opened in 1906, complete with dance pavilion and amusement park. Doomed by the automobile and new patterns of travel, however, it lasted only until 1927.

For several decades, the island was only used for picnicking and hunting. But its natural protection and easy water access spurred a rash of development proposals for a deepwater port. They included nuclear and fossil fuel power plants and the construction of a cargo port. A causeway to the mainland was built and initial construction began before previously overlooked wetlands were discovered on the island.

The project has been mired in legalities ever since. The State bought the island, and now a Sears Island Planning Initiative is trying, once and for all, to craft a vision for the future of the Maine coast's largest undeveloped island. On the table are two core recommendations: that the entire island should remain in its present, undeveloped, natural condition;

and that the cargo port on nearby Mack Point be expanded as demand warrants into an environmentally integrated inter-modal transportation hub. Perhaps someday you will be able to come to Mack Point by boat from Rockland to visit Sears Island and then catch a train to Baxter State Park.

APPROACHES. The approach, west of Sears Island, is deep and well buoyed for the use of ocean-going vessels. High, wooded Sears Island has a radio tower near its southern end.

ANCHORAGES, MOORINGS. You can anchor in the middle of Long Cove, in 10 to 20 feet at low.

STOCKTON HARBOR

 ★

44° 28.20'N 068° 51.85'W

CHARTS: 13309, 13302   CHART KIT: 67, 20

Sears Island and Cape Jellison form a large, very well-protected harbor at the mouth of the Penobscot River. The undeveloped shores of Sears Island to the west and the sparsely populated Cape are marred only by the Delta Chemical Company on the mainland shore. For cruising sailors, it provides a good harbor of refuge or a convenient place to spend the night before running up the Penobscot River.

APPROACHES. The approach is easy and well buoyed for large ship traffic. Sears Island is high and wooded, with a tall radio tower near the southern end. As you enter between Sears Island and Cape Jellison, you will see the tanks, sheds, and stacks of the chemical plant on the west side of the harbor.

The only danger is a Delta Chemical offshore loading platform standing in the water in the left-hand side of the entrance as shown on the chart. Often a barge is alongside. Leave nun "6" to starboard and then the loading facility to port. There is ample room to pass safely.

Pilings and ruined wharves line the eastern side of the harbor, but they present no dangers.

ANCHORAGES, MOORINGS. The harbor shoals gradually toward its head, leaving room for a fleet to anchor in 9 to 16 feet of water at low. Holding ground is good mud, and protection is good in every direction except due south. Several dozen lobsterboats, draggers, and small pleasure craft moor just past the ruined wharves on Cape Jellison.

It is a bit eerie here at night—not a single light on the Cape Jellison shore, just the blinking red signal on the Sears Island radio tower and the low hum of the chemical plant to lull you to sleep.

GETTING ASHORE. Land at the floats by the launching ramp on Cape Jellison, opposite the mooring field.

NORUMBEGA

Norumbega was a magnificent city on the banks of the Penobscot, three-quarters of a mile long, with splendid houses and streets as broad as London's. The Indians there wore furs and pearl-studded ornaments of gold and silver.

"Aranbega" first appears on Verrazano's map drawn from a 1524 voyage, and many later explorers sought the fabled city. One adventure that gave it credence was the extraordinary journey of an Englishman named David Ingram in 1568. A sailor who was abandoned with a hundred companions on the shores of the Gulf of Mexico for lack of provisions, Ingram walked along Indian trails with two companions all the way up the east coast through Maine to the Saint John River, where he embarked on a French ship that took him back to Europe. Remarkable though such a journey may have been, Ingram made an even better story of it, providing just what his audience wanted to hear and probably insuring himself free drinks for the rest of his life. His report, printed in 1589 by Hakluyt, chronicler of many early voyages of exploration, did much to embellish the reputation of Norumbega. According to historian Benjamin De Costa (*The Lost City of New England*, 1884), Ingram claimed to have seen "the spectacle of native monarchs borne to public audiences on sumptuous chairs of silver and crystal, adorned with precious stones. In all the houses pearls were common, and in some cottages were seen a peck or more."

Later explorers searched for Norumbega on the Penobscot River, and several reported finding it. The legend grew. More honest or more politic explorers maintained a discreet silence. In 1583, Sir Humphrey Gilbert sailed to establish a colony at Norumbega, but his ship was lost en route and Gilbert drowned. It was not until after his 1604 voyage that Samuel de Champlain reported he could find no trace of the fabulous city on the Penobscot. Even then, the myth faded slowly and reluctantly. People wanted to believe in Norumbega.

Later maps applied the name to a wider region, and all of New England was known as Norumbega until well into the 17th century. The name still appears from time to time, recently as a splendid turreted estate in Camden that is now converted into a bed and breakfast.

PENOBSCOT RIVER

The Penobscot is a grand, wide river, 24 miles long from the entrance at Fort Point to the head of navigation at Bangor. It is the second-longest river in Maine, rising near Mount Katahdin, coursing 240 miles to the sea, and draining one-third of the state.

Somewhere on the Penobscot was the fabled Indian city of Norumbega, reported by Verrazano in 1524 (see sidebar, page 241). Legends of its opulence slowly faded, but by the second half of the 19th century, the Penobscot was a prosperous thoroughfare for ships that brought goods from every nation and returned downriver loaded with lumber, granite, and ice. Steamships were seen daily on the Boston-to-Bangor run. The Penobscot was big time, and Bangor was the lumber capital of the world.

Today there are splendid miles of river with scarcely a house, almost unchanged since Verrazano's time. Most of the river traffic is small powerboats and occasional tankers. But it has recently been discovered by small cruise ships that visit the off-the-beaten-path ports of Bucksport and Bangor.

The beauty of the the river itself is reason enough to cruise up the Penobscot. You will pass under the new Penobscot Narrows Bridge, suspended by only one set of suspension cables. Just beyond, high on an escarpment opposite Bucksport, Fort Knox still commands this bend in the river. The prettiest stretch is Crosby Narrows, lined with cliffs and cedars. Ships returning from the West Indies loaded with rum dropped off a barrel here to keep the tax collector happy.

Once you have left Belfast or Castine, anchorages to the north are few and far between. Fuel is available in Bucksport, Winterport, and Hampden, and either Bucksport or Bangor are easy ports for provisioning.

Despite the river's Indian name which means "River with the rocks over there," going upriver in daylight and with the flood is easy for cruising boats. Navigation at night is dangerous, since there are few lighted buoys. There are a number of shoal areas, particularly off Lawrence and Luce coves and Frankfort Flats, but you will have no trouble if you follow the buoys and the chart carefully.

The wind usually blows straight up or downriver, and you will probably have to power. The Coast Pilot notes that currents of 5 knots are not unusual between Odom Ledge and Orrington. The ebb is stronger than the flood, and when the wind blows against the ebb, the river can be rough. The lumbering business has left the legacy of deadheads, waterlogged logs which have sunk to the bottom and periodically refloat to lurk dangerously at the surface.

There are two fixed bridges on the main channel of the Penobscot River, both with ample clearance. The new Penobscot Narrows Bridge replaces the old steel Waldo-Hancock Bridge and carries Route 1 across the river to Verona Island, just south of Bucksport. The second is a bridge routing I-95 just south of Bangor.

Turtle Cove is a burying ground for sailing ships. Sloops and brigs and schooners were run up here and scuttled when their useful days were done. Commodore Dudley Saltonstall sank his flagship, *Warren*, near Oak Point during the disastrous Penobscot Expedition in 1785 (see sidebar, page 251).

FORT POINT COVE

3

44° 28.35'N 068° 48.92'W

CHARTS: 13309   CHART KIT: 67

North of Fort Point, at the entrance to the Penobscot River, Fort Point Cove is a large, open bay with good anchorage. In the days when Bangor was the lumber capital of the world, the cove was full of ships and barges waiting out the ebb tide before heading upriver to pick up their cargoes, or pausing before setting out to sea. It still serves that purpose today, although you are likely to be the only boat in sight. Fort Point State Park and the lighthouse at the southern end of the cove are worth the stop here.

APPROACHES. Fort Point is wooded, with a bold cliff topped by the square white lighthouse tower and a substantial residence attached. Nearby is the white bell tower. Fort Point Ledge is marked by a granite structure with a red beacon on top. Leaving this to starboard and can "1" to port, curve left around the point and into the cove, well outside the long pier.

ANCHORAGES, MOORINGS. Anchor anywhere north and west of the pier in 8 to 29 feet of water at low. Holding ground is good, in mud. The cove is big enough so that there is considerable fetch if the wind is blowing.

GETTING ASHORE. Row in to the pier or land on the sand beach to the east.

THINGS TO DO. Walk up the wooded paths of the state park, working your way east toward the lighthouse. You will find the earthworks and foundations of old Fort Pownall, built in 1759.

The Coast Guard station adjacent to the park has a long history. Marking the entrance to the Penobscot River, it was established in 1836 during the administration of Andrew Jackson. The beautiful fourth-order brass-and-crystal Fresnel lens was made in France and mounted in 1857.

The bell tower northeast of the light originally housed a clockwork mechanism that struck the 1,200-pound bronze fog bell every 20 seconds. Survivor of a decade when similar towers were burned as useless structures, this is the only bell tower still owned by the government, and it is listed on the National Register of Historic Places.

MORSE COVE

1 ★★

44° 27.38′N 068° 47.33′W

CHARTS: 13309   CHART KIT: 67, 20

Morse Cove is a broad bight on the eastern shore of the Penobscot River, near its entrance, opposite Fort Point.

APPROACHES. The approach is straightforward. Morse Cove is easily identified by the rusting hulk, which shows on the chart and acts as a breakwater. It is shoal inside the moorings except near high tide.

ANCHORAGES, MOORINGS. Rental moorings, mostly 200-pound mushrooms, are available from Devereux Marine, or there is plenty of room to anchor. Morse Cove is exposed from the north to the southwest.

GETTING ASHORE. Take the dinghy in to the float behind the rusting *Squall*, bought for $1 from the Coast Guard and sunk here as a breakwater.

FOR THE BOAT. *Devereux Marine (207-326-4800).* Devereaux's floats are accessible only at higher tides, but they have water and electricity and pump-out facilities. The yard has a 30-ton boatlift and can perform hull, rigging, and engine repairs, or you can tackle them yourself. "And put in," say Bill Stephenson and Andrea Doyle, "that we're real nice."

FORT KNOX

44° 34.00′N 068° 48.00′W

CHARTS: 13309   CHART KIT: 67

During the American Revolution and the War of 1812, the British controlled the Penobscot River, so when a border dispute erupted far north in Fort Kent in 1839, there was a natural fear that the British might once more sail up the Penobscot and seize Bangor, a critical source of lumber.

Fort Knox was started in 1844 to forestall this possibility. Granite for the construction was quarried five miles upstream at Mount Waldo and barged downriver. The fort was manned during the Civil War and the Spanish-American War, but by then it was obsolete.

This is a fort for every taste, whether you like exploring dark, bespidered chambers and tunneled staircases or prefer the open air and grand vistas from the upper battlements. Some of the great 10-inch and 15-inch smoothbore cannons still command the river, and the hotshot furnaces remain.

This was the prototype of many granite forts built in Maine, including Popham, Gorges, Preble, and Scammell. It was named for Major General Henry Knox, who trained gun crews for George Washington during the Revolution and became the nation's first Secretary of War.

Most visitors arrive at Fort Knox State Park by car. If you are feeling bold, however, it is possible to anchor in the basin just north of the tunnel that emerges from the green embankment, north of the fort itself. The basin is about 120 feet long and has a flight of stone steps at the end.

At low, the basin is almost dry, so anchor just into the river from it, in about 30 feet. The bottom is rocky, and the river currents are strong, so it might be wise to leave someone onboard while you explore.

At high slack, it is possible to bring your boat in to the basin and tie up to the iron rings in the granite walls, though you have to be brave. Have your fenders ready and beware of any stray currents or eddies.

THINGS TO DO. In addition to exploring the fort, your can visit the observatory at the top of the nearest tower of the Penobscot Narrows Bridge and get a 420-foot-high, bird's-eye view of the river and upper reaches of Penobscot Bay.

BUCKSPORT

44° 34.24'N 068° 47.87'W
CHARTS: 13309   CHART KIT: 67

Bucksport is a small town located across the river from Fort Knox, at the fork above Verona Island. A less exciting view is of the huge paper mill next door. Supplies and provisions are available, and the town has created a large town landing to encourage visitors of the boating kind.

APPROACHES. There are no dangers in approaching the town dock except the 6-foot spot shown on the chart if you stray to the east.

ANCHORAGES, MOORINGS. It may be possible to berth on the town floats overnight. Otherwise, anchor outside the private moorings off the town dock, in 15 feet of water, mud bottom.

GETTING ASHORE. Row in to the town dock.

FOR THE BOAT. *Bucksport Town Landing (Harbormaster Michael Ormsby: 207-469-6616).* The town landing can be identified by its flagpole. The center two floats are for loading and unloading only. The maximum tie-up is limited to one hour, but the harbormaster may allow you to stay longer or overnight. The docks have 8 feet of depth at low, and they are plumbed for water.

Bucksport Marina (207-469-5902, 207-989-5840). Just to the east of the town floats, Bucksport Marina primarily rents seasonal slips, though when space is available, they will rent to transients hourly or overnight by boat length or slip length, whichever is greater. Gas, water, pump-outs, and electricity to 50 amps are available on the floats. Restrooms, showers, and ice are ashore.

FOR THE CREW. Bucksport is a classic river town, founded because of the river but built with its back turned to it. Recently, though, Bucksport has rediscovered its waterfront and its past. The town has built a spiffy waterfront park and walking path and the historical society is housed in an adjacent railroad station. Restrooms and a pay phone are at the park, and almost everything else you may need is within a short walk.

Turn right to find the Jed Prouty Tavern and Inn, now an assisted-living home, which dates back to 1798. Built entirely of hand-hewn timbers, it hosted Presidents Jackson, Van Buren, Harrison, and Tyler, not to mention Daniel Webster and Jefferson Davis.

In town you'll find a laundromat, a pharmacy, a bank, a bookstore, and several restaurants that run the gamut from the Dairy Port to MacLeod's (*469-3963*). Just where the bridge touches the mainland, there is a 24-hour convenience store. A Hannaford supermarket and liquor store is just beyond in the Bucksport Plaza.

THINGS TO DO. The Alamo Theatre, almost directly across the street from the waterfront park, is home to Northeast Historic Film (*www.oldfilm.org*), a nationally recognized archive of historic Maine movie footage. Video of their stock can be rented or purchased or seen at the theater.

Following the main street to the east, you will pass Buck Cemetery on the north side. The tall monument nearest to the street belongs to Colonel Jonathan Buck. Observe on the stone the dark outline of a woman's leg, which has resisted all efforts at removal. One story goes that before coming to Maine, Magistrate Buck sentenced a woman to be burned for witchcraft. The mark is her curse, which followed him to the grave. Locally, other stories allude to a mistress and a jealous wife.

WINTERPORT

3 ★★★

44° 38.15'N 068° 50.40'W

CHARTS: 13309   CHART KIT: 67

Winterport is an old river town, settled in 1766, 11 miles south of Bangor. It was the head of navigation during the winter months when the brackish water froze farther upstream. Many steamboat captains chose to settle here, and their stately homes are a reminder of the days when the Penobscot River was a major channel of commerce. More than a mile of Main Street and parts of many side streets are on the National Register of Historic Places, though sadly, Winterport now has the feeling of a town on the way to someplace else.

APPROACHES. From the south, Winterport can be identified by the long, flat-sided tin sheds once used for storing potatoes and frozen chicken. Continuing upriver a short way, you will notice Winterport Marine's buildings, docks, and floats on the west bank.

ANCHORAGES, MOORINGS. Winterport Marine has 4 heavy moorings, or you can come alongside. Mid Coast Marine, 2 miles farther upriver, also offers dockage and moorings. If you anchor, be aware that river current can run 3 knots or more.

GETTING ASHORE. Winterport Marine provides launch service, so you won't need to match your might against the river current. If you do use your own dinghy, take every precaution.

FOR THE BOAT. *Winterport Marine (Ch. 09, 16; 207-223-8885; www.winterportmarine.com).* This full-service marina was once a ship dock, and the floats have 13 feet of water alongside at low. Gas, diesel, water, ice, electricity, and pump-outs are available here. Winterport Marine has a full-time mechanic and can perform hull and engine repairs, with a 12-ton boatlift and a hydraulic trailer and ramp.

Mid Coast Marine (207-223-4781) lies 2 miles above Winterport. They have gas, water, ice, and electricity, dockage and moorings. They can haul by hydraulic trailer and perform most repairs.

FOR THE CREW. Winterport Marine has showers ashore. To reach civilization, walk straight up the road to Main Street (Route 1A), no more than .2 miles away. A large convenience store is at the south end of town. You will also pass a laundromat, Winterport Pizza, Rosie's Diner, and Winterport Winery, which specializes in fruit wines.

THINGS TO DO. Winterport is an ideal size for exploring on foot, and the Historical Association, on the main street, can give you guidance. Find the Union Meeting House, where a Revere Boston Bell hangs in the bell tower, or walk up the hill on Marine Street past the old homes for a view of the river from the crest.

Schooners being towed by steamer through the Penobscot River Narrows, 1912. (Frank Claes)

HAMPDEN

3 ★★

44° 45.77'N 068° 47.89'W

CHARTS: 13309   CHART KIT: 67

Hampden lies a couple of miles south of Bangor. Two marinas can take care of most boating needs, while Bangor is a short cab ride away.

ANCHORAGES, MOORINGS. Turtle Head Marina and Waterfront Marine both have rental moorings, and Waterfront has limited dock space.

FOR THE BOAT. *Turtle Head Marina (Ch. 16; 207-941-8619).* Gas, diesel, and water are available at the marina float, with ice and marine supplies ashore. The buoyed approach channel is dredged to 10 feet, but the channel is narrow. A public launching ramp is next to the marina.

Waterfront Marine (207-942-7029). This new marina is further upriver on the west bank in East Hampden, .75 miles south of the large Route 395 bridge. They occupy an old ax foundry that once forged the essential tools for north-woods lumberjacks. Gas, diesel, water, and limited marine supplies are available, and they service gasoline and diesel engines.

FOR THE CREW. Dana's Grill is in the Turtle Head building. A convenience store is half a mile from the parking lot, left on Route 1A.

BANGOR

3 ★★★

44° 46.79'N 068° 46.74'W

CHARTS: 13309 CHART KIT: 67 (B)

Bangor was once the lumber capital of the world, a great roaring, brawling frontier town full of lumberjacks, sailors, shipwrights, merchants, prostitutes, warehouses, sawmills, bars, and boardinghouses—its face turned toward the river and the seven seas. But with the decline of lumbering and a disastrous fire in 1911, the river lost its importance. Recently, Bangor has rediscovered that grand highway to the sea, investing millions into several public docks and a waterfront park.

During the lumbering heyday, an army of lumberjacks felled trees all winter long—first nearby and then deep in the northern woods as the forest became depleted. When the ice went out in April, the great river drives began, and the Penobscot was choked with logs on their way to the mills at Bangor and beyond. A hundred million board feet were shipped in 1842, and it was 200 million by mid-century. The records show that 3,376 ships cleared Bangor in 1860, and the river between Bangor and Brewer was crammed gunwale to gunwale.

Today Bangor is experiencing a revitalization. Although the view from the waterfront still encompasses oil tanks, railroad cars, and expanses of pavement, efforts are being made to improve the waterfront with new floats, landscaping, and riverside walks. The town is probably best known now as home to the horror writer Stephen King.

APPROACHES. The approach to Bangor is easy and clear, passing under the fixed bridge for I-395 (vertical clearance 78 feet) to the head of navigation just beyond at the fixed highway bridge between Bangor and Brewer (vertical clearance 22 feet).

Examining the chart carefully, you will discover a number of little squares scattered along each side of the river for a couple of miles south of Bangor. These are cribs—little green islands defined by cribwork—once used for mooring deep-draft vessels. The cribs are now dilapidated and dangerous.

The Bangor Landing is on the west side of the river, just below the second fixed bridge, north of the private floats of the Bangor Dock Facility.

ANCHORAGES, MOORINGS. Bangor has five town floats, numbered consecutively from upriver. The first float from the south, number five, is the heavy-vessel dock used by visiting cruise ships. The second float, number four, is rented seasonally. Floats two and three are available for transients, either for limited dockage or for the night by arrangement with the harbormaster. Both have about 15 feet of depth alongside at low, water, and free pump-out facilities. The last float, number one, is above the bridge and shallow, for dinghy use only.

Use caution in approaching the docks. Current generally floods at 3 or 4 knots and ebbs at several knots, with a tidal range of 10 to 12 feet.

10 rental moorings are planned. Check with the harbormaster for availability and reservations.

GETTING ASHORE. Dinghy to the public landing floats farthest north, just beyond the bridge.

FOR THE BOAT. *Bangor Landing (Harbormaster Jerry Ledwith: Ch. 09, 16; 207-947-5251).* There is water and a sewage pump-out station, at the floats. 50-amp electrical service is at the top of the ramp (you need a long cord). Restrooms, showers, and a phone are by the harbormaster's building in the parking lot.

FOR THE CREW. Initially, you may not need to go farther than the Sea Dog Brewery hard by floats two and three. The Muddy Rudder Restaurant is across the river. It has no dock for the dinghy, but it can be reached on foot by crossing the Route 1A bridge.

To get to the main part of the city, walk northward on the path along the waterfront, crossing the tracks and continuing up Broad and Water streets, to Main Street. It is about .3 miles from the waterfront. Turning right on Main Street, you will find a drugstore, bank, bookstore, deli, restaurants, and bakery.

The waterfront path leads southward and crosses the railroad tracks to a Shaw's plaza, visible from the waterfront. In addition to the huge supermarket, it has a liquor store and a full laundromat.

Bangor International Airport and the Eastern Maine Medical Center (*973-7000*) are nearby, and several large shopping malls are within easy taxi distance.

THINGS TO DO. Bangor has six historic districts with 20 landmark buildings, and there are stately Victorian mansions along West Broadway. One way to see the city is to take a bus or walking tour; inquire at the Tourist Information Bureau on Main Street. The Bangor Historical Society Museum is at 159 Union Street (*942-5766*).

The annual Bangor State Fair takes place at the end of July and beginning of August. In the last weekend of

August, the city hosts an annual Folk Festival. Dockage is extremely limited then, so call ahead.

The Bangor Municipal Golf Course is a mile away from the waterfront.

A walk south from the waterfront with a left on Main Street will bring you to Hollywood Slots, Bangor's first foray into casinos and gambling. If you would like to exercise your arms instead of your legs, they will come get you in their shuttle.

EAST PENOBSCOT BAY

CASTINE

Eaton's Boatyard.

2 ★★★★★

44° 23.19′N 068° 47.76′W

CHARTS: *13309, 13302*   CHART KIT: *66, 20*

Castine is a beautiful town on a high and almost isolated peninsula north of Cape Rosier at the mouth of the Bagaduce River. Lovely old homes grace streets still lined with elms, and the village is quiet and inviting. Castine's history is long and colorful, with believe-it-or-not military comings and goings that border on the bizarre.

Samuel de Champlain stopped here in 1604, on his way up the Penobscot River, and John Smith followed in 1614. A Plymouth Company trading post was established by 1629. From then on, Castine changed hands regularly between the French and English, and once, for a few days, the Dutch. Something like 16 different fortifications were built on the peninsula, of which Fort George and Fort Madison are still recognizable today.

One of the earliest, Fort Pentagoet, built by the French in 1635, has been excavated, revealing that Brittany slate, used as ships' ballast, was one of the building materials. Ceramic stew pots, soup ladles, and stemware for Rhennish wine uncovered here suggest that the French brought civilization with them to the wilderness. Baron Jean-Vincent d'Abbadie de Saint-Castin was a 22-year-old ensign at Fort Pentagoet when it was destroyed in 1674. Returning from France after his discharge, the baron married Mathilde, daughter of Madockawando, a Tarratine Indian sagamore, and rebuilt Pentagoet.

A century later, it was Britain's turn, and during the Revolution they started construction of Fort George—thus instigating the disastrous Penobscot Expedition (see sidebar, page 251). When the 1783 Treaty of Paris unexpectedly ceded Castine to the United States, a number of Tories who had settled there dismantled their houses and shipped them north to found the town of St. Andrews on Passamaquoddy Bay.

Castine was the last British post to be surrendered at the end of the American Revolution—only to be recaptured during the War of 1812. The British left at last in 1815.

APPROACHES. From the south, you can safely coast along quite close to Cape Rosier, which is high, wooded, and steep. From well south of Cape Rosier, you will be able to see the round, white tower of the abandoned lighthouse on Dice Head.

A surprising number of boats have discovered the rock shown on the chart on the northwestern corner of Nautilus Island. Do not cut the island close. Find red-and-white bell "CH" (*44° 22.50′N 068° 49.05′W*), marking the entrance to Castine Harbor and proceed up the middle of the river, leaving can "1" to port and red daybeacon "2," on its granite pile at Hosmer Ledge, well to starboard.

The Castine waterfront is opposite nun "2," which marks the western edge of Middle Ground. Be sure to locate the nun, well over on the north side of the river just outside the moored boats. Obvious landmarks are the large brick buildings of Maine Maritime Academy. Often the training ship *State of Maine* is docked alongside.

ANCHORAGES, MOORINGS. The Bagaduce River empties out past Castine, and the current is swift. This, and the extreme depth, make it impractical to anchor off Castine. Eaton's Boatyard rents moorings, as does the Castine Yacht Club and the Castine Harbor Lodge. ⇾ Dock space is available at Eaton's, Dennett's Wharf, the Harbor Lodge, the town docks, and sometimes at the yacht club.

Another strategy is to lie at the town dock or pick up a mooring for a couple of hours while exploring the town, then anchor for the night in nearby Smith Cove or Holbrook Island Harbor.

GETTING ASHORE. Land at the yacht club, Eaton's, or on the floats of the town dock.

FOR THE BOAT. *Castine Town Docks (Harbormaster George Plender: Ch. 09; 207-326-4502).* The town has two floats, both just east of the Maine Maritime Academy. ⇾ The larger, westernmost float allows dockage for up to two hours or overnight by arrangement with the harbormaster. It also has water and 30-amp power. The smaller, eastern dock is limited to 20 minutes.

Dennett's Wharf (207-326-9045). Dennett's is primarily a restaurant, but they have a rental mooring and ⇾ 80 feet available on their face dock and another 40 feet on their dock to the east. Depth is around 11 feet at low. Electricity, water, showers, and restrooms are available.

Eaton's Boatyard (Ch. 09 or 16; 207-326-8579). Eaton's dock and floats and big shingled shed are west of the yacht club and just east of Dennett's Wharf. Their fuel dock, with 16 feet alongside at low, has gas and diesel, pump-outs, ice, and water. They have a marine railway and can handle hull and engine repairs. Call the yard for a tow if you end up on a nearby ledge. Ken Eaton's workboat is said to be the hardest-working in Maine. The hull is 78 years young, and the engine is a mere 45.

Castine Yacht Club (pay phone: 207-326-9231). CYC is located in a modern gray building with a peaked roof beyond Eaton's and directly opposite nun "2." They have five well-marked guest moorings for rent to visitors. Boats over 35 feet should use the outer two moorings. ⇾ They also will rent overnight dockage on the west float when it is available. There is water at the float, with 20 feet of depth alongside at low.

Castine Harbor Lodge (Ch. 09; 207-326-4335; www.castinemaine.com). You can't miss this spectacular, bright yellow summer "cottage" as you enter the harbor. The lodge rents several moorings and ⇾ deep dock space on their large T-shaped wharf.

FOR THE CREW. If you are lucky enough to be on Eaton's docks, you'll find Ken a gracious host.

He'll even cook lobsters for you in his boathouse. The Castine Yacht Club has showers and a pay phone. Another shower is at the Castine Inn, up Main Street.

🍴 The Breeze takeout, restrooms, and a pay phone are at the Town Docks. If you have kids, or have a yen for a good cone or outstanding crab roll yourself, be sure to visit the Variety Store, an authentic soda fountain in the heart of town that also serves breakfasts.

For dinner in a charming room with a 360-degree mural of Castine, walk a short way up Main Street to the Castine Inn (*326-4365*). The graceful old Pentagoet Inn is across the street (*326-8616*). Or try the Bagaduce Oyster Bar in the Castine Harbor Lodge (*326-4335*). Reservations are advised.

Dennett's Wharf was built in the 1830s as a sail loft. Now it boasts "Maine's longest oyster bar." Dennett's also offers meals to go and lobsters cooked or uncooked, with a dining deck overlooking the harbor. Bring a dollar for the ceiling.

🛒 T&C Grocery (*326-4818*), at Main and Water streets, has good fresh meat, fish, and produce and is also the state liquor store. Bah's Bakehouse (*326-9510*) is located at the Village Inn. Just follow your nose.

A bank is on Main Street. So is The Four Flags gift shop, which carries boating supplies and charts. The Compass Rose Bookstore and Café is a Wi-Fi hotspot where you can lower your waterline with nautical titles. Or if you really want to go to sea, visit the Maine Maritime Academy Bookstore in Curtis Hall for charts, clothes, and books on everything you never knew you should know about shipping out.

The Maine Maritime Academy pool is open to the public for limited periods. Check at the academy (*326-4311*) for times. For the Castine Community Hospital, call *207-326-4348*.

THINGS TO DO. Most stores have the free pamphlet *Welcome to Castine*, which includes a map and suggested walking tours. The beautiful walk up Main Street to the earthworks at Fort George passes various old buildings and historical plaques scattered throughout the town. The steeple of the Unitarian Church, for example, was designed by Charles Bulfinch, the designer of the U.S. Capitol, and its bell is a genuine Paul Revere. One plaque, at small-boy eye-level, has a graphic description of torture by the Indians.

The *State of Maine* is usually open to the public for guided tours when she is in port. When she is not on an expedition, the schooner *Bowdoin* may be here too. She was commanded by onetime Castine resident Admiral Donald B. MacMillan on 26 voyages of exploration to the Arctic. She now sails as a training ship for Maine Maritime Academy.

Perkins Street runs west along the shorefront, passing a number of old homes, including the J.A. Webster House, once known as "The House of Sin" because the owner worked in a shipyard on Sundays. Colonial crafts and activities are demonstrated at the John Perkins House, built in 1665. Next door are a blacksmith shop and the Wilson Museum, containing prehistoric and local treasures.

Continue along Perkins Street, passing the earthworks of Fort Madison on the shore, to Battle Avenue and then Dice Head, with dramatic views of Penobscot Bay. The walk from the town dock to Dice Head is 1.5 miles. A fire in the spring of 1999 gutted the lighthouse keeper's quarters, but the tower was spared. Rebuilt, the lighthouse is leased by the town and is a private residence, but the grounds are open to the public. A short path leads down to the rocks.

The Maine Coast Heritage Trust has preserved a spectacular 132-acre forest with trails on bold Blockhouse Point, a short walk to the northwest.

The Castine Golf Club (*326-8844*), up Main Street, welcomes visitors to use its course and its tennis courts. Dennett's Wharf rents mountain bikes.

Each August, the Maine Maritime Academy sponsors the Retired Skippers Race, for skippers aged 65 to 92 in boats from 27 to 60 feet.

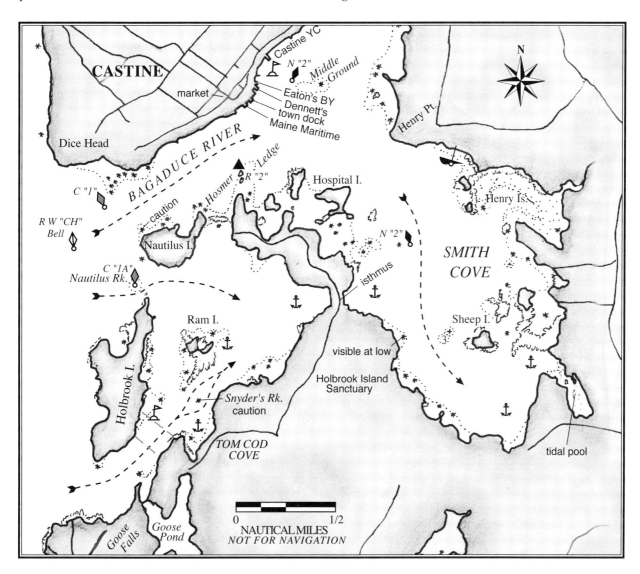

SMITH COVE

4 ★★★

E of Holbrook isthmus: 44° 22.27'N 068° 46.92'W
SW of Sheep Island: 44° 21.85'N 068° 46.30'W

CHARTS: 13309, 13302   CHART KIT: 66, 20

Due south of Castine, Smith Cove provides a convenient and secure anchorage under most conditions. Consider spending the day exploring Castine and then anchor in Smith Cove for the night. Most of Smith Cove south of nun "2" provides good depth for anchoring and is landlocked on all sides. It is also a mile across, though, so there is considerable fetch.

APPROACHES. Leaving Castine Harbor and observing nun "2" at the west end of Middle Ground, head for the opening between Hospital Island and Henry Point. The fleet of boats moored off Hospital Island belongs to Maine Maritime Academy. The Henry Islands are small, high, and wooded. At low tide, the islands are joined, and the wreck shown on the chart is visible. Leave nun "2" in Smith Cove to starboard.

Sheep Island is low and grassy with only a few trees and one small house on the south side. The deepwater entrance into the lower part of the cove is about 200 yards wide with a 3-foot spot near the middle. Estimate a point halfway between Sheep Island and the mainland to the west, and stay west of this halfway point. The rock off the western shore is visible for about an hour after low.

ANCHORAGES, MOORINGS. Anchor off the Holbrook isthmus on the cove's west shore or work deeper into the southeast corner of the cove beyond Sheep Island. Holding ground is good mud.

THINGS TO DO. The upper portion of Smith Cove makes a pleasant daystop. Anchor off the little beach at the narrow neck of land on the west side of the cove, in 15 to 20 feet, and row in to the beach. The headquarters of the Holbrook Island Sanctuary is a short distance to the left, and from there you can explore a network of beautiful trails (see Holbrook Island Harbor entry).

The tidal pool at the southeast end of Smith Cove is a favorite picnic and swimming spot.

BAGADUCE RIVER

The Narrows: 44° 25.55'N 068° 45.39'W

CHARTS: 13309, 13302   CHART KIT: 66, 20

The Bagaduce River drains a large area of land at the northeastern end of Penobscot Bay and forms Castine Harbor. Its currents are swift, particularly at the Narrows about three miles upriver. The river above Castine is best explored in a small powerboat. Beyond the Narrows, at Jones Point, the passage is full of rocks and shoals, most of which are unmarked.

On the passage en route to the Narrows, be sure to find nun "2" just off the docks of Castine, marking the western end of Middle Ground. After leaving nun "2" to starboard, cross to the other side of the river, leaving can "3" to port. Head upriver toward Negro Island, which is high and wooded, then around the bend, leaving can "5" to port, and proceed up the middle of the river. This section is quite wide. Summer cottages line the banks between stands of deciduous trees, and occasional fields run past farmhouses down to the river's edge. The river appears to end about a mile ahead where the Narrows turns eastward. From this point, you can see can "7" at the far end of the Narrows.

Beyond the Narrows, the river branches to the north and south around Youngs Island. Today most of Northern Bay dries out at low, but in the 1800s the town of Penobscot on its shores was a thriving port, where schooners took on lumber and bricks and navigated the Bagaduce under sail alone.

A deeper channel runs through South Bay and snakes between ledges toward the fixed bridge at Brooksville. This can make an exciting exploration in a dinghy with enough power to work against the current. Your reward is served up at the ¶ Bagaduce Lunch takeout by the bridge.

HOLBROOK ISLAND HARBOR

4 ★★★★

W of Holbrook isthmus: 44° 22.22′N 068° 47.42′W
CHARTS: 13309, 13302   CHART KIT: 66, 20

Holbrook and Nautilus Island form a beautiful harbor cradled in the northern tip of Cape Rosier just south of Castine. Not only is there good protection, but you will be treated to glimpses of the Camden Hills, Islesboro, Castine, and Dice Head Light. Most of its shores are preserved as part of the Holbrook Island Sanctuary.

APPROACHES. The entrance is fairly narrow but easy.

From the west, enter between Nautilus and Holbrook Island, leaving can "1A" at Nautilus Rock to port. Holbrook Island appears with a high wooded point at each end. Nautilus Island, also high and wooded, has a large house and a flagpole on its north end. Once past the can, jog to the left to avoid the north tip of Holbrook Island.

It is also possible to approach this harbor through the entrance south of Holbrook Island, though this approach can be hazardous for strangers. The isolated ledge in midchannel, south of Ram Island, shows only 2 to 3 feet above water at low, which makes it highly dangerous at midtide or above. "Snyder's Rock" has been located the hard way by many yachts—a dozen or more in recent memory.

ANCHORAGES, MOORINGS. The best anchorage is at the eastern edge of the harbor off the narrow isthmus. Anchor just past the headlands in about

---

THE PENOBSCOT EXPEDITION

On August 14, 1779, the fledgling American Republic suffered its worst naval defeat until Pearl Harbor, 162 years later. The history books generally omit this embarrassing episode—one in which we lost 17 armed vessels, 24 transports, and 474 men to a much smaller British force.

Castine was a village of about 20 houses in June of 1779, when the British landed 750 men and began construction of Fort George. The fort would secure the Penobscot River as a gateway for masts and lumber vital to the Royal Navy. It would also establish Castine as a refuge for loyalists as the Revolutionary War drew to a close.

In less than a month, the Commonwealth of Massachusetts assembled a fleet of militia and marines commanded by Commodore Dudley Saltonstall. Generals Solomon Lovell and Peleg Wadsworth assumed charge of the troops, and Paul Revere commanded the artillery.

On July 19, the fleet headed northeastward, bearing a sorry lot. Many were old men or young boys entirely lacking in military training. Sighting the fleet sailing up Penobscot Bay, the British hastily armed their partly finished fort and anchored three warships across the mouth of Castine Harbor. On July 25, the American fleet tried to enter the harbor but was repelled. They landed instead on Nautilus Island, where they managed to capture the British battery and regroup. A few days later, under British fire, the marines scaled Dice Head and captured the battery there, opening the way to Fort George. General Lovell's journal noted: "It struck me with admiration to see what a Precipice we had ascended...it is at least where we landed three hundred feet high, and almost perpendicular & the men were obliged to pull themselves up by the twigs and trees."

At this point the British, by their own account, were on the verge of surrender. Yet the American officers spent more than two weeks deliberating the merits of launching an attack by sea. Colonel Brewer, after having visited Fort George on routine business, noted that the British force was small and its defenses were light. He reported to Saltonstall, "I thought that as the wind breezed up he might go in with his shipping, silence the two vessels and the six gun battery, and in half an hour make everything his own. In reply to which," Brewer wrote, "he hove up his long chin, and said, 'You seem to be d-n knowing about the matter! I am not going to risk my shipping in that d-n hole.'"

Finally, on August 13, Lovell took command of a 400-man army reinforcement and prepared to storm the fort, a condition imposed by Saltonstall before he would agree to involve his ships. By noon, five of Saltonstall's ships anchored in line to await the tide. But their time had run out. By 5 PM, a British relief squadron of five ships was sighted entering Penobscot Bay. The American ships and Lovell's men retreated. Ignoring Lovell's pleas, Saltonstall led his fleet up the Penobscot River, pursued by the British. As the British closed in, the American crews scuttled or burned most of their ships, but the British captured two warships and nine transports, losing only 70 men and no ships.

General Lovell's journal entry for August 14 reports the sorry occasion: "...to attempt to give a description of this terrible Day is out of my Power...To see four Ships pursuing seventeen Sail of Armed Vessels, nine of which were stout Ships, Transports on fire, Men of War blowing up, Provisions of all kinds, & every kind of Stores on Shore...and as much confusion as can possibly be conceived."

The American survivors melted into the woods and walked home. The Penobscot Expedition was over..

16 feet at low. You can also anchor northeast or east of Ram Island in slightly deeper water.

⚓ Holbrook Island Sanctuary has a heavy guest mooring off the southeast flank of Holbrook Island, at the northern edge of the cable area shown on the chart. Use the southern entrance, and beware of the rock farther to the north. The mooring ball is white, marked "HIS Guest," with dayglow pickup floats.

GETTING ASHORE. Land on the isthmus beach or dinghy to the Holbrook Island dock.

THINGS TO DO. Turn right after landing at the beach and walk toward the headquarters of the Holbrook Island Sanctuary (*Phillip Farr, manager, 326-4012*). The sanctuary has an extensive system of trails leading inland and along the shore, and a trail map is available at the headquarters.

The wildlife sanctuary, covering 1,230 acres, was donated to the state in 1971 by Anita Harris "to preserve for the future a piece of the unspoiled Maine that I used to know." She lived in the big house on Holbrook Island until she died in 1985, leaving Holbrook Island and a trust fund to maintain it as a wildlife refuge and research station. The sanctuary can also be approached from Smith Cove, on the other side of the narrow neck (see previous entry).

Ram Island, long a local favorite for picnics and beachcombing, has been acquired by the Castine Conservation Trust to be preserved in its natural state for enjoyment by the public.

TOM COD COVE

3 ★★★

44° 21.63'N 068° 48.20'W

CHARTS: 13309, 13302   CHART KIT: 66, 20

On the north shore of Cape Rosier, due south of Ram Island, is a spot known locally as Tom Cod Cove. A handful of moorings occupy the cove, but you may be able to use one of them in this lovely and well-protected anchorage.

APPROACHES. Use extreme caution if you are approaching from Holbrook Island Harbor to the north. Snyder's Rock lurks south of Ram Island, smack dab in the middle of the channel. Instead, use the wide and clear entrance south of Holbrook Island. If you are coasting north along Cape Rosier, it will be the first opening. On the south side of the entrance are a gazebo, a breakwater, and a boathouse. Leaving these well to starboard, proceed up the middle of the channel. Leave to starboard Goose Falls, with its little bridges, and continue past the next rocky point to starboard, turning right into Tom Cod Cove.

ANCHORAGES, MOORINGS. ⚓ The Holbrook Island Sanctuary maintains several white moorings with dayglow floats, marked "HIS," available free to the public. Other moorings belong to the Castine Yacht Club. If the moorings are occupied and there is room, the holding ground is good mud, in 15 to 20 feet at low. The anchorage, pretty and unspoiled, is protected from prevailing southerlies but somewhat exposed to strong northwesterlies.

The rock charted off the west side of the cove is just south of the dock on the west shore and a bit farther out.

THINGS TO DO. Row to the sanctuary dock on the west side of the cove, to Goose Falls, or to the dock on Holbrook Island to explore the trails of Holbrook Island Sanctuary.

WEIR COVE

1 ★★★

44° 19.10'N 068° 47.10'W

CHARTS: 13309, 13302   CHART KIT: 66, 20

Although it offers less protection than neighboring harbors to the east, Weir Cove, on the southeast side of Cape Rosier, is a lovely anchorage in settled summer weather and a good harbor if the wind is from the north.

Abrupt and rocky Weir Cove Mountain at the head of the harbor is protected by the Maine Coast Heritage Trust.

APPROACHES. Run between Blake Point and the Spectacle Islands. When far enough past the Spectacles to clear the 7-foot spot, head for low-lying and rocky Buck Island (which has a little grass on top). Before reaching Buck Island, head north into Weir Cove. The ledges on the west side of the entrance, opposite Buck Island, are visible at halftide.

ANCHORAGES, MOORINGS. Anchor just outside the moored lobsterboats in 15 to 20 feet at low. There is exposure from the southeast to southwest, but holding ground is good.

HORSESHOE COVE

■ ★★★

44° 20.13'N 068° 46.10'W

CHARTS: 13309, 13302   CHART KIT: 66, 20

Horseshoe Cove is a slot on the eastern side of Cape Rosier, not far from the entrance to Eggemoggin Reach. On the chart, it looks impossible to enter. But once inside, this beautiful, unspoiled harbor has almost perfect protection.

The cove is named for the horseshoe crabs that migrate here annually to lay their eggs.

APPROACHES. The easiest approach is from the east, from the direction of Bucks Harbor and Eggemoggin Reach, passing north of low, grassy Thrumcap Island. Leave to starboard the small and hard-to-see red daybeacon "2" (privately maintained) at the end of the ledge marking the right side of the entrance to Horseshoe Cove. Continue westward until the cove opens up, then turn northward up the middle of the channel.

Pass a dock extending from the west shore and a series of mooring stakes in midchannel. Leave these stakes to starboard and aim for the small red daybeacon some distance ahead.

Green can "3" will nudge you over toward the east shore and align you to pass through the channel between red daybeacon "6" and green daybeacon "5." The red beacon is set on top of Cowpens Ledge, a large, rounded rock (covered about an hour before high tide) at the right-hand side of the entrance. The red beacon, though, is not at the edge of the ledge. Favor the green daybeacon slightly, staying at least 30 feet west of the red beacon as you pass into the mooring area.

Stay close in among the moorings. Joe Davis reports a rock less than 50 feet outside the westernmost moorings and a mere 4½-feet deep on drain tides.

ANCHORAGES, MOORINGS. 500- to 2,000-pound granite-block moorings are set out by Seal Cove Boatyard. Their sheds are just out of sight farther up the cove on the west side. Hang on a vacant mooring and check with the yard.

There is no room to anchor in the mooring area, but you may find some farther south in the approach channel. Protection is extremely good. The only exposure is to the south-southwest, but the tiny, wooded island at the harbor's mouth blocks the swells.

GETTING ASHORE. Take your dinghy in to the floats at Seal Cove Boatyard.

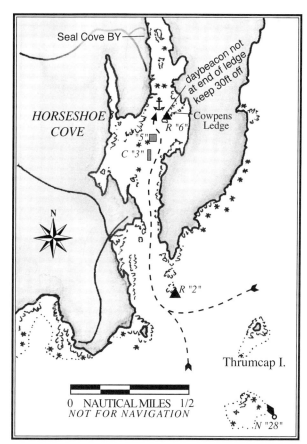

FOR THE BOAT. *Seal Cove Boatyard, Inc. (Ch. 09; 207-326-4422; www.sealcoveboatyard.com).* Owner Bob Vaughan specializes in rebuilding and restoring wooden boats and handles the usual storage and repairs. You are likely to see some great old classics at the moorings and up in the yard.

The route from the mooring area to the yard is narrow and crooked. Don't try it without someone from the yard to guide you. There is no fuel or ice, but water is available at the floats. The yard has mechanics for engine repairs, a large marine railway, and a 30-ton boatlift. Repairs can be made on wood and glass hulls.

THINGS TO DO. Past the boatyard, the river turns west and then north, running a long way back into Cape Rosier, and at midtide the water comes rushing in or out of the narrow passage between granite banks. The current runs through a salt pond and two sets of reversing falls, reaching 5 or 6 knots at its maximum and whipping up whitewater that attracts intrepid kayakers and even occasional yachtsmen in their dinghies. To explore this fascinating channel, Bob Vaughan suggests going upstream about an hour before high slack tide and returning with the first of the ebb.

ORCUTT HARBOR

4 ★★★

44° 20.70'N 068° 45.20'W

Charts: 13309, 13302   *Chart Kit: 66, 20*

Orcutt Harbor, on the east side of Cape Rosier, is just west of Bucks Harbor. The harbor is attractive, easy to enter, and well protected except from the southwest. In a northerly it would be a wonderful refuge. Orcutt is little used by yachtsmen, who perhaps prefer the amenities and security of Bucks Harbor next door.

APPROACHES. Go right up to the head of the harbor, favoring the eastern side to avoid two 5-foot spots to the west. There is good water almost to the rock at the head of the harbor.

ANCHORAGES, MOORINGS. Anchor in mid-channel in 7 to 14 feet at low. The bottom holds well.

BUCKS HARBOR and LEM'S COVE

■ ★★★★

Bucks Harbor: 44° 20.35'N 068° 44.25'W
Lem's Cove: 44° 20.02'N 068° 43.76'W

Charts: 13309, 13302   *Chart Kit: 66, 20*

Bucks Harbor is one of the best protected and most attractive harbors on Penobscot Bay. It is also one of the busiest. The American Coast Pilot of 1806 called it a hurricane hole, and, if you are in the right spots, it still is.

Bucks Harbor is nestled behind Harbor Island, just east of Cape Rosier and west of the entrance to Eggemoggin Reach. Bucks Harbor Yacht Club, the third oldest in Maine, is home to a substantial fleet of yachts. The best-protected portions of Bucks Harbor are behind Harbor Island and in the bight at the southeastern end called Lem's Cove. The harbor is the setting for many of Robert McCloskey's famous children's books, including *One Morning in Maine, Time of Wonder,* and *Blueberries for Sal.*

APPROACHES. Northwest of Little Deer Isle, there are several rocks and ledges, but most are marked. Find red-and-white bell "ER" *(44° 17.99'N 068° 46.48'W)* and steer for red-and-white bell "EG" *(44° 19.21'N 068° 44.56'W).* This leg is charted as an extension of Eggemoggin Reach, so you will pass between nun "32" to *port* and cans "33," "31," "29,"

Bucks Harbor from the Yacht Club.

and "27" to *starboard*. From "EG," enter Bucks Harbor either east or west of Harbor Island.

Green can "1" marks a ledge inside the harbor, but it may be hard to see among the moored boats. The ledge is visible during extremely low tides.

ANCHORAGES, MOORINGS. Bucks Harbor Marine rents moorings, and the Bucks Harbor Yacht Club maintains a couple of guest moorings and keeps track of members who are away.

Overnight dockage may be available at the Bucks Harbor Marine floats, with 22 feet of depth alongside.

Protection in Bucks Harbor generally is superb, depending somewhat on where you are located.

The best shelter is north of Harbor Island, where you are completely landlocked, but there is not likely to be space for visitors. The protection way east in Lem's Cove is also extremely good.

There is room to anchor outside the moored boats on either the east or the west side of the harbor, in good sand bottom, although the depth may run from 24 to 30 feet at low, and the protection is not as good.

GETTING ASHORE. Row your dinghy in to the yacht club floats or, if you are a customer, to Bucks Harbor Marine.

FOR THE BOAT. *Bucks Harbor Marine (Ch. 09 or 16; 207-326-8839).* Bucks Harbor Marine is east of the yacht club and easily recognizable by its white fuel tanks. Its "daily membership" fee includes a mooring, dinghy space, garbage disposal, and showers. They have gas, diesel, water, 30-amp electricity, ice, and some marine supplies, but they will charge for water if you are not a "day member." A diver is on site, and mechanics are on call for emergency repairs.

Bucks Harbor Yacht Club (pay phone: 207-326-9265). Water is available for filling tanks only, with a depth of 17 feet at the floats. Please don't wash down your boat.

Condon's Garage (207-326-4964). Condon's Garage, just up the road from the yacht club, can handle welding and some engine repairs. They are a Mercury outboard sales and service dealer.

FOR THE CREW. Bucks Harbor Marine has basic provisions—beer, soda, ice cream, and lobsters—and basic necessities, like showers. Their specialties include wines, cheeses, and a "famous" key lime pie. They can arrange long and short-term parking, and mail can be dropped off. They carry local and regional newspapers and books.

Visiting yachtsmen are welcome to use the yacht club's clay tennis court and clubhouse. On Thursday evenings there is joyful contra dancing for children of all ages in the clubhouse.

The Buck's Harbor Market (*326-8683*) sits at the first and only intersection in South Brooksville. In addition to basic groceries, they carry baked goods, produce, cheeses and deli meats, wine and beer, and *The New York Times*. The kitchen serves breakfast and lunch. In the evening it is taken over by the Rendezvous Restaurant (*326-8531*), which bills its fare as "Downeast Mediterranean."

THINGS TO DO. The pace is slow in South Brooksville. Stroll just to the west of town and marvel at the stepped granite seawall at the head of the harbor. Before the seawall was built, the road used to drop to the beach and was

impassable during extra high tides. The seawall's construction was funded as a WPA project, but it was still expensive enough that the locals dubbed it The Golden Stairs. Its western end has been rebuilt as a launching ramp.

Once your walk is done, try sitting in a rocking chair on the porch of the yacht club and observing the yachting scene below. Or wander through the clubhouse, which was built around 1912, and note the "first private burgee taken through the Panama Canal; also the Cape Cod Canal." One club member summed it up: "Nothing's changed. Change is all around us, but nothing's changed here."

POND ISLAND
44° 17.50'N 068° 48.00'W

CHARTS: 13309, 13302   CHART KIT: 66, 20

Pond Island, just south of Cape Rosier, was once part of a glacial river delta, composed entirely of sand and gravel, with characteristics more typical of islands around Cape Cod than the staunch rocky islands of Maine. A wide variety of seabirds have discovered the 30-acre island's loose, sandy soil for nesting, and its beaches have attracted generations of local picnickers, both modern and native. Clam shell middens date Indian feasts here to more than 2,000 years ago.

This is sensitive territory. In 1980, with the island facing a private sale and certain development, Betty Eberhart, a local resident and wife of Pulitzer-Prize-winning poet Richard Eberhart, persuaded the seller to delay the sale until she had time to raise funds to buy it. From her kitchen table, she solicited contributions from anyone she knew who had ever landed on or enjoyed Pond. Her remarkable effort lasted four years and included the formation of Friends of Pond Island, which ultimately presented the hard-won donations to the Nature Conservancy for the island's purchase, preservation and management. The Conservancy eventually transferred it to the Maine Coast Heritage Trust.

Thanks to Betty Eberhart's inspiration and dedication, Pond remains undeveloped and open to the public. Camping is not allowed, but caring visitors are welcome. Please tread lightly and disturb nothing.

Eiders build their camouflaged nests on the ground at the northwest end of the island, and they are easy to miss and easy to step on. Visitors should avoid the area until after mid-July. A delightful walk around the beaches of the southeastern end takes in the varied shoreline, wildflowers, and the marshy pond that gives the island its name.

On a nice summer day you might find a small armada anchored off the beach. The western end of Pond Island is high and wooded, but the southeastern end is long, low, and sandy, with a single clump of trees. Come in from the north and head for that single clump. Work your way in to a comfortable depth of 15 to 20 feet and anchor off the beach in sand bottom. Check the depth carefully and the state of the tide, since the beach shoals rapidly. Many boats ground out here while their owners relax ashore.

HOG ISLAND and FIDDLE HEAD

3 ★★★

44° 17.19′N   068° 47.27′W

CHARTS: 13309, 13302   CHART KIT: 66, 20

Hog Island lies southeast of Pond Island and is almost connected to it. No doubt it came by the name honestly, but "Hog" scarcely describes one of the prettiest islands in Penobscot Bay. Partially wooded with some open meadow, privately owned Hog Island is connected by a sand bar to rocky little Fiddle Head.

APPROACHES. Come in from the north.

ANCHORAGES, MOORINGS. Several private moorings are set off a dock on the northeastern end of Hog Island. Anchor outside these moorings, roughly on a line between the dock and Fiddle Head Island, in 15 to 25 feet at low. The bottom is sandy, but it seems to hold well. You will be well protected here during normal summer conditions. When the sandy bar is exposed, it creates the feeling of a peaceful lagoon.

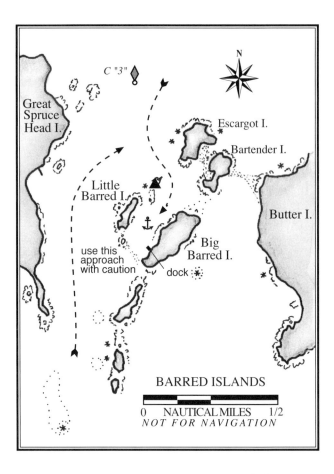

BARRED ISLANDS

PICKERING ISLAND

3 ★★★

Western Cove: 44° 16.06′N   068° 44.76′W
Eastern Cove: 44° 15.94′N   068° 43.89′W

CHARTS: 13305, 13309, 13302   CHART KIT: 66, 20

High, wooded Pickering Island lies at the northeastern end of Penobscot Bay, just south of Little Deer Isle. Two pleasant coves on the north side of the island provide good protection from prevailing winds.

The island is privately owned and frequently occupied during the summer months. It is also a preserve of The Nature Conservancy. You are welcome to go ashore at the western cove, but not at the eastern cove, where the owner has experienced "a constant intrusion of visitors."

APPROACHES-Western Cove. The easiest approach to either cove is from the west. Sailing up East Penobscot Bay, stay west of a line drawn between the western tips of Bradbury and Pickering Island to avoid the foul ground between them. Make your turn around the western tip of Pickering and coast along the north shore until the western cove opens up. Run up the middle of the cove toward the sand beach at the head.

APPROACHES-Eastern Cove. To reach the eastern cove, continue coasting fairly close to the north shore of Pickering. Little Eaton Island has a high, rocky bluff at the western end and a clump of spruce trees in the middle. Eaton Island is mostly wooded, with several beaches. A third, unnamed island is connected to the eastern tip of Pickering by a bar. Enter the middle of the cove formed by Pickering and the offlying islands.

ANCHORAGES, MOORINGS. In the western cove anchor in 10 to 12 feet at low. Protection is good except from the northeast. Lovely birch trees line the shore, and no houses are in sight.

In the eastern cove drop the hook in 12 to 15 feet at low along the southern or eastern perimeter of this cove. The bottom is rocky. Pickering's north shore is shoal, so favor the east end, by the little unnamed island.

This is the cove where the owner of the island lives. Respect his privacy and do not land here.

BRADBURY and CROW ISLAND
off Crow Island: 44° 14.85'N 068° 44.45'W

CHARTS: *13305, 13302*   CHART KIT: *68B, 20*

Small Crow Island, just east of Bradbury Island, off Deer Isle, is part woods and part meadows abundant with wildflowers. The state-owned island is enjoyed as a daystop by local residents, windjammers, and visiting yachtsmen. Under normal summer conditions, the anchorage is on the north side of the island. Crow Island's somewhat exposed location, combined with a dangerous ledge at each end of the island, makes this a dubious overnight stop.

Bigger Bradbury Island is owned and managed as a preserve by the Island Heritage Trust. This also makes a pleasant daystop.

APPROACHES. The approach to Crow is easiest from the east. A solitary boulder sits on the eastern ledge, but the ledge continues under water well beyond the boulder. Give it a wide berth.

From the west, coast along the north shore of Bradbury Island to avoid the rocks between Bradbury and Pickering Island. To continue to Crow Island you will need to turn northward at the east end of Bradbury to avoid the nasty ledge at the west end of Crow Island. This ledge extends some distance north of the two rocks shown on the chart, which are visible at high tide.

ANCHORAGES, MOORINGS. The north shore of Bradbury is steep-to except for the 21-foot shoulder just east of the middle of the shore and due west of the rock shown on the chart.

At Crow, anchor off the second little shingle beach from the east. The fringing ledge stops at the east side of this beach. The cove shoals rapidly, and you will have to get in quite close to find adequate anchoring depths. The bottom apparently is rocky.

THINGS TO DO. Crow Island is used by sea birds for nesting, so it is closed to human visitors from April 1 through August 15.

Bradbury is open for respectful day use. Please leave no trace. No camping is allowed, and no fires are permitted except below the high-water mark. Contact Island Heritage Trust (*348-2455; www.islandheritagetrust.org*) if you spot anything troubling.

GREAT SPRUCE HEAD ISLAND
44° 14.06'N 068° 49.10'W

CHARTS: *13305, 13302*   CHART KIT: *68B, 20*

Many lovers of Maine were introduced to the state by Eliot Porter's wonderful photographs in *Summer Island: Penobscot Country*, published by the Sierra Club in 1966. "Summer Island" is Great Spruce Head, still owned by Porter family.

The houses, barn, boathouse, and enormous bell are all on the east side of the island, opposite the Barred Islands. There is a float and dock, and a number of private moorings. Please do not go ashore on this private island.

BARRED ISLANDS

44° 13.95'N 068° 48.56'W

CHARTS: *13305, 13302*   CHART KIT: *68B, 20*

Three islands and some rocks form a tiny archipelago just southeast of Great Spruce Head Island and west of Butter Island. The little lagoon between them is one of the prettiest anchorages in Penobscot Bay. It is also one of the most popular.

The northernmost island is Escargot, with Bartender eastward toward Butter. Big Barred, the largest island, forms the eastern side of the harbor, and Little Barred forms the western side.

Once the hook is set, the beauty here is a beguiling temptress. No houses are visible on Big Barred Island, only a dock. It is hard to resist the urge to go ashore. The islands, however, are very much private, and landing is not allowed.

APPROACHES. The entrance to the Barred Islands is from the northwest. Run south past can "3" off Great Spruce Head Island and enter between Escargot, the northernmost island, and Little Barred, at the southwest corner.

Use caution at the entrance. A ledge, shown on the chart and visible at midtide, makes out a long way northeast of Little Barred. It is marked with a privately maintained daybeacon at its highest point—not at its extremity near the entrance. Don't cut the beacon close. The ledge extends underwater a considerable distance beyond it. You will find good water about halfway between the beacon and Escargot.

A second ledge lies east of the tip of Little Barred. When past this second ledge, turn right and run southward down the middle of the harbor, between low, wooded Little Barred to starboard and high, wooded Big Barred to port.

The approach from the south, between Great Spruce Head Island and the Barred Islands, is very narrow and lined with rocks. Use it with caution, on a rising tide.

ANCHORAGES, MOORINGS. Once past the ledge east of the tip of Little Barred, anchor anywhere in 20 to 34 feet at low, on the midline of the harbor or toward Little Barred. There is good water almost to the gravel bar that connects Little and Big Barred, at the south end of the harbor, and the holding ground is good.

From just opposite the dock you can see across the way to the houses, flagpole, and enormous bronze bell on Great Spruce Head Island.

You will be comfortable here in settled summer weather, but the anchorage can be rough and untenable with strong winds from the southwest, north, or northwest, especially at high tide when the bars are covered. If these conditions are expected, leave.

On most summer nights you will probably be sharing this anchorage with at least several other boats. Arrive early and respect the tranquillity of this special place. Off season, you may be lucky enough to have it all to yourself.

GETTING ASHORE. All of the Barred islands are private, and the owners request that you do not land. Bartender and Escargot are owned by the Cabot family, the owners of Butter Island. They remind cruisers that private property rights extend to the low tide line (see *Public and Private Property* in this book's introduction). The rules for their generous sharing of Butter Island are below.

BUTTER ISLAND

Orchard Beach: 44° 14.00'N 068° 47.17'W
Nubble Beach: 44° 13.67'N 068° 46.95'W

CHARTS: 13305, 13302   **CHART KIT: 68B, 20**

Butter Island, west of Deer Isle and north of Eagle Island, has long been a popular destination for cruising yachtsmen and local residents. There are lovely beaches to explore, abundant berries, and a 186-foot peak to climb. The shoreline forms no true harbors, however, so Butter is recommended only as a daystop.

Butter was owned by the late Tom Cabot of Boston, who for more than 50 years played a leading role in the preservation of Maine islands. For years the Cabot family generously welcomed visitors ashore, but an ever-increasing number of boaters, kayakers, windjammers, and excursion boats has forced them to establish rules and limitations to where the public is allowed on the island. Please read and respect the rules outlined below.

Like so many islands in Maine, Butter was settled by fishermen and farmers, beginning in 1785. In 1895, however, the island caught the attention of George and Emory Harriman of Boston, who envisioned an "Arabic-like town of tents and cottages" for genteel Bostonians of good social standing. Butter was renamed Dirigo, after Maine's state motto, and a hotel was built to accommodate as many as 100 guests. The New England Tent Club provided comfortable quarters and vigorous outdoor activities for their enthusiastic guests, who arrived by steamer from Boston. The round trip cost $5.50, leaving Boston at 5PM and arriving at Dirigo at 7AM, after a change at Rockland.

Like other Maine island resorts, the New England Tent Club succumbed to the advent of the automobile and the decline of steamships. The end came in 1916 when the ferry service was suspended. Today there is hardly a trace of the resort.

The lower, more eastern, of the two hills shown on the chart is Montserrat Hill, and the broad bights either north or south of it offer the only allowed access to the island. With a southwesterly wind, anchor off Nubble Beach in the southern bight or off Orchard Beach, in the northern bight. Exposure in either anchorage is from the north through the southeast. If the wind is from other quarters, anchor east or west of the peninsula extending from the southwest corner of Butter.

A less preferable anchorage is at the northwest corner of Butter, south of the sandbar connecting Butter and Bartender and east of the 64-foot spot. The mailboat plies this cove four times a day, and this is where the Cabot family has their dock and moorings, which under no circumstances should be used by visitors.

Approach any of these anchorages by working your way in to a comfortable depth and dropping the hook. Note that there are ledges making out a short way from shore near the southwest peninsula. The bottom in most locations is mud, providing good holding ground, but there is also some kelp.

THINGS TO DO. Rob Cabot, grandson of Tom Cabot, has forwarded us their latest public access policy. In addition to establishing where cruisers can and cannot go, it is a reminder of the huge debt of gratitude cruisers owe to private landowners who have managed, at great personal expense and effort, to keep the Maine coast as beautiful as it is. Please, when you visit islands such as Butter, keep your presence to a minimum—keep your group small, talk softly, and tread lightly. And please respect the following rules:

Visiting private boat owners may only use Nubble Beach, south of Montserrat Hill on the southeast side of the island, and Orchard Beach, north of Montserrat Hill at the northeast end, and may only hike the trail leading up Montserrat Hill to a granite bench overlooking these two coves. Fires and dogs are not allowed. No one is welcome on any other part of the island except by special invitation of the Cabot family. Commercial enterprises, including windjammers and kayak outfitters, are specifically not welcome. Use your holding tank, not the island, and remove all trash, including human waste. Two caretakers live on the island and patrol it on foot and by boat to ensure compliance. Limited camping for Maine Island Trail Association members and certain nonprofit groups is allowed at specific sites by advance reservation only (*446-4147*).

DEER ISLE—West Coast

NORTHWEST HARBOR

[4] ★★★

44° 13.95'N   068° 41.65'W

CHARTS: 13305, 13302   CHART KIT: 68B, 20

On the chart, Deer Isle is like a butterfly, and Northwest Harbor is the deep indentation between its wings. This little-used harbor is wide, free of dangers, and easy to enter in normal visibility. Its shores are tree-lined with only a few summer homes. The cluster of houses at the far end is the village of Deer Isle.

APPROACHES. Can "1" and nun "2" mark the ledges to the south, and unmarked Gull Ledge lies between. At high tide, Gull Ledge is a small, flat rocky island. Low tide reveals it to be a sizable ledge to be respected.

Coming from the south, use as your landmark Hardhead Island, which is high and barren with impressive cliffs. Then, leaving green can "1" well to starboard, head for Heart Island, north of the entrance. Heart is a small, steep island, with a cleared section facing southwest.

After leaving Gull Ledge and nun "2" to starboard, turn right and run down the middle of the harbor.

ANCHORAGES, MOORINGS. Anchor wherever you wish in 10 to 15 feet at low. You can go in comfortably as far as the large dock on the northeast shore, but beyond that it shoals rapidly.

Holding ground is good, in mud, and protection is excellent except from the northwest. Since so few boats use Northwest Harbor, there is as much swinging room as you could want.

GETTING ASHORE. It is a long way from the anchorage to the village, which is inaccessible by water at low tide. When there is water at the head of the harbor, land at one of the tiny beaches on either side of the causeway.

FOR THE CREW. Deer Isle Village has a post office, a library, and a bank, but like so many small towns, it has lost its market and hardware store. So you'll have to eat out. The Pilgrim's Inn (*348-6615*) serves dinner in their Whale's Rib Tavern (*348-5222*). In addition to the regular menu, they serve takeout and kids' meals.

SYLVESTER COVE

1 ★★★

44° 12.73′N   068° 43.41′W

CHARTS: 13305, 13302   CHART KIT: 68B, 20

Pleasant, friendly little Sylvester Cove, on the west side of Deer Isle, is shared by working fishermen, a summer colony, the Deer Isle Yacht Club, and the mailboats that transport the residents to Eagle and Great Spruce Head Island. It is a pretty place but not a well-protected harbor.

APPROACHES. Several clues will help you pick out Sylvester Cove from a distance. It is due east of the lighthouse on the northeast tip of Eagle Island. Hardhead Island, to the north, is easy to recognize, being high, treeless, and grassy, with high cliffs. And you will usually see a little cluster of masts in Sylvester Cove.

As you approach Sylvester Cove on an easterly course, leave nun "2" to starboard. This marks the end of a long ledge to the southeast that is not visible at half tide. You will see the yacht club dock on the north side of the cove and a miniature flagpole on the dock itself.

From the nun, you can run in among the moored boats if you wish, but use caution. Do not go beyond the floats at the dock. Local yachtsmen say that staying south of a line between the lighthouse on Eagle Island and the yacht club float will keep you in good water.

ANCHORAGES, MOORINGS. The Deer Isle Yacht Club maintains a couple of guest moorings outside the fleet near nun "2." You can also anchor in the same vicinity.

GETTING ASHORE. Row in to the back of the yacht club floats.

FOR THE BOAT. The Deer Isle Yacht Club has no clubhouse. Aside from parking and use of the floats (with 5 to 6 feet alongside at low), the only facilities are the guest moorings.

FOR THE CREW. Pilgrim's Inn (*348-5222*) in Deer Isle serves dinner. If you want to eat before their dinner rush, they may come pick you up. Or you can call a taxi at *968-3534*.

CROCKETT and GOOSE COVE

1 ★★★

Crockett Cove: 44° 10.23′N   068° 42.42′W
Goose Cove: 44° 10.09′N   068° 43.00′W

CHARTS: 13305, 13313   CHART KIT: 68B, 20, 21

Crockett Cove is a pretty, relatively unused anchorage on the southwest corner of Deer Isle, near Burnt Cove. It is wide open to the southwest, but offers good protection from the north with bold shores. There are a number of cottages and docks along the south shore, but no commercial activity.

In fair weather, Goose Cove, to the west, makes a pleasant daystop for exploring the Barred Island Preserve or for dinner at the Goose Cove Lodge.

APPROACHES. Coming from the west, use the steeple in West Stonington at the head of Burnt Cove as a mark. Once you have identified Burnt Cove, Crockett is the next one north. Run up the middle of the entrance.

Several huge rocks block the center of Goose Cove. The only place for a boat is on the 14-foot sounding, due east of the bar between Barred Island and Stinson Point.

ANCHORAGES, MOORINGS. There is plenty of room to anchor in Crockett Cove. Go in as far as the little peninsula to port. Anchor in 15 feet or so at low, in mud bottom.

Goose Cove Lodge has a heavy mooring with a white float in 10 feet of depth at low, or you can anchor outside the mooring.

GETTING ASHORE. Dinghy in to the crescent sand beach at the head of Goose Cove and land at the left end. Beware of the rocks lurking in the cove.

FOR THE CREW. Goose Cove Lodge (*348-2508*) welcomes cruisers for elegant dinners at their "casual bistro." Reservations are required.

THINGS TO DO. Trails for The Nature Conservancy's Barred Island Preserve leave from Goose Cove Lodge. The sandy bar to Barred Island is exposed 3 hours before and after low.

BURNT COVE

4 ★★

44° 09.81′N   068° 42.05′W

CHARTS: 13305, 13315, 13313   CHART KIT: 68B, 20, 21

Burnt Cove is on the southwestern corner of Deer Isle, about a mile north of the entrance to the Deer Island Thorofare. This is primarily a working harbor, with only a few pleasure boats, but it is friendly, easy to enter, and offers good protection.

APPROACHES. From the Fox Islands Thorofare, steer for the white church spire in West Stonington. As you get closer, check off the islands—Mark Island has the low lighthouse, Andrews and Second are both wooded, and, farthest north, The Fort is a huge granite ledge with just a few trees. A cleft in the rock on the end of Fifield Point suggests a missing front tooth. Stay in the middle as you enter Burnt Cove to avoid rocks along the south shore.

ANCHORAGES, MOORINGS. The inner part of the cove is full of moored boats. Anchor west of the moorings in a good mud bottom 9 feet deep at low. Billings Marine (*367-2328*), in Stonington, maintains a number of stout rental moorings here that may be available to transients. Strong southwesterlies will bring swells in from the west, but from all other directions the harbor is well protected.

GETTING ASHORE. Row in to the Fifield Lobster Company float. Or, if the tide is high and your destination is the market, take your dinghy to the head of the harbor and beach it near the red boathouse.

FOR THE BOAT. *Fifield Lobster Company (207-367-2313).* Fifield is the first commercial wharf on the south shore. They sell gas and diesel, with about 6 feet of low-tide depth at their floats.

FOR THE CREW. Try Fifield for lobsters. Walk to the head of the cove to find the big, fully stocked Burnt Cove Market (*348-2681*) and liquor store that also has ice and a pay phone. A hardware store (*367-5570*) and pharmacy (*367-5107*) are next door. The Island Medical Center (*367-2311*) is 2 miles away.

THINGS TO DO. The 100-acre Crockett Cove Woods Preserve was given to The Nature Conservancy by the well-known artist and architect Emily Muir. At the head of the harbor turn left on Whitman Road, which leads to the entrance of the preserve. There is a self-guiding nature trail through the quiet woods, with deep mosses and old man's beard lichens hanging from the trees.

ISLE AU HAUT

Lying five miles off Deer Isle, beautiful Isle au Haut is remote and hard to reach—probably the reason why the year-round population has always been low and why this rugged island looks little different now from when it was named by Samuel de Champlain in 1604.

"High Island," he called it, and so it impresses the visitor today—rising to 556 feet, six miles long, two miles wide, and heavily forested. A variety of birds live here, and so do deer. The deer are protected, and the islanders have built high wire fences around their gardens to keep them out.

More than half the island is part of Acadia National Park. Millions of people a year visit Mount Desert Island, but park visitors to Isle au Haut are limited to 50 a day. As a result, Isle au Haut remains a wonderful place to explore, with 30 miles of hiking trails, rustic campsites, fishing for salmon and trout in Long Pond, mountains to climb with splendid vistas, and rugged shorelines to walk.

The harbors on the west and northwest coasts are small and only marginally sheltered, but they are fine for settled summer weather. The two most useful ones are Isle au Haut Thorofare, adjoining the little village of Isle au Haut, and Duck Harbor, providing access to the park and trails.

The east coast of Isle au Haut is a forbidding place, where great Atlantic rollers crash on high cliffs. York Island, at the northeast corner, is private, inhabited, and difficult to approach. Head Harbor, at the south end of the island, is attractive but very much exposed to prevailing winds and ocean swells.

Electricity came to the island around 1970, but there were no telephones until 1988. Cars are expensive to transport by barge to Isle au Haut, so their owners keep them going as long as possible. As a result, the few "island cars" are in an extraordinary state of life-after-death. About 90 people live on the island all winter. Lobstering is the main occupation.

A ferry runs from Stonington to the village of Isle au Haut, and in summer it continues on to Duck Harbor. The *Miss Lizzie* was named for a much-loved lady who was postmistress for more than 50 years.

The best hiking trails are accessible from Duck Harbor. Trail maps are available at the information board ashore. The most popular hikes are around

Western Head, up Duck Harbor Mountain, and along the high ridge of Jerusalem Mountain. See *Acadia*, page 301 and 302.

Fishermen settled the island first in 1772, and the population grew gradually to a peak of 274 in 1880. By the mid-1800s, lobstering had become the main occupation, and a lobster cannery was established here in 1860, shipping delicacies as far as London. The cannery closed in 1873, but live lobsters then were shipped from Isle au Haut to customers in Boston and New York.

In 1879 a young Bostonian named Ernest Bowditch (grandson of *American Practical Navigator* author Nathaniel Bowditch) saw the island across the sea one summer day and was enchanted by its wild beauty. With a few friends, he formed the Point Look-out Club, and eventually they bought up most of the available acreage on the island. It was a bachelor club at first—no children, no dogs, and no women. But the bachelors went the way of all flesh, and it developed into a family resort.

Meals were served in a large clubhouse, and each member had a little cabin nearby, connected by a boardwalk. Entertainment was simple: sailing, deep-sea fishing, hiking, and tennis. There were church fairs, and it was the custom on Sunday nights to go to someone's cabin and sing hymns.

For half a century this was an exclusive paradise for a few families. Then, in 1945-46, the same summer families donated the land to Acadia National Park. Most of the 300 or so summer residents on the island today are descendants of the original club members.

How is the island's name pronounced? This may help: "Says the summer man, when the fog hangs low, 'There's a bridal wreath on Isle au Haut.' But the fisherman says as he loads his boat, 'It's thick-a-fog on Isle au Haut.'"

WINDJAMMERS

Few sights can rival the unequivocal beauty of a Maine windjammer under a stiff press of sail, barreling past the craggy coasts and spruce-studded islands. The seamanship is obvious, history seems to break the bonds of time, and all the elements converge in a moment of perfection—the wind, the sea, the following birds, the blurry background, the passengers, the crew, and the helmsman, leaning forward, loving it.

Maine's windjammer fleet represents the country's largest collection of actively sailed historic vessels. They have been lovingly restored, rebuilt, or even built new for the same reason they were originally built—profit. They once were fishing schooners, coasting freighters, oyster dredgers, and pilot schooners. By the mid-1930s, however, the profitability of working sailing ships had dwindled to the point where many were converted to power, laid up, or retired altogether.

The idea of Maine's "headboats," as they were sometimes called, can be credited to Captain Frank Swift of Camden. In 1936 he began taking passengers out on a pair of chartered fishing schooners. Twelve years later he had a fleet of nine. He also had some competition. Today, more than 20 windjammers work the Maine coast, most of the largest centered around Penobscot Bay. But instead of carrying lumber or oysters or fish, their cargo is paying passengers.

The passengers experience the Maine coast in a way like no other, under the tall masts and topsails of a working ship, guided by exceptional seamen with encyclopedic knowledge of the local waters. In turn, the passengers have created a market that has enabled these fine ships to be saved from certain collapse into the mudbanks of forgotten backwaters all along the Atlantic seaboard. In other cases, it has led to the birth of brand new schooners that continue the evolving tradition of working sailing ships.

Most of Maine's windjammers sail from Rockport, Camden, or Boothbay Harbor, but you are likely to encounter them almost anywhere along the coast. They include veterans *Stephen Taber*, the oldest U.S. sailing vessel in continuous service (now more than 135 years) and *Lewis R. French*, built at Christmas Cove that same year; *Grace Bailey* and *Mercantile*, both in Captain Swift's original fleet; the *Isaac H. Evans*, incongruously designated a National Historic Landmark; *American Eagle*, a true Gloucester fishing schooner; the William Hand-designed *Bowdoin*, Donald MacMillan's arctic research ship; the schooner yacht *Nathaniel Bowditch*; and newcomers *Mary Day*, the first schooner designed and built for the passenger trade, to be followed by Herb Smith's *Appledores*, *Anglique*, and *Heritage*.

HEAD HARBOR

1 ★★★

44° 01.21′N 068° 37.12′W

CHARTS: 13303, 13313, 13302, 13312   CHART KIT: 69, 22

This pretty little harbor is tucked inside Eastern Head at the south end of Isle au Haut, making it a handy for boats making more offshore passages.

It is exposed both to ocean swells and prevailing southwest winds but a lot less so than you might think by looking at the chart. If the wind is from the north or east, it is fine.

APPROACHES. Both Western Ear Ledge and Roaring Bull Ledge break, even on a calm day. After identifying them, run in for Head Harbor, staying well clear of the ledges that make out a long way from Eastern Head. The harbor entrance is wide and easy.

ANCHORAGES, MOORINGS. Anchor in the middle of the harbor before reaching the meadow on the eastern side, in 15 to 20 feet at low. The holding ground is good in muddy sand.

GETTING ASHORE. Land on the beach at the head of the harbor where there are a half dozen houses.

THINGS TO DO. A paved road leading west toward Duck Harbor intersects the hiking trails of Acadia National Park. Another road leads half a mile north to Long Pond.

DUCK HARBOR

3 ★★★★

44° 01.74′N 068° 39.30′W

CHARTS: 13303, 13305, 13302, 13312   CHART KIT: 69, 21

Duck Harbor is a peaceful little paradise at the southwestern end of Isle au Haut and the easiest place to access the trails of Acadia National Park. This is where the ferry from Stonington drops off and picks up campers and hikers.

The harbor itself is small and exposed to the west, and the bottom is rocky. Only a few boats will fit in the harbor, so arrive early. In very settled weather, latecomers will often anchor in the mouth of the harbor, or even just outside it, but in any kind of a blow from the southwest to northwest they will feel like, well, sitting ducks.

APPROACHES. It's hard to see Duck Harbor from a distance. Head in the general direction of Duck Harbor Mountain, the highest bump near the southern end of the island. It is just south of the harbor. As you get closer, look for a bare, conical rock at the north side of the entrance.

Run a straight line into the entrance of Duck Harbor from 200 or 300 yards north of The Brandies. They are marked with nun "4" and usually visible. Leave to port Duck Harbor Ledge, which can be identified by the lobster buoys floating over it, and look to starboard for an extended family of seals basking on Haddock Ledge.

Another, and perhaps easier, approach is to pass north of nun "6" at Rock T, heading toward Moores Harbor. From the entrance of Moores Harbor, turn south inside Moores Harbor Ledge (visible except at the highest tide) and run fairly close along the shoreline until you reach Duck Harbor.

ANCHORAGES, MOORINGS. Anchor in midchannel, in 10 to 15 feet at low, just before or just past the dock where the ferry lands on the south side of the harbor. Do not go as far as the large, sloping ledge that leads up into the trees on the south side.

The bottom is rocky, so be sure to get the hook set well. Some boats put out stern anchors so they don't swing. The anchorage is wide open to the west, but otherwise it offers good protection.

If you end up anchoring tenuously in the mouth of the harbor, be sure to have a backup plan if the wind should pipe up from the west. Isle au Haut Thorofare might work. Plot your course and program your waypoints ahead of time to steer you clear of the outer ledges in case you have to move unexpectedly in the night.

GETTING ASHORE. Row in to the float, leaving the outer face clear for the ferry.

THINGS TO DO. This is the least-visited part of Acadia National Park (*288-3338; ferry 367-5193*) and one of the most beautiful. The park bulletin board has a trail map and other information. A short hike leads to the top of Duck Harbor Mountain (314 feet), where there are great views of Penobscot Bay and glimpses of your boat down below.

A much longer walk will take you around Western Head, passing cobble beaches one after the other, each more beautiful than the next, and eventually back to the main trail. Another possibility is to cut straight across the island for a swim in Long Pond. The trails are likely to be rough and wet, and perhaps longer than you might expect, but they are well worth the effort.

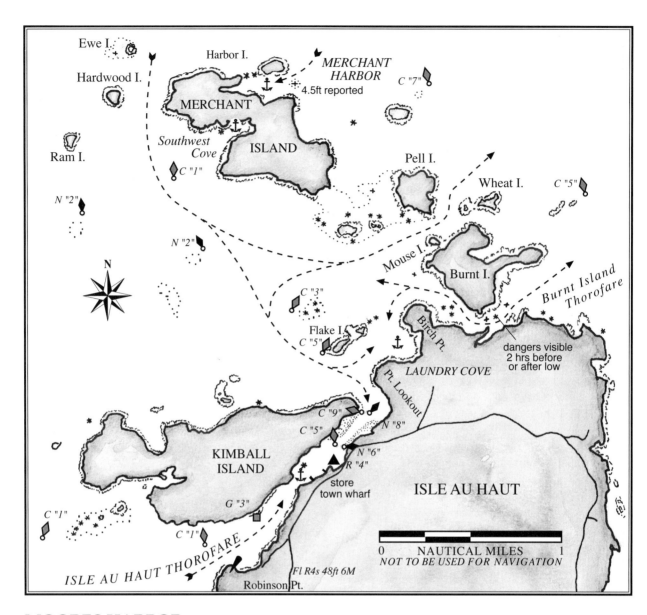

MOORES HARBOR

1 ★★★

44° 03.13′N 068° 38.70′W

CHARTS: *13303, 13305, 13302, 13312*   CHART KIT: *69, 21*

Moores Harbor, on the west side of Isle au Haut, is large and wide and open to prevailing southwesterlies, so it is not generally appealing as an anchorage. It does provide good protection, however, when the wind is from other quarters. In settled weather it is an easy alternative if Duck Harbor is full.

For years, a weir in the northwest corner was the westernmost active weir on the coast of Maine. Dwindling inshore runs of herring, however, have forced the Barter family to discontinue this family tradition.

APPROACHES. Approaching from the west, pass north of nun "6" at Rock T. Continue past Moores Harbor Ledge (visible except at the highest tide) and run into Moores Harbor on either side of the large ledge in midharbor. This ledge is awash 9 feet at low and visible except at dead high tide.

Coming from Duck Harbor, run fairly close along the shore, inside Moores Harbor Ledge.

ANCHORAGES, MOORINGS. Anchor north and east of the ledge in the mud bottom at 20 to 25 feet at low. Note the 3-foot spot.

GETTING ASHORE. Pull your dinghy up on the pebble beach southeast of the 3-foot spot.

THINGS TO DO. Musseling is great here. Acadia National Park trails run close to the shore.

ISLE AU HAUT THOROFARE

3 ★★★

44° 04.43'N 068° 38.48'W
CHARTS: 13313, 13302, 13305
CHART KIT: 69, 21

The narrow channel between Kimball Island and Isle au Haut is known as Isle au Haut Thorofare. It offers reasonable protection and access to the village of Isle au Haut, but the anchorage is not an easy one. The funnel-shaped harbor is deep, and it is open to the southwest. A dredged channel makes it possible to continue northward to Merchant Row or Jericho Bay.

Isle au Haut village is a pretty little community—small, inaccessible, and surprisingly remote. A resident-owned ferry serves it from Stonington. Isle au Haut is its own town, settled partly by fishermen and partly by summer people, governed by town meeting, where all decisions must be approved by a quorum. "Quorum," one resident says, "that's a big word around here."

Isle au Haut has had a resettlement program similar to Frenchboro's to attempt to infuse the island with new blood. The program offers low-interest loans to self-sufficient and self-employed people for the purchase of houses.

Needless to say, the pace is slow.

APPROACHES. Approaching from the west, pick up Kimball Rock, awash 10 feet at low, and can "1" off Marsh Cove Ledges, passing outside of both. The lighthouse on Robinson Point is clearly visible. It has a round granite base and white brick tower.

Approaching from Merchant Row and the north, pass between Hardwood and Merchant Island, following the cable track shown on the chart. Leave to port the two cans marking the ledges off Flake Island, turn to starboard, and enter the dredged channel leading through the Thorofare.

Avoid passing through the dredged channel at dead low, and preferably wait for a rising tide. The buoyed channel was originally dredged to 6 feet at low, but it has silted in. Now it barely accommodates the mailboat's 3.5-foot draft at low, but that's about it. As one local put it, "She's dry on a drainer." When pressed, another said, "There's plenty of water. It's just spread out."

Some stakes may be along the Kimball Island side of the channel to help keep you lined up. Proceed slowly, steer straight, and you should have no trouble. Once out of the dredged channel, be careful of the ledge, marked by red daybeacon "4," which extends at least 30 feet north of the beacon.

ANCHORAGES, MOORINGS. There is not much room in the Thorofare. Anchor opposite the village, outside the moored boats on the Kimball Island side, in 15 to 21 feet at low. Right across from the town dock, there is an open space among the moored boats, enticing you to anchor. Don't. The rock shown on the chart is really there.

Payson Barter has a couple of honor-system rental moorings on the north side of the Thorofare across from the town dock. The floats are marked "rent." Just place your money in the waterproof canister—otherwise known as a Pepsi bottle—attached to the pendant.

The current through the Thorofare is not very strong (it floods northeast and ebbs southwest), but the combination of wind and current is likely to swing you around in every direction during the night.

If there is no room to anchor in the Thorofare, consider Laundry Cove, nearby to the north.

GETTING ASHORE. Take your dinghy in to the float at the Isle au Haut town landing, the westernmost dock on the south side of the Thorofare. Do not add your trash to the island's burden. Kimball Island is private.

FOR THE CREW. From the wharf, walk left along the road a short distance. The post office is in a small outbuilding by the road.

🛒 The Island Store (*335-5211*) is farther along on the left. They are well-stocked with most groceries, some fresh produce, wine, beer, and ice cream. Owner Marie Meyer will procure special orders with a few days' notice.

An islander farther to the north may be selling produce, or, several days a week, 🍴 their brick oven may be fired up for fresh pizza.

THINGS TO DO. It is possible to explore trails of Acadia National Park from the village. Turn right from the town dock and walk about .3 miles to the ranger station and to the trails that depart from there.

Another trail leads to Mount Champlain. Continue north past the market for another quarter mile. On the right, immediately beyond the Point Lookout road, you'll see the trailhead. It's a fairly gentle half-hour hike to the summit, with fine views east over Jericho Bay to Swan's Island.

LAUNDRY COVE

3 ★★★

44° 05.16'N 068° 37.70'W

Charts: 13313, 13312   *Chart Kit:* 69, 21

Laundry Cove, between Point Lookout and Birch Point, north of the Isle au Haut Thorofare, has splendid views over the islands of Penobscot Bay, west to the Camden Hills and north to Merchant Row. Ashore are a few handsome cottages, some from the island's era as an exclusive Bostonian summer colony. The beauty of this spot attests to the appreciative eyes of 19th-century summer people.

APPROACHES. From the dredged channel of Isle au Haut Thorofare, round Point Lookout, noting the rocks that make out a long way from Flake Island. From north or west, pass south of Flake Island.

From the Burnt Island Thorofare, proceed southward to Flake Island, keeping it to port, and turn northeastward into Laundry Cove.

ANCHORAGES, MOORINGS. Work your way in midway between Birch Point and Point Lookout, anchoring in 10 to 14 feet at low. Holding ground is good. The anchorage is exposed to north and west but fine for summer weather. The stone wharf on Point Lookout is private.

BURNT ISLAND THOROFARE

44° 05.26'N 068° 37.10'W

Charts: 13313   *Chart Kit:* 69, 21

Burnt Island Thorofare is a rock-strewn east-west passage between Isle au Haut and Burnt Island to the north. With extreme care, visitors can thread through it to reach Jericho Bay and points east, but they should weigh this option against well-marked Merchant Row or the passage between Burnt and Pell Island, just to the north.

Most of the dangers in Burnt Island Thorofare are visible an hour or two before or after low tide. Refer carefully and frequently to your charts, your depth sounder, and the sketch chart on page 264.

It is possible to anchor halfway through the Thorofare, but the bottom is rocky and the location is not particularly attractive. Better choices are Laundry Cove or the Isle au Haut Thorofare.

BURNT-PELL PASSAGE

44° 05.78'N 068° 37.49'W

Charts: 13313   *Chart Kit:* 69, 21

To head east from Isle au Haut, the cleanest and most well-marked passage is between Merchant Island and Hardwood Island, then eastward through Merchant Row, leaving can "9" to starboard.

If you insist on shortcuts, a more direct route runs between Burnt Island and Pell Island, as shown on the sketch chart on page 264, but this passage requires careful piloting. The western tip of Burnt has long, sloping shelves of bare rock, behind which hides little Mouse Island. Leaving Burnt and Mouse Island close aboard to starboard, head for the western end of Wheat Island to avoid the rocks that make out a long way from the southern end of Pell. Wheat Island is low and grassy, with trees on the eastern end and a beach near its western tip.

When you are reasonably close to Wheat, turn north and pass between Wheat and Pell into Jericho Bay. Dangerous Channel Rock, close by to the northeast, is covered at high tide and breaks only occasionally, but it is awash 8 feet at low.

Mount Desert
Isle au Haut to Schoodic Point

A CRUISING GUIDE TO THE MAINE COAST

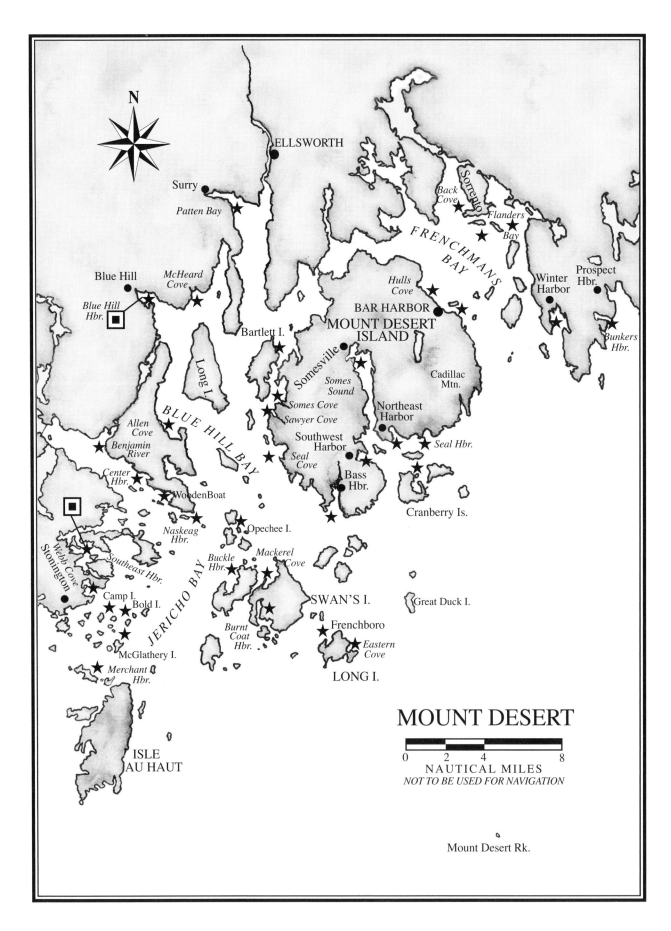

MOUNT DESERT

SOME of the most staggeringly beautiful cruising on the coast of Maine lies off the island of Mount Desert. Off Deer Isle, intricate islands pepper Merchant Row and meld together as Jericho Bay expands. Swan's Island and staunch Long Island lie offshore. Broad and deep Blue Hill Bay sweeps inland to the sophisticated town of Blue Hill. Frenchman Bay embraces the enclave of Sorrento, the bustle of Bar Harbor, and lonely Schoodic Point. And in the middle of it all, the age-old mountains of Mount Desert rise from the sea in craggy grandeur.

The sailor bound east from Penobscot Bay has a variety of routes for entering Jericho Bay on his way to Blue Hill and Mount Desert. Farthest north is the wide, open passage of Eggemoggin Reach. South of Deer Isle, the Deer Island Thorofare takes you past the fishing town of Stonington and north of an archipelago laced with granite and spruce islands. The broader passage of Merchant Row leads south of these same islands. Some sailors pass offshore, south of Isle au Haut, to avoid all complications, but they also miss the best cruising.

Jericho Bay is wide and deep and a beautiful place to sail. Mount Desert looms high to the northeast, Isle au Haut to the southwest, and Blue Hill to the north, with the Camden Hills still visible to the northwest. In the foreground are the myriad granite-shored islands of Merchants Row.

Swan's Island and Long Island are special among the outlying islands, with delightful harbors, thriving communities, and distinct personalities. Fifteen miles to sea lies lonely Mount Desert Rock, where adventurous cruisers may find great humpback and finback whales.

Through York Narrows, Casco Passage, or Pond Island Passage, the route leads to beautiful Blue Hill Bay, with its miles of open water and snug little coves.

The spectacular island of Mount Desert is set like a jewel between Blue Hill Bay and Frenchman Bay, and its mountains and dramatic headlands define the character of the entire region. The most scenic portions of Mount Desert Island have been preserved as part of Acadia National Park, enjoyed each year by millions. There are miles of hiking trails, cliffs and beaches, views from mountain tops, lakes, glacial formations, gardens, museums, and carriage roads. Much of the park can be visited from a cruising boat. The great sailing centers of Southwest Harbor and Northeast Harbor bustle with boating activity at the foot of the mountains.

On the eastern side of Mount Desert Island, bold granite islands—the Porcupines and Ironbound—rise from the deep waters of Frenchman Bay to guard the gilded town of Bar Harbor, once the playground of the rich, whose grand "cottages" still line the shores. To the east, summer and working communities coexist at Winter Harbor. Lonely, rugged Schoodic Peninsula juts into the ocean, forming the eastern shore of Frenchman Bay and the beginning of the raw beauty of points farther Down East.

MERCHANTS ROW
and Deer Island Thorofare

DEER ISLAND THOROFARE
western entrance between R"2" and Mark Island:
44° 08.10'N 068° 42.65'W
eastern entrance between Long and Potato Ledge:
44° 10.63'N 068° 32.92'W

CHARTS: *13315, 13305, 13313*   CHART KIT: *69, 22*

Deer Island Thorofare and Merchant Row, running to the north and south respectively of the splendid Merchants Row archipelago, are two of the three major passages from East Penobscot Bay to Jericho Bay. The third major passage, Eggemoggin Reach, is discussed later in this chapter.

The Thorofare is a narrow, well-marked channel no wider than 100 yards in several places. On the north side is the busy fishing village of Stonington. On the south side are the granite quarries of Crotch Island. Farther south lie some three dozen spruce-clad islands, largely uninhabited, with the shelving white-granite shores characteristic of this area.

The current through Deer Island Thorofare is not very strong, and it is influenced by the wind. With no wind, the tide floods eastward and ebbs westward. With westerly winds, both flood and ebb set eastward. With strong easterlies, both set westward.

APPROACHES from the west. Emerging from Fox Islands Thorofare in clear weather, you will often find the east side of Penobscot Bay shrouded by fog. Assuming good visibility, however, look for the standpipe north of Stonington and the tall crane, derricks, and granite tailings on Crotch Island. As you cross East Penobscot Bay, Mount Desert will be visible in the distance ahead, Blue Hill to the north-northeast, and high Isle au Haut to the southeast.

Approaching the Thorofare, check to make sure you are clear of the Brown Cow, an isolated ledge barely visible at high. The Thorofare's western end begins between West Mark Island Ledge to the north, marked by nun "2," and high Mark Island to the south, marked by a small lighthouse standing no taller than the trees. Its foghorn matches its size, only a decibel or two louder than the waves on the rocks.

After passing Mark Island, find green can "27," critical for passage between the outer ledges. Green marks are left to *starboard*, red marks to *port*. Thereafter, the passage is easy and well marked.

APPROACHES from the east. Coming from Casco Passage, north of Swan's Island, the easiest route is north or south of Egg Rock and then between Long Ledge and Potato Ledge, all of which are well marked. To the north, you will see the distinctive Lazygut Islands—one large wooded island and two small ones to the west, like a dash and two dots.

Deer Island Thorofare starts at the green can north of Eastern Mark Island, which is distinguished by the bold, shelving shores on its eastern side. In this direction, green marks are left to port, red marks to starboard.

ANCHORAGES, MOORINGS. If you are overtaken by fog or darkness on your way through Deer Island Thorofare, there are several alternatives to consider. You can anchor in Webb Cove, or among the islands of Merchants Row, or on the north side of the Thorofare. You may also be able to get a mooring or tie up at Billings Marine, on the north side, west of Stonington.

MERCHANT ROW
N"10": 44° 07.11'N 068° 39.71'W

CHARTS: *13315, 13305, 13313*   CHART KIT: *69, 22*

Merchant Row is the southern passage that links East Penobscot Bay and Jericho Bay. It passes north of Merchant Island and south of most of the islands off Stonington. The archipelago of islands themselves is known as Merchants Row, with an "s."

Merchant Row is far wider and less obstructed than Deer Island Thorofare, though it is a little longer. It is easier to navigate in the fog. The only unmarked danger in the passage is Channel Rock, at the eastern end, awash 8 feet at low but completely submerged at high.

A CRUISING GUIDE TO THE MAINE COAST

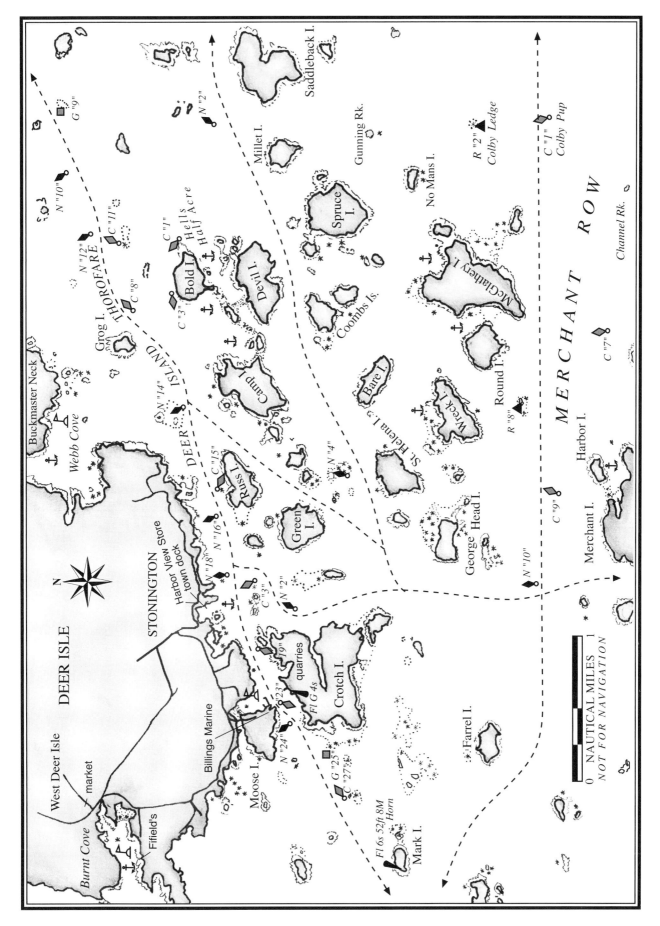

MOUNT DESERT

STONINGTON

2 ★★★★

44° 09.15'N 068° 39.75'W

CHARTS: 13315, 13305, 13313   CHART KIT: 69, 22

Stonington, at the southern tip of Deer Isle, is Maine's frontier town. It's a working harbor and it wants to stay that way. A town ordinance gives commercial fishing boats first choice for moorings, passenger vessels are next in line, and pleasure boats last. But slip in and don't get in the way, and everything is colorful and interesting. The derricks and granite quarries of Crotch Island operate across the way, and the islands of Merchants Row stretch south toward the blue heights of Isle au Haut.

Although fishing has always been part of Stonington's heritage, the town flourished with the granite industry in the late 19th century. Between 1870 and 1925, enormous quantities of granite were produced from quarries here and on Crotch Island, and the town, originally known as Green's Landing, rightfully earned its new name. Stonington granite built parts of Rockefeller Center, the Smithsonian Institution, Boston's Museum of Fine Arts, and several New York City bridges, including the George Washington and the Triboro. But just when permanent prosperity seemed to have come to Stonington, the granite industry declined and the quarries closed, returning the town to the fishermen. In the early 1980s, however, new technology spurred the reopening of Crotch Island to quarrying, and granite has once again become competitive as a building material.

If you arrive in Stonington on the third or fourth Saturday in July expecting a quiet coastal hamlet, beware. From 10AM until 4PM, high-speed lobsterboats will be drag racing down the Deer Island Thorofare. The annual Stonington Lobster Boat Race has gotten so big and so fast that the Coast Guard proposed a permanent special local regulation "to protect the boating public from the hazards associated with high-speed powerboat racing in confined waters."

These may not be the only hazards, though. The frontier-town attitude is often cultivated on the water, too. Without a doubt and not by accident, Stonington's lobsterboats are the loudest on the coast, and the men at their helms seem to take perverse pleasure in buzzing nearby yachts at predawn hours or cutting across their bows. As one cruiser put it, it can be a "very disgruntling experience."

Whether these are acts of defiance or envy or intolerance or independence or simply a chuckle in a long, hard workday, the message is clear: "A pleasure boat is a visitor here. Remember that."

But remember also, that were you to ever get into trouble on the water around here, you would strain to hear the bellowing approach of one of these boats, and it would be sweet music to your ears.

APPROACHES. The approach to Stonington is straightforward from east or west. See *Deer Island Thorofare*, page 270.

ANCHORAGES, MOORINGS. Consider two possible places to lie for the night. Moorings and dockage are available from Billings Marine, on Moose Island, but here you will be some distance west of the center of town.

Usually no moorings are available off the town itself, but there is good anchorage right off the moored boats and the town, on the north side of the Thorofare, west of nun "16" at Staple Point, in 15 to 20 feet of water. This is where the windjammers often lie, and it provides the best access to the public landing and the rest of town. Be aware, though, that there is not much protection from the south, and strong southwest winds will funnel past Crotch Island. Make sure your anchor is well set and that your anchor light works.

GETTING ASHORE. Stonington's waterfront can be as confusing as it is busy. Be careful of the constant crisscross of lobsterboats, cruisers' dinghies, lobsterboats, windjammer yawlboats, lobsterboats, kayaks, and more lobsterboats. The broad harbor is divided by a large stone fish pier and two little islets piled high with lobster gear. The floats at the fish pier are commercial only. The public floats are on the eastern shore of the eastern half of the harbor, along the granite seawall by the large, light-blue firehouse building. If a strong afternoon southwesterly is building, be prepared for a long row against it to get back to your boat.

FOR THE BOAT. *Island Fishing Gear and Auto Parts (207-367-5959)* at the head of the fish pier sells fuel filters, rubber boots, and plenty of whatchamacallits.

Billings Diesel and Marine Service (Ch. 09, 16; 207-367-2328, or 348-6980 nights). Billings is a very large yard with four marine railways capable of handling up to 425 tons, a 35-ton boatlift, and a 20-ton mobile crane. They can make hull and engine repairs of all kinds to wood, fiberglass, steel, and aluminum vessels. They have an extensive machine shop, an electronics shop, and a large chandlery. Curiously, they won't touch outboards (do they know something we don't?). A word of caution: a good part of the business at Billings comes from repairs to boats that have found Merchants Row the hard way—with their keels or props.

The fuel float is located north of the large finger pier. At low tide, give the end of the pier a good berth as you enter. Gas, diesel, pump-outs, and ice are available, with 5 or 6 feet alongside at low. Stonington water has been reported to be not good, but we have found it tolerable. Taste it before you tank.

Greenhead Lobster (207-367-0950) is a commercial wharf in the southeast corner of the harbor. They have gas and diesel but little time for yachts. Depth at low is 4.5 feet.

Blackmore Electronics (207-367-2703). Blackmore sits overlooking the water at the east end of town. They can repair your navigation equipment or rewind your starter or alternator.

New England Marine and Industrial, Inc. (207-367-2692; www.newenglandmarine.com), located just out of town, on Cemetery Road, stocks an extensive inventory of commercial fishing gear, including lines and ground tackle.

FOR THE CREW. Billings has phones, showers, and a weather-beaten coin-operated laundry machine and dryer that are regularly overtaxed by soggy yachtsmen. Don't let the smoking belts of the dryer alarm you. Billings sells ice and often has live lobsters for sale on the dock or from a floating buying station tied up nearby. From the yard, it is a mile in to town.

The public landing is right in town, next door to the Harbor View Store (*367-2532*), which sells ice, wine, beer, and essential groceries. For more extensive provisioning in Stonington, you will need to bum a ride to the Burnt Cove Market (*348-2681*) a couple of miles away. Every Friday morning, local farmers sell their produce in the parking lot of the Island Community Center, up the hill at the west end of Main Street.

Main Street wraps around the harbor, lined with several restaurants, galleries, and bookstores. The town float and small library are near the eastern end. The west is anchored by the large commercial wharf and the venerable Stonington Opera House. Dockside Books, at the far western end of Main, has an amazing array of nautical titles in an old fish shack at the water's edge.

The Harbor Cafe is the place to gam at dawn. They are open at 6AM for breakfast and also serve lunch and dinner. On Fridays they have an all-you-can-eat fish fry. The Fisherman's Friend, another local favorite, has moved and expanded into a location right at the town landing. For more elegant lunches and dinners, try the Maritime Cafe (*367-2600*), on the harbor side of Main.

THINGS TO DO. The Deer Isle Granite Museum (*367-6331*), located in the Webb building on Main, depicts the history of Maine's granite quarries and the men who worked them.

Boats leave several times daily for Isle au Haut (*367-5193*) from the docks next to the Bayview Restaurant. They take you to Duck Harbor, inside Acadia National Park, where there is spectacular day-hiking and lean-tos and shelters for camping (by reservation only, usually long in advance). Get in touch with Acadia National Park, Box 177, Bar Harbor, Maine 04609 (*207-288-3338*) for information or reservations. Daily sightseeing excursions also run out of Stonington among the islands and along the coast of Isle au Haut.

To stretch your legs, walk east out of town on Indian Point Road to visit the Lily Pond, about three quarters of a mile each way. Go early, before the lilies close for the day.

MERCHANTS ROW

An archipelago of 30 or 40 islands lies between Deer Isle and Isle au Haut. Merchants Row, as they are know collectively, is unsurpassed for beauty anywhere in Maine. The islands are all darkly wooded and fringed with white and pink sloping granite shores. Tight passages wind between them, and harbors are everywhere. Even the names of the islands are evocative—Sprout and Potato, Enchanted and Grog, Round, Bare, and Green. Other names remind you that for every island, there is rock—Hells Half Acre and Devil and Wreck Island.

Some islands have old quarries and stone wharves. Crotch Island is once again a working quarry and one of the few places where this fascinating operation can be observed.

Most of the islands in Merchants Row are privately owned, but six—Wreck, Round, McGlathery, Millet, Russ, and Sand—belong to environmental groups. Five more are owned by the Bureau of Parks and Lands and are part of the Maine Island Trail: Wheat, Harbor, Steves, Hell's Half Acre, and Little Sheep.

The most frequently used anchorages are at McGlathery, between Round and McGlathery, between Camp and Bold, between Bold and Devil, and at Merchant Harbor. There are many other places to drop the hook for a short time or for the night. The major anchorages are described here. The pleasure of discovering the gunkholes is left to you.

You can be dazzled by these islands. As you wander among them, it's easy to be distracted by their beauty. But the distances are small, and there are lots of hard spots. Keep the chart in hand and keep track of where you are at all times. Navigation is much more difficult in fog, so save this magnificent cruising ground for a sparkling day. It may be the most beautiful place in the world.

CROTCH ISLAND

off Mill Pond: 44° 08.58'N 068° 40.08'W
off wharf: 44° 08.85'N 068° 40.40'W
CHARTS: 13315, 13313   CHART KIT: 69, 22

Ever since the first quarry opened in 1870, Crotch Island has been renowned for the color and workability of its granite. It was Crotch Island's "Sherwood Pink" that Jacqueline Kennedy chose for the late president's memorial in Arlington National Cemetery. Crotch Island granite is still being quarried and shipped in large blocks to build museums and office buildings in Boston, Washington, and New York.

The original Goss Quarry drew workers to Crotch Island from Scandinavia, Italy, and the British Isles. Skilled workmen finished and polished the stones right on the island. The company store was run by the *padrone,* and even during Prohibition, he imported wine directly from Italy.

In 1983, New England Stone Industries of Smithfield, Rhode Island reopened the quarries. Currently the rough stone is shipped away to be cut and polished, but someday they may use the old cutting shed to once again finish the granite on site. A visit to the island makes a fascinating daystop.

APPROACHES. There are two stone wharves on Crotch Island—one on Deer Island Thorofare and the second on the island's eastern side at the entrance to the tidal Mill Pond. Granite is shipped from both sites, but the Mill Pond wharf is closer to the active quarry on Thurlow Head.

From the Thorofare, pass between Scott Island and can "3" east of Two Bush Island. Then head southwest for Mill Pond, leaving nun "2" to port.

ANCHORAGES, MOORINGS. There is good anchorage off the entrance to Mill Pond, west of the cable area marked on the chart, in 7 feet of water at low. The mud bottom holds well.

THINGS TO DO. From the anchorage, you can see the tops of the giant derricks and watch the heavy-duty front-loaders piling massive granite blocks on the wharf for shipment. The dimension of each stone is marked on it, and it's interesting to speculate on their weights.

It is fun to row a dinghy all the way up Mill Pond, past huge piles of granite blocks near the entrance. The whole north shore of the inlet is strewn with granite rubble, while the south shore remains in its natural state. At low there is not much water, and the midtide current can be strong.

A CRUISING GUIDE TO THE MAINE COAST

Loading granite at the John L. Goss quarry on Crotch Island, 1908. (Frank Claes)

RUSS ISLAND
off S. shore: 44° 08.90'N 068° 38.60'W
CHARTS: 13315, 13313   CHART KIT: 69, 22

Beautiful Russ Island, just east of Stonington, rises to a height of 100 feet and offers dramatic views of the surrounding islands and shores. This uninhabited island is covered by a mature spruce forest except for a low saddle across the central portion, which was once cleared as pasture for a small flock of sheep. Russ was quarried for granite, and the remains of two stone wharves may still be seen on the north side.

Russ was acquired by the Island Institute from painter and architect Emily Muir, whose family owned it for half a century. The Institute protected the island with conservation easements and sold it to the Chewonki Foundation, which will use it for their coastal kayaking and sailing programs. The public is welcome during the day for picnicking and hiking.

There is a small protected cove on the north side, but the usual anchorage is off the sandy beach on the south shore. The anchorage is exposed and recommended only as a daystop.

CAMP ISLAND
3 ★★★★
44° 09.01'N 068° 37.57'W
CHARTS: 13315, 13313   CHART KIT: 69, 22

One of the most pleasant anchorages in Merchants Row is just east of Camp Island, about 1.5 miles east of Stonington. The anchorage is encircled by Bold and Devil Island, both high and wooded. To the east, Hells Half Acre and the two little Coot Islands add beauty and protection.

APPROACHES. The easiest approach is from Deer Island Thorofare from the north, between Camp and Bold Island. There is also a nice little back door between Devil Island and the Coots that is fun to sail through, favoring the eastern side.

ANCHORAGES, MOORINGS. The best place to anchor is about where the 11-foot spot is shown on the chart, north of the Coot Islands, which provide protection from the prevailing southwesterlies. Or anchor eastward in depths of 17 to 24 feet at low. The bottom is excellent mud holding ground. In choosing your spot, stay well clear of the ledge extending north and east from the Coot Islands.

Boats anchored on the eastern side, toward Hells Half Acre, do not have the full protection of the Coot Islands. You will be perfectly comfortable under normal conditions, but in a strong southwesterly you will be exposed to the wind through the slot between Devil and the Coot Islands.

THINGS TO DO. It's fun to explore the gentle, ledgy shores of beautiful Hells Half Acre, which is state-owned but heavily used as part of the Maine Island Trail. Camp and Devil Island are private.

A lobster shack overlooking Mark Island and the western entrance to Deer Island Thorofare.

GRANITE

Think of Maine and you think of granite—pink and white granite from Penobscot Bay, shades of gray from Somes Sound, bright red from Jonesport, rose from Deer Isle.

Granite from Maine built the Library of Congress, the House of Representatives, and the Treasury Building in Washington. It went into the Museum of Fine Arts in Boston, the Customs House in Atlanta, Union Station in Chicago, the St. Louis Post Office, and the Cape Cod Canal. Maine granite was used for the Central Park Reservoir, the Metropolitan Museum, the Brooklyn Bridge, the George Washington Bridge, and Grant's Tomb in New York. More recently, it helped rebuild the Statue of Liberty.

There was soaring granite for the great cathedrals; humble granite for the curbstones and horse troughs of a nation; granite headstones to mark graves and granite blocks to moor boats; granite stones by the millions to pave muddy city streets; granite slabs to sheathe the office buildings of corporate America.

A wonderful substance, this—hard and lasting and beautiful. There may be flecks of shiny mica in it, crystals of quartz, chunks of rosy feldspar, and perhaps black specks of biotite. Granite of different colors and textures can be cut and polished to show an endless variety of patterns.

The largest pieces of granite ever cut in Maine were the huge shafts quarried on Vinalhaven for eight columns at the Cathedral of Saint John the Divine in New York City. In the rough, these shafts were 64 feet long and eight feet thick, each weighing about 120 tons when put on the lathe. The heavy columns broke while being finished, and eventually each was made in two separate pieces.

In most quarries, the granite lies in horizontal sheets several feet thick, with natural seams, called "lifts," between the layers. The vertical cleavage plane along the grain of the rock is the "rift," usually running east and west. The cut across the grain is known as the "hardway." Granite was quarried by drilling a series of vertical holes a few inches apart down to a lift, packing the holes with black powder or dynamite, and exploding off large chunks of rock.

The pieces of granite were swung from the quarry face by great derricks and booms. In some quarries, they could be rolled downhill by gravity on rail cars or slung under the high axles of galamanders, carriages with wheels eight feet in diameter, drawn by teams of horses or oxen. The granite was shaped and polished in cutting sheds and loaded onto granite schooners or barges for transportation to the project site.

The glamorous part of the industry supplied stone for famous monuments and buildings, but the tonnage came from paving blocks, hundreds of thousands of them, cut by hand from larger stones. Starting with a block four feet by four feet wide, and perhaps three feet thick, the stonecutter used a 32-pound hammer and chisel to split the stone to a thickness of 18 inches, then 9 inches, then 4½ inches. The final paving block was 4½ by 4 by 12 inches. A good man could cut 500 or more paving blocks in a working day, with clean, straight edges. The record was 786.

Accidents in the quarries were frequent. Sometimes a charge of powder failed to ignite, and a new charge was tamped down right on top of the smoldering embers—a good way to lose a leg or a life. Quarrymen wore no goggles, so stone chips frequently got in their eyes.

The cutters and polishers who labored inside sheds faced more insidious hazards. The cutting sheds trapped the deadly granite dust. Eventually, silicosis clogged the lungs of the exposed workers, and many died young of "consumption," long before the days of workers' compensation.

Sometimes accidents were caused by the inability to communicate. A quarry was a Tower of Babel. Quarrymen and stonecutters and polishers came from everywhere in Europe. They bunked in company barracks (the Shamrock and the Aberdeen on Dix Island could house 500 men apiece), and they bought whatever they needed in the company store. If they had cash left over, they spent it on hard living in Rockland or Stonington. They were a tough and rowdy bunch, these quarrymen, and there was a Wild West atmosphere wherever they went on Saturday nights.

If the workmen in this rough-and-tumble industry were colorful, so were their bosses. One of the pioneers was General Davis Tillson, who returned from the Civil War with only one leg, but with his other faculties intact. He became the dictator of Hurricane Island, and woe to anyone rash enough to cross him. The Italians called him the "Bombasto Furioso."

continued on page 278

BOLD and DEVIL ISLAND

3 ★★★★

44° 09.15'N   068° 36.80'W

CHARTS: 13315, 13313   CHART KIT: 69, 22

Bold and Devil Island lie about two miles east of Stonington, and the tongue of deep water between them makes a fine anchorage. The distant skyline of Mount Desert lies to the east, and all around are the beautiful spruce and granite islands of Merchants Row.

APPROACHES. From Deer Island Thorofare, approach close northward of Bold Island, leaving cans "3" and "1" to starboard and leaving Bold Island Ledges, a spot favored by seals, to port. After rounding can "1" off the eastern tip of Bold, head into the slot between Bold and Devil. Note the two islets to port. The easternmost one is smooth ledge with a clump of conifers. The westerly one is grassy on top. To starboard, ledges make out from Bold Island, tiny scraps of which are visible at all tides.

ANCHORAGES, MOORINGS. The protection and views are better in the eastern end of the anchorage, and the holding ground is good, in mud. Strong southwesterlies will funnel between Devil Island and Hells Half Acre.

THINGS TO DO. Explore the islets east of Devil Island and state-owned Hells Half Acre. The smooth granite ledges are beautiful, and the clamming reputedly is good.

McGLATHERY ISLAND

3 ★★★★

44° 07.73'N   068° 36.77'W

CHARTS: 13315, 13313   CHART KIT: 69, 22

Probably the best-known anchorage in Merchants Row is at McGlathery Island, one of the largest in the group. Windjammers use it often, as do cruising yachtsmen. Beautiful granite ledges and a sandy bar lead to a small offlying island. Ashore you might meet the woolliest of wild sheep.

The island is owned by a conservation group called Friends of Nature, which was formed in 1953 by visionary environmentalist Rudy Haase to purchase the island to prevent it from being clear-cut for pulpwood.

APPROACHES. The anchorage is at the northern end of McGlathery. Approach from the north. The little, wooded, unnamed island at the northeastern tip of McGlathery Island is conspicuous by a large boulder resting on the sloping ledge at its northern end. No Mans Island to the east is small and wooded, with a long, sloping ledge at its western end.

ANCHORAGES, MOORINGS. The eastern end of the anchorage is marked by a large, conical rock at the edge of the trees. A sandy, boulder-strewn beach lies below it. Come just inside the northern knob of McGlathery and anchor west of the conical rock in a mud bottom, 21 feet deep at low. Protection is good except from the northeast.

THINGS TO DO. Take a swim off the beach and bask on the sun-warmed ledges. Go across the sandbar to the adjoining little island and walk around it on the high ledges overlooking the sea. The walk is mostly easy going, with just an occasional leap across a chasm. On McGlathery, trails lead inland to where you may meet the sheep, which are likely to be just as startled as you are.

This lovely island is visited by sea kayakers, daytrippers, windjammers, and cruising boats. Remember that we are all here for the same reasons. Respect this fragile place and its tranquillity and show your appreciation for the efforts of the Friends of Nature by leaving McGlathery cleaner than you found it.

ROUND ISLAND

3 ★★★★

44° 07.55'N   068° 37.35'W

CHARTS: 13315, 13313   CHART KIT: 69, 22

Round Island is an Island Heritage Trust preserve between McGlathery Island and Wreck Island to the west. The bight between Round and McGlathery makes a delightful anchorage. There are views past the island to Stonington and a glimpse southward of high Isle au Haut.

APPROACHES. Coast down the western shore of McGlathery Island until you come to the bump of land that extends westward, then ease into the bight between Round and McGlathery Island. If you are coming from the anchorage at McGlathery, stay well offshore to avoid the unmarked rock off the northern point of McGlathery.

Coming from Merchant Row, identify the passage between Wreck Island and Round Island just north of Barter Island Ledges. There is a 5-foot spot southwest of Round Island, but otherwise its shores are steep-to. Keep the Round Island shore close aboard to starboard and follow it into the anchorage.

ANCHORAGES, MOORINGS. Opposite the bump of land, anchor in mud halfway between Round Island and McGlathery in 10 feet at low. Protection is reasonably good, although a strong southwester will funnel through.

THINGS TO DO. From the anchorage, you can land conveniently on the beach at McGlathery or on the beach on the eastern tip of Round Island to explore these preserves, both open to the public. Because Round Island is so heavily wooded, you probably will want to stick to the shoreline. No fires or camping are permitted.

WRECK ISLAND

3 ★★★★

44° 07.80'N 068° 38.13'W

CHARTS: 13315, 13313   CHART KIT: 69, 22

Just west of Round Island, Wreck is also an Island Heritage Trust preserve. It has the great advantage, however, of being partially open. Much of the interior is grassy fields with patches of raspberries.

APPROACHES. The easiest approach is from the east, south of distinctive Bare Island with its western end marked by sloping granite outcroppings interspersed with tufts of grass. A house and dock are at Bare's eastern end. Neighboring St. Helena Island has large, open spaces with granite monoliths and an unobtrusive house and stone dock on the south shore.

ANCHORAGES, MOORINGS. While it is possible to anchor off the eastern end of Wreck and land on

(continued from page 276)

The granite industry was organized in the decade after the Civil War, starting about 1867. Government contracts for post offices and public buildings were the lifeblood of the industry, with a guaranteed 15 percent profit. By 1877 (when it was declared illegal), the "Granite Ring" had established a monopoly on these lucrative contracts: Bodwell of Vinalhaven, Tillson of Hurricane Island, St. John of Clark Island, and Dixon of Dix Island.

But even during its palmy days, the industry was never a consistent one, dependent as it was on the uncertainties of bidding for each job and the political pressures involved in the awarding of contracts. No two jobs were alike, so it was impossible to standardize the product or build an inventory between contracts. Granite was susceptible to the periodic booms and busts of the building trade. The industry reached its peak quickly and faded out within 50 years, the victim of new materials, competition from other states, and the high cost of shipping.

In recent years, the natural beauty and durability of Maine granite has begun to be appreciated again after a generation or two of aluminum and glass buildings. New techniques allow granite to be sliced in sheets as thin as a pane of glass, creating new architectural possiblities. Granite today is moved by large front-load tractors and loaded on barges for shipment. Drilling and dynamite blasting have been replaced by a high-temperature gas torch used to cut a groove in the stone. That's the way it is done these days on Crotch Island, off Stonington. It is a noisy business, and the roar is audible a long way downwind—not very popular with the summer folk who live nearby.

Crotch Island is about the only place you can see a working quarry today on the coast of Maine, but there are lots of old quarries worth visiting. They are usually recognizable by the piles of grout and tailings and sometimes by great abandoned derricks and booms still in place. A number of old quarries have wonderful freshwater pools where you can swim—for example, Wildcat near Tenants Harbor, Hall Quarry in Somes Sound, several quarries on Vinalhaven near Carvers Harbor, and the quarry in Burnt Coat Harbor on Swan's Island. These spots serve as reminders of an industry that forever marked the face of Maine yet lasted just one lifetime.

the nearby beaches, the best protection is on the north side. Drop the hook where Chart 13315 shows 9 to 14 feet of water, east of the large knob of land on the north side of the island. The anchorage is small, but the holding ground is good mud.

THINGS TO DO. The eastern half of Wreck Island was kept open by a century of grazing, and sheep were reintroduced by the U.S. Forest Service in 1980 to maintain this open quality. As you explore, be careful of the old fencing and cellar holes, which are the only remaining evidence of the several farm families who lived on Wreck. No dogs, camping, or fires are allowed.

MERCHANT HARBOR

3 ★★★★

44° 06.59'N 068° 38.61'W

CHARTS: **13315, 13313**   CHART KIT: **69, 22**

One of the best and most attractive anchorages in the area is Merchant Harbor, on the north side of Merchant Island. Harbor Island, with beautiful, open meadows and conifers, protects the anchorage from the north and adds a sense of serenity and security. A private dock and float are at the head of the harbor, along with a few small craft on moorings.

APPROACHES. The approach is easy from east or west, following the marked channel of Merchant Row. As you enter, stay reasonably close to the Harbor Island side to avoid the 4½-foot spot shown on the chart.

ANCHORAGES, MOORINGS. In good weather anchor past Harbor Island toward the head of the harbor in 16 feet at low. If winds from the north are expected, anchor in the lee of Harbor Island in 27 feet at low. Protection is excellent from west to south. The bottom is mostly mud, but some of it may be rocky. Rig a tripline and be sure your hook is well set.

THINGS TO DO. State-owned Harbor Island has a pleasing combination of trees and raspberry thickets, open meadows, and ledges. It is also extremely popular with kayakers.

SOUTHWEST COVE

1 ★★★

mouth: 44° 06.26'N 068° 39.12'W

CHARTS: **13315, 13313**   CHART KIT: **69, 22**

A lovely little cove lies on the southwest side of Merchant Island, exposed to prevailing southwesterly winds but offering excellent protection if the wind is in the north or east. There is a gravel beach at the head of the cove and a house or two set well back in the trees. The land is private, and visitors are not welcome ashore.

APPROACHES. From the north, run along the cable area shown on the chart, starting west of nun "10" and leaving Ewe and Hardwood Island to starboard. Both of these islands are heavily wooded, with sloping granite ledges.

Find can "1" at the 8-foot spot off Merchant Island and enter the cove north of the can. Favor the south side of the entrance, which is all sloping ledge, to avoid the rocks on the north side of the entrance, which are visible two hours either side of low tide.

ANCHORAGES, MOORINGS. Anchor all the way up in the pool at the head of the cove, in 12 to 13 feet at low. The bottom is mud with some grass.

DEER ISLE—Eastern Coast

WEBB COVE

3 4 (inner cove) ★★★

off Ocean Adventures: 44° 10.21'N 068° 38.27'W
CHARTS: 13315, 13305, 13313   CHART KIT: 69, 22

On the southeastern side of Deer Isle, Webb Cove is a sizable bay protected to the north by Buckmaster Neck. Protection is reasonable in the outer portion of the cove and even better in the inner section, although it is a bit ticklish to enter and full of lobsterboats. The smooth dome of granite that forms the southwest height of Buckmaster Neck was named Settlement Quarry.

APPROACHES. From Deer Island Thorofare, pass west of Humpkins Ledge or between Humpkins Ledge and Grog Island. Identify Channel Rock, which appears as two separate rocks, in the middle of the entrance, and pass to either side. Then head straight for the quarry derricks and dock. Webb Cove is broad, but shallow. Watch your depthsounder.

ANCHORAGES, MOORINGS. Anchor south or west of the granite quarry dock in 8 to 10 feet at low, in mud. Exposure here is only to the southeast.

⚓ The dock and float to the south of the granite quarry belong to Old Quarry Ocean Adventures, which rents moorings to boats with a draft of less than 7 feet.

The inner harbor is shallow and full of boats, but if you can get in and find room the protection is excellent. Stay close to the quarry dock and make a gradual turn to port, running about halfway between the peninsula of Deer Isle to the south and the little island to the north, using the mooring stakes as your guide. Enter cautiously. The deep water is narrow, and the rocks shown on the chart are difficult to locate. Anchor in 6 to 8 feet at low in soft mud bottom and consider a stern anchor. Here you will be landlocked in all directions except for a very small opening to the southeast.

GETTING ASHORE. Take your dinghy in to the float off the quarry dock or to the dock of Old Quarry Ocean Adventures.

FOR THE BOAT. *Old Quarry Ocean Adventures (Ch. 16; 207-367-8977; www.oldquarry.com)* has four heavy rental moorings for boats with a maximum draft of 7 feet.

FOR THE CREW. Old Quarry Ocean Adventures has hot showers and a small camp store with basic supplies. They also can arrange lobster bakes.

THINGS TO DO. Settlement Quarry is interesting to explore. Equipment and great, smooth slabs of stone were left behind when the quarry closed, and it is still used as a landing place for huge stones cut from Crotch Island.

The Island Heritage Trust maintains trails that lead up to the dome of the quarry. One trail starts at Old Quarry Campground. Another is off Route 15.

Old Quarry Ocean Adventures rents small boats and kayaks, and they have campsites and a small swimming hole. Owner Bill Baker can arrange long-term parking, making this a convenient place to connect with crew.

PICKERING COVE

3 ★★★

44° 12.22'N 068° 37.19'W
CHARTS: 13313, 13316   CHART KIT: 69, 22

On the east side of Deer Isle, south of Stinson Neck, lie several coves that offer pleasant sailing in a little-used area. There are some interesting gunkholes in the deeper recesses. Pickering Cove is formed by Freese Island and a smaller peninsula of Deer Isle, just east of the entrance to Southeast Harbor. The anchorage is peaceful and attractive, easy to enter, and provides good protection except to the southeast.

APPROACHES. As you approach from Deer Island Thorofare, you will pass the internationally known Haystack Mountain School of Crafts in Western Cove on Stinson Neck. Billings Cove and the bay west of Stinson Neck offer good anchorages and a pleasing shoreline, but they are too wide and open for overnight protection. Pickering Cove, to the west, is more secure.

The rocks outside the southwestern corner of the entrance are visible 2 or 3 feet above water at high tide. Run up the middle of the cove toward the hunk of rock crowned by trees at the head of the deep water. On the mainland behind this little island are several houses.

ANCHORAGES, MOORINGS. Anchor anywhere up to the island itself in 14 to 15 feet at low in good mud holding ground.

A CRUISING GUIDE TO THE MAINE COAST

SOUTHEAST HARBOR

 ★★★

44° 12.05'N   068° 38.77'W

CHARTS: *13313, 13316*   CHART KIT: *69, 22*

Southeast Harbor offers excellent protection and an attractive setting on the east side of Deer Isle, deep between Stinson and Whitmore Neck.

APPROACHES. The entrance to Southeast Harbor is north of Whitmore Neck. It is easier to navigate than it appears on the chart. Pass north of the rocks in midchannel, bits of which are visible at all tides. Favor the north side, where the channel at low water is roughly 75 yards wide and 40 feet deep.

ANCHORAGES, MOORINGS. Continue heading westward and anchor in the basin northeast of Warren Point, in 22 to 30 feet at low. The "rk rpt" shown on the chart north of Warren Point is reported to be visible at low.

INNER HARBOR (Southeast Harbor)

■ ★★★

44° 11.52'N   068° 39.07'W

CHARTS: *13313, 13316*   CHART KIT: *69*

Inner Harbor is a challenging gunkhole west of Whitmore Neck on Deer Isle and south of Warren Point. Protection is excellent, but there is not much swinging room.

APPROACHES. Because of the narrow channel and generous supply of rocks, Inner Harbor is easiest to enter near low tide.

First observe the line of rocks running north from Whitmore Neck, which should be left to port. The second important guidepost is the tiny island with a cluster of spruces (next to the "H" in "Inner Harbor" on chart 13313), also left to port. Some of the shoal areas are marked by stakes.

ANCHORAGES, MOORINGS. Anchor outside the moored lobsterboats in 16 to 18 feet at low.

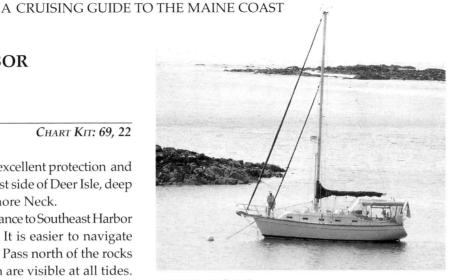

DEEP HOLE

 ★★

44° 12.33'N   068° 39.17'W

CHARTS: *13313, 13316*   CHART KIT: *69*

A small, wooded island lies north of Warren Point on Deer Isle, on the west side of the arm of water leading to the pocket called Deep Hole (Deep Hole itself is not shown on the Chart Kits). Tucked in behind the island is a cluster of sailboats used by the Deer Isle Sailing Center, whose dock and float are on the opposite shore. The anchorage is pleasant and reasonably well protected.

APPROACHES. Stay in midchannel, leaving a ledge to starboard and rounding the little island to port.

ANCHORAGES, MOORINGS. Anchor north and west of the island in midchannel, in 15 to 20 feet at low. Protection is good, but there are usually 1 or 2 knots of current.

THINGS TO DO. Use your dinghy to explore northward into intriguing Deep Hole. Author and sailor Roger Taylor reports that he once lowered a leadline here, and it really is deep—102 feet!

GREENLAW COVE

between Oak Point and Campbell Island:
44° 13.56'N   068° 36.52'W

CHARTS: *13313, 13316*   CHART KIT: *71*

The long arm of Greenlaw Cove, at the northeastern corner of Deer Isle, looks intriguing on the chart, but it has little scenic beauty or interest. The protection and holding ground are good, however, between Oak Point and Campbell Island.

EGGEMOGGIN REACH

Eggemoggin Reach is a great body of water between Deer Isle and the mainland that connects East Penobscot Bay with Jericho Bay. It is about 10 miles long and averages a mile in width. With summer winds, it can be an exhilarating sail in either direction. As Robert Carter wrote in his 1858 *Summer Cruise on the Coast of New England*, "There cannot be a finer sheet of water in the world than this Reach, which is bounded on every side by superb views."

While there are a number of ledges in the Reach, they are all well buoyed. The western end is marked by red-and-white bell "EG" (44°19.21'N 068°44.56'W). Halfway through, a graceful suspension bridge connects Little Deer Isle to the mainland, with a vertical clearance of 85 feet at the center. Red-and-white bell "EE" (44°12.74'N 068°32.30'W) marks the eastern end. Current in the Reach floods northwest and ebbs southeast, but it is not very strong.

Winds sometimes funnel unexpectedly down the Reach, so be cautious on gusty days. In September 1984, the 65-foot schooner *Isaac H. Evans* was knocked down by squally winds just south of Grays Point near Bucks Harbor, at the western end of the Reach. She sank in 60 feet of water, showing only the tip of her mainmast and the peak of her gaff. Most of the passengers and crew swam ashore and the rest were saved by local rescue groups.

Early each August, about 120 of the fastest and most beautiful wooden boats gather for the Eggemoggin Reach Regatta, a spectacle to watch or win.

The banks of Eggemoggin Reach are wooded, mostly with deciduous trees. At the eastern end, the islands become crowned with conifers, and as you approach, there is a definite feeling of arrival in some new and mysterious place. Here the channel passes south of the Torrey Islands and the ledge at Torrey Castle.

Several good harbors line the Reach. By far the best and easiest to enter is Bucks Harbor, at the western end, in Penobscot Bay. Another good refuge is Center Harbor, near the Torrey Islands, although it is small and crowded. Benjamin River, east of the bridge, offers extremely good protection. *WoodenBoat* magazine and the WoodenBoat School, at the eastern end of the Reach, welcome visitors and provide guest moorings.

The suspension bridge over Eggemoggin Reach. Clearance is 85 feet at its center.

BILLINGS COVE

44° 17.95'N 068° 40.75'W

CHARTS: 13309, 13316   CHART KIT: 66, 20

Billings Cove is easy to find on the north side of Eggemoggin Reach, just east of the bridge. It is easy to enter and protected except from the south and southeast. Other nearby harbors, however, have more to offer.

APPROACHES. The cove is wide open to the southeast, with no obstructions other than the shoreside ledges. Go down the middle of the cove.

ANCHORAGES, MOORINGS. Anchor in 10 to 20 feet at low on the midline of the cove.

BENJAMIN RIVER

entrance channel at ledge: 44° 17.46'N 068° 37.72'W

CHARTS: 13316   CHART KIT: 71

Benjamin River is a remarkable, 50-foot-deep hole surrounded by mussel flats and guarded at the entrance by an awesome ledge. Once inside, there is excellent anchorage and protection.

APPROACHES. The key to entering Benjamin River is the long, low ledge that makes out from the east shore almost all the way across the channel, leaving an entrance only about 100 yards wide. Two nuns and a can mark the channel and make the entrance relatively easy, but don't cut sharply around the last

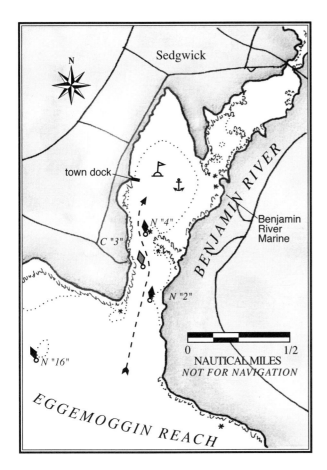

red nun. Even at dead low, you can't see the end of the ledge, which extends underwater an additional 40 feet or so.

Here is the old rule of thumb used by local sailors when the channel was unbuoyed. Just as you enter from the Reach, you will see on your right a whitish rock slab climbing up into the trees on shore. Run a straight line from the slab of rock to the landward end of the town dock, and it will clear the ledge that obstructs the entrance. The town dock is shown on the chart on the west side of the cove.

ANCHORAGES, MOORINGS. Benjamin River Marine maintains about 13 rental moorings. There usually is plenty of room to anchor in 7 to 18 feet at low along the eastern side of the cove. Holding ground is excellent, in mud. The cove provides protection in every direction except south-southwest, and the ledge at the entrance prevents most seas from entering.

GETTING ASHORE. Land at the Sedgwick town dock, on the west side of the harbor, or—except at the lowest tides—at the Benjamin River Marine float.

FOR THE BOAT. *Benjamin River Marine (207-359-2244).* This small, serious repair yard is on the east side of the cove. The yard hauls boats on a marine railway and provides storage and limited marine supplies. No fuel is available. There is water and electricity at the float, which dries out at extreme low tides.

FOR THE CREW. The nearest provisions are in Sedgwick at the Eggemoggin Country Store (*359-2125*), about 2.5 miles from the town dock. Head south on Carter Point Road, take a right on Old Steamboat Road and then a left on Reach Road. The store has what general stores generally have—everything: groceries, liquor, baked goods, and pizza.

CENTER HARBOR

3 ★★★

44° 15.70′N   068° 35.00′W

CHARTS: **13316**   CHART KIT: **71**

Near the eastern end of Eggemoggin Reach, north of the Torrey Islands, Center Harbor is an idyllic little place full of boats, with room for just one boatyard and a yacht club. The harbor is bounded on the south by Chatto Island and its rock ledges. A fleet of lovely gaff-rigged Beetle Cats is moored just inside the entrance. Astoundingly beautiful bigger boats are moored deeper in the small harbor—one of the most concentrated pockets of classic yachts on the coast.

APPROACHES. Leave to starboard nun "2," off the rocky shore of Chatto Island (it's hard to see), and leave to port green daybeacon "3," which stands on a rock in the harbor.

ANCHORAGES, MOORINGS. You can anchor outside the moored boats in about 16 feet at low, but there is not apt to be room to anchor inside the harbor. The Center Harbor Yacht Club has a mooring for overnight visitors, and the Brooklin Boat Yard has a few moorings that may be available for rental. Protection in the harbor is good except from the northwest.

GETTING ASHORE. Row in to the dinghy dock at the Brooklin Boat Yard or at the Center Harbor Yacht Club. The yard is closer to the market.

FOR THE BOAT. *Brooklin Boat Yard (207-359-2236; www.brooklinboatyard.com).* Brooklin Boat Yard was started in 1960 by Joel White, the son of author E.B. White. It is now run by Joel's son Steve. It has gained a reputation for designing and building boats of refined simplicity and classic grace in the aesthetic tradition of Herreshoff. In addition, the yard maintains or services the classics in the harbor. The yard has marine railways and a 35-ton boatlift and can handle most repairs. There is water at the float, with 4 to 5 feet alongside at low, but no fuel.

Brooklin Marine Supply (207-359-5030). This chandlery is located in a former school building in Brooklin. It serves both the public and as the main stock room for Brooklin Boatyard. It carries fittings, tools, and hardware.

Doyle Center Harbor (207-359-2003). This sailmaker's loft is located in the Old School Building with Brooklin Marine Supply.

Center Harbor Yacht Club (pay phone: 207-359-8868). You will see the gray-roofed, dark red clubhouse and the flagpole, dock, and floats of CHYC on the north shore near the harbor's entrance. Fresh water is at the floats, with 4 feet alongside at low.

FOR THE CREW. From the boatyard, walk up to the main road and turn right. It's .7 miles to the post office and the Brooklin General Store (*359-8817*), which in addition to a reasonable selection of groceries carries *The New York Times* and exchanges propane canisters. If you are loaded with groceries, they probably will give you a ride back to the harbor.

Across the street is the Morning Moon Café (*359-2373*), which has baked goods and serves breakfasts, lunches, and dinners. Another dinner option is the Brooklin Inn's pub or dining room (*359-2777*).

THINGS TO DO. Resident E.B. White donated the original *Stuart Little* illustrations to the mouse-sized Friend Library, right downtown.

It is a long but pleasant walk from the General Store to the stunning headquarters of *WoodenBoat* magazine and school, or you can visit by boat (see below).

The harbor and some of the fleet at *WoodenBoat* Magazine.

WOODENBOAT SCHOOL

3 ★★★★

44° 14.63'N 068° 33.36'W

CHARTS: 13316   CHART KIT: 71, 22

Among aficionados of wooden boats, *WoodenBoat* magazine and its WoodenBoat School is a Mecca that is well worth a pilgrimage.

Near the eastern end of Eggemoggin Reach, on the north side, is the rambling, whitewashed-brick estate that serves as the headquarters of the magazine. The large, red-brick stables nearby house the boatbuilding school. There are some moorings out front, and visitors are welcome.

WoodenBoat magazine was founded on a shoestring in 1974 by Jon Wilson. Against all odds, including a disastrous fire, this seductive magazine became the bellwether of a resurgent interest in wooden boats. The publishing empire now includes a book division and the trade magazine *Professional Boatbuilder*.

The associated summer school (*359-4651, www. woodenboat.com*) provides short courses in boatbuilding skills of all kinds, bringing in as instructors master boatbuilders, surveyors, and sailmakers. The atmosphere in the school is a wonderful blend of wood shavings, modern epoxy glues, practical work, and fun. Stop here for a visit if you want to learn about wooden boats or simply watch how these beautiful objects are built. The school also offers courses on the water, teaching seamanship in a variety of traditional craft from schooners to Friendship sloops and small open boats.

APPROACHES. WoodenBoat's dock is located north of Babson Island. The approach may be made either west or east of the two Babson Islands. Small Shoofly Ledge, between the Babsons and to the north, is just about covered at high tide, and seals like to haul out here. An old green-and-gray boathouse with a fieldstone base graces the head of the dock.

ANCHORAGES, MOORINGS. Anchor in good mud 10 to 12-feet deep or check with the school to see whether a mooring is available. Hard south or southeasterlies make the anchorage pretty lumpy.

GETTING ASHORE. Row in to the dinghy float and walk up the dock past the green boathouse.

FOR THE CREW. The magazine has a pay phone and a small store displaying publications. It's a 1.5-mile walk to Brooklin, which has a post office, general store, and the Morning Moon Café.

Coasting eastward on Eggemoggin Reach.

NASKEAG HARBOR

3 ★★

44° 13.62'N 068° 31.80'W
CHARTS: 13316   CHART KIT: 71, 22

Naskeag Harbor shelters a small fleet of fishing boats at the eastern end of Eggemoggin Reach, but it is not very useful for cruising yachts. A bar running nearly all the way from the mainland to Harbor Island separates Naskeag into an eastern and western part. The fishing boats are moored in the eastern part. There is room to anchor outside the moored boats in 15 to 20 feet at low, but the bottom is rocky and uncertain, and low tide reveals boulders scattered along the shore like a marble game among giants.

The western portion is difficult to enter because of the unmarked Triangles, which block the center of the harbor and are invisible even at low tide.

CASCO PASSAGE and YORK NARROWS

Casco Passage between N"10" and C"9":
44° 11.60'N 068° 29.35'W
York Narrows between N"8" and C"7":
44° 11.19'N 068° 28.80'W
RW"CP": 44° 11.80'N 068° 26.51'W
CHARTS: 13315 (INSET), 13313, 13316
CHART KIT: 70, 71, 22

Casco Passage joins Jericho Bay and Blue Hill Bay north of Swan's Island. From Jericho Bay, the islands north of the passage—Johns, Opechee, and Black—are low and dark against the background of Mount Desert. The opening between Swan's and Black is visible from a long distance away.

The northern channel of Casco Passage is almost a straight shot and a bit easier than York Narrows, the southern fork. Running from west to east, leave red marks to port, green marks to starboard. The current through the passage ebbs west and floods east with considerable force.

From the west, starting at red bell "8," the York Narrows passage is deep but narrow, and it turns sharply north at Orono Island. Be sure to find green beacon "lA," marking a separate little treeless island and ledge north of Orono. York Narrows is one of the approaches to Buckle Harbor and to Mackerel Cove on Swan's Island (see page 287).

OPECHEE ISLAND

2 ★★★

44° 12.50'N 068° 28.10'W
CHARTS: 13313, 13316, 13312   CHART KIT: 70, 71, 22

At the southwest corner of Blue Hill Bay, north of Casco Passage, there is a small archipelago whose principal islands are Black, Pond, and Opechee. A very pretty little anchorage lies east of Opechee surrounded by wooded islands and views of Mount Desert and Isle au Haut.

APPROACHES. Approaching from Blue Hill Bay, Pond Island is clearly separated from the others of the group, while Black and Opechee merge at a distance. There is a rocky promontory near the north tip of Black. Enter the anchorage north of Black Island, leaving Eagle and Sheep to starboard. Small Eagle Island is entirely wooded, while Sheep Island beyond it has been cleared. Head for a sloping granite shelf on the eastern shore of Opechee.

ANCHORAGES, MOORINGS. Anchor in the large 9-foot area shown on the chart, about equidistant from Black, Sheep, and Opechee Island. The ledges between Black and Opechee break the seas even at high tide, but the anchorage is exposed to winds from several directions.

SWAN'S ISLAND

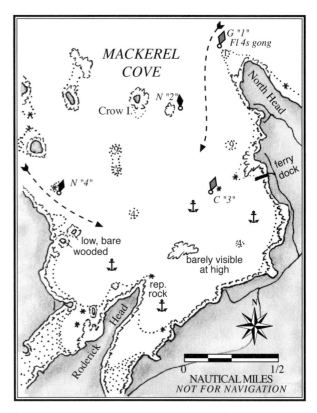

Although served by several daily ferry trips from the mainland at Bass Harbor, Swan's Island is still remote. There is only one small store, no liquor sales, and no amusements except those the islanders make for themselves. The pace is slow—and Swan's Islanders like it that way. The year-round population is about 350 people, and it triples in the summertime. Fishing is still the main occupation.

Swan's is big, nearly 7,000 acres. It is hilly, but not particularly high. The island has three little villages— Atlantic in Mackerel Cove on the north coast, where the ferry comes in, and Minturn and Swan's Island, on the shores of Burnt Coat Harbor in the south.

Swan's irregular shoreline creates three good harbors. Burnt Coat is nearly landlocked and has the most facilities. Mackerel Cove is big, but good in prevailing southwesterlies. Buckle Harbor, on the northwest coast, is a little gem.

It must have been a dry year in 1603; perhaps lightning started a fire in the forests of Swan's Island. When French explorer Samuel de Champlain visited the area in 1604, the story goes, he named it *Brule Cote*, meaning "burnt coast." This later was Anglicized to Burnt Coat.

Thomas Kench was the first white settler. He came to Burnt Coat in 1786 and lived as a hermit for 14 years. He was followed by "King David" Smith, who built his cabin on the same islet in Burnt Coat Harbor. Smith was not of hermit stock. He and his three wives produced and raised some two dozen children, practically populating Swan's by themselves.

The same year that Thomas Kench took up residence, a colorful land speculator named Colonel James Swan bought the island along with several others from the Commonwealth of Massachusetts. Swan had participated in the Boston Tea Party and was wounded at Bunker Hill. This bright and energetic young man was noticed early by those in power, and he moved easily among the leaders of the day. George Washington was a friend, and so was Lafayette. Swan's portrait was painted by Gilbert Stuart.

With inherited money, Swan became a speculator on a grand scale, buying up land confiscated from the Tories and also commissioning privateers. The "Burnt Coat Group" of islands, which included Swan's, was to become "an island empire," and indeed, an enormous mansion was built on the island as his headquarters. To populate his empire, Swan offered 100 acres to any homesteader who stayed seven years.

In 1788, Swan's speculations turned sour. He escaped his creditors by moving to France, where his fortunes rose again until he was charged with embezzlement and thrown into debtors' prison. Although he could have paid his debts and regained his freedom, Swan chose to protest his innocence and spent 22 years in a Paris jail, where he died in 1830. Swan never lived in the mansion built for him on the island.

Since 1986, honoring their island's bicentennial, Swan's Islanders have made a deliberate effort—in defiance of the nautical charts and the U.S. Postal Service—to restore the grammatically and historically correct apostrophe to the name Swans, a fitting tribute to their colorful founding father.

BUCKLE HARBOR

4 ★★★★

44° 10.85′N   068° 28.35′W

CHARTS: 13315 (INSET), 13312, 13313   CHART KIT: 70, 22

Buckle Harbor, at the northwest corner of Swan's Island, is a geographically convenient anchorage, an easy place to stop on your way east or west. It is also lovely and unspoiled, but not undiscovered. You will rarely have Buckle to yourself on a summer evening. Not far away is the spectacular skyline of Mount Desert, and a procession of boats passes the harbor entrance on their way through York Narrows.

APPROACHES. From the west, stay in the deep tongue of water north of Buckle Island and south of York Narrows, being sure to give a wide berth to the extensive ledges along Buckle's northwest shore. From the east, leave York Narrows by heading south just east of can "5" and continue south to the harbor.

ANCHORAGES, MOORINGS. Anchor off the east shore of Buckle Island in 7 to 9 feet at low. Holding ground is good, in mud. The harbor provides protection all around except from the north and northeast.

GETTING ASHORE. Take your dinghy to any of the little beaches on Buckle Island. Lovely trails lead around and through the island.

MACKEREL COVE

3 ★★★

SW of C"3": 44° 10.65′N   068° 25.85′W
W of Roderick Head: 44° 10.35′N   068° 26.60′W
E of Roderick Head: 44° 10.25′N   068° 26.10′W

CHARTS: 13313 (INSET), 13312, 13315   CHART KIT: 70, 22

Mackerel Cove is a large bay on the north side of Swan's Island. The ferry from Bass Harbor docks here just east of can "3," and the tiny fishing community of Atlantic is in the southeast corner. The cove is large enough to handle the New York Yacht Club cruise fleet with space left over, and it offers good protection, particularly in the anchorages east and west of Roderick Head.

APPROACHES. From the north, the approach is wide and easy. Start at green gong "1" at North Point, left to port, and keep the nun off the ledge at Crow Island to starboard.

Mackerel Cove may also be entered from the west. From York Narrows (inset on chart 13315), turn to starboard after passing Buckle Harbor and green can "5." There is ample room between Swan's Island and Orono Island, which is marked by a granite wharf at the southern end. This is a remarkably pretty passage through which the current floods eastward and ebbs westward. After passing Orono Island and the two islets to starboard, continue along the shore of Swan's, with its sloping granite slabs, and leave nun "4" close aboard to port. The outermost of the two islets south of nun "4" is low and bare. The inner one is wooded.

The key to Mackerel Cove is the enormous ledge off Roderick Head. At high tide, only a tiny portion is visible, even though the rest extends a long way east and southwest, as you will observe at low. Boats frequently underestimate the extent of the ledge and anchor too close for comfortable swinging, or they actually find it with their keels. Anchor a long way from the ledge, just to be sure.

There is room to pass between the ledge and Roderick Head, but a nasty rock is reported about 250 feet offshore on a line between Roderick Head and the ledge. Proceed with caution about midway between the two, but avoid the passage altogether within two hours either side of low.

ANCHORAGES, MOORINGS. There are several places to anchor in Mackerel Cove, all with good holding ground and beautiful views of Mount Desert Island. When a large fleet is in, they usually cluster west and south of can "3," but the protection is better farther south, on either side of Roderick Head. The anchorage east of Roderick Head is small and unobstructed. Anchor in 7 to 10 feet at low.

West of Roderick Head, the first small ledge on the east side is visible until about halftide. Deep water extends southward to a tiny islet that has two tanks on top of it, and there are small craft moored in the shoal cove beyond. Anchor in 15 feet at low.

It is also possible to anchor east of can "3," along the shoreline south of the ferry landing, in 10 to 13 feet at low, but the protection is not as good.

GETTING ASHORE. The public landing and float are on the south side of the ferry terminal, with good depth alongside. Yachts often make crew changes here. It can be a long dinghy ride from the more distant anchorages, particularly if you are rowing, and the afternoon southwesterlies can make a job of getting back to your boat.

FOR THE CREW. The Island Bake Shoppe (*526-4153*) is located in the second house from the ferry terminal.

A 20-minute walk into Atlantic will bring you to a remote laundromat and Claire's Kitchen (*526-4425*), a takeout and small grocery store—the only one on Swan's. Follow the ferry road to its intersection with the main road. A left and left again will put you on the Atlantic Loop Road, where you will find the small Atlantic Laundromat (*526-4478*). Continuing straight on the main road or a right at the intersection past the laundromat will bring you to the village center and the store.

THINGS TO DO. The Lobster and Marine Museum (*526-4423*) celebrates the history of the lobster industry and fishing heritage. It is 100 yards from the ferry landing in the Captain Henry Lee House, which formerly housed the ferryboat captain.

The Seaside Hall Museum, about a mile away, documents Swan's Island history. Take the road from the ferry and turn right on the main road. The museum is on the left. The island library (*526-4330*) is in a former schoolhouse on the way to Atlantic.

BURNT COAT HARBOR

4 ★★★★

entrance: 44° 07.97'N 068° 26.74'W
inner channel: 44° 08.45'N 068° 26.90'W

CHARTS: 13313, 13312   **CHART KIT: 70, 22**

Burnt Coat Harbor, on the south shore of Swan's Island, is an attractive, well-protected harbor with one of the island's largest communities. It remains a serious fishing town—lobsterboats still leave the harbor at first light, rocking you gently in your bunk, and roar back in at dusk.

APPROACHES from the west. Burnt Coat Harbor is approached from the west through well-marked Toothacher Bay. The barren rock of High Sheriff merges at first with grassy Gooseberry Island behind it, and in good visibility you will see the white lighthouse on Hockamock Head. Keep can "3," off Gooseberry Ledge, to port and head for the narrow entrance just south of the lighthouse. The square white tower and white building of the lighthouse on Hockamock Head are conspicuous. Enter the harbor between red beacon "4" off Harbor Island and gong "5." There is plenty of room to drop your sails after entering.

APPROACHES from the east. The "back door" to Burnt Coat Harbor is a neat little tickle between Stanley Point and Harbor Island, marked by two green cans left to port. Use caution here; the scale of the paper chart is woefully inadequate, and many of the electronic charts are completely misleading.

From red gong "2," run inside the Baker Islands, leaving the first can, can "3," well to port. It marks the western end of a very mean ledge, visible only at dead low.

Curving right around Stanley Point, you will see the inner can, can "5," which marks the southern end of another ledge, which is visible at midtide or lower. The problem with chart 13313 is that the orientation of the symbol for can "5" obscures the narrow channel to the east of the can, and it obscures the fact that the ledge continues to the northnorthwest, well beyond the can. Leave both the can and the ledge to port. To clear the ledge that continues beyond the can, you need to stay close to the two little wooded islands, The Hubs, which are left to starboard.

Once you are in Burnt Coat, be sure to note the ledges off Long Cove and the little island (Potato), left to starboard.

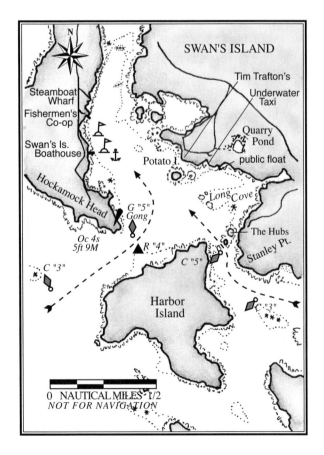

ANCHORAGES, MOORINGS. Fishermen here have discovered that cruisers will pay for the convenience and security of a mooring. Nearly thirty heavy transient moorings are available to visiting boats. The moorings are owned by various fishermen and managed by Kevin Staples, with a 50/50 split of the proceeds. They have green polyball floats and pickup buoys. The mooring field fills most of the harbor from the Swan's Island Boathouse, on the west side of the harbor, to the Fishermen's Co-op.

There is also plenty of room to anchor along the western side of the harbor south of the moorings, in 17 to 25 feet at low. The bottom is mud, but we've had reports of anchors fouled by seaweed and boats dragging. Be sure your anchor is well set. It is also possible to anchor north of Harbor Island, but the swells may give you an uncomfortable night.

GETTING ASHORE. Land at the float at the Swan's Island Boathouse or, heading north, at the Fishermen's Co-op or at Steamboat Wharf. At the Co-op, use the float to the north of the green bait shed.

FOR THE BOAT. *Swan's Island Boathouse (207-526-4201).* This operation is run by the Staples family. They manage the moorings, sell cube ice, and provide dinghy docking. They also have a takeout (see *For the Crew*, below) and gift shop.

Fishermen's Co-op (207-526-4327). The first long wharf to the north of the Boathouse has gas and diesel at the southern float (4 feet at low) and some marine supplies. The green Co-op is strictly a "fishermen first" operation, but it is very friendly. The manager requests that visiting yachtsmen not buy fuel after 1PM, when wharf space is needed for returning lobsterboats. If you come ashore here by dinghy, please use the float north of the green bait shed.

Steamboat Wharf (Ch. 68; 207-526-4186). The next wharf to the north is occupied by Swan's Island Lobster, a wholesale buyer. When it is not busy, they welcome cruising boats to their float, which has 6 feet of depth at low. They have gas and diesel and accept checks or credit cards.

FOR THE CREW. Beginning at 11AM, the Boathouse (526-4201) has a takeout, ice cream, sandwiches, and local produce in season. They also will cook lobster and chowder dinners and deliver them to your boat. Undoubtedly, they could sell you live lobsters as well, or you can buy them at the Co-op.

Across the harbor, the large, dark wharf north of Potato Island is Tim Trafton's wharf (526-4427), a takeout that sells cooked lobsters, freshly dug clams, and scallops. Further south, Underwater Taxi (526-4204) has a seafood store with another takeout. They sell lobsters, steamers, scallops, crabmeat, chowder, and fresh fish. For a modest upcharge, they will cook any fish they sell.

The Co-op has a credit-card phone outside the office. The post office is about a quarter mile up the hill to the right.

THINGS TO DO. It is a short walk south down the road to the lighthouse at Hockamock Head, with great views of the harbor and bay. The square white tower was built in 1872 and served as a useful mark for fishermen in setting their traps and finding their way home. In a late-1970s cost-cutting move, the Coast Guard closed the lighthouse and replaced it with a 20-foot beacon whose light could be seen only half as far. Under pressure from Swan's Islanders, the light in the tower was restored.

In 1982, the Coast Guard stripped the white paint off the brick tower to reduce repainting costs. This made the tower almost invisible against the background. The islanders protested again, and the tower was finally restored to its brilliant white.

In 1994, as part of the Maine Lights Program, the Coast Guard transferred ownership of the light to the town of Swan's Island as part of a public park.

For a long walk, go past the narrow "Carrying Place" to Fine Sand Beach. This is a lovely pocket beach reached by a trail through the woods from the parking area. Do not be surprised if you see deer along the way. Or you can rent bikes or kayaks from the Harbor Watch Motel (526-4563, 800-532-7929) in Minturn and explore the island's roads. If you plan ahead, the motel may be able to deliver them to one of the wharves.

To swim at freshwater Quarry Pond, take your dinghy eastward from the anchorage, passing inside Potato Island. Go beyond the old stone wharf and tie up at a public landing float. Then walk a short distance up to the right to the quarry. Beware of hidden rocks in the pond.

SAND and POPPLESTONE COVE

Sand Cove: 44° 06.50'N 068° 30.00'W

CHARTS: *13313, 13313*   CHART KIT: *69, 22*

High, wooded Marshall Island lies southwest of Swan's Island. Sand Cove, at the southeast side of Marshall, has a beautiful little white sand beach in the western corner. On a sunny summer day, you are likely to find several boats anchored and people ashore enjoying the beach. The rest of the cove has rocky shores, and the pink granite of Devils Head is striped with black basaltic dikes.

In strong southwesterlies, it is hard to get far enough in for full protection, and there's an entering swell, but at other times, this is a fine daystop.

In good visibility, the approach is easy. Stay well clear of Yellow Ledge, the southern part of which is always visible.

Popplestone Cove, north of Devils Head, looks good on the chart, but try it only if you have a yachtsman or fisherman anchor. It is aptly named for the smooth, round stones that cover the bottom, which make for very poor holding ground.

THOSE ENTREPRENEURIAL LONG ISLANDERS

Massachusetts land speculator Colonel James Swan first acquired Long Island in 1786, together with Swan's Island and 23 others. From debtors' prison in Paris (see *Swan's Island* entry), Swan continued to wheel and deal, eventually losing Long Island to Michael O'Maley in 1812. O'Maley offered land to settlers, and by 1820 the census showed three households and a population of 19. In 1835, Israel B. Lunt bought 1,132 acres for $600. Lunt was truly the progenitor of Frenchboro. A third or more of today's residents have his surname, and his descendants still live in the Israel B. Lunt homestead.

Showing early signs of the opportunism that still exists today, the islanders agreed to name part of Long Island Plantation for lawyer Webster French if he would arrange for a post office on the island. French used his political connections successfully—thus Frenchboro.

Fishing provided the livelihood of the early settlers, who went after groundfish such as haddock and cod. Today the fishing centers on lobstering, seining for herring, and dragging for scallops. For a while, pulpwood was harvested on Long Island and shipped by barge and coaster to the St. Regis Paper mill in Bucksport.

The island population reached a peak of 197 in 1910 and then started a steady decline. A schoolhouse built in 1907 to hold 60 students was serving half that number by 1930, and only two children attended school in 1964. Threatened with the loss of the school, the islanders rose to the occasion by inviting 14 foster children to live in Frenchboro. The school was saved, and most of the children stayed on the island until it was time to leave for high school on Mount Desert. Two have returned to live in Frenchboro.

Modern times have come slowly to Frenchboro. A cable brought electricity in the 1950s, but telephone service didn't arrive until 1982, and many families still move off the island so their children can attend high school. As the population continued to decline, it became obvious to some that Frenchboro was in danger of extinction, like many island communities before it.

With impetus from the women—"a matriarchal society," as Town Manager James Haskell noted—Frenchboro obtained a $360,000 block grant from the state. Some of the money was used to build a new community hall and dump and to upgrade housing. Most of the money was used toward a more long-sighted goal—to allow Frenchboro to offer inexpensive land parcels and low-interest housing loans to attract young residents and increase the population. Fifty-five acres were donated by the Rockefeller family to this end.

Articles about the low-cost housing project appeared in such far-flung places as *The New York Times* and *USA Today*, and a tabloid dubbed it a "homesteading" project on "Fantasy Island." Applicants came from all walks of life, yet, to date, the results have been bittersweet. Since the inception of the program, a dozen or more families have tried to adapt to the rigors and solitude of island life, but only one of the original families has remained.

FRENCHBORO (Long Island)

4 ★★★★★

Lunt Harbor: 44° 07.26'N 068° 21.81'W

CHARTS: 13313, 13312   CHART KIT: 70, 22

Cruising and working boats share the harbor of Frenchboro off Lunt's wharf.

Long Island is remote, very remote, and a wonderful place to visit. About 70 islanders live in this lobstering community about 8 miles from Mount Desert Island, clinging to their centuries-old way of life with great determination and considerable ingenuity.

The town of Frenchboro consists of twelve islands—Long, Crow, Harbor, Mount Desert Rock, Great Duck, Little Duck, Black and Placentia, the two Green islands, Pond and Drum—but the center of this staunch community rings Lunt Harbor on Long Island. Here there is a seasonal summer restaurant at Lunt's wharf, a post office, the one-room schoolhouse, the island church, and a library and small museum run by the Historical Society. Visitors can explore by hiking the well-maintained trails. A few cellar holes are all that remain to tell the story of former settlers.

At the height of summer, Long Island is not as isolated as it once was. Frenchboro has become a popular destination for boats from all over the Mount Desert Island area, and the succulent lobsters steamed up at Lunt's Deli attract yacht club cruises that pack the small harbor. It's also on the itineraries of several day-charter boats. Lunt Harbor is probably busiest on the second Saturday in August, when the islanders hold their annual Lobster Festival. In addition to untold quantities of chicken salad and homemade pies, 500 lobsters or more are boiled in the church, like an offering of thanks, and devoured in the field next door.

APPROACHES. Long Island's shores are bold, and the outlying dangers are well marked. Lunt Harbor may be entered from the west by passing south of green bell "1" and Harbor Island, or from the northeast from red-and-white gong "LI" (*44° 08.33'N 068° 20.52'W*) passing east of Crow and Harbor Island. Watch out for the occasional ferry that plies this route to and from Bass Harbor.

The ferry docks at the first large stone wharf on the eastern side of the harbor, with a green steel ramp. The first large wharf on the opposite, western side of the harbor is Lunt & Lunt.

ANCHORAGES, MOORINGS. Lunt & Lunt puts down about 15 large granite-block moorings in 10 feet of water north of their wharf. These are not specifically marked. Pick up any mooring that isn't marked with the name of a fishing boat or that does not have a punt on it and row ashore to confirm and pay at the deli.

You can also anchor outside the moorings in 25 to 35 feet, although you will be more exposed. Protection inside the harbor is extremely good except for northeasters, which occasionally wreak havoc here.

Another possibility is to anchor on the 15-foot bar east of Harbor Island. While the inner harbor to the south is dredged to 6 or 7 feet along the western shore, there is no room to anchor.

GETTING ASHORE. Row in to the dinghy float at Lunt & Lunt.

FOR THE BOAT. *Lunt & Lunt (Ch. 80; 207-334-2902, 207-334-2922).* Lunt & Lunt has gas, diesel, ice, and water at their wharf, with 3 or 4 feet alongside at low. Please do not bring trash ashore.

FOR THE CREW. Lobsters are available fresh or cooked at Lunt's Dockside Deli, which is open for lunch, dinner, and Sunday breakfast. Somehow, everything tastes sweeter here—the lobster, clam rolls, steamers, scallops, and the blueberry pie. The shed behind Lunt's dock is where Frenchboro Island Seafood once packed mason jars of lobster delicacies for Bloomingdale's gourmet market in New York City.

THINGS TO DO. Frenchboro is an island of simple pleasures and unspoiled natural beauty. Most of the island was held privately for generations. In recent years, however, a 900-acre tract of undeveloped land—more than half the island—was put on the market, threatening major social and economic changes for the small island fishing community. The Maine Coast Heritage Trust, the Island Institute, and the Maine Sea Coast Mission coordinated an emergency fundraising effort, and individual and corporate donors did the rest, raising $3 million to buy the land for what at the time was the single biggest conservation effort in Maine's history.

Beautiful trails lead in all directions to the various coves and beaches. For example, walk to the head of the harbor and continue south along the dirt road. Then branch off on a footpath through the woods to Little Beach, which you can reach in about 10 or 15 minutes. There you can gather beach peas, look for orchids in the nearby bog, or sit on a driftwood log while you picnic, watching lobsterboats work offshore.

For a longer walk of about 1½ hours, take the footpath from the ferry landing around Northeast Point to Eastern Beach, past green mosses, pink and green crab carapaces, and bleached skeletons of sea urchins, and then head back on the dirt road. If you are feeling energetic, walk from the ferry landing or Eastern Beach to Richs Head and the double-crescent beach at Eastern Cove.

Walking down the western side of the harbor from Lunt's, you will pass family graveyards set in the steep slopes between houses. At the head of the harbor is the fire pond, with an extended family of ducks. Several deer may come wandering up for a handout. The deer are loved and protected by the islanders and almost tame. They stand on their hind legs to pick apples off the trees right in town.

The little building of the Frenchboro Historical Museum is on the hill above the fire pond. The Historical Society opens the museum most afternoons to show their interesting displays of island memorabilia. It also contains the library and a small gift shop. Across the street is the playground and the school.

The school's enrollment is a good gauge of the strength of the community. It currently has 13 scholars, up from one in the late 1970s.

Blessings from above and below in Lunt Harbor.

EASTERN COVE

3 ★★★★

44° 06.68'N 068° 20.40'W

CHARTS: 13313, 13312   CHART KIT: 70, 22

Remote, unspoiled, and beautiful, Eastern Cove lies between Richs Head and the main part of Long Island. Richs Head is connected by a graceful tombolo with beaches on both sides.

Richs Head was settled by farmer William Rich in the early 1820s, and by 1850 the population had grown enough to warrant a school. All that remains today are seven cellar holes and stone walls.

APPROACHES. Easily identified by Richs Head, the cove is wide open, with no obstructions.

ANCHORAGES, MOORINGS. Run in toward the center of the beach and anchor in 24 to 30 feet at low. The chart shows the bottom as rocky, but a plow anchor holds well here. Protection is good except from north to east, with little swell.

GETTING ASHORE. Row in to the cobble beach.

THINGS TO DO. The southern shore of Eastern Cove has open meadows with blueberry plants everywhere, but the deer will probably beat you to them. There are cranberries, too. Delightful!

It's a one-hour walk to Lunt Harbor, through mossy woods. The trail starts among the trees at the western end of the southern beach. Another trail leads along the shore from the beach to the cliffs on the eastern side of Richs Head.

MOUNT DESERT ROCK
43° 58.22′N   068° 07.70′W

CHARTS: *13313, 13312*   CHART KIT: *22*

One of the most likely places to see whales along the coast is off Mount Desert Rock, where the great, gentle creatures come to feed in the upwelling waters. This is a lonely place indeed, almost 15 miles from the nearest land. The lighthouse here stands on three acres of low-lying rock, often shrouded in fog. The light is automated, but you will still find human inhabitants here: researchers and volunteers of Allied Whale (*www.coa.edu/alliedwhale*), a project of College of the Atlantic in Bar Harbor. In 1996, after decades of using the Rock as a research station, COA acquired the island and buildings from the Coast Guard as part of the Maine Lights Program. Often they will answer VHF calls to "*Petrel.*" A heavy Coast Guard mooring, labeled "CG," makes it possible to tie up and go ashore, landing your dinghy on the ways.

Pick a day with good visibility and a reasonable weather forecast. Any of the harbors on Mount Desert Island are good points of departure, but it's a shorter run from Burnt Coat Harbor or from Frenchboro, Long Island.

When you set out toward the empty horizon, Frenchboro will drop astern, and you will feel as if you have left the world behind. Before long, all land will disappear. On a typical hazy day, you may be within 3 miles of Mount Desert Rock before you glimpse its lighthouse.

Be prepared to be disappointed. After all, whales are not sighted every day. But also be prepared to see a great, black glistening shape rise suddenly from the water close aboard and pass down your side without a sound. Even pilot whales are impressive enough, and the great whales may dwarf your boat. Commonly sighted whale species are humpback (*Megaptera novaeangliae*), finback (*Balaenoptera physalus*), and northern right (*Eubalaena glacialis*) whales.

Whale-watching vessels may be near the rock, and perhaps some smaller boats. Listen for the chatter about whales on various VHF channels. Watch for a sudden plume of mist shooting skyward and a great humpback or finback whale rising majestically to the surface. Followed by a little fleet of observers, the whale will swim in leisurely pursuit of food, sounding occasionally and showing the white undersides of its flukes, then surfacing and blowing again. See *Whale-watching Guidelines,* page 295.

BLUE HILL BAY

On the west side of Mount Desert Island, Blue Hill Bay has long stretches of sparkling, wide-open waters, protected from ocean swells by the many islands at its southern entrance. Although less dramatic than neighboring Frenchman Bay, it offers great sailing, with alluring Blue Hill ahead and the hills of Mount Desert to the east.

This magnificent sailing territory has relatively few good harbors. Blue Hill Harbor, at the northwest corner, is the jewel of the bay and home to the Kollegewidgwok Yacht Club. In the center, Long Island is the largest island off the coast of Maine without a year-round population. At the southeastern end of the bay, Bass Harbor on Mount Desert is the base for a substantial fishing fleet and also offers complete services for yachtsmen. Pretty Marsh Harbor and several little coves lie along the west coast of Mount Desert Island and the eastern shore of Bartlett Island.

Much of the beauty of Blue Hill Bay will be forever preserved. The string of islands dividing the southern end of the bay—Bar, Trumpet, Ship, and the Barges—have important nesting colonies of eiders and are owned in part by The Nature Conservancy. Maine Coast Heritage Trust conservation easements protect 430-Acre Tinker Island. Bartlett Island, off the west shore of Mount Desert, was bought by the Rockefeller family to preserve it from development. The Penobscot Indian word *kollegewidgwok* says it all. It means "blue hill on shining green water."

ALLEN COVE
3 ★★

44° 17.77′N   068° 32.70′W

CHARTS: *13316, 13312*   CHART KIT: *71*

Allen Cove is on the west side of Blue Hill Bay between Tinker and Long Island. It is a large, open bay offering good protection from prevailing winds.

APPROACHES. There are no obstructions except the ledges at Harriman Point.

ANCHORAGES, MOORINGS. The cove shoals gradually. Anchor near the middle in mud and grass.

WHALES AND WHALE-WATCHING

The great whales are the largest mammals that have ever lived—land creatures that returned to the sea eons ago. Their brains are larger than man's, and one species sings a song whose eerie notes may be channeled for thousands of miles along deep acoustic passages in the sea.

Cetaceans (from the Greek word meaning "sea monster") include whales, dolphins, and porpoises. Out of more than 85 species, six are often seen in the Gulf of Maine, including three of the great whales—the finback, the humpback, and the minke. The smaller pilot whales are also common, as are harbor porpoises and white-sided dolphins.

Four other species are also seen occasionally: the rare right whale, the killer whale (orca), and common and white-beaked dolphins.

The great whales are usually seen near the fishing banks, where upwelling waters bring food in the form of tiny crustaceans, krill, squid, and small fish. These are filtered from enormous gulps of seawater by baleen sieves that hang from the upper jaws of most large whales. (Of the great whales, only the sperm whale has teeth).

Not unlike the human species, some whales go south in winter, to breed in Florida and the Antilles. They come north again in the spring to their feeding grounds in the Gulf of Maine.

Man has hunted whales for thousands of years, eaten their blubber, and used their oil to light his lamps. For a long time, it was a fair fight, and men in their small boats risked their lives to catch a few of the great leviathans. Then the demand for whale oil, the growth of fishing fleets, and technological developments such as engines and explosive harpoons began to turn the game into a slaughter.

Many species of whales were hunted to the edge of extinction. The North Atlantic right whale is the most endangered large whale in the world. It got its name because these were the "right" whales to hunt—they were easy to approach and floated after they were killed. Only about 350 of these whales survive.

Finally, the discovery of petroleum sources for oil and the advent of the electric light reduced the market for whale oil. And a growing awareness of the intelligence, gentleness, and beauty of these great creatures brought environmentalists and finally the public to the defense of the whale. The major countries involved in hunting whales agreed to establish quotas, and whaling came under regulation by the International Whaling Commission in 1946. Progress has been slow and frustrating, but today the factory ships of Russia and the United States are gone. However, Japan, Norway, and Korea still hunt whales in spite of the growing strength of public opinion. In the powerful words of Chief Seattle, spoken in 1884, "What happens to the beasts will happen to man. All things are connected. If the great beasts are gone, man would surely die of a great loneliness of spirit."

Six whales may be sighted in the Gulf of Maine, shown in the same scale as a 43-foot cruising ketch. Waterlines indicate the part of the whale usually seen above water. From left to right: humpback, orca (killer), minke, finback, right, and pilot whales.

A whale surfaces off lonely Mount Desert Rock. (Keary Nichols, MPBN)

In Maine, you are most likely to spot whales off Matinicus Rock, Seal Island (southeast of Isle au Haut), and Mount Desert Rock. Canada's Grand Manan also is a likely whale-watching site. From the shore, whales can be sighted from lighthouses such as Bass Harbor (on Mount Desert Island) and West Quoddy Head, though you might be surprised anywhere. In 1993 a True's beaked whale, one of only 24 ever sighted, beached itself on Curtis Island at the mouth of Camden Harbor.

There are lots of clues to use in identifying whales. Size is important, of course, as is silhouette. The shape of the spout may give you positive identification, or if you get close enough, distinctive behavior may be conclusive. Experienced whale-watchers can recognize individuals by details such as the color pattern, shape of a fluke, or even by the callosities on a right whale's head.

In his *Field Guide to the Whales, Porpoises and Seals*, Steven Katona notes that whales sometimes sleep on the surface, but there is little likelihood of hitting one at night or in the fog. Whales occasionally approach boats, but according to Katona, "We have never heard of a case where an undisturbed whale approached and damaged a vessel in our waters." On the other hand, he points out, "Vessels coming too close to a feeding or jumping whale have been struck, resulting in narrow escapes. Give the animals a wide berth."

WHALE-WATCHING GUIDELINES

Because of the growing interest in whale-watching and the increasing number of boats following whales, these guidelines have been developed by the National Marine Fisheries Service (NMFS) and others to prevent harassment and damage to whales and danger to the boating public.

· When operating within a quarter of a mile of whales, avoid excessive speed or sudden changes in speed or direction. Don't rev your engines.

· When close to whales (within 300 feet) don't approach stationary whales at more than idle speed, don't approach moving or resting whales head-on, and don't get between a mother and her calf. Instead, parallel the course and speed of moving whales. Right whales are particularly slow moving, sometimes even stationary at the surface, so they are particularly prone to being hit by boats and ships.

· Do not intentionally approach within 100 feet of whales. If they come within 100 feet of your vessel, put engines in neutral and don't reengage props until whales are observed at the surface, clear of the vessel.

· Active whales require ample space. Breaching, lobtailing, and flipper slapping may endanger a vessel. Feeding whales often emit subsurface bubbles to concentrate their prey before rising to feed at the surface. Stand clear of light-green bubble patches.

· Diving in the vicinity of whales is not advised because of their active and unpredictable behavior. Divers should not approach within 100 feet of whales.

· In all cases, do not restrict the normal movement or behavior of whales or take action that may evoke a reaction from them or result in physical contact.

STRANDINGS

If you sight any marine mammal in distress, including seals and whales, please report it to the Maine Marine Animal hotline at 800-532-9551.

WHALE-WATCHING EXPEDITIONS

If you want professional help in locating whales, whale-watching expeditions depart regularly in season from Portsmouth, Kennebunkport, Portland, Boothbay Harbor, New Harbor, Port Clyde, Rockland, Camden, Bass Harbor, Southwest Harbor, Northeast Harbor, Bar Harbor, Jonesport, Cutler, and Lubec, as well as Grand Manan.

BLUE HILL HARBOR

4 (outer harbor) ■ (inner harbor)

★★★★

outer harbor: 44° 24.49′N 068° 33.92′W
inner harbor: 44° 24.42′N 068° 34.67′W

CHARTS: 13316 (INSET), 13312 CHART KIT: 72 (A)

Year-round residents and summer people have cared about Blue Hill for a long time. The old clapboard houses are freshly painted, the lawns are tended, the market and the hardware stores are housed in handsome old buildings, the imposing town hall rises high on a grassy slope, and the public library stands on a quiet block. A pleasing selection of restaurants and attractive little shops offer quilts and antiques.

Blue Hill is also the peninsula's cultural center, with theater, dance, art, and music.

Impressive elm trees form the backdrop for this little town. Many are enormous, and all of them are healthy. They were saved by the efforts of a dedicated warden and a community united in its determination to preserve these magnificent elms.

Blue Hill was settled in 1762. The first industry was shipbuilding, which was later replaced by the mining of copper and the quarrying of granite.

In summertime, Blue Hill is jammed with people, threatening the very tranquillity and charm they seek, but it remains a most delightful town to visit.

For boaters, this requires a little doing. The center of town is at the very western end of the inner harbor, which dries out at low. You can dinghy to the town dock—or from it—only two hours either side of high tide. Otherwise, you will be exercising your legs from the yacht club or the rugged landing on Peters Point (see *Getting Ashore*, below).

APPROACHES. See chart next page. Blue Hill Harbor has inner and outer portions, both inside Sculpin and Parker points. Once past the ledges of Closson Point, run along the northern shore and north of the large ledge near the entrance, leaving cans "1" and "3" well to port. Then pass through the narrow entrance at Sculpin Point between can "5" and nun "6." There is a substantial current here at midtide.

The Kollegewidgwok Yacht Club is in the outer harbor, on the eastern shore, opposite Peters Point.

The Penobscot Indian word kollegewidgwok *says it all. It means "blue hill on shining green water."*

After leaving nun "6" to starboard, head for the moored boats and the gray, flat-roofed clubhouse.

ANCHORAGES, MOORINGS—Outer Harbor. The yacht club has fifteen reservable guest moorings for rent. If none is available, there may be room to anchor south of the moored boats in 17 to 24 feet, or try the inner harbor.

APPROACHES—Inner Harbor. Beyond Peters Point, the inner harbor provides even better protection and gets you considerably closer to the town of Blue Hill. There are a number of small craft and fishing boats moored here, but you might find room to anchor. Enter for the first time at midtide or below, when all the dangers are visible. It looks a bit daunting, but it's easy if you trust the buoys. Leaving can "7" close aboard to port and watching for crosscurrents, head for nun "8." Turn sharply around nun "8," leaving it to starboard, and head for the northernmost spire in Blue Hill.

ANCHORAGES, MOORINGS—Inner Harbor. Continue until you reach the moored boats, then turn westward among them. Anchor in 14 to 20 feet at low, in a good mud bottom. This part of the inner harbor is completely landlocked and makes a good hurricane hole. You are surrounded by rocky islets, with the masts of the yacht club to the east and the steeples of Blue Hill to the northwest—a lovely spot. Seals bask here undisturbed by your approach, and there are ospreys.

GETTING ASHORE. If you are at the yacht club, row in to the dinghy floats. From the inner harbor, dinghy ashore to the muddy beach just north of the private dock and stone wharf, the old steamer landing, on Peters Point. This land is private but may be used to access the mooring field. If you time the tides, you can take the dinghy all the way into town and land at the ramp next to the town wharf. There is adequate depth two hours either side of high.

FOR THE BOAT. *Kollegewidgwok Yacht Club (Ch. 09; 207-374-5581).* Gas, diesel, water, and ice may be obtained at the floats, with 10 feet alongside at low.

Raynes Marine Works (207-374-2877). Located in a cove just west of Peters Point, this small yard can haul boats up to 32 feet for storage and for hull, rigging, and engine repairs. The cove dries out at low, so call from the yacht club.

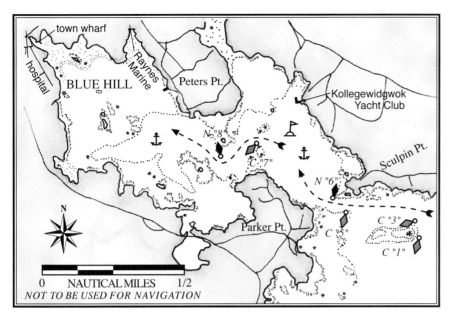

A lumberyard and hardware store are also tucked in along Main Street. A FedEx box is diagonally across from Merrill & Hinckley, near Blue Hill Travel.

Directly across from the town wharf is the Blue Hill Memorial Hospital (*374-2836*), which, along with St. Andrews Hospital in Boothbay Harbor, must rank as one of the most water-accessible medical facilities anywhere.

THINGS TO DO. A pleasant little grassy park is on the water, right next to the town wharf, with picnic tables and a small beach. A visitors' center, also by the wharf, is the perfect stop to discover local events.

The hill for which the town and harbor are named is 940 feet high, with open blueberry fields, spruces, and granite ledges. The magnificent views are well worth the climb. The distance from the post office to the base of the trail is 1.25 miles. Walk up Route 15 (north) about .75 miles and turn right on Mountain Road. You'll see the watchtower from the intersection. Continue half a mile to a sign that says, "Trail to Tower." It's a steep woodland trail, often masquerading as a stream bed. Allow 1¼ hours to get to the top and back to the base of the trail.

As you near the summit, a clearing provides wonderful vistas west toward the northern end of Penobscot Bay and southwest to the Camden Hills. There are also clearings on top, from which you can see south to Isle au Haut, southeast down Blue Hill Bay, and east to Mount Desert. If you are feeling adventurous, climb the stairway of the watchtower whose sign reads "Visitors Welcome—At Your Own Risk."

On most Saturdays, the Blue Hill Heritage Trust (*374-5118*) offers guided hikes, farm tours, and nature walks.

The Jonathan Fisher Memorial, a lovely early-19th-century home built by one of the town's most illustrious residents, is worth a visit. Among the building's attractions are fascinating inventions attributed to the multi-talented Fisher.

Blue Hill supports a rich and varied music scene. Even before you arrive you can tune into Blue Hill's

FOR THE CREW. The Kollegewidgwok Yacht Club has a pay phone. To get into Blue Hill, walk up the yacht club road and turn left. It's 1.7 miles from the club into town, but you probably can arrange a ride.

🛒 The Merrill & Hinckley grocery and fish market (*374-2821*) is close to the town wharf, up the hill on Union Street, which is Route 177. Children love to shop here, pushing their miniature shopping carts down the aisles. Grown-ups do too—the market is also a liquor store and has an ATM. For a small fee, orders can be taken over the phone and delivered to the yacht club.

Fresh breads and whole-meal takeout can be picked up at Pain de Famille, on Main. The Blue Hill Food Co-op, north along 177N, sells organic grains and cereals and fresh produce. A laundromat is around the corner from the Co-op. The Tradewinds (*374-5137*) supermarket and a Rite Aid, .6 miles out of town at the intersection of Routes 15 and 175, might be better for major provisioning.

Turn right from the town wharf onto Main Street to reach the post office and several restaurants. 🍴 Fishnet Seafood on Main, heading north, fries up classic takeout. The Wescott Forge (*374-9909*) has more formal dining and a casual bisto and bar in an old forge building on Main Street by a small stream. The Blue Moose (*374-3274*) offers family dining for lunch and dinner and even hosts cooking classes. For elegant dining, try The Arborvine (*374-2119*), on Main up Tenny Hill as you head toward the Tradewinds Market.

wonderful community radio station WERU, 89.9 FM (*469-6600*) to hear listings of local events or to simply enjoy great music. The station began broadcasting in 1988 from a studio known as the "henhouse" in a converted chicken coop of Noel Paul Stookey, "Paul" of Peter, Paul, and Mary fame.

The renowned Kneisel Hall School for String and Ensemble Music, based here each summer, presents twice-weekly concerts on its campus at the edge of town. Call *374-2811* for the schedule. The unique Bagaduce Music Lending Library is a short walk out of town on Main Street. It lends vocal and instrumental music to music lovers worldwide. Flash in the Pan, a community steel drum band, is often playing nearby.

The Blue Hill Fair has been an annual event for a century or so, and it draws people from all over the state in early September. This is a real country fair, reputed to be the inspiration for E.B. White's *Charlotte's Web*, with oxen straining to pull the heaviest loads, harness racing, sheep trials, and many other traditional events. The fairground is 1.3 miles east of the post office, on Route 172.

The brick Blue Hill Public Library is a couple of blocks south. Hanging on a wooden stand by the fireplace is an extraordinary relic—a chain-and-plate vest reputedly worn by Ferdinand Magellan when he was killed in the Philippines in 1521 during the first circumnavigation of the world.

The yellowed exhibit card in the library reports that a General Rumbough, with the U.S. force that conquered the Philippines during the Spanish-American War, searched the beach where Magellan was killed and found numerous artifacts. Most were given to New York's Metropolitan Museum of Art, but the fanciest piece of armor was given by General Rumbough to the Blue Hill Public Library.

Carroll Connard, married to General Rumbough's granddaughter, has written us to clarify a few points. David Rumbough, having graduated from West Point in 1917, acquired the vest while serving as Chief of Staff to the commanding general in the Phillipines in the early 1930s, not during the Spanish-American War. Mr. Connard has not been able to verify its authenticity or the existence of the companion pieces. Beyond those facts, the waters get murky. "Exactly where and how he acquired it," Mr. Connard writes, "and with what provenance is not known."

The gift of Blue Hill Bay's Long Island was accepted by Acadia National Park with one stipulation: the buffalos had to go.

McHEARD COVE

3 ★★★

44° 24.72′N 068° 31.15′W

CHARTS: *13316, 13312*   CHART KIT: *72*

About two miles east of Blue Hill, McHeard Cove runs north to the village of East Blue Hill. The inner half of the cove dries out at low tide.

APPROACHES. After passing east of Darling Island, enter McHeard Cove, staying on the western side. The ledges and islands on the east side are generally obvious and easy to avoid.

ANCHORAGES, MOORINGS. Webber's Cove Boatyard maintains several moorings near the 9-foot spot on the chart next to a little white granite island. There is exposure to the south and southeast.

GETTING ASHORE. Land at the boatyard floats if there is enough water. Otherwise, land at the ramp at the playground just before you reach the yard.

FOR THE BOAT. *Webber's Cove Boatyard (207-374-2841).* At the head of McHeard Cove, the floats of the boatyard are dry at low water. There is a 15-ton boatlift, and repairs can be made. Limited marine supplies are available, but no fuel or water.

Longfield Dory Co. (207-374-5656). Just west of the boatyard, Longfield builds dinghies and can also handle wood and fiberglass repairs.

FOR THE CREW. Webber's Cove Boatyard can supply lobsters. The post office and public library in this peaceful village are at the entrance to the boatyard. The nearest groceries and shops are in Blue Hill, about 3 miles away.

MORGAN BAY

44° 26.12'N   068° 28.98'W
CHARTS: 13316, 13312   CHART KIT: 72

Morgan Bay lies west of Newbury Neck at the northern end of Blue Hill Bay. A confusion of islands and ledges pepper the entrance, but once past them, the bay becomes wide and open all the way to its head. To enter Morgan Bay, leave Conary Nub to starboard.

Newbury Neck, to the east, is largely wooded and undeveloped, and there are attractive houses and meadows along the western shore. At least one bald eagle makes its home here.

UNION RIVER BAY

44° 27.00'N   068° 26.60'W
CHARTS: 13316, 13312   CHART KIT: 72

Above Bartlett Island, Union River Bay runs six miles northward toward the large town of Ellsworth. More than a mile wide and unobstructed, the bay is a pleasant diversion if you like to sail with no particular destination in mind. Attractive saltwater farms line the shores along Newbury Neck, and the two radio towers shown on the chart are farther north.

Union River Bay can get rough when strong southwesterlies blow against an ebb tide. At the northern end of the bay, the narrow and shoal Union River, dredged in 2001, leads to busy Ellsworth. There is a reasonable anchorage in Patten Bay.

PATTEN BAY

3 ★★

44° 29.47'N   068° 28.91'W
CHARTS: 13316, 13312   CHART KIT: 72

Patten Bay opens off the northwestern corner of Union River Bay and leads to the town of Surry. The handsome south shore is steep and wooded, but there are a number of houses along the north shore.

APPROACHES. Leave nun "2" at the northern end of Union River Bay to starboard and favor the deeper water along the north shore of Patten Bay, heading for the spire in Surry.

ANCHORAGES, MOORINGS. Anchor in 9 to 14 feet at low along the north shore. Protection is good, although southwest winds may curl around the corner.

THINGS TO DO. Music lovers should check the schedule of the Surry Opera House (*667-2629*) for an eclectic performance.

CONTENTION COVE

44° 29.45'N   068° 28.15'W
CHARTS: 13316, 13312   CHART KIT: 72

On the northern side of the entrance to Patten Bay, little Contention Cove makes a nice daystop and provides access to the 🍴 Surry Inn for dinner. Anchor outside the moored boats in mud bottom. The Surry Inn (*667-5091, 800-742-3414*), built in 1834, is the second white house back from the road on the east side of the cove, with a large expanse of lawn in front.

The cove earned its name from an incident of typical Yankee stubbornness. It seems the farmers wanted the road to follow the most direct route from Blue Hill to Ellsworth, while the fishermen, naturally, wanted it to run along the head of the cove—hence, Contention Cove.

NORTHWEST COVE

44° 23.65'N   068° 21.85'W
CHARTS: 13316, 13312   CHART KIT: 72, 22

Western Bay forms the approach to Mount Desert Narrows. The shores are relatively low-lying and developed. Ahead, a stream of traffic crosses the bridge to Mount Desert Island. The unmarked shoals surrounding Alley Island lurk to the north.

There is very little here to attract the cruising boat, but in a pinch, Northwest Cove is easy to enter, reasonably protected, and the bottom holds well.

BARTLETT NARROWS

at Ledges Point: 44° 20.80'N   068° 25.08'W
CHARTS: 13316, 13312   CHART KIT: 71, 72, 23

Bartlett Island lies close to Mount Desert's western shore. With 2,500 acres, Bartlett is the second largest island in Blue Hill Bay. In 1973, David and Peggy Rockefeller bought the island to preserve it from resort development.

Three existing homes have been remodeled, and two new ones have been built. A herd of Simmental cattle grazes on the pastures coming down to the water near Birch Cove. Bartlett is one of many examples of the efforts of the Rockefellers to preserve the beauties of the Maine coast and to restore traditional island occupations. No camping is allowed on Bartlett without permission, and no fires are allowed above the high tide mark. The Hub at the northern tip of Bartlett is state owned.

From the south, run west of Folly and John Island, past West Point, into the Narrows. The rocks make out a long way from Ledges Point, so get well over to the Mount Desert shore as you approach Great Cove. The ledges are partially visible at all tides. Once through the narrows, stay clear of the shoal off Goose Marsh Point.

Boats are moored in the bight north of West Point, on the east shore of the Narrows. The water, however, is too deep for anchoring. Outfitters from Bar Harbor often use the public ramp on the north shore of the bight to launch scores of kayakers, some with little or no experience. Beware of kayaks throughout Bartlett Narrows, particularly in the fog.

GREAT COVE

3 ★★★

44° 20.95'N 068° 25.45'W

CHARTS: 13316, 13312   CHART KIT: 71, 72, 23

Great Cove is on the eastern side of Bartlett Island halfway through the Narrows. It's a good anchorage and easy to enter, with views of the hills across the way on Mount Desert.

APPROACHES. From the south, avoid the fish pens that may be moored off the Bartlett Island Farm and run up the middle of the wide cove. If you are coming from the north, be sure to stay well clear of the ledges off—what else?—Ledges Point before making your turn into the cove.

ANCHORAGES, MOORINGS. Anchor wherever you find swinging room in 15 to 20 feet at low. Holding ground is excellent, in mud, and exposure is only to the south and southeast.

GALLEY COVE

3 ★★★

44° 21.73'N 068° 25.15'W

CHARTS: 13316, 13312   CHART KIT: 71, 72, 23

This beautiful little cove is tucked into the eastern shore of Bartlett Island, just south of Galley Point. A small gravel beach graces its head, and it is crowned with views of the hills on Mount Desert.

APPROACHES. From either the south or north, run along the eastern shore of Bartlett Island toward Galley Point, which is mostly wooded with some clearings at the end. Enter in the middle of the cove.

ANCHORAGES, MOORINGS. There are private moorings in the cove close to shore. Anchor in the middle of the cove in 18 to 20 feet at low, in a mud bottom. There is protection here from all directions except the east and northeast.

MOUNT DESERT ISLAND

The Abenaki Indians called it *Pemetic*, "the sloping land." In 1524, Giovanni da Verrazano, a Florentine working for French and Italian bankers, was the first European to record the sighting of Mount Desert Island. Eighty years later, it was visited by Samuel de Champlain, representing Pierre du Guast, Sieur de Monts, who had been designated lieutenant general of New France by King Henry IV and granted a 10-year fur monopoly. Coasting down from a French settlement in Nova Scotia, Champlain named it *L'isle des Monts Deserts*, or "Island of the Barren Mounts." He struck a shoal off Otter Point and landed September 5, 1604 to repair his ship and to explore.

With the assassination of Henry IV, de Monts lost royal favor, and his patent was bought by Antoinette de Pons, Marquise de Guercheville. This rich and virtuous lady founded the short-lived Jesuit mission on Fernald Point in Somes Sound in 1613. It was destroyed by Captain Samuel Argall, under British orders to clear the coast of Frenchmen.

In 1688, Antoine Laumet successfully petitioned Louis XIV for an enormous grant of land in New France, including Mount Desert Island. Adopting the title of Sieur Antoine de la Mothe Cadillac, he brought his bride to Mount Desert to establish a feudal estate. This romantic scheme failed for unknown reasons, and Cadillac went on to found Detroit, but his name lives on here at Cadillac Mountain, the tallest mountain on the Atlantic coast north of Brazil.

With the conquest of Quebec by General James Wolfe in 1759, the British finally won their long struggle with the French, and the lands of New France were up for grabs. Governor Francis Bernard of Massachusetts coveted Mount Desert and persuaded Abraham Somes and other families to colonize the island in 1761, but the Revolution nullified Bernard's claims. After the Revolution, the General Court of Massachusetts split the island between two claimants. The western half went to Bernard's son, the eastern half to Marie Theresa de Gregoire, granddaughter of Cadillac.

In the mid-19th century, Mount Desert became fashionable. Hotels sprang up as bases for campers, naturalists, and artists. Rusticators, as the summer people were labeled, built camps, and the wealthy started to build imposing "cottages." "The Briars" was built in 1881 for J. Montgomery Sears of Boston, "Stanwood" in 1885 for James G. Blaine of Maine, "Casa Far Niente" for S. Weir Mitchell of Philadelphia, "Chatwold" in 1894 for publisher Joseph Pulitzer of New York, and so on, each trying to outdo the others. The roster of people who summered on Mount Desert included Vanderbilts and Fords, Carnegies and Astors, Rockefellers and Morgans. But by the time E.J. Stotesbury built his 80-room mansion in 1925, the Great Depression was just over the horizon,

---

ACADIA— FOR THE ENJOYMENT OF ALL PEOPLE

In an old photograph, George B. Dorr stands on the shore of Jordan Pond, dressed in baggy tweeds with a three-button jacket, bow tie, and cap. A canoe is drawn up, and Dorr is talking to Charles W. Eliot, former president of Harvard. These men and others like them—wealthy, educated, and public-spirited—had the vision of preserving this grand island from woodsmen and developers for the enjoyment of all people.

Dorr was a doer as well as dreamer, and he persuaded others to share his dream and donate land and money. He was also a politician, able to facilitate the sticky process of donating private lands for public use. The first association was Hancock County Trustees of Public Reservations. When its tax-free status was challenged, Dorr decided to aim for a national park. At a time when too many such proposals were overwhelming Congress, the first 6,000 acres became Sieur de Monts National Monument in 1916, requiring only the president's approval.

With further pressure from Dorr, it finally became Lafayette National Park in 1919. When much of Schoodic Peninsula was added 10 years later, potential donors of land objected to the name, and in 1929 Congress was persuaded to change it to Acadia National Park.

One of the major land donors was John D. Rockefeller, Jr., who gave almost one-third of the park's present acreage. He is also remembered as the man who conceived the network of carriage roads crisscrossing the island, so horses and carriages and hikers might be insulated from the automobiles that threatened to invade the island. (They were not allowed until 1913.) He did it right, inviting Frederick Olmstead (son of the designer of New York City's Central Park) to lay out the roads and design the rustic stone bridges. Today, they've been discovered by cross-country skiers and mountain-bikers, who can easily climb their modest grades to discover breathtaking vistas of the mountains, hills, and ocean.

EXPLORING THE PARK BY BOAT

Acadia National Park is easy to explore from any of the major harbors. A free Island Explorer shuttle bus can take you anywhere in the park from Bass Harbor, Southwest Harbor, Somesville, Northeast Harbor, Seal Harbor, and Bar Harbor.

The following is a list of the harbors nearest major park attractions. For more details, see the individual harbor or contact: Superintendent, Acadia National Park, Bar Harbor, Maine 04609; 207-288-3338; www.acadiainfo.com or www.nps.gov/acad.

MOUNT DESERT ISLAND
Bass Harbor: Bass Harbor Head Lighthouse, Ship Harbor Nature Trail.

Valley Cove: Flying Mountain, Man o'War Brook, St. Sauveur Mountain.

Northeast Harbor: Asticou Terraces, Thuya Lodge and Gardens, Eliot Mountain.

Seal Harbor: Jordon Pond House, Jordan Pond Trails (Sargent Mountain, Pemetic Mountain, The Bubbles), carriage roads.

Bar Harbor: Cadillac Mountain, Sieur de Monts Spring, The Abbe Museum, Wild Gardens of Acadia, Thunder Hole, Park Loop Road, Natural History Museum, Jackson Laboratory, Bar Island, Sand Beach, Otter Cliffs.

CRANBERRY ISLANDS
Islesford: Islesford Historical Museum.

Baker Island: fields and trails, lighthouse.

SCHOODIC PENINSULA
Pond Island: Schoodic Head, Schoodic Point.

ISLE AU HAUT
Duck Harbor: hiking trails, Duck Harbor Mountain.

the automobile was opening up Mount Desert to the public, and the era of grand "cottages" was coming to a close.

The *coup de grace* came less than two decades later. A sign in Acadia National Park relates the sad tale: "In October 1947, a series of fires lasting 26 days ravaged more than 25 square miles of Mount Desert Island. Bar Harbor was severely threatened, and most of the landscape in front of you was transformed into an apparent wasteland. The fire consumed 170 homes of year-round residents. Over 60 summer mansions burned, leaving only chimneys and garden statues standing. One third of the park woodlands burned before the flames died at the ocean's edge."

ACADIA NATIONAL PARK

Acadia National Park is visited by nearly three million people each year. It is one of the smallest national parks, but one of the most popular. The park covers large areas of Mount Desert Island, Schoodic Peninsula, Isle au Haut, and parts of the Cranberry Islands.

Thanks to the efforts and foresight of some determined individuals, the 45,000 acres of the park cover some of the most spectacular scenery in Maine. The first rays of the morning sun to touch this country brighten the summit of Cadillac Mountain. There are coastal drives, sheer cliffs, calm lakes between wooded hills, natural springs, wild and formal gardens, and even a tea house at Jordan Pond. In addition to a museum of stone-age artifacts, there are natural walks, woods and wildlife, sea birds and ocean beaches, roads leading to the mountaintops, and 45 miles of carriage roads winding everywhere for hiker, bicyclist, horse-drawn carriage, and rider.

People come from everywhere to see this beautiful island, this granite-hard, fog-softened, primeval meeting place of land and sea. It is a landscape that appeals to many; Acadia is one of the nation's most visited parks.

The park's Visitors Center is located near the village of Hulls Cove, in the northeastern corner of Mount Desert Island. They have maps, information on naturalist-guided tours, and a short film about the park. Also at the information center is a trail head to get on the majestic carriage trails.

PRETTY MARSH HARBOR

3 ★★★

44° 20.15'N 068° 24.70'W

CHARTS: 13316, 13312   CHART KIT: 71, 23

On the west side of Mount Desert Island, inside Bartlett Island, Pretty Marsh Harbor provides good protection in a tranquil setting. To the southwest there are islands beyond islands, with handsome buildings and meadows on Hardwood Island in view.

APPROACHES. Approaching from the south, run inside Hardwood Island. Hardwood is 113 feet high and wooded. Leave Folly Island, which is low and bushy with a few trees, well to port. The ledge about 250 yards east of Folly is covered 3 feet at low. Entering Pretty Marsh Harbor, favor the eastern side to avoid the shoal extending 350 yards southeast of West Point.

ANCHORAGES, MOORINGS. Anchor in the middle of the harbor in 8 to 21 feet at low. Holding ground varies from hard to good mud. Make sure the hook is in. Protection is good all around except in a heavy southwesterly.

THINGS TO DO. On the east shore of the entrance to the harbor, opposite the end of West Point, stairs lead up from the water to a rustic gazebo. You can land the dinghy there and walk through a stately grove of spruce and cedar, a national park picnic area.

The little tickle of water leading off to the northeast is fun to explore by dinghy.

SOMES COVE

4 ★★★

44° 19.21'N 068° 25.03'W

CHARTS: 13316, 13312   CHART KIT: 71, 23

Somes is a tiny cove on the west side of Mount Desert Island, opposite the southern tip of Bartlett Island. Anchorage and protection are good.

APPROACHES. Two islets sit at the southern corner of the entrance, the outer one low and rocky, the inner one bushy. Enter north of the islets, down the middle of the cove.

ANCHORAGES, MOORINGS. Anchor just west of a line between the dock on the north shore and the dock on the south shore, in 12 to 20 feet at low. The only exposure is to the north. The bottom is mud and grass, with good holding.

SAWYER COVE

4 ★★★

44° 18.80′N 068° 25.47′W

CHARTS: 13316, 13312   CHART KIT: 71, 23

Sawyer Cove is an attractive small anchorage opposite the northern tip of Hardwood Island, with excellent protection from prevailing winds. A bald eagle may sometimes be seen soaring above the 100-foot bluff that forms the southern entrance.

APPROACHES. The large ledge in the mouth of the entrance is visible almost until high water, and some kind soul has built a small rock cairn on its eastern part. It can be passed to either side, but the approach to the south is wider.

ANCHORAGES, MOORINGS. Anchor in 8 to 21 feet at low, with good mud holding ground. The only exposure is to the north.

SEAL COVE

1 ★★★

44° 16.87′N 068° 24.58′W

CHARTS: 13316, 13312   CHART KIT: 71, 23

Seal Cove is on the west side of Mount Desert opposite Tinker Island. The farther into Seal Cove you get, the prettier it becomes. Flanked by Murphy Hill and Robbins Hill, with Bernard Mountain in the background, the innermost section is a delight.

APPROACHES. Dodge Point can be identified by the conspicuous white boulder on the shore just south of the point. Moose Island is wooded in the middle, with meadows around the southern shore. Run up the middle of Seal Cove toward the estuary at the head of the cove, staying well clear of the ledges east of Dodge Point.

ANCHORAGES, MOORINGS. Anchor in 6½ to 14 feet at low among the moored boats. Holding ground is good, in mud. The cove is exposed to the west and southwest, but its protection is excellent from the north around to east and from the south.

GETTING ASHORE. Land at the public launching ramp on the north shore.

THINGS TO DO. Explore the estuary with your dinghy toward Seal Cove Pond, or walk along Seal Cove Road, which skirts the harbor.

GOOSE COVE

1 ★★

44° 15.28′N 068° 23.65′W

CHARTS: 13316, 13312   CHART KIT: 71, 23

Goose Cove lies a couple of miles north of Bass Harbor on the west shore of Mount Desert. The cove is open and easy to enter but exposed to the prevailing southwesterlies.

APPROACHES. The best way to identify Goose Cove is by the spire in West Tremont. The cove is south of Goose Cove Rock, which is whitish and grassy, and the entrance is between Dix and Nutter Point, both of which are wooded and extend a long way. Run up the middle.

ANCHORAGES, MOORINGS. Anchor outside the private moorings on the south side in 10 to 15 feet at low. Despite the "Rky" designation on the chart, there is mud bottom with good holding.

BASS HARBOR

1 (outer harbor)   4 (inner harbor)

★★

Outer harbor, off Morris: 44° 14.00′N 068° 20.95′W

CHARTS: 13316, 13318, 13312, 13313
CHART KIT: 71, 74, 23

Bass Harbor, at the southwestern corner of Mount Desert Island, is the home port for Mount Desert's largest lobstering fleet. The harbor's few yachting amenities are centered around Morris Yachts, at the harbor's mouth, and Up Harbor Marine, inside.

In 1990, a developer proposed building an 84-slip marina next to Bass Harbor's biggest lobster buyer. The fishermen were afraid they would lose another increasingly scarce piece of working waterfront. The Bureau of Public Lands, however, denied the developer the required subtidal leasing permit on the grounds that the marina impacted the established fisheries use, and that decision was upheld even after the developer sued the bureau.

The harbor is flanked by the towns of Bernard on the west shore and Bass Harbor on the east. The outer harbor is open to the south, but there is better protection inside. The ferry to Swan's Island and Frenchboro leaves from Bass Harbor on the east side. This is where Ruth Moore lived for 40 years and wrote most of her novels about life on the Maine coast, now popular once more. Among the best known are *Speak to the Winds* and *The Weir*.

APPROACHES. Bass Harbor Head Light marks the harbor's eastern shore, with a small white tower and connected building.

Coming from the west, there are no problems entering Bass Harbor. Weaver Ledge, at the entrance, is marked by a nun and a can at either end. From the east, the usual approach is across Bass Harbor Bar.

ANCHORAGES, MOORINGS. Morris Yachts maintains a number of rental moorings south of their dock on the eastern shore, identifiable by their large white polyball floats and tall pickup buoys. You can anchor in about 23 feet near the moorings in good holding ground, but it is exposed and likely to be rolly.

Morris also manages slips and moorings way up inside the harbor near nun "6." Up Harbor Marine, at the bend in the channel, has five 40-foot transient slips with depths of at least 8 feet. The dredged channel shown on the chart is deep enough but narrow, and full of moored lobsterboats. The rest of the inner harbor is ledgy and a lot of it dries out at low.

GETTING ASHORE. In Bass Harbor, take the dinghy to the Morris floats or to the public landing just north of the docks of the Swan's Island ferry. In Bernard land at Thurston's takeout.

FOR THE BOAT—Bass Harbor. *Morris Yachts (Ch. 09; 207-244-5511).* Morris Yachts is located near the mouth of the harbor, on the eastern, Bass Harbor side. Morris builds their yachts inland, but this is their full-service repair yard and the headquarters for their charter fleet. They haul with a 30-ton boatlift and can handle repairs of all kinds. Diesel, water, electricity, and pump-outs are available at the floats, with 15 feet alongside at low.

Bass Harbor Public Landing (Harbormaster Tim Butler: 207-244-4564). The floats at the public landing are on the east side, just north of the ferry dock, with 6 to 7 feet alongside at low and a two-hour maximum tie-up. Pay phones and a dumpster are available at the ferry terminal.

C.H. Rich Co. (207-244-3485). North of the ferry landing is the red brick crab cannery—now condos—and just north of that is the blue shingled building and lobster wharf of C.H. Rich. Gas and diesel are available alongside, but this is a busy commercial wharf.

Harbor Divers (Ch. 07, 16, 71; 207-244-5751). May you never need them.

FOR THE BOAT—Bernard. *F.W. Thurston Co. (Ch. 16; 207-244-3320).* The green-and-gray buildings of F.W. Thurston are on the Bernard side, part way along the dredged channel and just south of the big red fish plant. They are just past the yellow awnings on the wharf of the Thurston takeout. They cater primarily to commercial fishermen, selling gas and diesel at their fuel float, with 7 feet alongside at low.

Up Harbor Marine (207-244-0270; www.up-harbor.com). This little operation's gray-and-white buildings occupy the last dock and floats northward, on the Bernard side near nun "6." Water, electricity, and pump-outs are available at the floats, with 8 feet of depth at low.

FOR THE CREW. Morris Yachts has showers, a laundromat, and a phone.

The Maine-ly Delights takeout is next to Morris, hard by the Swan's Island ferry terminal. Be sure to save room for the famous, fried Oh-Boy Doughboy dessert. A quarter-mile walk north brings you to The Seafood Ketch restaurant (244-7463), which serves a full menu of seafood. Across the harbor in Bernard, you can buy lobsters live or to eat in the rough at F.W. Thurston (244-7600), and there is a fish market a short distance south of Up Harbor Marine.

THINGS TO DO. There are several antique shops in Bernard, on the west side of the harbor. The easiest way to get there is to row.

To visit the Bass Harbor Head Light, turn right after the post office onto Route 102A and continue until you see signs. The round trip is 3 miles, a delightful walk.

A short trail to the left takes you down to the pink granite ledges and provides a good view of the lighthouse, which is now privately owned. A paved path to the right goes to the lighthouse itself, with a great bronze bell mounted next to it, marked "U.S. Lighthouse Establishment 1891."

Both children and adults might like to visit Ravenswood (*244-9621*), which sells ship models and kits. Walk left from the ferry and turn right on McMullen Avenue.

East of Bass Harbor is Ship Harbor, which is part of Acadia National Park and has an interesting nature trail. To get there, walk out Route 102A toward the lighthouse but continue on 102A instead of turning off toward the light. After .7 miles, you will find the beginning of this self-guiding, 1.3-mile trail along shoreline paths, through the spruce, and past bogs with insectivorous pitcher plants.

Tired of all this walking? The free Island Explorer shuttle bus stops at the ferry terminal and will take you nearly anywhere on Mount Desert Island.

BASS HARBOR BAR

"WB": 44° 13.08'N 068° 20.47'W
"EB": 44° 13.23'N 068° 20.01'W

CHARTS: *13316, 13318, 13312, 13313*
CHART KIT: *71, 74, 23*

Bass Harbor Bar extends south from Bass Harbor Head on Mount Desert Island to Great Gott Island. A dredged east-west channel runs from red-and-white gong "WB," close to Bass Harbor Head Light, to red-and-white bell "EB" at the eastern end. Run a straight line the short distance between them.

The current floods west and ebbs east across the bar at an angle to the channel, so it tends to set you to one side or the other of the channel or right out of it altogether. Keep track of a rough bearing on the mark behind you so you can gauge it. Normally this is an easy passage, but a nasty chop can build up when the wind is against the current. With conditions quite benign just a short distance away, even 50-footers can pitch violently and dip their bowsprits as they cross the bar. Stow the lunch dishes ahead of time.

SOUTHWEST HARBOR

44° 16.40'N 068° 18.85'W

CHARTS: *13321, 13312, 13313, 13318*
CHART KIT: *74, 23*

Coast Guard: 207-244-5121

The entrance to Somes Sound is flanked by the great boating centers of Mount Desert— Southwest Harbor and Northeast Harbor. Both are working harbors, with an interesting mixture of fishing boats and handsome yachts.

There are two concentrations of boats in Southwest Harbor. Along the southern shore, in Manset, the well-known Hinckley yard provides a multitude of services for yachtsmen. On this side of the harbor, however, you will be a long way from the town of Southwest Harbor, but it can be reached via the Island Explorer shuttle bus. Manset will be considered separately, in the following entry.

The working fleet is concentrated along the north shore of Southwest Harbor on Clark Point, near the Coast Guard, the town dock, and many other facilities, a mile or less from the center of town. The extensive floats of Great Harbor Marina lie right at the head of the harbor.

Except for a channel along the northern shore, the harbor is full of moorings, and anchoring is not allowed. Southwest Harbor offers good protection except for winds out of the east.

Although Hinckley is probably the best-known name around here, half-a-dozen other builders are also based in Southwest Harbor, including two Maine boatbuilding legends, Jarvis Newman and Ralph Stanley. A broad range of marine services are also based here, from supplies to repairs to charters to custom fabrication.

APPROACHES. Southwest Harbor is easy to approach by either Western Way or Eastern Way.

Coming from the west, start at green gong "1" off Long Ledge and run north through Western Way, between Great Cranberry Island and Mount Desert Island. Keep South Bunker Ledge, marked by red daybeacon "2," to starboard. The start of the narrower portion is marked by nun "4" and can "5" marking shoal areas on either side. Thereafter, the passage is clear except for the nun on Cow Ledge, which you leave to starboard.

Eastern Way runs along the north shore of Sutton Island. Coming from the east, pass north or south of East Bunker Ledge, and then follow Eastern Way past several red marks and the lighthouse on Bear Island. Leave them all to starboard.

ANCHORAGES, MOORINGS. The town of Southwest Harbor puts out 6 moorings with white floats marked "rental." Call Harbormaster Gene Thurston (*Ch. 09, 16; 244-7913*) for details. Beal's Lobster Pier, indentifiable by the red buildings next to the Coast Guard station, also has a few rental moorings. Southwest Harbor's lower town dock lies next to Beal's, for temporary dockage. Adjacent Southwest Boat may have moorings or dockage available.

There may also be room to anchor among the moorings, although technically this is not allowed. Make sure you are not in the main channel to the north, and be prepared to move if asked to do so. Holding ground is good in about 10 feet at low. Exposure is to the east.

Great Harbor Marina sits at the head of the harbor. They rent slips to transients but have no moorings. Reservations are advisable. The Hinckley yard in Manset also has dock space and moorings (see *Manset*, page 308).

GETTING ASHORE. Dinghy to the lower town dock next to Beal's Lobster Pier. If you want to land closer to town, use the upper town dock, near the head of the harbor, next to Ralph W. Stanley's boatshop or the dinghy dock at Great Harbor Marina.

FOR THE BOAT. The Coast Guard station is at the end of Clark Point (*244-5517; emergencies: Ch. 16, 244-5121*).

Beal's Lobster Pier (Ch. 16; 207-244-3202). Next to the Coast Guard, Beal's has water, gas, and diesel at their floats with 12 feet alongside at low, as well as ice, marine hardware, and lobsters crawling or on the plate.

Southwest Harbor Town Dock (Harbormaster Gene Thurston: 207-244-7913), next to Beal's, has floats for limited, 2-hour tie-ups. The town also has several rental moorings.

Downeast Diesel & Marine (207-244-5145), lies beyond the lower town dock. Their floats are for repair work only, but they are the place to bring your boat to make your engine behave.

When we put in here with a jury-rigged fuel pump, owner John Spofford pulled one of the semi-exotic pumps from his spare-parts shelf and lent it to us for our trip home. He would take neither money nor my name. "But you'll need mine," he added, "so you can send it back."

Southwest Boat (207-244-5525). Identifiable by its large marine railway tucked next to Downeast Diesel, Southwest Boat specializes in steel boat work. Depending on their workload, they may have moorings or dockage, with depths of 10 to 12 feet, and pump-outs.

Hamilton Marine (207-244-7870). This branch of Maine's discount marine chandlery is located on Clark Point Road, across from the lower town dock. They stock or can order you just about any part you need.

Ralph W. Stanley (207-244-3795). Stanley, a well-known builder of wooden boats, is located near the head of the harbor. He has three marine railways that can handle boats to 15 tons, and the yard can perform complete hull repairs. In recent years, Ralph was flown to the White House to be named a "National Historic Treasure." Do you say thank you when somebody calls you that?

Great Harbor Marina (207-244-0117). Built on the site of one of Maine's sardine canneries, this large facility at the head of the harbor has numerous slips with big power, pump-out facilities, and plenty of depth. The fuel dock, around to

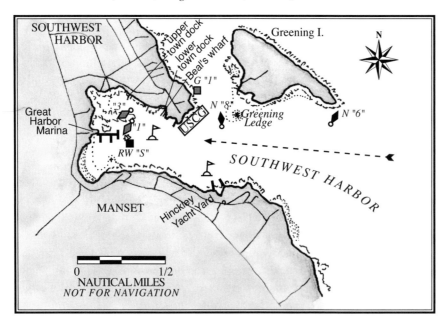

the left, pumps diesel. Showers and laundry and associated marine businesses (see below) are ashore.

Acadia Sails (207-244-5722). Located at Great Harbor Marina, this branch of Yarmouth's Maine Sailing Partners is a full-service loft, providing new sails, sail repair, and canvas work.

West Marine (207-244-0300). This branch of West Marine is on the left, just up from the docks at Great Harbor Marina.

McEachem & Hutchins (207-244-7243). This hardware store, at the main intersection in Southwest Harbor, carries a variety of marine items.

FOR THE CREW. Dumpsters and restrooms are located at the town docks.

You can eat outside on the deck at Beal's Lobster Pier, next to the lower town dock. Walk up Clark Point Road toward the town of Southwest Harbor, and you will pass several inns that serve meals. There are restaurants in town, within a mile or so from either town dock. Red Sky Restaurant (*244-0476*), on Clark Point Road as you enter town, serves an excellent menu of contemporary American cuisine. If you are at Great Harbor Marina, the Deck House Restaurant (*244-5044*) serves up everything from takeout to dinner theater.

It is also possible to sail around the peninsula that forms Southwest Harbor and anchor off the docks and floats of the Claremont Hotel (*244-5036*), opposite Greening Island. The hotel maintains guest moorings, and there is 8 feet alongside the floats. Lunch and cocktails are served in the informal waterfront Boathouse, and dinner is in the formal dining room. For dinner, coats and ties are required, and reservations are recommended.

Lobster and fresh fish are available at Beal's. The excellent, hardwood-floored Sawyer's Market, in the center of town, can take care of the rest if Little Notch Cafe and Bakery, next door, doesn't tempt you away. Sawyer's Specialties, across the street, can provide the extras, with wines, cheeses, fancy cigars, and a dozen varieties of olives. Sawyer's (*244-3315*) will deliver to the waterfront. So will Southwest Foodmart (*244-5601*), located about half a mile north of the center of town on Route 102. The Village Washtub laundromat (*244-7228*) is slightly closer, .2 miles north of town.

On Main Street in town, you will find the library, which has Wi-Fi, banks, galleries, art supplies, a hardware store, a natural foods store, and even an espresso bar. The post office has a FedEx box outside, and Tom Cat, Inc., across the street, has an excellent selection of newspapers. For children, there is a wonderful playground at the school, behind the recreational center.

The Southwest Harbor Medical Center (*244-5513*), a division of Ellsworth's Maine Coast Memorial Hospital, provides emergency care at all hours. Walk up Clark Point Road and turn right on Herrick Road at the Medical Center sign. The distance from the town dock is .8 miles.

The Southwest Harbor Chamber of Commerce can be reached at *244-9264*.

THINGS TO DO. The Mount Desert Oceanarium (*244-7330*) is right on the waterfront, west of the lower town dock, marked by a blue lobsterboat. Its building once was a "coal and vittlin' station" for coastal schooners. It's well worth a visit, especially for the children. Founder David Mills gives lectures on lobsters, and there are fish tanks and exhibits on commercial fishing, lobstering, and the causes of tides, and there is a touch tank where you can pick up sea cucumbers, clams, and other shellfish.

The superb Wendell Gilley Museum is north of town on Route 102, about 1.3 miles from the upper town dock. The late Wendell Gilley earned his living as a plumber, but while fixing oil burners and thawing pipes, he started carving birds. Completely self-taught, he carved barn owls and ospreys, ducks and herons. Samples sent to Abercrombie and Fitch in New York were well received, and Gilley put his son through college with the proceeds. After 27 years in the plumbing business, he turned to carving full time.

The handsome museum has an extensive collection of his work as well as a complete set of Audubon prints. They also show some good films and hold frequent classes in bird-carving. Even if you aren't a dedicated bird-lover, this is a fascinating place.

Bikes can be rented at Southwest Cycle on Main Street (*244-5856*). The ferry to the Cranberry Islands (*244-5882*) leaves from the upper town dock, and so does the water taxi (*460-3977*).

Free Island Explorer shuttle buses leave from Southwest Harbor and roam all over Mount Desert. This is an excellent way to get to places like Bar Harbor if you don't plan on putting in there by boat. They can also take you to trailheads, so you can explore Acadia by foot. If you are waiting, be patient. As our driver explained, "We're allowed to be late, we just can't be early."

MANSET (Southwest Harbor)
44° 16.23'N 068° 18.44'W
(For Charts, Harbor Ratings, and Approaches, see Southwest Harbor, above.)

Manset is on the southern shore of Southwest Harbor. On this side of the harbor, the well-known Hinckley yard provides a multitude of services for yachtsmen, although by land you will be a long way from the activities centered across the way.

Call them a cult or call them a group of sailors who *know* they own the best boats in the world, Hinckley owners are a breed apart. The company was started by Henry Hinckley in 1932, building wooden boats including Sou'westers and Pilots. By the late 1940s, Hinckley had become the nation's largest builder of cruising sailboats.

Then came fiberglass. Hinckley was one of the first manufacturers to learn how to use the new material. Their first glass model, starting in 1959, was the classic Bermuda 40.

Henry Hinckley sold the company to Canadians in the late 1970s, but in 1983 it was purchased back by Shep McKenney and Bob Hinckley, Henry's son. So the family tradition continues. The yard today limits production to 10 or 15 boats a year, including the B-40, the popular Sou'wester series, and, more recently, Talaria jet-powered picnic boats.

Still on the leading edge of technology, Hinckley is now building their hulls with a resin infusion process. In addition to molding hulls, Hinckley fabricates all its own spars, electrical panels, and metal parts—everything except the engine—with the emphasis, always, on quality.

ANCHORAGES, MOORINGS. Hinckley has the first very long dock on the south side of the harbor, marked with a little gray building at the end. The yard maintains a large number of moorings, and dockage for boats to 100 feet. There is also plenty of room to anchor off the moored boats in 20 to 30 feet of water at low.

The next dock to the west belongs to Manset Yacht Services. Their dockage is for water or loading only, but they do have rental moorings.

The Manset town dock is west of Manset Yacht, with temporary tie-ups for boats to 45 feet, plus water and electricity.

GETTING ASHORE. Take your dinghy in to the floats at Hinckley, Manset Yacht, or the Manset town dock. Hinckley has a launch service that responds to channel 9 or three toots.

FOR THE BOAT. *The Hinckley Company (Ch. 09 or 16; 207-244-5531; www.thehinckleyco.com).* This major boatyard provides diesel, water, ice, pump-outs, and electricity at the floats with 10 feet alongside at low. Hinckley has a marine railway that can handle boats to 65 feet, a 20-ton crane, and a 70-ton lift, and they still are expanding their service facilities, which include hull, rigging, electronics, metalworking, and engine repairs of all kinds. There is a well-stocked ship's store (*244-2553*).

Manset Yacht Services (207-244-4040). Manset Yacht has rental moorings and water at their floats for transients. They haul with a 40-ton boatlift and can handle general repairs, and they are a dealer for J-boats. Their real specialty is storage, and lots of it, in a huge, heated indoor storage shed.

Manset Town Dock (harbormaster: Ch. 09) allows limited tie-ups and has water and electricity.

Fortune, Inc. (207-244-9104). This loft of the large Falmouth sailmaker is at Manset Yacht Services.

Manset Dive Service (Ch. 09, 10, 16; 207-244-0469). Let's hope you don't need them.

FOR THE CREW. Showers, laundry, and a pay phone are right at the head of Hinckley's dock.

🍴 The Moorings Restaurant (*244-7070*) is right next door to Hinckley, and The Dockside Restaurant (*244-5221*) is directly opposite the town dock, just down Shore Road. XYZ Restaurant (*244-5221*), .5 miles from the town landing, serves superb regional Mexican food and a mean margarita. Follow Shore Road westward to Alder Lane, which tees with Route 102A (Seawall Road). Take a right a short distance to Bennet Lane, on the left side, and follow Bennet up to the restaurant.

🛒 Walk south about a quarter of a mile from Hinckley to the Double J Grocery & Deli (*244-5544*), at the intersection of Route 102 and look for their fresh-baked muffins.

From Hinckley to the center of Southwest Harbor via the shore road (Route 102A and then 102), it is 1.7 miles. If you don't want to walk, you can hop on the free Island Explorer shuttle bus at Hinckley. Or you can work up an appetite by rowing directly across the harbor.

THINGS TO DO. Once you've picked up the jaw you dropped when you were walking around the Hinckley yard, you might want to stroll over to the boatshop of Lee S. Wilbur, where perfection comes in the form of beautiful Downeast lobsterboat hulls, both for work and pleasure. His shop is across the street from Manset Yacht Services on Seawall Road.

With your boat secure on a mooring, the Island Explorer can take you from Manset all over Mount Desert Island and Acadia.

GILPATRICK COVE
44° 17.19′N   068° 17.53′W

CHARTS: 13321, 13312, 13318   **CHART KIT:** 74, 23

Gilpatrick Cove, at the entrance to Somes Sound, and around the corner from Northeast Harbor, is home to the Northeast Harbor Fleet (*276-5101*). There is a handsome little clubhouse on the east shore, with a launch, water, and 4.5 feet at the floats at low. The rare visiting yachtsman is well received, but there are no guest moorings or other facilities. The club exists primarily for racing and has a fleet of Internationals and J-24's.

SOMES SOUND

Somes Sound is a glacial river valley drowned by the ocean among the geologically old, rounded mountains of Mount Desert Island. Technically, this is the only fjord on the eastern Atlantic seaboard, though it is a modest example compared with the majestic fjords of Norway or New Zealand. Still, a sail up Somes Sound is exhilarating. Offshore islands drop astern and shear cliffs plunge into the deep water. Gradually the shoreline levels. At its head lies graceful Somes Harbor.

Sailing in Somes Sound is variable and challenging. The wind usually funnels straight up or straight down the sound, and it is likely to shift rapidly. Downdrafts come off the mountains, cat's paws spread in all directions, and a strong current runs in The Narrows. With many of the cliffs falling straight into the sea, you can hold your course till the last minute and tack under the looming rocks. In one or two spots, you can lay your boat right up against the cliffs. Occasionally you will see porpoises.

There are few dangers in Somes Sound. Middle Rock is marked by can "5" (*Fl G 2.5s*) just before The Narrows. The 4-foot spot off the eastern shore just past The Narrows claims more than its share of unsuspecting yachts. So does the shoaling area off Sand Point.

Valley Cove, under Eagle Cliffs, is a wonderful spot to anchor and climb Flying Mountain. Around the next headland, Man o'War Brook still tumbles into the sound as it did when English ships filled their water casks here. If you venture close to the western shore, you'll find a bronze plaque set in the vertical stone in memory of Reverend Cornelius Smith and his wife, Mary Wheeler. These pioneers of the summer colony of Northeast Harbor (1886-1913) gave Acadia Mountain to the public.

The silent granite faces of Hall Quarry lie just beyond. Five miles up, at the head of the sound, is the narrow entrance to Somes Harbor and the peaceful village of Somesville.

MOUNT DESERT

ATLANTIC NEPTUNE

Cruising the coast of Maine is challenging enough with today's excellent and detailed charts. Think what it must have been like with no charts, or—perhaps worse—with inaccurate charts. When Joseph F.W. DesBarres began his survey of the eastern coast of Canada and the United States in 1764, the English charts (even those of the British Isles) were full of mistakes. The Hydrographer in Ordinary to Charles II referred to "the common scandal of their badness."

When DesBarres published the first of the *Atlantic Neptune* series in 1774, they showed a marked advance in accuracy and detail over what had been previously available to mariners. Called "the most splendid collection of charts, plans and views ever published," the *Atlantic Neptune* charts remain fascinating to this day. Strikingly colored, they are works of art in themselves.

The foreword to the 1781 edition, kept in the Map Room of the British Museum, says this about the utility of the *Atlantic Neptune*: "The Northern Coast and Harbours of His American dominion were unexplored, and very partially and imperfectly known, and that only to a few fishermen. The Isle of Sable itself was a terror to all Navigators at large going to America, or returning thence to Europe, and shipwrecks were innumerable. In the progressive execution of the Royal Commands, every master of a vessel became himself a Pilot of the district exhibited in the part of the work which fell into his hands, and no shipwrecks are heard of, excepting from insurmountable stress of weather or negligence. In the safety, abridgment, confidence, and frequency of Voyages, the notion has saved millions…"

The *Atlantic Neptune* became the standard charts for almost a century, and they were still the best when DesBarres died in Halifax at age 102.

Part of an *Atlantic Neptune* chart (1772) showing Mount Desert Island but carefully omitting Northeast Harbor, where the British fleet lay. (Maine Historical Society)

The *Atlantic Neptune* shows a number of intriguing and amusing differences in nomenclature from our present charts. Rockport is "Goose River," Rockland is "Owls Head Bay," Camden is "Meguanticook Harbor," and the Camden Hills are the "Penobscot Mountains." Going east, you pass through "Edgemogin River" and reach "Mount Desart Island," beyond which is "Moose a Becky's Beach," leading to "Great Island," which is now Roque.

There is a fascinating footnote. If you compare the *Atlantic Neptune* chart of Mount Desert to your own chart, you may notice a strange omission. The southeast coast of Mount Desert is devoid of harbors. British surveyor Samuel Holland, a meticulous cartographer known for the accuracy of his charts, has somehow missed Northeast Harbor and Seal Harbor altogether. Historian Samuel Eliot Morison surmises that "fog must have concealed the harbor entrance" on the day the surveyor sailed by. More likely, this was where the British fleet was based, and the chart was an early example of "disinformation."

DesBarres' methods are interesting to those of us who wonder how such a thing can be done. In one of his reports, he explains that he measured a base of 350 fathoms along the shore of a harbor, then sighted the angles from the end of the base line to objects placed on the opposite shore, calculated the other sides of the triangles by trigonometry, and drew the results on paper fixed on a plain table.

Then, he said, "from points as were most commodiously situated on those islands, and head lands, I observed the distant head lands, bays, islands, points, and other remarkable objects, as far as they could be distinguished. Next I went along shore, and reexamined the accuracy of every intersected object, delineated the true shape of every head land, island, point, bay, rock above water, etc. and every winding and irregularity of the coast; and, with boats sent around the shoals, rocks and breakers, determined from observations on shore, their positions and extent, as perfectly as I could.

"When the map of any part of the coast was completed in this manner, I provided immediately each craft with copys of it: the sloop was employed in beating off and on, upon the coast, to the distance of ten and twelve miles in the offing, laying down the soundings in their proper bearings and distance, remarking every where the quality of the bottom. The shallop was, in the meantime, kept busy in sounding, and remarking around the headlands, islands, and rocks in the offing; and the boats within the indraught, upwards, to the heads of bays, harbors etc."

The *Atlantic Neptune* charts the coast in detail from the St. Lawrence to the Mississippi—quite an accomplishment!

VALLEY COVE

3 ★★★★

44° 18.55'N 068° 19.00'W

CHARTS: 13318, 13312   CHART KIT: 74, 23

Valley Cove is one of the grandest spots on Mount Desert for the cruising sailor. The French and British and perhaps even Vikings have moored here before, just inside the entrance to Somes Sound on the western shore. Below the great cliffs and spruce forest, a boat seems small against the backdrop of nature and history.

A short but vigorous hike from Valley Cove leads to the top of Flying Mountain, with marvelous views.

APPROACHES. The only dangers at the entrance to Somes Sound are marked by can "5" (*Fl G 2.5s*) which is left to port, just before The Narrows. Thereafter, the sound is deep to Valley Cove, but favor the west shore to avoid the shoal areas on the east. It's easy to assume that all sides of Somes Sound are steep-to and forget these spots only to be suddenly reminded in the worst way.

ANCHORAGES, MOORINGS. Feel your way in toward the southern shore and anchor in 25 to 35 feet at low. Be sure your anchor is well set. The bottom is deep and may be rocky.

Several huge, steel-balled moorings are set in the cove, and Hinckley maintains a few others. All of them are used by visiting boats at their own risk on a first-come, first-served basis. Although the wind can funnel up or down Somes Sound and gusts sometimes blow down from the mountains, protection here is good.

GETTING ASHORE. Land on the pebble beach on the southern shore, near a small footbridge.

THINGS TO DO. To the left, the trail leads southward through a spruce forest to the summit of Flying Mountain. The Appalachian Mountain Club calls this trail "the greatest reward on the island for a small effort." It is rough and steep in places, but well-blazed with orange marks on the rocks. There are two lookouts from which you can gaze up and down Somes Sound, but don't stop. Continue southward along the crest of the mountain until the trail leads you to the bare rocks on top, with U.S. Coast and Geodetic Survey marks that call it "Fernald Hill." From there, look south to the meadows of Jesuit Spring, past Greening and Sutton Island to the Cranberry Islands and the great Atlantic. The distance is .3 miles to the crest and about an equal distance along the crest to the southern summit. Allow three-quarters of an hour to go up and back, or take a picnic and linger on top.

If you follow the trail north for a mile, you will reach Man o'War Brook. This is a difficult hike. The first part leads over the tumbled rock scree just under the cliff face, and the footing is tricky. When you round the headland at the north end of Valley Cove, the path passes through cedar, pine, and spruce forest and is much easier.

After one mile, turn right at the signpost and take the path down toward the sound of rushing waters and the overlook. Man o'War Brook tumbles down steep ledges and plummets into Somes Sound. The brook runs into the tiny indentation just north of Valley Cove, and it is easier to reach by water than by land. British ships took on drinking water from the stream, and it served naval vessels as late as 1878, when *Cimbria* of the Imperial Russian Navy topped off here.

HALL QUARRY

2 ★★

44° 20.35'N 068° 19.15'W

CHARTS: 13318, 13312   CHART KIT: 74, 23

Hall Quarry lies beyond Valley Cove, on the west shore of Somes Sound. The site of the old granite quarry is now occupied by the custom boatbuilder and repair yard of the John M. Williams Company.

APPROACHES. Once past The Narrows of Somes Sound, the west shore is bold. Hall Quarry is labeled on the chart by the 69-foot sounding.

ANCHORAGES, MOORINGS. The Williams yard has several moorings for transients. Don't try to anchor here, depths drop off to over 100 feet! Bar Harbor Boating, on the opposite shore, may also have moorings. Winds tend to funnel up and down the Sound.

FOR THE BOAT. *John M. Williams Co. (Ch. 09; 207-244-7854; www.stanleyboats.com).* Williams builds the Stanley line of boats based on the lobsterboat hull form in various sizes. They also haul with a 30-ton boatlift and can make engine, hull, refrigeration, metalworking, and rigging repairs, or you can tackle them yourself.

In addition to their rental moorings, they have a float with water. In a pinch, fuel can be delivered by truck.

Bar Harbor Boating (Ch. 09 or 10; 207-276-5838). This small yard is a division of Bar Harbor Boating in Hulls Cove, north of Bar Harbor. Their big red shed with gray doors sits on a little cove on the east side, directly opposite Hall Quarry. They have a couple of rental moorings, floats (6 feet alongside), and a 20-ton marine railway, and they can make hull and engine repairs. Water and propane are available, but no fuel.

THINGS TO DO. From the Williams yard, walk up the tarred road until you are even with the top of the quarry, then follow the dirt road to the left, past the old crane shack, where the gears and levers have rusted in place. There is a dizzying view down a vertical face of stone into the quarry pool. Pink granite quarried here was used for the Franklin Mint in Philadelphia, and the quarry was reopened in the 1950s when an exact match was needed.

From the yard, walk up to the paved road and turn left, continuing to Route 102. Turn left again in about 100 yards at the sign for Ike's Point. There you will discover a tiny beach on beautiful Echo Lake. The swimming is great. The distance is 1.3 miles each way.

SOMES HARBOR

4 ★★★★★

44° 21.68'N 068° 19.60'W
Abel's Yacht Yard: 44° 21.57'N 068° 18.58'W

Charts: 13318, 13312   *Chart Kit:* 74, 23

At the northwest end of Somes Sound, the little village of Somesville is a quiet backwater that seems to have changed little since a cooper, Abraham Somes, brought his family here from Gloucester in 1761, starting the first permanent settlement on Mount Desert Island. In its heyday, seven mills, five shipyards, and four blacksmith shops crowded the little town. Attractive old white clapboard homes still line the main street; then there is a market, a gas station, the Masonic Hall, and suddenly there's the blinking light and you're out of town.

Somes Harbor is snug and well protected, open only through the narrow entrance to the south. There is no commercial activity in the harbor, and it is thoroughly restful and charming.

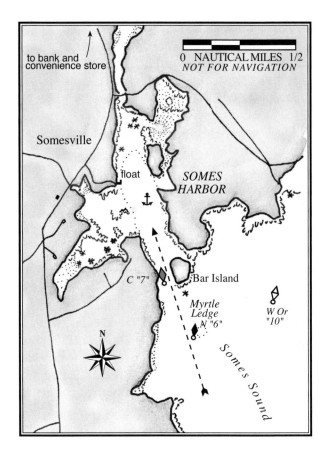

APPROACHES. Somes Sound is deep and clear almost until you reach the end. Leave nun "6," marking Myrtle Ledge, to starboard. Then pass through the narrow entrance west of Bar Island, leaving can "7" to port.

ANCHORAGES, MOORINGS. Anchor in 15 to 20 feet of water at low in the main slot of the harbor, north and east of the ledge shown in the middle of the shoal area along the harbor's west side. Holding ground is good. Or somebody might direct you to a vacant mooring. Bar Island breaks the wind and gives Somes Harbor excellent protection.

⚓ Henry R. Abel Yacht Yard has rental moorings outside Somes Harbor, at the head of Somes Sound. These are heavy granite blocks with orange balls inside the ledge marked by the can.

GETTING ASHORE. Take your dinghy in to the float and dock on the west side of the harbor. This private landing is maintained by the nonprofit Somesville Landing Corp. *(288-4661)* to provide access to the harbor for Somesville residents. They welcome contributions, which are tax deductible.

FOR THE BOAT. *Henry R. Abel Yacht Yard (207-276-5777).* Though not directly in Somes Harbor, the Abel yard is located nearby, at the head of Somes

Sound, on the northeast shore where the chart reads "Marine railway." In addition to their moorings, the yard has several floats, with 7 feet reported alongside. Water, electricity, and ice are available, but no fuel. This spotless yard handles hull repairs and painting in its large sheds, including one for spray-painting. It has two boatlifts, of 35 and 50 tons. Engine repairs can be arranged.

FOR THE CREW. The gravel road leads to Route 102. Turn left to reach the pretty little village of Somesville, with a white-clapboarded library, the historical society, and the wonderful Port in a Storm bookstore clustered around an old mill pond.

If you turn right on Route 102 instead, a half-mile walk will bring you a Mobil gas (and diesel) station and 🛒 major convenience store with essential groceries, a liquor store, and Subway sandwiches. A pay phone, propane exchange, and FedEx and UPS boxes are outside. A bank and the post office are across the street.

🍴 If you have a fast dinghy or your boat is on an Abel mooring, Abel's Lobster Pound restaurant (*276-5827*) is next to the yard, with tables outside in a peaceful pine grove.

THINGS TO DO. The free Island Explorer shuttle buses stop here and can take you anywhere on Mount Desert Island. Check at Port in a Storm (*244-4114*) for stops and times. The bookstore often has readings or talks by authors in the evenings. If the tide is high, you can bring your dinghy right to the back of the bookstore.

On rainy days, people often pack the little library in Somesville, open Wednesdays and Saturdays, or the even tinier Mount Desert Museum next to the millpond.

If it is sunny, head for a dip in Somes Pond. Cross the street south of Port in a Storm and follow Oak Hill Road for half a mile.

In July and August, the Acadia Repertory Theatre (*244-7260*) produces thrillers, drama, and farce in the Masonic Hall at the south end of town.

To reach the national park trails from Abel's, walk up from the yard to the main road and turn right. About three-quarters of a mile south you will find a sign for Giant Slide Trail, leading to Sargent Mountain. A shorter walk in the same direction will bring you to the Mount Desert Island Historical Society's Sound School House Museum.

NORTHEAST HARBOR

4 ★★★★★

44° 17.75′N 068° 16.90′W

CHARTS: 13321, 13318, 13312   **CHART KIT: 74, 23**

At the eastern side of the entrance to Somes Sound, in a setting of great natural beauty, Northeast Harbor is one of Maine's major yachting centers. Still reflecting its history as the playground of affluent society, the harbor has the elegant Asticou Inn and a wonderful armada of pleasure craft, with a working fleet of lobsterboats mixed in. The best large harbor in the Mount Desert area, Northeast makes it very easy for the yachtsman: services are concentrated on the waterfront, stores deliver, and the town is water-oriented.

This is the homeport of the missionary vessel *Sunbeam V*, affectionately dubbed by some as "God's tugboat," which brings religious services and practical help to many of the isolated islands of the coast. The ferry to the Cranberry Islands operates from here, and so do several natural-history and whale-watching boats.

One of the treasures of Northeast Harbor is Asticou Terraces, gardens that are easy to reach by dinghy and not to be missed.

APPROACHES. Coming from west and south, approach through Western Way, between Great Cranberry Island and Mount Desert Island. Coming from the east, pass north or south of East Bunker Ledge and run through Eastern Way, along the north shore of Sutton Island. Leave to starboard several red marks, the lighthouse on Bear Island, and red bell "2" at the entrance, and head north into the harbor.

ANCHORAGES, MOORINGS. A large town dock and marina is on the west side of the harbor, near the southern end. The marina has dock space and moorings and mooring floats in the harbor for visiting boats.

⚓ The town marina's inner portion is reserved for commercial use, but the outer finger floats are available to visiting boats, with 10 feet alongside at low. Dock space can be reserved in advance, way in advance; they start taking reservations (*276-5737*) on the first of January for the upcoming season.

⚓ The town's 50 moorings and numerous mooring floats are available on a first-come, first served basis. The moorings have bright green pickup buoys and three-digit numbers on the floats. They vary in weight. Boats 40-50 feet should use 400-series

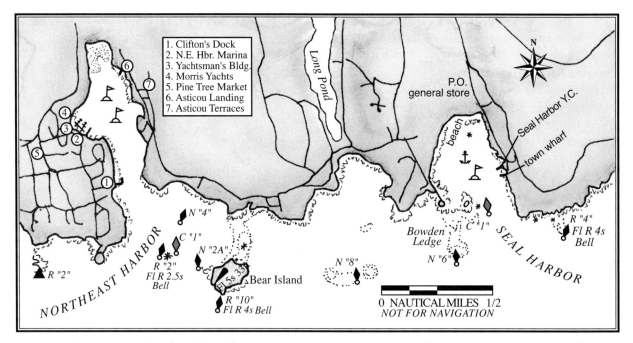

moorings, boats 30-40 feet should use the 300 series, boats 20-30 feet should use the 200 series, and boats less than 20 feet should use the 100-series moorings. The mooring floats are obvious. Contact Northeast Harbor Moorings on VHF Ch. 09 for assignment as you approach or when you arrive.

If the town moorings are all taken, additional moorings are sometimes available through Clifton's Dock (*276-5308*) or the MDI Water Taxi (*Ch. 16, 68; 244-7312*).

All of the deep water in the inner harbor is full of moorings, and anchoring is not allowed. There appears to be space on the east side of the entrance, but it is exposed, and the bottom is hard clay.

GETTING ASHORE. Yachtsmen should come in to the dinghy floats farthest north and parallel to the shore. The public float south of the town pier is intended for fishermen's dinghies.

FOR THE BOAT. *Northeast Harbor Marina (harbormaster: Ch. 09, 16 or 68; 207-276-5737).* The town dock here is one of the largest facilities of its kind, with many floats, parking spaces, open greens, and tennis courts. The dock is used by tour boats, ferries, fishing boats, and yachts, so traffic is heavy. Water and electricity are available at the public float (farthest south), where you can tie up for two hours, with 8 or 9 feet at low, and water is also piped to float #300, moored near the head of the harbor, which has a bright green band painted on each of its ends. You may need to BYOH (bring your own hose). The marina also has holding tank pump-out facilities. Packages will be held for cruisers. Address them to Northeast Harbor Harbormaster, P.O. Box 237, 18 Harbor Drive, Northeast Harbor, ME 04662.

Clifton Dock (207-276-5308). On the west side of the entrance to Northeast Harbor, Clifton Dock is a convenient spot to take on gas, diesel, water, or ice, with 22 feet alongside the fuel float. Pump-outs are also available.

Morris Yachts (207-276-5300). Morris Yachts has taken over what was formerly Mount Desert Yacht Yard, at the northwestern end of the harbor, tucked in beyond the town marina. They run this facility as another service center (in addition to their yard in Bass Harbor), with repairs of all kinds, storage, and a brokerage.

F.T. Brown (207-276-3329) is a large hardware store on Main Street. Their serious boating department stocks hardware, propane, charts, line, and marine supplies.

FOR THE CREW. Pay phones are right at the town dock. The Mount Desert Chamber of Commerce's shingled "Yachtsman's Building" (*276-5040*) is right by the marina, with showers available 24 hours a day (the key will be in the harbormaster's office after hours), a reading room with internet connections, a paperback book swap, and recent copies of *The New York Times*. They will even rent you towels and a hair dryer. The Chamber of Commerce handles reservations (and rents rackets and balls) for the public tennis courts near the marina, and they have literature available describing the nearby attractions.

Sea Street leads from the waterfront up to the main street in town. Turn right on Main Street and you will find the post office, the full-service Shirt Off Your Back Laundry (*276-5611*), Brown's Hardware, and a bank with an ATM.

⛵ The Pine Tree Market (*276-3335*) is to the left on Main. Pine Tree Market is everything a good market should be—creaking floors and sweet smells from the bakery, a good butcher, wine and fine cheese, fresh produce, liquor, and ice. And they will deliver to the dock. The coin-op Downtown Laundry Cellar is beneath the market.

McGrath's newstand, beyond the market, sells fresh papers, including *The New York Times*, and Sherman's Books stocks bestsellers. Wikhegan Books, on Main at the head of Sea Street, carries rare and out-of-print Mount Desert Island and nautical titles.

🍴 The Docksider takeout (*276-3965*) on the way up Sea Street, has block ice and ice cream—something our kids don't let us forget—and a surprising wine list, something we always remember. Full Belly Deli, also on Sea Street, can take care of sandwiches and Edy's ice cream. Colonel's Bakery, pizzeria, and restaurant (*276-5147*), on Main, dishes up breakfast, lunch and dinner. The House of M (*276-9898*), beyond Colonel's Bakery on Main, cooks Asian-inspired dinners. Main Sail Restaurant at the Kimball Terrace Inn (*276-3383*) serves three meals a day overlooking the town docks. The restaurant at the Asticou Inn (*276-3344*), overlooking the harbor from above the east shore, is open to the public for breakfast and formal dinners. The inn is a 10-minute walk from their landing on the harbor. Be sure to bring flashlights for the return trip.

Mount Desert Medical Center (*276-3331*) is on Kimball Road at the south end of town.

THINGS TO DO. The small but elegant Great Harbor Maritime Museum (*276-5262, closed Mondays*) is right in town in the old fire house. Their exhibits feature local maritime history and boatbuilding greats.

If you are socked in by bad weather, visit the charming public library. The Milliken Room is devoted entirely to books about Maine.

Check for cultural events at the Neighborhood House at the south end of Main. They host the Mount Desert Festival of Chamber Music, bike parades, and ice-cream socials (after which, you may want to work out in their fitness room).

A visit to Thuya Lodge and Gardens in Asticou Terraces is unforgettable. Take your dinghy in to the float of the pink-granite Asticou Landing, at the northeastern part of the harbor. Follow the path to the main road and cross on the crosswalk.

Rustic paths lead upward past terraces and gazebos designed by Joseph Henry Curtis (1841-1928) "for the quiet recreation of the people of this town and their summer guests." The small formal garden is a delight for botanist and gardener alike, and there is a rare botanical book collection in the lodge. From the gardens, a leisurely trail leads to Eliot Mountain, or you can follow the park trails east a couple of miles to Jordan Pond House.

Bikes can be rented at a couple of locations on Main Street, and the trails of Acadia begin nearby. The free Island Explorer shuttle bus stops right at the harbormaster's office for exploring the rest of Mount Desert. For taxis call Airport Taxi (*667-5995*).

Beal & Bunker, Inc. offers a variety of sightseeing cruises starting from the waterfront, including an evening cocktail cruise and another with an Acadia National Park naturalist on board. Sign up at the waterfront booth (*244-3575*). They also run the mailboat trips to the Cranberry Islands. The *Delight*, a 32-foot antique open launch, provides water-taxi service to the outlying islands (*244-5724*).

Maine Whalewatch (*276-5803*) makes a full-day trip out to Mount Desert Rock with a naturalist, looking for finback and humpback whales, porpoises, and many species of seabirds.

CRANBERRY ISLANDS

The Cranberries are beautiful and peaceful islands, with miles of walks and spectacular views of sea and mountains. Named for their extensive cranberry bogs and known for their mosquitoes, the islands undertook a program of drainage in the 1920s. As islander Wilfred Bunker says, "The mosquitoes came back, but the cranberries didn't."

The Cranberry Islands are in transition. After two centuries of a fishing economy, their small population now includes more summer people than fishermen. Of the five islands considered part of the Cranberries, the two closest to the mainland, Bear and Sutton, are now private. Baker Island is uninhabited, and half of it belongs to Acadia National Park.

Only Great and Little Cranberry still have the post offices and stores that make them viable year-round communities; only Little Cranberry has a school. About 75 people live on Little Cranberry year round, and about 40-50 hunker down in the winter on Great Cranberry.

A privately run ferry service connects the Cranberries to Northeast Harbor and Southwest Harbor, taking islanders to the mainland for modern necessities and taking tourists to the islands for simple pleasures.

GREAT CRANBERRY ISLAND

2 ★★★

Spurling Cove: 44° 15.55'N 068° 16.10'W

CHARTS: 13321, 13312, 13318   CHART KIT: 74, 23

Great Cranberry is larger but seems more remote than the other Cranberry islands, despite the fact that it lies just two miles away from the hum of Northeast Harbor. The lobstermen share the island with summer people, many of whom have come here for generations, but the year-round population is dwindling. The school's two students have moved on, and the school has closed. This is a delightful place to enjoy peaceful walks, island beauty, friendly people, and deer that are almost tame.

The year-rounders have other concerns. After policing themselves since the founding of the Cranberry Islands in the early 19th century, they are considering a police force. The long winters have become longer with an increase in drunk driving, break-ins, and recreational gunfire.

APPROACHES. The approach to Spurling Cove, on the north shore, is open and unobstructed.

ANCHORAGES, MOORINGS. If it's not spring or fall, ⚓ moorings may be available from Newman and Gray. You can anchor outside the moored boats, in 16 to 25 feet at low. There is considerable exposure from northwest to northeast.

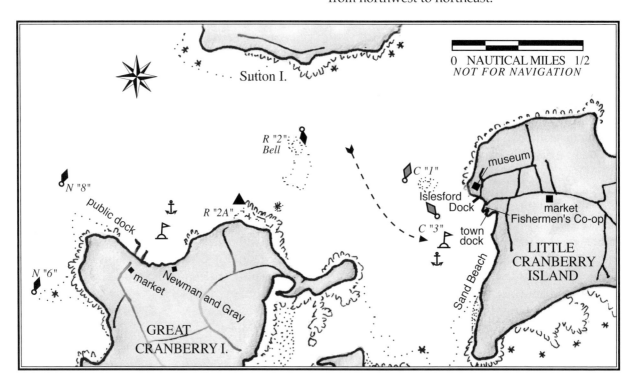

GETTING ASHORE. Land at the dinghy float of the public dock, farthest to the west. Do not block the two public docks.

FOR THE BOAT. *Newman and Gray Boatyard (207-244-0575).* Originally from Southwest Harbor, Jarvis Newman and Ed Gray have built a year-round yard with sheds, a marine railway, and boatlift to the east of the town wharf. They build new boats and do all kinds of repairs.

Cranberry Island Boatyard (207-244-7316). Cranberry Island Boatyard is located on the east side of the island on the tricky cove known as The Pool. They maintain the local fleet and build the elegant Western Way dinghies and a jet boat and they can probably help you if you need repairs. Ask for their help before entering The Pool.

FOR THE CREW. Ferries and water taxis use the town docks, so please don't block them. The post office and a pay phone are at the head of the next dock to the east, which is private. Restrooms are in the parking lot. A general store by the ferry landing has groceries, and the Seawitch lunch counter specializes in gourmet paninis and crabrolls.

THINGS TO DO. A single paved road, dubbed "I-95" by the islanders, runs the length of the island, about 2 miles. Well-kept cottages sit in flower-filled meadows, and side roads beckon for exploration. One of these leads eastward to The Pool, where the Cranberry Island Boatyard builds custom boats. The Pool is almost completely landlocked and can be entered only at high tide and only with local advice. The large heath to the west was once bogs, long since dried up, that supported the cranberries for which the islands are named. The many high fences around the gardens are to protect them from the deer. You will pass the new Historical Museum and Cultural Center and the Whale's Rib gift shop, which offers a selection of seeds of island plants and flowers—lupine, beach pea, beach rose, and wild iris.

Eventually the main road leads you to beautiful rocks on the southern shore and out to The Gut between Great Cranberry and Little Cranberry Island. The mountains of Mount Desert Island range high to the north, and to the south lies Baker Island and the open Atlantic.

The Islesford Post Office sells more stamps than any other post office in Maine.

LITTLE CRANBERRY ISLAND
(Islesford)

2 ★★★★

44° 15.60'N 068° 14.45'W

CHARTS: *13321, 13312, 13318*   CHART KIT: *74, 23*

Little Cranberry Island lies less than two miles to seaward of Mount Desert. Its views are framed to the north by the mountains of Mount Desert Island and the twinkling lights of Southwest Harbor to the west. This small, quiet island has pleasant walks, an interesting little museum, and a waterfront restaurant. Transportation is furnished by little ferries from Northeast Harbor and Manset and by the constant coming and going of water taxis, fishing boats, and yachts in the anchorage.

Little Cranberry, Great Cranberry, Bear, Baker, and Sutton Island are all part of the town of Cranberry Isles. In 1884, tired of having Little Cranberry Island mail accidentally end up at the Cranberry Isles post office on Great Cranberry, the residents of Little Cranberry came up with the more distinctive village name of Islesford.

APPROACHES. The harbor in Hadlock Cove can be approached from the west or north. From the west, leave red bell "2" marking Spurling Rock in Gilley Thorofare to starboard. Leave cans "1" and "3" to port.

As you look toward land from the harbor, there are three docks. The one to the left is Islesford Dock, the Fishermen's Co-op is in the center, and the town dock is to the right. Be careful of the granite remains of a former pier (shown on the chart) just left of the Islesford Dock.

ANCHORAGES, MOORINGS. All of Cranberry Harbor can be used as an anchorage, with depths of 14 to 23 feet at low, though it is broad and quite exposed. The best protection is found in Hadlock Cove, the wide bight where the docks are located.

⚓ The Little Cranberry Yacht Club, whose one-room clubhouse is on Islesford Dock, has two memorial guest moorings. Islesford Dock Restaurant also manages several rental moorings. The moorings are actually owned by restaurant patrons, and, as one insider put it, "the rentals pay off the bar tabs."

If no mooring is available, there is plenty of room to anchor outside the moored boats. Some people have reported difficulty with kelp on the bottom, so be sure your hook is well set before you leave your boat.

GETTING ASHORE. Land your dinghy at the town dock or at Islesford Dock.

FOR THE BOAT. *Fishermen's Co-op.* If you are stuck, 🛢 limited amounts of gas and diesel are available at the floats of the Fishermen's Co-op, with 9 feet at low. It is much more convenient to fuel up at Northeast or Southwest Harbor or Manset.

Islesford Dock Restaurant (207-244-7494). The restaurant has ⛵ slips for patrons at the end of Islesford Dock and several rental moorings.

FOR THE CREW. 🍴 The Islesford Dock Restaurant (*244-7494*) serves meals on the harbor with greenhouse-grown produce and occasional live music.

Public restrooms are at the head of the Islesford Dock building. A short walk eastward on the main road will bring you to the 🛒 Islesford Market (*244-7667*), where local notices are thumbtacked to the bulletin board and the screen door creaks and slams. The post office cubby inside sells $58,000 of stamps a year, more than any other post office in Maine. The trick? Mail order.

The Fishermen's Co-op sells lobsters.

THINGS TO DO. Be sure to visit the Islesford Historic Museum (*244-9224*), part of Acadia National Park, in the handsome Georgian brick-and-granite building close to the docks. It houses a collection of

The Gilley graveyard on Baker Island.

nautical memorabilia, old documents, tools, charts, photos, and domestic furnishings depicting the history of the Cranberry Islands. The museum has a small but interesting collection of books about the history and natural history of the Maine coast for sale.

Islesford women sell their "winter work" in a shop on Islesford Dock, and a potter has a gallery there. The Islesford Artists gallery, a short walk "uptown," displays the works of island artists.

A short walk leads down Sand Beach to the Maypole, site of the island's 18th-century spring celebrations. According to Ted Spurling, "this custom was brought by a beautiful Frenchwoman, Margarite La Croix Stanley, wife of John Stanley, the first permanent settler."

On the way you will pass the workshop of Islesford Boatworks (*www.islesfordboatworks.org*), a program started by Brendan Ravenhill and his siblings Amanda and Geoffrey to teach island kids boatbuilding skills and the confidence of creation. Visitors are welcome to poke their heads in the door and sniff the sawdust.

It's a half-hour walk from the waterfront to the old Coast Guard station at Bar Point. The road that leads straight to the shore is public, but the Coast Guard station and the road that leads to it to the right are private. On July 4, 1944, a month after D-Day, a patrol boat from this station towed ashore the collapsed remains of a Navy dirigible known as K-14. The dirigible had been on patrol looking for enemy submarines when it plummeted into the sea under somewhat mysterious circumstances. The Navy insists that no coastal patrol dirigible was ever lost to enemy fire, but rumors abounded on Islesford of sub sightings, bullet holes, and gunfire.

BAKER ISLAND

44° 14.77′N 068° 12.15′W

CHARTS: *13321, 13312, 13318* CHART KIT: *74, 23*

In 1899, Charles W. Eliot, then the president of Harvard, wrote a history about pioneer life on Baker Island, *John Gilley of Baker's Island*. William Gilley and his wife, Hannah, arrived on Baker Island in 1806. "There it lay in the sea, unoccupied and unclaimed," wrote Eliot, "and they simply took possession of it."

Not much has changed since then. Today, half of Baker Island is privately owned, the other half is a part of Acadia National Park, and a lighthouse is in the middle. Sightseeing boats come to Baker Island from Northeast Harbor with a park ranger aboard, or you can make a daystop here.

A sign in the park portion of the island summarizes Eliot's history of the Gilleys: "William and Hannah Gilley raised 12 children to maturity. Forest and sea shaped their world; farming, fishing and hunting sustained them. They worked hard. Calloused hands wielded an ax, guided a plow, trimmed a sail, treadled a spinningwheel, or held a child. To their skilled labor, the land produced vegetables, forage for stock and a little wheat. They caught herring and mackerel and picked up lobsters in the shallows. Like us they knew joy and sorrow, labor and rest, adversity and success."

APPROACHES. As shown on chart 13318 or Chart Kit 74, a line of cans and a whistle guard the eastern shores of Little Cranberry and Baker Island. Run outside this line until you reach can "3," at Baker Island. Coast along the north shore of Baker toward the bight in the ledge south of Gravel Nobble.

ANCHORAGES, MOORINGS. Anchor off the north shore of Baker, in 10 to 17 feet at low, west of the cable area. The anchorage is exposed to prevailing southwesterlies, so be sure your hook is well set.

GETTING ASHORE. Row in to the rock beach where the path hits the shore, as shown on the chart.

THINGS TO DO. Walk along the beach to where the path leads through a thick growth of beach roses, then follow it through a beautiful field past an old island cape and another red house. At the lighthouse, the path heads through the woods and leads to the storm beach and "Dance Hall Floor," an aptly named spot. Cranberry Islanders used to bring food, spirits, and windup Victrolas and while away time on these stunning flat ledges.

SEAL HARBOR

2 ★★★

44° 17.48′N 068° 14.30′W

CHARTS: *13321, 13312, 13318* CHART KIT: *74, 23*

The first home in Seal Harbor was built by John Clement in 1809, but Rockefeller is the name more commonly associated with this harbor. Peggy and David Rockefeller, who summer in Seal Harbor, helped found the Maine Coast Heritage Trust, which has been influential in preserving and protecting many of the state's islands and shores that yachtsmen enjoy today.

The harbor itself is exposed to the south. During normal summer weather, you will be safe here, but there is likely to be a roll even when it's calm. During strong southerlies or southeasterlies, the harbor would be unpleasant.

Other than the town wharf and the Seal Harbor Yacht Club, there are no boat facilities. The big gray shed at the western end of the harbor is the Rockefeller boathouse.

An entrance to Acadia National Park is nearby, with access to Jordan Pond House and to Wildwood Stables.

APPROACHES. The approach to Seal Harbor is easy (see sketch chart on page 314). Nun "6" is an Eastern Way buoy that marks Bowden Ledge, off the entrance. Can "1" marks a ledge on the left side of the entrance. Leave can "1" to port as you enter. From the east, the white pyramid on East Bunker Ledge is a good landmark.

ANCHORAGES, MOORINGS. The Seal Harbor Yacht Club has some guest moorings, marked "SHYC." Pick one up and check at the club. Or look for the harbormaster at the town dock and ask if there is a vacant mooring. Otherwise anchor in 9 to 18 feet at low anywhere convenient except to the north of the yacht club floats, where a ledge is reported. Holding ground is good.

GETTING ASHORE. Tie your dinghy behind the float at the town wharf or at the yacht club docks.

SAILING WITH CHILDREN

Note: for our family, these lessons are dated. Does anyone have suggestions for cruising with teenagers?

The anchor is set. The water is still. There isn't a house or a person in sight on the tranquil harbor. The sun begins to settle toward the horizon. All is peaceful after a long day's sail. Or is it? A slight storm seems to be brewing down below. One of your children is demanding food. The other is idly lobbing crayons across the dinette. A flailing elbow sends a cup of juice flying to the floor, and a shouting match erupts about whose fault it is. The bucket in the lazarette where you usually keep the big sponge is now full of an odd assortment of gritty treasures—pebbles and shells and small sea creatures in various stages of decay. You recoil from the smell, and the lazarette drops on your head. And from somewhere in the distance, you hear a meek voice saying, "Daddy, the head is overflowing again…"

The very idea of children on boats seems, to some sailors, like their worst nightmare. Already cruisers have to navigate and sail the boat, keep the crew happy, safe, and well fed, establish the cruising itinerary, provision the galley, and maintain the boat. Why add the massive responsibility of keeping small children safe around the cold water and the near impossibility of keeping them entertained onboard? Is it fair to subject them to the confinement of a boat, the stresses of their parents, or the inherent risks of sailing? Is that what you call a vacation?

Still, many cruising families have young children aboard. We have sailed in Maine with our children, Nathaniel and Hannah, since they were born, and we have discovered that they have shaped our cruises in myriad marvelous ways—by sharing their squeals of excitement and laughter, by inventing imaginative games, by proudly displaying their collections of beachcombings or berries, by making friends, and by helping their parents make friends. What more valuable treasure could you give your child than the joys of discovery? And you, if you are lucky, might discover the world through a child's eyes.

Here are the basics of what we have learned:

Safety. The first order of business is keeping children safe. Lifejackets are a must. PFD's must be properly sized and, if they have straps, properly adjusted to fit snuggly so they won't ride up over the child's head. Most importantly, they must be comfortable. If the jacket digs into the child's neck or armpits or if it makes the wearer feel like the Michelin Man, it simply doesn't get worn, at least not without a fight. Lifejackets should also be habitual. Small children, especially nonswimmers, should have theirs on whenever they are even near the water—on the dock, in the dinghy, in the cockpit. They should learn as early as possible how to put on and zip or snap up their own jackets.

Maine has a child lifejacket law that requires children under 10 to wear lifejackets at all times when aboard. The law, the Marine Patrol is quick to point out, was developed as a result of lake boating accidents and is not yet refined to cover situations where children live, eat, and sleep on cruising vessels, so you will still have to use your own judgment.

When Carol and I are sailing with Nathaniel and Hannah, we try to make realistic lifejacket rules and to enforce them consistently. Our children must wear lifejackets in the cockpit when we are under way except when one adult can devote his or her total attention to them. Down below, lifejackets aren't needed. But if the children are coming up and down the companionway, they either have to keep putting on their lifejackets or keep them on. You can form your own rules based on your child's age and disposition, your boat's configuration, and the number of attentive adult eyes you have onboard.

As an alternative to the lifejacket on substantial cruising boats, we sometimes use a lifeline and harness on our children. Climbing webbing makes a good tether because it is flat and won't roll underfoot, and it is strong. We attach it to the steering pedestal in the cockpit where it allows the most movement and the least tangling, and we limit the tether's length so the child cannot reach the edge of the boat. The actual harness doesn't have to be built like an adult safety harness, designed to stop the kinetic force of a full-grown body being hurled across a boat or into the water. In any kind of heavy weather, our children will be down below, not on their tether. A simple harness of rope or webbing works well.

Likewise, extreme strength of the tether clip is not as important as its operation. It should not be easy for the child to undo, yet you should be able to undo it quickly. We use an aluminum climbing carabiner with a twisting lock mechanism. Both Nathaniel and Hannah quickly learned the limits of the tether and how to keep from getting tangled in it, and on hot days, they begged for it instead of their lifejackets.

Infants can be secured in the boat in a number of ways. Hannah's favorite was swinging in the seat from her home infant swing, which we tied to the interior grab rails of the cabin. Here, secure in her swing, she loved every wave, the bigger the better. Swings with trays help hold toys and snacks. Another option is to tie an infant car seat securely to the boat and buckle the child right in. And if your dinette table is more secure than ours, you can use one of the small cantilevered infant chairs (known as a Sassy Seat) as a highchair. At anchor, as a treat for young children, we lashed a Johnny Jump-Up seat, a bouncy seat hung from a spring, to the boom and let them bounce away their energy in the cockpit during the cocktail hour. Nathaniel is now old enough to prefer the bosun's chair.

Place. Once the basics of safety are covered, consider place. Every child needs a sense of place, afloat as well as ashore. Each child should have a personal place where he can keep his books and toys and collections of feathers and shells. He also needs his own place to play and sleep. On our boat, this place is his berth and the lockers near it. More importantly, the child needs a sense of other places—what can and can't be done in other places and at other times. Our kids are free to run around on deck when we are at anchor and watching, but they aren't allowed out of the cockpit otherwise. They are free to climb in and out of the companionway, but they aren't allowed to play on it. We encourage them to paint at the dinette when we are anchored, but when we are beating to weather, the watercolors tend to paint by themselves.

Purpose. Along with place, we try to give our children purpose. Just like adults, they love to help if it is fun, but they hate it if it is work. When I am rowing the dinghy, my pilot Nathaniel in the bow tells me which way to steer, and I try to trust him and not look over my shoulder. I hand the wet dishes to them to dry or stack. Nathaniel likes to kill the engine, and Hannah tries to be in charge of the sail ties and the figure-eight knots.

For basic maneuvers, we establish routines and try to develop habits. We try to get into the dinghy in the same order each time and sit in the same places. We remind them to keep fingers in each time we come alongside, and we disembark in the same order. Nathaniel knows to go into the cockpit and cross over to the far seat so there is room for the rest of us to board.

Activity. Once aboard, what do children do all day in the confines of a boat? Rest assured that entertainment is everywhere. Even with a library full of books, lockers full of art supplies, and crates of small toys, the most interesting things aboard will be the tools you use—the parallel rules that transform themselves into a gun, the charts that become treasure maps, the courtesy flags that become the Jolly Roger, the dinette table that becomes a fort with a blanket draped over it. We inflate a small swimming pool in the cockpit for warm baths and water play, both at anchor and on easy sails.

One year our children discovered the radios. They would punch in a tape, turn up the volume, and straddle the benches down below and dance away their energy even while the boat was on its ear. And when the VHF was off—and once when it wasn't—they would grab the mike and say, "Do you copy? Do you copy? Do you copy copycat?"

At an anchorage, they are tireless explorers and prodigious collectors. We made a checkers game using different colored periwinkles as playing pieces. Gull feathers become masks. Sticks look like cutlasses, and pirate treasure might be buried anywhere. Smooth round stones become pirate money, and sea glass is pirates' gold. Windjammers, of course, are the pirate ships, ghosting by on the horizon or dropping anchor at close range. We've collected a rainbow of lobster bands on our "lobster stick." And we clean up after careless mermaids by collecting their elegant sea china. One beachcombing expedition even turned up a soaked and slightly singed stuffed teddy bear wearing handmade boxer shorts who, when dried off, became "Boat Bear," a favorite mascot.

Schedule. Our children change our sailing habits too. We try to sail when our children are the most tired in the hopes that they will nap. Likewise we try to be at anchor or exploring ashore when they have the most energy. This reduces the sailing to shorter legs, but we end up seeing more coast more intimately this way. Often it backfires—the children don't take naps, and they are exhausted when we finally get off the boat.

We get exhausted too. Sometimes it seems as though there simply is no rest. If the children don't need attending to, the boat does. Clothespins mysteriously find their way into the sockets of the winches. Superballs jam the cockpit scuppers. Crayons appear in your berth. The head gets clogged, and the beds get wet. One mess is being created even while you clean up the others. Moments of silence are treasured rarities.

Yet the real treasure lies in all these things. Here we are, a family coping with family things, yet we are sailing, exploring, discovering. Onboard, children learn to entertain themselves and to discover their constantly changing surroundings. They learn to make friends quickly and earnestly. They learn to live in close quarters and to respect others. They learn to help with chores that need to get done and to obey orders that need to be obeyed. They learn to keep themselves safe and to be careful. And above all, they learn to love the new and the unfamiliar, to embrace the unknown. Look at your children's faces as they peer into a tidal pool or skip up the dock at a new port of call, and you will be glad you didn't leave them at home.

Children's books about boating are bountiful. Here is a partial list of the books our children have loved while sailing:

Andre, Lew Dietz
Bert Dow, Deepwater Man,
   Robert McClosky
Boris and Amos, William Steig
Captain Pugwash, John Ryan
Going Lobstering, Jerry Pallotta
Keep the Lights Burning Abbie,
   Peter and Connie Roop
Mr. Bear's Boat, Thomas Graham
One Morning in Maine,
   Robert McClosky
Salty Dog, Salty Sails North,
   Gloria Rand
Sea Story, Jill Barklem
Where the Wild Things Are,
   Maurice Sendak
Sarah's Boat, Douglas Alvord
Scuppers the Sailor Dog,
   Margaret Wise Brown

FOR THE BOAT. *Seal Harbor Yacht Club (207-276-5888).* The Seal Harbor Yacht Club is perched on the rocks on the east side of the harbor with stairs and an elevator leading down from the road above. Water and electricity are available at the floats, with about 6 feet alongside at low, and there are restrooms in the beautiful clubhouse. Contradances are held every Wednesday night.

Town Wharf. The town wharf is a substantial granite dock on the east side of the harbor, just south of the yacht club. The two 40-foot floats have 10 to 13 feet alongside at low, drinking water, and electricity. Dockage is limited to two hours on the outer float and 15 minutes on the inner float.

FOR THE CREW. Seal Harbor is half a mile from the town wharf, past the head of the harbor. The crossroad village has a post office, the Lighthouse Restaurant (*276-3958*), a gas station, and the Village Market (*276-4094*) with essential groceries, beer, wine, soda, ice, and ice cream.

THINGS TO DO. A highlight for many visitors to Acadia is Jordan Pond House, a unique restaurant that serves luncheon, tea, and dinner (*276-3316*). It is about a 2-mile walk from the Stanley Brook entrance to the park, at the head of Seal Harbor, but you can probably hitch a ride. A sign at the restaurant reveals something of its history: "In the late 1800's and early 1900's fashionable cottage dwellers and wealthy visitors dined in rustic style at the Jordan Pond House. Built as a simple farmhouse in 1847, it later expanded into a restaurant with several wings. Fire ravaged the entire structure on June 21, 1979. Rising from the ashes a new tea house and restaurant has replaced the old, continuing a tradition of gracious dining in an atmosphere of genial surroundings—luncheon on the porch, tea on the lawn, dinner by the fireside. Now, as then, specialties include chicken, lobster, homemade ice cream, and fresh popovers—big, brown, and featherweight." Tea on the lawn has been a special event for generations of patrons. Jordan Pond House is not a secret, however. Make reservations well in advance.

North of the restaurant, a number of lovely trails skirt Jordan Pond to the Bubbles, Asticou, Sargent Mountain, and Pemetic Mountain.

One of the most delightful ways to see Acadia National Park is by horse-drawn carriage over the 45 miles of carriage roads that meander through the eastern part of the park. Wildwood Stables (*276-3622*) will pick you up for one of their scheduled tours.

FRENCHMAN BAY

Frenchman Bay commemorates Samuel de Champlain, the early French explorer who visited Mount Desert in 1604. It is Maine's most dramatic bay, dominated by the pink granite of Cadillac Mountain, the highest point on the eastern seaboard. Champlain Mountain rises from the ocean, then Schooner Head, Great Head, and Otter Cliffs. Rolling, wooded Schoodic Peninsula frames the bay to the east.

The smooth ledges of Egg Rock and its welcoming lighthouse mark the middle of the entrance to Frenchman Bay. A procession of bold and cliffy islands and peninsulas lies to the north—the Porcupines, Ironbound, and Grindstone Neck. There are caves in the seaward cliffs of these islands and narrow passages that are a pleasure to thread under sail—Halibut Hole between Ironbound and Jordan, the slot between Rum Key and Long Porcupine, and the passage west of Sheep Porcupine.

Only a few good harbors grace the Bay: Bar Harbor, Winter Harbor, and Sorrento, but the sailing couldn't be finer. Below the Porcupines, the bay is open to the ocean swells. North of the Porcupines, the bay is calmer and the shores less dramatic.

Much of the natural beauty of Frenchman Bay is protected. Acadia National Park encompasses the mountains of Mount Desert and most of Schoodic Peninsula. The Nature Conservancy preserves Turtle Island in the south, Long Porcupine, and Dram and Preble Island near Sorrento.

BAR HARBOR

3 ★★★

44° 10.63′N 068° 32.92′W

CHARTS: 13323, 13312, 13318   CHART KIT: 73, 23

Bar Harbor was called Eden when it appeared on the map in 1796. Set against the awesome mass of Cadillac Mountain and with the bold islands and Schoodic Peninsula lying offshore in the sparkling waters of Frenchman Bay, it is easy to see why. But Eden was not named for its staggering beauty. Instead, Massachusetts, which at the time controlled all of Maine, named it after an obscure Englishman. The locals, being more practical, called it Bar Harbor for the long gravel bar that extends across the head of the harbor, and after 125 years of their stubbornness, the name of the town was officially changed.

In the mid-1800s, with the advent of direct steamer service from Boston, Bar Harbor developed into a summer Eden for the very wealthy, who built the vast hotels and great "cottages" that still line the streets. But now, Bar Harbor is an Eden for the everyman. The cottages have new lives as colleges, museums, nursing homes, and inns, and the streets are jammed with the festive bustle of tourism from dawn until late at night.

As a yachting center, Bar Harbor is small compared to Southwest or Northeast Harbor, but the town has a variety of attractions. There is a constant coming-and-going of boats of every size and description—from traditional schooners to excursion boats and cruise ships. Sightseers line the wharf and scrutinize every move, and a few fishermen actually manage to make a living amid the fray.

As a port of entry, Bar Harbor is serviced by a seasonal, local Customs Office (*207-288-4675*), but private vessels checking into the U.S. should call the Small Boat Reporting Center at *207-532-2131 ext. 255*.

APPROACHES. All of the moorings and shore facilities of Bar Harbor are east of the bar connecting Bar Island to Mount Desert, and the approach is without dangers except in fog. The usual path from the south is east of bold Bald Porcupine Island, which has cliffs at its southern end. It is also possible to run between the breakwater west of Porcupine Island and the Mount Desert shore. The breakwater is covered at high, but its western end is marked by a white beacon.

In fog, the real dangers lie in the amount of relatively high-speed vessel traffic coming and going from Bar Harbor—the ferries, the excursion boats, and the cruise ships. Monitor your radio and place a *securité* call for peace of mind.

ANCHORAGES, MOORINGS. Bar Harbor maintains a number of rental moorings east of the large town pier, separated from the private moorings to the north by an approach channel. A mooring can be reserved by calling Harbormaster Charlie Phippen (*Ch. 09, 16; 288-5571, cell 266-2110*). The harbormaster also rents overnight dockage on the two town floats east of the pier with 8 to 9 feet at low. Additional dockage can be found at the Harborside Hotel's marina or at the Bar Harbor Regency north of the bar.

You can anchor outside of the moorings in 15 to 30 feet at low with good holding ground and fairly good protection from the breakwater to the south and the surrounding islands. Allow plenty of room to swing, since the first part of the ebb runs over the bar to the north and the current can hold boats at different angles to the wind.

A mile north of town, just beyond the ferry terminal, the Bar Harbor Regency Hotel marina can

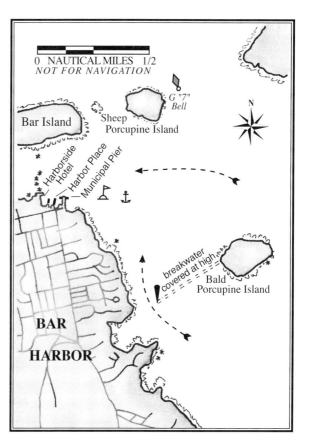

accommodate yachts up to 120 feet, with deep water along the outer floats.

FOR THE BOAT. *Municipal Pier (Harbormaster Charlie Phippen: Ch. 09, 16; 207-288-5571).* Floats on the east side of the pier are for pleasure boats; floats on the west side are for fishermen. Overnight dockage is rented by the harbormaster, whose office is on the pier. Water and electricity are available.

Harbor Place (207-288-3322). Harbor Place is the building next to the municipal pier, owned by and home to Bar Harbor Whale Watch. Their fuel dock, however, is open to the public, with gas, diesel, water, and pump-outs. Owner Bob Collier also has five or six rental moorings. Reservations should be made in advance at 288-5410.

Harborside Hotel and Marina (207-288-5033; www.theharborsidehotel.com). Harborside's marina is west of the town pier. Their docks can accomodate vessels to 160 feet and have 30 and 50-amp power and water. Dockage fees include the full use of the hotel amenities, including an outdoor pool, showers, Jacuzzi, and fitness room.

Bar Harbor Regency Hotel (Ch. 09, 16; 207-288-9723) is located north of town and the Porcupine Islands. They have slips for transients to 160 feet and one mooring. Water, ice, electricity, showers, and laundry are available, but no fuel. Visitors may eat at the hotel or have food delivered to the floats.

FOR THE CREW. Restrooms and phones are located on the town pier by the harbormaster's office. From there, Main Street (with all the lampposts) heads up the hill, and Cottage Street leads off to the right.

The well-stocked Bayside Liquors is on Main, on the right as you head uphill. A large Hannaford is a short walk down Cottage Street. Natural foods can be found on Mount Desert Street, farther up Main and to the right.

On Cottage, you'll find other essentials: the post office, a Rite Aid pharmacy, a FedEx box, fax and copy services, a hardware store, a small Radio Shack, a laundromat, and—at the far end—an Irving gas station that sells block ice (apparently the only source in town). Also on Cottage is the 1930s art deco Criterion Movie Theater, listed on the National Register of Historic Places and still showing movies.

The nearest public showers are at the YWCA (*288-5008*) at 38 Mount Desert Street or at the new YMCA (*288-3511*) farther out Main and then to the right on Park Street.

Bar Harbor probably has more restaurants per square mile than anywhere on the Maine coast, and by walking the downtown streets you'll discover most of them. If you take it black, you can still get a cup of coffee for 20 cents at the old-fashioned soda fountain at West End Drug, but you'll be bumped up to 25 cents if you want cream, and the Governor will hit you for another two cents.

On the waterfront, Fisherman's Landing and Stewman's both offer lobsters in the rough and waterfront bars. Rosalie's, on Cottage Street, makes superb pizzas while you watch. Nakorn Thai, on Cottage Street, serves excellent Thai food. Head to the end of Cottage Street to Cafe Bluefish (*288-3696*) for exquisite fish. Elsewhere you'll find superb French, Italian, Mexican, and Downeast fare.

For a Mount Desert dessert, stop in at Ben and Bill's Chocolate Emporium and try the lobster ice cream. Inventor Jeff Young says that the salty flavor of the lobster goes well with sweet vanilla. "But," he adds, "I've also got a few things in there that kind of beat the fishy taste."

Bar Harbor has no shortage of taxis, and it is the hub of the free Island Explorer shuttle for exploring the rest of the island. The Mount Desert Island hospital can be reached at *207-288-5081*.

THINGS TO DO. There is no end to the exercise you can get around Bar Harbor. At halftide or lower, you can cross the gravel bar at the head of the harbor and explore the western half of Bar Island, part of Acadia National Park. The Shore Path is another pleasant walk. It starts in front of the venerable Bar Harbor Inn, east of the town pier, and leads south along the shore for half a mile to Grant Park, with its balancing rock.

Most bookstores sell guidebooks to the hikes and climbs in Acadia National Park, and the free shuttle can get you to their trailheads. Even rock climbing instruction can be arranged through Acadian Mountain Guides (*288-0342, 288-8186*). Lightweight mountain bikes can be rented on Cottage Street for touring the miles of carriage roads that wind through the interior of Acadia National Park. The moderate grades meander through spectacular scenery, and the bikes are so well suited to the terrain that it almost seems as if the roads were built for the bikes. Take lunch and liquids with you. Sea kayak tours and rentals can be arranged through several outfitters on Cottage Street.

If you've gotten your fill of fitness on the boat, tour buses depart several times a day from the town green above the town pier to circle Acadia's Loop Road and climb Cadillac Mountain. Call National Park Tours (*288-3327*) or Olie's Trolley (*288-9899*). Numerous nature, whale, and sightseeing cruises offer daily trips into the bay. The three-masted schooner *Natalie Todd* and the converted sardine carrier *Francis Todd* sail from the dock at the Bar Harbor Inn. Or for some real sailing, gliders leave from—and hopefully return to—the airport (*667-SOAR*).

Mount Desert Street, off Main, is one-stop shopping for Bar Harbor history. The Abbe Museum, in a renovated shingle-style building, celebrates the history of Mount Desert's earliest inhabitants. The Jesup Memorial Library picks up the story with an exhibit on the "rusticators" and their summer cottages, visitors, hotels, steamers, and the Green Mountain Railway. And "Ledgelawn," one of the last great cottages, built in 1904 by a wealthy Bostonian, is open to the public. The sweeping staircase, the grand fireplace, and the original furnishings are monuments to the elegance of old Bar Harbor, much of which was lost in the disastrous fire of 1947.

For natural history, visit the Natural History Museum at College of the Atlantic (*Ch. 16, 288-5015*). Less than a mile north of the town pier, the waterfront campus can be reached by land or sea. The college has several large guest moorings and 11 feet of water at low on the end of the float. You can also anchor off in 15 to 30 feet. By land, walk out West Street to Route 3.

The campus features a former summer estate appropriately named "The Turrets," and you can enjoy the lovely, historic terraced gardens, waterfront landscape, and pebble beach for a quiet stroll or picnic. The museum displays birds, mammals, and the huge skull of a finback whale and has interpretive programs and a natural history speaker program. The college is headquarters to the Allied Whale Project, which studies whales in the Gulf of Maine. The program is funded in part by the Bar Harbor Whale Museum, on West Street opposite the waterfront.

A more touristy but still interesting stop is the Mount Desert Oceanarium (*244-5005*), 5 miles north of town on Route 3, with exhibits on marine life that include whales and live harbor seals. Whale-watching is big business in Bar Harbor. The 92-foot whale-watching catamaran *Friendship IV* cost an estimated 1.5 million dollars, and it can take you to the whales at about 26 knots.

Large and larger on Bar Harbor's town floats.

For mammals of a different kind—and certainly this is why you have sailed so far!—Jackson Laboratory is the world's largest breeder of genetically pure mice for mammalian genetic research. It is also Bar Harbor's largest employer. More than 700 pure mouse strains have been bred for decades, and each year two million mice are shipped worldwide to labs for research on cell biology, immunology, genetics, pathology, and drugs. The lab (*288-3371*) is located south on Route 3, 1.7 miles from the waterfront, and it offers a fascinating, hour-long presentation for visitors several afternoons a week.

Throughout the summer season, the Bar Harbor Festival and the Arcady Music Festival hold regularly scheduled concerts, and many of the bars have live music. There are two dinner theaters in town. In July and August, evening band concerts are performed on the Village Green twice a week.

For a real side trip, the *Cat*, a high-speed catamaran ferry, leaves Bar Harbor three or four times a week and blasts to Yarmouth, Nova Scotia in less than 3 hours and returns the same day (*877-359-3760; www.catferry.com*).

HULLS COVE

3 ★★

44° 24.59'N 068° 14.33'W

CHARTS: 13323, 13312, 13318   CHART KIT: 73, 23

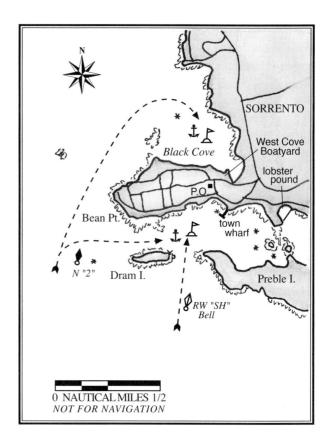

Hulls Cove is a couple of miles north of the bustle of Bar Harbor. Here the pace is slower and more peaceful. The cove, however, is exposed and offers little protection. A cluster of boats is moored off the small clubhouse of the Bar Harbor Yacht Club.

APPROACHES. From Bar Harbor, round the bell off Sheep Porcupine Island and run alongshore past the *Cat* ferry dock. The Bar Harbor Yacht Club is near Canoe Point, inside nun "4."

ANCHORAGES, MOORINGS. The BHYC has a guest mooring, or you can anchor off in 24 feet at low. The location is quite exposed.

FOR THE BOAT. *Bar Harbor Yacht Club (207-288-3275).* This modest club has a guest mooring and water at the floats, with 12 feet alongside at low. The club has few other facilities, but the visiting yachtsman will find plenty of kindred spirits.

FOR THE CREW. Hulls Cove Village, about .7 miles away to the north, has a pizza parlor and a general store that sells groceries, beer, and ice.

MOUNT DESERT NARROWS

1 ★

44° 26.60'N 068° 20.00'W

CHARTS: 13318, 13312, 13316   CHART KIT: 73, 23

Mount Desert was once an island that could be circumnavigated, going from Eastern Bay into Western Bay through Mount Desert Narrows. While this is still possible for smaller boats, the Narrows is shoal and unmarked, and the fixed bridge has a vertical clearance of only 35 feet.

West of Sorrento and Back Cove, there are few anchorages or facilities, but the open waters are perfect for pleasant sailing against the magnificent backdrop of Mount Desert.

Sullivan Harbor leads to Taunton Bay, but do not plan to go above the reversing falls at Falls Point, which are turbulent and dangerous (see the *Coast Pilot*). The swing bridge just above the falls is normally closed.

The Skillings River degenerates rapidly into a narrow, crooked passage with many unmarked dangers, as does the Jordan River, farther west.

Past Old Point, just beyond the entrance to the Jordan River, you will see the large sheds of Morris Yachts on the mainland. Note the rocks and shoals west of the yard.

ANCHORAGES, MOORINGS. Morris may have moorings available, with 10 feet of water at low. The depths here are somewhat greater than shown on the chart.

FOR THE BOAT. *Morris Yachts (207-244-5509).* Morris builds rugged cruising boats with the beautiful lines of Chuck Paine's designs. This is their production facility. Their service and charter departments are in Bass Harbor and Northeast Harbor.

FOR THE CREW. A short walk from the yard will take you west to Route 3 and a selection of restaurants. You can rent a car at the neighboring airport to explore Mount Desert and Acadia National Park.

SORRENTO HARBOR

2 ★★★★

44° 28.17′N 068° 11.11′W

CHARTS: 13318, 13312   CHART KIT: 73

Beautiful Sorrento Harbor nestles between wooded islands and a graceful summer community on the mainland. The sun sets behind the vista of the Cadillac Mountain range on Mount Desert Island and washes Sorrento with sweet light. Sorrento was founded as Waukeag Neck, a resort in the 1880s. The atmosphere was one of low-key elegance. In the early days, a steamship sailed from Bar Harbor to the docks of the great hotel at Sorrento. Nowadays residents take their Boston Whalers.

Fifteen-foot Wee Scots were the first one-design class raced by the Sorrento Yacht Club, in 1926. Despite the later introduction of S-boats and Mercurys, a number of the stable little Wee Scots still exist, some owned by their original families.

Part of Sorrento's beauty is the view of Mount Desert between unspoiled Preble and Dram Island. In the mid-1960s, when it appeared that Dram Island might be clear-cut for pulpwood, summer resident Bayard Ewing made an extensive search to locate the owner, who turned out to be a minister living in Wyoming. The minister needed an organ for his church. Ewing swapped an organ for the island and then donated the island to The Nature Conservancy. The protection of the harbor is now complete with the donation of Preble Island as well to The Nature Conservancy by the Ewing family.

APPROACHES. Run up Frenchman Bay for red-and-white bell "SH" (*44° 27.84′N 068° 11.16′W*), which marks the entrance, and enter halfway between Dram and Preble Island, both high and wooded. Ledges make out from both sides. From the west, it is also possible to enter between Dram Island and Bean Point. Since ledges extend almost halfway across from the north, favor the Dram Island side.

ANCHORAGES, MOORINGS. The Sorrento Yacht Club puts out three guest moorings at the west end of the harbor. Two are memorial moorings with wooden pickup floats. One is marked, "Robert M. Lewis, 1886-1958. He cruised." The other is for Robert E. Montgomery. What a unique way to be remembered.

Anchor anywhere you can find room outside the moored boats, in 6 to 15 feet at low. Holding ground is good mud. Do not approach the two little islands (each with a clump of trees) marking the ledge that separates Sorrento Harbor from Eastern Point Harbor.

One pays a price for the view. It's likely to be a bit rolly, and there is considerable exposure to the west and southeast or southwest, depending on where you are in the harbor. If there is no room or if strong southerlies are expected, head to Black Cove.

GETTING ASHORE. Dinghy to the town wharf.

FOR THE BOAT. *Town Wharf.* Water and electricity are available at the town wharf, with 4 feet alongside at low. Rocks lurk off the west end, so at low tide, approach the floats from the east.

Sorrento Yacht Club. The yacht club uses the town wharf and has no facilities except the guest moorings. Their office is in the Sorrento library.

FOR THE CREW. The town wharf has a phone.

THINGS TO DO. Take a beautiful walk westward along the harbor and back through the streets of this appealing community. Visit the Sorrento Public Library, in an 1892 shingled building.

BLACK COVE

4 ★★

44° 28.66′N 068° 11.10′W

CHARTS: 13318, 13312   CHART KIT: 73

What you gain on the swings, you lose on the roundabouts. Just around Bean Point from Sorrento, Black Cove is much better protected, but you lose the spectacular view.

APPROACHES. Approach around Bean Point. There is a long halftide ledge running north from the southern side of the entrance to Black Cove. A rock in the middle of the entrance is barely visible at low tide. To avoid these dangers, come in near the northern side of the entrance and run down the eastern shore.

ANCHORAGES, MOORINGS. The West Cove Boat Yard has several moorings. Anchorage is good in the middle of the cove in 7 to 9 feet at low, in a mud bottom.

FOR THE BOAT. *West Cove Boat Yard (Ch. 07; 207-422-3137).* The yard is a custom builder of wooden boats. It has a 20-ton marine railway plus a 15-ton crane and can do hull and engine repairs.

EASTERN POINT HARBOR

44° 28.10'N 068° 10.17'W

Charts: 13318, 13312   Chart Kit: 73

Eastern Point is primarily a working harbor. A lobster pound is penned off in the northwest corner (wholesale only), and many lobsterboats are moored on stakes here.

APPROACHES. The entrance between Calf and Preble Island is open and easy. Head for the small, wooded island on the bar that separates Sorrento from Eastern Point.

ANCHORAGES, MOORINGS. Anchor outside the moored boats wherever you can find room in 8 to 15 feet at low, mud bottom. Protection is not as good as it appears on the chart, since southerly winds are funneled in by Calf Island.

FLANDERS BAY

between Schieffelin Point and Ash Island:
44° 28.63'N 068° 08.63'W

Charts: 13318, 13312   Chart Kit: 73

Flanders is a large, round bay in the northeastern corner of Frenchman Bay, big enough to hold a whole cruising club. Reached by a zigzag route, it is almost entirely landlocked and would be an excellent hurricane hole if it were not so large. Seals are often on Halftide Ledge, and sometimes a bald eagle soars overhead.

APPROACHES. There are two possible approaches, either north or south of Calf Island. The most direct route is south of Calf, leaving can "1" close to port. Note that there is a 3-foot spot to the east. Continue around can "3" at Halftide Ledge and northwest again past Schieffelin Point (with several houses, trees, and large meadows) into Flanders Bay. The route north of Calf Island is more open but longer.

ANCHORAGES, MOORINGS. The bay shoals gradually. Anchor in whatever section provides the best protection from the wind in 8 to 20 feet of water at low. The mud bottom is good holding ground.

STAVE ISLAND HARBOR

Myrick Cove: 44° 25.02'N 068° 06.71'W

Charts: 13318, 13312   Chart Kit: 73

On the east side of Frenchman Bay, Stave and Jordan Island form Stave Island Harbor. While this would be a good harbor for small ships, it is much too large and open to provide a comfortable anchorage for cruising boats. Reasonable protection can be found in Myrick Cove, east of Jordan Island.

WINTER HARBOR

(For Protection and Facility ratings, see individual harbor descriptions below.)

Charts: 13322, 13312, 13318   Chart Kit: 74 (A), 73, 23

Winter Harbor is tucked between rugged Schoodic Peninsula and the handsome summer community of Grindstone Neck. The town's population is equally distinct: well-heeled summer folk and staunch fishermen, retirees, and, until 2002, navy personel.

In 1935, John D. Rockefeller, feeling that the Navy radio installation on Otter Cliffs on Mount Desert was unsightly, paid out of his own pocket to have it moved to nearby Big Moose Island, at the end of Schoodic Peninsula. For decades, this was the east coast's best site for radio reception. In 2002, however, the technology was deemed obsolete, and the station was closed.

The town lost nearly half of its population, from 988 to 492. After the closing, the 180-student school only had 27 students. Much of the Navy housing was in the town, and the town was able to sell all of it for a nearly $5 million windfall. But the buyers weren't like the young families with children that had left. Instead, they were retirees or middle-aged summer people. The base itself, located on National Park Service land, reverted to the NPS, which plans to convert this complete village into a learning and educational center. Another 405 acres will be deeded to the U.S. Fish and Wildlife Service as a refuge.

Every August, the Winter Harbor Lobster Festival draws a large crowd of spectators, especially for the main event, the annual lobsterboat races, with 13 classes rated by size and power. Speed is the objective. Not only do the races generate intense rivalries and devoted fans, but they teach fishermen and boatbuilders how to improve hull design and engines.

The harbor provides varying degrees of protection depending on which part you choose. There are three coves. To the west is Sand Cove, where the Winter Harbor Yacht Club is located. It is quite deep and wide open, useful for a summer fleet, but not for boats working in the winter. The fishermen, who were here first, sensibly chose Inner Winter Harbor, a tight, almost landlocked hole in the middle. Henry Cove, to the east, is exposed to south and southwest winds and is the least used. The anchorages and facilities of these three coves are considered separately, below.

APPROACHES. There are several good landmarks as you approach Winter Harbor—the big, white lighthouse building on Egg Rock to the west, the white abandoned lighthouse on Mark Island at the entrance to the harbor (visible a long way), and a green water tank on Big Moose Island, at the end of Schoodic Peninsula.

Starting with red-and-green bell "MI" (*Fl (2+1) G 6s, 44° 21.48'N 068° 05.11'W*) off Mark Island, the approach to Winter Harbor is wide and easy. Follow Grindstone Neck around into Sand Cove, or follow the buoys to Inner Winter Harbor and Henry Cove.

SAND COVE (Winter Harbor)

2 ★★★

44° 23.00'N 068° 05.40'W
CHARTS: *13322, 13312, 13318* CHART KIT: *74 (A), 73, 23*

Grindstone Neck is no more than five miles east of Bar Harbor by water. Inspired by its proximity, a group of New Yorkers and Philadelphians bought it in 1889, "for development into a cottage colony of the character, for instance, of Tuxedo, of Llewellyn Park, or of North East Harbor."

In time, the great cottages of Bar Harbor have been recycled as nursing homes, museums, and motels, but the summer colony on Grindstone Neck has maintained its character. Though socially exclusive, the people of Grindstone Neck have always believed that the beauties of the land should be open to everyone, so there is access to the shoreline, and "the rocks are free."

The handsome Winter Harbor Yacht Club is on the west side of Sand Cove. Proudly moored in front of the club is their fleet of all nine Winter Harbor

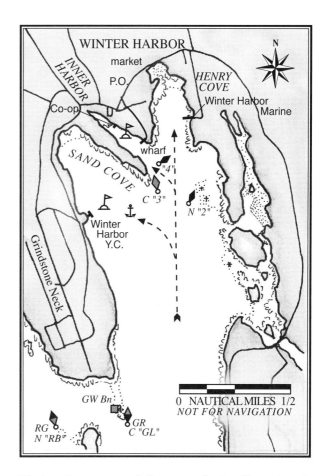

21s in existence, carefully researched, collected, and meticulously restored.

APPROACHES. From red-and-green gong "MI" (*Fl (2+1) G 6s, 44° 21.48'N 068° 05.11'W*) off Mark Island, follow Grindstone Neck around into the cove, staying well clear of Grindstone Ledge to port.

ANCHORAGES, MOORINGS. For a mooring rental, ask the launch attendant or at the Winter Harbor Yacht Club clubhouse, the large, gray-shingled building on the west side of the cove whose cupola is shown on the chart. The mooring floats are mostly white lobster buoys, marked "Guest."

You can anchor off to the north or outside the moored boats, as many yachts do. Holding ground is reported to be good in the vicinity of the club but questionable north of the mooring field. The anchorage is open and exposed to the south and likely to be quite rolly.

GETTING ASHORE. Land at the yacht club floats or call the launch.

FOR THE BOAT. *Winter Harbor Yacht Club (Ch. 16 or 71; 207-963-2275).* The yacht club floats have water and electricity, with 20 feet alongside. Stay clear of the southern floats, which are private.

FOR THE CREW. The yacht club is hospitable to visiting yachtsmen and offers launch service and showers. Lunch is served in the comfortable clubhouse, where a fire burns on cold days. Look for the grindstone set in the fieldstone chimney and an old print of the original land division on Grindstone Neck. There is a pay phone.

Walk north on the shore road for just over a mile, past the golf course, to the town of Winter Harbor. Once past the fieldstone-and-shingle library, you will find J.M. Gerrish Provisions, a gourmet coffee, sandwich, and ice cream shop. A classic 5&10¢ store is across the street, the last of its breed. The school, with its playground, is around the corner.

A good IGA market (with ice) is down the hill and so are the post office and several restaurants, including Chase's (*963-7171*), an affordable local hangout, and the Fisherman's Inn Restaurant (*963-5585*). Often a yacht club member will spot you lugging groceries and offer a ride back to the clubhouse.

THINGS TO DO. It is a pleasant walk of about a mile from the yacht club to the end of Grindstone Point, with its great smooth whalebacks of pink granite striped by dikes of black basalt. The mountains of Mount Desert lie on the horizon, beyond the islands scattered in the bay.

TERNS

Terns are the acrobats of the air. With their lithe, quick-winged bodies, they hover above the sea, then dive straight down into the water with a delicate splash to seize tiny fish, and then return effortlessly to the air with a beat or two of their strong wings. Terns are often called sea swallows. Their bodies are white, with a long, white forked tail. The tops of their swept-back wings are gray, and their heads wear a black cap.

Terns and seagulls are natural enemies. In the places where puffins and auks thrive, such as Matinicus Rock and Machias Seal Island, there is usually a large population of scrappy terns. They make a lot of fuss and chase off intruders, including gulls. But given any opportunity, the gulls will eat tern eggs and chicks and sometimes will drive terns from a nesting site.

The resident naturalist on Machias Seal Island tells this story: Being in mortal competition with gulls, terns react with fury to the orange spot under the bill of a herring gull. One day a cruising sloop arrived at the island with a white mainsail and orange spinnaker. As they dropped the spinnaker, it made a crescent of orange under the white mainsail, forming the world's most enormous seagull. The whole tern population went wild.

Near the end of the 19th century, terns were hunted for their feathers to grace milady's hat, and by 1900, they were almost wiped out on the coast of Maine. One group of feather hunters reported a kill of 1,200 birds in a single day, and as many as 100,000 were killed in a season. Although long since protected from such deliberate slaughter, terns have come back slowly.

Terns and gulls and people are involved in a story with confusing moral issues. Part of the proliferation of the herring gull and consequent pressure on terns was caused by the free food found in open dumps and by fish wastes thrown over the side from fishing boats. Not long ago, gulls took over Maine's largest nesting sites for arctic and common terns. Feeling responsible for redressing the original balance, the U.S. Fish and Wildlife Service, with the reluctant support of the Maine Audubon Society and other wildlife groups, poisoned the gulls on Petit Manan Island and neighboring Green Island.

The terns returned. The experiment was such an unqualified success that it has been repeated at other sites along the coast, the latest being Pond Island, in the mouth of the Kennebec. Today, 88 percent of Maine's terns live on managed islands.

Different kinds of terns are difficult to distinguish. The tern most frequently seen along the coast of Maine is the common tern, which breeds from Labrador to the Caribbean. The least tern is found along Maine's southern coast.

Three other species of terns are found in Maine during the summer. The roseate tern is the rarest, and probably most endangered. The arctic tern, aside from its grace and beauty, is extraordinary for its annual migration of some 22,000 miles—almost a complete circumnavigation of the earth. It breeds in the Arctic and down the U.S. east coast as far as Massachusetts. Then these terns cross the Atlantic and fly down the coasts of Europe and Africa.

The handsome clubhouse of the Winter Harbor Yacht Club is on the west side of Sand Cove, on Grindstone Neck.

INNER HARBOR (Winter Harbor)

4 ★★★

44° 23.30'N 068° 05.20'W

CHARTS: 13322, 13312, 13318   CHART KIT: 74 (A), 73, 23

This tight little harbor is home to Winter Harbor's fleet of lobsterboats. There is a purposeful bustle, starting early in the morning, and much to observe. It's close to town and an interesting place to spend the night if you can find a mooring.

APPROACHES. The entrance is marked by a can left to port and a nun left to starboard. Note the red daybeacon marking the ledges to starboard.

ANCHORAGES, MOORINGS. The inner harbor is jammed with lobsterboats, floats, and dinghies, and there is no room to anchor. On a given night, there may be two or three moorings available. Ask a lobsterman. Sometimes eight or nine yachts will be here, nested three on a mooring. This little harbor is extremely well protected.

GETTING ASHORE. Land at the town wharf.

FOR THE BOAT. *Town Wharf (Harbormaster Dale Torrey, 207-963-2235).* The town wharf is on the east side as you enter, just past the daybeacon. The float has 15 feet of depth at low. Water is available.

Winter Harbor Co-op (Ch. 06; 207-963-5857). The Co-op has a dock and a float near the head of the harbor on the east side, with water, electricity, gas, and diesel.

FOR THE CREW. The town of Winter Harbor is near the head of the cove, up the street from the town wharf (see *Sand Cove*, page 329).

HENRY COVE (Winter Harbor)

2 ★★

44° 23.45'N 068° 04.95'W

CHARTS: 13322, 13312, 13318   CHART KIT: 74 (A), 73, 23

Though somewhat exposed to south and southwest winds, Henry Cove is convenient to town and to the Winter Harbor Marina.

APPROACHES. Leave the nun marking the ledge off Sargents Point to starboard. Pass Inner Harbor and its entrance buoys, off to port, and run up the middle into Henry Cove.

ANCHORAGES, MOORINGS. Winter Harbor Marine has a number of rental moorings. The cove is exposed to the south and likely to be rolly.

GETTING ASHORE. Land at the marina floats.

FOR THE BOAT. *Winter Harbor Marine (207-963-7449).* Warren Pettegrow has transformed Winter Harbor Marine into a base for his Bar Harbor Ferry, but the yard still serves visitors by boats. The floats, with 6 feet at low, have water and diesel on tap. Although primarily serving the local lobster fleet, they could haul you with their hydraulic trailer in an emergency and orchestrate most repairs. They can also arrange long-term parking.

FOR THE CREW. WHM's Bar Harbor Ferry *(244-5882, 460-1981)* can run you into Bar Harbor in the morning and back in the afternoon—a nice way to take in the town without taking in your boat. From WHM it is half a mile into the town of Winter Harbor.

SCHOODIC POINT

Schoodic is a very special place. Once it belonged to one man, John Moore, who loved it so well that he opened it to the public in 1897. Now most of it is part of Acadia National Park.

You can stand here at the end of the world and sense the great forces that have formed the coast of Maine—upwellings of granite eons ago, then the intrusion of dramatic black basaltic dikes into the pink granite, then the glaciers carving and grinding down the rocks, and the ceaseless working of the wind and sea. Waves roll unchecked across the Atlantic and crash in dazzling spray upon the pink granite ledges.

Schoodic is difficult for the cruising boat to visit. The harbors on either side of the point—Pond Island, on the west shore, and Wonsqueak and Bunkers Harbor, on the east (see Down East)—are tricky at best. Of the three, Pond Island is the best choice for temporary anchorage and access to Schoodic Head.

A safer way to explore Schoodic is to leave your boat in Winter Harbor and hitch a ride or find a garage there willing to rent you a car.

is rocky and the holding ground dubious. Consider leaving someone aboard while you explore.

Land near the causeway that crosses the head of the road, which will put you right next to the gravel road leading up to Schoodic Head. It is a half-hour hike to the top of Schoodic Head and well worth it. From the turnaround, you can look west across Frenchman Bay to Mount Desert, the lighthouse on Egg Rock, and your own boat riding at anchor below.

A short trail continues to the granite ledges on top of Schoodic Head, with views east along the coast to the lighthouse of Petit Manan. Another vantage point overlooks Schoodic Island.

From your landing opposite Pond Island, it is a 1.3-mile walk to the end of Schoodic Peninsula, where great Atlantic swells break on the pink granite ledges.

For a less strenuous outing, explore West Pond in your dinghy.

POND ISLAND
44° 20.70'N 068° 04.15'W

CHARTS: *13322, 13312, 13318*   CHART KIT: *74, 23*

The anchorage north of wooded Pond Island is a great gunkhole and provides access for the cruising boat to the beauties of Schoodic Peninsula, most of which is part of Acadia National Park. Although the cove is reasonably easy to enter, the bottom is rocky and there is some exposure to the ocean, so it is not recommended as an overnight anchorage.

Big Moose Island is easily identified at the end of Schoodic Peninsula by the high green water tank. Wooded Pond Island lies close by to the north, and 440-foot-high Schoodic Head is prominent to the east. Enter north of Pond Island, favoring the northern side of the cove to avoid the ledges along the southern shore. The westernmost of these ledges is just visible at low tide.

Anchor beyond the ledges north of the eastern tip of Pond Island, in 11 to 16 feet at low. The bottom

Down East
Schoodic Point to West Quoddy Head

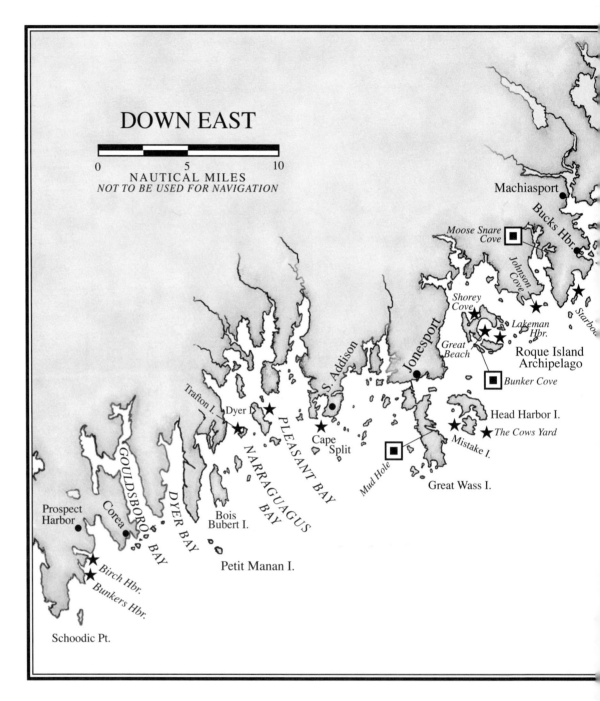

As you pass Schoodic Point heading east, civilization falls behind, and you enter a more primitive world—one where fishing and lobstering are all-important, and the affairs of Boston and New York seem far away and insignificant. All the coves are working harbors filled with lobsterboats and trawlers, where the docks are piled high with traps and bait barrels. Yachts are rare, and facilities for yachts are almost nonexistent. The last big marina lies astern, so now you must depend on your own resources more than ever.

This is a land of weirs and blueberry barrens, of uninhabited islands, narrow bays, and countless estuaries. Wildlife is abundant here, including seals and ospreys and bald eagles. You can see puffins and razorbills at close range on Machias Seal Island. There are lovely white granite and spruce islands in Eastern and Western Bay, and unusual harbors and a remarkable Nature Conservancy preserve in the area of Great Wass. Just beyond is fabled Roque Island, with its mile-long white sand beach, a goal for generations of cruising sailors.

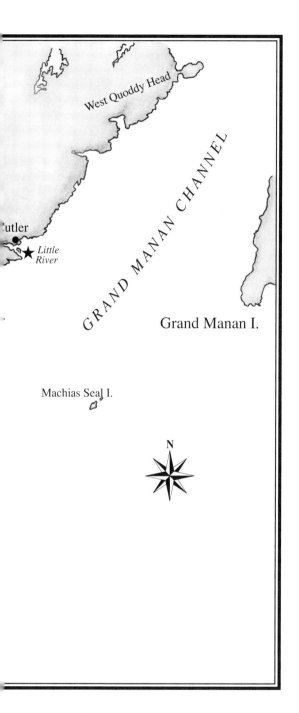

between the ocean and the air can build thick fog. Fog is most prevalent in July and August, when the air is warmest and the humidity highest. Radar and GPS, while not essential, certainly add confidence to navigating this part of the coast.

The few good harbors are spread over greater distances that the cruising grounds to the west, and the water between them is bold ocean. Before starting, it is wise to identify on the chart a number of possible harbors of refuge along the way, including Winter Harbor, Prospect Harbor, Trafton Island, Eastern Harbor, Mistake Island Harbor, and The Cows Yard—all of which are easy to enter and not far out of your way.

Running east from Schoodic, you will go progressively farther out of shore-based VHF range, and don't expect coverage for your cellphone. The U.S. Coast Guard station at Jonesport monitors channel 16 for emergencies.

Fuel for recreational boats is not readily available along this stretch of the coast. While many harbors have diesel or gas for their resident lobster fleets, it is generally sold on account at the commercial buying stations. In a pinch you can certainly tank up at the commercial wharves, but retail business is not encouraged. Plan your fuel use and range carefully, round Schoodic with full tanks, and keep topped off when you can.

Sailing Down East is very different from the typical cruising experience. This is where true voyaging begins. If you arrive in June or September, you will hardly see another cruising boat, and even in the summer months you'll probably be glad for the company you find. When you return to more populated cruising grounds, be prepared for a bit of a culture shock. You will feel like you have really been somewhere, and you have.

This is also a region of dense fogs, strong currents, and steadily higher tides as you approach Canada and the Bay of Fundy. Anchoring and docking in an area with tidal ranges of 20 feet or more require different techniques from those used with smaller tides. Starting with Cutler, the mean and large spring tidal ranges are given for every anchorage. Time your passage to take advantage of the currents, which flood east along the coast and ebb west.

The huge tides move enormous volumes of cold ocean water inshore, and the temperature difference

SCHOODIC PENINSULA
to Petit Manan and Cape Split

WONSQUEAK HARBOR
44° 21.69′N 068° 02.19′W
CHARTS: 13324, 13312, 13318   CHART KIT: 74, 75, 23

Wonsqueak is a crack in the rocks at the northern end of Schoodic Harbor. This narrow estuary has just enough room for half a dozen lobsterboats, moored fore-and-aft. The harbor is shoal, exposed to the south, and of no value to cruising boats.

BUNKERS HARBOR
3 ★★
44° 22.20′N 068° 01.75′W
CHARTS: 13324, 13312   CHART KIT: 75, 23

Bunkers is a snug harbor for a dozen or more lobsterboats on the east side of Schoodic Peninsula. It has an outer harbor with room for anchoring, but it is more exposed. The entrance is tricky. Unmarked ledges obstruct the entrance, and they are visible only within two hours of low tide.

APPROACHES. Enter from the south. From the green-and-red whistle buoy "BC" (44° 22.95′N 068° 01.43′W) off Spruce Point, run north along the shoreline at a comfortable distance. Dangerous Bunkers Ledge is marked by can "1" at its eastern, outside end. Some of the ledge is visible two or three hours after low. Come in close along the shore and pass inside the ledge, staying between the large reddish rock near the southern tip of the harbor and Bunkers Ledge to starboard. Once past the southern point and beyond the 7-foot sounding shown on the chart, turn to port and run down the midline of the harbor.

ANCHORAGES, MOORINGS. The inner harbor has depths of 6 to 7 feet at low, but it is jammed with lobsterboats. The southernmost of the moorings is owned by Bunker's Wharf Restaurant and is available on a first-come, first-served basis.

Alternatively, anchor in the outer harbor in 23 to 25 feet at low. Protection is reasonably good although some swells enter, even in settled weather, and you may feel like you are in a fish-bowl view of a large house on the shore.

GETTING ASHORE. Row in to the floats near the head of the harbor or to Bunker's Wharf Restaurant.

FOR THE CREW. Bill Osgood runs the Bunker's Wharf Restaurant (963-2244) on the southernmost wharf on the west side of the harbor. The wharf has a float that can handle powerboats to 48 feet. The deepest water is at the southern end of the float. Reservations are strongly recommended for dinner, and they are not a bad idea for lunch either.

It is about a mile walk northward to a variety of stores at the intersection of Route 186.

BIRCH HARBOR
44° 22.95′N 068° 01.40′W
CHARTS: 13324, 13312   CHART KIT: 75, 23

Birch Harbor is mostly shoal, unmarked, and exposed to ocean swells. It would make an uncomfortable anchorage. Add to that a difficult entrance, where Roaring Bull breaks at low water.

The intrepid can enter along the southwestern shore. At the head of the harbor, a complex of stores includes a market, laundromat, liquor store, post office, and takeout.

PROSPECT HARBOR
3 ★
44° 24.10′N 068° 01.28′W
CHARTS: 13324, 13312   CHART KIT: 75, 23

The westerly bight of Prospect Harbor is easy to enter and provides reasonably good protection in summer weather, although it is exposed to the south and southeast. A dozen lobsterboats moor in the inner harbor. There are no facilities for visitors.

The harbor is dominated by the Stinson Canning factory, on the west side, and there are houses and a road all around the head of the cove. The radio antenna on Cranberry Point and the white dome on Prospect Harbor Point further detract from the scenery.

APPROACHES. Gong "3" and green daybeacon "5" are both left to port. The lighthouse on Prospect Harbor Point marks the eastern shore.

ANCHORAGES, MOORINGS. Anchor past the cannery, outside the moored boats, in 14 to 18 feet at low. The mud bottom holds well.

A CRUISING GUIDE TO THE MAINE COAST

COREA

between "REP" and Western Island:
44° 23.50'N 067° 58.12'W
off white ledges: 44° 23.68'N 067° 57.97'W

CHARTS: *13324, 13312*   CHART KIT: *75, 23*

Corea is the quintessential lobstering port, remote, well protected, and charming, with room, perhaps, for one or two visiting yachts. It is home to a fleet of about 40 lobsterboats, and the surrounding docks are piled high with traps. The summer community is relatively small, and there is no doubt that the fishermen are in charge. This is the home of Young Brothers, well-known builders of hundreds of lobsterboats.

Corea, however, is a difficult place to visit by boat. The harbor is shallow and crammed with working boats. If you are lucky, you may be directed to a vacant mooring for the night. Arrive early, and have a backup plan in case there is no room.

APPROACHES. The *Coast Pilot* says that in May 1984, a depth of 7½ feet could be carried to the anchorage, where 6 feet is available, except for shoaling along the edges. Count on at least a foot less water now.

From a distance, you can see a white church steeple at the head of the harbor and a huge circular antenna west of Corea on the same peninsula. The antenna sinks below the trees as you enter the harbor. On radar, lobstermen say, it looks just like a miniature doughnut.

Western Island is higher and far more wooded than Outer Bar Island. Western Island has a red-roofed cottage on the south side and smooth, sloping reddish rocks.

Because several unmarked rocks and ledges are covered at high tide, this is not an easy harbor to enter. Avoid coming in at dead high, dead low, or in poor visibility. Start at red-and-white whistle buoy "CE" (*44° 22.92'N 067° 57.79'W*). Pass west of Western Island, fairly close aboard, curving eastward around the northern tip of the island. Then head for the sloping white granite ledges on the eastern side of the entrance to Corea Harbor. When you have almost reached these ledges, turn to port and head up into the harbor, staying toward the right-hand side of the channel as you pass the gray buildings of the Corea Lobster Co-op.

This will keep you clear of the nasty rock ("REP" on the chart) and the two ledges in the mouth of the entrance. At high tide the inner and higher of these two ledges may show a slight ripple on the surface, or it may not. These ledges are shown on the chart, but they are very deceiving, and many boats have found them with their keels. They are on a line from the middle of Western Island to the center of the harbor entrance, and you must be east of this line while approaching the entrance.

ANCHORAGES, MOORINGS. There appears to be room to anchor opposite the Co-op, but there

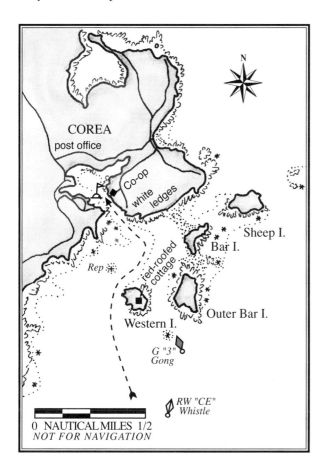

are hidden rocks. Elsewhere, the harbor is too full to anchor. You will need to ask a lobsterman if there is an available mooring. Try the Co-op or Young Brothers (*963-7467*), the second pier on the left.

GETTING ASHORE. Land at the Co-op or at the low-tide beach across the harbor.

FOR THE BOAT. *Corea Lobster Co-op (207-963-7936).* Gas, diesel, and lobsters can be bought at the floats below the red buildings of the Co-op, at the right side of the entrance, with 8 to 10 feet alongside.

FOR THE CREW. The post office is at the head of the harbor.

GOULDSBORO and DYER BAYS

Gouldsboro and Dyer Bays have little to offer the cruising sailor. Both are five or six miles long, but they hold no significant harbors, and their northern ends dwindle to narrow, unmarked estuaries.

Gouldsboro Bay is probably the more attractive of the two. The handsome Sally Islands lie across the entrance, and the bay itself is wide and open for several miles, with a miniature range of beautiful hills at its head. There is an active weir on the west side of the bay near Newman Cove. West Bay is choked with unmarked ledges and shoal areas, as is Joy Bay, leading to Steuben.

Of the several entrances to Gouldsboro Bay, the easiest is Western Passage. The current can be two or three knots, but it runs straight through the channel, and Sheep Island is bold on the eastern side. Be sure to identify the Sally Islands correctly to make sure you are in the right place. Outer Bar Island and Bar Island are both low, with just a few trees. Sheep Island is heavily wooded. Leave Sheep Island close to port. The channel is about 100 yards wide.

In the early 1980s, Gouldsboro Bay was selected for a detailed study by the Smithsonian Institution as a representative water body of the Maine coast. The resulting Washington exhibit includes mudflat, rocky shore, and salt marsh, complete with rockweed, crabs, lobsters, and a three-foot tide.

Dyer Bay, next door to Gouldsboro, flanks the west side of low-lying Petit Manan Point. It is marked by sandy bluffs at Yellow Birch Head and by Eagle Hill near the head of the bay.

The 123-foot lighthouse on 'tit Manan marks one of the foggiest places on the coast of Maine. (Christopher Ayers)

PETIT MANAN ISLAND
South of R "2": 44° 21.50'N 067° 51.75'W
CHARTS: 13324, 13312   CHART KIT: 75, 23, 24

Petit Manan Island (known familiarly and universally as 'tit Manan) is an important landmark on the direct passage from Mount Desert cruising waters to Roque Island. On a clear day, the 123-foot-tall pencil of the 'tit Manan lighthouse can be seen from as far away as Frenchman Bay, beckoning enticingly over Schoodic Point.

But clear days are rare here. The coast from Schoodic to Head Harbor Island is one of the foggiest areas in Maine. At 'tit Manan there is an average of 250 hours of fog a month during July and August. Unless you're prepared to depend on radar and GPS, you could be holed up in a harbor here for days, waiting to head east. Or home.

Green Island and 'tit Manan are low-lying, treeless, and surrounded by ledges, and in thick fog you could easily be up on the rocks before seeing the light. In a southwesterly breeze, the foghorn to leeward is often inaudible until you have left it astern.

Waves coming in from the Gulf of Maine encounter relatively shoal water as they approach 'tit Manan, and often a very rough sea builds up. The current floods east along the coast and ebbs west. It also floods north into the bays and ebbs south, resulting in turbulent waters off 'tit Manan, where the currents meet. This area is especially rough when a southwest wind blows across an ebbing tide. To avoid the worst of it, pass about a mile offshore to a point halfway between nun "2," south of 'tit Manan, and can "1," off Simms Rock. In thick fog or heavy weather, you would do well to run even farther offshore, outside red whistle "6A" off Southeast Rock.

PETIT MANAN BAR
RW "WB": 44° 22.98'N   067° 53.20'W
RW "EB": 44° 23.11'N   067° 52.84'W

CHARTS: 13324, 13312   CHART KIT: 75, 23, 24

A long bar runs between 'tit Manan Point and Green Island. The safest passage around the bar is outside Petit Manan Island, described above. A marked channel, however, crosses the bar midway along its length, making a convenient shortcut for boats hailing from Corea or Pigeon Hill Bay.

The western end of the short passage is marked by red-and-white bell "WB" and the eastern end by red-and-white gong "EB." Run a straight line between the buoys, leaving them close aboard on either side. The more shoal water is to the north.

A very ugly chop can build up on this bar, particularly when the wind blows against the current. In poor visibility the small buoys are very hard to find, and in heavy weather the passage will certainly be unpleasant or even dangerous. It's usually easier and safer to take the longer route outside 'tit Manan. On occasion, however, the passage over the bar is clear when it's thick-a-fog outside.

PIGEON HILL BAY
CHARTS: 13324, 13312   CHART KIT: 75, 23, 24

Pigeon Hill Bay lies between Bois Bubert Island and the eastern low-lying shore of Petit Manan Point. The bay is open and relatively unobstructed and makes a pleasant sail south of Pigeon Hill, a local landmark. The entrance north of Pigeon Hill and Bois Bubert is a rock garden.

BOIS BUBERT
Seal Cove: 44° 26.15'N   067° 51.40'W
CHARTS: 13324, 13312   CHART KIT: 75, 23, 24

Bois Bubert is a large island of 1,000 acres that parallels Petit Manan Point. The island preserves for history the name of a Frenchman otherwise unknown and locally pronounced "Bo-Bare." Monsieur Bubert's woods now shelter ruffed grouse, woodcock, and white-tailed deer.

Most of the island has been acquired by The Nature Conservancy and transferred to the U.S. Fish and Wildlife Service, complementing the refuge on 'tit Manan Point, as part of the ever-growing Maine Coastal Islands National Wildlife Refuge. The seven-acre freshwater pond at the northern end of Bois Bubert is surrounded by a quaking bog.

Shoal-draft boats can use Little Bois Bubert Harbor, at the south end, to get ashore, but for deep-keel boats the best bet is Seal Cove, on the east coast. It is sufficiently exposed to ocean swells to make it an uncomfortable anchorage for the night, but it is a pleasant daystop.

There is an old weir leading from Bois Bubert opposite the north end of Seal Cove Ledge, so the easiest entrance is from the south. Seal Cove Ledge extends a long way southward, so favor the southern shore. There is reasonable protection near the beach at the southwest corner, but the bottom is rocky. There are views of the handsome Douglas Islands and an audience of curious seals usually watches from the ledge.

NARRAGUAGUS, HARRINGTON, and PLEASANT BAYS

The complex and sparsely settled coastline between Petit Manan and Cape Split has been formed by four small rivers and many more creeks and brooks which in turn broaden out to Narraguagus, Harrington, and Pleasant Bay.

Pleasant Bay is by far the most open of these bodies of water, and it offers delightful sailing among spectacular islands. These bays are often free of fog when it is thick farther out. Many islands in this region have distinctive shapes or formations, including Ladle Ledges, Pot Rock, Shipstern Island, Jordans Delight, and the Douglas Islands. Old weirs extend from the shores, some still actively fished.

Secure harbors are few and far between. Eastern Harbor, between Cape Split and Moose Neck, is the best in heavy weather, although it is exposed to the southwest. With winds from south to east, the anchorage north of Trafton Island is comfortable.

This part of the coast is far more important to ospreys and bald eagles, fishermen and blueberry growers than it is to cruising sailors. Enormous numbers of fish run up the estuaries to spawn, among them alewives, herring, striped bass, and shad. At certain high tides in the spring, every local resident

BALD EAGLES

The bald eagle is a magnificent wild creature with a wingspan of as much as six or eight feet, awesome talons and beak, and the fierce, independent gaze that made it the national emblem of the United States. The adults are easily recognizable by their white feathered heads, white tails, and enormous size. More eagles nest near Pleasant Bay, east of Petit Manan, than any other place on the coast of Maine because of the ample supply of fish nearby. Cobscook Bay is another favorite eagle haunt.

Not surprisingly, eagles are at the top of the pecking order as far as birds are concerned (with the exception of the great horned owl). They frequently steal fish from ospreys, and sometimes they prey on ducks and other seabirds. Smaller birds cease their chatter and seem to melt into the trees when an eagle is near. For a bird with such an intimidating reputation, eagles are indifferent fishermen, subsisting largely on dead or dying fish at the surface and the occasional live fish stolen from another bird.

In Indian and colonial times, eagles were common along the coast of Maine. Numerous Eagle islands and Eagle points are still named on the chart. Eagles have returned to *Swango*, the place the Indians called "eagle island," what we now call Swan Island in the Kennebec River.

Eagles mate for life and use the same nests year after year. Eagle nests, or aeries, are impressive structures of sticks and branches, floored by grasses, moss, and feathers, and held together by a rich layer of guano. Every spring, the nest is repaired with more sticks, feathers, and grass. Older nests can reach an enormous size. One record nest described in a National Audubon Society publication was 10 feet across and 20 feet deep.

Each pair usually maintains alternate nesting sites, and during a given year they will abandon one nest for another nearby. This almost human desire for change of scenery is apparently related to survival of the young, which are thus protected from the parasites that thrive in the nest guano. The nesting season lasts from March through August.

There is an active eagle nest on wonderfully inaccessible Shipstern Island (a Nature Conservancy preserve), off Pleasant Bay. A decade ago, the owners of this nest moved over to neighboring Flint Island (also protected by The Nature Conservancy) because their large nest became so heavy that it broke the branch on which it rested and fell to the ground. Apparently this is the only eagle pair in Maine to claim two islands for their nesting sites.

Eagles do not tolerate human disturbance. As the human population grew, their range shrank. Far worse was the genocidal effect of pesticides such as DDT, which accumulated in the food chain and interfered with the reproduction of eagles, ospreys, and other birds. By 1965, only four nesting pairs of eagles in Maine succeeded in producing eaglets.

With the banning of DDT and the conservation of islands and other areas suitable for eagle nests, this trend has gradually been reversed. Maine's bald eagle population has rebounded from a mere 25 pairs in the 1970s to more that 370 pairs in 2005.

Still, the eagles struggle. Maine is at the northern limit of the bald eagle's breeding range. Living and breeding conditions are harsher here, habitat is in decline, and contamination from pollution is still a suspect.

With the tentative success of the eagle population, the federal government changed the bald eagle's status from "endangered" to "threatened," but eagles are still far from common in Maine. A sighting is always a surprise—and always awe inspiring. Few thrills can match the sight of one of these great creatures soaring on the wind.

---

who can walk and who owns a net or a scoop is standing up to his knees in alewives, feverishly collecting fish for the larder.

Two of the rivers are buoyed, the Narraguagus and the Pleasant, so it is possible to reach the towns of Milbridge and Addison. However, both routes are difficult and are more suited to a salmon than a cruising boat. Attempt them only on a rising tide. The town of Milbridge has expanded its launching ramp and floats into a town "marina" to encourage visitors.

The Narraguagus River, in particular, has been famous for its runs of the endangered Atlantic salmon. Herons based in the Douglas Islands and ospreys on Trafton Island also appreciate the fishing, and for the same reason there are more bald eagles in Pleasant Bay than anywhere else on the Maine coast.

The 1806 edition of *The American Coast Pilot* dismisses Pleasant River as "too difficult to describe." Two centuries later, the *Coast Pilot* is slightly less ominous: "the river is seldom used except by fishermen, and the once-extensive trade in lumber ceased many years ago. Passage up the river is suitable for small craft only, except with local knowledge, as the river is reported to have shoaled in many places."

Along the shores of Washington County from Cherryfield to Machias Bay lie the great blueberry barrens—open slopes and heaths often dotted with boulders left by glaciers. In June the blueberry fields are covered with delicate white blossoms, and by late July or early August the harvesters are raking the blueberries. If the harvest is late, the opening of schools is delayed, so the schoolchildren can keep helping with the harvest. With the first frost, the fields take on an extraordinary reddish hue. Each year some 30 to 40 million pounds of blueberries are shipped fresh or frozen from the fields of Maine.

TRAFTON ISLAND

3 ★★★

44° 29.35'N   067° 49.80'W

CHARTS: *13324, 13325*   CHART KIT: *76, 24*

Trafton Island has been a favorite of yachtsmen for generations. Lying just east of Petit Manan and northward, at the mouth of Narraguagus Bay, the island has one of the few harbors in the area. The anchorage at the northern end provides good protection from the east around to the south and southwest. From here you can explore the nearby bays or depart for Great Wass and Roque Island. There might be half a dozen yachts anchored here on a summer afternoon, or you might be lucky enough to have it to yourself.

This is a lovely anchorage and one of the best places in Maine to watch the ospreys, which thrive on the alewives, bass, and smelt that run up the nearby rivers to spawn. At least two osprey families live near the anchorage. The adults, white-breasted and alert, perch by the nest or fly home chirping with fish in their talons or with branches to make repairs. The young birds in the nest tentatively lift their wings and are carried aloft on the breeze. Sometimes six or seven ospreys are soaring over the island.

Trafton has been privately owned by one family for decades, and the owners are in residence much of the summer. The owners request that you do not land unless invited.

APPROACHES. From the west, the steep little Douglas Islands are easy to identify. Then you'll see the abandoned lighthouse on the east coast of large Pond Island. As you pass Jordans Delight to starboard, admire the unusual sea arch buttressing the cliffs.

The easiest approach is around the east side of the island. Dramatic Shipstern Island has a rocky southern point shaped like the stern of a galleon. Keep to the west of Western Reef, marked by a nun, and west of partially wooded Tommy Island, which has multicolored bands of reddish, gray, black, and white rock. Then follow the bold eastern shore of Trafton Island. You will see the red beacon on Trafton Halftide Ledge half a mile to the north. Round the northern tip of Trafton and head back into the anchorage.

When Trafton is familiar to you, it is also easy to approach around the west side of the island, focusing on the outlying ledge on the west side of the anchorage.

ANCHORAGES, MOORINGS. Anchor in 8 to 10 feet at low, eastward of the little wooded island and the outlying ledge, which is visible at all tides. The water shoals beyond a line that runs from the northern tip of the little island to another ledge making out from the south shore about 100 yards eastward, to a big white rock with dark lichen on the eastern shore of the anchorage. Anchor well north of this line in good mud.

DYER ISLAND (Northeast Cove)

3 ★★★

44° 30.56'N   067° 48.34'W

CHARTS: *13324, 13325*   CHART KIT: *76, 24*

Dyer is a long, wooded island that separates Narraguagus from Pleasant Bay. Of the several coves on the island, only Northeast Cove offers a good overnight anchorage. Southwest Cove has an active lobster pound and a fishing camp at its head, but it is exposed to the southwest and has dubious holding ground. Northwest Cove is lined with the rustic cottages and buildings of a boys' camp. There are docks and floats and several small sailboats on moorings.

Northeast Cove is a charmer, with two little islands on the north side and a secret little inlet in the corner to explore. There once was a lobster pound here, but today there is no sign of civilization.

APPROACHES. Run up the east side of Dyer Island and enter south of Otter Island.

ANCHORAGES, MOORINGS. Do not go west of the smaller island. Anchor in 9 feet at low. The cove has good holding ground and is well protected from prevailing summer winds, although it is exposed to the north and southeast.

The bridge spanning Moosabec Reach between Beals Island and Jonesport.

EASTERN HARBOR (Cape Split)

3 ★★★

Otter Cove: 44° 30.25'N 067° 43.55'W
Inner Harbor: 44° 30.51'N 067° 43.51'W

CHARTS: **13324, 13325**   CHART KIT: **76, 24**

Eastern Harbor, between Moose Neck and Cape Split, is one of the best harbors between Schoodic Point and Head Harbor Island and one of the easiest to enter. It can be bouncy in a strong southwesterly, but it offers excellent protection in every other direction. The harbor entrance frames the beautiful view of the bright-green, rounded forms of Ladle Ledges. The small fishing community of South Addison has considerable atmosphere but not much to offer in the way of facilities.

APPROACHES. Enter halfway between the nun and the can off Eastern Pitch and Marsh Island and continue between the next pair of red and green buoys. The remains of a large weir may be visible at the western side of the entrance to the inner harbor.

ANCHORAGES, MOORINGS. You can anchor either in Otter Cove, which has lots of room, or farther north in the inner harbor, where the lobsterboats lie at their moorings. Both anchorages have about the same amount of exposure to the southwest.

In Otter Cove, anchor in about 8 to 9 feet at low, outside the few moored boats. It is hard to get in far enough to avoid the current, so you will skate around a bit between wind and tide. Holding ground is good mud.

In the inner harbor the lobsterboats are moored between the small dock on the eastern side and the ledge at the northern end of the harbor. There may be room to anchor outside the moored boats in 7 to 8 feet at low, but be sure you know where the ledge starts.

GETTING ASHORE. Row in to the dinghy floats east of the small dock.

FOR THE BOAT. *North Atlantic Lobster Sales (207-483-2908 dock, 483-2888 office).* Gas and diesel are available at this lobster dock in the inner harbor, but there is only about 4 feet alongside at low. Water can be obtained from a nearby spring if you are willing to lug it. There is chipped ice at the ice plant on the dock.

Tyler's Marine (207-483-2886). Tyler's, located right at the intersection of the town, can help you with any engine problems.

FOR THE CREW. You'll have to rough it here and settle for lobsters from North Atlantic Lobster or one of the local boats.

MOOSABEC REACH, Western and Eastern Bays

Separated by Moosabec Reach and protected by the complex of islands to seaward, Jonesport and Beals Island are busy fishing communities, with the largest fleet of lobsterboats in eastern Maine. Passage all the way through the Reach is barred to many cruising sailboats by the Beals Island Bridge, built in 1959, with a vertical clearance of only 39 feet. The voyage from West Jonesport around Great Wass Island, Head Harbor Island, and back up to Jonesport covers some 25 miles, including a sea passage.

Moosabec Reach was known as "Moose a' Becky's Beach" on the 1770s British charts of the area. By 1896 the *American Coast Pilot* was referring to it as "Moosepeek Reach."

Great Wass, Steele Harbor, and Head Harbor Island stand boldly out in the Gulf of Maine, taking the brunt of the ocean's attack. Both on the chart and in the appearance of their weathered cliffs, these great rock bastions show ample evidence of the ceaseless pounding of the waves. This is one of the foggiest parts of Maine. The combination of fog, swift currents, complicated channels, and ledges demand respect from the cruising sailor.

There are, however, two excellent harbors in the area, both easy to enter under most conditions. The first is The Cows Yard at Head Harbor Island, and the second is Mistake Island Harbor. A third anchorage, the Mud Hole on the east side of Great Wass Island, provides absolute security, but it is much harder to enter.

The islands of Eastern and Western Bay are spectacular. Sloping shores of white and pink granite contrast with dark spruce and fir trees and the white foam and froth of the ocean swells breaking on the ledges.

There are seabirds of every kind here. Great blue herons wade along the shores, and bald eagles soar silently overhead. In recent years razorbills have returned, and you may even see an occasional puffin. Much of Great Wass Island has been preserved by The Nature Conservancy as a treasury of rare, endangered species.

WESTERN BAY

Tibbett Narrows: 44° 29.73'N 067° 42.25'W
CHARTS: *13326, 13324, 13325*   CHART KIT: *76, 77, 24*

Western Bay is easily approached through the wide southern entrances or from the west through Tibbett Narrows. The well-marked Narrows squeezes between low, partially wooded Tibbett Island, with gravel beaches, and higher Ram Island, with shores of shelving white granite.

There is pleasant sailing in the open stretches of Western Bay, particularly around the beautiful western islands like Drisko, Toms, and Stevens Island, which is owned by the state. The bay has no good harbors, but you can find some temporary anchorages in settled weather.

WEST JONESPORT

1 ★★   🛒 🍴

44° 31.57'N 067° 36.95'W
CHARTS: *13326, 13325*   CHART KIT: *77, 24*

Coast Guard: 207-497-2200

Ashore, West Jonesport and Jonesport are part of the same community. However, since the bridge was built across Moosabec Reach to Beals Island, West Jonesport has been completely separated from Jonesport as far as the cruising sailor is concerned (unless your mast can fit under the bridge).

The U.S. Coast Guard has a major installation just west of the bridge—the last one until you reach Eastport. Most services, however, are in Jonesport, within easy walking distance east of the bridge.

APPROACHES. The western part of Moosabec Reach is well buoyed and deeper and wider than the eastern section. Since the current floods eastward and ebbs westward, a southwest wind against the ebb can kick up chop in the Reach.

After passing Shabbit Island, be alert for crosscurrents from Western Bay that may push you out of the channel. The shoreline along the north side from Wohoa Bay to Jonesport is low and dark in color with

gravel beaches—in startling contrast to the white granite islands of Western and Eastern Bay. One striking landmark on privately owned Hardwood Island is the rusting red sheds, stone wharf, tailings, and machinery of a former granite quarry.

Lobster docks line the north shore as you approach West Jonesport, and the Coast Guard station's long pier and large boat shed are clearly visible at the northern end of the bridge causeway.

The causeway narrows the Reach and greatly increases the current there. Sometimes it reaches 6 to 8 knots. If your mast is too high to fit under the bridge (vertical clearance is 39 feet; mean tidal range is 11.5 feet), stay well away from the bridge and the currents there, particularly at midtide, and allow for strong eddies near the north side.

ANCHORAGES, MOORINGS. You can anchor in 16 to 24 feet at low off the northern shore, near the end of the Coast Guard pier. Holding ground is good, and you will be out of most of the current.

⇒ It's possible to come alongside the Coast Guard pier, except near low tide. You must call in advance on channel 16 or *497-2200* to ask permission so you don't interfere with their operations. Tie up on either side of the east-west finger or to the west side of the pier itself. The inside of the east-west finger has about 4 feet at low, and the pier has less.

FOR THE CREW. There is a pay phone inside the Coast Guard building or across the street on the outside wall of the firehouse. The IGA supermarket and the post office are a mile eastward, and you can usually get a ride back with your groceries. Tall Barney's restaurant is a local favorite, right at the foot of the Beals Island Bridge. For other facilities, see the Jonesport entry, page 348.

THINGS TO DO. Walk across the bridge to explore Beals Island.

BEALS ISLAND
east of the bridge: 44° 31.38′N 067° 36.79′W
CHARTS: 13326, 13325   CHART KIT: 77, 24

Beals Island is connected to the mainland by the bridge over Moosabec Reach (39-foot vertical clearance). It is an island of fishermen, lobstermen, and boatbuilders, and the land of the lobsterboat. The boats built here are famous for their speed and seaworthiness.

The first Beals Island settler in 1764 was an unusual man in many ways. Manwarren Beal was enormous, smart, and tough. He was also a prolific poet.

Other unusual people came to Beals, among them George Washington Adams, an itinerant preacher. In 1865, he launched the Palestine Emigration Association and persuaded 156 presumably sensible citizens of Beals and Jonesport to emigrate to Jaffa (now Israel) to establish a colony to carry on the work of the Lord. They sailed from Jonesport and arrived in the Holy Land. But funds ran out, most of the colonists fell ill, and many died. Most never got back to Maine. "One old lady," Maine author Louise Rich reports, "stayed in order to spite herself for having been such a fool as to come in the first place."

Before the bridge was built in 1959, Beals was an isolated community that even had its own peculiarities of speech—a quaint Elizabethan English that has now disappeared.

To visit Beals Island, walk across the bridge from Jonesport or anchor in the first cove west of the bridge, outside the moored boats, and row ashore to a convenient float.

EASTERN BAY
east of Mud Hole Point: 44° 28.95′N 067° 33.95′W
CHARTS: 13326, 13325   CHART KIT: 77, 24

Ledge-spangled Eastern Bay lies between Great Wass Island and Head Harbor Island. Strangers should attempt to navigate its northern portion only in good visibility and near low tide when most of the dangers are visible.

The usual approach to Eastern Bay is from the south, through the deep and clear passages of Mud Hole Channel or Main Channel Way. It is also possible, but considerably more difficult, to approach from the north from Moosabec Reach, through the winding buoyed passage just west of Head Harbor Island.

In addition to Mistake Island Harbor and the Mud Hole, Eastern Bay offers other anchorages to the adventurous sailor, including Sand Cove North on the east shore of Great Wass Island and the cove north of Middle Hardwood known as Sealand. In good years, naturalist Philip Conkling reports, Middle Hardwood has an enormous crop of blueberries and huckleberries. "The island looks like a sea of blue."

MUD HOLE (Great Wass Island)

■ ★★★★★

entrance: 44° 29.04'N 067° 34.70'W

CHARTS: 13326, 13325	CHART KIT: 77, 24

Mud Hole is a wonderfully private and well-protected anchorage on the east coast of Great Wass Island. It is difficult to enter, but once inside, you are in a completely natural little fjord lined with granite and spruce, with herons, seagulls, and bald eagles for company. Mud Hole provides easy access to The Nature Conservancy's Great Wass Island Preserve and some wonderful trails through mossy forests and stands of rare, gnarled jack pines, across sunny—or foggy—granite ledges, and out along the cliffs of Eastern Bay.

Don't be surprised to find yourself sharing Mud Hole with several million clamlets growing in incubator cars. They are part of an experiment by the University of Maine at Machias underwritten by a group of nearby towns and shellfish dealers. Spawn from the Beals Island hatchery (*497-5769*) are raised in the protected water of the Mud Hole and then distributed to participating towns to seed local clam flats.

APPROACHES. Approach through Mud Hole Channel, paralleling and favoring the line of islands and ledges south and west of Mistake Island. Black Ledges to port are usually breaking, but note the position of their northernmost rocks. Green Island is the first little island with trees to starboard. Mink Island is the second. Well before reaching Mink Island, bear westward and head in north of Mud Hole Point, giving it a wide berth.

For years an old weir was rotting along the south shore. It was located just north of the word "Hole" in "Mud Hole Pt." on the chart. To avoid any remains, approach near the northern entrance point where the chart shows 14 feet (see sketch chart).

A ledge sits smack dab in the middle of Mud Hole's narrow entrance. It covers at about halftide, but its dark shape and rockweed are still discernible until a couple of hours before high—the ideal time to enter.

Use the channel to the south of the ledge. After approaching the northern entrance point to avoid the weir, turn sharply to port and head directly over to the south shore. Creep along this shore until you have safely passed the midchannel ledge to starboard. The lobster buoys will usually give you an indication of the deep water which runs close by the southern shore to port. The kelp in the channel will fool your depthsounder and make your heart leap into your mouth.

Cruiser Sterling Blake conducted a thorough leadline survey of the entrance by dinghy at dead low during a new moon. He reports, "The 2-foot sounding reported for the entrance is accurate. The best water to be had was 2'4" and most of it is 1'6". The bottom is very soft."

After passing the entrance ledge, continue to favor the south shore until you have passed a second ledge extending southward from the north shore. After the ledge, the anchorage becomes wider.

ANCHORAGES, MOORINGS. The deep water runs a relatively short distance west of the second ledge. Feel your way in slowly and anchor in 12 to 20 feet at low. Sometimes, as part of the aquaculture operation, a floating shed is moored where the deep water begins to shoal. You may find it useful to put out a stern anchor to keep from swinging with the tide into shallower water. The bottom—you guessed it—is mud.

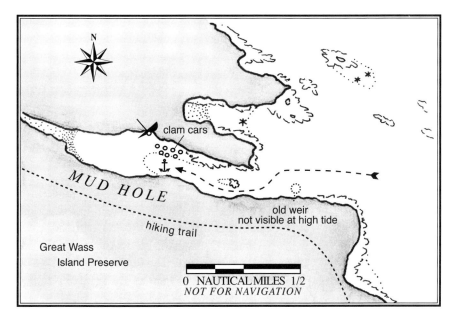

GETTING ASHORE. Row to the south shore of the anchorage and tie up to a convenient tree.

THINGS TO DO. The Mud Hole Trail runs along the south shore of the harbor. Scramble up the hill 50 yards or so until you reach the well-defined trail. For a long, three or four-hour walk, head west to the parking lot and then back via the Little Cape Cove Trail, closing the loop by walking along the cliffs and beaches and northward back to Mud Hole Point.

For a shorter walk, head east from the anchorage on Mud Hole Trail to the cliffs at Mud Hole Point. Then walk south along the shore to Little Cape Point.

Just south of Mud Hole Point, after a good rain, a gurgling stream runs out of the woods and over the shelving granite to form Taft's Bath—a delightful series of little pools, basins, and cascades of fresh water just right for sitting and splashing. Dry off in the sun on the warm rocks.

MISTAKE ISLAND HARBOR

44° 28.60'N 067° 32.50'W

Charts: 13326, 13325   *Chart Kit:* 77, 24

One of the most beautiful harbors in Maine is hidden away among the ledges behind Mistake and Knight Island. Don't mistake it for the "Mistake Harbor" shown on the chart farther north. Once inside, you are surrounded by a pink and white granite shoreline and protected from the ocean swells that crash on the rocks only a few hundred yards away. Over this tranquil scene of natural beauty stands Moose Peak Light. You may be lulled to sleep by its sonorous double note.

APPROACHES. Most boats enter from offshore through Main Channel Way, leaving Moose Peak Light to port and the sloping granite cliffs of Steele Harbor Island to starboard. As you pass the lighthouse, there is a glimpse of open water at high tide between treeless Mistake Island and heavily wooded Knight Island. Do not be tempted, however. Continue past Knight Island, turning to port around the end of the island and southward back into the harbor.

A line between the northern tips of Knight and Mistake crosses a tiny and dangerous ledge that should be left to port. The rockweed on this ledge is

The Coast Guard boathouse on Mistake Island.

just barely visible at high tide, and some kind soul has placed a short metal pipe on it. The ledge is just opposite a small green cottage on privately owned Knight Island. The larger ledge to the west is visible at all tides. Leave it to starboard.

If you are coming from Mud Hole or elsewhere in Eastern Bay, Mistake Island Harbor is easily approached from the north. Pass south of can "1," then head for the northern end of Knight Island.

ANCHORAGES, MOORINGS. Note that the entire anchorage is a cable area. We've had no reports of problems, but recommend a tripline on your anchor. Drop the hook in 10 feet at low off the northern end of Mistake Island. The best protection is in the slot between Mistake and Knight.

Mistake Island Harbor is fairly well protected from ocean swells by the outlying ledges, though strong southwesterlies will blow into the anchorage. It is also exposed to the north and northwest, and boats may drag if a strong front comes through.

GETTING ASHORE. Row into the ways of the old Coast Guard boathouse at the northern end of Mistake Island.

THINGS TO DO. A boardwalk leads from the boathouse to Moose Peak Light, automated in 1970. Stay on the boardwalk or granite outcroppings to avoid damage to the rare and delicate plants found on Mistake Island. These are mostly members of one plant family notable for its ability to survive in the acidic, peaty soil and the cold, moist, and salty air. They include lush blueberry, crowberry, leatherleaf, lambkill, and Labrador tea. You will also find abundant raspberries and the beautiful beachhead iris.

Most of Mistake Island was acquired by The Nature Conservancy from a grandson of the man who was lighthouse keeper from 1890 to 1910. The remaining six acres are still owned by the Coast Guard.

HEAD HARBOR and THE COWS YARD

4 ★★★

Cows Yard entrance: 44° 29.83′N 067° 31.64′W

CHARTS: 13326, 13325   CHART KIT: 77, 24

The Cows Yard is a secure anchorage, attractive and easy to enter under any conditions from the south through Head Harbor. It's one of the best harbors between Mount Desert and Roque Island. If you are caught out in fog or bad weather, this is a good refuge. The anchorage is largely undeveloped, with only a few summer cottages.

APPROACHES. This coast is steep-to all the way from Crumple Island, past Great Wass, Mistake Island, and Head Harbor. Even in thick fog you can approach within visual range without getting into trouble and run along the 150-foot bottom contour with your fathometer, paralleling the coast. Or you can tack in toward the cliffs, seeing and hearing the breaking surf in plenty of time to come about for the next outward leg.

Two excellent clues help identify the entrance. First is the lonely lighthouse at Moose Peak, whose foghorn has warned mariners since 1827. The second is a dramatic difference between the light-colored rocks on the west side of the entrance and the dark rocks of Head Harbor Island and Man Island on the east side of the entrance.

The western ledge at the entrance to Head Harbor normally breaks. The first ledge to starboard, off the little point with trees, is visible above halftide. Run up the middle of Head Harbor, aiming for a rock that forms the entrance to The Cows Yard, identifiable by a long, gray-shingled cottage to the right of it. Enter The Cows Yard halfway between the rock to starboard, which shows at all tides, and the island to port.

ANCHORAGES, MOORINGS. The Cows Yard is much shallower than shown on the chart but still big enough to provide anchorage for several boats. Where the chart shows 20 feet, the actual depth is only 9 or 10 feet at low. Do not go past the second island to port. Best protection is found toward the east side of the deep water, behind the big rock that forms the entrance. There is considerable kelp on the bottom, so make sure your anchor is set.

Even with a strong southwest wind, hardly any ocean swell penetrates into The Cows Yard. This is generally a serene anchorage, often with the reassuring double note of the Moose Peak foghorn.

The same cannot be said of the outer Head Harbor, which has a rocky bottom and is exposed to ocean swells.

JONESPORT

2 ★★★

44° 31.80'N 067° 35.67'W

CHARTS: 13326, 13325   CHART KIT: 77, 24

Coast Guard: 207-497-2200

Jonesport, like Beals Island across Moosabec Reach, is a fishing town. The shores are lined with docks piled with lobster gear, and lobsterboats are moored in every cove.

This is the best place in the area to pick up supplies, most of which are found east of the bridge in town. The Coast Guard station, with search and rescue capabilities, is in West Jonesport, just west of the bridge (*Ch. 16; 497-2200*).

Each year, the Fourth of July Races are held here to determine the "World's Fastest Lobsterboat." The races began in dories around the time of the Civil War. Dories were succeeded by Friendship sloops, then by boats powered with naphtha engines, and finally by the handsome, powerful, and seaworthy Jonesport lobsterboats of today.

APPROACHES. Moosabec Reach is a narrow thoroughfare, with lots of traffic and considerable current, flooding east and ebbing west. The eastern entrance is well marked, starting with green bell "1" north of Mark Island, but the passage quickly narrows with little leeway between the ledges. It is difficult in low visibility. If caught in the fog, favor the north side or anchor in mud just outside the channel.

ANCHORAGES, MOORINGS. Jonesport's harbor lies in Sawyer Cove behind a 1,200-foot steel and stone breakwater that extends across the harbor's mouth from the east. It provides good shelter from southeast winds but less from the prevailing southwesterlies.

Enter the anchorage at the west end of the breakwater. The harbor is tightly packed with moored boats, and it quickly becomes ledgy toward its head. It is unlikely that you will find room to anchor, but you may be lucky enough to be directed to a vacant mooring, or you may be able to rent one from Jonesport Shipyard. Otherwise, temporary dockage can be had on the town marina floats (the easternmost of the docks) with 3 feet at low. Ask at the floats, at the Jonesport Shipyard, at Look's, or contact the harbormaster.

GETTING ASHORE. Row into the town marina.

FOR THE BOAT. *Town Marina (Harbormaster Russell Batson: 207-497-5931; town hall: 497-5926).* The town marina consists of a large town wharf and floats and a launching ramp and parking. The first, southernmost float is for commercial vessels only. The other float, to the north, is for the limited docking of recreational boats. Despite the longer wharf, both floats only have about 3 feet of depth at low tide.

Jonesport Shipyard (Ch. 09, 16; 207-497-2701, 800-544-3708). Jonesport Shipyard is located east of the town floats near the head of the harbor. They can haul and repair boats up to 17 tons or 45 feet, and they provide showers, laundry, and ice. Don't be tempted to reach the shipyard by boat, or you'll find the bottom instead.

Look Lobster (Ch. 77; 207-497-2353). Look's floats and buildings are opposite the west end of the breakwater. Gas and diesel are available at the floats, but no water. Come at halftide or better—there is only 3 feet of depth off the floats at low, and a ledge lurks a couple of boat-lengths to the east.

Moosabec Marine (207-497-2196; www.moosabecmarine.com). Moosabec Marine is located in town, just before the IGA. They specialize in outboard sales, service, and repairs, with a full Mercury and Yamaha parts department.

FOR THE CREW. King's Hardware store (*497-2274*), right up from the landing, carries marine hardware, U.S. charts, and lumber. They also have a copy and fax machine. Walk left into town to find the post office, several takeouts, and Church's Hardware, which is a propane and U.S. chart dealer. The street signs here are all in the shape of the Jonesport lobsterboats. The IGA is just beyond, then Tall Barney's Restaurant, a local favorite. There is a pay phone all the way past the Beals Island bridge, outside the firehouse.

THINGS TO DO. Captain Barna Norton, who lived in West Jonesport, claimed direct ownership of Machias Seal Island through his great-grandfather Barna Beal, known as Tall Barney. Machias Seal Island, however, is claimed by the Canadians. In 1940, Barna Norton began running charter trips to the surrounding islands, and, in season, to his beloved Machias Seal Island, home of arctic terns, puffins, and razorbills. His son, Captain John Norton, now carries on the tradition (*497-2560, 497-5933; www.machiassealisland.com*).

The bridge to Beals Island makes an interesting walk with a nice view of the working waterfront.

ROQUE ISLAND ARCHIPELAGO

For a long time, Roque Island has been the ultimate goal of sailors cruising Down East—perhaps because of its beautiful, mile-long white sand beach, perhaps because there is the sense of something special about this island, perhaps because it takes determination to sail east of Schoodic and Petit Manan.

Roque is the centerpiece of an archipelago that includes Great and Little Spruce, Lakeman, Marsh and Bar, Double Shot, Anguilla, and Halifax Island—all set in a body of water called Chandler Bay to the west and Englishman Bay to the east (see sketch chart on page 351). Shaped roughly like an H, the Roque archipelago offers a delightful variety of anchorages. The great southern beach on Roque Harbor is the most familiar. Lakeman Harbor to the east, surrounded by Lakeman, Marsh, and Bar Island, provides a secure anchorage. Tiny, landlocked Bunker Cove is to the west. Another sand beach curves around Shorey Cove, on the north side of the island.

Indians were the earliest-known summer inhabitants of Roque, and numerous shell heaps have been studied here by archaeologists. Joseph Peabody acquired Roque in 1806, and for almost two centuries the island has served as a resort and retreat for his descendants, the Gardner and Monks families. The old family buildings and farmhouses, red and yellow, are on the eastern side of Squire Point, overlooking Shorey Cove. There is a private boatyard, with a dock, metal and woodworking shops, and a small fleet at the moorings. Boats are hauled by attaching a farm tractor to a great old anchor half-buried in the ground and winching them up the ways.

Roque is a working farm, and it is almost self-sufficient, with cattle, sheep, pigs, geese, chickens, pigeons, and other animals. Products of the island include milk, butter, eggs, wool, beef, pork, squab, raspberries, rhubarb, herbs, and vegetables. The caretaker and the owners take pride in both frugality and ingenuity, and the farm has some marvelous Rube Goldberg contraptions, including a continuous conveyor belt built from scrap material to cut, carry, and split firewood.

In summertime, it is not unusual to see family members traveling in a horse-drawn carriage. During the winter, the resident Clydesdales haul guests on old blue sledges. As John Peabody Monks said in his book *Roque Island, Maine—A History*: "The visitor to Roque forgets the urgencies of time and place."

Perroquet is the French word for parrot, and it seems plausible that, as Samuel Eliot Morison suggests, Roque was named by the French explorers for the puffin, or sea parrot. Or perhaps the origin was "rogue," considering the pirates who were based nearby.

When Joseph Peabody bought Roque, he made good use of it. First he built a tidal dam across Paradise Cove to power a gristmill and a sawmill. Several large vessels were built for him at a shipyard in the little bight at the mouth of Paradise Cove, just west of Point Olga. At one time, Peabody owned 63 ships and employed more than 3,000 men in his various shipping and trading enterprises.

In 1868, John and Catharine Gardner sold Roque Island for reasons unknown. Ten years later, two of their sons bought it back for double the price, and the island has been in the family ever since. Shortly thereafter, the Gardners bought Great and Little Spruce, Lakeman, Anguilla, and the little Bar Islands. Double Shot was acquired in the 1930s.

A lumber mill once stood at the head of Patten Cove. Between 1926 and 1928, three million of the mill's laths were shipped on coastal steamers to New York.

There are lots of stories of life on Roque, both human and animal. Pigs were once allowed to roam free on the island, and it is reported that a particularly bright pig learned how to dig clams on the shore. Then there is the story of lobsterman Horace Dunbar, who enjoyed a party ashore with some visiting yachtsmen and had his share of liquor. Finally it was time to go, so Horace got in his skiff and started to row home. "Two hours later, he was found by more sober members of the party, still rowing steadily, with his boat's painter attached to the wharf."

> "Two hours later,
> he was found by more sober
> members of the party,
> still rowing steadily,
> with his boat's painter
> attached to the wharf."

THE THOROFARE
at "REP": 44° 33.74'N 067° 31.09'W

CHARTS: 13326, 13325   CHART KIT: 77, 78, 24

The Thorofare is a narrow, winding, hidden, and altogether delightful passage from Chandler Bay to Roque Island Harbor. From the west, it is hard to find. One good way is to run a compass course from the eastern end of the Jonesport channel until the Thorofare starts to open up. You will see small, wooded West Bar Island southeast of Bonney Point. Pass midway between this island and the western end of Little Spruce Island and turn southeast toward the slot between Little and Great Spruce Island. Then, leaving East Bar Island to port, turn eastward through the Thorofare.

With a southwesterly wind, it's a pleasure to sail the Thorofare. You will be blanketed by Little Spruce but probably get enough puffs to tiptoe through the passage. There is not much current.

The chart shows a rock ("REP") right next to Easterly Bar Island, but many people have sailed through the Thorofare over the years without finding a trace of it. Perhaps the prudent thing to do is avoid the passage near dead low. The shallowest portions of the Thorofare are farther east, where there may be 7 or 8 feet or less.

BUNKER COVE

44° 33.63'N 067° 31.18'W

CHARTS: 13326, 13325   CHART KIT: 77, 78, 24

A splendid anchorage known as Bunker Cove lies at the western end of the Thorofare in the gut between Great and Little Spruce Island, secluded and secure. In his history of Roque Island, Monks calls it "an almost landlocked small anchorage, considered by some yachtsmen the most beautiful on the Atlantic Coast."

The cove itself, however, has silted in. At most, there is only room for two or three boats to anchor in deep water without being forced into the Thorofare.

Bunker Cove, like neighboring Bunker Hole, is named for Jack Bunker of Somes Sound. During the Revolution he heard about a British ship on the Sheepscot River collecting food, so he and a friend canoed down the coast and found the ship at anchor in Wiscasset, unguarded and full of provisions. They cut her cable, hoisted sail, and headed east to Somes Sound, where they distributed the provisions to their starving neighbors.

Then, chased by another British vessel, they fled with their prize toward Roque and ran her into Bunker Hole, east of Little Spruce Island. Without local knowledge, the British were afraid to follow, and by the time they rowed past Little Spruce, the ship's masts had been cut away and she was camouflaged with branches and trees. They never saw a thing.

APPROACHES. From the west, enter the Thorofare between Little Spruce Island and West Bar Island, which guards Patten Cove. Swing southward, leaving East Bar Island to port, and enter Bunker Cove, in the narrow slot between Great and Little Spruce. The water shoals rapidly after you pass the cliffs on Great Spruce.

ANCHORAGES, MOORINGS. The only deep water in Bunker Cove is at its mouth. Cruising boats of moderate draft shouldn't go in much beyond a range between the first point on Great Spruce Island and the obvious shoulder on Little Spruce. The bottom is mud, with some seaweed.

One effective way to anchor here is to drop the hook in the middle and take a stern line to a tree on the cliffs of Great Spruce.

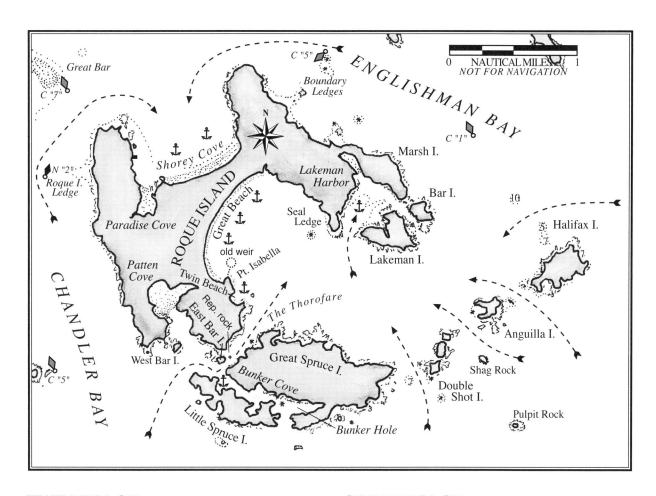

TWIN BEACH

3 ★★★

44° 34.20′N 067° 30.95′W

CHARTS: *13326, 13325*   CHART KIT: *77, 78, 24*

Near the southern end of Great Beach, south of Point Isabella, is a little cove between rocky headlands with a tiny beach of its own. There is room for one boat to lie here in relative privacy, with good protection from prevailing winds.

APPROACHES. There are no dangers. Head straight into the cove.

ANCHORAGES, MOORINGS. Anchor in 10 to 11 feet at low, just inside the headlands. The bottom is mud but there is also some kelp.

GREAT BEACH

3 ★★★★★

44° 34.75′N 067° 31.05′W

CHARTS: *13326, 13325*   CHART KIT: *77, 78, 24*

There are no signs of civilization on the great southern beach of Roque—just the sweeping, mile-long arc of white sand flanked by a meadow and framed by the dark spruces of fabled Roque Island.

Signs posted on the beach explain the owners' wishes, and all visiting yachtsmen should respect them. In particular, avoid the southern end of Great Beach, which is reserved for the family, and refrain from walking inland beyond the beach.

APPROACHES. Great Beach is usually approached from the west through the Thorofare (see page 350). From the east, the approach is wide and easy, north of Halifax Island into Roque Island Harbor. It is also possible, with care, to run through the slots between Halifax, Anguilla, Double Shot, and Great Spruce.

The only danger in Roque Island Harbor is Seal Ledge, visible for about two hours either side of low tide. Remnants of an old weir may still extend about 350 feet from the rocky point at the south end of the beach, but the weir is more of a nuisance than a real danger. The sketch chart will help you steer clear.

ANCHORAGES, MOORINGS. Anchor anywhere along the beach in a comfortable depth. The best protection from prevailing winds is toward the southern end. The holding ground is fine, but a slight sea swell can make this a rolly anchorage, in which case you may be happier for the night somewhere else.

On most days in the summer, you will find six or eight boats anchored here. Occasionally a yacht club cruise will make it seem downright congested. In September, Great Beach returns to its natural state.

The sweeping sands of Roque Island's Great Beach.

LAKEMAN HARBOR

44° 34.75′N 067° 29.75′W

Charts: 13326, 13325   *Chart Kit:* 77, 78, 24

Lakeman is an excellent harbor formed by Marsh Island, Lakeman Island, and the southeastern tip of Roque Island. Ocean swells do not enter, and there is good protection in every direction except southwest. Except for a fishing camp or two, the islands have remained in their natural state, and you will be lulled by the chirps of ospreys, the chuckles of eiders, and the baas of sheep on the islands.

APPROACHES. Coming from Great Beach, be cautious of Seal Ledge, which is visible for a couple of hours after low tide. Otherwise, the approach is obvious between Roque and Lakeman Island. Favor the Lakeman side to avoid the ledge that makes out from Roque. From the entrance, head toward the southern end of the cliffs on Marsh Island. The deep water ends about 300 yards from Marsh.

ANCHORAGES, MOORINGS. Anchor in 7 or 8 feet of water at low. Holding ground is good, in mud, although there is some kelp. Lakeman is open to the southwest, and the afternoon breeze comes in across Roque Harbor. This usually calms down in the evening. Otherwise, the anchorage is entirely secure.

THINGS TO DO. Row up between Marsh Island and Roque at high tide, or out between Bar and Lakeman. Do not land ashore.

HALIFAX ISLAND

44° 34.37′N 067° 27.84′W

Charts: 13326, 13325   *Chart Kit:* 77, 78, 24

Uninhabited Halifax Island lies at the eastern tip of the Roque archipelago, next to Anguilla. Gently rolling meadows rise to a 60-foot cliff at the western end, where guillemots nest. At one time, Halifax was probably cleared by fire or logging or for use as sheep pasture.

The island is protected by a Nature Conservancy easement because of its freshwater bog and several uncommon plants, including the beachhead iris and roseroot stonecrop. Blueberries, raspberries, and cranberries abound. Beware of the fierce, insectivorous sundew. It is also part of the Maine Coastal Islands National Wildlife Refuge and the Maine Island Trail. Halifax Island is used by Hurricane Island Outward Bound and the Chewonki Foundation, so don't be surprised to find it occupied by castaway sea kayakers.

There is a partially protected anchorage at the northwestern tip, toward Roque, and landing is easy on the cobble beaches. To protect the bog vegetation, please keep to the shore in the northeastern area.

SHOREY COVE

4 ★★★

44° 35.30'N 067° 31.45'W

CHARTS: 13326, 13325   CHART KIT: 77, 78, 24

Roque's Great Beach will spoil you. The beach at Shorey Cove on the north side of Roque is neither as long nor as beautiful. Nevertheless, it runs almost three-quarters of a mile from Paradise Cove on the west to the high cliffs of Great Head on the east. The red and yellow buildings of the Monks and Gardner clans, plus their dock, boathouses, and barns, grace the eastern shore of Squire Point.

John Shorey, shipwright and onetime owner of Roque Island, lived here and built several ships for his extensive fleet on the shore of Paradise Cove. The 19th-century island owner, Joseph Peabody, had constructed a tidal dam across the cove, and it powered the sawmill.

APPROACHES. Coming up Chandler Bay against the current is a slow proposition. Look for small nun "2" at Roque Island Ledge and green can "7" marking Great Bar. You will see green open meadows on Squire Point and horses, sheep, and cattle grazing. Give Squire Point a wide berth to clear the ledge at its northeastern end.

Approaching through Englishman Bay, note in particular can "5" at Boundary Ledges, left to port.

ANCHORAGES, MOORINGS. Shorey Cove provides a secure anchorage anywhere along the beach. A small fleet of boats is moored in the western part of the cove, and the owners prefer that you do not anchor "right out in front." Anchor anywhere else along the sand beach, in 7 to 10 feet at low. If winds from the east are expected, excellent protection can be found in the lee of Great Head.

THINGS TO DO. You are welcome to walk on the beach, but fires or camping are not allowed.

SPAR ISLAND

44° 37.12'N 067° 33.66'W

CHARTS: 13326, 13325   CHART KIT: 77, 24

Mason Bay lies to the northwest of Roque, and tiny Spar Island is west of Dunn Island at the mouth of the bay. The 9-foot hole just north of Spar Island makes an interesting anchorage. The approach between Flake Point Bar and Dunn Island is narrow but deep.

Local fishermen use Flake Point Bar as though it were a pier, pulling the bows of their boats up on the hard gravel spit to land and load their gear.

JOHNSON COVE

3 ★★★

44° 36.20'N 067° 26.80'W

CHARTS: 13326, 13325   CHART KIT: 77, 24

Johnson Cove is near the western entrance to Little Kennebec Bay, tucked behind Calf Point and Calf Island. The protection is good, but there are still pretty vistas out to sea. A gravel beach rings the head of the cove, where there are several cottages. Calf Island is rocky and wooded. The cove is a favorite feeding area for eiders.

APPROACHES. Approach from either side of Calf Island. The wider eastern entrance is easier.

If you use the western approach, favor the Calf Point side as you enter. The ledge north of Calf Island is very large and visible until high tide.

ANCHORAGES, MOORINGS. Anchor where Calf Island provides the best protection from the wind, in 16 feet at low. Holding ground is good, thick mud.

Fueling up at Look Lobster, Jonesport.

LITTLE KENNEBEC BAY

between Hope Is. and Yoho Head:
44° 37.60'N 067° 26.00'W

CHARTS: **13326, 13325**   CHART KIT: **77, 24**

Little Kennebec Bay beckons for exploration northward from Englishman Bay. With a summer southwesterly, you can sail gently up the sparsely settled shores of the bay. It's eyeball navigation all the way, with flats on either side of the channel—a gunkholer's dream. The shores are low-lying, and in places the bay widens to a mile or more.

The western side of the entrance is easily distinguished by the high, wooded headlands east of sandy Roque Bluffs. Point of Main flanks the eastern side of the entrance, high and rocky. Two white domes crown Howard Mountain to the north. Hickey Island is barren at its southern end, and scraggly trees cling to the north. Leave it to either side with plenty of room. Fan Island is easy to identify—high, rocky, and wooded.

After leaving Fan Island to port, head for Grays Beach, at the eastern end of the cable area marked on the chart, to avoid the ledge making out from Sea Wall Point. Once clear of Sea Wall Point, turn northwestward to avoid the rock lying off Grays Beach, visible for two hours after low. Give Yoho Head ample clearance. The ledges off Yoho are visible at halftide.

From there on, finding the deep channel between the flats is a matter of informed guesswork, made easier when the tide is low and the sun behind you. Caution and a rising tide are the key elements. It's possible to work your way up Collins Branch into Moose Snare Cove or up the West Branch past Marston Point. Beware, however, of silting. We've had reports that the west branch could be as much as 10 feet shallower than the soundings on the chart, with 6 feet of depth at one point.

MOOSE SNARE COVE

44° 38.79'N 067° 25.85'W

CHARTS: **13326, 13325**   CHART KIT: **77, 24**

At the very end of the Collins Branch of Little Kennebec Bay lies intriguing Moose Snare Cove. How do you snare a moose? The cove provides the perfect gunkhole, hidden way up around and beyond Hog Island. The north shore is high and partially cleared for a cabin or two, and a number of clamming skiffs are moored nearby.

APPROACHES. Like the western branch of Little Kennebec Bay, Collins Branch has silted considerably. It is best to work your way up into Moose Snare on a low tide when it is rising. Tiny, wooded Porcupine Island lies off the grassy meadows of Johnson Point. Work your way slowly past Hog Island and then stay close to the northern shore of the cove.

At low tide you may ground out, but no harm will come to you in the mud. The cove continues around into Mill Pond where there may be a cluster of yurts, but the pond is too shoal except for small craft.

ANCHORAGES, MOORINGS. Anchor along the north shore of Moose Snare Cove in the mud bottom, which is 5 to 8-feet deep at low. A stern anchor will keep you lined up in the narrow channel and off the mudflats to the south. This would be a perfect spot to hide during a hurricane, and it is unlikely you will find other boats here.

MACHIAS BAY

Machias Bay is neither grand nor extensive, but it is blessed with wonderful scenery, wildlife sanctuaries, and serenity. The barren Libby Islands and the Libby Island Light guard the entrance.

Cross Island is to the east. It is one of the largest undeveloped islands on the coast, home to deer, bears, seals, eiders, bald eagles, razorbills, and Outward Bound students. North of Cross, the skyline is dominated by the huge radio towers on the Cutler peninsula.

The western shore of Machias Bay is particularly beautiful. Dramatic Yellow Head Island lies like a dragon in the water, and high and handsome Bare and Bar islands stand in attendance beyond. Part of Salt Island is a preserve of The Nature Conservancy and a home to bald eagles.

At the entrance to Machias Bay, the 90-foot cliffs of Stone Island drop straight into deep water. In the 1960s, there were plans to off-load supertankers here, abandoned only after a decade of public debate. Stone Island eventually became part of The Nature Conservancy, and now great blue herons, ospreys, and bald eagles nest on these forbidding shores. The island is closed to human visitors until mid-August, after nesting season.

In Howard Cove, there is a rare jasper beach and a gravel bar, which is exposed at low tide.

Looming over Howard Cove—and hard to miss anywhere in Machias Bay—are the huge, white radomes on Howard Mountain. Bucks Harbor is an excellent anchorage with good protection and a large and active lobster fleet.

Machias is the Indian word for "bad little falls." The Machias River empties into the northwestern corner of the bay, passing on its way the town of Machias, home of the best strawberry pie in Maine at Helen's Restaurant (*255-8423*) and also the home of Downeast Hospital (*255-3356*). In the past, there were lumber mills here and shipyards and extensive fisheries based in Machiasport. In the early 18th century, pirates such as John Rhoades and Captain Bellamy kept their ships in this remote area.

Machias Bay is open and generally free of obstacles in its southern reaches. High Avery Rock marks the middle of the bay with its beacon and adjacent bell. Sailing is clear as far north as Round and Salt Island, but beyond that the bottom degenerates into the flats and shoals of Holmes Bay. A good chop can build up in Machias Bay, especially with the southwest wind blowing against the ebb, but generally it provides a lovely sail.

The Machias area is full of historical footnotes. At Burnham's Tavern, now a museum, in Machias, patriots met to discuss the British demand for lumber to build barracks. The discussion continued on June 11, 1775 near a little stream now known as "Foster's Rubicon." Calling for resistance to the British, Colonel Benjamin Foster leapt across, followed by the men of Machias. The next day saw the first naval battle of the Revolution and the capture of the armed British schooner *Margaretta* by the patriots' sloop *Unity,* under the command of Captain Jeremiah O'Brien.

STARBOARD COVE

2 ★★★

44° 36.35'N 067° 23.50'W

CHARTS: 13326, 13325   **CHART KIT: 77, 24, 25**

Starboard Cove is formed between the mainland and Starboard Island to the south. Residents of Starboard Island reach home by driving their pickup trucks across the gravel bar at low tide.

APPROACHES. Enter north of Starboard Island. Beware of the extensive fish pens that are moored off the island.

ANCHORAGES, MOORINGS. Anchor anywhere in the cove in 13 to 20 feet at low. The bottom is mud. A line between the northern ends of Stone and Starboard Island marks the start of shoaling near the gravel bar and by an old weir. Anchor north of this line. Or check with Pettegrow Boat for a mooring.

GETTING ASHORE. Row in to the beach.

FOR THE BOAT. *Pettegrow Boat Yard (207-255-8740).* This unpretentious yard has rental moorings, limited marine supplies, charts, and a 90-ton marine railway, but no fuel. They are capable of hull and engine repairs. This is the last boatyard until you reach Eastport and one of the least expensive in Maine.

THINGS TO DO. Explore Jasper Beach in Howard Cove or walk across the bar to Starboard Island at low tide.

HOWARD COVE

44° 37.45'N 067° 23.20'W

CHARTS: 13326, 13325
CHART KIT: 78

In the southwestern corner of Machias Bay, Howard Cove is easy to enter but exposed to south and southeast winds. Holding ground is poor, and there are no facilities.

Jasper Beach, at the head of the cove, is a half-mile crescent gravel beach, backed by a marsh, with a tidal inlet at the eastern end. The beach pebbles are a mixture of multicolored jasper quartz and smooth, dark, volcanic rhyolites. The town of Machiasport, with help from The Nature Conservancy, has preserved public access to this unique beach, which has been registered by the state as a Critical Area. To visit the beach, anchor off and row your dinghy in. The western half of the cove is a cable area, so be sure to drop your hook in the eastern half.

The Pettegrow Boat Yard on Starboard Cove.

BUCKS HARBOR

3 ★★★

44° 38.30'N 067° 22.10'W

CHARTS: 13326, 13325   CHART KIT: 78, 25

Bucks Harbor is the most useful harbor in Machias Bay, and it is used by 40 or more lobsterboats, some scallopers, and several draggers. The harbor is on the western side of the bay, open to the east, but otherwise it is well protected, with an attractive and authentic fishing village at its head.

APPROACHES. Enter halfway between Bar Island and Bucks Head.

ANCHORAGES, MOORINGS. Anchor outside the moored boats, west or southwest of Bar Island, in thick mud 8 to 15 feet down at low.

GETTING ASHORE. Row in either to the Bucks Harbor town pier or to the beach south of the lobster pound.

FOR THE BOAT. *Bucks Harbor Town Pier.* This new, large, concrete wharf is the second wharf on the bump of shore northwest of Bucks Head. A small metal float for dinghies or temporary dockage is on its northwest side, with 4 feet of depth at low.

BBS Lobster Company (207-255-8888). In a pinch, you can obtain gas or diesel at the dock and float of BBS, just east of the town pier.

FOR THE CREW. Lobsters may be bought at BBS Lobster Company or from any obliging lobsterman. There are no amenities in the town.

MACHIASPORT and MACHIAS RIVER

off Machiasport: 44° 41.80'N 067° 23.51'W

CHARTS: *13326, 13325* CHART KIT: *78*

Machiasport, two miles up the Machias River, has little to offer the cruising sailor. The channel through the mudflats is deep but very narrow, and the buoys are quite far apart. The only remaining wharf belongs to Machiasport Packing. If it's empty, you might be able to get permission to tie up to the wharf.

The river is unbuoyed above Machiasport, and the fixed bridge at Crocker Point has a vertical clearance of only 25 feet.

Some eroding Indian petroglyphs are carved on a ledge close to shore near Birch Point, south of Machiasport. The carvings, showing deer and moose, human figures, and images from Indian legend, are thought to be more than 3,000 years old.

European settlers built sawmills on the river, and ships were built here in the 19th century. Great booms of logs were collected in Whitneyville, way upstream from Machias, and floated downriver for shipment from Machiasport. This was the site of the last long river drive on the eastern seaboard.

In modern times, little Machiasport became known for its valiant and successful fight against the oil refineries that tried to locate here and in Eastport, attracted by deep water and easy access. Today the economy is fueled by blueberries.

CROSS ISLAND (Northwest Harbor)

44° 37.00'N 067° 19.10'W

CHARTS: *13326, 13325* CHART KIT: *78, 25*

Northwest Harbor is on Cross Island at, as you might guess, its northwestern corner. It is easy to enter and beautiful. In the early morning, deer might be browsing along its shores, and the outlet of a saltwater pond might be sparkling in the sun as it runs across the small beach. The radio towers on the Cutler peninsula are almost out of sight.

But this is not an ideal overnight anchorage. Rafted fish pens fill most of the western portion of the cove, the holding ground is dubious, and rockweed and kelp obscure most of the sand bottom.

Northwest Harbor is best enjoyed as a daystop. Try to find a sandy patch for your anchor, and don't leave your boat for long.

CROSS ISLAND (Northeast Harbor)

3 ★★★

44° 37.13'N 067° 17.26'W

CHARTS: *13392, 13325, 13326* CHART KIT: *25* (INSET)

Looking from Cross Island Narrows, Northeast Harbor's shores appear natural and unspoiled. But from the harbor itself, the view is dominated by the enormous radio towers on the Cutler peninsula. These towers, some almost 1,000 feet tall, form one of the world's most powerful radio transmitters and are used by the U.S. Navy to communicate with their submerged submarines (see *Little Machias Bay*, next). Here, across the Narrows, contrasting worlds seem to collide. At night, red lights silently pulse on each tower, and low-lying mists come alive with an eerie and beautiful alien glow.

In contrast, there have been signs of bears on Cross Island and there are an abundance of deer. Bald eagles have nested on Mink Island. Cross Island is uninhabited except for the Hurricane Island Outward Bound School (HIOBS) base at the old Coast Guard station, just east of the harbor. The island was formerly owned by Tom Cabot of Boston, who gave the Coast Guard buildings to HIOBS and the rest of the island to The Nature Conservancy, which passed it on to the U.S. Fish and Wildlife Service.

APPROACHES. Approach Northeast Harbor through Cross Island Narrows, either from the west or east. At the western end of the Narrows, leave green can "7" off Dogfish Rocks to starboard. Run through Cross Island Narrows almost to green can "5" before turning south toward Northeast Harbor. Leave Mink Island to port and favor it slightly.

From the east, pass on either side of green, rocky Old Man Island. If you are downwind, you can guess why the island was originally called Old Man's Ass. One-hundred-and-forty pairs of razorbills nest in a cleft in the middle of the island—one of their few nesting sites on the east coast of the United States. When eiders faced extinction in Maine, the last pair nested on Old Man. Seals bask on the ledges east of Mink Island.

Find flashing green bell "3" (*Fl G 4s*) close to the end of the Cutler peninsula and leave it to port. Follow the narrow buoyed channel, leaving red beacon "4" to starboard and green can "5" to port. Turn sharply to port around can "5" to head south into Northeast Harbor. Leave Mink Island to port and favor it slightly.

ANCHORAGES, MOORINGS. Anchor to the southwest of Mink Island in 10 to 12 feet at low. This should be far enough in to get you out of the current of Cross Island Narrows. Outward Bound has several moorings here, but they are small 100-pound mushrooms.

THINGS TO DO. Trails lead all around this spectacular island. Row in to the south shore and walk around the shoreline to the HIOBS base.

LITTLE MACHIAS BAY
44° 38.40'N   067° 15.20'W

CHARTS: 13392, 13325   CHART KIT: 25 (INSET), 25

As the *Coast Pilot* says, Little Machias Bay "is not used for an anchorage as it is exposed to southerly and southeasterly winds and is close to Little River and Machias Bay, both excellent anchorages." Another good reason not to anchor here is the overwhelming presence of the radio towers on the Cutler peninsula at the western side of the bay, creating one of the most powerful radio stations in the world.

The two-million-watt radio transmitter was established at Cutler in 1961 to broadcast low-frequency radio communication to submerged submarines in the North Atlantic, the Arctic Ocean, and the Mediterranean. Its 24 kHz signal is so powerful that it can be detected across the continental United States, and an interesting side effect of the signal has been discovered. Because the signal travels close to the ground during daylight hours, it is particularly sensitive to the state of the earth's ionosphere. A flare on the sun produces x-rays that affect the ionosphere, which, in turn, enhance Cutler's signal. The strength of the signal, then, can be used to measure sun flare activity.

Activity on the U.S. Navy base in Cutler, however, has declined in recent years. Nearly all of the personnel have been relocated, and most of the base will be transferred to the public. The towers are now operated by civilian subcontractors, and messages once encoded at Cutler are now sent from Virginia before being broadcast by the transmitters.

Perhaps the only redeeming feature—other than the importance to our worldwide defense system—is the pleasure of trying to count the towers as you sail by. Sometimes it's 21, sometimes 25, sometimes 26. You try it.

CUTLER (Little River)
4 ★★★★

44° 39.33'N   067° 12.35'W

CHARTS: 13392, 13325   CHART KIT: 25 (INSET), 25

Tidal Range: *Mean* 13.5 feet; *Spring* 15.4 feet

Cutler is a working harbor with a small fleet of lobsterboats and a pleasant little town of 400 residents. The island at its entrance provides a warm and prehistoric feeling of security, like the rock at the mouth of your cave.

This is the last good harbor in Maine before Eastport—easy to enter and well protected in almost all conditions. The southern tip of Canada's Grand Manan Island is visible out the entrance, and Cutler is the best place from which to sail for Machias Seal Island or Grand Manan, six miles offshore.

On a return trip from Canada, you can check in here with U.S. Customs by phoning their Lubec office (*207-733-4331*). For more on entering the United States, see page 368.

Like many other Maine coastal towns, the history of Cutler includes a golden era of shipbuilding, from 1845 to 1855. Excursion vessels docked here in the 1880s, but since then, it has been pretty quiet.

Unfortunately, Cutler is better known elsewhere in the world. Because the VLF radio towers on the Cutler peninsula communicate with submerged U.S. subs, Cutler was once grouped with the Pentagon, the White House, and Strategic Air Command headquarters in Omaha as likely first targets in a nuclear war.

The entrance to Cutler Harbor looking past Look's Wharf, Little River Island, and the salmon pens.

APPROACHES. From the west, the enormous Cutler peninsula radio towers near Little River are visible a long way up the coast, popping in and out from behind the headlands. Then Western Head appears around the corner, looking at first like a little wooded island.

Find red-and-white bell "LR" (*44° 39.18'N 068° 10.89'W*) outside the entrance. The harbor is not immediately obvious. Enter north of Little River Island, with its 56-foot light tower, observing nun "2," which marks the ledge just beyond. As you approach the nun, the entrance opens up, comfortably wide.

Past the nun, be careful of an old weir that extends a long way from the right shore. Salmon pens are usually moored along both shores, but they are well out of the channel.

ANCHORAGES, MOORINGS. Lobsterboats are moored in two groups, opposite each of the two working docks. The first group is in the little indentation to the right, opposite Look's Wharf. The second is toward the head of the harbor, off Deano's Wharf. There is room to anchor on the south side of the harbor between the two groups of moorings in a good mud bottom 14 to 16-feet deep at low. Protection is excellent, with only slight exposure to the southeast. Note the cable area farther to the east.

Be sure to put out enough scope. Lobstermen report that yachts drag and go aground here every year.

GETTING ASHORE. Both lobster docks have vertical ladders from their floats. Land your dinghy on the beach in the middle of town to the right of the launching ramp.

FOR THE BOAT. In an emergency, gas and diesel may be obtained at the two lobster docks. Normally these commercial operations would rather not cater to cruising boats.

A.M. Look Canning Co. (207-259-7712). Look's dock and red shack is closest to the entrance, in the indentation to the right. There is 3 to 4 feet at low alongside the small float.

Deano's Wharf, Little River Lobster (207-259-7704). Deano's dock is about 300 yards farther into the harbor, with 7 to 8 feet alongside the float.

FOR THE CREW. There are few amenities for visitors in this working village. Walk up the hill to the left to find the Little River Lodge (*259-4437*), across the street from the town hall. They can provide yachtsmen with showers and a phone and can be a convenient base for crew changes. The library is at the top of the hill.

THINGS TO DO. This is a fascinating place to observe the workings of a Downeast fishing community. Watch the boats bringing in their catch and visit with local residents, who couldn't be friendlier or more outgoing to visitors on boats. There is a peaceful back road east along the coast past Money Cove, about a 20-minute walk.

Both of the beautiful headlands at the entrance to Little River are now open to the public. To explore the more than three miles of wild and unspoiled coastline from Western Head to neighboring Great Head, owned by Maine Coast Heritage Trust, land your dinghy somewhere convenient on the western shore and follow the road and path toward the sea.

Eastern Knubble is protected by a conservation easement. Walk eastward along the road toward Money Cove (shown on chart 13392). Just after the last house on the right before Money Cove, look for a tiny path to the right with a "Money Cove" sign. One branch of this woodland path leads to the headland, another to the cove itself.

Farther east, just south of Holmes Cove, the Bureau of Public Lands has acquired a 2,174-acre parcel along 4.5 miles of the "Bold Coast." Hitch a ride east on Route 191 for several miles to reach the trailhead of the Cutler Coast Reserve. Once there, you will be rewarded by carefully crafted trails and breathtaking vistas.

The Canadian lighthouse on Machias Seal Island.

MACHIAS SEAL ISLAND

44° 30.15′N 067° 05.92′W

Charts: 13392, 13325   *Chart Kit: 25 (inset), 25*
Canadian Charts: 4340

Remote, fascinating Machias Seal Island lies 10 miles southeast of Cutler and almost equidistant from Grand Manan Island to the northeast. The island is tiny, a mere 15-acre speck on the chart, but both the United States and Canada claim it as theirs, and they have disputed it for more than a century and a half.

It is hard to imagine why the sovereignty of this scrap of land is in contention, but it is, and the outcome probably will be decided at the International Court of Justice in The Hague several thousand miles to the east. If it was simply a question of which flag should fly at the lighthouse, it would be fairly ludicrous, but nowadays, with the 200-mile limit and fishing rights at stake, the issues are more serious.

Apparently, American sovereignty of the island was confirmed in both the 1783 Treaty of Paris and in the Treaty of Ghent, which ended the War of 1812. But by the time of the Civil War, it was in dispute again. The American known as "Tall Barney" claimed Machias Seal and the surrounding waters in 1865 and single-handedly defended them against a landing of Canadian officers in the spring of that year. But Canadian merchants from Saint John with shipping interests had already established a lighthouse here in 1832, and eventually the lighthouse was taken over by the Canadian government. They now claim that the operation of the lighthouse established "effective territorial occupation."

What is not in dispute is who the island really belongs to—the seabirds. As noted in the Canadian *Sailing Directions*, "Machias Seal Island is home to five species of breeding seabirds: puffins, razorbills, petrels, arctic and common terns. It is one of the largest known colonies of arctic terns on the east coast of North America and the largest razorbill and puffin colony south of Newfoundland." If you want to see the spectacular bird population, come early in the summer, because they will have migrated by mid-August.

The Canadian authorities feel strongly that human intrusion adversely affects the nesting birds, so they impose a strict limit of 30 visitors per day. If you are number 31, you are not allowed to land.

Several commercial skippers make predawn departures and call ahead to be sure their customers are included in the quota. Andy Patterson sails out of Cutler (*259-4484*) and Preston Wilcox operates out of Seal Cove on Grand Manan Island (*506-662-8552*). The Maine Audubon Society also sponsors occasional trips.

Perhaps the best known skipper was Captain Barna Norton who operated out of Jonesport. Captain Norton championed American ownership—and even his ownership—of the island for most of his life. His great-grandfather, Barna Beal, was none other than the one they called "Tall Barney." Once when Captain Norton rowed ashore brandishing the stars and stripes, a Royal Canadian Mounted Police helicopter swooped down and buzzed the island in retaliation. His son, Captain John Norton, now runs charters to the island (*497-2560; www.machias-sealisland.com*).

You can get to Machias Seal Island in your own boat, too, but the visitor quota is almost exclusively filled by the commercial charters. Their operators note with wry satisfaction that yachts tend to arrive in the afternoon. If you want to go ashore, hail Machias Seal Island on VHF well in advance of your arrival to be sure to be included in the quota. Even if you can't get ashore, you will see plenty of puffins. Cutler is the most convenient point of departure. Set your course directly for Machias Seal. The 82-foot lighthouse (*Fl 3s*) is visible for 17 miles.

On the way out you will occasionally see buoys marking bottom trawls, sometimes with a radar reflector at one end. You may also see fluorescent floats marking the nets beneath. All of these things are to be avoided. Be sure to pick up the flashing red bell that guards North Shoal and North Rock 2 miles north of Machias Seal Island.

In July of 1964 the fog horn on Machias Seal sounded continuously for 31 days, but don't let that discourage you. The same month the following year there was only one day of fog.

The moorings off the island are private and lightweight, and usually used by the tour boats. Anchor due east of the lighthouse, in the lee of the island, in 9 to 16 feet. If a heavy swell is running, anchoring will be difficult. The bottom is rocky and holding uncertain.

If you are allowed to land, use extreme caution. There are no docks (the keepers arrive by helicopter), so you will need to scramble up the seaweed-covered ledges. If a swell is up, don't attempt to land. As the sign says, *"Debarcadere Dangereux."*

Be sure to bring your binoculars, your camera, and a lot of film or batteries. Also bring a boat hook or oar to hold over your head to discourage the protective terns from dive-bombing you. Even if you can't get ashore, you'll see lots of puffins, auks, and terns flying and fishing all around.

PUFFINS

Surely the puffin is one of the world's most endearing creatures. Seen up close, this is a small, chunky bird that walks as though wearing galoshes. Its cheeks are puffy, its eyes are marked like the tuft of a sofa, and it flies like a buzz bomb, with rapidly beating wings. Part of the puffin's charm is the contrast between its sober black morning coat and earnest expression, with its orange feet and brilliantly colored beak. It looks rather like a clergyman on a binge.

Puffins and razorbills are both members of the alcid family, whose characteristics include black-and-white plumage and short, stubby wings that propel them underwater for fishing. Puffins and razorbills are usually found in the same nesting area, often along with terns, whose noisy and aggressive tactics help drive off potential predators, primarily gulls.

Historically, puffins nested on six of the outlying islands of the coast of Maine, as far west as Muscongus Bay. But the harvesting of eggs by farmers and fishermen and the killing of adult birds for their feathers destroyed their populations. By 1900, they were gone from all the Maine islands except Matinicus Rock.

Then, in the mid-1970s, Steven Kress, then the director of Audubon's Ecology Camp on Hog Island, launched an imaginative program to reintroduce the puffins to their old nesting sites. Burrows were dug by hand in the tuff of Eastern Egg Rock, in Muscongus Bay, and young birds were brought from the abundant puffin colonies of Newfoundland.

Naturalists became foster parents, handfeeding puffin chicks for months. The chicks were allowed to fledge naturally and leave to spend the fall and winter in the North Atlantic. Then the waiting began. Would they return to nest? To encourage them, the naturalists stood puffin decoys on the rocks and broadcast seductive mating calls out over Muscongus Bay. After years of effort, the puffins reared on Eastern Egg did, indeed, return to mate and nest, and the colony has been successfully reestablished.

Today you can see puffins in three places along the coast of Maine if you arrive before August 15—Eastern Egg Rock, in Muscongus Bay; Matinicus Rock, at the entrance to Penobscot Bay; and Machias Seal Island, way Down East. At Eastern Egg and Matinicus Rock, you can see them from your boat, but it is extremely difficult to land, and you would be likely to disturb the birds. At Machias Seal Island, you can land and watch the birds from blinds, only a few feet away.

Maine is at the southern end of the puffin breeding range. The puffins arrive at Machias Seal in late March but remain in the water around the island for some time. In the shelter and concealment of the large granite boulders, they make simple nests from seaweed and grasses. A single egg is incubated by both parents. The boulders also serve as convenient launching pads for these short-winged birds, which frequently have difficulty achieving flight.

The peak hatching time is mid-June, and for the next six weeks the demands of the offspring totally occupy the parents. Often one parent will spend all day away from the island diving for small herring, their principal food. Finally the young refuse food, and the parents leave them to fend for themselves. For three days to a week, they fast in their burrows before leaving their nests in the middle of the night for the surrounding waters. Before learning how to fly, the young become expert swimmers and divers. By mid-August, all the puffins have left the island to spend the fall and winter in the North Atlantic.

SANDY COVE
44° 44.96'N 067° 04.41'W

CHARTS: 13392, 13325   CHART KIT: 25

Sandy Cove is a delightful and altogether unspoiled little cove between the bright green hill of Eastern Head and Haycock Harbor. The cove is separated from Haycock by a spine of rocks. Eastern Head is owned by Maine's Bureau of Parks and Lands. On a clear day, the romantic cliffs of Grand Manan stretch along the eastern horizon.

A weir used to project from the north shore of the cove, and a line of rocks extends from the west side. There is just room to pass between these hazards and anchor off the pretty sand beach in 15 to 22 feet at low. Holding is uncertain in sand and rock.

Sandy Cove provides protection from the southwest, but it is open to ocean swells and not sheltered enough for safe overnight anchorage. This is recommended only as a daystop.

HAYCOCK HARBOR
off entrance: 44° 45.09'N 067° 04.08'W

CHARTS: 13392, 13325   CHART KIT: 25

This little cleft in the western shore of Grand Manan Channel was a favorite of yachtsman and island collector Tom Cabot. In the past, it was possible to bring a deep-draft cruising boat over the shallow entrance and into The Pool, and it may still be. Haycock, however, is not a gunkhole for the faint-of-heart.

The scale of the chart is woefully inadequate, and those who attempt to feel their way in do so, I suspect, as much for bragging rights as for beauty. And they do so at their own risk. Even Cabot managed to get his 50-foot *Avelinda* "hung by the tail" when her skeg got caught in a crevice of a cliff on a falling tide. You would do better in a kayak.

BAILEYS MISTAKE
44° 45.85'N 067° 03.50'W

CHARTS: 13392, 13325   CHART KIT: 25

Baileys Mistake is a beautiful spot with an evocative name. Only a few houses ring the cove, and the grassy shores roll to a gravel beach. And there are hardly any boats at all.

There are good reasons for this. The harbor is exposed from the southwest to southeast, and the holding ground is poor, making it a poor choice as an overnight anchorage. It is a pleasant daystop, though, for lunch or to wait for the tide.

The harbor is easily entered from red whistle "2BM." Favor high, wooded Jims Head to avoid the ledges at the entrance, which are marked with a green can "1." Look for the old weir close to the eastern shore.

One foggy day Captain Bailey sailed his four-masted schooner with a load of lumber into Quoddy Narrows, headed toward Lubec—or so he thought. His dead reckoning was about six miles off, and the ship ran hard aground. Rather than face the music with the ship's Boston owners, so the story goes, Bailey and his crew unloaded the lumber and used it to build a settlement on the shores of this lovely little harbor.

Passamaquoddy Bay
West Quoddy Head to Calais

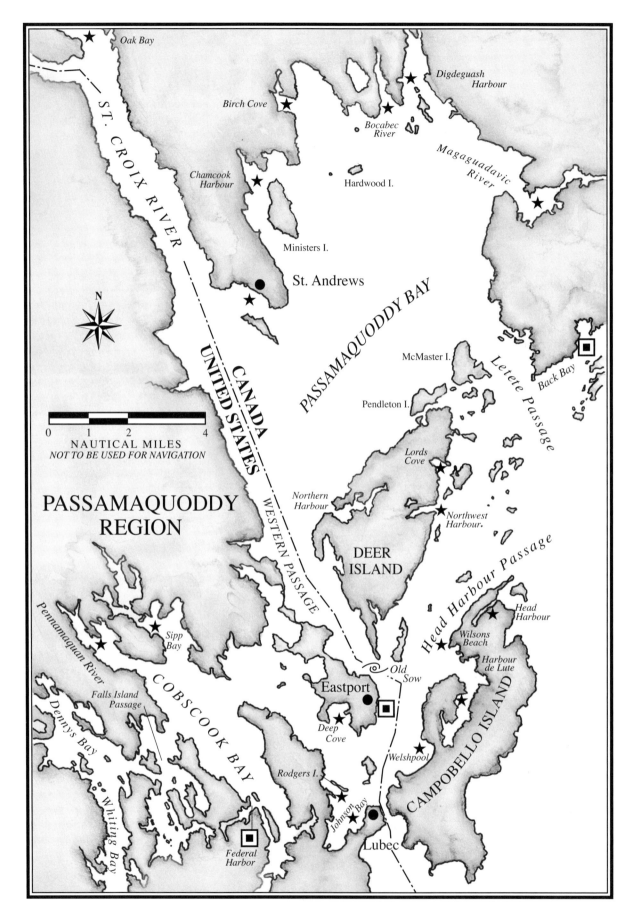

CANADA lies before you. Just a few miles off rugged West Quoddy Head lies New Brunswick's Grand Manan Island, with foggy cliffs, snug harbors, brightly colored fishing boats, and friendly people.

Across Lubec Narrows lies Campobello, best know for Franklin Delano Roosevelt's summer cottage of the same name, now an international park.

Due north is the splendor of Passamaquoddy Bay, teeming with fish and birds, guarded by swift currents and whirlpools. The St. Croix River flows into Passamaquoddy Bay from the north, marking the border between the United States and Canada. It passes St. Croix Island, where Samuel de Champlain established the first European settlement, in 1604. And it empties into the bay at the resort of St. Andrews, founded by Loyalists who barged their houses from Castine when they fled from the United States in the aftermath of the American Revolution.

This is where Maine and the United States trickle to an end in the struggling towns of Lubec and Eastport, once bustling fishing communities, now fighting to regain their economic health. Both communities flank the entrance to isolated Cobscook Bay, a haunt of eagles and seals but very few people.

NAVIGATION

Navigation can be challenging in this region, with its huge tides, strong currents, and frequent heavy fog. Average tidal ranges run between 18 and 20 feet, and they can reach 27 feet during the monthly spring tides.

Before arriving, study the Canadian charts and the buoyage system described in the Canadian *Sailing Directions*. The Chart Kit or similar commercial chart packages usually do not include any Canadian waters that are not mapped on NOAA charts, since the Canadian charts are copyright protected. For information on obtaining Canadian charts and publications, see *Introduction*, page 21. Canadian *Notices to Mariners* can be obtained at *www.notmar.com*.

Although based on the same general principle as in the United States (red, right, returning), Canada's navigation system is based on the IALA Maritime Buoyage System, with buoys that differ in size and shape, coloring, numbering, and chart symbols. Notice that most—but not all—the Canadian charts show depth in meters, not feet. Often there is a conversion scale in the margins. One meter equals 3.28 feet, or you can go metric and convert your draft from feet to meters by multiplying your draft in feet by 0.305.

The Canadian Coast Guard operates a Vessel Traffic Services (call name "Fundy Traffic") based in Saint John, New Brunswick. Fundy Traffic can be of great assistance to you when you are in the Bay of Fundy, especially in fog or heavy weather. Hail them on channel 14, 12, or 16, or call them at *506-636-4696*. By radar or triangulation on your VHF transmissions, they can pinpoint your location

The red-and-white candy-striped lighthouse at West Quoddy Head, the easternmost point in the continental United States. The original lighthouse was built in 1808 by order of President Thomas Jefferson. (Boutilier)

and warn you of nearby ship traffic much like an air-traffic controller. Fundy Traffic will also inform you of weather conditions, times of high and low slack tides, the location of endangered right whales, and other useful information. You should have no trouble reaching them by VHF from Grand Manan or Campobello.

Endangered right whales congregate east of Grand Manan for feeding and raising calves, and a conservation area has been established there for their protection. These whales are unaware of and will not move out of the way for ships and boats, so please treat them as if they were another vessel and steer well clear to avoid collisions.

If right whales are spotted, please report their location to Fundy Traffic on VHF (Ch. 14, 12) so that they can guide other ships around them.

Weirs are still common here, though many have fallen into disrepair and are visible only at lower tides. If you remember that they normally extend out from the land to trap herring swimming alongshore, they won't cause you much trouble. When in doubt, approach new shorelines at lower tides.

Lobstering season ends the last Friday in June in this part of the world, so the risk of fouling your prop in lobster buoys is greatly reduced.

ANCHORING and DOCKING

If this is your first visit, you are probably worried about how to deal with the large tides, how to anchor, and how to tie up to the high bulkheads of the many man-made harbors. These imposing structures have been built by the Canadian government at many locations to provide convenient shelter for the fishing fleets. The large wharves are designed with one or more right-angle turns, and once inside of them, boats find almost perfect protection. Often they have electricity, waste oil disposal bins, and sometimes cranes with electric hoists, which coupled with the height of the tides can handle serious rigging tasks.

With a tidal range of 20 feet, a depth of 16 feet at low becomes 36 feet at high. At high tide with a scope-to-depth ratio of 6:1, you would need 216 feet of scope. Low tide translates the same amount of scope into a ratio of more than 13:1, and that requires a lot of swinging room. If the anchorages here were as crowded as they are near Martha's Vineyard, this amount of swinging room would be unavailable. Fortunately, in this uncrowded part of the world, there is still plenty of room in most harbors to swing with all the scope you need.

In the crowded man-made harbors there are other solutions. Here the answer is to tie up to something afloat—another boat, a mooring, a lobster car, a floating pile driver. It is unlikely that you ever will be the inside boat, right next to the wharf.

Probably your boat will be rafted outside five or six assorted fishing vessels. This is the accepted practice. Only the inner boat has to worry very much about the rise and fall of the tide. When the boats inside leave at 4:30 AM, you won't even be disturbed. The fishermen will quietly shift your lines and slip out while you sleep.

Generally, you can avoid being the inside boat. If you need to be up against the wharf, your bow and stern lines should be at least 100 feet long, and you'll be glad you remembered to bring a couple of fenderboards. You will need some way to hold yourself close to the wharf as the tide rises to keep your boat and the whole line of boats tied outside of you from drifting out. The usual solution is a breastline looped around a vertical rope or wire on the wharf itself. The local fishermen are pros. Watch how they do it.

When five or 10 boats are nested, wind and current will bend the line of boats, like crack-the-whip. To counteract this problem, the inner three or four boats often run long bow and stern lines to the wharf. Be careful to avoid them as you approach. A final consideration is the tall, slippery ladders that scale the faces of the wharves at lower tides. If you have weak knees—or if heights make them weak—you may be somewhat limited when there isn't some kind of a ramp available.

CURRENTS and WHIRLPOOLS

The huge tides pull huge volumes of water through some tight spots, and the resulting currents are a force to be reckoned with. Plan your departures to spend a maximum amount of time with the current and the least possible time bucking it, especially on long passages like Grand Manan Channel. Study the approaches to Passamaquoddy Bay—Head Harbour, Letete, and Lubec Narrows— and time your passages for the best conditions of tide and visibility.

The eddies, boils, and whirlpools in the approaches to Passamaquoddy Bay are unlike anything encountered elsewhere on the coast of Maine. At full strength, they may make your boat yaw violently or even spin you around in a circle. These conditions are definitely dangerous to small craft but merely disconcerting for a stout cruising boat. One whirlpool off Eastport is called the Old Sow for the little piglet whirlpools she spins free (see *Western Passage*, page 392).

MONEY, TIME ZONES, BOATYARDS, and SUPPLIES

You do not have to worry about getting Canadian currency; U.S. dollars are accepted everywhere in the region, and most stores give an appropriate discount. With daily rate fluctuations, another strategy is to use your credit card and let your bank do the rate exchanging.

Canada adds a Goods and Services Tax (GST) to almost everything, but in an effort to encourage tourism, some of the GST is refundable to visitors. Save all of your original receipts for purchases and accommodations. Applications for the refund are available at customs offices or tourist information offices, but the problem for cruisers is that the receipts must be validated by customs as you leave Canada, a near impossibility on the high seas. For more information on the rebate program, call *902-432-5608* or visit *www.ccra-adrc.gc.ca*.

New Brunswick is in the Atlantic Time Zone, so the time is one hour later than in Maine. However, dates for switching to Daylight Savings Time may vary between the two countries.

This is an area of small towns, so most of the harbors offer no supplies or services. Plan to stock up in Eastport, Lubec, or St. Andrews. The only boatyards in the area are in Eastport, which is also the base for the easternmost station of the U.S. Coast Guard.

FATHOMS	1	2	3	4	5	6	7	8	9	10	11	12	13	14	15	16	17
FEET	6	12	18	24	30	36	42	48	54	60	66	72	78	84	90	96	102
METERS	1 2	3 4	5 6	7	8 9	10 11	12 13	14 15	16 17	18 19	20 21	22 23	24 25	26 27	28 29	30 31	

BORDER CROSSINGS

The once easygoing border-crossing formalities between these two friendly countries have become a lot more serious because of increased drug trafficking (especially by boat) and the threat of terrorism. Canadian and American customs officials are still friendly and courteous to cruising sailors, but in border-crossing matters, you'd be wise to go by the book. Penalties for customs violations are severe, including possible seizure of your boat.

This is also a time of flux for border security. Canadian and U.S. Customs regulations may change from the time of this writing.

Entering Canada. As you enter Canadian waters you should be flying a Canadian courtesy flag (ensign) at the starboard spreader. The yellow "Q" flag is not in general use.

Plan to arrive during regular, Monday-through-Friday business hours (remember the time change, one hour later than in Maine) at a designated port of entry: North Head or Seal Cove on Grand Manan; Head Harbour, Wilsons Beach or Welshpool on Campobello; Lord's Cove or Leonardsville Wharf on Deer Island; St. Andrews in Passamaquoddy Bay; St. Stephens on the St. Croix River; or Blacks Harbour, Beaver Harbour, or Dipper Harbour in the Bay of Fundy (see *Fundy and Saint John*, page 404).

You can try contacting Canadian Customs on VHF channel 87 when you enter Canadian waters or as you approach your destination. Otherwise, call *888-CANPASS (888-226-7277)* by cellphone up to four hours before your arrival or immediately once you have docked. Note that not many of the harbors detailed in this book have convenient pay phones. Only one member of the crew should go ashore to make this call; all others should remain aboard until the boat is cleared.

If you arrive outside of business hours, inspection is mandatory, and you will be charged a fee for the custom officer's overtime and mileage.

CANPASS applies only to boats entering from the U.S. with U.S. and/or Canadian residents aboard. CANPASS will need the following information: your time and place of arrival, your expected stay in Canada and its purpose, the boat's documentation or registration, a complete crew list with identification and citizenship documentation, the exact amount of tobacco and alcohol aboard, and, if you have pets, rabies vaccination certificates that are at least 30 days old from a licensed veterinarian.

Ship's papers can be either your documentation or registration. Passports are ideal for crew identification, but American and Canadian crew do not need passports. Instead, they must have proof of citizenship, such as a certified birth certificate, and photo identification such as a driver's license.

The tobacco and alcohol question almost seems silly, but the Canadians take it very seriously. Don't sail into Canada with the boat's liquor locker—or bilge—provisioned for the whole summer (or for a big party). Canada allows each nonresident to bring 40 ounces of alcohol or 1.5 liters of wine or 24 beers tax and duty free into the country. Beyond that, they will impose stiff duties which are doubly painful if you don't end up drinking your booze in Canadian waters. When you are asked if you have any alcoholic beverages, be prepared to say more than "some beers." Count your bottles. Ditto for cigarettes.

If you sound believable, you will be issued a clearance number. If you don't, or if you are entering Canada from another country, sit tight until a customs officer can come and inspect you and your vessel. Later, if you've enjoyed yourself so much that you want to extend your Canadian visit, you can call CANPASS again and extend your dates.

Boats that cross the international border regularly can register in advance with Canadian Customs and receive an actual Canpass, which expedites the check-in procedure. For more information on duties, exemptions, and entry by a private boat into Canada (including regulations on firearms, alcohol, and tobacco) call *506-636-5064* or visit *www.cbsa-asfc.gc.ca*.

Entering or Reentering the United States. When a pleasure boat or yacht arrives in the United States, the first landing must be at a Customs port or designated place where Customs and Border Patrol (CBP) service is available. Those ports are where Customs has a presence for the arrival of ships or cruise ships: Portsmouth, NH, Portland, Belfast, Bar Harbor, Eastport, and Lubec.

It is often useful to call ahead and tell CBP your estimated time of arrival. Private vessels arriving in Maine should report arrival by calling *207-532-2131 x255*. Boats might then be directed to report to a staffed location for inspection. No person or baggage may leave a vessel until it has been cleared by CBP, with the exception of the captain, who may leave the

vessel to call CBP. Once the vessel's arrival has been reported, the captain is required to return to the vessel until cleared. CBP does not charge overtime fees, but try to avoid arriving outside of regular business hours.

The captain will need his boat's registration or certification documents and crew identification and proof of citizenship. Passports are ideal, but Americans and Canadians can use other proof of citizenship such as a certified birth certificate along with photo identification such as a driver's license. As part of the Homeland Security Program, nonresidents who apply for entry with nonimmigrant visas may be subject to "biometric procedures," which means they may be digitally fingerprinted and photographed.

Be sure to declare all foodstuffs on board, particularly fresh fruit, vegetables, and meat. If you have firearms you will need to have a pre-approved form (No. 6) from Alcohol, Tobacco, and Firearms.

Non-U.S. pleasure boats entering the United States must pay a navigation fee and apply for and receive a Cruising License, which is valid for one year.

An alternate program called I-68 is the American equivalent of the Canadian Canpass. It requires a boat to be inspected once each year, but then allows it to enter and reenter the U.S. and its waters from Canada without reporting to Customs for each entry. This program would make cruising in Passamaquoddy Bay considerably easier; you could stop at Campobello, for instance, then visit Lubec or Eastport, then St. Andrews, and then return to the U.S. without repeatedly checking in with customs. Unfortunately, the program has stringent requirements that make it practical only for local boats that are constantly back and forth between the two countries: everyone onboard needs to apply in person ahead of time, each needing three photographs and a fingerprint, and each paying a processing fee.

More information on duties, exemptions, and entry by a private boat into the United States is at *www.cbp.gov* or call *202-354-1000*.

GRAND MANAN

Sailing to Grand Manan is a grand adventure for cruising boats. There is a definite feeling of setting out to sea toward a foreign land, and the sunlit cliffs of the western shore are infinitely alluring on a clear day. When the fog shuts in, the voyage is an anxious one, but the bold shores of the island make it reasonably easy to approach in safety.

Because U.S. maps and charts usually eliminate Grand Manan, a great many Americans are unaware that it exists, only six miles off our shores. Simple and old-world in feeling, Grand Manan is absorbed with fishing and the sea. Tides are more important than the clock, a man's boat more sacred than his car.

The Passamaquoddy Indians called it *Mun-a-nook* which means, simply, "island place." Settlement was spurred by British Loyalists fleeing the American Revolutionary War. Now there is charter-plane service to an airstrip on the island, but most of the inhabitants are linked to the outside world by ferry service from Blacks Harbour on the New Brunswick mainland, a 20-mile trip. With only a handful of lodgings and restaurants, Grand Manan has avoided being too touristy, though ferry service six times a day in oceanliner-sized boats is beginning to change island life. Still, tourists are given the wonderful name of "strangers-from-away," and people come here to enjoy the simple pleasures.

It seems that the colder and harsher the environment, the warmer and more hospitable its inhabitants. And so it is here. The people of Grand Manan are extremely friendly and courteous. "If you need help," they say, "just stop anyone and ask."

The large draggers and seiners of Grand Manan are freshly painted in brilliant greens and reds and blues, almost as though the paint salesman had just come through town—typical of islands where the predominant color is fog gray. The fishing fleet appears remarkably sturdy, new, and well maintained. The big trawlers head north to fish off Nova Scotia, and their catch is sold to factory ships at sea. Cod and halibut are taken in the waters off Grand Manan, though they are increasingly scarce. Lobster fishing is limited to the winter, beginning in November and lasting until June, and to protect the stocks, each lobsterman is limited to 375 traps. Draggers and divers catch scallops and urchins, and herring are seined from weirs.

Like boats, each weir has a name: *Admiral, Jubilee, Hard Luck, Grit,* and *Try Again.* The enormous Bay of Fundy tides require that the weirs have very long poles, driven into the bottom with floating pile drivers. The Grand Manan herring weir fishery is the largest in the Bay of Fundy, and much of the catch is sold directly to factory ships for export. Some is still smoked traditionally, with spruce or hemlock in weathered shingle smokehouses with red doors, although this picturesque industry is on the decline.

Dulse, a kind of edible seaweed raked from the rocks of Dark Harbour on the west side of the island, is used to make emulsifiers for ice cream and other products. It was also valued in the 1800s as a source of iodine, to prevent goiter.

More than 300 species of birds have been identified on Grand Manan, and its walking trails are a paradise for hikers, birders, and photographers. The trails along the rugged shoreline and on top of the high cliffs of the northern and western side are particularly spectacular, providing views of interesting geological formations like "Hole in the Wall." Near the south end of the island, an aptly named cluster of white glacial erratics is called "Flock of Sheep."

Since lobstering season in this part of Canada ends on the last Friday in June, the waters around Grand Manan are pleasantly free of lobster buoys during the summer. In late June, fishermen drag loads of lobster traps ashore on homemade sledges and stack them in colorful piles. On July 1, Canada Day, there are water sports and other events, including a greasy-pole contest.

The two best harbors for boats drawing 6 feet or more are North Head and Seal Cove, at either end of the island. Both are man-made and easy to reach.

Most of the east coast of Grand Manan is strewn with islands, ledges, and shoals, among which the strong tidal currents swirl and tumble. The passages between Ross, Cheney, and White Head Island dry out at low. Local fishermen use these passages and know the signs that indicate when time is running out, but strangers shouldn't try them. If you see what little is left of these passages at low tides, you'll be glad you left them to the locals.

The shoals that stretch southeastward from White Head Island—Bulkhead Rip, Clarks Ground, and Old Proprietor Shoal—are even more dangerous. Here the current runs 4 to 6 knots, and heavy rips build up on the ebb. Heading north on a flood tide with good visibility, you might have a lovely sail. In the fog or against the ebb among the shoals, it could be a nightmare. The most conservative approach is to avoid the east coast of Grand Manan altogether or to stay outside the 50-fathom curve, well away from its dangers.

There are various ways to get around Grand Manan and see the sights. The island is 14 miles long, so walking the whole way is impractical. Bikes can be rented, there is a small car rental operation, and you will often be asked if you need a ride.

PASSAGES TO GRAND MANAN

If conditions are right, you can reach Grand Manan comfortably on your third day out of Penobscot Bay, spending the first night near Mount Desert and the second at Roque Island. You would be wise to allow at least one more day for the trip to provide for the usual surprises.

The last leg of the voyage can begin at Roque Island, Cutler, or Head Harbour (Campobello Island). Roque is probably as far away as you want to start, about 40 miles from Grand Manan's North Head. If the weather turns bad, Cutler can be a fallback.

The trick is to pick a day with reasonable visibility and to maximize the hours you can ride the flood current up Grand Manan Channel. The current runs 1.5 to 2.5 knots, flooding northeast and ebbing southwest. Along the west coast of Grand Manan, the current reaches 3 knots. Grand Manan Channel can be extremely rough when the wind is against the current, creating short, steep waves. When strong southeasterlies blow up the Grand Manan Channel against the ebb, rough and sometimes breaking seas can be found in the approaches to Quoddy Narrows and off Nancy Head on Campobello.

Grand Manan is split on U.S. charts, so you will have to do some switching back and forth between chart 13392, the southern half, and 13394, which shows the northern half.

Approach the island along the bold western shore, rounding the southern tip of the island to reach Seal Cove or the northern tip to reach North Head.

To avoid a substantial fee, it is desirable to arrive at Customs during business hours Monday through Friday, preferably in the morning on Monday, Wednesday, or Friday (see *Entering Canada,* page 368). That takes a bit of planning. If you have to make a choice, opt for visibility and favorable current and let Customs fall where it may.

NORTH HEAD

4 ★★★★★

44° 45.65′N   066° 45.00′W
CANADIAN CHARTS: **4340, 4342**   U.S. CHARTS: **13394**
Tidal Range: *Mean* 17.9 feet; *Large* 24 feet

Customs: 888-CANPASS, office: 506-662-3232
Emergency: 911, 506-662-8484, 800-665-6663
Ferry information: 506-662-3606, tickets: 506-662-3724

North Head is a fishing village on the northeast corner of Grand Manan, home to a fleet of brightly painted seiners and trawlers. The harbor is always bustling with activity, a fascinating place to spend a few days and meet the friendly residents.

The village, a man-made harbor, and the ferry docks are on the south side of the peninsula of North Head. This is a port of entry for Canadian Customs, an excellent refuge, and the best place to start your exploration of the island.

APPROACHES. Coming from the coast of Maine, by far the safest approach is along the western coast of Grand Manan. The current in Grand Manan Channel averages 1.5 to 2.5 knots, flooding northeast and

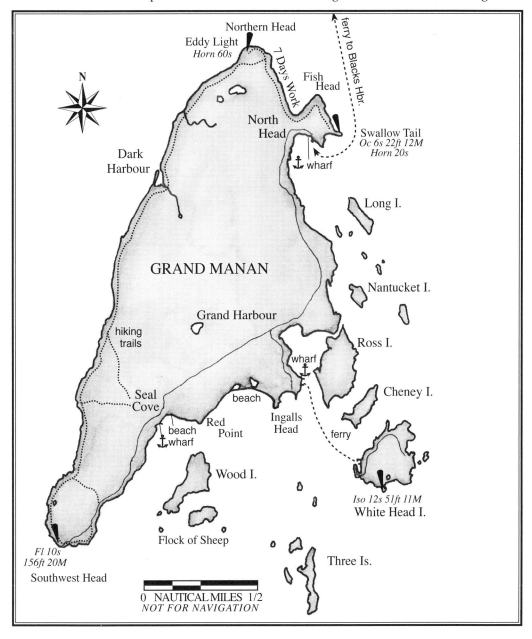

A CRUISING GUIDE TO THE MAINE COAST

PASSAMAQUODDY BAY

ebbing southwest, and near the west coast of Grand Manan, the current runs about 3 knots, parallel to the shore.

Grand Manan's west coast is wooded, with steep cliffs 300 feet high. North of Dark Harbour they rise to 400 feet. In clear weather, the cliffs are visible from a long way away. Head for a point off the northwestern tip of Grand Manan, keeping good dead reckoning and GPS fixes against the possibility that fog may close in. The coast is bold, with no outlying dangers. Even in thick weather, you would probably be aware of the crash of the surf and loom of the cliffs before getting into trouble.

As you near the northern tip of the island, you will see the lighthouse at Northern Head on a plateau halfway up the cliffs: a square white tower and several separate houses with red roofs. Rounding the northern point, with its dramatic scalloped cliffs, note the multiple layers at Ashburton Head known as "Seven Days Work." There may be turbulence and tide rips here, and you are likely to be blanketed for a while by the island. Keep a lookout for the ferry, which runs several times a day between North Head and Blacks Harbour to the north. In fog, you might want to call the ferry on channel 16 for their position or to call Fundy Traffic on channel 14 or 12 to report yours. At this point, there is still some distance to go to reach the harbor, especially if the tide is flooding and the current against you.

As you round Fish Head, the distinctive Swallow Tail lighthouse will appear as a white octagonal tower with a red top, guyed all around. Be on the lookout for whales here, particularly the endangered right whale. If you see these rare creatures, steer clear—their numbers have been decimated because they aren't aware of vessels—and report their location to Fundy Traffic on VHF channel 14 or 12.

Continue south past Swallow Tail, leaving the red bell off Net Point to starboard as you make the turn westward toward the harbor. The weirs close to shore are typical of those found all around Grand Manan.

The Canadian courtesy flag should now be flying at the starboard spreader, and you are in the Atlantic Time Zone. The time here is one hour later than in the eastern United States.

ANCHORAGES, MOORINGS. There are three large wharves in Flagg Cove, off North Head. The more easterly wharf, marked with a fixed red light at the outer corner, is for the large ferries from Blacks Harbour. The middle wharf, to the west, is the old ferry wharf. Just west of that is Fishermen's Wharf, in the shape of a bent T, with a flashing red light at the outer corner (see inset, Canadian chart 4342). Fish pens moored in the western part of the cove are far enough from the wharves to present little danger, although they may not be lit at night.

⚓ Both ends of Fishermen's Wharf have deep water, and you can tie up inside either end of the crosspiece of the T. The easterly part is exposed to the east and southeast. The most convenient place to lie is inside the westerly part where there is more room and better access to a float for landing. The inner corner is reserved for large herring seiners. Along the western portion of the wharf, you can probably find a boat about your size to tie alongside. Boats here usually moor heading westward. If there are fishermen around, they will help you tend the lines.

The harbormaster will come by to collect a $25 fee for dockage here, even when you are rafted to the fishing boats. He might also shepherd you into his pickup and give you a lift if you need one.

GETTING ASHORE. If this is your port of entry, call *888-CANPASS* or report to Customs (*506-662-3232*) in the brick house—also the post office—near the wharf (see *Entering Canada*, page 368). Row in to the "pontoon" and ramp near the head of the wharf. Another possibility is to clamber across the neighboring boats and climb a vertical ladder.

FOR THE BOAT. *Harbour Authority of Grand Manan (506-662-8482).* There is a trash bin at the head of the wharf. ⛽ Fuel can be obtained by truck from

Irving (*506-662-3433*), located left down the road a short distance from the wharf. Electricity can be arranged at the dock by permit, through the harbormaster.

Ice can be bought at most stores. Water is harder to come by. Talk to the captains of the large seiners docked at the wharf. They have installed—at their own expense—running water to the outer portion of the wharf. Or use a five-gallon container and ask for water at a nearby store. Charts and publications are sold at Island Home Hardware (*506-662-8441*) in Grand Harbour.

Daggett Agencies (506-662-3652) in Grand Harbour has some marine supplies and can steer you to mechanics.

M.G. Fisheries (506-662-3696), also in Grand Harbour, carries limited marine supplies.

General Marine Service (506-662-3288). General Marine is centrally located in Grand Harbour. Wiley McDowell can help you with repairs of all kinds.

FOR THE CREW. Laundry facilities, showers, and toilets are available at Hole-in-the-Wall Campground (*506-662-3152*) about .25 miles away. Take a right on the ferry landing road, then a left on Old Airport Road.

Several takeouts, dairy bars, and restaurants cluster near the ferry landing, within walking distance west of the wharf. The Compass Rose (*506-662-8570*), on the harbor, serves three meals a day and—remember, we're in Canada—tea.

On summer Saturdays the village of North Head has a farmers' market featuring local produce and crafts. The Grandisle grocery and drug store (*506-662-3439*) is about a mile west of the docks, where the main road turns south. A bakery and the Grand Manan Hospital (*506-662-8411*) are just beyond. If you arrange transportation, there is a market, the Grand Manan Save Easy (*506-662-8152*), and a laundromat in Woodwards Cove.

Pay phones can be found near the ferry dock and at the Customs House where the post office is also located. Bikes can be rented and kayak tours planned at Adventure High (*506-662-3563*), a short distance from the ferry landing. At present, there is no taxi service on the island.

Grand Harbour has the island's only bank, a hardware store with charts (*506-662-8441*), and the fabulous, old-style Newton Market (*506-662-8166*). The only liquor store is in Castalia, about 2 miles away.

THINGS TO DO. Tourist pamphlets and maps of the walking trails are available at markets and stores.

One nearby walk is a must. Turn right from the wharf and follow the road and signs .7 miles out to Swallow Tail Light. The path leads down and across a bridge over a chasm to the spectacular little peninsula on which its white-shingled, octagonal tower stands firmly anchored against the winter winds by guywires. The former keeper's house and a boathouse are still there. Notice the angle of the ways down which the boat was launched. Look out to sea, and you might see whales.

For a spectacular but much longer hike, take the trail that starts near the Swallow Tail parking lot. It heads through the woods and along the cliffs past Whale Cove and Hole-in-the-Wall and out to Fish Point and back. The trail is rough and often emerges on high cliffs. Wear long pants, take insect repellent, and allow 2½ hours.

If you are sailing with children, take them to Stanley Beach at the west end of the North Head bight to look for sand dollars at low tide.

The main road, Route 776, runs down the eastern shore of Grand Manan, past several small coves. At Woodwards Cove, take a look at the smokehouses, with their weathered shingles and red doors, still used to smoke herring.

A free ferry runs from the man-made harbor at Ingalls Head to White Head Island, which is fun to explore by bike. Just make sure you catch the last ferry back at 4:30 PM (double-check that you have set your watches ahead by one hour).

At Grand Harbour, farther south via Route 776, there is a wonderful little museum with an impressive Fresnel lens from a lighthouse, marine and geological exhibits, and a large collection of island birds. The Tourist Information Centre (*506-662-3442*) is located beneath the museum.

Several operators offer charter excursions to watch seals, whales, or puffins. Or you can take to the skies with Atlantic Charters (*506-662-8525*) for sightseeing by air from Seal Cove.

Don't miss the social phenomenon of loading the ferry, lifeline of all islands: cars line up on the main street hours before departure time, large trucks inch down the ramp into the ferry's maw, and families walk aboard for the long ride to the mainland.

(Frank Claes)

WEIRS

A weir (pronounced "ware") is a large trap for herring, caught mostly for bait. Slice the head off a small herring, and it becomes a sardine. Sardines used to be big business in Maine and Canada. A century ago, there were more than 50 canneries along the coast of Maine and annual shipments of 500 million cans of sardines.

Herring are caught by surrounding a school of fish with nets (purse-seining) or by closing off a cove when the fish are inside (stop-seining). But one of the oldest methods of catching herring is by building weirs.

With the advent of an offshore herring fishery, however, fewer herring are running close inshore, and weirs are fast becoming a thing of the past. You won't see any east of Schoodic Peninsula anymore, but you may begin to see them—or the rotting remains of abandoned weirs—as you head Downeast. When you cross over into Canada, in Passamaquoddy Bay and around Grand Manan, there are still weirs everywhere, though most aren't active.

Fortunately, weirs are not a great danger to alert yachtsmen. Remember that weirs usually extend from the shoreline, so do not try to pass inside them. Weirs are generally visible at high tide or are marked by a high pole at the outside. Old weirs may lurk under the surface until midtide or below. Even so, hitting an old weir pole is not like hitting a rock.

Weirs have been around for a long time, and their heart-shaped design hasn't changed much. They are usually made from materials found locally, and repaired each spring after the ravages of winter. Poles are driven into the mud leading from the shore into the top of the heart-shaped "pocket." In the Bay of Fundy, enormous stakes are required because of the huge tides. You may see 75-foot Douglas fir stakes piled in Grand Manan waiting to be driven. Homemade, floating pile drivers are a common sight way Downeast, and small cruising boats often moor alongside one of them for the night.

After the long weir stakes are driven, shorter top poles are attached. The structure is tied together with a rope rib, lashed horizontally. In earlier days, the space between the poles was filled with brush. Today, the more effective netting, or "twine," is used. The "top twine" is hung on the rib around the pocket and reaches to high water level, and divers secure the "bottom twine" on the sea floor.

Herring come inshore at night, usually with the flooding tide. The long "leader" of poles and netting from the shore diverts schools of fish into the pocket where they become disoriented and swim around in circles, unable to find the exit.

When the fishermen become aware that there are herring playing in the weir (often by the dark of the moon), they bring in a dory or two and drop a purse seine beneath the herring. Drawing the fish together as they close the seine, the fishermen then dip or pump their catch into the hold of a lobsterboat or herring carrier.

A weir is a semipermanent and very valuable installation. "My father fished it before me, and my grandfather before him," a fisherman may tell you. In Canada, weirs have names, and there are weirs such as *King George*, *Victoria*, *Gamble and Ruin*, *Bread 'n Butter*, *Last Ditch*, *Iron Maiden*, and *Cora Belle*.

Weir fishing is passive, and weir fishermen become experts on the habits, needs, and whims of the elusive silvery fish, analyzing their cyclical behavior and debating how their movements are affected by tides, moon, and weather. As one old-timer says, "You got to think like a herring."

SEAL COVE

★★★★

44° 38.85′N 066° 50.30′W

CANADIAN CHARTS: *4124, 4340, 4342* U.S. CHARTS: *13392*
Tidal Range: *Mean* 14.3 feet; *Large* 19.6 feet

Seal Cove is a small and pleasant fishing community on the eastern side of Grand Manan near the southern end and a possible port of entry into Canada. The approach from the south is easy, and a man-made harbor provides excellent protection. Like most Grand Mananers, the people of Seal Cove are extremely friendly and helpful.

This harbor is used mostly by lobsterboats. The lobster season officially ends in June to allow lobsters to molt and breed, so the boats gear up for other fisheries. Often nets are laid out in the fields and among the neat little houses to dry. In the spring, Douglas fir poles 60 to 80 feet long are piled on the docks. They are sharpened at one end, ready to be driven by the floating pile drivers that build or repair the island's weirs.

The old harbor at the head of the cove dries out, so it is not useful as an anchorage. It is still used, though, by herring seiners and dories when the herring are running. The catch is unloaded and racked and hung in the weathered shingle smokehouses that cram the shore and perch over the water on old pilings.

Lying secure in Seal Cove on a foggy day, there is a great feeling of peace, gently interrupted by occasional raucous seagull cries, the rumble of a lobsterboat's diesel starting up, or the rise and fall of voices on the wharf as fishermen discuss the catch, the market, and the weather.

With prevailing summer winds, there will be fog at Grand Manan if the mainland is warm. "It can be desperate foggy in Seal Cove," the fishermen say, "but burned off by the time you reach North Head."

APPROACHES. From the south or west, make for high Southwest Head and the lighthouse (with two radio towers nearby). Then pick up red-and-white bell "XAD," southeast of the head. From the bell, coast up between low and partially wooded Wood Island and the bolder east shore of Grand Manan.

Note green can "XA3" off Buck Rock, left to port, and look for the "Flock of Sheep" near Pat's Cove—glacial erratics left browsing 10,000 years ago. Stay in midchannel to clear the weirs along the Grand Manan shoreline. Otherwise, this channel is wide and open with the few dangers clearly marked. Ahead you will see the dark bulk of the wharves and the cluster of houses in Seal Cove. The fish pens moored south of the harbor are clearly visible, but they may not be lit at night.

ANCHORAGES, MOORINGS. There are two man-made harbors in Seal Cove. The northern one dries out at low water.

The southern harbor, which you come to first, has good, deep water and excellent protection. A breakwater and wharf that are not shown on the charts extend from the north, perpendicular to the main wharf, and form a narrow entrance. Be careful as you enter that someone isn't coming out.

Tie up to a boat along the main wharf or, after June 30, to a lobster car in midharbor. An average low tide leaves about 12 to 14 feet of water along the inner wall of the south wharf and 9 to 10 feet at the southern edge of both rows of lobster cars and along the inner edge of the north wharf. The harbormaster will stop by to lighten your pockets by $25 for dockage.

Use caution if you decide to dock to a lobster car. Get permission to stay alongside and check how well it is moored. Lobster cars float around quite a lot. If you are tied to one, don't be surprised to find yourself bumping one of the boats nested at the wharf. Fenders are good insurance.

GETTING ASHORE. Take the dinghy in to the float near the land end of the south wharf. If you are entering Canada here, call Customs at *888-CANPASS* from the pay phone near the head of the dock.

FOR THE BOAT. *Harbour Authority of Grand Manan (506-662-8482).* The Harbour Authority is responsible for all dockage. There is a trash can on the dock but no water. If you need water, ask around. Someone may give you water at his home, but you will have to lug it in containers. Irving (*506-662-3433*) will truck fuel to the wharf. Small amounts of ice are sold at the markets.

General Marine Service (506-662-3288), in Grand Harbour, usually works on commercial boats, but they can help you with mechanical trouble.

Daggett Agencies (506-662-3652) in Grand Harbour has some marine supplies.

M.G. Fisheries (506-662-3696), also in Grand Harbour, carries limited marine supplies.

FOR THE CREW. A half-mile walk into town will bring you to a small convenience store.

THINGS TO DO. Visit the handsome, shingled smokehouses around the northern harbor. Their owner, New York architect Michael Zimmer, runs a fun, free museum of art and fish, including a Herring Hall of Fame. Bring a picnic. A nice sandy beach with good sun and protected water extends farther to the north.

The public Brookside Golf Course is an easy walk away, and it rents clubs. Windsor Park, on the hill above the town, has tennis courts.

For a long hike, walk about 4.5 miles down the road to Southwest Head. You will be rewarded by wonderful views southward to Machias Seal Island and southwestward to the mainland.

Preston Wilcox makes daily trips aboard *Day's Catch* to Machias Seal Island out of Seal Cove (*506-662-8552*).

On Canada Day, on July 1, Seal Cove hosts an annual Canada Days Celebration with a children's fancy dress parade, village suppers, and scallop-shucking and greasy-pole contests.

INGALLS HEAD

44° 39.73′N 066° 45.22′W

CANADIAN CHARTS: 4340, 4342, 4011   U.S. CHARTS: 13392
Tidal Range: *Mean* 14.3 feet; *Large* 19.6 feet

Ingalls Head is a man-made harbor at the southeast corner of Grand Manan. It serves a substantial fishing fleet that unloads its catch here for processing in shoreside plants.

The harbor provides excellent protection and has been dredged to depths of about 10 feet. The ferry to White Head Island leaves from here.

APPROACHES. From Seal Cove, follow the buoyed channel north of Wood Island. Leave the green bell off Ox Head to port, then head north for Ingalls, leaving the two green spar buoys to port. Watch for the sudden departure or arrival of the red ferry from White Head Island.

ANCHORAGES, MOORINGS. The man-made harbor is heart-shaped, divided by a central pier. The southern section is more convenient because it has a float for landing. Tie up in any free space along the sides of the harbor or, preferably, alongside another boat. The finger piers in the northern section are used by Fundy Marine's boatlift.

FOR THE BOAT. Gas and diesel might be available from lobster cars in the harbor. Otherwise, Irving (*506-662-3433*) can deliver it.

Fundy Marine Service Center (Ch. 69; 506-662-8481). This storage and hauling facility is just north of the harbor. It is owned by the province and managed by the Grand Manan Fishermen's Association as a do-it-yourself yard, and it is available to transients. The yard has an enormous 150-ton boatlift and electricity, but no other facilities.

GETTING ASHORE. Row in to the float in the southern section of the harbor.

FOR THE CREW. The nearest amenities are in Grand Harbour, a pretty walk of about .75 miles. There you will find a bank, groceries, hardware, and restaurants.

THINGS TO DO. For a pleasant outing to a remote island, take the free ferry from the concrete ramp in the southern section of the harbor to White Head Island. Fishing families have lived on White Head for over two hundred years. It's a friendly and hospitable place and a haven for wildlife. The main occupations of the island are fishing, tending weirs, and raking dulse.

Stroll along Battle Beach to the lighthouse on Long Point past ponds full of ducks and other birds. For a long walk, take the main road along the northwest coast and around to Gull Cove on the eastern side of the island. There's a grocery store at the docks. Make sure you have set your watches ahead an hour to Atlantic time, so you won't miss the last ferry back to Ingalls Head at 4:30 PM.

Weathered sheds and smokehouses cram the shores of Old Harbour, Seal Cove.

APPROACHES to PASSAMAQUODDY BAY

Campobello Island, Deer Island, and a chain of smaller islands overlap to guard the entrances of Passamaquoddy Bay, tucked away to the north. Campobello also protects the Maine towns of Eastport and Lubec, which flank the entrance to the wild and empty waters of Cobscook Bay.

As the enormous Bay of Fundy tides pour through these complex passages, they form strong and turbulent currents, which, together with the frequent fogs, make navigation difficult in the area. The three major approaches to Passamaquoddy Bay are Lubec Narrows, Head Harbour Passage, and Letete Passage—each with very strong currents. Head Harbour Passage is the main channel and by far the easiest.

U.S. chart 13396 is particularly useful in this area. Based on Canadian data, its soundings are in meters, and it includes details such as the location of fish weirs.

Lubec Narrows. The International Bridge joining Lubec and Campobello Island over Lubec Narrows is fixed and has a vertical clearance of only 47 feet (at high tide). This passage should be used by strangers only at high slack, if they can pass comfortably under the bridge. Currents at maximum strength run 7 or 8 knots with substantial eddies and turbulence, and there is danger of colliding with one of the bridge piers. At low tide the approach channel is narrow and crooked. It's far better and safer to make the longer trip around Campobello Island and through Head Harbour Passage. For sailing directions for Lubec Narrows, see the Lubec approaches on page 383.

Head Harbour Passage. Head Harbour Passage runs between Campobello and Deer Island toward Eastport and Lubec. It is several hundred yards wide and clear of obstructions. Although you may encounter tide rips and boils, the passage is basically safe and easy. See page 380.

Letete Passage. Letete Passage cuts north of Deer and McMaster Island and is the most direct route to St. Andrews. It is wide and full of obstructions, and the current can rip along at 8 knots, with strong eddies and boils. It is well marked, though, and in good visibility near high or low slack it is easy. Passing with the flood is like having an extra engine. Unless you know it well, avoid Letete Passage in poor visibility. See page 394.

Western Passage. From Eastport, the usual approach to St. Andrews and Passamaquoddy Bay is through Western Passage. Time your passage to avoid the worst currents. Near slack water, Western Passage is usually calm and benign. But the full force of the flood or the ebb can bring to life "Old Sow," one of the world's largest whirlpools! See page 392.

CAMPOBELLO ISLAND

Campobello is a large Canadian island that protects the entrances to Cobscook and Passamaquoddy Bay, joined to the mainland by the International Bridge over Lubec Narrows. It is a lovely island, 7.5 miles long and 2.5 miles wide, with hard-working fishing communities and the famous summer home of President Franklin D. Roosevelt on Friars Bay. His cottage has now become part of the Roosevelt Campobello International Park and can be reached by docking in the man-made harbor of Welshpool.

Campobello's southern and eastern shores front on Grand Manan Channel. Head Harbour Passage runs along its northern and western shores, with depths of 200 and 300 feet. The island has two excellent natural harbors, Head Harbour at the northern end and Harbour de Lute on the western side, as well as several small man-made harbors that can provide refuge in a pinch.

A Canadian customs office is at the bridge to Campobello, but there is no place for a boat to dock. The easiest harbors to clear customs are Head Harbour, Wilsons Beach, or Welshpool, but none have payphones.

HEAD HARBOUR

4 ★★★★★

W of Head Harbour Is.: 44° 57.10'N 066° 54.50'W
CANADIAN CHARTS: 4373
U.S. CHARTS: 13396, 13328, 13329, 13394, 13398
Tidal Range: *Mean* 18.1 feet; *Large* 25.2 feet

A picturesque lighthouse at the end of a series of bold, rocky islets gives the mariner a dramatic welcome to Head Harbour, at the northeastern end of Campobello Island. On early French charts, this was *Port aux Coquilles*. Though less euphoniously named nowadays, it is still one of the best harbors in the Passamaquoddy region, a good place to spend the night or to wait out the tide before entering Head Harbour or Letete Passage. The entrance is easy, the harbor is snug, and there is always fascinating fishing activity.

The accepted practice, even for pleasure boats, is to tie alongside one of the fishing boats at the large, concrete, government wharf. To get ashore you need to clamber across industrial decks and up iron ladders. If you want a less gritty experience, however, there is a small marina on the opposite shore.

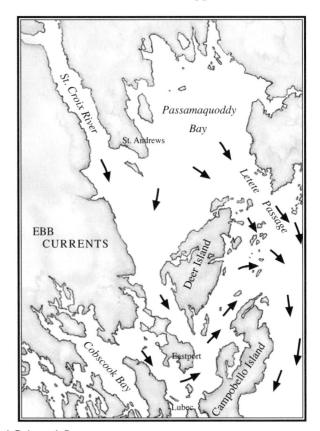

Flood and ebb currents in the approaches to Passamaquoddy and Cobscook Bay.

Aerial view of the entrance to Head Harbour, at the north end of Campobello Island. East Quoddy Head lighthouse is in the foreground. Head Harbour Passage and Deer Island are in the background. (Christopher Ayres)

If this is your first port of call in Canada, it can be difficult to call customs from here (*888-CANPASS*). Surprisingly, there is no public phone at the wharf, and there is no cellular coverage. In season, a small shingled shed by the wharf is occupied by Island Cruises, and they will let you use their phone during their regular 9-5 business hours.

Likewise, if you are at the marina of boatbuilders S.E. Newman and Son, opposite the fishing wharf, you can use their shop phone when they are open. Otherwise, it is a 1.3-mile walk to a store in Wilsons Beach.

Also remember that if you plan to arrive back in any U.S. anchorage the next day, you will have to reenter the U.S. at Eastport or Lubec.

APPROACHES. Nancy Head, on the east coast of Campobello Island, is high and prominent. The distinctive lighthouse at East Quoddy Head, with its white buildings and red roofs, is visible from a long distance. The white tower sports an enormous red cross, as though wrapped in red ribbon. Islands and headlands range away to the northeast along this beautiful coast.

There are two entrances to Head Harbour, one on each side of Head Harbour Island. The northern entrance is better. Run down the middle of the northern entrance, passing weirs on both sides. You'll also pass moorings and a long private dock near the western end of Head Harbour Island.

Three green buoys are grouped close together beyond the island. Run north of them, leaving them to port. Favor the north shore at Cubs Point, then proceed down the middle of the channel. From just past Cubs Point to the L-shaped public wharf, the water in the channel is more than 30 feet deep at low. Salmon pens may be moored along the edges of the channel.

ANCHORAGES, MOORINGS. Fishing boats are tied up along every wall of the public wharf, with the largest nested along the outer edges of the L. Depth is good inside as well, where there is about 15 feet at low. Pick a convenient boat to come alongside.

Alternatively, S.E. Newman's marina (*506-752-2300*), directly opposite the wharf, has plenty of dock space with depths at low to 25 feet. The floats have water and electricity.

There may be room to anchor among the line of stake moorings that extend westward to shoal water. The rocky southern shore west of the wharf is steep-to, with depths nearby of 12 to 15 feet at low and good mud holding ground.

The halogen lights on the wharf will cast an eerie yellow glow at night, especially in the fog.

GETTING ASHORE. Clamber over the adjoining boats and climb the ladder to the top of the wharf. Or, if you are less athletically inclined, dinghy to the float on the inside of the L and use the ramp.

FOR THE BOAT. Inquire at the wharf about water. There are waste-oil containers ashore.

S.E. Newman and Sons, Inc. (506-752-2620, 506-752-2300). Steve Newman and his crew build new boats, but they can probably help you with any emergency repairs.

FOR THE CREW. From the dock, turn left and walk 1.3 miles to the small market and post office at Wilsons Beach.

THINGS TO DO. East Quoddy Head Light stands on the most seaward of a number of small islets. After a challenging walk up and down steel ladders and across treacherous seaweed and sharp rocks, you will reach a sign that reads, "Danger—Rising Tides can get you stranded on this island." Proceed to the lighthouse only if it is near low tide.

During July and August you may be able to spot whales from East Quoddy Head. Often a pod will come in on the flood, spouting frequently. The walk from the Head Harbour wharf to the parking lot at the lighthouse is 1.6 miles. If you want to get closer, Mac Green's Island Cruises depart from the wharf for whale-watching.

The head of Head Harbour has numerous wrecks, derelict vessels, and a sagging cannery on its banks. It feels like another era, well worth a little time-travel by dinghy.

HEAD HARBOUR PASSAGE
off Brown Head: 44° 56.60'N 066° 56.10'W

CANADIAN CHARTS: 4373, 4011
U.S. CHARTS: 13396, 13394

Head Harbour Passage is the main approach to Eastport, Lubec, and Cobscook Bay. The entrance is at the northern end of Campobello Island, marked by East Quoddy Head Light, with its distinctive white tower and large red cross.

This deep and unobstructed passage follows the steep north shore of Campobello Island for about 4 miles until it meets Friar Roads and Western Passage south of Deer Island, a spot marked by the lighthouse on Cherry Island.

Ideally, come through Head Harbour Passage with the current, which floods southwest and ebbs northeast at a maximum strength of about 5 knots. At the entrance, strong east or northeast winds against an ebb current may create a heavy breaking sea off East Quoddy Head. The flood tide will set you hard toward Spruce Island and Black Rock. In the passage, you may encounter boils, eddies, and turbulence strong enough to make your boat yaw—unsettling, but not dangerous except for small craft.

The old boathouses of Wilsons Beach.
Opposite page: aquaculture tenders at Welshpool.

WILSONS BEACH
4 ★★★
44° 55.88'N 066° 56.50'W

CANADIAN CHARTS: 4343, 4340, 4373
U.S. CHARTS: 13396, 13328, 13394   CHART KIT: 79
Tidal Range: *Mean* 18.1 feet; *Large* 25.2 feet

Just north of Windmill Point on Head Harbour Passage, Wilsons Beach is a small fishing community with a man-made harbor. It can be identified on the chart by the light shown at the end of the wharf, and by the large red wharf buildings that have collapsed in recent years.

You can enter Canada at Wilsons Beach by calling Canadian Customs (*888-CANPASS*).

APPROACHES. The approach from Head Harbour Passage is straightforward. The long wharf is angled toward the land to provide a protective pocket. Several boats will probably be rafted inside the face of the wharf, with barely room to squeeze by between them and the land.

ANCHORAGES, MOORINGS. There is about 6 feet of depth at low along the inner face of the long wharf. Tie up to one of the resident boats.

GETTING ASHORE. There are no floats in the harbor, so you will have to climb a vertical ladder to get ashore.

FOR THE CREW. You will find a small variety store and a post office near the docks. Family Fisheries takeout and fish market (*506-752-2470*) is about a mile south.

HARBOUR DE LUTE

4 ★★★★

44° 54.48′N 066° 55.83′W

CANADIAN CHARTS: 4343, 4340, 4373
U.S. CHARTS: 13396, 13328, 13394   CHART KIT: 79
Tidal Range: *Mean* 18.1 feet; *Large* 25.2 feet

This excellent and seldom-used refuge lies on the northwest side of Campobello Island, opposite Eastport. Early charts called it *Harbour de L'Outre*. Several man-made harbors have been built on either side of the entrance, and presumably the fishermen find these more convenient than the deeper indent past Man of War Head. For a cruising yacht, however, Harbour de Lute provides a serene and private anchorage.

Because of the many decaying weirs in the harbor, it is preferable to enter below midtide, when most of the stakes are visible.

APPROACHES. The entrance to Harbour de Lute, between Windmill Point and Man of War Head, is wide and open except for salmon pens on either side. Favor Man of War Head to stay clear of the 9-foot spot and head for green spar "UX1." Two weirs extend from the western shore, one before the green spar and one beyond it. The second weir has lost its top poles and is not visible after halftide.

Leave the spar fairly close to port to clear the weirs and then make a wide turn to starboard toward Bunker Hill, favoring the eastern shore. There are several more weirs in the inner portion of the harbor. Continue to the area east of Seal Rock, which is visible except at very high tides, lying off the western shore among a series of pilings.

ANCHORAGES, MOORINGS. Anchor in midharbor, setting the hook in good mud 11 to 20 feet deep at low. There should be ample swinging room. Protection is good from every direction, although strong northerlies can funnel into the harbor.

After anchoring, pick out a range on the northeast shore that will take you clear of the weir on the western shore in case you want to leave at high tide.

GETTING ASHORE. Row in to the beach at a convenient spot on the eastern shore.

THINGS TO DO. If you land at the meadow north of Bunker Hill, a brief struggle through the bushes will bring you to a dirt road shown on the chart. There are lots of wildflowers, and you may see a resident bald eagle.

WELSHPOOL

4 ★★★

44° 53.30′N 066° 57.45′W

CANADIAN CHARTS: 4343, 4340, 4373
U.S. CHARTS: 13396, 13328, 13394   CHART KIT: 79
Tidal Range: *Mean* 18.4 feet; *Large* 24.7 feet

Welshpool is a man-made harbor at the northern end of Friars Bay on the west side of Campobello. It offers the cruising sailor a way to visit President Franklin Roosevelt's summer cottage, the 2,600-acre Roosevelt Campobello International Park, and other parts of the island. Welshpool is the home of a small aquaculture operation that maintains the salmon pens inside Friars Bay.

In 1769, Campobello was granted to a Captain Owen, and his family ran the island as a feudal estate for almost a century. In 1829, Captain Owen's son, Admiral William Fitzwilliam Owen, built a large colonial house on the prominent northwest point of Friars Bay. The home was once a small private yacht club and is now run as a bed and breakfast. It can be seen to the left of the wharf as you approach.

APPROACHES. The harbor is wide open. The only dangers are two large weirs and some fish pens.

ANCHORAGES, MOORINGS. As shown on the chart, the man-made harbor at Welshpool has the shape of an L, with the arm pointing southeast. Boats and barges for tending the salmon pens will usually be rafted two or three deep against the inner face of the wharf, where the depths were originally 9 or 10 feet at low. Some silting has occurred, and now the deepest water is found one or two boats out from the wharf. If no more than three or four boats are nested, there should be room to squeeze through between the outer boat and the land, with 5 or 6 feet of water at low.

⇒ Tie up to one of the fishing boats, and check the depth. Depending on the tide and how close you are rafted to the wharf, you can probably lie over here for a few hours or even overnight.

You can also anchor in the northeast corner of Friars Bay in 25 to 30 feet of water, but protection is limited, and there are several salmon pens to avoid. Check the locations of the weirs and fish pens and notice the two piles indicated on the chart.

GETTING ASHORE. Land at the float at the inner corner of the harbor.

FOR THE BOAT. There is electricity at the wharf and a one-ton crane, but no water or fuel.

FOR THE CREW. A small convenience store is .6 miles away. Follow the road from the wharf, turn right at the first intersection, and then go straight through the next and up the hill. Family Fisheries takeout and fish market is 1.25 miles in the same direction.

THINGS TO DO. From the wharf, take the first left to find the small, shingled library and museum, built in 1898. The post office and a church playground are up the road from the wharf and to the right.

Franklin Roosevelt's cottage is about three-quarters of the way down Friars Bay, a walk of 1.5 miles from the wharf. The building is unmistakable—red with a green roof—and signs point the way. There are films about Roosevelt and Campobello at the Reception Centre, and furnishings used by the Roosevelt family are on display. A plaque at the home commemorates the former president: "In happy memory of Franklin Delano Roosevelt (1882-1945) who, during many years of his eventful life, found, in this tranquil island, rest, refreshment, and freedom from care. To him it was always the 'Beloved Island.'"

Halfway up the hill toward the Roosevelt Cottage, a trail leads off to the left, through the woods 1.8 miles to Herring Cove Provincial Park, on Campobello's east coast. Here there is a lovely, mile-long crescent beach, with dark sand, grassy borders, driftwood, and a great view of Grand Manan.

Friars Head lies a bit beyond Roosevelt's house. Nature trails lead to a lookout station with a spectacular view of Cobscook Bay, Lubec, Eastport, and the islands. Three hours before high tide, you may also get a distant look at "Old Sow," one of the world's largest whirlpools, which forms between Eastport and Deer Island.

LUBEC

3 ★★

44° 51.85'N 066° 59.05'W

U.S. Charts: 13396, 13328, 13394   **Chart Kit: 79**
Canadian Charts: 4340, 4373
Tidal Range: *Mean 17.5 feet; Spring 20 feet*

U.S. Customs: 207-532-2131 x255; office: 207-733-4331

Lubec is a small town on the west side of Lubec Narrows, through which pour the great tides of Cobscook and Passamaquoddy Bay. Lubec was once the sardine-packing capital of the world, with 20 or more canneries and a can factory in operation and streets lined with elegant homes.

Now the canneries are gone and with them, most of the town's economic base. Recent efforts to revive the struggling downtown have included the building of a large municipal marina, and a fundraising campaign is under way to turn the last sardine plant into a museum.

The marina, however, suffered from winter weather and lack of visitors and had to be discontinued. Even during the height of summer, there is a ghostly feeling to this place. Many of the false-fronted stores are dusty and boarded, and quite a few of the homes are for sale. The homes, though, are being cared for. When the sun comes out, so do the paint brushes, and the air rings with the sounds of hammers and saws.

In 1897, Lubec was the site of an infamous fraud concocted by the local Baptist minister and a partner. Together they convinced the townspeople and many out-of-state investors that gold could be profitably extracted from seawater.

The last smokehouse complex in the U.S. perches on the shore of Lubec Narrows opposite Campobello Island.

They formed the Electrolytic Marine Salts Company and built a secretive plant in North Lubec. The company claimed profits of over $100 per day, their stock soared, and construction was begun for a second plant. But less than a year after the company's incorporation, a New York bank became suspicious of the extracted gold's consistency and investigated the Lubec operation. The minister fled to France and then to the South Seas. He was never prosecuted.

The renovated storefronts on Water Street.

Lubec is a Customs port of entry for the United States. The International Bridge, spanning Lubec Narrows, leads to Canada's Campobello Island. The fixed bridge has a vertical clearance of 47 feet at high tide.

APPROACHES. The easiest and safest approach to Lubec is from Friar Roads, after having first rounded Campobello Island and negotiated Head Harbour Passage. The water is deep and unobstructed to the Lubec town docks.

APPROACHES via LUBEC NARROWS. From the south, the most direct approach to Lubec and the Passamaquoddy Bay area is through Lubec Narrows, but it is certainly not the easiest or safest.

Lubec Narrows is spanned by a fixed bridge to Campobello Island with a minimum vertical clearance of 47 feet. Add to that Lubec's mean tidal range of 17.5 feet, and most masts should be able to clear the bridge. But at low the dredged channel south of the bridge is narrow, crooked, and "dicey," as some local sailors put it.

Since the passage is barely navigable at low tide, the only time to attempt this passage is close to high slack. Therefore, if your mast is too tall to pass under the bridge at high slack, use Head Harbour Passage instead. Remember that high water will be higher than normal during spring tides and low water will be lower.

Passing through the narrows at anything other than high or low slack poses another problem—current, and plenty of it. The maximum flood runs 6 knots in toward Lubec, and the ebb surges out at as much as 8 knots. The current is at its strongest near Mulholland Point.

Fighting your way under the bridge against the ebb would be difficult. Even going through with the flood is likely to be uncomfortable and dangerous because of the risk of yawing into the bridge piers. The period of high slack water in the Narrows is only five to 15 minutes.

Leave flashing green whistle "1" off West Quoddy Head well to port, and then run to red-and-white fairway bell "WQ." If you are waiting for the tide, anchor in Quoddy Narrows, in 12 to 25 feet, north of West Quoddy Head.

When the tide is favorable, follow buoyed Lubec Channel, running between three sets of red buoys and green buoys. After leaving green can "5" to port, head for the center span of the bridge, sheathed in riprap. Note that the line of small crosses on the chart marks the international boundary, not the channel.

After passing under the bridge, look for the breakwater to port at the northern end of the Narrows. The breakwater is normally easy to see, and even when it is covered by extreme high water, it is marked by a white pyramid midway along its length.

To head south through Lubec Narrows, wait for the tide at the northern end of the Narrows or dock at the Lubec town floats. You can also wait at anchor in nearby Johnson Bay. When the tide is favorable, start the passage near green can "7" and head for the center span of the bridge.

ANCHORAGES, MOORINGS. The Lubec town floats are at the northern end of town, just past the breakwater and to the west. This is a convenient temporary dockage for exploring Lubec or checking in with Customs, but it can be very uncomfortable here in a north or northwest wind.

You can also anchor to the west in Johnson Bay while waiting for the tide, in 15 to 25 feet at low, but this is not convenient to the town of Lubec.

GETTING ASHORE. Land at the Lubec town dock at the northern end of town.

To check in with U.S. Customs, call *207-733-4331* or walk a few blocks to the brick customs building at the International Bridge. They are open 24 hours a day.

FOR THE BOAT. *Lubec town floats and wharf.* These floats are right in the center of town, but they are exposed. Temporary dockage is allowed for a couple of hours. A concrete town wharf lies farther to the west. It has a crane for use by commercial fishermen, but it could be useful for rigging repairs.

FOR THE CREW. On Water Street, the Mulholland Brothers Market has been restored as the first phase of preserving the MuCurdy Fish Plant, the last United States smokehouse complex.

It's about a half mile from the town dock to a group of markets. Turn right up the hill on Main Street, and continue past the bandstand and down the other side to a Quik-Stop convenience store (ice and pay phone) and a Lyon's Market (*733-0925*), which is open every day. A bank with an ATM is just up the road toward Campobello. Bayside Chocolates, on Water Street, can sweeten your galley with chocolates or live lobster or both.

There are several restaurants in town, including the Chowder House, overlooking the town floats, and Home Port Inn and Restaurant (*733-2077*), at the top of the hill. The Atlantic House coffee and deli on Water Street can start your day, and Annabell's Pub, a few doors away, can finish it.

The post office is at the bridge at the far end of Water Street. The Regional Medical Center (*733-5541, 733-4321*) is on the road leading to Quoddy Head.

THINGS TO DO. Lubec's heyday and hard times are jumbled together in the little Sardine Village Museum (*733-2822*) outside of town.

Although there is no pedestrian path on the International Bridge, it is easy enough to walk across to Canada. American and Canadian Customs officials guard the road at either end.

For a fee, certain locals will drive you to the lighthouse at West Quoddy Head and Quoddy Head State Park. Ask at the restaurants for their names. The candy-striped lighthouse was built in 1791 on the easternmost point of land in the United States. There are trails all along the cliffs, picnicking sites, and wonderful views of Grand Manan.

Herring racked for smoking at Eastport, 1920. (Frank Claes)

JOHNSON BAY

3 ★★

44° 51.35′N 067° 00.25′W

U.S. CHARTS: 13396, 13328, 13394   **CHART KIT:** 79
CANADIAN CHARTS: 4340, 4373

Tidal Range: *Mean* 17.5 feet; *Spring* 20 feet

West of Lubec, Johnson Bay is more useful than attractive. It is easy to enter and provides good anchorage and protection for boats waiting out the tide at Lubec Narrows. It is also a more comfortable place to spend the night than the town docks at Lubec if the wind is from the north.

For aesthetics, Johnson Bay offers a gravel pit, a trailer park, and considerable development. Once you are anchored, however, the views north and east are most attractive. From this distance, Lubec is transformed into a quaint little town, its white houses clustered on the hill that is topped with two steeples. Beyond handsome Popes Folly Island, you can see Friars Head and Campobello.

APPROACHES. The approach is wide and straightforward on either side of Dudley and Treat Island. A large group of fish pens is centered in the bay, marked by privately maintained buoys.

ANCHORAGES, MOORINGS. Continue in until you reach depths of 13 to 14 feet at low, between the little peninsulas on either side of the bay. Favor the south or the north side, depending on the wind. Holding ground is good, in mud.

RODGERS ISLAND

3 ★★★

44° 52.05′N 067° 00.40′W

U.S. Charts: 13396, 13328, 13394   Chart Kit: 79
Tidal Range: *Mean* 17.5 feet; *Spring* 20 feet

Rodgers is a small, wooded island at the northwest corner of Johnson Bay, with rocky headlands and several attractive little beaches. A pleasant anchorage lies between it and the mainland.

APPROACHES. Approach from Johnson Bay. Aim halfway between the eastern end of Rodgers Island to starboard and the mainland peninsula to port.

ANCHORAGES, MOORINGS. The passage between Rodgers Island and the mainland is narrow and unmarked. Ledges extend south from the eastern tip of Rodgers and northeast from the mainland shore.

It is possible to get past the mainland ledge, which is partly visible at halftide, but it is safer to anchor before you reach it, halfway between the eastern tip of Rodgers Island and the eastern tip of the mainland peninsula. Depths are 9 to 15 feet at low, and the bottom is good mud.

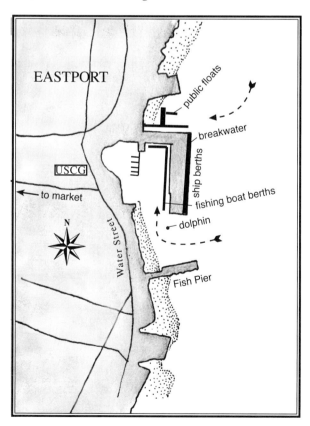

EASTPORT

■ ★★★★

44° 54.40′N 066° 58.95′W

U.S. Charts: 13396 (INSET), 13328, 13394
Chart Kit: 79 (INSET)
Tidal Range: *Mean* 18.4 feet; *Spring* 20.9 feet

Coast Guard: 207-853-2845
U.S. Customs: 207-532-2131 x255; office: 207-853-4313

Eastport is the easternmost deepwater port in the United States and one of the most isolated. It is dubbed "Island City" because it is built on Moose Island and connected to the mainland by causeways. It forms the northern entrance to Cobscook Bay and looks out across the border to Deer Island and Campobello.

Eastport is a particularly interesting fishing town. Since the mid-1970s, it has attracted millions of dollars in state and federal funds to rebuild its waterfront area. This has resulted in a large cargo pier from which freighters load lumber and paper for shipment overseas. Eastport has also been the center for American efforts at aquaculture. Salmon smolt are raised here in pens until they reach market size.

However, it's still a toss-up which will win out—the periodic renovation efforts or the process of decay. Eastport will always be remote, and the town remains in a depressed state, with little industry and high unemployment. Eastporters look to Calais, 28 miles away, for entertainment.

Water Street is the main street of Eastport, where everything happens. The old stone post office is here, and the Customs House, and the huge cargo wharf which forms a man-made harbor as good as those in Canada. The wharf proved so successful that a new, larger terminal was built off Estes Head.

Twice every day, 70 billion cubic feet of water enter and leave Passamaquoddy Bay. Underlying the causeway that links Eastport to the mainland is part of a dam built in the 1930s as the first step in harnessing this enormous source of free power. The project languished during the Depression and was finally killed, but the concept of tidal power on a huge scale is still intriguing, and schemes for reviving the project surface periodically.

From a beach just north of town, a ferry-scow leaves for Deer Island, in Canada. Another ferry runs from Deer Island to Campobello.

SALMON AQUACULTURE

Salmon have always been prized as a delicacy. Returning from the ocean and fighting their way up ancestral rivers to spawn, they are hunted by eagles, bears, and fishermen.

Salmon raised in pens taste wonderful, too.

For decades now, salmon have been raised all over the world, from Chile to Japan, from Ireland to Iceland. Norway, where the industry is heavily subsidized, is the world's leading exporter. Canadian researchers successfully raised salmon off Deer Island in 1974, and salmon aquaculture rapidly became a major enterprise. In some places on the New Brunswick coast, you can almost walk across the water from pen to pen. By 1988, the industry in New Brunswick was producing more dollars than the province's traditional fisheries.

To encourage the development of aquaculture in this country, Maine passed a law in 1973 that provides exclusive leases of bottom acreage. Oysters, mussels, coho salmon, and rainbow trout were among the early aquaculture crops. Yet there was strong resistance among fishermen, and there still is. As one of them puts it, "Most fishermen I know don't want to be farmers."

Lobstermen complain that good lobstering bottom is being taken away. Others worry about the spread of disease from the farmed fish to salmon in the wild and about genetic intermingling when farm-raised salmon escape and breed with wild fish. The introduction of antibiotics in fish feed is hotly debated for its effect on other species. There is fear of the buildup of feces and feed under the cages, and of the degradation of water quality. There are complaints by summer people about fish pens spoiling the view and reducing land values, as well as concerns about interference with navigation.

Maine was late joining the game. The overtures of a Norwegian company were rejected by Vinalhaven in 1987, but the following year Swan's Island fishermen entered a joint venture with an American company to raise salmon in Toothacher Cove. Now fish pens have blossomed in many Maine locations, including Cobscook Bay, Eastport, Cutler, and Cross Island.

Growing salmon is an art. Eggs are removed from females and artificially inseminated. The eggs are hatched and raised in hatcheries until they become smolts and are ready to enter salt water. The smolt, or juvenile salmon, are about five inches long and weigh only a few ounces when they are introduced to their saltwater pens, where they swim around endlessly, occasionally breaking the surface in a flash of spray.

Mussels may be encouraged to grow in the cages to filter the water. If need be, antibiotics may be added to the feed, or the fish may be inoculated directly. The feed may be various mixtures of fish meal, krill to provide pinkness, plus vitamins and other additives.

The salmon are fed two or three times a day. Tons of fish meal are needed daily for large operations. The rule of thumb says that two pounds of feed produce one pound of fish and one pound of waste.

Divers keep careful watch of the bottom under salmon pens, monitoring any buildup of wastes and checking for "morts," fish that have died from natural causes or been killed by predators such as seals.

The water temperature in winter is critical. When it falls just below 32 degrees Fahrenheit—"super-chill" conditions—ice crystals form in the tissues of the salmon and they die in great numbers. Swan's Island lost 30 percent of its crop to super chill the first winter.

For the 18 to 24 months it takes to grow the fish to market size (four to 12 pounds), it's all expense and high risk for the fish farmers. Finally, when it is time to bring the fish to market, the salmon are bled in the water, chilled, eviscerated, and packed in ice for shipment.

Small, individually owned fish farms typically include one or two round pens, 30 feet in diameter, each containing a few thousand salmon, but the risks and economics seem to favor larger corporate operations with 16 to 64 pens, great rectangular rafts moored to the bottom, with 100,000 or more fish. An averge pen holds $500,000 of salmon.

In more recent years, many of the smaller Maine operations have been bought and consolidated by Canadian companies. Still, Maine remains the United States's largest producer of farmed salmon, grossing over $100 million annually.

The most easterly U.S. Coast Guard station is in Eastport. It is a division of the USCG facility in Jonesport, with search-and-rescue capability (*emergencies: Ch. 16, 207-853-2845; non-emergencies: 207-853-4544*). The Border Patrol also operates from here.

APPROACHES. Eastport is easily approached through the deepwater entrances of Head Harbour Passage and Western Passage, which converge into Friar Roads. Aim for the south end of the L-shaped breakwater, which dominates the town. The pink granite seawall lining the shore behind the breakwater is visible from a long distance.

To enter the harbor, go between the southern tip of the breakwater and the Lower Pier, also known as the Fish Pier, which extends eastward from the shore to form the southern end of the anchorage.

This entrance is wide enough, but there is a large dolphin with a bollard standing in the middle. Sometimes boats at the end of the dock warp barely-visible docking lines to the dophin—or the lines may be in the water—so pass to the left of it and turn northward into the anchorage. There is often a tugboat tied up at the southern tip of the breakwater, and someone may be coming out, so enter cautiously.

Another series of floats, reserved for pleasure craft, lies outside the breakwater to the north of it. Obviously, it is much more exposed to the north and east, but depending on the state of the inner harbor and the state of the wind, it may be easier to lie here. Depth is ample except on the inner face of the inner float.

When leaving Eastport for St. Andrews through Western Passage (see page 392), time your departure to avoid maximum turbulence in Old Sow, one of the world's largest whirlpools, between Eastport and the southern tip of Deer Island. Old Sow is most active during flood tide, about three hours before high.

ANCHORAGES, MOORINGS. Inside the breakwater. Although space inside the breakwater is reserved primarily for commercial fishing boats, transient cruising boats may still be able to find space if foul weather is approaching. ⇒ You probably will need to tie up alongside a large fishing boat or dragger on the inside face of the breakwater, remembering that the fishermen may be getting under way at dawn. The floats along the landward side of the harbor are rented seasonally for small craft. A second option is to tie up to the town float at the north end, with permission from the harbormaster. Next to the town float is the Coast Guard float. There is 13 feet of water in the southern end of the harbor and 9 feet in the northern end, but no room to anchor.

Outside the breakwater. ⇒ Visiting boats are requested to come alongside the floats lining the northern, outer face of the breakwater and the finger heading north, in depths of 12 feet at low. The town charges a dockage fee. Protection here is excellent from prevailing winds, but you would have to move inside the breakwater in the event of severe weather from the north or east.

Additional public dock space may also be available on the south side of the Fish Pier.

⚓ Three town moorings are also available to rent. Of the handful of moorings to the north of the breakwater, the most southerly and the most easterly are the town rentals. The third town mooring is south of the Fish Pier.

GETTING ASHORE. Ramps lead from all floats in the harbor to the top of the breakwater.

For yachts returning from Canada, Eastport is the most convenient port of entry. Check in by calling U.S. Customs (*207-532-2131 x255*). The Eastport office is beneath the granite post office, right next to the breakwater. Try to arrive during the business hours of 8AM - 4PM.

FOR THE BOAT. Currently, fuel is not available for boats except by truck.

Eastport town docks (City Hall, 207-853-2300; Harbormaster Eric Voisine: 207-853-4614). Water and electricity are available at the pleasure craft floats on the northern, outer face of the breakwater. Ice can be purchased at the Moose Island General Store across the street from the wharf.

Moose Island Marine, Inc. (Ch. 09, 10, 11, 16; 207-853-6058; www.mooseislandmarine.com). Across the street from the head of the breakwater, Dean Pike's Moose Island Marine handles all sorts of dockside service, particularly mechanical repairs. The chandlery offers the largest selection of marine supplies and fasteners east of Mount Desert Island, including charts, courtesy flags, outboards, and Volvo parts. They also have a larger repair yard in Deep Cove.

S.L. Wadsworth & Son (207-853-4343). This classic hardware store and chandlery on Water Street claims to be the nation's oldest ship chandlery, with charts, maps, fishing gear, and hardware.

Eastport's breakwater and inner harbor.

FOR THE CREW. There is a pay phone on the town wharf, next to Rosie's hot dog stand, but you won't need it to check in with customs. Walk around the corner to the left and across the street to their office.

A number of restaurants, takeouts, and diners line Water Street, along with banks, a hardware store, a drugstore, galleries, and bookstores. A laundromat is two blocks up the hill from Water Street.

🛒 The big IGA supermarket (*853-6138*) is four blocks up from the post office on Washington Street. It is also a liquor store. Deliveries or rides can be arranged if you stock up.

If your crew is jumping ship, a coastal bus route runs southward down Route 1 and then connects to buses in Bangor. The nearest stop is at the intersection of Route 190 and U.S. Route 1, several miles out of town. For a taxi call *853-6162*. If you or your crew happen to be driving the stretch of Route 190 between Eastport and Route 1, be especially careful of the speed limit as you pass through Sipayik, the Passamaquoddy Reservation. Tickets here are notoriously steep.

THINGS TO DO. Stroll one way along Water Street and back along the attractive waterfront path. 🍴 If it is dawn or later, you can swill coffee with the locals at the Wa-co (pronounced "whacko") Diner, where the regulars have running tabs, the eggs fly, and the signs read, "Don't ask for credit if your bill is over $50." and "Bills due yesterday." Renovations replaced some of the old-world character but added an outdoor deck.

The office of the *Quoddy Tides* newspaper, next to the wharf, also houses a little marine library, shop, and aquarium. And if, having come this far, you've begun to feel like an old salt, you'll find DownEast Tattoos behind the IGA or Jaded Ink on Dana Street.

Two windjammers sail daily from Eastport. One, the *Halie Matthew*, becomes a floating schooner bar on Thursdays, Fridays, and Saturdays. She's docked alongside the Fish Pier, which forms the southern end of the harbor.

Celebrating the Fourth of July has been a big deal in Eastport for more than a century, and it is lots of fun. Festivities last for several days, involving everything from beauty contests and parades to water sports and a codfish relay. Events may include the Caledonian Bag Pipe Band, in kilts, a torchlight parade, and a blueberry pie-eating contest. Parachutists come floating down out of the sky to land at the wharf, and sometimes a U.S. Navy ship makes an appearance. The finale is a grand display of fireworks over the bay, watched by half of the population of Washington County and New Brunswick. Even on Grand Manan people head to the west coast to watch. Firecrackers continue to go off unofficially until 2 AM, and sleep is scarce.

BROAD COVE
44° 53.95′N 067° 00.40′W

U.S. CHARTS: 13396, 13328, 13394   **CHART KIT: 79**

Broad Cove is a large commercial bay on the south shore of Moose Island, next to Eastport. Although anchorage here is good, the cove is wide open to the south, obstructed by a very large salmon aquaculture operation, and home to lumber mills, two large fish-reduction plants, and a cargo terminal off Estes Head. Fish reduction produces the pearlescence derived from fish scales which is used to add luster to eye shadow and nail polish. It also produces fish meal and fertilizer and a certain odor. Find another place to spend the night.

Cobscook is an Abenaki word meaning "boiling tides," and the Abenakis had it right.

DEEP COVE

2 ★★

44° 54.38′N 067° 01.20′W

U.S. Charts: 13396, 13328, 13394   **Chart Kit:** 79
Tidal Range: *Mean* 18.7 feet; *Spring* 21.3 feet

Deep Cove is on the west side of Moose Island, next to the Eastport airport. Fish pens are moored in the cove, and a long dock extends from near the big blue building of the Washington County Marine Trades Center, which offers nationally-renowned courses in commercial fisheries, marine mechanics, painting, and boatbuilding.

APPROACHES. Approach through the entrance to Cobscook Bay, where the current runs up to 4 or 5 knots. Favor the Shackford Head side of the channel to avoid Cooper Island Ledge. The entrance to Deep Cove is wide and unobstructed, except for a large flotilla of fish pens.

ANCHORAGES, MOORINGS. Tie up near the end of the long dock, which has 12 feet at low along the outer face, or pick up one of the several rental moorings that Moose Island Marine maintains here. The cove is very exposed from northwest to southwest.

GETTING ASHORE. A float and ladder are on the south side of the dock.

FOR THE BOAT. *Moose Island Marine (yard 207-853-9692; www.mooseislandmarine.com).* This repair yard of Moose Island Marine can haul, store and repair most boats, and they have a shower. They operate the 60-ton boatlift for the Marine Trades Center.

Washington County Marine Trades Center **(207-853-2518).** The marine center has a 60-ton boatlift for emergencies. You can also make below-waterline repairs by careening at the specially reinforced dock, letting the falling tide settle your keel on the granite ramp along the northern side of the dock.

FOR THE CREW. A walk of 1.5 miles takes you to the stores and waterfront of Eastport.

COBSCOOK BAY

Cobscook Bay is the most northerly bay on the east coast of the United States, and it still remains isolated and undeveloped. Tremendous currents carry rich nutrients and abundant fish, saltwater farms dot the shores, and bald eagles glide overhead. And there is hardly a boat to be seen.

Areas attractive to most cruising sailors are limited. There are pleasant anchorages in South Bay and the Pennamaquan River, but the bays beyond Falls Island—Dennys and Whiting—are inaccessible except through narrow passages with extremely strong currents. *Cobscook* is an Abenaki word meaning "boiling tides," and the Abenakis had it right.

APPROACHES. After passing Broad Cove and Deep Cove to starboard, leave Cooper Island Ledge and cans "5" and "7" to port and nun "8" marking Birch Point Ledge to starboard. This nun lies well off Birch Point but is known for towing under. If you don't see it, hug the end of Grove Point to avoid this dangerous ledge. Then head for little Red Island—which really is red—and leave it to starboard. Use caution to avoid the unmarked, submerged ledge (4.6 feet at low) south of Red Island.

SIPP BAY

4 ★★

44° 56.30′N 067° 07.00′W

U.S. Charts: 13328, 13394   **Chart Kit:** 79
Tidal Range: *Mean* 19.1 feet; *Spring* 21.8 feet

At the northern end of Cobscook Bay, East Bay has two branches, with Sipp Bay stretching to the northwest. At the head of Sipp Bay is a trailer park.

APPROACHES. Leave Red Island to starboard and bear northward into East Bay. Then turn west into Sipp Bay and favor the east side until you pass the ledge off Rogers Point. Stop short of the little island on the north shore.

ANCHORAGES, MOORINGS. Anchor in the middle of Sipp Bay southeast of the little island in 8 to 12 feet at low, mud bottom. Protection is good all around except for some exposure to the south.

THINGS TO DO. Two small, state-owned islands, Cedar and Virgin, lie farther up in the bay and are accessible by dinghy.

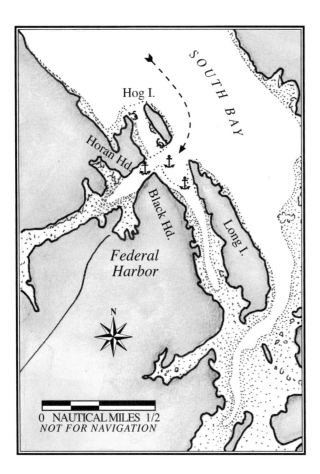

PENNAMAQUAN RIVER

3 ★★★

44° 55.70'N 067° 08.70'W

U.S. Charts: 13328, 13394 **Chart Kit:** 79
Tidal Range: *Mean* 19.1 feet; *Spring* 21.8 feet

Pennamaquan River flows into the northwest corner of Cobscook Bay, between Leighton Neck and Hersey Neck. This pleasant river is rich in wildlife, including seals, loons, and bald eagles.

APPROACHES. After leaving Red Island to starboard, head for green can "11" at Gangway Ledge, south of Hersey Neck. The river is comfortably wide. Follow the buoyed channel, leaving the green cans to port. Can "15" brings you way over to the northern shore. You will pass a wharf and oil tanks to starboard, and you might see ospreys nesting there.

ANCHORAGES, MOORINGS. Anchor southwest of Hersey Point before it becomes too shoal. Note the rocks off the point. The mud-and-weed bottom holds adequately. Protection is reasonably good, but there is considerable fetch from the northwest and some southeastern exposure.

FEDERAL HARBOR

■ ★★★★

44° 51.85'N 067° 03.40'W

U.S. Charts: 13328, 13394 **Chart Kit:** 79
Tidal Range: *Mean* 19.2 feet; *Spring* 21.9 feet

Snug little Federal Harbor lies amid four rocky headlands in South Bay. It is the best and most attractive harbor in Cobscook Bay. Bald eagles nest nearby, and traces of civilization can only be detected with binoculars.

APPROACHES. Leaving Red Island to starboard, make your turn down into South Bay. Stay well clear of Razor Island, where the underlying ledge extends a long way east and west. Favor the Denbow Neck side of the bay. Horan Head is conspicuously high, while Hog Island is an even higher bump. Enter the anchorage between Hog Island and Long Island.

ANCHORAGES, MOORINGS. Anchor dead center among Hog Island, Long Island, Black Head, and Horan Head, in 8 to 12 feet at low. There is ample swinging room for one boat. The rocky islet between Hog Island and Horan Head provides considerable protection to the north, and you are landlocked in every other direction. Holding ground is good mud.

For an even more secure spot, continue westward between Black Head and Horan Head and anchor in 8 feet at low. It's narrow enough that you may want to set a stern anchor. There may be a mooring in this area. The inner part of Federal Harbor dries out at low.

A third possibility is the long, narrow slot between Long Island and the mainland, where you can anchor in 10 feet at low.

THINGS TO DO. Eagle-watching from your boat is good sport. Long Island, donated to The Nature Conservancy by Robert Rimoldi, is the largest undeveloped island in Cobscook Bay and a sanctuary for bald eagles. A series of looped trails lead around Horan Head and South Bay, and tiny Federal Island, southwest of Horan Head, is owned by the state.

FALLS ISLAND PASSAGE

entrance between Fox and Falls Island:
44° 53.18'N 067° 06.98'W

U.S. Charts: *13328, 13394* **Chart Kit:** *79*

On the west shore of Cobscook Bay, a narrow slot leads westward between Leighton and Denbow Point to Dennys and Whiting Bay. In the middle of this entrance, like a cork in a bottle, Falls Island obstructs the great flow of water that rushes by on either side.

Between Falls Island and Leighton Neck are the well-known reversing Cobscook Falls. The current ebbing from the inner bays, after passing the ledge south of Mahar Point, turns sharply north, then plunges southeastward around the ledge at the western corner of Falls Island, forming an S-pattern. At full strength the current is probably 8 knots or more of pure whitewater. The ledges are covered at high tide.

Although the passage south of Falls Island is far easier, the tidal stream is still extremely strong, and there are dangerous unmarked ledges. The current around Falls Island is strongest during periods of spring tides, when the area becomes very turbulent. A wind against the current can produce large standing waves in the passages.

Falls Island Passage is not for the faint of heart. It is probably impossible for a sailboat with a low-powered auxiliary engine to fight its way in against the current. The best time to make the passage south of Falls Island is with the last of the flood or at high slack, which lasts only a short time.

Entering between Fox Island and Falls Island, stay east of the ledge northeast of Ruth Point (the southernmost point of Falls Island, opposite Race Point). After making your turn westward between Race Point and Ruth Point, favor the Crow Neck side of the passage to avoid the dangerous ledge southwest of Falls Island. This ledge shows about 5 feet at midtide and is covered at high.

Cruising yachts are at substantial risk here. Intrepid yachtsmen take their boats through south of Falls Island, and some have run Cobscook Falls to the north, but unless you have a great need for white-knuckle adventure, leave Whiting and Dennys Bay for the eagles—or make the passage in someone else's boat.

Mending nets for fish pens along the shores of Western Passage, with Deer Island on the horizon. The stakes of an old weir are to the right. The sailboat in the distance is heading north (to the left) toward Passamaquoddy Bay.

WESTERN PASSAGE and OLD SOW
Old Sow: 44° 55.30'N 066° 59.20'W
Canadian Charts: **4373**
U.S. Charts: **13396, 13328, 13394, 13398**
Chart Kit: **79, 80**

Western Passage runs west of Deer Island between Friar Roads to the south and the St. Croix River and Passamaquoddy Bay to the north. This is the usual route between Eastport and St. Andrews. Currents are strong in the passage, averaging 3 knots for both flood and ebb. A rate of 7 knots has been measured off Deer Island Point, diminishing northward. Strong eddies and countercurrents form near the shores. Also, be alert for the ferry-scows which make frequent crossings between Eastport and the eastern tip of Deer Island.

Old Sow. The flood that streams down through Head Harbour Passage and passes either side of Indian Island turns sharply north into Western Passage, producing "Old Sow," often touted as the world's largest whirlpool, between Deer Island Point and Dog Island. Old Sow gets her name from her offspring, the little piglet whirlpools that spin away from their mother.

Time your passage to avoid Old Sow at its most turbulent, about three hours before high, when it is a dangerous place for small craft. It may even be difficult for substantial vessels, causing them to lose control, yaw back and forth, lose headway, or even spin around. While this is not the Edgar Allan Poe kind of maelstrom that sucks you down into the depths, it can be awesome enough to produce white knuckles and high anxiety. At other times the passage may be quite benign.

The lighthouse on Deer Island Point is a small white tower. The lighthouse on Cherry Island is also a small, round white tower, but with a red top. At a distance, both look like they belong in a child's train set.

The large Passamaquoddy Indian Reservation is on the west side of the passage opposite Clam Cove, on the promontory of Pleasant Point, surrounded on three sides by the ocean. A monument in the cemetery honors 200 members of the tribe who assisted the early colonists during the American Revolution. Pleasant Point administers Passamaquoddy lands and holdings throughout eastern Maine, acquired as a result of the historic Indian Land Claims Settlement of 1981.

DEER ISLAND

Large, beautiful Deer Island sits smack in the middle of the Passamaquoddy region, almost equidistant from Eastport, Campobello, and Letete. It is bounded by Passamaquoddy Bay to the north and on its other sides by the turbulent currents of the Western, Head Harbour, and Letete passages.

A number of small fishing villages are scattered along Deer Island's wooded shores, while its interior is almost a wilderness. Ferries run to Deer Island from Eastport, Campobello, and from Letete to the north, but much of the traffic they bring to the island just uses the island as a stepping stone to cross to mainland New Brunswick or back without having to drive all the way around Passamaquoddy Bay.

The east coast of Deer Island is quite delightful, with sparsely settled rocky shores, weirs, jutting promontories, and handsome farmhouses. Starting at the southern end, keep your eyes open for the ferry-scows from Eastport and Campobello, which land on the eastern side of Deer Island Point. The sailing is great between Indian Island and Deer Island, especially on the flood. The current here is strong, with some small tide rips, boils, and eddies that weaken as you go farther north. You pass the tiny hole of Chocolate Cove and then the busy fishing community of Leonardville. Casco Bay Island, Popes, and many other lovely islands dot the eastern horizon. Northwest Harbour presents a possible refuge, and north of it is intriguing Lords Cove, whose entrance channel is barely wider than your boat. The northeast side of the island looks on the chart like a wine glass dropped on the kitchen floor—shards everywhere with liquid in-between. This area is wonderful to explore in good visibility but dicey for a stranger in the fog.

Great Chief Glooscap was paddling his canoe one day along the shores of Passamaquoddy, so Indian legend tells, when he saw a deer and a moose pursued by a pack of wolves. With a wave of his hand, Glooscap changed them all to islands. The Wolves still slaver north of Grand Manan, but Deer and Moose lie safe in Passamaquoddy Bay.

NORTHERN HARBOUR

44° 59.03′N 067° 00.60′W

CANADIAN CHARTS: **4331, 4373**

U.S. CHARTS: **13396, 13398**   CHART KIT: **79, 80**

The intriguing gunkhole of Northern Harbour is on the western side of Deer Island, north of Clam Cove Head. It is a pretty spot with a rocky shoreline all around and only two or three houses visible.

The holding ground is poor, however, and the harbor is exposed to the northwest. Northern Harbour is only recommended as a daystop.

A group of fish pens flank the entrance like gatekeepers, and one of the world's largest lobster pounds occupies the harbor's northern branches.

NORTHWEST HARBOUR

44° 59.01′N 066° 57.55′W

CANADIAN CHARTS: **4331, 4373**

U.S. CHARTS: **13396, 13398**   CHART KIT: **80**

Tidal Range: *Mean* 18.1 feet; *Large* 25.2 feet

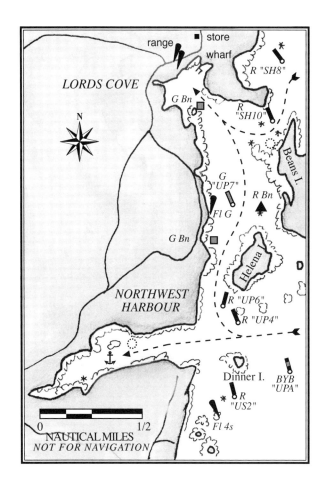

Northwest Harbour is a delightful surprise midway along the *east* coast of Deer Island. High wooded banks make this a tiny fjord, deep and well protected, shared only with a few houses. This is not a recommended harbor for entry into Canada since there is no public access ashore. Instead, head to Lords Cove.

APPROACHES. From the south, run up the Quoddy River—not really a river, but at maximum ebb and flood it might feel like it. Pass Bar Island to port and find the black-yellow-black spar "UPA."

From the east, round East Quoddy Head and cut across Head Harbour Passage. Pass to either side of Black Rock, with a green daybeacon, and turn northward in the Quoddy River.

At spar "UPA" turn to the west and pass between small Dinner Island well to port and red spar "UP4" close to starboard. Stay in the middle of the entrance, and once in the harbor, favor the southern side to avoid the ledge and weir along the north shore. Do not go past the point at the western end of the cove.

ANCHORAGES, MOORINGS. The middle of the cove's western end has about 21 feet at low and good holding ground. Southwest winds tend to funnel down the length of the harbor from its head.

LORDS COVE

4 ★★★   ⚓ 🛒

entrance outside daymark:
45° 00.15′N 066° 56.70′W

CANADIAN CHARTS: **4331**

U.S. CHARTS: **13398**   CHART KIT: **80**

Tidal Range: *Mean* 18.1 feet; *Large* 25.2 feet

Lords Cove is nestled inside Beans and St. Helena Island at the northeastern corner of Deer Island. The approach channel is extremely narrow but well marked. Once inside, there is a tight little working harbor with excellent protection. This is a designated harbor for entering Canada, though an unusual one. Call *888-CANPASS* to clear customs.

APPROACHES. Lords Cove may be approached from the north or south. From the south, the buoyed channel west of St. Helena Island is easier, since the approach east of St. Helena requires some eyeball navigation. Follow the general approach description in the Northwest Harbour entry to reach red spar "UP4." Heading north, leave red spars "UP4" and

"UP6" to starboard. Favor the St. Helena side until you are past the green daybeacon, the green light, and the black-yellow-black spar "UPC," leaving them all to port.

From the north, pass between Deer Island and Beans Island by keeping red "SH10" to starboard. Give it a wide berth as you turn west toward Lords Cove. Stay about midway between the mark and the fish weir and ledge that jut from Beans Island.

A green daybeacon stands on a ledge at the narrow entrance to Lords Cove. Leave it close aboard to port.

Proceed slowly, close by the green daybeacon, and then in a line from there to the far end of the wharf. Fish pens flank either side of the narrow channel and a sunken vessel lies on the bank to starboard.

ANCHORAGES, MOORINGS. The several moorings in the small turning basin leave little room to anchor. Tie up to the floats off the western end of the wharf. There is good water on the inside faces of the floats, but don't try the outer, western side. Do not enter the little cove on the west side of the wharf, which is shoal.

FOR THE BOAT. An ice plant is across the harbor.

FOR THE CREW. A small convenience store, the post office, and a credit union are on the coastal road a short distance to the right.

LETETE PASSAGE

CANADIAN CHARTS: *4111, 4313, 4331*
U.S. CHARTS: *13398*   CHART KIT: *80*

Letete Passage is the main entrance to Passamaquoddy Bay, passing north of Deer Island, between McMaster Island and the mainland to the north. It is the most direct route to St. Andrews after rounding East Quoddy Head.

Letete's currents are strong. It should be avoided two hours either side of the maximum ebb or flood, when the current swirls along at 8 knots with eddies and boils.

Sailing through Letete with a strong flood current is an exhilarating experience. The shore moves by at an alarming rate and quick decisions are required. It is less exciting but more comfortable to make the passage near high slack, preferably 45 minutes before or after high water. Fighting against the ebb is a losing proposition.

Navigating Letete Passage is complicated by many rocks and ledges and by ferries and overhead power cables that cross it. All the fixed dangers, however, are well marked. Canadian chart 4111 is much larger in scale than the U.S. chart and is more likely to show the latest changes in buoyage. The power lines cross at Dry Ledges with a vertical clearance of 128 feet. The blue ferries run between Butler Point on Deer Island and Matthews Cove, and they can be hailed on channel 69. Good visibility is extremely important, so you should avoid this passage in fog.

The passage is quite sheltered, except from the east and southeast. It can become very rough or even impassable, however, when strong southeasterlies blow against the ebb and swells reach the passage. Conversely, northwest winds against the flood will produce choppy conditions, though they are less serious without the accompanying swell.

Letete Passage can be identified at a distance by the three high towers carrying overhead power lines. These are very useful in orienting yourself during the passage. The middle tower stands on Dry Ledge, which you can pass to either side. Wooded Mohawk Island is another useful landmark. Greens Point is marked by a white, octagonal lighthouse with a cluster of buildings and a radio tower. A light on a skeleton tower marks Morgan Ledge.

By far the easiest route from the south is to pass between Greens Point and Mohawk Island, dead center in the passage. Leave the series of red buoys along the eastern shore to starboard and pass to either side of Dry Ledge. Note Little Dry Ledge to the south, which appears merely as a rock on chart 13398. Exit south or north of Thumb Island, which is grassy with some scraggly trees.

Another route takes you between Parker Island and Black Ledge, marked with a green daybeacon, then west of Mohawk Island. This passage is much narrower than the route to the north, with a confusion of buoys and more turbulent current.

As you emerge from Letete, you will see the great red roof of the Algonquin Hotel in St. Andrews and the circling mountains of Passamaquoddy Bay.

Coming from the north, reverse the route described above. Enter Letete Passage between the high northern tip of McMaster Island and scruffy little Thumb Island. Use the center tower on Dry Ledge to guide you, passing to either side, then go over to the eastern shore, leaving the line of red marks to port and exiting between Mohawk Island and Greens Point.

PASSAMAQUODDY BAY

There is magic in the name of Passamaquoddy Bay—perhaps because it is so remote from the areas where most sailors usually cruise; perhaps because it shares with the Bay of Fundy the image of enormous tides, swift currents, and impenetrable fogs.

In the 6.2 hours between low and high tides, more than two billion tons of water swirl into the bay. The deep, cold, upwelling waters bring nutrients to the surface and make this an immensely rich feeding area for birds, fish, and even whales. Millions of phalaropes, Bonaparte's gulls, and birds of other species congregate in Passamaquoddy Bay during the late summer. *Passamaquoddy* itself means "pollock plenty place."

This is a beautiful open bay of no great extent—merely five miles across and a bit more north and south. The bay is ringed by a bold shoreline and well protected by the islands across the entrance. There are hardly any dangers except for a few islands. Hardwood Island, in the middle of the bay, provides a convenient reference point.

Passamaquoddy's northern and eastern parts are prettier than its western portion. The encircling hills are much less developed and almost in a natural state. Because of the protecting islands, ocean swells do not penetrate the bay, making for calm anchorages. The bay's air temperature is warmer than elsewhere on the coast, so the bay is often clear when thick fog banks lie outside.

Though it is small, there are several good harbors in Passamaquoddy Bay. The only facilities are at St. Andrews, at the southern end. Good protection may be found in Chamcook Harbour, Birch Cove, the Bocabec River, the Digdeguash River, and the Magaguadavic River. Of these, the three river anchorages are the most isolated and beautiful. Cruising sailboats are still a rarity here.

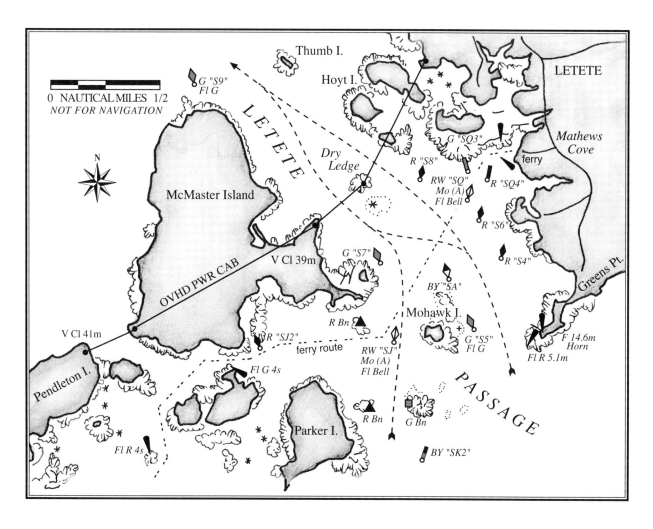

PENDLETON ISLAND

45° 02.15′N 066° 57.05′W

CANADIAN CHARTS: *4331*
U.S. CHARTS: *13398*   CHART KIT: *80*

High, wooded Pendleton Island lies north of Deer Island. A nice beach that is a favorite daystop of local yachtsmen lies in the bight on the north shore of Pendleton. They report that with a stern anchor out, you can put your bow right up on the sand.

MAGAGUADAVIC RIVER

4 ★★★   🛒 🍴

45° 06.55′N 066° 53.65′W

CANADIAN CHARTS: *4331*
U.S. CHARTS: *13398*   CHART KIT: *80*
Tidal Range: *Mean* 19.6 feet; *Large* 27.2 feet

The Magaguadavic River, on the east side of Passamaquoddy Bay, is fed by the waters of Lake Utopia. The river offers very good protection in a peaceful setting. This is one of the nicest harbors on the bay.

Shoal-draft boats or dinghies can follow the buoyed channel upriver, preferably on a rising tide, as far as the little town of St. George.

APPROACHES. Use high Midjik Bluff as your landmark, rounding its steep cliffs into the wide entrance of the river. Favor the red buoys on the starboard side of the channel to avoid the flats that extend from the low point to the north. Fish pens are moored in mid-harbor.

ANCHORAGES, MOORINGS. Anchor just south of a line between red buoys "SM4" and "SM6," in 6 to 12 feet at low. Protection is good except from the north.

Harder to reach but better protected is a spot northwest of red buoy "SM8," just west of a brown rock bluff with a house above it. There is 7 to 12 feet at low, and the bottom is partly rock, partly mud. There are shoal spots nearby, so feel your way in cautiously.

GETTING ASHORE. Dinghy in to the floats by the town ramp or to the Eagle's Nest Lookout, just below the green steel bridge.

FOR THE CREW. The small village of St. George has 🍴 several restaurants, a bank, a liquor store, and a 🛒 country store with fresh baked goods.

THINGS TO DO. The Magaguadavic River is squeezed over a dam and through a tight gorge right in town. A small walkway leads to an overview of the gorge and fishway. Above the dam, another trail leads along the river's southern bank.

DIGDEGUASH HARBOUR

3 ★★★★

45° 09.45′N 066° 57.95′W

CANADIAN CHARTS: *4331*
U.S. CHARTS: *13398*   CHART KIT: *80*
Tidal Range: *Mean* 19.6 feet; *Large* 27.2 feet

A little group of beautiful islands clusters in the mouth of the Digdeguash River at the northern end of Passamaquoddy Bay. Their rocky shores are tinged with orange lichen, and they provide good protection. The only signs of civilization are one or two houses along the forested banks of the river and a highway bridge that crosses well up at the head of the harbor. At night, the cries of loons echo across the water.

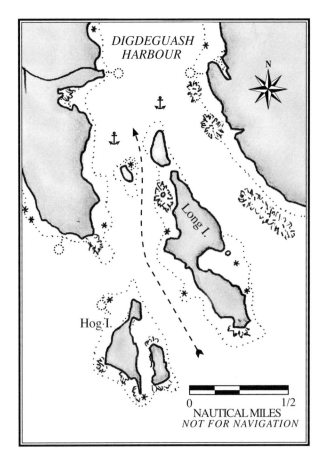

Digdeguash Harbour, looking south.

APPROACHES. The eastern side of the entrance to the Digdeguash River can be identified at a distance by Oven Head on the mainland, with its distinctive rounded brown bluffs and a meadow.

Run up the tongue of deep water between Long Island to starboard and a group of three smaller islands (including Hog) to port. Having passed the western tip of Long Island turn gradually to starboard, favoring the western side of the channel to avoid the ledges shown on the chart, and continue northward between two small islands.

ANCHORAGES, MOORINGS. Anchor to either side of the deep tongue of water north of the two small islands in 9 to 10 feet at low. Under most conditions, the western part of the anchorage offers the best protection. There is some exposure through the islands to the south or southwest at high tide, but at low tide you will be entirely landlocked. The holding ground is good, in mud.

THINGS TO DO. The three parts of Hog Island are connected at low tide and pleasant to explore. There are small meadows filled with ferns and abundant raspberries in July. The islands at the mouth of the Digdeguash are nesting areas for herring gulls.

BOCABEC RIVER

 ★★★

45° 09.61′N 067° 58.92′W

CANADIAN CHARTS: 4331
U.S. CHARTS: 13398   CHART KIT: 80
Tidal Range: *Mean* 19.6 feet; *Large* 27.2 feet

The attractive Bocabec River runs into Passamquoddy Bay from the north, just west of Digdeguash. A small fleet of fishing boats is based here, and the anchorage has good protection except in a southerly gale.

West of Bocabec River, Dicks Island was taken over by cormorants a couple of decades ago. The deciduous trees have all been killed by shag droppings, and the remaining skeletons are full of cormorant nests—sometimes six or more on a branch.

APPROACHES. Starting west of Hog Island, run up the middle of the entrance to Bocabec Cove. Stay in the deep water just off the ends of the two weirs extending from the western shore to clear a weir and a rock near the eastern shore. Continue to favor the western shore as you run past another weir extending from the squarish promontory on the eastern shore and past several offlying rocks.

ANCHORAGES, MOORINGS. The water deepens just beyond the promontory on the eastern shore. Anchor west of the moored boats in 15 to 20 feet at low. The water dries out north of the moorings.

GETTING ASHORE. Land on the eastern shore next to the white house by the public road.

BIRCH COVE

4 ★★

45° 09.00′N 067° 02.36′W

CANADIAN CHARTS: 4331
U.S. CHARTS: 13398   CHART KIT: 80
Tidal Range: *Mean* 19.6 feet; *Large* 27.2 feet

At the northwestern corner of Passamaquoddy Bay, Birch Cove extends west of Bocabec Bay. It offers good protection except from the southeast, but the western shore is crowded with camps.

APPROACHES. The approach from Bocabec Bay is wide open except at the entrance, where a large weir extends from the south shore of the harbor.

ANCHORAGES, MOORINGS. Anchor near the western shore in 9 to 14 feet at low in good mud.

CHAMCOOK HARBOUR

3 ★★★

entrance channel: 45° 07.23'N 067° 02.78'W

CANADIAN CHARTS: **4331**
U.S. CHARTS: **13398** (INSET)   CHART KIT: **80**
Tidal Range: *Mean* 19.6 feet; *Large* 27.2 feet

Large Chamcook Harbour is formed behind Ministers Island, north of St. Andrews. Both sections of Chamcook Harbour are well protected, so you can choose between the views of the ruins of a cannery and wharf on the west shore or a magnificent barn on Ministers Island to the east.

APPROACHES. Ministers Island can be identified from a distance by the round stone tower with a white conical roof at its southern end. Enter through the narrow buoyed channel north of Ministers. The current is moderately strong in the channel but not bothersome inside the harbor.

ANCHORAGES, MOORINGS. Anchor either in the northern portion of the harbor, west of Craig Point, in 18 to 21 feet at low, or in the center of the bay, where the anchor is shown on the charts, in 33 feet or more. A campground is on the north shore.

A third, more daring option is to feel your way into the southern portion of the harbor. The inlet over the bar leading to the southern part of the harbor has 6 to 12 feet at low, but it is hard to locate. Cross on a rising tide. Once across the bar, anchor in 24 to 30 feet at low. Some of the available space is taken up by aquaculture operations. The harbor is partially protected to the south by a bar between Ministers Island and the mainland (cars drive over it at low tide), but this protection disappears at high tide.

FOR THE CREW. At higher tides, it is possible to take the dinghy up Chamcook Creek to the northwest and scramble up the shore to an inconvenient convenience store and gas station on Route 127. Don't let the tide leave you high and dry.

THINGS TO DO. Near low tide, walk across the gravel bar to Ministers Island and Covenhoven, the magnificent estate of Sir William Van Horne, who built the Canadian Pacific railway from Montreal to Vancouver. The island is now owned by the province and run as an Historic Site, with tours available (*506-529-5081*). *Caution:* The bar is exposed about two hours either side of low tide, but don't cut it close. People have lost their lives here in the rapidly rising Passamaquoddy tides.

ST. ANDREWS

3 ★★★★★

off wharf: 45° 04.23'N 067° 03.30'W

CANADIAN CHARTS: **4331, 4332**
U.S. CHARTS: **13398** (INSET)   CHART KIT: **80**
Tidal Range: *Mean* 19.6 feet; *Large* 27.2 feet

St. Andrews-by-the-Sea is a pleasant resort town at the southern end of the peninsula between Passamaquoddy Bay and the entrance to the St. Croix River. It is the largest Canadian settlement in Passamaquoddy Bay and a convenient port of entry.

The wide streets and handsome houses of St. Andrews may remind you of the Maine town of Castine—not surprising since it was founded in 1783 by British loyalists from Castine who moved their houses here by barge. They prospered in the lumber and shipbuilding trades with Britian and in dried and salted fish with the West Indies. St. Andrews never suffered a major fire, and it was overlooked by 19th century industrialists, so most of the original architecture has survived. On top of the hill looms the great old Algonquin Hotel.

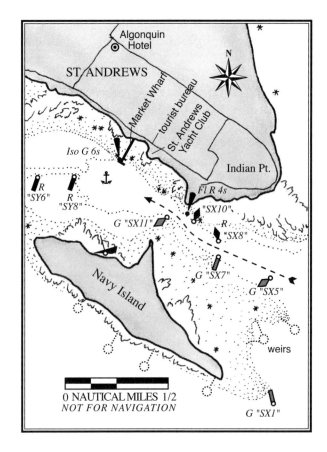

St. Andrews Harbour is partially protected by Navy Island and relatively easy to enter. The focus of the harbor for both working and pleasure boats is the huge Market Wharf. In 1994 a cigarette fell between the creosoted timbers, and in 45 seconds eight cars and three boats were consumed by the inferno. The rebuilt wharf opened four years later.

APPROACHES. From the East or South. Enter Passamaquoddy Bay either through Letete Passage (page 394) or Western Passage (page 392). As you approach St. Andrews, some of the buildings can be seen partially obscured by Navy Island, and the long red roof of the Algonquin Hotel emerges above the trees. Although the port can be entered from either side of Navy Island, the eastern approach is deeper and preferable.

The southeastern tip of Navy Island is a sand bluff with offlying dangers marked by spar buoy "SX1." Pick up green bell "SX3" (*45° 03.50'N 067° 01.70'W*) off the eastern entrance and follow the series of channel buoys through the flats into the harbor. It is almost a straight channel. To starboard, a flashing red light on a skeleton tower (between buoys "SX10" and "SX11") marks the end of what was once a wharf.

From the North. When approaching the eastern entrance to St. Andrews from the north, or leaving St. Andrews for northern Passamaquoddy Bay, be sure to clear red "SX2," which marks the eastern extremity of Tongue Shoal.

From the West. Western Channel is well marked but narrow. Avoid it near low tide.

ANCHORAGES, MOORINGS. Overnight dockage on Market Wharf can be arranged in advance through the wharfinger (see "For the Boat"), or he may be able to direct you to a vacant mooring. Members of the St. Andrews Yacht Club are also helpful.

If no moorings are available or if you prefer to anchor, drop the hook south of the wharf in 8 to 14 feet at low. The harbor is well protected at low tides by the flats surrounding Navy Island, but as the tide rises, they disappear. Current floods west and ebbs east strongly through the harbor, so you may hang askew. You will need lots of scope, and if the wind is contrary to the current, you may spend a rolly night.

GETTING ASHORE. Dinghy to the wharf.

St. Andrews is a convenient port of entry. Call *888-CANPASS* for customs clearance. See *Border Crossings* on page 368 for more details.

On Market Wharf, with Navy Island in the background.

FOR THE BOAT. *Market Wharf (Wharfinger B.B. Chamberlain: Ch. 78; 506-529-5170).* Overnight dockage on Market Wharf can be arranged in advance by contacting the wharfinger. Water, pump-outs, power, and a pay phone are available on the wharf. Ice can be bought at the Save-Easy supermarket, a block away.

St. Andrews Yacht Club. The club has a small clubhouse not far east of Market Wharf, a hundred yards from the water at low tide. There are no facilities for visiting yachtsmen, but you will find some kindred spirits there.

Navy Island Dive Co. (506-529-4555). This dive shop, steps from Market Wharf, is a Canadian chart dealer. Turn left on Water Street and left again on William Street.

FOR THE CREW. Most of the streets of St. Andrews are named after the royal family except Water Street, which runs along the harbor. On Water Street you will find almost everything you might need within a short walking distance—banks, restaurants, a hardware store, a liquor store, a pharmacy, the post office, bike rentals, and a tourist information office which has maps of the town.

Pay phones and public restrooms are across the street from the post office. The Save-Easy supermarket is only a block away to the right. Washboard Laundry is in the same direction. Showers—free to boaters—are available at the Kiwanis Campground (*506-529-3439*), a short walk farther to the east end of town.

There are several places to eat in town. For a good dinner try L'Europe (*506-529-3818*) on King Street or Windsor House (*506-529-3330*) two blocks to the left from the wharf. For a fabulous Sunday brunch, walk up the hill to the Algonquin Hotel (*506-529-8823*). For morning coffee and muffins—or midnight music—visit the Fulcrum Cafe to the right on Water Street.

The Seafarers' Internet Cafe is past the Fulcrum. The Ross Memorial Library (*506-529-5125*) is two blocks up from the wharf.

The St. Andrews Health Centre (medical, *506-529-8881*; dental, *506-529-8814*) is several blocks to the northeast of the wharf. For a taxi call *506-529-3371*.

THINGS TO DO. St. Andrews is famous for its fine woolens and craft shops. The woolens, tweeds, and hand-knit sweaters at Cottage Craft, in the red building opposite the Tourist Bureau, are exceptional.

St. Andrews is unique in Canada because of its high proportion of buildings that are more than 100 years old. Many of these attractive old structures are listed in the *Walking Guide for Historic St. Andrews*, available at the Tourist Bureau. Also noteworthy are the shingle-style buildings of Canadian architect Edward S. Maxwell.

In town, historic plaques show where the Loyalists landed, and the old Loyalist Cemetery still guards their graves.

It's an invigorating walk up the hill to the great five-story Algonquin Hotel (*506-529-8823, 800-563-4299*). The hotel has a heated pool, tennis courts, and an 18-hole golf course (rackets and clubs can be rented). They also run a daily activity program for children aged five to 12. Children might also enjoy the vast playground at the Vincent Massey Elementary School, four blocks up from the wharf.

Walk half a mile west along the waterfront to the Blockhouse, with a pleasant little park and three 18-pounders still commanding the harbor. This is the sole survivor of the 12 blockhouses built to defend New Brunswick during the War of 1812 between the United States and Britain. Don't forget to check the machicolations.

RAZORBILLS

Razor-billed auks, or razorbills, are the rarest nesting birds on the coast of Maine. In contrast to the gaudy colors of their smaller cousins, the puffins, auks are as elegant as Victorian dandies—jet black with modish white piping on back and bill and snowy white shirtfronts.

Razorbills are rare in Maine and seldom seen from the shore. About 100 pairs nest on Machias Seal Island, coexisting comfortably with puffins and terns. A smaller colony of about 40 pairs nests on Matinicus Rock, also shared with terns and puffins, and two even smaller colonies nest on Old Man Island and Freeman Rock. They also breed in Greenland, Iceland, and northeastern Russia.

If you take your boat to Machias Seal Island or Matinicus Rock, it's not hard to spot these handsome birds. They have little fear of man, and they are actually curious enough to be attracted by loud noises or arm-waving.

Like the puffin, the razorbill dives below the surface for fish and crustaceans, flying underwater with its short, stubby wings. Its nest is often on an inaccessible cliff or in a crevice.

A pair of razorbills produces only a single egg each year. The eggs, resting precariously on tiny ledges high above the sea, would be vulnerable to the slightest misstep were it not for an ingenious solution of nature. The eggs are radically tapered at one end, so when disturbed they merely roll around in a small circle.

Until well into the 19th century, the razorbill had a larger relative in the North Atlantic, the great auk. Standing 30 inches high, this ungainly fowl was flightless. Having evolved in conditions similar to those in Antarctica, the great auk resembled the penguin, and it occupied a similar ecological niche. The auks and the penguins, however, are unrelated. Their similarities are an example of convergent evolution.

Naturalist Phil Conkling in *Islands in Time* says the Great Auks were "the fastest and most powerful diving bird in the evolutionary record of life. They used their short, tapered wings to propel themselves down to 40 fathoms in pursuit of schools of smelt, herring, or capelin. The few observations of their fishing techniques made by naturalists before the auks disappeared described rafts of 20 to 50 birds that would dive to surround a large school of herring, actually driving the fish to the surface."

Early European explorers and fishermen found enormous concentrations of great auks on this side of the Atlantic, in colonies of hundreds of thousands. The birds were easy prey for a man on foot, providing badly needed meat and oil, and later a European market developed for these products. For a few seasons, commercial hunters prospered, wiping out whole populations at a time. By 1844, the great auk was extinct.

A walk of 1.8 miles to the west, or a short cab ride, will take you to Huntsman Marine Centre in Brandy Cove (*506-529-8895*), where you can watch seals being fed twice a day or dabble in a touch tank to pick up starfish, sea cucumbers, and other marine creatures in your own hands. Nearby is the St. Andrews Marine Biological Station (*506-529-8854*), which researches herring and salmon biology and aquaculture.

The extensive beaches uncovered at low tides are rich beachcombing grounds. Pieces of china, Dover flint, and West Indian coral once used as ship ballast occasionally turn up.

For those who ballast their own boats with ice skates and bowling balls, there are facilities for both at the W. C. O'Neill Arena (*506-529-4578*) several blocks from the waterfront. Skating is open to the public most evenings, year-round. Public tennis courts and the Chamber of Commerce (*506-529-3555*) are next door. Sea kayaking tours are available through Bruce Smith at Seascape (*506-529-4866*), at the head of the wharf.

ST. CROIX RIVER

CALAIS and ST. STEPHEN
SE of "S34": 45° 11.40′N 067° 16.00′W
CANADIAN CHARTS: **4331**   U.S. CHARTS: **13398** (INSET)
Tidal Range: *Mean* 20 feet; *Spring* 22.8 feet

The St. Croix River, at the northwest corner of Passamaquoddy Bay, marks the boundary between the United States and Canada. The lower section of the river, from St. Andrews northward, is deep and almost a mile wide, more like a bay than a river, with plenty of room for good sailing. Currents in this part of the river run about 2 knots.

The shores of the river are relatively low and mostly wooded, with occasional houses and large meadows. Hills and mountains inland frame the peaceful scene. The big, dark pyramid of Devils Head is visible in the distance upriver, where the river narrows, turns westward, and forms part of the cross for which it was named.

St. Croix Island was the site of the ill-fated first attempt at European settlement in the New World.

Don't let this broad and peaceful river lull you into forgetting about the shoals south of St. Croix Island. The first European settlement in the New World was on this island, where Sieur de Monts, Samuel de Champlain, and 79 other Frenchmen landed in 1604. Right in the middle of the river, it was relatively safe from Indians, and a cannon at the south end could command the approaches.

But it was an unusually severe winter. By the time a supply ship arrived in June, two months late, 35 of the Frenchmen had died of scurvy. The settlement was abandoned, and the French retreated to Port Royal, Nova Scotia.

The five-acre island is now an international park. Among its attractions are an inviting beach on the southeastern end, a freshwater spring, and clumps of trees. There is a grassy plateau with room for an acre or two of farmland on top of a sandy bluff. It was more extensive in Champlain's day. His men cut trees to build their fort and houses, and subsequent erosion washed away much of the original island, creating today's shoals.

Opposite Devils Head, you may see a large freighter tied up at the Sand Point Marine Terminal loading lumber and potatoes. Devils Head is high and wooded and sometimes creates downdrafts that can be dangerous to small craft.

From Devils Head, the river runs west about six miles to the International Bridge linking Calais (United States) and St. Stephen (Canada). Both towns are ports of entry. This section of the river should be navigated on a rising tide. Above Spruce Point the river is narrow and winding, with a channel that was dredged but is not maintained. The maximum current runs 3 to 4 knots and sometimes tows under the buoys on the north side of the channel at The Narrows, near Whitlocks Mill.

It is possible to anchor with excellent protection on the west side of the channel, just above Whitlocks Mill Light, in 14 feet at low.

The river west of Devils Head is lined with small camps interspersed with occasional stretches of pristine riverbank. There remains a great deal of wildlife to enjoy, including herons, seals, loons, an occasional eagle, and a constant procession of ospreys carrying fish back to their nests.

The scene suddenly becomes suburban and then urban as you arrive at the head of navigation in Calais and St. Stephen. There are wharves and floats on both sides of the river. The wharves dry out at low tide, but it is possible to anchor in midriver during average tides if you draw no more than 4 feet. The mean tidal range in Calais is 20 feet—more than any other town in the continental U.S.

For most cruising boats there is not enough water to anchor at Calais/St. Stephen, but it's possible to make a brief visit by powering upriver on a rising tide and coming down with the ebb. Allow an hour or so each way from Devils Head. Clearance from both U.S. and Canadian customs can be arranged here.

Calais and St. Stephen were once prosperous ports and lumber towns, their banks lined with wharves and sailing ships. Many of the handsome old houses and churches still remain.

Calais. Calais (rhymes with Alice) is a small city, with supermarkets, drugstores, restaurants, a bank, and a movie theater, but no facilities for boats other than the town landing. There is a tourist information center and the Downeast Heritage Museum just above the landing and a duty-free shop next to the bridge. The Calais Regional Hospital can be reached at *207-454-7521*.

St. Stephen. St. Stephen is also convenient, with nearby restaurants, shops, a liquor store, a laundromat, a supermarket, and pay phones. Washrooms (hey, we're in Canada!) and a tourist information center are located in an old train station directly opposite the town floats.

Bring your sweet tooth to the Ganong Chocolate store (with a brown awning) for sinfully delicious chocolate by the piece or the pound. Ganong has been in business since 1809, and there are interesting displays of early candy making and equipment.

OAK BAY

45° 12.00'N   067° 10.65'W

CANADIAN CHARTS: **4331**   U.S. CHARTS: **13398**

Tidal Range: *Mean* 19.6 feet; *Large* 27.2 feet

Oak Bay lies north of the entrance to the narrow portion of the St. Croix River. The wide, encircling shore of this attractive bay has a number of small summer camps, and St. Croix Mountain forms a backdrop across the way.

Anchorage is good in much of the shallow bay, and this would be a convenient spot to wait out the tide in the St. Croix River.

APPROACHES. The approach from the southern portion of the St. Croix River is wide open and without dangers except for possible strong downdrafts off Devils Head.

ANCHORAGES, MOORINGS. Anchor in 9 to 12 feet along the western shore south of Spoon Island, the pointed little island that lies near the head of the bay. Holding ground is good, in mud bottom. You will be comfortable here in settled summer weather, but the bay is too wide to offer protection if it blows.

Fundy and Saint John

Passamaquoddy Bay to Fredericton

FUNDY AND SAINT JOHN

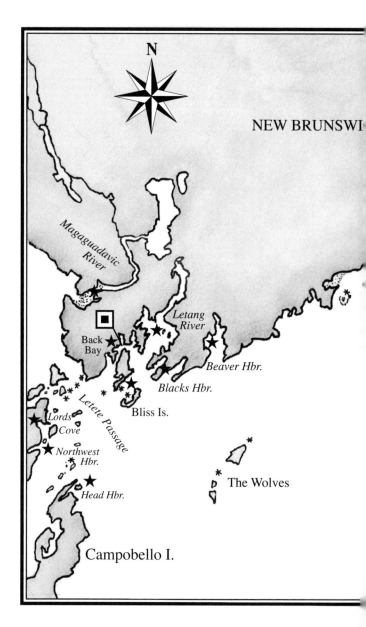

ONCE through the Grand Manan Channel, the wide-open and temperamental Bay of Fundy lies before you. Huge tides of icy water move in and out of the bay and turn warm, moist air into concrete-thick fog. When the wind blows against the strong currents, the bay is whipped into steep, unruly waves. At other times, Fundy lies under a glass surface, lulling, warm, and beguiling.

Fog is most prevalent in spring and early summer in light southerly winds of less than 10 knots. Ship reports indicate fog 30 percent of the time in July, the worst month, but this is an average. Between foggy days, other days can sparkle.

The last of the good harbors are near Letete Passage, tucked behind the Bliss Islands or on either side of Frye Island and the Letang Peninsula. Farther east, the coast of New Brunswick is bold, with only one or two harbors for refuge until you reach the shipping port of Saint John. This small metropolis lies at the mouth of the Saint John River, fortified along its waterfront by huge concrete container-ship terminals, oil tanks, a ferry terminal, cruise-ship docks, and the Canadian Coast Guard. It is not an easy place for a yacht to lie.

Beyond the city, the dramatic Reversing Falls is the gatekeeper to the inland wonders of the Saint John River. Two irrefutable forces of nature meet at the falls—the enormous Fundy tides and the ceaseless seaward flow of the river. At low tides, the river cascades out to sea. At high tides, the sea surges upriver. And during the short slack periods, the falls are navigable by boat.

With surprising suddenness, you enter a different world, where the warm, fresh waters of the Saint John River stretch inland, tinged brown and smelling of rich earth. The fog is dissolved by the sun, the tides are measured in inches instead of meters, and salt is found only on your food. Languid arms and bays flood the valleys between the

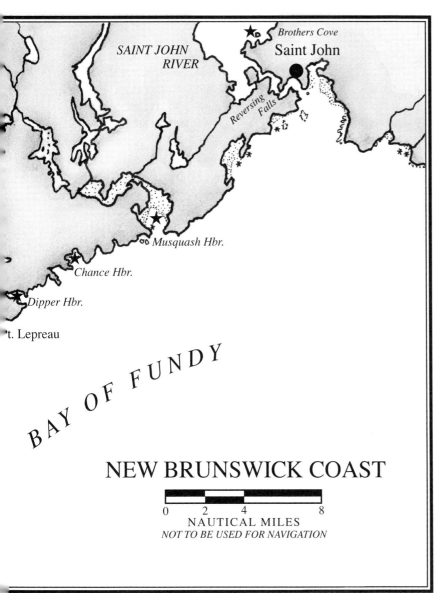

BAY OF FUNDY to SAINT JOHN

The 40-mile, day-long passage to Saint John from the area around Letete Passage is more complicated in timing than in navigation. There are several good anchorages from which to depart, but almost no good harbors en route. The shores are bold, with few offlying dangers. Once at Saint John, it is difficult to enter the harbor against the tide and river current, and it can be an uncomfortable place to spend the night. Ideally, try to time your arrival so you can pass directly up the Reversing Falls and into the Saint John River.

The Bay of Fundy is host to a large amount of shipping traffic, most of it converging on Saint John. The Marine Communications Traffic Center, based in Saint John, maintains a radar surveillance of the entire bay and helps direct the traffic, much like an air traffic controller. Also known as Fundy Traffic, they can be reached or monitored on channel 14 west of The Wolves or on channel 12 east of The Wolves, or at *506-636-4696*. In limited visibility, you should identify yourself and your position so they can monitor your progress in relation to other ships nearby. Be particularly alert for the ferry running between Grand Manan and Blacks Harbour and for shipping traffic in the lanes and anchorages outside Saint John.

At Saint John, the unique phenomenon of the Reversing Falls delays the upriver flow of current until slack water at the falls, about 3 hours and 50 minutes after low tide. During the delay, even while the tide is *flooding*, the current in the river and in the harbor is *out-flowing*. After slack water, the current flows upriver. It continues flowing *upriver* even during the first 2 hours and 25 minutes of the *ebb*, until the next slack water at the falls (see *Reversing Falls* on page 414 for more information).

The delayed current, however, can work to your advantage. Plan to arrive in Saint John at least 4 hours after low when the current will be flowing upriver in your favor. Allow plenty of time for

rolling hills, and wavelets lap at wooded shores and beaches of river sand.

Farther upriver, storybook cable ferries cross from one hamlet to another. Farms and pasture spill to the banks, and cows are barged to the low, grassy islands for grazing. The narrow tributary of the Jemseg River opens into wide Grand Lake, where the inland sailing is unobstructed and unsurpassed on the warm breezes. Fredericton, the capital of New Brunswick and the head of navigation, flanks the river just below the Mactaquac Dam.

In emergencies, the Canadian Coast Guard can be reached on VHF channel 16, or Vessel Traffic Services, known as Fundy Traffic, can be reached on channels 12 and 14.

customs clearance if you are entering Canada here. You can start sailing on the last of the ebb in the Bay of Fundy, catch the beginning of the flood around Point Lepreau where the current is strong, and ride the flood to Saint John, arriving at the Reversing Falls just before slack water, when the current begins to flow upriver, and you can pass through the falls. If you have arrived in Saint John after slack at the Falls, you can dock in the heart of downtown at Market Slip and explore this interesting city while you wait for the next slack water.

BACK BAY

◼ ★★★ ⇌

off wharf: 45° 03.25'N 066° 51.55'W

CANADIAN CHARTS: **4124, 4331**   U.S. CHARTS: **13398**
Tidal Range: *Mean* 18.1 feet (5.5m); *Large* 25.2 feet (7.7m)

Back Bay is a broad bay to the east of Greens Point, just to the east of Letete Passage. The fishing community of Back Bay is on the bay's north shore. Most of the homes in the town are newly built, lacking in character, and dominated by the fish plant at the wharf. There is a post office is just up the hill from the wharf, but no other amenities are close by. The harbor, with its large breakwater, provides good protection from the south and easy access to Letete Passage, but it has little in the way of ambiance.

The real gem of Back Bay is the incredibly deep gut at the north end of the bay that leads to an almost completely landlocked, three-armed gunkhole, home to herons, eagles, and seals.

APPROACHES. At the mouth of the bay, favor the eastern side with the three red buoys to avoid the large shoals jutting from the northwest shore, halfway between Greens Point and the town. Once past the last nun, head for the prominent point of land at the south end of the town and then run for the green can at the tip of the breakwater.

To continue up the bay, keep the green can to port and stay centered in the gut to the north. You will have to pass several fish cages, moored to the west and dead center in the narrows. This passage does not require timing with the tides, and there will always be plenty of water.

ANCHORAGES, MOORINGS. ⇌ Tie alongside the wharf, or you may find room to anchor just to the north of the breakwater. Do not anchor in the large, shoal part of the harbor to the north of the wharf.

At the head of the bay, anchor in the mouth of any of the three arms in 12 to 18 feet at low. Don't try to get too deep into the arms, which shoal quickly.

FOR THE CREW. 🍴 From the wharf, a quater-mile walk will bring you to the little Bay View Variety.

THINGS TO DO. In town, head up the hill to the enormous Seaview Gospel Church that overlooks the town. As the sign outside says, "Destroy the sin in your life or it will destroy you."

Or contemplate nature in the arms at the head of the bay by dinghy. The small tidal pond to the southwest looks as though it was once dammed for a lobster pound. On a high tide, you might be able to pass to the north of Frye Island and into the head of Letang Harbour, but make sure you get back before the tide dries it out.

BLISS HARBOUR

45° 01.65'N 066° 50.25'W

CANADIAN CHARTS: **4124, 4331**   U.S. CHARTS: **13398**
Tidal Range: *Mean* 18.1 feet (5.5m); *Large* 25.2 feet (7.7m)

Bliss Harbour is a wide, open body of water formed by Frye Island to the north and west and by the Bliss Islands to the south. The chart shows an anchorage in Bliss Harbour, but this would be a naked spot to lie in anything but the fairest weather, even if you could find room among the fish pens.

The Bliss Islands make a good daystop, though, particularly when coming from the east, for exploring and for waiting for favorable tides for nearby Letete Passage. The Bliss Islands form an amphitheater around a small but wide cove which offers good protection from the prevailing southwesterlies. Several cottages lie along this cove, and a fish-farming operation somewhat limits anchoring. The approach is easiest from the south, leaving cans "SA3" and "SA5" to port and rounding to starboard into the cove.

BLACKS HARBOUR

3 ★★ 🍴 🛒

off ferry wharf: 45° 02.88'N 066° 48.55'W

CANADIAN CHARTS: *4124*   U.S. CHARTS: *13398*

Tidal Range: *Mean* 18.1 feet (5.5m); *Large* 25.2 feet (7.7m)

Blacks Harbour is well protected but industrial. This is the headquarters of Conners Brothers Seafoods, the world's largest sardine packers. Their operations include a huge fish plant and shipyard. The large ferry departs and arrives here from Grand Manan several times a day.

Under normal circumstances, Blacks Harbour would hold little appeal for cruising sailors, but with its easy access from the Bay of Fundy, it could offer protection at a time when immediate shelter is more important than ambiance. Otherwise take the time to head into Letang Harbour and up the river.

The harbor is open to the prevailing southwesterlies but snug in blows from any other quarter.

APPROACHES. This is an easy harbor to enter, even in fog, but you must use extreme caution when approaching Blacks to avoid the Grand Manan ferries, which are constantly leaving and entering the harbor at cruising speed.

From the Bay of Fundy, find the lighthouse and horn on the point south of Blacks. Keep flashing red bell "KA2" close by to starboard. Then pass between can "KA3" to port and the flashing red daybeacon on the ledge to starboard. Once clear, head for the high bluff due north and follow it northeast into the harbor.

ANCHORAGES, MOORINGS. It may be possible to find a mooring or swinging room just past the ferry wharf on the south side of the harbor. A second possibility is just past nun "KA4," inside the harbor along the south shore, near the government wharf. A third option is to anchor near the head of the harbor, on the north side just past the cannery.

FOR THE CREW. Blacks is a Canadian port of entry. Call *888-CANPASS* to clear customs.

Restrooms and a pay phone are on the ferry wharf. 🍴 During the day several kiosks and takeouts may be operating out of the parking lot. From there, the town of Blacks Harbour is about a mile walk.

The small village has a hardware store, a convenience store, an ATM, a liquor store, 🛒 a large market, and—in honor of the sardines—🍴 the Silver King Restaurant (*506-456-3836*), which delivers.

A buoyed net from shore leads schools of fish into the weir.

LETANG RIVER

4 ★★★★

45° 04.50'N 066° 47.65'W

CANADIAN CHARTS: *4124*   U.S. CHARTS: *13398*

Tidal Range: *Mean* 18.1 feet (5.5m); *Large* 25.2 feet (7.7m)

Just north of Blacks Harbour, the Letang River, like Back Bay, runs to the northeast, feeling ever more like an inland lake the farther upriver you go.

Letang Harbour is the wide, lower portion east of Frye Island, exposed and full of fish cages. Several boats are moored near a wharf at the end of the road on the Letang Peninsula, but this would be a fair-weather anchorage at best.

Instead, follow the river farther upstream, past the fish cages, until it widens among numerous wooded islands and points. You can find protection from almost any wind in any one of the numerous anchoring spots. The solitude is broken only by the cry of loons, a farmhouse in the distance, and the faint whish of the world going by on a road to the northwest.

APPROACHES. From the south, pass Blacks Harbour to starboard and head between the high headland to starboard and the small island to port, both of which are bold and would be relatively easy to find in the fog. Round the headland and follow the arm of the harbor to the northeast but be on the lookout for fish cages. Run down the long narrow section along the southeast side of Letang Peninsula until it opens to coves to the north and south and the Letang River dead ahead and tending to the north. All three options make wonderful anchorages.

If you choose to anchor in the northern cove, keep close to the islands to the east to avoid the ledges along the west shore, particularly the one that sits dead center in the cove's mouth.

If you choose to go farther upriver, keep those same islands close to port to clear a ledge that extends westward from the small island on the eastern side of the river. Unfortunately, a rather uninspired house sits high on a hill to the north of this otherwise magnificent spot.

ANCHORAGES, MOORINGS. Depending on the prevailing wind, anchor either in the cove to the south, the cove to the north, or follow the river farther upstream and anchor just off the cluster of islands to the west or in the elbow of the river. Holding ground everywhere is good mud.

BEAVER HARBOUR

3 ★★

off wharf: 45° 04.20'N 066° 44.20'W

CANADIAN CHARTS: **4116**   U.S. CHARTS: **13398 (INSET)**
Tidal Range: *Mean* 18.1 feet (5.5m); *Large* 25.2 feet (7.7m)

Beaver Harbour is wide and wide open to the south. It is also spangled by weirs and fish pens that make it difficult for strangers to enter at night or in the fog, despite the lighthouse at Drews Head and the flashing light at the end of the government wharf. Still, protection can be found to the west of the government wharf among the fleet of fishing boats. A processing plant of Heritage Salmon is ashore by the wharf.

APPROACHES. The Canadian chart shows additional buoys that are absent on the U.S. chart. Find red-and-white bell "KEA" (45° 03.10'N 066° 43.52'W) at the harbor entrance, and run in toward the lighthouse and fog signal at Lighthouse Point on Drews Head (*Fl 15s 14.2m 13M*). Pass between the lighthouse and nun "KE2" (not shown on U.S. chart), which marks the shoal area to the east.

Past the lighthouse, can "KE3" marks the easternmost point of the weir projecting from the west shore. Keeping the can to port, turn toward the end of the wharf, marked by a flashing red light, and keep it to port while turning behind it.

ANCHORAGES, MOORINGS. Anchor to the west of the wharf among the boats in 10 to 15 feet of water at low. Bottom is mud. Or come alongside the wharf or a fishing boat if there is space and you won't get in the way.

THE WOLVES

off Paul's Cove: 44° 58.65'N 066° 42.80'W

CANADIAN CHARTS: **4011, 4116**   U.S. CHARTS: **13398**

When Great Chief Glooscap, with a mere wave of his hand, changed a deer and a moose and the pack of wolves on their heels into islands, he granted Deer and Moose Island eternal safety in the protected embrace of Passamaquoddy Bay. But he doomed the wolves to be outcasts forever. And so they are—a small pack of islands, huddling together in the broad expanse of the Bay of Fundy.

Southern Wolf has a prominent light (*Fl 10s 38.1m 15M*) at its western end. Eastern End Cove, a small bight on the northeast end of Eastern Wolf Island, provides a lee in normal southwest winds but is totally exposed in anything else. It would make an uncomfortable overnight anchorage despite one or two heavy moorings that may be unoccupied. The moorings, though, make Eastern Wolf a convenient daystop.

The safest approach is from the northeast. Note the rock, due north of the cove, which is exposed at low. You can land and explore the island, where there are a few fishermen's cottages.

POINT LEPREAU

S. of Point Lepreau: 45° 03.00'N 066° 27.50'W

CANADIAN CHARTS: **4116**

On a clear day, Point Lepreau can be seen from almost anywhere in the Bay of Fundy, though there are many sailors who have passed this way and never seen it at all. More prominent than the 85-foot lighthouse (*Fl 5s 26m 16M, horn*) on the point is the nuclear power plant on the point slightly to the north, the first nuclear generator in Canada.

When sailing from Letete Passage to Saint John, Point Lepreau marks the halfway point. Offshore, the red-and-white Mo(A) whistle lies in an area of strong current and tide rips that can produce nasty chop when the wind is opposing the current. Under these conditions, it is best to stay farther offshore and give Point Lepreau a wide berth. Cutting inside the whistle and going through the rips is not recommended.

DIPPER HARBOUR

3 ★★

outside breakwater: 45° 05.50'N 066° 24.80'W

CANADIAN CHARTS: **4116** (INSET)
Tidal Range: *Mean* 21.9 feet (6.7m); *Spring* 30 feet (8.8m)

Good harbors along the north coast of the Bay of Fundy past Point Lepreau are a scarce commodity until you reach Saint John. The exception is Dipper Harbour, which qualifies as an ideal refuge. The harbor's enlarged breakwater and wharf provide good shelter from all directions.

APPROACHES. Dipper Harbour is easy to enter in any conditions. Steer for the tip of the breakwater, marked by a flashing green light. Keep off Fishing Point and the weir that extends from the west, and keep off nun "KN2" to the north. Round the breakwater into the harbor.

ANCHORAGES, MOORINGS. Dipper Harbour has plenty of room for anchoring with good holding in mud. It is also filled with heavy winter moorings for the fishing fleet. These are not used as much in the summer, so one might be available. The wharf extends north from the beginning of the breakwater with floating docks on both sides. Anchor, pick up a vacant mooring, or come alongside the docks or a fishing boat at the wharf.

THINGS TO DO. If holed-up by weather and looking for things to do, you might be able to find a ride to Point Lepreau to tour Canada's first nuclear power plant (*506-659-2220 ext. 6433*).

CHANCE HARBOUR

3 ★★

off wharf: 45° 07.00'N 066° 20.70'W

CANADIAN CHARTS: **4116**
Tidal Range: *Mean* 21.9 feet (6.7m); *Spring* 30 feet (8.8m)

Chance Harbour lies to the west of Dipper Harbour. In the winter of 1999/2000 a storm destroyed half the government wharf. It has since been rebuilt. In fair summer weather the bluffs on Lighthouse Point shelter the small bight to the north, and there is good anchorage off the wharf. The dramatic red cliffs of Cranberry Head flank the east shore.

APPROACHES. From red-and-white bell "KS," run toward the lighthouse on Lighthouse Point. Be sure to leave to port can "KS1," which marks the ledges making out from the point. A weir makes out from the north shore of Lighthouse Point. Round the point broadly and turn westward into the anchorage. Beware of fish pens opposite the end of the wharf.

ANCHORAGES, MOORINGS. Dock to a fishing boat or the wharf.

FOR THE CREW. Walk up the hill from the wharf and turn right on Route 790. The Mariner's Inn (*506-659-2619*), about half a mile away, serves lunch and dinner. Innkeepers Susan and Bill Postma also rent bikes and—more importantly, perhaps, after all your sailing—provide quality footcare, healing arts, and reflexology.

MUSQUASH HARBOUR

between Western Head and Musquash Head:
45° 09.55'N 066° 14.65'W

CANADIAN CHARTS: **4116** (INSET)
Tidal Range: *Mean* 21.9 feet (6.7m); *Spring* 30 feet (8.8m)

Despite its large size and the large inset devoted to it on the Canadian chart, Musquash Harbour is too broad and shallow to be of use to the cruising yachtsman. The anchorage shown to the west of Musquash Island in the mouth of the Musquash River has no protection at all.

A second, more secure anchorage known as Five Fathom Hole can be reached by following the buoys farther up the harbor. There is room to anchor off the government wharf, or you may find an available mooring. Deep-draft boats will probably need to enter at half tide or better, preferably with the flood, and time their departure similarly.

SAINT JOHN

1 ★★★★★ ⚓ 🛒 ⛽ 🍴

E. of Partridge Is.: 45° 14.20'N 066° 02.50'W

CANADIAN CHARTS: **4117, 4416**
Tidal Range: *Mean* 21.9 feet (6.7m); *Spring* 30 feet (8.8m)

Emergency: 911, upriver 800-442-9722
Hospital: 506-648-6000, 506-632-5555
Canadian Coast Guard: Ch. 16; 800-565-1582, *16
Fundy Traffic: Ch. 12, 14; 506-636-4696
Customs: 888-CANPASS
Marine forecast: 506-648-4986, 506-636-4991

The face modern Saint John presents to the water is a very different sight than the mouth of the mighty river that Samuel de Champlain named in 1604 in honor of St. John the Baptist Day. Huge freighters and expansive cruise ships come and go from wharves fortified against enormous tides with ranges taller than the masts of some cruising sailboats. The shores are lined with tank farms, dry docks, and the largest refinery in Canada. And the constant current either catapults a boat upriver or fights it every inch of the way.

Behind the industrial facade is a city of human proportions that is a testament to the perseverance of the Saint Johners who rebuilt most of it after it was razed by fire in 1877. A European sense of cleanliness and order prevails. The streets are laid out in neat rows, cars stop for pedestrians, and the crossing signals beep audible signals for the blind. Most of the core of the city can be reached on foot—the City Market, King's Square, the rows of historic houses, the fine dining and shopping. If the fog happens to be too thick to navigate the streets on foot, a network of indoor skywalks connects major portions of the city.

As a courtesy, notice that the "Saint" in Saint John is never abbreviated, and that there is only one "John." St. Johns is in Newfoundland, and you haven't sailed that far... yet.

APPROACHES. Use caution approaching the very busy port of Saint John. The Marine Communications Traffic Center (*Ch. 12; 506-636-4696*), also called "Fundy Traffic" on the radio, maintains radar surveillance of all vessel movements in the entire Bay of Fundy, much like an air traffic controller, but their primary reason for being is to direct the flow of traffic in and out of Saint John. Fundy Traffic should be contacted on channel 12 well before approaching the harbor, particularly during low visibility, and they will inform you of vessel movements nearby. Channel 12 should be monitored until you arrive safely.

Because the denser, heavier salt water tends to flood in a wedge beneath the lighter fresh water flowing out of the Saint John River, the brown, fresh water from the river meets the greenish-blue salt water along a foam line well outside the mouth of the harbor, even on a flood tide. Floating debris may be lurking in the foam line, so use caution when crossing it. South or southwesterly winds build short, steep waves in this area that are uncomfortable but not dangerous unless they are breaking in strong southerly winds. The waves become progressively shorter and steeper toward the mouth of the harbor.

When the visibility is good, avoid shipping traffic by approaching close to Partridge Island, which was the site of North America's first quarantine station and later the site of the first steam-powered fog whistle in the world. Pick up the green channel buoys and follow them up the west side of the channel. Once around the ledge extending eastward from Partridge, you can stay just outside the channel to the west if you need to avoid shipping traffic. At night, follow the lighted range markers. Proceed northward toward the first wharf on the west side of the harbor where the ferry departs for Digby, Nova Scotia.

In pea soup, the red-and-white Mo(A) whistle south of Partridge might be reassuring, but it marks the center of the shipping lanes. Most shipping in the harbor occurs 2½ hours before or after high water, but count on traffic at all times regardless of visibility. Note on your charts that a ship anchorage lies off Manawagonish Island, known locally as Mahogany Island. Fundy Traffic can inform you if any ships are lying there.

Proceed up the western fork of the channel. A strong current with complex eddies runs in the harbor as the tides interact with the outflow from the Saint John River. The unique phenomena of the Reversing Falls delays the upriver flow of current until slack water at the falls, about 3 hours and 50 minutes after low tide. During the delay, even while the tide is *flooding*, the current is *out-flowing*. After slack water, the current flows upriver. It continues flowing *upriver* even during the first 2 hours and 25 minutes of the *ebb*, until the next slack water at the falls (see *Reversing Falls* on page 414). Plan to arrive at least 4 hours after low water while the current is flowing upriver. In an ideal world, the current will be with you, or you will have a big auxiliary, or both.

ANCHORAGES, MOORINGS. There is only one option for docking in the harbor, and it is temporary at best. Most boats try to navigate the Reversing Falls at the first opportunity. ⚓ Tie to the floating dock at Market Slip, on the east side at the head of the harbor just past the Coast Guard wharf. Dock anywhere but the innermost boatlength where it shoals to a reported depth of only 2 or 3 feet. When approaching the float, use extreme care in navigating the eddies that swirl around the mouth of the slip and the outer end of the dock.

Market Slip was the site of the first Loyalist landing in 1783, and it once extended to the foot of King Street. Later, it was lined with warehouses, and merchant ships of all kinds unloaded their wares. It still is the heart of the city, now renovated and known as the Boardwalk at Market Square. The city could not be more accessible, but you and your boat could not be more public. The dock is exposed to the swells which roll into the harbor everywhere, and the current can be strong enough to press boats against the dock or pull them away with remarkable force, particularly at the outer end of the dock. The dock is about three boatlengths long, and boats often raft off the middle section of the dock to avoid the eddies at the outer end and the shoal water farther in.

Customs clearance (*888-CANPASS, regular hours: M-F 8AM - 9PM*) can be gained from Market Slip (see page 368 for more customs information).

FOR THE BOAT. Most amenities will be found in the boating Mecca beyond the Reversing Falls. A schedule of the tides and the high and low slack waters for the Reversing Falls can be obtained at the tourist information center next to Market Slip. The schedule, printed by the Irving Oil Company, clearly indicates that it is not intended for navigational purposes, but you will find a copy on almost every boat in the area.

FOR THE CREW. There is a laundromat on Princess Street, three blocks away, and a variety of restaurants nearby. As part of revitalization efforts, the city has constructed an artificial beach at the end of Market Slip, an incongruous piece of California planted in the middle of the city. Here, at almost any time of day and often in the drizzle or fog, league and pick-up games of volleyball are played in bare feet and bathing suits. The Canada Games Aquatic Centre, one block away, is open most afternoons for public swimming (they have showers, too), and the Market Square complex, located right next to Market Slip, houses the city library.

THINGS TO DO. The exploring, dining, and shopping possibilities here are unlimited. The tourist bureau, located at Loyalist Plaza next to Market Slip, has pamphlets describing walking tours of the city. Wander the shops and restaurants in Market Square, a mall in restored buildings that were once warehouses for the clipper trade. The New Brunswick Museum, next to Market Square, can fill you in on the history.

Walk up the hill to King's Square, where the paths are laid out like the stripes of the Union Jack and an Edwardian bandstand is elevated over a fountain. Nearby is the Imperial Theatre and the Loyalist Burying Grounds. 🛒 City Market, off one corner of the square, is one of the oldest farmers' markets in the country with fresh fruits, vegetables, seafood, and meats. It opens and closes with a ringing of a bell at 8:30 AM and 5:30 PM six days a week, and it is housed in an 1876 building spanned by timber-framed trusses on cast-iron columns. To the south and east, walk the lovely streets of townhouses and captains' homes near Queens Square.

The city has an urban transit system that can considerably broaden your explorations. Routes and schedules are available at the tourist bureau or by calling *506-658-4700*. One awesome spectacle that many cruisers never see from their boat (hopefully!) is the Reversing Falls at their maximum strength. A bus passes near the Falls View Park, or you can take a cab. You can also take a jetboat tour through the Falls, leaving from Market Slip (*506-634-8987*). Do this after you have already navigated the falls at slack. Otherwise, you might not believe that it is possible, or be a little less inclined to find out.

THE SAINT JOHN RIVER

The Saint John River above the Reversing Falls is one of the most extensive and interesting freshwater cruising areas anywhere. After the Bay of Fundy's extreme tides, currents, and fogs, the Saint John River is a benevolent world with practically no fog, 20-inch tides, warm water, sandy spits, and unlimited gunkholing. The water is tinged brown and has the earthy smell of rich dirt and leaves. The wind is warm and carries the breath of conifers. The Saint John River branches out into three long arms—the Kennebecasis, Long Reach, and Belleisle Bay—and an interconnected network of lakes that includes Grand Lake, Maquapit Lake, Washademoak Lake, and French and Indian Lake.

The Saint John River has always been a main thoroughfare, an artery through the heart of New Brunswick. Even the smallest cluster of farms still has its own wharf, left from the days when steamships were the main means of transportation to the markets of Saint John. Farmhouses face the river, cable ferries ply back and forth between the banks, and each crossroads has the feel of an isolated junction town, connected to others only by the ceaseless water flowing by.

At its southern end near Saint John, the river cuts through bold hills and penetrates deep valleys to form Kennebecasis and Belleisle bays. Farther upriver, the steep hillsides and bluffs give way to rolling hills, where the river meanders through bottom lands and flat silt islands and spreads into broad shallow lakes. The head of navigation is at Fredericton, 60 miles from the Bay of Fundy.

One of the oldest promoters of sailing this labyrinth is the Royal Kennebeccasis Yacht Club at the mouth of Kennebecasis Bay. It is the largest of several yacht and boat clubs in the area, and its active boating community is renowned for its hospitality to guests.

The level of water in the navigable portions of the Saint John River is a balance between the water running through the hydroelectric dam at Mactaquac above Fredericton and the heights of the tides that back it up at the Reversing Falls in Saint John. During the periods of heavy spring meltdown or rain, known in river parlance as the freshet, the volume of water flowing down the river is greater than the volume of water that can pass through the falls during a tide cycle. The level of the river's bays and lakes rises dramatically, flooding many of the low islands and wharves. In extreme cases in early spring, the level of the river exceeds the height of the high tide at the falls, and the falls don't reverse themselves, so they become impassable to upriver-bound boats. The clubhouse at the Royal Kennebeccasis Yacht Club is set atop high footings, yet the freshet has still lapped at its floor joists.

During the summer cruising months, the water level returns to normal. At the Reversing Falls, low tides in the Bay of Fundy allow the river to run out twice a day, and the high tides back the river up twice a day. The result is a tidal cycle on the river that is identical to the cycle in the Bay of Fundy, a high and a low approximately every 12 hours and 40 minutes, but with a range that is only 2 feet in Kennebecasis Bay, 11 inches in Gagetown, and even less on Grand Lake. Though the tidal height is felt far upriver and salt water can reach all the way to Oak Point, the current runs ceaselessly downriver except at the Reversing Falls themselves. Free of dramatic tides, boats can moor with a stern anchor out and bow tied to trees, nudged onto sand spits, or pulled to shear rocks.

The river itself presents few navigational obstacles, and the odd rock or shoal areas are clearly buoyed. In early summer, occasional fish and eel traps are staked just offshore, but they rarely occupy navigable water. The remains of old logging operations, from the days when logs were driven down the river and boomed in coves, present more unique obstacles. Old cribbings and pilings from the booms still foul the mouths of some coves and render others useless as anchorages, but they are well charted. Waterlogged sticks and branches on the bottom can foul anchors, so triplines are a wise precaution.

Less exact and more ubiquitous are deadheads—waterlogged logs which sank during the logging days and which now, years later, have become buoyant enough to lurk just at the surface. Areas particularly prone to deadheads are marked on the charts, but deadheads can surface anywhere, particularly near the banks of the river and in the eddies of the Reversing Falls, where they accumulate.

References aboard should include the Canadian chart packets 4141 (which has two sheets) and 4142 (four sheets) as well as the current *Sailing Directions*

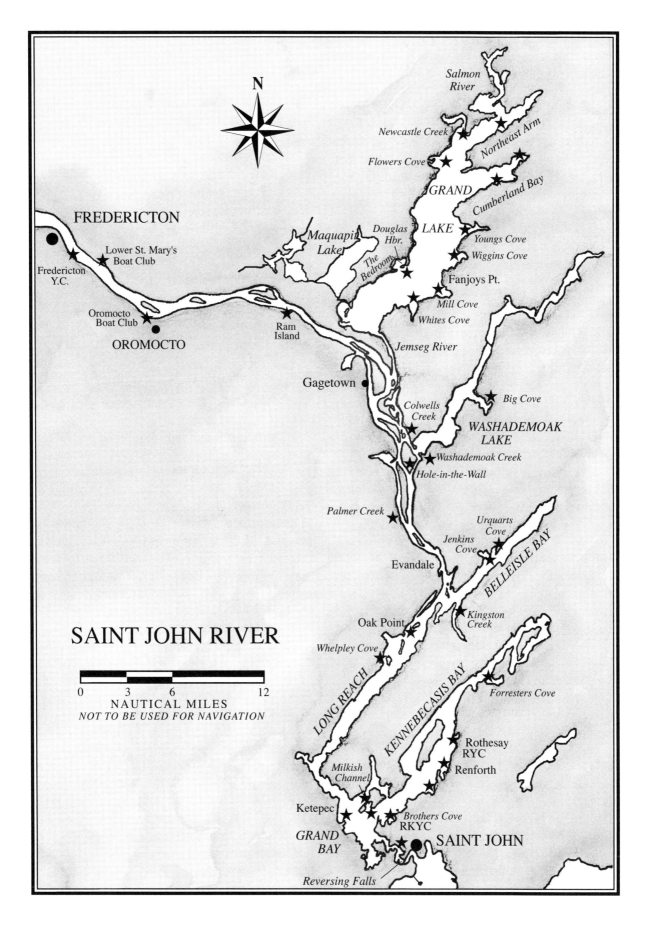

for the Saint John River, publication ATL 107, available at local chart dealers or through the Hydrographic Chart Distribution Office, Department of Fisheries and Oceans, 1675 Russell Road, P.O. Box 8080, Ottawa, Ontario, Canada K1G 3H6; *800-668-5222.* These charts should be corrected for river buoys that were overhauled in 1995. Note that the Canadian charts for the river system still give depths in feet, and that the charts are not oriented with north at the top but rather follow the direction of the river.

THE REVERSING FALLS

The Pots, between Upper and Lower Falls:
45° 15.75'N   066° 05.30'W

CANADIAN CHARTS: **4141-1, 4117, 4116**

The Reversing Falls is the key that unlocks the inland cruising treasures of the Saint John River. The Saint John River meets Saint John Harbour at a narrow gorge partially dammed by a submerged ledge. This damming, along with the difference in water levels between the river and the tremendous Fundy tides, produces the unique phenomenon of the Reversing Falls.

When the tide is lower than the river, the river behaves as any normal river should, cascading seaward through the gorge. An average low tide is about 14 feet lower than the river level, and millions of gallons crash through the falls—an awesome sight. But when the rising tide exceeds the level of the river, the flow reverses itself, and sea water flows upstream.

Samuel de Champlain wrote that the river, "having turned a point...narrows once more and makes a kind of fall between two great cliffs, where the water rushes with so much power that any wood thrown there is drawn under and seen no more. Awaiting, however, the (half) tide, one can pass this strait very easily..."

The moment of the reversal, the slack water, occurs two times every tide cycle. Low slack occurs roughly 3 hours and 50 minutes after low water in Saint John when the tide is *flooding* and the upstream flow will begin. High slack occurs when the tide is *ebbing,* about 2 hours and 25 minutes after high water in Saint John, when the downstream flow will resume. More intuitive terms for low and high slack might be ebb and flood slack. The period of slack water lasts all of 10 minutes, but the falls are generally navigable for 20 minutes either side of slack.

The falls are divided into the Upper Falls and the Lower Falls. The difference between the water levels of the tide and the river is most pronounced at the Upper Falls, where the water cascades over a submerged ledge that stretches between Union Point and Crow Island to the east. The Lower Falls are created by the turbulent water being squeezed through a narrow gorge spanned above by a highway and railroad bridge with a clearance of 80 feet. Union Point projects between the Upper Falls and the Lower Falls, adorned with a huge frothing and belching pulp mill.

To those from afar, passing through the Reversing Falls represents a certain rite of passage. Stories of hair-raising adventures in the falls abound, though local sailors will pass through them and back for a day sail or navigate them at night. The entire trick is in the timing. Time the falls correctly, and the current is no worse than the strong currents you might encounter in places like Passamaquoddy Bay, but time them incorrectly, and your yacht will behave as if it is on a roller coaster. If you want local knowledge, call the Kennebeccasis Yacht Club at *506-652-9430.* Here are some suggestions from the pros:

SLACK WATER. The exact time of slack water may vary depending on the water level of the Saint

The Reversing Falls near slack water, looking upriver. The Lower Falls are beneath the bridges. The Upper Falls are between the pulp mill on Union Point and the islands. Split Rock, in the foreground, is a good reference for gauging the current. (Canadian Hydrographic Service)

John River, which fluctuates seasonally and with the amount of recent rainfall. If the river is high, a low slack will be slightly later, and a high slack will be slightly earlier. Conversely, if the river level is low, a low slack will be slightly earlier, and a high slack will be slightly later. Always arrive at the falls early to allow for these discrepancies and to wait for slack.

Irving Oil Company publishes a table of slack water predictions for the Reversing Falls which most locals keep on board and find reliable despite its printed warning that it is not intended for navigation. The table can be picked up at the tourist information center at Market Slip and many other locations. The most accurate slack predictions are found at the Canadian Hydrographic Service's website (*www.charts.gc.ca/pub/en*), then follow the water levels links.

DEPTH. Depth through the falls is ample except for the charted 5-foot area just off Union Point. The soundings charted at the Reversing Falls were taken at slack water, but depth at slack will also vary with the height of the river at any given time.

APPROACHES heading upriver. When heading upriver, pass under the Harbour Bridge (76-foot clearance), the first bridge at the head of the harbor, and arrive at the Reversing Falls early. Circle in the elbow of the river to the west, just below the Reversing Falls bridge, while waiting for slack water. Watch the water between the east shore and freestanding Split Rock to judge the water levels and the direction and force of the current. Take your time and wait for the tide. You will see the force of the current passing between Split Rock gradually diminish.

Heading Upriver at Low Slack. Before slack, the current will be against you; after slack, the current will be with you. Arrive early and circle clockwise while waiting. The water flowing between Split Rock is a good gauge of slack water. Before slack, the water will be flowing downriver, cascading in a mini falls through the split. When circling, head into the current and feel its strength against you.

As slack approaches, the current's force will diminish and eventually reverse and sweep you through the falls. It is better to go slightly early, against the last of the current, rather than later when there might be so much current in your favor that you lose steerageway. When it is time to go through the falls, approach the Lower Falls from the elbow in the river south of the bridges rather than trying to turn into the current from behind Split Rock.

Heading Upriver at High Slack. Before slack, the current will be with you; after slack, it will be against you. Don't be tempted to go too soon with too much current heading upriver, or you will lose steerageway. Circle counterclockwise in the elbow of the river so you will be running with the current when you are closest to the middle of the river but can then turn to port into the eddy of the elbow. Be careful not to get too close to the falls, where the current could sweep you through.

Alternatively, you can wait until the current turns slightly against you, but don't wait too long, or the contrary current will become too strong to buck, particularly in the Upper Falls where it is stronger. Don't approach from behind Split Rock, particularly when the current is still running upriver—your leeway as you turn might put you onto the rock. Even when approaching in the straight-line course from the elbow, you may still be set to the west at Split Rock. Be careful to stay in midchannel.

Once you are past Split Rock and through the Lower Falls, the channel widens, and Union Point with its Kafkaesque pulp mill projects from the left. Steer slightly to starboard to stay in the middle of the river and pass between the pulp mill and a turbulent eddy to starboard called "The Pot." Stay midchannel through the Upper Falls.

APPROACHES heading downriver. When heading downriver, arrive early and circle in Lee Cove to the west, above the pulp mill on Union Point, near the log boom. If you are very early, good holding is reported off Indiantown. Watch the current at the submerged ledge and against the rocks at Union Point. The route through the falls will be the reverse of the upriver route.

The Reversing Falls during ebb flow. Lower Falls and Split Rock are in the foreground. Upper Falls and Crow Island are past the bridge.
(Canadian Hydrographic Service)

Heading Downriver at Low Slack. Before slack, the current will be with you; after slack, the current will be against you. When circling in Lee Cove, keep well away from the Upper Falls themselves so the current won't sweep you over them. Go just before slack water and let the current help you, but don't go too early or you'll lose steerageway. You can also go at slack or just after the current has turned against you, but don't wait too long or you won't be able to power against it.

Heading Downriver at High Slack. Before slack, the current will be against you; after slack, the current will be with you. Go through against the last of the current or wait until it turns in your favor. Don't wait too long, or you may lose your steerageway in the falls.

CAUTIONS. Logs from upriver tend to accumulate in the eddies of the falls and should be avoided. The logs will be floating on their sides or they could be waterlogged "deadheads" standing vertically in the water, with just one end near the surface. The logs can be anywhere, but most probably they will be lurking in the eddy lines, clearly defined by foam (created by the turbulence and the pulp mill's effluent) and other wrack. Use extra caution when crossing these eddy lines.

If through timing or temptation you run the falls with too much current in your favor, you may lose steerageway. Even though you are moving fast, the boat may hardly be moving relative to the water, and the rudder could have little or no effect. Control can be regained by throttling forward faster than the current, but you may be swapping one danger for another. Now the boat will be moving faster, and your decisions will need to keep up. The helm, meanwhile, might still be sluggish.

A more dramatic emergency technique to regain control when steerageway is lost is to throttle hard forward or in reverse to turn the boat 180 degrees into the current, but this must be done quickly and decisively. Now, powering forward against the current, the flow of water over the rudder is the sum of the current and your speed, so steerageway is excellent. If your engine is more powerful than the current, you can either power against the current and out of it the way you came, or you can hold the boat stationary in the current. Or, by powering *forward* at a speed slightly less than that of the current, you can let the current carry the boat slowly *backward* in your original direction but with much more steering and speed control.

Another mistake is to head upriver too late after high slack, when the current has turned against you. Since the current is less at the Lower Falls than the Upper Falls, a late arrival might be able to power against it through the Lower Falls, only to find that the boat can't overcome the current in the Upper Falls. And with every passing moment, the strength of the contrary current increases.

A skipper caught in this situation has a couple of nerve-wracking options. He can attempt to spin the boat around and shoot out the Lower Falls like a cork out of a Champagne bottle, but his 180-degree turn will need to be quick to minimize the boat's leeway, and his steerage may be marginal once he has turned unless he powers forward even faster than the current on his way out.

Or, perhaps more prudently, he can throttle back slightly while heading forward into the current and let the current carry the boat backwards slowly while he maneuvers toward the small cove south of the pulp mill on Union Point. If necessary, you can stay in this relatively calm water until the next slack.

If you are caught in the falls and feel you need help, call the Coast Guard (Ch. 16) or Fundy Traffic (Ch. 12). They may send the powerful jetboat of the Reversing Falls Jetboat Tours (*506-634-8987*) to tow you to safety.

Upper Falls at maximum ebb, looking upriver past Goat, Middle, and Crow Island. The pulp mill on Union Point is on the left.

It is possible to navigate the Reversing Falls at night or in limited visibility, but this is not recommended for strangers. A tourist bureau up by the highway shines several blinding lights on the Lower Falls, and their glare makes gauging slack water difficult. Orange halogen lights from the highway and pulp mill provide surreal illumination through the Upper Falls, but by then your night vision will be nonexistent. The darkness will seem particularly inky, and there are almost no illuminated buoys from there to Grand Bay.

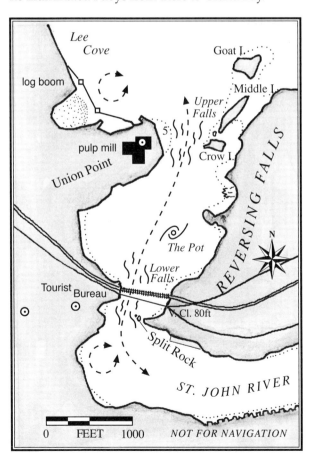

MARBLE COVE

off channel entrance: 45° 16.30'N 066° 05.40'W

CANADIAN CHARTS: **4141-1**

Marble Cove lies just above the Reversing Falls on the east bank of the river in the midst of what used to be the industrial heart of Saint John. The cove is broad and shallow, and most of it dries out at low tide. Areas that do not dry out are occupied by sunken ruins from the time when the cove was used as a timber pond, so anchoring within the cove is impossible. A dredged, 6-foot-deep channel runs through the cove to the Saint John Power Boat Club in the northeast corner, where cruisers can spend the night or wait for slack water before navigating the Reversing Falls.

APPROACHES. The western entrance to the dredged channel is marked first by two red spars and then by a flashing red light. The light makes Marble Cove easy to spot if you've come up the Reversing Falls at night. Once in, the channel is clearly marked and 6 feet deep. If you do touch bottom, it is soft mud.

ANCHORAGES, MOORINGS. Slips are available at the Saint John Power Boat Club for powerboats and shallow-draft sailboats.

Do not anchor in the cove. It is shallow and foul. The *Sailing Directions* indicate good holding ground outside the cove north of Rowans Point or in the mouth of Lee Cove farther upriver.

FOR THE BOAT. *Saint John Power Boat Club (manager, 506-642-5233)* has water and a pay phone. They can haul boats with up to a 5.5-foot draft on their 20-ton marine railway, and they have an example of what must be one of the world's first Travel Lifts.

GRAND BAY

Grand Bay is the first wide bay on the Saint John River above the Reversing Falls. Kennebecasis Bay flows into its midpoint from the east, and narrow Long Reach flows into it from the north.

True anchorages are few. The most convenient might be McCormack Cove, on the south side of Kennebecasis Island, or the back. A breakwater forms a marina at Ketepec, on the west shore, and the Royal Kennebeccasis Yacht Club lies just inside the Kennebecasis. South Bay is boomed off for pulp logs for the mill.

SAND COVE

2 ★★

45° 16.20'N 066° 07.60'W

CANADIAN CHARTS: 4141-1

Just north of large South Bay and tucked north of Sand Point lies the small bight of Sand Cove. Protection is adequate from the north and south, but the cove is wide open to the west. The beach at the head of the cove is the Dominion Park swimming area.

APPROACHES. Pass under the power lines (82 feet) that run from Arthurs Point to Green Head and run down the east shore into Sand Cove. Log booms and their concrete piers cross the mouth of South Bay, so you shouldn't.

ANCHORAGES, MOORINGS. There are only one or two permanent moorings in the cove. Anchor anywhere in 7 to 9 feet of water.

THINGS TO DO. The park puts up volleyball nets on the beach so you can join a pick-up game.

KETEPEC

4 ★★★

entrance: 45° 17.00'N 066° 09.40'W

CANADIAN CHARTS: 4141-1

A harbor is formed behind a breakwater between the suburban towns of Ketepec and Belmont on the west shore of Grand Bay. The Saint John Marina, located in Ketepec, is one of the largest on the Saint John River and can function as a port of entry by calling *888-CANPASS*.

APPROACHES. The northwest end of the breakwater is marked by flashing green spar "JK1." Pass north of the spar, leaving it to port, and follow the buoyed channel to the marina. Note the rocks shown on the chart in the middle of the anchorage.

ANCHORAGES, MOORINGS. The Saint John Marina has dockage with 8 feet of water available for visitors, or you can anchor inside the breakwater to the right of the buoyed approach channel in 10 to 12 feet of water.

FOR THE BOAT. *Saint John Marina (506-738-8484).* This is a full-service facility. They pump gas and diesel and have water, electricity, pump-outs, and ice at the floats with 8 feet of water alongside. They haul boats with a crane and do engine, hull, rigging, fiberglass, and outboard repairs. Their store sells marine hardware and charts, and their Pierside Marine is an outboard dealer and parts supplier.

FOR THE CREW. Showers and a pay phone are ashore, as well as a licensed restaurant and lounge, often with live music.

THINGS TO DO. The marina has a volleyball court for some great fun and exercise. Bus 25 runs to shopping centers nearby and into the city of Saint John. Cabs are also available.

McCORMACK COVE

 ★★★

45° 19.10'N   066° 08.45'W

CANADIAN CHARTS: 4141-1

McCormack Cove is large and beautiful, lying on the southwest side of Kennebecasis Island, near the northern end of Grand Bay. The inner end of the cove is divided into two smaller coves, each beautiful and unspoiled, with only occasional glimpses of the few houses ashore. Protection is good from all directions except the south, where it is wide open to fetch from across Grand Bay. In southerlies, seek protection to the north of Kennebecasis Island or in Brothers Cove to the south.

APPROACHES. The broad mouth of the cove is wide and clear, but two sunken log cribs lie about halfway in, shown as small squares on the chart. Beyond them the cove is divided by a small island. If you plan to anchor in the western portion, keep close to the western shore to avoid the cribs. The northeast portion can be reached by following the eastern shore and then following the 6-foot contour past the eastern crib. Or you can estimate the width of the cove and keep about a third of the way off the eastern shore to pass between the two cribs.

ANCHORAGES, MOORINGS. Anchor anywhere in the northeastern end of the cove in a comfortable depth. In the western portion, anchor just inside the mouth before it shoals. Note the rocks just to the south of the small island.

KENNEBECASIS ISLAND

 ★★★

45° 19.70'N   066° 08.60'W

CANADIAN CHARTS: 4141-1

The Milkish Channel curves around Kennebecasis Island to the northwest, north, and northeast. At the west end of the channel, just off Grand Bay, the northwest shore of Kennebecasis Island bulges to the north and forms a broad bight, making a beautiful, though somewhat exposed, anchorage. Grazing cows might moo you to sleep.

APPROACHES. From Grand Bay to the west, pass to either side of Man of War Rock, marked by a black-red-black isolated danger spar. Deeper water lies to the north. Run in to the shore of Kennebecasis Island toward the bight above the word "Kennebecasis" on the chart.

ANCHORAGES, MOORINGS. Note the extent of the shoal area by the shore and anchor in a comfortable depth. The island provides a good lee in winds from the south and east but no protection if it blows from the west or north.

KENNEBECASIS BAY

The Indian word *Kennebecasis* means "Little Water" and has the same derivation as the word Kennebec, meaning "Big Water." Sure enough, the Kennebecasis River is not much more than a creek meandering through the Hampton Marsh. But the name hardly applies to the beautiful Kennebecasis Bay, which stretches from Grand Bay northeastward about 16 miles. Its shores are steep and mostly forested, and even though it is close to Saint John and the center of the river's boating activity, it is already off the beaten path of upriver-bound boats. Its mouth is flanked by the Royal Kennebeccasis Yacht Club to the south in Brothers Cove and Kennebecasis Island to the north. Bold Long Island lies dead center, with its brooding cliffs known as the "Minister's Face." At the head of navigation at The Neck, the bay gives way to the extensive Hampton Marsh.

A cable ferry crosses Kennebecasis Bay from Turners Flats, just east of Brothers Cove, to the Kingston Peninsula at McColgan Point. Another crosses between Gondola and Reeds Point. *Do not cross close ahead of cable ferries.* See *Westfield Beach* on page 424 for more details.

BROTHERS COVE

3 ★★★★

45° 18.25'N 066° 06.40'W

Canadian Charts: 4141-1

The clubhouse of the Royal Kennebeccasis Yacht Club.

Brothers Cove lies just east of Boars Head at the mouth of Kennebecasis Bay and harbors the Royal Kennebeccasis Yacht Club and its fleet. The Cove is broad and relatively shallow, but it is sheltered to the northeast by Indian and Goat Island. The prominent feature ashore is the yacht club, with its 70-foot-high observation tower that acts as one end of the starting line for club races. The club garnered an extra "c" in its name by virtue of a spelling error on its Royal Warrant in 1898. Guests are warmly welcomed here and are free to use the facilities. Fellow sailors will help visitors find moorings or slips and gladly offer pointers on favorite anchorages or navigating the river or the Reversing Falls. They may even offer you a ride to nearby markets or over to Saint John.

If you haven't checked in with customs, this is a convenient place to call *888-CANPASS*.

APPROACHES. From Grand Bay, turn northwestward into Kennebecasis Bay, rounding Boars Head and Ragged Point. Pass westward of Indian Island and follow the channel buoys in toward the club.

ANCHORAGES, MOORINGS. Anchoring in Brothers Cove is not recommended since the cove was once used for log booms, and the bottom is still fouled with sunken logs. The RKYC will happily help you find a mooring (no charge) or a slip.

FOR THE BOAT. *Royal Kennebeccasis Yacht Club (506-652-9430; www.rkyc.nb.ca).* Gas, diesel, water, ice, and pump-out facilities are available at the club's wharf. They accept credit cards. Electricity is available at their slips. They have a small marine railway, a hydraulic trailer, and a mast crane.

Millidgeville Marine (506-633-1454), right next to the club, stocks a large inventory of boat hardware, rigging, charts, and nautical books.

R.W. Ring Marine Services (506-633-0044). This boatyard is located directly across the street from the yacht club.

FOR THE CREW. The club's bar serves a light dinner menu. Crew are also welcome to use the showers in the restrooms. A pay phone is at the clubhouse. Most importantly, the club is full of kindred spirits.

A ten-minute, half-mile walk up Millidge Avenue brings you to a small shopping plaza with a bank and an elaborate convenience store at an Irving station with an on-site bakery and limited fresh fruits, vegetables, and groceries.

A club member may offer you a ride or you can take a bus to a large supermarket on Somerset Avenue. Head up Millidgeville Avenue then take a left on Somerset.

THINGS TO DO. Bus 15 leaves from the stop just past the intersection of Millidge Avenue and University Avenue, steps away from the plaza and the convenience store. The bus runs every hour at 20 minutes past the hour and will take you to Kings Square in Saint John. It also passes a shopping center with a supermarket and laundromat. Call *506-658-4700* for more transit information.

The huge Rockwood Park is a little over two miles away. Hiking and riding trails crisscross its woodsy interior, and it includes a zoo and an aquatic driving range. Get a ride from a club member or follow Millidge Avenue to a left on University Avenue. The park is dead ahead when the road comes to a T.

MILKISH CHANNEL

 ★★★

off Polly Sams Pt.: 45° 20.45'N 066° 07.40'W
CANADIAN CHARTS: **4141-1**

Milkish Channel wraps around Kennebecasis Island. From the southeastern entrance at McColgan Point all the way to the southwestern end at Grand Bay, good anchorages can be found along its unspoiled shores. The anchorages, though, are not where you would expect them. Keith Cove is shoal, Sea Dog Cove is fouled with logging debris, and the mouth of Milkish Creek at Bayswater is silted.

APPROACHES. Enter Milkish either at the southwest end from Grand Bay or at the southeast end from Kennebecasis Bay. Beware of Man of War Rock, marked by an obstruction buoy, and the cable ferry crossing from McColgan Point to Kennebecasis Island. See *Westfield Beach* on page 424 for more on cable ferries.

ANCHORAGES, MOORINGS. With the exceptions mentioned above, anchor anywhere along the shores where you can find protection from the prevailing winds. Depths are comfortable everywhere, and the bottom is mud and holds well. See also *Kennebecasis Island* entry on page 419.

DRURY COVE

 ★★★

outer cove: 45° 20.20'N 066° 02.40'W
CANADIAN CHARTS: **4141-1**

Drury Cove is a hole-in-the-wall with excellent protection on the southern shore of Kennebecasis Bay, just opposite the mouth of North Channel. The sides of the cove are steep and wooded, and the narrow entrance is well marked and easy.

Once inside, there is a surprise pocket to the west which is wonderfully snug but quite shallow, depending on the river level. A house and garage lie at the head of the inner cove, so you might feel a little like you've anchored in their swimming pool.

APPROACHES. The west side of the mouth of the cove is shoal, and fish pens may be moored off the western point of the entrance. Enter the inner harbor by passing between the red and green spar that mark the channel over a limestone bar. The reported depth is 8 feet.

ANCHORAGES, MOORINGS. Both parts of the cove are well protected. In fair weather, anchor in the outer part of the cove. Exposure is to the north.

If something foul is brewing, it might be worth trying to find the depth or the room to anchor in the inner cove.

VANCES BEACH

45° 21.80'N 066° 03.90'W
CANADIAN CHARTS: **4141-1**

At the southwestern end of Long Island, just before North Channel meets Kennebecasis Bay, there is a sandy beach along a pristine shore. This makes a wonderful daystop, particularly in the afternoon when the beach gets the sun. Anchor off the beach, or drop a stern anchor and take a bow line ashore.

The anchorage offers very little protection, and the water is deep right up to the shore, so it isn't recommended as an overnight anchorage in anything but the fairest weather.

CATHLINE COVE

 ★★★

45° 21.90'N 066° 03.10'W
CANADIAN CHARTS: **4141-1**

Cathline Cove lies at the southern end of Long Island. The cove has a broad mouth, open to the south, but the high flanks of the island give it good protection from the north and west. The shores are pristine.

APPROACHES. Don't go in too deep where the cove shoals.

ANCHORAGES, MOORINGS. Anchor anywhere in the cove in 8 to 18 feet.

RENFORTH and TORRYBURN COVE

2 ★★

45° 21.50'N 066° 01.00'W

CANADIAN CHARTS: 4141-1

Torryburn Cove and broader Renforth Cove lie on the southern shore of Kennebecasis Bay west of Torryburn Point at the town of Renforth. Torryburn Cove is filled with residential moorings. Kennebecasis Park, shown on the chart, is a housing development, not a recreational park. Renforth is an older bedroom community of Saint John, and its waterfront has a small yacht club, a granite wharf, a sandy beach, and an elegant small park with a lighthouse and a playground.

APPROACHES. Identify the little red-and-white lighthouse at the head of the cove. The approach is wide and free of obstructions.

ANCHORAGES, MOORINGS. There is plenty of room to anchor among the moored boats in Renforth Cove in depths from 10 to 20 feet with good holding in mud. Protection is good from the southerly quarters, but wide open to the north. The Renforth Boat Club may be able to direct you to a vacant mooring.

FOR THE BOAT. *Renforth Boat Club.* This club has no facilities other than a launching ramp and a small marine railway, but if you need help—or need help finding help—you will find it here.

FOR THE CREW. A small market with a pay phone is up the hill and across the railroad tracks.

THINGS TO DO. A community center and tennis courts are steps away from the waterfront to the right.

ROTHESAY

3 ★★★

45° 23.60'N 066° 00.20'W

CANADIAN CHARTS: 4141-1

Rothesay is directly opposite the brooding cliffs on Long Island known as the Minister's Face. Kennebecasis Bay tends to the north here, and Rothesay Cove is tucked into the east shore, filled with the fleet of the Rothesay Yacht Club. The town of Rothesay is nearby.

APPROACHES. From the south, pass outside Troop Reef, marked with a red conical buoy. Find lighted green buoy "AP1" and pass between it and red spar "AP2" to starboard.

ANCHORAGES, MOORINGS. Anchor among or outside the moored boats in 10 to 15 feet of water. There might be temporary dock space along the club float, to the left of the concrete public wharf, or a member might direct you to a vacant mooring. The unpleasant-sounding sewage lagoon shown on the chart at the head of the cove is not at all noticeable in the harbor. Protection is good from the east, but both southerlies and northerlies will find their way in, and westerlies make themselves right at home.

FOR THE BOAT. *Rothesay Yacht Club (506-847-7245)* has no guest moorings or any facilities for visiting yachtsmen other than its small clubhouse and pump-outs. They are somewhat off the beaten path of long-distance cruisers heading upriver, so visitors are welcomed with open arms. You will find no end of help and hospitality here. Ice, water, and power are available.

FOR THE CREW. The town of Rothesay, which has a market, banks, and a post office, is less than a mile away—a pleasant walk.

MATHER ISLAND

45° 25.00'N 066° 01.40'W
CANADIAN CHARTS: **4141-1**

Mather Island is almost a miniature replica of splendid Long Island, lying off Long's northeastern end and practically connected by a rock breakwater. The bight formed by the west side of Mather Island and the breakwater makes a nice anchorage with a backdrop of Mather's long sand beach and Long's wooded slopes. The island is private, and landing is not allowed.

APPROACHES. Round the northern end of Mather Island and follow its west shore into the bight. Or run down the fjord-like North Channel to the end of Long Island. There are no obstructions.

ANCHORAGES, MOORINGS. Anchor in the lee of Long and Mather Island in a comfortable depth.

MATTHEWS COVE

45° 27.10'N 065° 58.40'W
CANADIAN CHARTS: **4141-1**

Kennebecasis Bay narrows dramatically between Gondola and Reeds Point. Matthews Cove lies just beyond Gondola Point, extending almost half a mile southward. Most of it is too shallow to use, and a cable area occupies much of the rest. However, it is possible to get some shelter at the mouth of the cove, hard against the western shore.

APPROACHES. Pass Gondola Point and run toward Matthews Cove along the shore. Round the small nub on the west side of the entrance.

ANCHORAGES, MOORINGS. Anchor immediately past the nub as close to the western shore as possible. Depth will be about 7 feet.

FORRESTERS COVE

45° 27.45'N 065° 57.25'W
CANADIAN CHARTS: **4141-1**

Forresters Cove, just beyond Forresters Bluff, is the most popular anchorage at the northeastern end of Kennebecasis Bay. It is beautiful and large enough to hold an entire fleet, yet it's well protected from all directions but north.

APPROACHES. There are no obstructions. Round Forresters Bluff into the cove.

ANCHORAGES, MOORINGS. Work your way shoreward to a comfortable depth anywhere along the head of the cove. Holding is good in mud.

MEEHANS COVE

3 ★★

45° 27.50'N 065° 56.40'W
CANADIAN CHARTS: **4141-1**

Meehans Cove lies to the east of Forresters Cove, with equally good protection. It is less popular among cruisers, though, because of the campground and boat-launching ramp on the east shore.

APPROACHES. The entrance is broad and unobstructed.

ANCHORAGES, MOORINGS. Anchor anywhere the depth and protection from prevailing winds are adequate. The inner part of the cove at the head is too shallow to use.

KENNEBECASIS RIVER

off Olive Pt.: 45° 28.40'N 065° 57.10'W
CANADIAN CHARTS: **4141-1**

Beyond Perry Point, at the northeastern extremity of Kennebecasis Bay, the depths of the bay give way to the broad shallows of the Hampton Marsh. The Kennebecasis River winds through the shallows, with a very narrow buoyed channel and charted depths of 6 feet or greater, but it is subject to shifting bars and not recommended for cruising boats.

The Minister's Face on Long Island forms the backdrop to the wharf of the Renforth Yacht Club.

LONG REACH

Above Grand Bay, the Saint John River wraps around the end of the Kingston Peninsula. From Lands End to Westfield, the river tends northwest until it turns abruptly around Gregory Point. From there, Long Reach stretches almost 10 miles northeast in a straight and river-like fashion. Rarely, though, is Long Reach a reach. It is either a tacking marathon or a magnificent run or a lulling chug under power.

Here the river behaves like a river, with downriver currents of up to a couple of knots, moderated slightly by the flood through the Reversing Falls and amplified by the ebb. The intensity of the current can be predicted by the shape of the shoreline. Where a point projects into the river, the current will be greater as it is forced around it. Likewise, where the river is constricted, the current will be greater. Where the river broadens, the current will diminish, often dropping its suspended sediment and making the river more shoal.

Southerly winds will funnel up the Reach against the current, sometimes producing a steep chop or standing waves. Under these conditions, the east side of the Reach offers a little more lee. Turbulence is greatest off Oak Point and between Purdys and Belyeas Point.

Real harbors are a rarity here, though you can anchor almost anywhere along the riverbanks, choosing your location for protection from prevailing winds. The standard stopover is Whelpley Cove, behind Cantons Island. The indentation just upriver from Purdys Point, on the west shore, makes a nice fair-weather anchorage or daystop, with a beach and a high tree-lined bluff with a farm field beyond.

Upriver from the beautiful farm on Carters Point, the east shore becomes high and much less developed with small beaches at the toes of steep, forested hills. Once past Cantons Island, the main channel of the Reach becomes a labyrinth through low silt-islands—Grassy Island, Mistake Intervale, and Hog Island at the mouth of Belleisle Bay.

Watch for the spar buoys carefully, particularly the green spar at the submerged tip of Mistake Intervale, which is almost invisible against the trees along the east bank and surprisingly far—more than half a mile when the river is up—from the visible tip itself.

WESTFIELD BEACH
45° 21.00'N 066° 13.30'W

CANADIAN CHARTS: **4141-2**

Cable Ferries. A cable ferry crosses the river between Westfield Beach and Hardings Point a couple of miles upriver from Grand Bay and Lands End. If this is the first time you have crossed paths with a cable ferry on the Saint John River, use extra care. The ferry pulls itself along a submerged cable which spans the river. The cable ahead of the ferry is under tension and rises close to the surface of the water a long distance ahead of the ferry. To avoid being snagged, never pass close ahead of a cable ferry. Aft of the ferry the slack cable sinks back to the bottom.

The ferries also dock, load, and return across the river surprisingly quickly. This particular ferry averages a crossing every 5½ minutes.

The Canadian Coast Guard keeps a high-speed search-and-rescue boat in Westfield Beach or across the river at Hardings Point for rescues and assistance on the Saint John River. They can be reached on VHF channel 16. A general store is located just up from the western ferry landing.

GREGORYS POINT
45° 21.65'N 066° 13.40'W

CANADIAN CHARTS: **4141-2**

Gregorys Point marks an almost 90-degree turn in the Saint John River, above which Long Reach tends straight to the northeast. Minor turbulence may be found when the river current flows between Gregorys Point and Woodmans Point and then is forced to turn southeastward, particularly with a southerly wind. Note the large shoal area that extends from the mouth of the Nerepis River, marked by green spars "J29" and "J31," and the shoal area upriver from Gregorys Point, marked by red spars "J34" and "J36."

PURDYS AND BELYEAS POINT

 ★★

45° 22.75'N 066° 12.50'W

CANADIAN CHARTS: *4141-2*

At the southern end of Long Reach, about a mile above Gregorys Point, the Saint John River is constricted between Purdys and Belyeas Point, each marked by a flashing light.

The current increases here, particularly on the Belyeas Point side, so this area is prone to chop from opposing southerly winds.

Pleasant anchorages can be found on either side of either point, providing good protection in most winds.

APPROACHES. The river is broad and free of obstructions on both sides of the points.

ANCHORAGES, MOORINGS. The anchorage upriver from Purdys Point will provide the best protection from southerly winds. The point is low and sandy, but it quickly rises to a bluff of pines.

Work your way in to a comfortable depth. Holding ground is good in sand and mud. From here you can watch the passing river traffic and the setting sun.

CARTERS COVE

 ★★

45° 23.40'N 066° 11.40'W

CANADIAN CHARTS: *4141-2*

A beautiful farmhouse sits on Carters Point, on the eastern shore of Long Reach. Carters Cove and Mill Cove, two of the few coves on Long Reach, lie to either side of the point.

Despite their names, neither is much of a cove. Both are broad bights with protection only from the direction of the shore, and boulders are reported on the bottom of Mill Cove. Carters Cove, however, would make a pleasant daystop or fair-weather anchorage.

A wonderful photograph in *Rags and Strings,* the history of the Royal Kennebeccasis Yacht Club, shows the yacht club fleet at anchor in Carters Cove during the annual cruise of 1912, with a stern-wheeled steamer docked at the wharf.

WHELPLEY COVE

3 ★★★

45° 28.75'N 066° 07.25'W

CANADIAN CHARTS: *4141-2*

Whelpley Cove is a favorite anchorage, more because of its convenience and protection than its beauty. It can usually be reached as a first-night layover for boats bound upriver from Saint John, and it is the only true anchorage along the entire stretch of Long Reach—a fact that did not escape early settlers. The cove is protected by a high landmass to the west and by Cantons Island to the east. The west shore was the site of the first European settlement in New Brunswick, and the island was the site of the first religious services held along the Saint John River, in 1611.

Its beauty is marred somewhat by several buildings on its west shore and a gravel parking lot for the active camp on the island.

APPROACHES. Pass the small retreat community of Beulah Camp and follow the green spars north between the west shore and Cantons Island.

ANCHORAGES, MOORINGS. The usual anchorage is in the cove in the west shore. Avoid the underwater cables that lie north of the sandy spit on Cantons. Otherwise, anchor anywhere, in about 10 feet in good mud. Some bottom grass grows nearer the shore, so be sure your hook is well set.

OAK POINT

2 ★★

45° 30.65'N 066° 05.00'W

CANADIAN CHARTS: *4141-2*

Oak Point is a provincial park and campsite. There is a sandy beach on the northern side of the point, protected from the prevailing southerly winds. The lee also makes a good anchorage, out of the wind and current.

APPROACHES. One of the unique light towers of the Saint John River sits on the point itself. Due to the topography of the west shore, Long Reach is exposed to northwesterlies between Cantons Island and Oak Point. The river current is strong here, particularly on the ebb, and opposing southerlies can make the waves stand tall.

At the Isle of Pines, the river bifurcates around large, low Grassy Island and the extensive grassy bar downstream of it. Note that the small channel to the east of the bar is no longer buoyed. The main channel passes close by Oak Point and west of Grassy Island.

Follow the buoys of the main channel keeping Isle of Pines, Rush Island, and the extremity of Oak Point, marked by green spar "J55" (formerly "J57"), to port. Stay close to spar "JA" to keep off the shoals farther to the north, then turn westward just north of Oak Point. Be on the lookout for a buoyed rope that defines a swimming area off the campground or for any bathers in the water.

ANCHORAGES, MOORINGS. Anchor anywhere in the bight north of Oak Point, in about 10 feet.

THINGS TO DO. If you have children aboard, this is a perfect swimming beach, sandy and gradual, and there is an extensive playground ashore. Plan for a morning swim when it is warmed by the sun. For a cleansing of a different kind, visit the simple Anglican church.

MISTAKE COVE
45° 32.60'N 066° 03.00'W
CANADIAN CHARTS: **4141-2**

Mistake Cove is long, broad, and shallow, and Mistake Intervale, which forms the cove, is low and grassy, affording no protection from the prevailing winds. It is of little use to the cruising sailor.

Long, low Mistake Intervale is sometimes difficult to see. Occasionally a herd of black-and-white cows are grazing on it. Once past Oak Point, look for green spar "J63," which marks the shallow southern extremity of Mistake Intervale. It is difficult to see the thin green buoy against the green shore, and it lies almost a mile from the visual tip of the Intervale. Find it and keep it to port. Otherwise Mistake Intervale might be named for yours.

At anchor in Whelpley Cove.

BELLEISLE BAY

Belleisle Bay is a beautiful arm off Long Reach running 10 miles northeast through rolling hills of forest and farmland. The first half contains Kingston Creek and Jenkins Cove, two of the best anchorages on the entire Saint John River. A cable ferry crosses the midpoint of Belleisle Bay at Long Point, and the communities of Keirsteadville and Hatfield Point flank its head.

Belleisle Bay can be entered through shallow Shampers Cove, though this is not recommended, or through the narrow but deeper passage between Hog and Pig Island.

Low, grassy Hog Island is easily identified by three high poles supporting osprey nests. The Shampers Cove route is more difficult because the entire cove is shoal and grassy.

A stranger to this area can't help wondering which island is (or was) Belle Isle?

HOG ISLAND PASSAGE
N. of "C2": 45° 33.90'N 066° 01.10'W
CANADIAN CHARTS: **4141-2**

Hog Island Passage is the easiest approach to Belleisle Bay, though it is somewhat longer and considerably narrower than approaching through Shampers Cove. Its advantage is depth—something Shampers Cove is noticeably lacking—charted to be at least 11 feet in 1991.

Hog Island is low, flat, and grassy, with three large poles supporting osprey nests. Continue up the Saint John River west of Hog Island to the red buoy "C2." Turn hard to the east and then south, keeping the buoy to starboard and then following the buoyed channel between Hog and Pig Island. The buoys here have been changed to conform to red-right-returning as you approach Belleisle Bay.

SHAMPERS COVE

2 ★★

entrance off Gorhams Bluff:
45° 32.85'N 066° 01.60'W
CANADIAN CHARTS: 4141-2

Shampers Cove is almost uniformly shallow and broad and exposed. Protection from the southwest can be found in the lee of the northeast shore of Gorhams Bluff in front of an active riverfront farm where mooing and the rich smell of manure float across the water. You will be vulnerable from other directions.

As a passage to Belleisle Bay, Shampers Cove is not recommended. The buoyage in Shampers Cove changes with the bottom, and the latest version of the charts may not be correct. The passage north of Hog Island is preferable.

APPROACHES. There are still green spars, which you still keep to port, but the red spars may be missing. Once past the farm inside Gorhams Bluff, try to spot the green spar against the far bank to the southeast. It is almost invisible against the shore. This, paired with a red spar, marks your turn to the northeast and out of the cove. Note that this buoyage is likely to change with the shifting channel.

ANCHORAGES, MOORINGS. Unfortunately, Gorhams Creek is too shoal for most boats to use. Otherwise it would provide excellent shelter. Shampers Cove is shallow enough to anchor almost anywhere—if you don't run aground first. Gorhams Bluff provides protection from the southwest, and Shampers Bluff provides a slight lee to winds from the south.

KINGSTON CREEK

4 ★★★★

45° 32.15'N 066° 00.20'W
CANADIAN CHARTS: 4141-2

This is one of the best harbors on the entire Saint John River. Kingston Creek lies to the east of high and steep Shampers Bluff near the mouth of Belleisle Bay.

Protection is almost complete, and the anchorage is convenient both to boats in Belleisle Bay and to boats bound up or down the Saint John River. There is room for numerous boats, yet it still feels snug. The shores are mostly undeveloped.

APPROACHES. The entrance is easy. Head south into the cove. Fish traps are sometimes staked off the eastern shore, just across from high Shampers Bluff, but otherwise there are no obstructions.

Kingston Creek is navigable for several miles, but the upper portion is notorious for deadheads, as indicated on the chart, and not recommended for boats with deep drafts.

ANCHORAGES, MOORINGS. Anchor in the first bight to the west against the steep slopes of Shampers Bluff, before a fish trap off the west shore and before the first major narrowing of the creek. You'll find plenty of room, even if you find other boats. The water, however, is quite deep.

Protection is complete except from north winds or strong southerlies, which tend to funnel down the creek.

THINGS TO DO. From the cockpit, watch for ospreys riding the thermals above Shampers Bluff. Take the dingy up the creek or around to a small beach on the northeast tip of Shampers Bluff.

FAIRWINDS

1 ★★

45° 33.35'N 065° 58.70'W
CANADIAN CHARTS: 4141-2

The full-service Belleisle Bay Marina (*506-832-7373*) is about .75 miles east of Andersons Point on the opposite shore. Dockage and moorings are available, or you can anchor, but there is no protection here. They offer gas, diesel, water, showers, a pay phone, and pump-outs. Their Fairwinds Restaurant serves a full lunch and dinner menu.

GHOST ISLAND

45° 34.55'N 065° 58.30'W
CANADIAN CHARTS: *4141-2*

The bight behind Ghost Island on the east side of Belleisle Bay is less appealing than its name. The anchorage is exposed, and several cottages with swimming floats dominate the mainland shore.

A vertical, head-shaped rock stands on the southwest shore of Ghost Island, presumably giving the island its name. In case you can't pick out the face, somebody has painted eyes and a mouth on the rock, looking northeast down the length of the bay.

APPROACHES. Enter from either side of Ghost Island.

ANCHORAGES, MOORINGS. Anchor anywhere behind Ghost Island. The water is shallower to the west of the island and deeper to the north.

JENKINS COVE

45° 35.60'N 065° 57.50'W
CANADIAN CHARTS: *4141-2 (INSET B)*

Jenkins Cove is the jewel of Belleisle Bay. It is large, well protected, and unspoiled. The northern shores are the verdant slopes of working farms, the nearest being the Jenkins family farm for which the cove is named. The southern shores are rocky and wooded. There is ample room here for a fleet of boats, yet a sense of serenity and privacy remains.

Once in the cove, there is a marvelous pocket in the southern shore with perfect protection and room for just about one boat. Unfortunately, two cottages sit at the head of this inner cove, giving it the unwelcome feeling of private property.

APPROACHES. The entrance to Jenkins is wide and easy. Keep the red spar "CC2" to starboard and round westward into the cove.

ANCHORAGES, MOORINGS. Depth and holding ground are reasonable throughout the cove except near the two marshy areas along the north shore. The usual anchorage in the prevailing southerlies is along the southern shore, just off the round nub to the west of the inner pocket.

URQUHARTS COVE

45° 36.00'N 065° 56.70'W
CANADIAN CHARTS: *4141-2*

Urquharts Cove lies just to the east of Jenkins Cove. It is deeper and narrower than Jenkins Cove, but it is wide open to the south and ringed by cottages.

APPROACHES. Keep red spar "CC2" to port and head straight into the cove.

ANCHORAGES. Anchor at the head of the cove in a comfortable depth.

ERBS COVE

2 ★

45° 35.25'N 065° 55.70'W
CANADIAN CHARTS: *4141-2*

Erbs Cove lies on the opposite shore of Belleisle Bay from Jenkins and Urquharts Cove. The head of Erbs Cove is shallow, so the anchorage is more of a broad bight than a cove. Protection is adequate in southerly winds, but otherwise the anchorage is exposed.

APPROACHES. There are no obstructions.

ANCHORAGES, MOORINGS. The best protection will be found close to the southern shore, well to the right of the bridge that crosses Peters Brook at the head of the cove.

LONG POINT to HATFIELD POINT

45° 36.75'N 065° 54.50'W
CANADIAN CHARTS: *4141-2*

A cable ferry crosses Belleisle Bay between Long Point and the shore to the east of Earle Cove. Beyond the ferry, there are several miles of broad sailing before the bay becomes shoal at its head. The hamlet of Hatfield Point lies at the head of navigation at the end of a short and shallow buoyed channel. It features the 🛒 Bayview Grocery Store and 🍴 Brenda's Kitchen, in back, which serves burgers and platters of fries that will lower your waterline.

There are no specific anchorages in this end of the bay, only spectacular sailing.

EVANDALE to GAGETOWN

Above Hog Island, from Evandale to Gagetown, the Saint John River winds its way through rich farmland, past barns and small hamlets. Shrouded in a morning mist, it takes on a timeless, painterly quality. Graceful deciduous trees grow to the river's edge. The river divides around low, grassy islands and reconverges in a labyrinth of small channels and enchanting creeks.

Here, the definition of an anchorage takes on a new meaning. Good anchorages can be measured by how close to the cows you can tie your boat. In little Hole-in-the-Wall or in Colwells Creek, you can drop out a stern anchor and tie to the trees.

Narrow Washademoak Creek flows out of long and narrow Washademoak Lake at Lower Musquash Island. Historic Gagetown was founded by Loyalists on the banks of deep, narrow, and well-protected Gagetown Creek.

Note that the river charts are oriented in the direction of the river, not in a north-south direction. The river runs up chart 4142-1 from right to left.

TENNANTS COVE
45° 34.20'N 066° 00.70'W
CANADIAN CHARTS: 4141-2

Just north of Hog Island at the mouth of Belleisle Bay is a long cove between bold shores. Unfortunately the bottom contours aren't bold at all. There is hardly enough water in the inner part of the cove for a skiff, and entrance to it is blocked by a bar 3 feet under the surface. The outer part is broad and exposed and of little use to cruisers.

If you are determined to explore this cove, the best approach is via the channel leading off the channel to the north of Hog Island.

PALMER CREEK

4 ★★★

entrance: 45° 37.90'N 066° 05.40'W
CANADIAN CHARTS: 4142-1

Little River and Palmer Creek run into the west channel of the Saint John about 3.5 miles north of Evandale, just past the southern tip of Long Island. Palmer Creek makes a beautiful and well-protected anchorage with plenty of depth, but the entrance is overgrown and tricky and only has enough water for boats with 5 feet of draft of less. Local advice is recommended.

APPROACHES. The channel at the entrance lies slightly toward the south bank of the creek. Water on the north side of the entrance is foul with reeds. Proceed cautiously.

ANCHORAGES. The deep water in Palmer Creek is about 50 feet wide and runs upriver as far as the point where the creek bends to the north. Anchor anywhere in the deep water.

HOLE-IN-THE-WALL

3 ★★★★

45° 41.40'N 066° 04.65'W
CANADIAN CHARTS: 4142-1

Hole-in-the-Wall is just that—a slot not much wider than a couple of boat widths that cuts through Lower Musquash Island to Musquash Lake. Several boats can squeeze in here and tie to the trees along the banks in the kind of serene and peaceful setting that is unique to the Saint John River. Current of a knot or so runs out the creek, and it is reported at times to be buggy. Protection is good, though not as good from the south as the chart might indicate. The wind is only broken by a couple of stands of deciduous trees along the banks of that part of Lower Musquash.

APPROACHES. Approaching from the river channel east of Long Island, find green spar "D1" at the junction of Washademoak Creek and keep it to starboard. Once Lower Musquash Island is abeam to starboard, look for a small treed islet against the island's shore. Head in to Hole-in-the-Wall upriver of the islet, favoring the Lower Musquash Island shore and crossing a shoal area about 9 feet deep. Depth increases beyond the island.

Approaching from the river channel west of Long Island, continue upriver past Long and make a U-turn around the bifurcation buoy "JTE." Head downriver between Long and Lower Musquash Island until you see the small islet against the shore of Lower Musquash.

ANCHORAGES. Tuck the boat into the branches on the south shore and tie the bow and stern off to the trees.

WASHADEMOAK CREEK

45° 42.20'N 066° 03.70'W

CANADIAN CHARTS: **4142-1**

Washademoak Creek leads from the main channel of the Saint John northeastward into Washademoak Lake. Washademoak Creek widens slightly by Hog Island, at the intersection of Colwells Creek. A classic farmhouse commands the southern shore, and a couple of boats are moored in front.

APPROACHES. At Long Island, take the eastern channel to green spar "D1" at Washademoak Creek. Notice the stunning farm on the east bank and the small cattle barge they have to move cows to pasture on the islands. Keep the spar to port and follow Washademoak Creek to the intersection with Colwells Creek.

ANCHORAGES, MOORINGS. Anchor at the intersection of Colwells and Washademoak creeks near the moored boats. Protection is good in all directions except the north.

COLWELLS CREEK

4 ★★★★

45° 42.85'N 066° 04.40'W

CANADIAN CHARTS: **4142-1**

Colwells Creek is about as rural as you can get with your saltwater cruising boat. You will be anchored between the fields, farmhouse, and outbuildings of the Colwell farm and the treed banks of Lower Musquash Island. Depending on the river level, there may be small sandy riverbanks for swimming.

APPROACHES. From Washademoak Creek, turn left into Colwells. Pass the light tower on Lower Musquash and the entrance to Washademoak Lake. Then squeeze between spars "DA2" and "DA5."

It is also possible to pass between Lower and Upper Musquash islands and turn into Colwells Creek around black-and-yellow spar "JTK," though care should be exercised here due to shoaling. If coming downriver, do not attempt to approach through Lawson Passage, which is shoal and unbuoyed.

ANCHORAGES, MOORINGS. You can anchor anywhere in Colwells Creek. In the middle of the creek, the river current will keep you heading upriver. Along shore the current is diminished, and it is possible to tie off to the trees.

WASHADEMOAK LAKE

Washademoak Lake is long and narrow, beautiful and relatively unspoiled. The shores are rolling and wooded, cleared here and there for farms, with a few cottages along the shores.

From the entrance at Lower Musquash Island to the bridge that spans the lake at Cambridge Narrows (vertical clearance of 45 feet), the lake is more than 10 miles long. It is also narrow, with only one or two good anchorages. Past Appleby Point, the depth doesn't get below 15 feet, which can be somewhat disconcerting when you are running before a fresh southerly. Note that almost every buoy in the lake has been positioned to keep you in what little water there is.

Washademoak is for those who love to explore without a particular destination. For that reason, it is not on the itinerary of most upriver-bound boats, and it is left for those that do.

APPROACHES. Washademoak Lake is entered through a channel between Colwells Point and low Hog Island. The channel through the grassy, shallow western end of the lake is marked with red and green spar buoys and a range between the lights on Lower Musquash Island and on the lake's north shore opposite MacDonalds Point. Keep the red buoys to starboard as you enter, green to port.

CRAFTS COVE

45° 44.25'N 066° 00.15'W

CANADIAN CHARTS: *4142-3*

Crafts Cove is the first real cove on the east shore of Washademoak Lake. Belyeas Cove farther south is too shoal to be useful as an anchorage. Even Crafts Cove is shoal at its head and along its shores, making it difficult to tuck in deep enough for good protection from the south.

APPROACHES. Work your way slowly into the cove from the middle. The head shoals quickly.

ANCHORAGES, MOORINGS. Anchor in a comfortable depth as far in the cove as possible. Holding ground is good mud.

BIG COVE

45° 46.00'N 065° 57.70'W

CANADIAN CHARTS: *4142-3*

Big Cove is the largest feature of Washademoak Lake, lying midway down its length on the eastern shore. Like the name says, Big Cove is big—more than half a mile wide and a mile long, but it is shallow, giving way to marsh at its head beyond Birch Island. The shores are forested, with only a few cottages.

Good protection can be found in the bight behind Pine Island formed by The Bluff, although the tranquillity of the unspoiled shoreline is broken by a cottage there. During high waters, a single boat can work carefully in to the 8-foot spot off the eastern shore of The Bluff and anchor in a wilderness.

APPROACHES. Appleby Point can be recognized by its beautiful farm. Round the point and head to Barnes Point on the opposite shore of the cove. Favor that shore as you run into Big Cove, past Pine Island, before turning to the south and west.

There may be fish traps off the Barnes Point shore, and deep-draft vessels should proceed slowly with an eye on the depthsounder. Depths throughout the navigable part of Big Cove are around 10 feet.

To get to the east side of The Bluff, hug The Bluff and keep a careful lookout for deadheads.

At anchor in rural Colwells Creek.

ANCHORAGES, MOORINGS. The bottom in the bight is mud, with good holding and good protection from the south. Pine Island is too small to give much shelter from the north.

To the east of The Bluff, the bottom is mud, but it may be foul. Be sure to rig a tripline and watch for deadheads. Swinging room is limited by depth. Protection is good except from the northeast.

CAMBRIDGE NARROWS

45° 49.70'N 065° 57.15'W

CANADIAN CHARTS: *4142-3*

Cambridge Narrows is the head of navigation on Washademoak Lake for boats that need more vertical clearance than the 45 feet provided by the bridge that spans the lake there. Boats with shorter masts and less draft can explore a couple more miles of Washademoak Lake past the bridge or poke their way up through Narrow Piece, which just about says it all.

ANCHORAGES, MOORINGS. The *Sailing Directions* report good anchorage on the west bank, south of the bridge and the concrete wharf, though there is little protection from southerlies running down the lake. Coxs Cove, to the east, is broad and shallow and of limited use.

FOR THE CREW. The town of Cambridge Narrows has a pay phone, a general store, and the Village Inn (*506-488-3105*) with licensed dining. There is a liquor store at a gas station a half mile to the east of the bridge.

GAGETOWN

③ ★★★★★

45° 47.05′N 066° 08.56′W

CANADIAN CHARTS: 4142-1

Gagetown is a quintessential river town, laid out straight and true along the straight part of Gagetown Creek. Its modest Front Street parallels the river, with small shops and fine homes and an inn at the old steamer landing. French Acadians settled here in 1691 and called the town *Grimross*, Malecite for "place of settlement," only to be bodily unsettled by British forces led by Colonel Robert Monckton in their 1758 effort to drive French settlers from the lower Saint John River Valley. Renamed Gagetown after Colonel Thomas Gage, the original grantee, it was transformed again with the arrival of Empire Loyalists in 1783, who laid out the town on a block grid as other Loyalists had done in Saint John and Saint Andrews. A small and welcoming marina in the center of town has floats along the banks and moorings in the creek itself, so it is easy to land and explore.

The southern half of Gagetown Island, across the creek, is the Mount Ararat Wildlife Management Area.

APPROACHES. Approaching from downriver, be careful of the cable ferry that crosses from Scovil Point. Find the green-red-green bifurcation buoy "JU" at Gagetown Creek near Scovil and keep it to starboard as you enter Gagetown Creek. You will see the floats and moored boats of the marina ahead.

Current in the creek can run a couple of knots, and southerly winds can make it dance. Tides at Gagetown have a range of about 11 inches, and they are felt about 2¼ hours after the tide at Saint John.

ANCHORAGES, MOORINGS. Gagetown Marina has dockage and moorings. The marina floats are on the west bank, visible as soon as you are in Gagetown Creek. Use caution in maneuvering to the docks in the strong current. Dock port side to, heading into the current.

If you anchor, drop the hook above the marina to avoid snagging their ground tackle. Protection is good from east or west, but southerlies and northerlies funnel up and down Gagetown Creek.

FOR THE BOAT. *Gagetown Marina (Ch. 68; 506-488-1992; www.gagetownmarina.ca).* The marina has premium and regular gas, diesel, propane, water, pump-outs, and ice. Owner Ross Wetmore will pick up or deliver crew to the Fredericton airport, and he can arrange help in delivering your boat to the Saint John River or through the Reversing Falls from as far away as Lubec.

FOR THE CREW. The marina has showers and a small laundromat. K&W Quality Meats, a well-stocked market with good meat, beer, and liquor departments, is at the north end of town by a large Irving station.

Front Street, parrallel to the river, has a bank, a post office and a couple of eateries. Try the Tidewalker for lunches and the Old Boot Pub for beer and dinner.

THINGS TO DO. Gagetown publishes a nice map of the town that highlights the town's historical heritage. There is a beautiful Greek Revival courthouse and the Tilley House of 1786, now a museum, and there are numerous craftspeople in the area.

The Mount Ararat Wildlife Management Area has several hiking trails open to the public. Take your dinghy to one of the boat landings south of town and on the opposite bank.

The rural Gagetown courthouse overlooks a field and the town.

On the docks at Gagetown Marina with the Mount Ararat Wildlife Management Area on the opposite bank.

GRAND LAKE

Grand Lake is just that. Approached through the narrow little Jemseg River, the lake opens and expands before you, broad and inspiring. Grand Lake is the widest and largest of all the freshwater appendages to the Saint John River, created by a broad depression between subtle hills. It is distinguished not so much by its shoreline and anchorages as simply by water—open expanses of it—made all the grander in contrast to the rivers and creeks that lead to it.

For somebody arriving from the river, it is sheer joy to crank up the halyards, feel the press of wind on the sails, and let the boat run off. The lake's width of up to four miles allows for long lazy tacks on a warm summer afternoon, and its varied shoreline offers plenty to explore. Grand indeed.

Grand Lake is entered through the Jemseg River and then through the long buoyed channel known as the "Bush Track" at the lake's shallow southwest end. Douglas Harbour and the famed pocket called "The Bedroom" lie four miles away on the west shore. Grand Point Bar projects a surprising distance into the otherwise deep lake, but it is well marked if you don't forget to look for it. Three wide coves—Whites, Mill, and Wiggins Cove—and deeper Youngs Cove make up the east shore.

Past Cumberland Point, the lake divides into Cumberland Bay and Northeast Arm, each about 6 miles in length, tending, again, to the northeast. Feeding the lake are Newcastle Creek and the Salmon River from the north and the Indian, French, and Maquapit lakes through the microscopic Lower Thoroughfare near the Marsh Path.

The lake's charms have not gone unrecognized. Cottages rim the lake's shores, two provincial parks flank its waist, and pleasure boats of the inland type abound. But on Grand Lake, there's plenty of pleasure to go around.

The lake is notorious for thundershowers and summer squalls precipitated by its water mass. The accompanying winds kick up a steep chop. The relative shallowness of the lake enables large seas of 6 or 7 feet to build up quickly, particularly in strong northeasterlies, which blow down the lake and break the waves in the lake's shallow end by the Jemseg River. Under these conditions, navigating the Bush Track either to or from Grand Lake is not recommended.

JEMSEG RIVER
at bridge: 45° 49.60'N 066° 06.97'W
CANADIAN CHARTS: **4142-1, 4142-4**

It is almost inconceivable that the entire body of Grand Lake is drained by the tiny Jemseg River, or that the approach to the lake is through such a tiny back door. The Jemseg is typical of other creeks off the Saint John River, with trees overhanging the banks of low, flat islands and current of not much more than a couple of knots.

When heading to Grand Lake, enter the Jemseg west of Gagetown Island, between Huestis Island and Nevers Island. The entrance is marked by a bold daybeacon and light on the bank of Huestis Island and by green spar "E1" left to port. Deep-draft boats should not attempt approaching the Jemseg through shoal Raft Channel farther north.

The river is constricted at Dykeman Shoals but well marked. Here, the current is most pronounced and can slow progress. Keep green spars "E3," "E5," and "E7" to port, then red "E8" to starboard.

At Jemseg, the river is crossed by Route 2, the Trans-Canada Highway, on a bridge with 75 feet of vertical clearance. Sailors report that, as fate would usually have it, it is here, in the 66 feet of space between the pilings, that you will most likely encounter the "Chip Barge," loaded with ground pulp chips being towed downriver from—where else?—Chipman, at the head of Grand Lake, to the pulp mill at the Reversing Falls.

Just past the bridge, there is a small marina on the east bank with a fuel dock and a nearby restaurant off the highway.

The northern end of the Jemseg River breaks into the broad expanse of Grand Lake, but don't be deceived. This end of the lake is shoal, and the well-buoyed channel through the shallows known as the "Bush Track" must be followed before you can ease off on a run down the lake. The Bush Track ends with flashing green buoy "E43."

DOUGLAS HARBOUR and THE BEDROOM

4 ★★★★

45° 54.82'N 066° 05.75'W

CANADIAN CHARTS: 4142-4 (INSET)

Douglas Harbour is the first harbor on the west shore, and Grand Lake doesn't save the best for last. This is the center of boating activity on Grand Lake, yet Douglas Harbour remains remarkably low-key. Boats are quietly moored in the harbor and in the lovely arm to the east known as The Bedroom, but there is none of the bustle of a marina or a club. Most of these boats, in fact, belong to members of the Fredericton Yacht Club, who leave them on Grand Lake for the season. The club even owns the wharf, which, like so many others on the river, is a concrete pier edged with wood at a branch of a rural crossroads. Dinghies are tied to nearby trees, and you are as likely see a tractor drive by as a car.

The arm of Douglas Harbour to the west would offer absolute protection in a snug cove ringed by trees were it not for the sandbar one foot beneath the surface across its mouth.

The beautiful eastern arm, though, squeezed between an island and Earles Meadows, is as cozy as a bedroom—just as its name implies. The first half is occupied by moored boats, but you might find anchoring room at its head.

APPROACHES. The channel into Douglas Harbour is well marked and buoyed. At the mouth, the red spar off Earles Meadows is shown on the inset of chart 4142-4, but not on the chart itself.

The body of water to the west of the wharf, between the wharf and a sandy spit, has a charted depth of 10 feet. It is actually only 5 or less, but the bottom is soft. That, I'm embarrassed to say, is firsthand information.

When maneuvering in The Bedroom, pass between the moored boats rather than outside of them, or you may find the bottom.

ANCHORAGES. Do not attempt to anchor to the west of the wharf. Instead, anchor among the moored boats between the last two red buoys to the east of the channel to the wharf. Or anchor at the head of The Bedroom, past the mooring field, favoring the Earles Meadows side. Holding ground is sand and mud.

FOR THE CREW. From the wharf, take the road to the right to the local Irving station and Hunters One Stop convenience store for fuel (you'll need to lug it), ice (cubes only), limited groceries, a takeout, a pay phone, and a small laundromat.

WHITES COVE

2 ★★★

45° 51.90'N 066° 04.40'W

CANADIAN CHARTS: 4142-4

Whites Cove is a large cove on the east shore of Grand Lake, due south of Douglas Harbour. It is broad but pretty, ringed by scattered summer cottages. Shoaling near its head prevents getting close in to shore, so it is still exposed to southerly gusts, and it is wide open to the north.

APPROACHES. The mouth of the cove is defined by Taylor and Ferris Points. Shoal areas off both points are marked with red spars. Otherwise, there are no obstructions.

ANCHORAGES, MOORINGS. Anchor anywhere in the cove that provides shelter. The spot shown on the chart by an anchor provides the best protection from southerly winds.

MILL COVE

2 ★★

45° 52.85'N 066° 01.50'W

CANADIAN CHARTS: 4142-4

Mill Cove is the elbow of Grand Lake, on the east shore opposite Grand Point Bar. It is a broad and exposed anchorage. Lakeside Provincial Park is on the east shore of the cove, near its mouth.

APPROACHES. The approach is unobstructed.

ANCHORAGES, MOORINGS. Shoaling around the perimeter of the cove prevents anchoring close in to shore. Anchor anywhere in a comfortable depth for the best protection from prevailing winds. The most exposure is from the north and west.

FANJOYS POINT

2 ★★ 🍴

45° 54.50'N 066° 00.80'W

CANADIAN CHARTS: *4142-4*

Between Fanjoys Point and Branscombe Point to the east, there is a small bight by a public wharf that provides adequate shelter in settled weather. The Trans-Canada Highway runs right along its southwestern shore, so it feels a little more like a truck stop than an anchorage.

APPROACHES. Enter between the shoal area off Fanjoys Point, marked by red spars "EC2" and "EC4," and the shoal area off Branscombe Point. The water is shoal from the end of the dock westward toward Fanjoys Point, as shown on the chart. Do not attempt to dock at the wharf. An obstruction is reported a couple of feet off its face.

ANCHORAGES, MOORINGS. Anchor anywhere in the cove. Protection is adequate in light breezes from the south or in fair weather.

FOR THE CREW. 🍴 A restaurant with a pay phone is reported .1 miles up the highway to the left.

WIGGINS COVE

2 ★★

45° 55.30'N 065° 59.70'W

CANADIAN CHARTS: *4142-4*

Wiggins lies wide open on the east shore of Grand Lake. The head is shoal a good distance out, protection is minimal, and the Trans-Canada Highway is within earshot.

APPROACHES. The shoal area off Wiggins Point is not marked. Otherwise, the approach is clear.

ANCHORAGES, MOORINGS. The best protection is at the southern end of the shoal water at the head of the cove, though this is not very good.

YOUNGS COVE

3 ★★★

45° 57.25'N 065° 58.10'W

CANADIAN CHARTS: *4142-4*

Youngs Cove is the best of the coves on the east shore of Grand Lake proper. It has a tranquil, undeveloped feel, even though several cottages have frontage on the cove, and the Trans-Canada Highway whishes nearby. The sun sets over the mouth of the cove.

APPROACHES. Wonderfully-named Holms Lump at the mouth of the cove can be passed to either side, though passing to the north is easier. Shoals off Foshay Beach are marked with green spar "EE3," which should be left to port. Be on the lookout for fish traps that may be set off the hooked point on the southern shore.

ANCHORAGES, MOORINGS. The anchor shown on the chart in the center of the cove is one option. A more popular spot is against the cove's southern shore, past the hooked point, as near the head of the cove as possible. Protection is better here, though it is closer to a pair of cottages ashore and to the highway. Another option is to work into the mini-cove east of Foshay Beach. The bottom is good mud.

CUMBERLAND BAY

3 ★★★

sharp cove: 46° 00.30'N 065° 56.20'W
Wasson Brook: 46° 00.40'N 065° 54.50'W
Cumberland Bay: 46° 01.80'N 065° 54.15'W

CANADIAN CHARTS: *4142-4*

Cumberland Bay and Northeast Arm make up the northern end of Grand Lake. Both are long and tend to the northeast. Cumberland Bay begins at Cumberland Point, which is like a two-faced coin. A beautiful farm sits high on the hill facing south, its metal roofs glinting in the sun, visible from almost anywhere in the southern half of Grand Lake. The north side of the point is lined with trailers.

The bay itself is wide and unobstructed enough for pleasant sailing, and its shores are largely undeveloped. Anchorages can be found in the small, sharp cove on the southern shore, in the cove by Wasson Brook, or at the head of the bay by Cumberland Bay Stream.

APPROACHES. Once around Cumberland Point, there are no obstructions in the Bay. When heading into the cove by Cumberland Bay Stream, favor the north shore to avoid the shoal water to the south.

ANCHORAGES, MOORINGS. The first, sharp cove on the southern shore is shared on either side by cottages and by loud bullfrogs that serenade at dusk. Anchor just inside the cove in 8 to 9 feet. Power lines that are not shown on the chart cross the shoal head of the cove.

The cove by Wasson Brook is broad, and there are several cottages on its smooth shores. Exposure is from the northwest. Anchor anywhere in a comfortable depth except in the shoals by the eastern shore.

The cove at the head of Cumberland Bay is less developed and offers the best protection from winds from the north. Anchor just inside the hook on the north shore or farther in toward the mouth of Cumberland Bay Stream. Don't be tempted to go in too far. The deepest water is toward the north shore.

GOAT ISLAND

46° 00.75′N 066° 00.80′W

CANADIAN CHARTS: **4142-4**

Sandy Goat Island lies in the center of Grand Lake at the north end, marked by a flashing light and ringed by sandy beaches. This is a popular day-stop for picnics and sunbathing. The beach is steep enough on the north side that deep-draft boats can approach it from the north, drop a stern anchor, and nose the bow onto the beach.

Note the unmarked, isolated boulder to the southwest of Goat Island. More rocks lie on the unmarked bar extending from the mainland toward Goat Island. While it is possible to pass between them in about 7 feet of water, it is not recommended for strangers. Instead, stay outside green lighted buoy "E75" off the east end of Goat.

FLOWERS COVE

46° 02.00′N 066° 02.10′W

CANADIAN CHARTS: **4142-4**

Flowers Cove lies on the west shore past Goat Island at the narrowest part of Grand Lake. It consists of two coves, one north of the other. Each offers excellent protection.

Cottages in the area called Wuhrs Beach line the southernmost cove (one boasts a possibly confusing ornamental lighthouse), but the cove has an inner pocket with deep water and absolute protection. The mouth of the pocket is spanned by almost invisible power lines, which must have been taken down regularly enough by masts to be raised from 25 feet of clearance to 54.

The entrance to the cove to the north is marked by prominent brown slag heaps from a now-defunct open-faced coal mining operation. The heaps sound like much more of an eyesore than they are. Inside is a totally unspoiled cove ringed by tall pines to the south and treed slopes to the north, with 10 to 12 feet of water and room for plenty of boats.

APPROACHES. Approach both inner harbors with care. The wires spanning the southern harbor run from a pole behind a brown cottage on the southern point to a pole by a white cottage on the northern point.

The chart shows a pair of red and green buoys at the mouth of the northern cove marking a pile of rocks jutting northward from the southern point. When we visited, the buoys weren't there, but we assumed the rocks still were. Make your entrance slowly, keeping off the southern point by about two thirds of the mouth's width, favoring the slag side to the north. Ten feet of depth is reported at the entrance, but it seems to be less, and it will vary with the height of the lake.

ANCHORAGES, MOORINGS. Once inside, anchor anywhere in 10 to 12 feet.

THINGS TO DO. Row the dingy to the slag piles, where there is an old set of wooden steps on the bank. Climb the piles for a wonderful bird's-eye view of the cove and out to Goat Island on the lake.

PRINTZ COVE

off entrance: 46° 02.35′N 066° 00.95′W

CANADIAN CHARTS: *4142-4*

Printz Cove, to the north of Flowers Cove, would be a perfect gunkhole—completely protected, with reported depths of 6 and 7 feet—were it not for the cottages that dominate the shore. Instead, it feels more like a driveway.

NEWCASTLE CREEK

46° 03.35′N 066° 00.25′W

CANADIAN CHARTS: *4142-4*

The cove at Newcastle Creek provides good shelter from northerly and westerly winds, and it even offers some protection from the south, but the otherwise inviting cove is overshadowed by a large power plant on the north shore.

APPROACHES. The power plant's red-and-white stack can be seen from a long distance away. Pass close to red spar "EM2," keeping it to starboard. Stay to the south of a line from the end of the southern power plant pier (which has a light) and the north face of the public wharf to avoid submerged anchor pilings.

ANCHORAGES, MOORINGS. Anchor anywhere in the cove south of the line mentioned above. The wharf has a reported depth of 10 feet at its face.

NORTHEAST ARM

behind Barton Is.: 46° 03.85′N 065° 58.00′W

CANADIAN CHARTS: *4142-4*

The Northeast Arm of Grand Lake is broad and shallow, and its shores are low. The Salmon River flows into its head, winding between the bushy gravel bars that make up Hawkes, Indian, and Baileys points. The best anchorage is behind Barton Island. Other possibilities are to anchor in the mouth of either of the two sharp indentations in the southern shore on either side of Barton, though these are shallow. The head of the arm becomes quite shoal, so protection is hard to find.

APPROACHES. Northeast Arm is free of obstructions despite its shallowness. Keep Barton Island and lighted red "E86" to starboard and round in behind the island.

To approach the indentation in the southern shore farther east, you will need to estimate the extent of the shoal area off Hawkes Point, since it is not buoyed. Keep well to the south and run into the cove.

ANCHORAGES, MOORINGS. Depths are good for anchoring everywhere, so find the best protection from the prevailing winds. Protection is good from the south and west behind Barton Island and in the eastern indentation. The western indentation has more exposure to the west.

CHIPMAN and the SALMON RIVER

entrance off Hawkes Pt.:
46° 04.70′N 065° 55.32′W

CANADIAN CHARTS: *4142-4*

The Salmon River is navigable to deep-draft cruising boats as far as Chipman, though the sight of one there is a rarity. For almost 10 miles, it passes through open-face coal mines and scrub shores. The river is used mostly by the "Chip Barge" that gets towed from a pulp chipper in Chipman all the way to the mill in Saint John at the Reversing Falls. Because of this traffic and the narrowness of the channel, you should not anchor in the Salmon River.

A relatively deep spur of the Salmon River (10 to 15 feet) winds through Salmon Bay, but it is unmarked and does not come close to the low shores, so the bay offers no protection as an anchorage.

Chipman, if you are intrepid enough to explore that far, has excellent protection at its public dock with as much as 14 feet alongside and pump-out facilities. The town has groceries, a bank, and a restaurant.

MAQUAPIT, FRENCH, and INDIAN LAKE

Thoroughfare entrance:
45° 52.17'N 066° 10.00'W

CANADIAN CHARTS: **4142-4**

Three interconnected lakes—Maquapit, French, and Indian—are connected to the southwestern corner of Grand Lake by the torturously narrow Lower Thoroughfare, beginning at Indian Point. All three lakes are shallow. Controlling depth for all of them appears to be 6 feet at the beginning of the Lower Thoroughfare. Channels marked by bushes lead through the shoal ends of the lakes and from one lake to another. It is conceivable that these lakes could be explored with a deep-draft cruising boat, though they are better suited to a canoe.

GAGETOWN to FREDERICTON

Above Gagetown, the character of the Saint John River changes again. The north shore is low and swampy, with names like The Intervale, Ash Swamp, Broad Meadows, and Beaver Heath. Most of these drain northward through the sprawl of Indian, French, and Maquapit Lake. The Trans-Canada Highway, Route 2, runs along most of the north bank.

Anchorages are limited to tucks behind the low, grassy river islands or to the boat clubs and marinas of Oromocto and Fredericton. Above Oromocto, the draft that boats can carry farther upriver is limited to less than 7 feet by Oromocto Shoals.

THATCH ISLAND

45° 49.50'N 066° 09.10'W

CANADIAN CHARTS: **4142-1**

Low, grassy Thatch Island lies north of Grimross Island, just upriver from Gagetown Island. The *Sailing Directions* suggest a good anchorage at its southeast end, though the approach is limited to boats with a draft of less than 5 feet by the shoals to the south.

RAM ISLAND

off entrance: 45° 51.62'N 066° 15.35'W

CANADIAN CHARTS: **4142-2**

Ram Island is long and narrow, lying close to the river's southern bank, with just a few trees at its eastern end. The narrow slot between the island and the mainland is one of the better upriver anchorages, though protection is not as good as it might appear because Ram is so low.

APPROACHES. The slot between Ram and the mainland must be approached from the east end, near the mouth of Swan Creek, since the opposite end is shoal. Controlling depth at the entrance is 7 feet. Water is shoal directly east of Ram's southern tip and due east of the mouth of the creek. Pass between the shoal areas by approaching from the east heading toward the west bank of the entrance to Swan Creek. Once past the tip of Ram, turn to the northwest and stay in midchannel.

ANCHORAGES, MOORINGS. Anchor anywhere in the slot in about 10 feet of water. You may want to use a stern anchor to keep aligned.

OX ISLAND CHANNEL

45° 52.32'N 066° 19.00'W

CANADIAN CHARTS: **4142-2**

The Ox Island Channel passes to the south of Ox and Gilbert Island. It is shallower than Sheffield Channel to the north, but more direct. Heading upriver, find the lighted red-green-red bifurcation buoy and keep it to starboard. Avoid the Ox Island Bar by staying close to the three red spars and keeping them to starboard. Look for a lone red spar past the western tip of Ox Island and also keep it to starboard.

OROMOCTO

4 ★★★

45° 51.32'N 066° 28.50'W

CANADIAN CHARTS: **4142-2** (INSET)

The town of Oromocto is the location of one of Canada's largest military training bases. The surrounding countryside is flat, partially wooded and partially farmed. A major shopping center is located right on the bank of the river next to the long floats of the Oromocto Boat Club, making this one of the most convenient ports for provisioning covered in this guide.

APPROACHES. The Oromocto Boat Club is located in the Oromocto River near its mouth. Pass under the bridge (65-foot clearance) at Burton, and proceed upriver south of Oromocto Island. A green-red-green bifurcation buoy marks shoals at the eastern end of Thatch Island. Keep it well to starboard and turn southward into the Oromocto River. The long club floats will be obvious along the left bank.

Water levels at Oromocto are influenced more by the releases at the Mactaquac Dam than by the tide. Random fluctuations have a range of about a foot.

ANCHORAGES, MOORINGS. The Oromocto Boat Club has room for visiting boats along its floats with 8 to 10 feet of depth.

FOR THE BOAT. *Oromocto Boat Club (Town Hall for information: 506-357-4400; Paul McCleod, 506-446-6135).* The boat club pumps gas and water in, and sewage out. They sell ice, and a pay phone is ashore.

FOR THE CREW. From the boat club, the shopping center is a short walk across the grassy Sir Douglas Hazen Park. It has a major grocery store and pharmacy, a liquor store, banks, and a restaurant and the Golden Arches. Buses run daily to Saint John and Fredericton, and to the Fredericton Airport.

FREDERICTON

2 ★★★★

Fredericton YC: 45° 55.95'N 066° 37.25'W

CANADIAN CHARTS: **4142-2** (INSET)

Emergency: 911
Hospital: 506-452-5400
Weather: 506-446-6244

Fredericton, lying some 63 miles up the Saint John River from the Bay of Fundy, is the capital of the province of New Brunswick. The city is built on both banks of the river, which is spanned by the Princess Margaret Bridge, with 84-foot clearance at its center, and by the railroad bridge, whose 25-foot clearance marks the head of navigation for most masted vessels. Unfortunately for sailors, the downtown area and the city's wharf lie upriver from the railroad bridge.

Fredericton came under English control in 1759. Fredrick Town was named after the son of King George III. Beginning in 1783, the United Empire Loyalists did their usual street-straightening and urban planning. Now it is home to the New Brunswick Legislature, the University of New Brunswick, St. Thomas University, a college of craft and design, and forestry and agriculture centers, not to mention the New Brunswick Country Music Hall of Fame.

APPROACHES. Once past the Oromocto Shoals, the upper Saint John is mostly free of obstacles and clearly marked. Power lines with a vertical clearance of 75 feet cross the river at Lower Saint Marys. From there to the Princess Margaret Bridge, shoals and logging cribs foul the north shore opposite the Fredericton Yacht Club, but they are well marked.

On the docks at the Oromocto Boat Club.

Fredericton's railroad bridge limits its waterfront to vessels with less than 25 feet of vertical clearance.

Princess Margaret Bridge is fixed with a center span that has 84 feet of clearance.

Due to Fredericton's proximity to the Mactaquac Dam, water levels can fluctuate by as much as 2 feet from the random opening of the dam.

ANCHORAGES, MOORINGS. Sailboats with masts taller than 25 feet will need to stop at the Fredericton Yacht Club, on the south bank about a half mile south of the Princess Margaret Bridge. The yacht club has moorings available for visitors. Boats that can pass beneath the railroad bridge (25 feet) can dock right downtown at the Regent Street Wharf.

FOR THE BOAT. *Fredericton Yacht Club (www.fyc.ca).* Identifiable by the fleet moored along the west bank, the club has a small float with water and can provide guests with a vacant mooring. Gas, groceries, a pay phone, and an excellent Chinese restaurant are nearby. Downtown Fredericton is about 2 miles away.

Regent Street Wharf (Port Warden, Ch. 68; 506-455-1445). The Regent Street wharf lies a block away from the downtown district of Fredericton on the west bank of the Saint John, but its access is limited to boats requiring less than the 25 feet of vertical clearance provided by the railroad bridge to the south. The Port Warden can be found here in the summer, and visitors should check in. The large public float has 10 feet of water alongside. Moorings, mostly seasonal, lie off the float.

FOR THE CREW. Fredericton has numerous markets, restaurants, and stores, as well as the large Dr. Everett Chalmers Hospital (*506-452-5400*). On Saturday mornings, farmers set up a market three blocks away from the wharf. Of the several laundromats in town, the best-named is the Spin and Grin on Smythe Street. A liquor store is located on King Street, a block up from the waterfront.

THINGS TO DO. Tourist information (*506-460-2041*) is available at City Hall, two blocks west of the wharf on Queen Street. The park by the Regent Street Wharf is called The Green. Children can climb to the top of a river lighthouse in the little lighthouse museum, which also rents bikes.

Next to The Green is Officers' Square, a military complex with a museum and a changing-of-the guard twice daily. Free open-air concerts are performed here on Tuesday and Thursday evenings in July and August. Behind the square is the large public library. The Fredericton Playhouse is one block from the wharf.

New Brunswick's largest collection of Canadian and British paintings can be seen daily at The Beaverbrook Art Gallery, across the street from the copper-roofed Legislative Assembly Building, a couple of blocks east on Queen Street. The Legislative Building holds, among other things, a complete set of hand-colored copper engravings of the *Birds of America* by John James Audubon.

If you would like a guided walking tour of the town, go to City Hall and ask for a Calithumpian.

APPENDIX A
EMERGENCY NUMBERS

SEARCH AND RESCUE	VHF	TELEPHONE
U.S COAST GUARD		
Boston	16	617-565-9200
Portland	16	207-799-1680
Southwest Harbor	16	207-244-5121
CANADIAN COAST GUARD		0 ask for search and rescue
Saint John, N.B.	16	800-565-1582, 902-427-8200
EMERGENCY		911
Maine State Police		800-482-0730
Poison Control Center		800-442-6305
COMMERCIAL TOWING		
TowBoat/U.S. (Portsmouth)	09, 16	603-436-0915
Sea Tow (Portland, midcoast)	09, 16	207-772-6724
TowBoat/U.S. (Portland, Boothbay, Castine)	09, 16	207-460-5866
OTHER		
Waterway Watch (to report suspicious activity)		877-24WATCH
Department of Marine Resources Marine Patrol		207-624-6571
Red tide hotline		800-232-4733, 207-633-9571
Marine mammal stranding hotline		800-532-9551

APPENDIX B
ENVIRONMENTAL ADVOCACY GROUPS

The preservation of the coast of Maine is no accident. Thousands of acres have been conserved by individuals, local land trusts, and environmental organizations. Often these efforts are collaborative. Some nonprofit organizations, such as The Nature Conservancy, purchase islands and shorelines outright. Others, such as the Maine Coast Heritage Trust, collaborate with federal or state agencies or local land trusts, by making initial purchases for them or creating conservation easements. If you would like to support the work of these groups or become more closely involved yourself, here are some of the major nonprofit environmental organizations active throughout the state.

ORGANIZATION	ADDRESS	WEBSITE	TELEPHONE
Maine Audubon Society	20 Gisland Farm Road, Falmouth, ME 04105	www.maineaudubon.org	207-781-2330
Maine Coast Heritage Trust	1 Bowdoin Mill Island, Suite 201, Topsham, ME 04662	www.mcht.org	207-276-5156
Maine Island Trail Association	58 Fore Street, Suite 30-3, Portland, ME 04101-4849	www.mita.org	207-761-8225
The Nature Conservancy	14 Maine Street, Suite 401, Brunswick, ME 04011	www.nature.org	207-729-5181
Island Institute	386 Main Street, Rockland, ME 04841	www.islandinstitute.org	207-594-9202

APPENDIX C
MARINE FACILITIES

PUMP-OUTS

Federal law prohibits the discharge of untreated sewage from boats with permanently installed heads within three miles of the coast. For guidelines on managing onboard human waste, see *Introduction*, page 9. In the past, considerable stretches of the Maine coast lacked pump-out facilities, but recent laws have declared Casco Bay a no-discharge zone and have mandated marinas of a certain size to offer pump-out service. New no-discharge zones will be created in subsequent years.

Use of holding tanks is also mandatory on the Saint John River, but nearly every marine facility there offers pump-out service, often for free.

Disposing of your waste, however, is not the primary business of marinas, boatyards, or fuel docks, so don't expect it to be their top priority. We recommend being a good customer first, then a customer in need of a pump-out. Plan on combining your fueling up with your pumping out, and call ahead for convenient times. If you discover

SOUTHERN COAST	FACILITY	VHF	TELEPHONE	⛽	⛽	🚿	⚓	🔧	🛠	🔩	🧺	🎣	PAGE
Little Harbor	Wentworth-by-the-Sea	09, 68	603-433-5050	g	d		•	•	•		•	•	35
Pepperrell Cove	Kittery Town Dock	09	603-439-0912	g	d	•		•					37
Back Channel	Kittery Point Yacht Yard	09	207-439-9582			•	•	•	•	•		•	38
Portsmouth, east	Portsmouth Yacht Club	09	603-436-9877	g	d	•		•				•	39
	Kittery Point Yacht Club	09	603-436-9303			•		•					39
	Prescott Park Wharf		603-431-8748					•					39
	Kittery Landing Marina	09	207-439-1661					•	•		•	•	39
	Fisherman's Coop			g	d								39
Portsmouth, west	Harbor Place Marina		603-436-0915					•	•				39
	Harbor Towing and Pumpout	09, 16	603-436-0915							•			40
	Badgers Island Marina	09	207-439-3820					•	•		•	•	39
	Island Marine Service	09	207-439-3810							•			40
Piscataqua River	Patten Yacht Yard	16, 68	207-439-3967							•			41
	Great Bay Marine	68	603-436-5299	g	d	•		•	•	•	•	•	41
York Harbor	Town of York, harbormaster	09, 16	207-363-0433				•						42
	Agamenticus Yacht Club		207-363-8510						•				42
	Donnell's Marine		207-363-4308				•						42
	York Harbor Marine Service	09, 16	207-363-3602	g	d	•	•	•		•	•	•	42
Perkins Cove, Ogunquit	Town dock, harbormaster	72	207-646-2136				•						46
Wells Harbor	Town dock, harbormaster	09, 16	207-646-3236				•	•	•				47
	Webhannet River Boatyard		207-646-9649							•			47
Kennebunk River	Nonantum Hotel and Marina		207-967-4050				•						48
	Chick's Marina	09, 68	207-967-2782	g	d		•	•	•	•	•	•	48
	Kennebunkport Marine	09	207-967-3411			o	•	•	•		•		48
	Yachtsman's Lodge	09, 16	207-967-2511	g	d	o	•	•			•		48
	Arundel Yacht Club	09	207-967-3060				•	•					49
	Performance Marine	09, 16	207-967-5550	g	d	o							49
Cape Porpoise	*Town dock, harbormaster*	*16*	*207-967-5040*	*g*	*d*		•						52
Biddeford Pool	Biddeford Pool Yacht Club	16, 68	207-282-0485	g	d	•		•					55
Saco River	Camp Ellis Wharf	16, 78	207-284-6288	g	d	1				•			57
	Marston's Marina	09	207-283-3727	g				•	•				57
	Saco Yacht Club	16, 69	207-282-9893			•	o	•					57
	Rumery's Boat Yard	09, 68	207-282-0408			o	o	•		•		•	57
Prouts Neck	Prouts Neck Yacht Club	09, 16	207-883-9362			•		•					59
SOUTHERN COAST	FACILITY	VHF	TELEPHONE	⛽	⛽	🚿	⚓	🔧	🛠	🔩	🧺	🎣	PAGE

A CRUISING GUIDE TO THE MAINE COAST

pump-out facilities that are not in working order, please report them to the boat pump-out coordinator of the Maine Department of Environmental Protection at *800-452-1942*.

KEY TO TABLES
g gas
d diesel
o when available
t temporary dockage
italics primarily commercial

CASCO BAY	FACILITY	VHF	TELEPHONE										PAGE
Portland	Town docks, harbormaster	09, 16	207-772-8121				t						66
	Portland Yacht Services	09, 16	207-774-1067			●		●	●	●	●		66
	DiMillo's Marina	09, 16	207-773-7632	g	d			●	●	●	●	●	66
	Gowen, Inc.	09, 16	207-773-1761							●			
	Maine Yacht Center	09	207-842-9000	g	d			●	●	●	●	●	66
	Baykeeper pump-out	09	207-776-0136							●			66
South Portland	Spring Point Marina	09	207-767-3254	g	d			●	●	●	●	●	69
	Sunset Marina	09	207-767-4729	g	d			●	●		●	●	69
	Aspasia Marina		207-767-1914							●			69
	Centerboard Yacht Club	68	207-799-7084			●					●		69
	South Port Marine	09	207-799-8191	g	d			●	●	●	●	●	69
	Reo Marine		207-767-5219							●			70
Peaks Island	Peaks Island Marina		207-766-5783	g		●		●			●		72
Great Diamond Island	Diamond Cove Marina	09	207-766-5850					●	●	●	●	●	73
Falmouth Foreside	Handy Boat Service	09	207-781-5110	g	d	●	o	●	●	●			75
	Portland Yacht Club	68	207-781-9820				o		●		●	●	75
	Town dock, harbormaster	09	207-7817317				t						75
	Falmouth Pumpout Boat	09	207-781-2300							●			75
Great Chebeague Island	Chebeague Island Boatyard		207-846-4146	g	d	●		●					78
Cliff Island	Fisherman's Wharf	72	207-766-2046	g	d								79
Royal River, Yarmouth	Royal River Boat Yard		207-846-9577	g	d			●	●	●	●		83
	Yankee Marina	09	207-846-4326					●	●	●		●	83
	Yarmouth Boat Yard	09	207-846-9050				o	●	●				83
Harraseeket River	Harraseeket Yacht Club		207-865-4949				o		●				85
	Strouts Point Wharf Co.	09, 16	207-865-3899	g	d	●		●	●	●		●	85
	Town landing, harbormaster	09	207-865-4546				t						85
	Brewer's S. Freeport Marine	09, 16	207-865-3181	g	d	●		●	●	●	●	●	85
Merepoint Bay	Paul's Marina	09	207-729-3067	g				●		●	●		86
Potts Harbor	Dolphin Marine Service		207-833-6000	g	d			●	●		●	●	88
	Finest Kind		207-833-6885							●			88
Orrs-Bailey Island	Orrs-Bailey Yacht Club							●					90
	Cook's Lobster House		207-833-2818	g	d			●	●				90
Orrs Cove, Quahog Bay	Great Island Boat Yard	09	207-729-1639	g	d	●	o	●	●	●		●	95
Cundys Harbor	Watson's General Store		207-725-7794	g	d								97
Sebasco	Casco Bay Boat Works		207-389-1302							●			98
New Meadows River	New Meadows Marina		207-443-6277	g				●	●	●	●		101
Sebasco Harbor	Sebasco Harbor Resort	09	207-389-1161	g			●		●	●	●	●	102
Small Point Harbor	Robert Stevens Boatbuilder		207-389-1794						●		●		104
	Hermit Island Campground		207-443-2101	g	d	o	o						104
CASCO BAY	FACILITY	VHF	TELEPHONE										PAGE

443

A CRUISING GUIDE TO THE MAINE COAST

MARINE FACILITIES

MIDCOAST	FACILITY	VHF	TELEPHONE	⛽	⛽	⚓	🚤	💧	🔌	🔧	🧺	🚿	PAGE
Bath	Percy and Small Shipyard		207-443-1316			•	•	•			•	•	113
	BFC Marine		207-443-3022	g					•		•		113
	Town landing, harbormaster		207-443-8339					t	•	•			113
	Kennebec Tavern Marina		207-443-9363	g		•		o	•				113
Richmond	Town landing							t					115
	Swan Island Yacht Club						o						115
	Ideal Auto and Marine		207-737-8401								•		115
Gardiner	Smithtown Marina		207-582-3153	g			o		•				117
	Gardiner town landing							t	•				117
Cape Harbor	Newagen Seaside Inn		207-633-5242			•							121
Harmon Harbor	Grey Havens Inn		207-371-2616			•	o						121
Cozy Harbor	Southport Yacht Club		207-633-5767				o		•				122
Five Islands	Town dock							t					123
	Five Islands Yacht Club					•							123
	Sheepscot Bay Boat Co.	09	207-371-2442	g	d	•	•	•			•		123
	Five Islands Boat Yard		207-371-2837								•		123
Riggs Cove	Robinhood Marine Center	09	207-371-2525	g	d	•	•	•	•	•	•	•	124
Maddock Cove	Boothbay Region Boatyard	09	207-633-2970	g	d	•	•	•	•	•	•	•	125
Wiscasset	Town pier, harbormaster							t					131
	Wiscasset Yacht Club	09	207-882-4058				o		•			•	131
	The Eddy Marina	16	207-882-7776	g	d	•	•	•				•	131
Boothbay Harbor	Boothbay Harbor Marina	09	207-633-6003				•	•			•	•	137
	Public landing, harbormaster	09, 16	207-633-3671					t					137
	Boothbay Harbor Pumpout Boat	09	207-633-4220						•				137
	Tugboat Inn Marina	09	207-633-4434			•	•	•			•	•	138
	Boothbay Harbor Shipyard	09, 68	207-633-3171			•				•			138
	Wotton's Wharf	09	207-633-7440	g	d	•	•	•					138
	Rocktide Motor Inn		207-633-4455				o						138
	Cap'n Fish's Motel and Marina	09	207-633-6605			•	•	•					138
	Brown's Wharf Marina	09	207-633-5440			•		•	•			•	139
	Carousel Marina	09	207-633-2922	g	d	•	•	•			•	•	139
	Boothbay Harbor Yacht Club	09	207-633-5750			•		•			•	•	139
	Blake's Boatyard	09	207-633-5040					•		•			139
Linekin Bay	Paul E. Luke, Inc.		207-633-4971			•				•			140
Little River	Spar Shed Marina		207-633-4389			•	o						142
	Little River Lobster		207-633-2648	g	d								142
Christmas Cove	Coveside Marina	09	207-644-8282	g	d	•	•	•	•			•	142
Soth Bristol	Gamage Shipbuilder, Inc.		207-644-8181	g	d	•	•	•		•		•	143
	Bittersweet Landing Boatyard		207-644-8731							•			151
East Boothbay	Ocean Point Marina	09	207-633-0773	g	d	•	•	•	•		•	•	144
Damariscotta	Riverside Boat Company		207-563-3398			•				•			149
	Schooner Landing		207-563-7447					•	•				149
	Newcastle Marine		207-563-5550							•			149
Pemaquid Harbor	Pemaquid Marine		207-677-2024							•			152
New Harbor	Shaw's Wharf		207-677-2200	g	d			•					156
	New Harbor Co-op		207-677-2791	g	d								156
	Gosnold Arms		207-677-3797			o	o						156
Round Pound	Muscongus Bay Lobster Co.		207-529-5528	g	d								157
	Padebco		207-529-5106			•				•			157
	Town landing							t					157
Oar Island Cove	Keene Narrows Lobster			g	d								159
Bremen	Broad Cove Marine Services		207-529-5186	g	d	•	•	•	•				160
Friendship	Town landing							t					163
Thomaston	Lyman-Morse Boatbuilding	09	207-354-6904			o	o			•			176
	Town dock, harbormaster							t					176
	Harbor View Rest. & Marina		207-354-8173			•	o						176
	Jeff's Marine Service		207-354-8777							•			176
Port Clyde	Port Clyde General Store	09, 28	207-372-6543	g	d	•	t	•				o	177
	Town landing							t					177
MIDCOAST	FACILITY	VHF	TELEPHONE	⛽	⛽	⚓	🚤	💧	🔌	🔧	🧺	🚿	PAGE

A CRUISING GUIDE TO THE MAINE COAST

PENOBSCOT BAY	FACILITY	VHF	TELEPHONE	⛽	⛽	🔧	⚓	💧	🚿	🔧	📦	♿	PAGE
Tenants Harbor	*Witham's Lobsters*			g	d								190
	Art's Lobsters		207-372-6265			●							190
	Lyman-Morse	09, 68	207-372-8063			●	●	●				●	190
	Cod End	09, 69	207-372-6782	g	d	●		●					190
	Town landing, harbormaster	09, 16	207-372-6597										190
Seal Harbor	*Spruce Head Fishermen's Co-op*			g	d								194
False Whitehead	Merchant's Landing		207-594-7459			o			●				194
	Spruce Head Marine		207-594-7545								●		194
Owls Head Harbor	*P.K. Reed & Sons*		207-594-4606	g	d								197
Rockland	Journey's End Marina	09, 18	207-594-4444	g	d	●	●	●	●	●		●	199
	Knight Marine Service	09	207-594-4068	g	d	●	●	●		●	●	●	200
	Beggar's Wharf	09	207-594-8500			●							200
	Rockland Harbor Boatyard		207-594-1766			●		●	●	●			200
	Ocean Pursuits		207-596-7357							●			200
	Samoset Resort		207-594-2511			●						●	201
	Rockland Landings Marina	09, 16	207-596-6573	g	d	●	●	●	●		●	●	201
	Municipal Dock, harbormaster	09	207-596-0312			●	●	●	●		●	●	201
Rockport	Marine Park, harbormaster	16	207-236-0676				●	●			●	●	204
	Rockport Marine		207-236-9651	g	d	●	●	●	●	●			204
	Rockport Boat Club					o		●					204
Camden	Camden Yacht Club	68	207-236-3014				t	●					207
	Wayfarer Marine	09, 71	207-236-4378	g	d	●	●	●	●	●	●	●	207
	Wiley Wharf		207-236-3256			●	●	●				●	207
	Camden-Rockport Pumpout	16	207-23697967							●			208
North Haven	J.O.Brown & Son	16	207-867-4621	g	d	●		●		●	●	●	215
	North Haven Casino		207-867-4696			●	t	●					215
	Thayer's Y-Knot Boatyard	09	207-867-4701			●		●				●	214
Hurricane Island	H.I. Outward Bound	09, 18				o							225
Vinalhaven	Hopkins Boatyard		207-863-2551							●			228
	Fishermen's Co-op			g	d								228
	Town dock						t						228
Islesboro	Pendleton Yacht Yard	09, 16	207-734-6728	g	d					●			232
	Tarratine Yacht Club		207-734-6994				t					●	232
	Grindel Point town dock						t						234
700-Acre Island	Dark Harbor Boat Yard	09	207-734-2246	g	d	●		●		●	●	●	233
Warren Island	State of Maine					●							233
Bayside	Town wharf, wharfmaster		207-338-1312				t	●					237
Belfast	Belfast Boatyard		207-338-5098			●	●	●	●	●			238
	City Landing, harbormaster	09	207-338-6222	g	d	●	●	●	●			●	238
Searsport	Public Landing, harbormaster	09, 10	207-548-2985			o	t	●					240
Morse Cove	Devereux Marine		207-326-4800					●	●	●			243
Bucksport	Town landing		207-469-6616				t	●					244
	Bucksport Marina		207-469-5902	g	d		●	●	●			●	244
Winterport	Winterport Marine	09	207-223-8885	g	d	●	●	●	●	●		●	245
	Mid Coast Marine		207-223-4781	g			●	●	●	●			245
Hampden	Turtle Head Marina	16	207-941-8619	g	d	●		●					245
	Waterfront Marine		207-942-7029	g	d	●	●	●		●			245
Bangor	Town landing, harbormaster	09	207-947-5251			●		●	●	●		●	246
Castine	Town docks, harbormaster	09	207-326-4502				t	●					248
	Dennett's Wharf		207-326-9045			●	●	●				●	248
	Eaton's Boatyard	09, 16	207-326-8579	g	d		●	●		●			248
	Castine Yacht Club		207-326-9231				●	●	●			●	248
	Castine Harbor Lodge	09	207-326-4335				●	●					248
Horseshoe Cove	Seal Cove Boatyard	09	207-326-4422			o				●			253
Bucks Harbor	Bucks Harbor Marine	09, 16	207-326-8839	g	d	●	●	●				●	254
	Bucks Harbor Yacht Club								●				255
Sylvester Cove	Deer Isle Yacht Club					o							260
Burnt Cove	*Fifield Lobster Co.*		207-367-2313	g	d								261
PENOBSCOT BAY	FACILITY	VHF	TELEPHONE	⛽	⛽	🔧	⚓	💧	🚿	🔧	📦	♿	PAGE

MARINE FACILITIES

MARINE FACILITIES

MOUNT DESERT	FACILITY	VHF	TELEPHONE	⛽	⛽	⚓	🚽	💧	🔌	🔧	🧺	🚿	PAGE
Stonington	Billings Diesel and Marine	09, 16	207-367-2328	g	d	●	●	●	●	●	●	●	272
	Town dock						t						272
	Greenhead Lobster		207-367-0950	g	d								273
Webb Cove	Old Quarry Ocean Adventures	16	207-367-8977			●						●	280
Benjamin River	Benjamin River Marine		207-359-2244			●		●		●			283
Center Harbor	Brooklin Boatyard		207-359-2236			o		●		●			283
	Center Harbor Yacht Club		207-359-8868			o		●					284
Swan's Island	Swan's Island Boathouse		207-526-4201			●							289
	Fishermen's Co-op		207-526-4327	g	d								289
	Steamboat Wharf	68	207-526-4186	g	d								289
Frenchboro	Lunt & Lunt	80	207-334-2902	g	d	●		●					291
Blue Hill	Kollegewidgwok Yacht Club	09	207-374-5581	g	d	●		●					296
	Raynes Marine Works		207-374-2877							●			296
McHeard Cove	Webber's Cove Boatyard		207-374-2841							●			298
Bass Harbor	Morris Yachts	09	207-244-5511		d	●		●	●	●	●	●	304
	Town dock, harbormaster		207-244-4564				t						304
	C.H. Rich Co.		207-244-3485	g	d								304
	F.W. Thurston	16	207-244-3320	g	d								304
	Up Harbor Marine		207-244-0270					●	●	●			304
Southwest Harbor	Town dock, harbormaster		207-244-7913			●	t						306
	Beal's Lobster Pier	16	207-244-3202	g	d								306
	Downeast Diesel and Marine		207-244-5145							●			306
	Southwest Boat		207-244-5525			o	o			●			306
	Ralph W. Stanley		207-244-3795							●			306
	Great Harbor Marina		207-244-0117		d		●	●	●		●	●	306
Manset	Hinckley Company	09, 16	207-244-5531	g	d	●	●	●	●	●	●	●	308
	Manset Yacht Services		207-244-4040			●		●	●	●			308
	Town dock, harbormaster	09					t	●					308
Somes Sound	John M. Williams Co.	09	207-244-7854			●		●		●			311
	Bar Harbor Boating	09, 10	207-276-5838			●		●					312
	Henry R. Abel Yacht Yard		207-276-5777			●			●	●			312
Northeast Harbor	NE Hbr Marina, harbormaster	09, 68	207-276-5737			●	●	●	●			●	314
	Clifton Dock		207-276-5308	g	d			●	●				314
	Morris Yachts, NE Harbor		207-276-5300							●			314
Great Cranberry Island	Newman and Gray Boatyard		207-244-0575			o				●			317
	Cranberry Island Boatyard		207-244-7316							●			317
Little Cranberry Island	Islesford Dock Restaurant		207-244-7494			●	o						318
	Little Cranberry Yacht Club					●							318
	Fishermen's Co-op			g	d								318
Seal Harbor	Seal Harbor Yacht Club		207-276-5888			●		●					322
	Town dock						t	●					322
Bar Harbor	Municipal Pier, harbormaster	09, 16	207-288-5571			●	●	●					324
	Harbor Place		207-288-3322	g	d	●			●	●			324
	Harborside Hotel and Marina		207-288-5033					●	●			●	324
	Bar Harbor Regency	09, 16	207-244-9723					●	●		●	●	324
Hulls Cove	Bar Harbor Yacht Club		207-244-3275			●							326
Sorrento	West Cove Boat Yard	07	207-422-3137			●				●			327
Winter Harbor	Winter Harbor Yacht Club	16, 71	207-963-2275			●						●	329
	Town dock, harbormaster		207-963-2235				t	●					331
	Winter Harbor Co-op	06	207-963-5857	g	d								331
	Winter Harbor Marine		207-963-7449		d	●			●	●			331
MOUNT DESERT	FACILITY	VHF	TELEPHONE	⛽	⛽	⚓	🚽	💧	🔌	🔧	🧺	🚿	PAGE

MARINE FACILITIES

DOWN EAST	FACILITY	VHF	TELEPHONE	gas	diesel	pump-out	launch	water	elec	repair	laundry	shower	PAGE
Bunker Harbor	Bunker's Wharf Restaurant		207-963-2244				•	t					336
Corea	Corea Lobster Co-op		207-963-7936	g	d								338
Eastern Harbor	North Atlantic Lobster Sales		207-483-2908	g	d								342
	Tyler's Marine		207-483-2886							•			342
Jonesport	Town dock, harbormaster		207-497-5931					t					348
	Jonesport Shipyard	09, 16	207-497-2701			o				•	•	•	348
	Look Lobster	77	207-497-2353	g	d								348
	Moosabec Marine		207-497-2196							•			348
Starboard Cove	Pettegrow Boat Yard		207-255-8740			o				•			355
Bucks Harbor	Bucks Harbor Town Pier							t					356
	BBS Lobster Company		207-255-8888	g	d								356
Cutler, Little River	A.M. Look Canning Co.		207-259-7712	g	d								359
	Deano's Wharf		207-259-7704	g	d								359
	Little River Lodge		207-259-4437									•	359
DOWN EAST	FACILITY	VHF	TELEPHONE										PAGE

PASSAMAQUODDY	FACILITY	VHF	TELEPHONE	gas	diesel	pump-out	launch	water	elec	repair	laundry	shower	PAGE
Grand Manan, N. Head	Harbour Authority		506-662-8482					•					372
Seal Cove	Harbour Authority		506-662-8482					•					375
	General Marine Service		506-662-3288							•			375
Ingalls Head	Harbour Authority		506-662-8482					•					376
	Fundy Marine Service Center		506-662-8481							•			376
Campobello, Head Hbr	Public wharf							•					379
	S.E. Newman and Sons		506-752-2620							•			379
Wilson's Beach	Public wharf							•					380
Welshpool	Public wharf							•					381
Lubec	Town dock							t					383
Eastport	Town dock, harbormaster		207-853-6414					•					387
	Moose Island Marine		207-853-6058							•			387
Deep Cove	Moose Island Marine (yard)		207-853-9692			•				•			389
	Marine Trades Center		207-853-2518					t					389
Lords Cove, Deer Island	Public dock							•					394
St. Andrews, N.B.	Market Wharf, wharfinger	78	506-529-5170					•	•				399
	St. Andrews Yacht Club					o							399
	Kiwanis Campground		506-529-3439									•	399
PASSAMAQUODDY	FACILITY	VHF	TELEPHONE										PAGE

FUNDY and SAINT JOHN	FACILITY	VHF	TELEPHONE	gas	diesel	pump-out	launch	water	elec	repair	laundry	shower	PAGE
Back Bay	Public Wharf							•					406
Beaver Harbour	Public Wharf							•					408
Dipper Harbour	Public Wharf							•					409
Chance Harbour	Public Wharf							•					409
Saint John	Market Slip							t					411
Marble Cove	Saint John Power Boat Club		506-642-5233					•	•		•		417
Ketepec	Saint John Marina		506-738-8484	g	d			•	•	•		•	418
Brothers Cove	Royal Kennebeccasis Yacht Club		506-652-9430	g	d	•		•	•			•	420
	Millidgeville Marine		506-633-1454							•			420
	R.W. Ring		506-633-0044							•			420
Rothesay	Rothesay Yacht Club		506-847-7245					•	•				422
Belleisle Bay	Belleisle Bay Marina		506-832-7373	g	d	•		•	•			•	427
Gagetown	Gagetown Marina	16	506-488-1992	g	d	•		•	•		•	•	432
	Steamer Stop Inn		506-488-2903					o					432
Salmon River	Chipman public dock							•	•				437
Fredericton	Fredericton Yacht Club						o	•					440
	Regent Street Wharf	68	506-455-1445					•					440
FUNDY and SAINT JOHN	FACILITY	VHF	TELEPHONE										PAGE

447

APPENDIX D
BOAT CHARTERS IN MAINE

You don't have to be a boat owner to cruise in Maine. The following companies charter everything from beautiful classics to ultra-light multihulls. They are listed by the location of the companies themselves, not the boats they manage, though typically a company located in one region will have more boats available in that region. The type of boats under management is subject to constant change.

CASCO BAY	WEBSITE	TELEPHONE	SAIL	POWER
Yacht North Group *Yarmouth*	yachtnorth.com	207-221-5285	Pearson, Sabre, Hinkley to 41'	Lowell, Lyman-Morse, Grand Banks to 43'
PENOBSCOT BAY				
Bay Island Yacht Charters *Rockland*	sailme.com	877-456-4267 207-596-7060	J, Sabre, Catalina, Palmer Johnson to 36'	Seaway
Johanson Boatworks *Rockland*	jboatworks.com	207-729-5181	Hunter, Beneteau, J, Tartan, C&C to 46'	
North Point Yacht Charters *Rockland*	northpointyachtcharters.com	207-875-2465	Freedom, Alden, Concordia, Hylas to 46'	
Bucks Harbor Marine and Charter *South Brooksville*	bucksharbor.com	207-348-5253	Columbia, Pearson, Hunter to 40'	Grand Banks to 36'
MOUNT DESERT				
Hinckley Yacht Charters *Manset*	hinckleyyachtcharters.com	800-492-7245 207-244-5008	Hinckley, Sabre to 51'	Grand Banks to 36'

APPENDIX E
PUBLIC TRANSPORTATION

The following public transportation options may help in planning crew changes or boat logistics.

BUS SERVICE	WEBSITE	TELEPHONE	ROUTES
Concord Trailways	concordtrailways.com	800-639-3317	Logan, Boston, Portland, Brunswick, Bath, Rockland, Camden, Belfast, Searsport, Bangor
Greyhound	greyhound.com	800-894-3355	Shuttle from Bangor to Ellsworth and Mount Desert Island
Bar Harbor Shuttle	barharborshuttle.com	207-479-5911	Bangor (airport, downtown) to Ellsworth and Bar Harbor
West's Coastal Connection	westbusservice.com	800-596-2823	Bangor, Ellsworth, Machias, Calais
SMT	smtbus.com	800-567-5151	Bangor, St. Stephen, St. Andrews, Saint John airport, Fredericton
LOCAL BUS SERVICE			
Portland Metro	gpmetrobus.com	207-774-0351	Greater Portland
Downeast Transportation	exploreacadia.com	207-667-5796	Mount Desert Island Explorer shuttle (free)
LIMO AND CHARTER			
Boothbay Stage Lines		207-633-7380	24 hr/day airport shuttles from Boothbay region
Mermaid Transportation	gomermaid.com	800-696-2463	Portland to Logan, Manchester airports
Mid-Coast Limo	midcoastlimo.com	800-937-2424	Coastal airport shuttle from Belfast south to Portland
Transportation Matters		207 374-2988	Blue Hill peninsula to Bar Harbor, Bangor, Portland airports
AIR SERVICE			
U.S. Airways Express (Colgan Air)	usairways.com	800-428-4322	Augusta, Bar Harbor
Penobscot Island Air	penobscotislandair.net	207-596-7500	Rockland, Belfast, Matinicus, Vinalhaven, North Haven, Islesboro, Stonington, Swan's Island, Bar Harbor

APPENDIX F
ISLAND FERRY SERVICE

Many of the larger islands along the coast are serviced by ferries, both publicly and privately owned. They enable the cruising yachtsman to explore places he doesn't want to go in his own boat or a way to arrange crew or repair logistics. The following is a list of ferry service along the coast.

ISLANDS SERVED	TERMINAL	OPERATOR	WEBSITE	TELEPHONE
SOUTHERN COAST				
Star Island, Isles of Shoals	Portsmouth, NH	Isles of Shoals Steamship Co.	islesofshoals.com	603-431-5500
CASCO BAY				
Peaks Island	Portland	Casco Bay Lines	cascobaylines.com	207-774-7871
Little Diamond Island	"	"	"	"
Great Diamond Island	"	"	"	"
Long Island	"	"	"	"
Chebeague Island	"	"	"	"
Cliff Island	"	"	"	"
Chebeague Island	Cousins Island	Chebeague Island Tranportation Co.	chebeaguetrans.com	207-846-3700
MIDCOAST				
Monhegan	Boothbay Harbor	*Balmy Days*	balmydayscruises.com	207-633-2284
	New Harbor	*Hardy*	hardyboat.com	800-278-3346
	Port Clyde	Monhegan Boat Lines	monheganboat.com	207-372-8848
PENOBSCOT BAY				
Islesboro	Lincolnville Beach	Maine State Ferry Service	maine.gov/mdot/opt/ferry/maine-ferry-service.php	207-596-2202
North Haven	Rockland	"		"
Vinalhaven	"	"		"
Matinicus	"	"		"
Isle au Haut	Stonington	Isle au Haut Boat Company	isleauhaut.com	207-367-6516
MOUNT DESERT				
Swan's Island	Bass Harbor	Maine State Ferry Service	*see above*	207-596-2202
Long Island, Frenchboro	Bass Harbor	"	"	"
Cranberry Islands	Northeast Harbor	Beal and Bunker		207-244-3575
Cranberry Islands	Manset	Cranberry Cove Boating, Co.		207-244-5882
DOWN EAST				
Grand Manan, NB	Blacks Harbour, NB	Coastal Transport Ltd.	coastaltransport.ca	506-662-3606
Campobello, NB	Eastport	East Coast Ferries, Ltd.	eastcoastferries.nb.ca	506-747-2159
Deer Island, NB	Eastport	"	"	"
Deer Island, NB	Campobello, NB	"	"	"
Deer Island, NB	Back Bay, NB	"	"	"
NOVA SCOTIA				
Yarmouth, NS	Portland	Bay Ferries	bayferries.com	207-761-4228
Yarmouth, NS	Bar Harbor	"	bayferries.com	207-288-3395
Digby, NS	Saint John, NB	"	"	506-649-7777

APPENDIX G
SALTWATER FISHING

Maine waters have numerous saltwater fish that are good eating and good sport. Licenses are not required, though size and bag limits do apply to some species, particularly striped bass. Regulations are subject to change, so please check in advance of fishing. For Canadian fishing regulations, contact any Canadian Tourist Information Center or the Fisheries and Oceans offices listed below.

It is unlawful to fish for striped bass in federal waters, which begin three miles offshore.

FISH	TIME IN MAINE WATERS	RESTRICTIONS, COMMENTS	
striped bass	May to October	1 fish, 20"-26" and >40", special regulations on some rivers	
bluefish	June through September	3 fish, average 8 - 18 lbs.	
mackerel	May through October	the kids' favorite, jig off docks, good striper bait	
shad	May through July	2 fish	
flounder	May through December	8 fish, 12" minimum	
eel	May through September		
smelt		no limit June 14 - March 14, otherwise 2-quart limit	
others: cunners (bergalls), menhaden (pogies)			
BOTTOM FISH			
cod		10 fish, 22" minimum, 2 hooks/line, 1 line/angler	
haddock		19" minimum, average 2 - 10 lbs., 2 hooks/line, 1 line/angler	
pollock		6 fish, 19" minimum, average 3 - 25 lbs.	
cusk		average 5 - 20 lbs.	
halibut		1 fish/boat, 36" minimum, can top 200 lbs.	
sturgeon		prohibited	
Atlantic salmon		prohibited, do not remove from water, release	
others: hake, wolffish, flounders, sculpins			
LARGE PELAGIC FISH			
Atlantic tuna, sharks, swordfish, billfish		permit required	
COMMON BAITS			
sandworms, sand eels, clams, mussels, squid, mackerel, shrimp, pogies			
OTHER REFERENCES			
Maine Department of Marine Resources		www.maine.gov/dmr	207-633-9500
Canadian Fisheries and Oceans		www.dfo-mpo.gc.ca/home-accueil_e.htm	506-529-5808

INDEX

Abagadasset River (Merrymeeting Bay), 114
Acadia National Park, 182, 269, 273, **301-302**, 305, 316, 319, 322, 324, 326
Adams Island (Ogunquit), 45
Addison (Pleasant River), 340
Agamenticus Yacht Club (York Harbor), 42, 43
U.S.S. *Albacore*, 41
Allen Cove (Blue Hill Bay), 293
Allen Island (Muscongus Bay), 154, 168, **171-172**, 178
Alley Island (Mount Desert Narrows), 299
Allied Whale Project, 170, 293, 325
America, 144
American Eagle, 202
American Revolution, 30, 33, 82, 301, 350, 355, 365, 369, 392
Ames Cove (Islesboro), 230, 231, **232**
Anchors, Anchoring; with large tidal range, 6, **366–367**
Androscoggin River, 114
Angel Gabriel, 152
Angelique, 262
Anguilla Island (Roque Island), 349, 351, 352
Antoinette, Marie, Queen of France, **132**
Appalachian Mountain Club, 119
Appledore, 262
Appledore Island (Isles of Shoals), 30, **32**
Aquaculture, 141, 148, 385, 386; clams, 345; hatcheries, 345, 386; research, 148; salmon, 388, 391, 401
Archangel, 82, 171
Arnold, Benedict, 116
Arrossac, Chief, 171
Arrowsic Bridge (Sasanoa River), 118
Arundel Yacht Club (Kennebunk River), 48, 49
Ash Island (Muscle Ridge Channel), 194
Aspinquid, 29
Atlantic Challenge, 202, 227
Atlantic Neptune (J.F.W. DesBarres), 310
Audubon, John James, 308, 440
Audubon, Maine Society, 11, 56, 86, 87, 330, 360, 441
Audubon, National Society, 58, 96, 101, 154, 170, 340, 361; Casco Bay, 104; Eastern Egg Rock, 165; Matinicus Rock, 188; midcoast, 165; Penobscot Bay, 170, 182; Ross Island, 129
Augusta (Kennebec River), 107; tidal range, 108
Auks, 188, 330, 360, **400**
Avelinda, 362

Babbidge Island (Fox Islands Thorofare), 218
Babson Islands (Eggemoggin Reach), 284
Back Bay, N.B. (Letete Passage), 406, 407
Back Channel (Kittery), 35, 37, 38
Back Cove (Casco Bay), 65
Back Cove (New Meadows River), 101
Back River (Inside Passage), 120
Back River (Sheepscot River), 128
Badgers Island (Piscataqua River), 33, 39
Bagaduce River (Castine), 181, 247, **250**
Baileys Mistake (Eastern Harbor), 362
Bailey Island (Casco Bay), 63, 90, 94

Baker Island (Cranberry Islands), 301, 316, 317, **319**
Bald eagles, 176, 334, 339, **340**, 343, 345, 355, 357, 365, 389, 390
Bald Porcupine Island (Frenchman Bay), 323
Balmy Days, 167, 168
Bangor, 181, 238, 242, 243, 245, **246-247**
Bangs Island (Casco Bay), 78
Bar Harbor (Frenchman Bay), 269, 295, 301, 302, 322, **323-325**, 326, 329
Bar Harbor Yacht Club (Hulls Cove), 326
Bar Island (Bar Harbor), 301, 323
Bar Island (Blue Hill Bay), 293
Bar Island (Deer Island, N.B.), 393
Bar Island (Gouldsboro Bay), 338
Bar Island (Machias Bay), **355**, 356
Bar Island (Muscongus Bay), 157, 172
Bar Island (Roque Island), **349**, 352
Bar Island (Somes Harbor), 312
Bareneck Island (Inside Passage), 119
Bare Island (Machias Bay), 355
Bare Island (Merchants Row), 274
Barges, The (Blue Hill Bay), 293
Barred Islands (Penobscot Bay), 182, **257**
Bartender Island (Barred Islands), 257, 258
Barters Island (Sheepscot River), 128, 130
Bartlett Harbor (North Haven Island), 212
Bartlett Island (Blue Hill Bay), 293, 299, **299–300**, 300-302
Bartlett Narrows (Blue Hill Bay), 299–332
Barton Island (Grand Lake), 437
Basin, The (New Meadows River), 64, **99–100**, 103, 104
Basin, The (Vinalhaven), 221, **223**
Basin Cove (Potts Harbor), 89
Basin Island (New Meadows River), 99
Basket Island (Casco Bay), 75
Basket Island (Wood Island Harbor), 55
Bass, striped, 108, **450**
Bass Harbor (Mount Desert Island), 286, 287, 293, 295, 301, **303-305**
Bass Harbor Bar (Mount Desert Island), 305; currents, 22
Bath (Kennebec River), 103, 107, 108, **111-114**; Inside Passage, 117, 118, 126, 139; tidal range, 108
Bath Iron Works, 111, 112
Bay of Fundy, 24, 335, 370, 374, 395, 404, 407, 408, 409, 410, 412, 414, 439; passages to Saint John, N.B., 405; tides and currents, 367, 377
Beal Island (Inside Passage), 119
Beals Island (Moosabec Reach), 342, 343, 344, 348
Bean, L.L., 68, 82, 85, 86
Beans Island (Deer Island, N.B.), 393, 394
Bear Island (New Meadows River), 97, 98
Bear Island (Cranberry Islands), 306, 316
Beaver Harbour (Bay of Fundy), 408
Beaver Island (Johns Bay), 152
Bedroom, The (Grand Lake), 433, **434**
Belfast (West Penobscot Bay), 112, 181, **238-239**
Bellamy, Captain, 355
Belleisle Bay (Saint John River), 412, 424, **426-428**, 429
Bellows, George, 169

451

Belyeas Cove (Washademoak Lake), 431
Belyeas Point (Long Reach), 424, 425
Ben Island (Quahog Bay), 95
Benjamin River (Eggemoggin Reach), 282–283
Benner Island (Muscongus Bay), 171
Bernard (Bass Harbor), 304
Bickford Island (Cape Porpoise Harbor), 52
Biddeford (Saco River), 57, 58
Biddeford Pool, 53, **54–56**
Biddeford Pool Yacht Club, 54, 55
Big Barred Island (Penobscot Bay), 257, 258
Big Cove (Washademoak Lake), 431
Big Garden Island (White Islands), 224, 225
Big Hen Island (Ridley Cove), 96, 103
Big Hen Island (Winter Harbor), 219
Big Moose Island (Schoodic Peninsula), 328, 329, 332
Big White Island (Hurricane Sound), 222, **224-225**
Billings Cove (Eggemoggin Reach), 282
Birch Cove (Bartlett Island), 300
Birch Cove (Passamaquoddy Bay), 395, 397
Birch Harbor (Schoodic Peninsula), 336
Birch Island (Middle Bay), 87
Birch Island (Muscle Ridge Channel), 195
Birch Island (Washademoak Lake), 431
Blackbeard (Edward Teach), 31, 79
Black Cove (Sorrento), 326, **327**
Black Island (Casco Passage), 285
Black Island (Muscongus Bay), 154
Blacks Harbour, N.B. (Bay of Fundy), 369, 405, **407**; ferry to Grand Manan Island, 372, 405, 407
Bliss Harbour, Bliss Islands (Bay of Fundy), 404, 406
Blue Hill, Blue Hill Harbor, 269, 293, **296-298**, 299
Blue Hill Bay, 6, 209, 269, 285, **293**, 297
Blueberries, 341, 357
Bluefish, 92, 108
Bluff Island (Saco Bay), 58, 59
Boat charters, 448
Bocabec Cove, Bocabec River (Passamaquoddy Bay), 395, 397
Bois Bubert Island (Pigeon Hill Bay), 339
Bold Island (Merchants Row), 274, 275, **277**
Bombazine Island (New Meadows River), 96, **101**, 104
Boon Island (southern coast), 34, 38, **41**, 170
Boothbay Harbor, 107, 108, 111, **136-139**, 140, 145, 167, 295; Inside Passage, 117–119, 126
Boothbay Harbor Yacht Club, 137
Bounty Cove (Islesboro), 236
Bowdoin, 112, 113, 144
Bradbury, Charles, 51
Bradbury Island (East Penobscot Bay), 257
Bragdon Island (New Meadows River), 100
Bragdon Island (York Harbor), 43
Bremen (Medomak River), 160
Bremen Long Island (Medomak River), 160, 161
Brewer (Penobscot River), 246
Brickyard Cove (Quahog Bay), 96
Brighams Cove (New Meadows River), 100
Brimstone Island (Penobscot Bay), 182, 229, **230-231**
Bristol, 145, 151
Broad Cove (Eastport), 388, 389
Broad Cove (Rockland), 198

Broad Cove (St. George River), 175
Broad Sound (Casco Bay), 63, **81**, 89
Brooklin (Eggemoggin Reach), 284
Brothers Cove (Kennebecasis Bay), 419, **420**
Buckle Harbor (Swan's Island), 285, 286, **287**
Bucksport (Penobscot River), 290
Bucks Harbor, Bucks Harbor Yacht Club (Eggemoggin Reach), 182, 253, **254-255**, 282
Bucks Harbor (Machias Bay), 355, 356
Bull, Dixie, 151
Bunker Cove (Roque Island), 349, **350**
Bunker Hole (Roque Island), 350
Bunkers Harbor (Schoodic Peninsula), 336
Buoys. *See* Lobster buoys; Navigation
Burgess, Abbie, 170, **189**
Burnt Coat Harbor (Swan's Island), 278, 286, **288-289**, 293
Burnt Coat Island (New Meadows River), 96
Burnt Cove (Deer Isle), 260, **261**
Burnt Island (Booth Bay), 136
Burnt Island (Fox Islands Thorofare), 218
Burnt Island (Isle au Haut), 266; Thorofare, 266
Burnt Island (Muscongus Bay), 50, 154, 168, 172
Burnt Island (Seal Harbor, Muscle Ridge), 193, 194
Burr, Aaron, 116
Bush, George H.W., 48
Bustins Island (Casco Bay), 64
Butter Island (Penobscot Bay), 182, 257, **258-259**

Cabbage Island (Linekin Bay), 140
Cabot Cove (Pulpit Harbor), 212
Cabot, Thomas D., 357, 362
Cadillac, Sieur Antione de la Mothe, 300
Cadillac Mountain (Mound Desert Island), 301, 322, 323, 325, 327
Calais (St. Croix River), 401, 402
Calderwood Island (Fox Islands Thorofare), 218
Caldwell Island (Muscongus Bay), 154, 172, 172–173
Calendar Islands (Casco Bay), 63
Calf Island (Flanders Bay), 328
Calf Island (Little Kennebec Bay), 353
Cambridge Narrows, N.B. (Washademoak Lake), 430, **431**
Camden (Penobscot Bay), 112, 175, 181, 204, **206-209**, 213, 295, 310; passages east from, 210
Camden Yacht Club, 207, 208
Camp Ellis (Saco River), 57, 58
Camp Island (Merchants Row), 274, 275
Campbell Island (Greenlaw Cove), 281
Camping, 9, 10
Campobello Island (New Brunswick), 365, 366, 376, 377, **378**, 379–402
Canada; entering, 7; customs, 372, 375, 378, 380, 384, 402, 411; duties, 368
Canadian Coast Guard, *See* Coast Guard, Canadian
CANPASS, **368**, 369, 372, 375, 378, 380, 393, 399
Cantons Island (Long Reach), 424, 425
Cape Elizabeth, 16, 24, 29, 60, 63, 170
Cape Harbor (Cape Newagen), 120, **121**
Cape Island (Cape Harbor), 121
Cape Island (Cape Porpoise), 53
Cape Neddick (southern coast), 44, **45**

Cape Newagen (Southport Island), 26, **121**
Cape Porpoise Harbor (southern coast), 29, 41, 50, **51–53**
Cape Rosier (East Penobscot Bay), 182, 247, 251-255
Cape Small (Casco Bay), 63, 103, 104, 107, 109
Cape Small Harbor (Casco Bay), 103, **104**
Cape Split, 339, **342**
Card Cove (Quahog Bay), 95
Carrying Place Cove (New Meadows River), 102
Carson, Rachel, 120, 176; Rachel Carson National Wildlife Refuge, 47
Carter, Robert, 31, 50, 126, 197, 282
Carters Cove, Carters Point (Long Reach), 424, 425
Carvers Harbor (Vinalhaven), 182, **227–229**, 278
Carver Cove (Fox Islands Thorofare), 218
Casco Bay Bridge, 69, 70
Casco Bay, Friends of, 70
Casco Bay Island (Head Harbour Passage, N.B.), 392
Casco Passage, 269, 270, **285**; currents, 22
Castin, Baron de, 151
Castine (East Penobscot Bay), 181, 235, 236, 237, 242, **247-249**, 250, 251, 252, 365, 398
Castine Yacht Club, **248**, 252
Castle Island (Hockomock Bay), 119
Cat, The, ferry to Nova Scotia, 325, 326
Cathance River (Merrymeeting Bay), 114
Cathline Cove (Kennebecasis Bay), 421
Cedar Island (Hurricane Sound), 221, 222, **224**
Cedar Island (Isles of Shoals), 30, **31**
Cedar Island (Sipp Bay), 389
Cellular phones, 19
Center Harbor (Eggemoggin Reach), 282, **283**
Center Harbor Yacht Club (Eggemoggin Reach), 283, 284
Center Island (Quahog Bay), 94, 95
Centerboard Yacht Club (South Portland), 64, 69
Chamcook Harbour (Passamaquoddy Bay), 395, **398**
Champlain, Samuel de, 107, 166, 171, 286, 300, 322, 365, 401; exploration by, 414; places named by, 60, 410
Chance Harbour, N.B. (Bay of Fundy), 409
Chandler Bay, 349, 350, 353
Chandler Cove (Great Chebeague Island), 76
Charters, *See* Boat charters
Charts, 18; *See also* Navigation; Atlantic Neptune, 310; Canadian, 365; harbor, 2; region overviews, 2
Chatto Island (Center Harbor), 283
Chauncey Creek (Pepperrell Cove), 37
Cheney Island (Grand Manan), 370
Cherry Island (Friar Roads), 380, 392
Children, sailing with, 320
Chipman, N.B. (Salmon River), 433, 437
Christmas Cove (Damariscotta River), 140, **142-143**, 144, 150
Cimbria, 311
Civil War, 67, 70, 72, 75, 110, 278, 348, 360
Clam Cove (Penobscot Bay), 203
Clam Island (Medomak River), 160
Clapboard Island (Casco Bay), 74, 75
Clark Island (Kittery), 37
Clark Island (Long Cove), 191, 278
Cliff Island (Casco Bay), 63, **78–79**, 80
Clough, Captain Samuel, 132
CNG, 13, 207

Coast Guard, Canadian, 23, 365, 404, 411, 416, 424; at Saint John, 410; **emergency numbers, 441**
Coast Guard, United States, 199, 272, 289, 305, 306, 318; and lighthouses, 170; boardings, 109; Boothbay Harbor, 136; Eastport, 367, 385, 387; Jonesport, 343, 344, 348; navigation recommendations, 18; Portland, 65; Portsmouth Harbor, 38; radio monitoring, 19; rescues, 109; Rockland, 198; **emergency numbers, 441**
Cobscook Bay, 340, 365, 377, 378, 380, 382, 385, 386, 389, 389–402; Cobscook Falls, 391
Cocktail Cove (Great Diamond Island), 73
Colwells Creek (Saint John River), 429, **430**, 431
Communications, *See* Cellular phones; VHF radio
Conary Nub (Morgan Bay), 299
Conkling, Philip, 344, 400
USS Constitution, 145
Contention Cove (Patten Bay), 299
Coot Islands (Merchants Row), 275
Cora F. Cressy, 159
Corea Harbor, 337–338
Cormorants, **129**, 397
Corsair, 111
Cousins Island (Casco Bay), 76, 81, 83
Cousins River (Royal River), 83
Cow Island (Great Diamond Island), 73
Cow Island (Medomak River), 161
Cow Island (Saco River), 57
Cows Yard, The (Head Harbor Island), 335, 343, **347**
Cowseagan Narrows (Back River), 120
Cozy Harbor (Sheepscot River), 120, **122**
Crab Island (Harraseeket River)), 85
Cradle Cove (Seven Hundred Acre Island), 231, **232-233**
Crafts Cove (Washademoak Lake), 431
Cranberry Island (Muscongus Bay), 164
Cranberry Islands (Mount Desert Island), 301, 311, 313, 315, **316-319**. *See also* Acadia National Park
Crane Island (Hurricane Sound), 221, 224
Crawford Island (Kennebec River), 114
Crescent Beach (Muscle Ridge Channel), 197
Crescent Beach (Cape Elizabeth), 60
Crie, Robert, 187
Criehaven, *See* Ragged Island (Criehaven)
Crockett Cove (Deer Isle), **260**, 261
Crockett Cove (Vinalhaven), **221**, 222
Cross Island (Machias Bay), **357**, 386
Cross River (Sheepscot River), 130
Crotch Island (Meduncook River), 163
Crotch Island (Merchants Row), 270, 272, **274**, 278
Crotch Island (Casco Bay) pinky, 78
Crotch Islands (Muscongus Bay), 159
Crow Cove, 230, **234**
Crow Island (East Penobscot Bay), 257
Crow Island (Frenchboro), 291
Crow Island (Great Chebeague Island), 78
Crow Island (Great Diamond Island), 73
Crow Island (Mackerel Cove), 287
Crow Island (Muscongus Bay), 159
Crow Island (Reversing Falls), 414
Crow Island (Thread of Life), 150
Crumple Island (Western Bay), 347

Cuckolds, The (Booth Bay), 109, 136; lighthouse, 26
Cumberland Bay (Grand Lake), 433, **435–436**
Cundys Harbor (New Meadows River), **97**, 99
Currents, *See* Tides and tidal currents
Curtis Island (Camden), 206, 295
Cushing Island (Casco Bay), 63, 71, 74
Customs, *See* Canada, entering; United States, entering
Cutler (Little River), 295, 335, **358–359**, 360, 370, 386

Dagger Island (North Haven Island), 219
Damarill, Humphrey, 133
Damariscotta, 141, 148, **149**
Damariscotta River, 52, 107, **140–149**, 150
Damariscove Island (Booth Bay), 107, **133–134**, 139
Dark Harbor (Islesboro), 182
Dark Harbour (Grand Manan), 370, 372
Darling, Ira C., Center, 52, 141, **148**
Darling Island (Blue Hill Bay), 298
Dash, 85
David Island (Cozy Harbor), 122
Davies, Captain Robert, 110
Davis Cove (Muscongus Bay), 154, **163–164**
Davis Island (Muscongus Bay), 154, 171, 178
Davis Strait (Muscongus Bay), 154, 178
Deadheads, 108, 412, 416, 427
Deadman Cove (Monhegan Island), 168
Deckers Cove (Townsend Gut), 127
Deep Cove (Eastport), 389
Deep Cove (Rockland), 197
Deep Hole (Deer Isle), 281
Deer Island (New Brunswick), 377, 379, 380, 382, 387, 391, 392, 393, 394, 396, 408
Deer Island Thorofare (Merchants Row), 182, 269, **270**, 272, 274, 275, 280; currents, 22
Deer Isle, 182, 209, 269, 274, 276, 280, 281, 282, 386; east coast, 280-281; west coast, 259-261
Deer Isle (village), 259, 260
Deer Isle Sailing Center (Deep Hole), 281
Deer Isle Yacht Club (Sylvester Cove), 260
Delight, 315
Dennetts Island (Piscataqua River), 33
Dennys Bay (Cobscook Bay), 389
DesBarres, Joseph F.W., *See* Atlantic Neptune
Devil Island (Merchants Row), 274, 275, **277**
Devils Head (Marshall Island), 290
Diamond Cove (Great Diamond Island), 71, **73–74**
Dick Island (Bocabec River), 397
Digdeguash Harbour, Digdeguash River (Passamaquoddy), 395, **396–397**
Dingley Cove, Island (New Meadows River), 99, 100
Dinner Island (Deer Island, N.B.), 393
Dipper Harbour, N.B. (Bay of Fundy), 409
Discoverer, 211
Dix Island (Muscle Ridge Channel), 182, **195–196**, 278
Dix Island Harbor (Muscle Ridge Channel), 195
Dodge Point (Damariscotta River), 141, **148**
Dog Island (Western Passage), 392
Dorr, George B., 301
Double Shot Island (Roque Island), 349, 351
Douglas Harbour, N.B. (Grand Lake), 433, 434

Douglas Islands (Narraguagus Bay), 339, 340
Down East Yacht Club (Boothbay Harbor), 138
Dram Island (Sorrento), 322, 327
Drisko Island (Western Bay), 343
Drury Cove (Kennebecasis Bay), 421
Duck Harbor (Isle au Haut), 261, **301**
Duck Island (Isles of Shoals), 30, **33**, 129
Dudley Island (Friar Roads), 384
Dumpling Islands (Fox Islands Thorofare), 214
Dunn Island (Mason Bay), 353
Dyer Bay, 338
Dyer Island (Pleasant Bay), 341–342
Dyon, 190

Eagle Island (Blue Hill Bay), 285
Eagle Island (Casco Bay), 64, **81**
Eagle Island (Penobscot Bay), 182
Eagles, *See* Bald eagles
East Bay (Cobscook Bay), 389
East Boothbay (Damariscotta River), 144, 146
East Quoddy Head (Campobello Island), 379, 380, 393; lighthouse, 379, 380
Easterly Bar Island (Roque Island), 350
Eastern Bay (Frenchman Bay), 326
Eastern Bay (Great Wass Island), 334, 343, 344, 346
Eastern Branch (Johns River), 150, 152, **153**
Eastern Cove (Frenchboro), 292
Eastern Egg Rock (Muscongus Bay), 156, 159, **165**, 361
Eastern Harbor (Cape Split), 335, 339, **342**
Eastern Mark Island (Deer Island Thorofare), 270
Eastern Point Harbor (Flanders Bay), 328
Eastern River (Kennebec River), 114
Eastern Wolf Island (The Wolves), 408
Eastport (Friar Roads), 50, 343, 358, 365, 367, 377, 379, 380, 382, 384, **385-388**, 389, 392
Eaton Island (East Penobscot Bay), 256
Ebenecook Harbor (Sheepscot River), 119, 120, **125–126**, 128
Edward J. Lawrence, 70
Egg Rock (Frenchman Bay), 322, 329, 332
Eggemoggin Reach, 182, 269, 270, 282, **282–285**, 310; currents, 22
Eiders, 134, 135, 165, 293, 352, 353, 355
Eldridge, Captain Asa, 200
Eliot, Charles W., 301, 319
Ellingwood Rock (Seguin Island), 26, 109
Elliot (Piscataqua River), 41
Ellsworth (Union River), 299
Emergency help, 22
EMERGENCY NUMBERS, 441
Emery Island (Muscle Ridge Channel), 197
Enchanted Island (Merchants Row), 274
Endeavour, 111
Englishman Bay, 349, 353, 354
Ensign Islands (Gilkey Harbor), 231
Environmental advocacy groups, 11, **441**
Erbs Cove, 428
Escargot Island (Barred Islands), 257, 258
Evandale, N.B. (Saint John River), 429
Ewe Island (Merchants Row), 279

Factory Island (Saco River), 57
Fairwinds (Belleisle Bay), 427
Falls Island (Cobscook Bay), 389, 391
Falls Island Passage (Cobscook Bay), 391
Falmouth Foreside (Casco Bay), 74–75
False Whitehead Harbor (Muscle Ridge Channel), 194
Fan Island (Little Kennebec Bay), 354
Fanjoys Point (Grand Lake), 435
Farnham Cove (Damariscotta River), 144
Farnsworth Art Museum (Rockland), 199, 202
Federal Harbor, Federal Island (Cobscook Bay), 390
Ferry Beach (Saco Bay), 57, 58
Ferry service, 449
Fiddle Head Island (East Penobscot Bay), 256
Fiddlehead Island (Hurricane Sound), 223
Fires; fire-danger broadcasts, 10; permits, 10
Fisherman's Cove (Cliff Island), 78, 79
Fishing, fishermen, 10, 11; cod, 30, 43; herring, 157, 158, 374, 384, 388; sport, 129, **450**; tuna, 90, 123; VHF channels, 19
Fishing Island (Portsmouth Harbor), 36
Five Fathom Hole (Musquash Harbour), 409
Five Islands (Sheepscot River), 120, 121, **123**
Five Islands Yacht Club, 123
Flake Island (Isle au Haut), 265, 266
Flanders Bay (Frenchman Bay), 328
Flash Island (Ridley Cove), 96
Flat Island (West Penobscot Bay), 234
Flea Island (Pleasant Point Gut), 173
Fletcher Neck (Biddeford Pool), 54
Flint Island (Narraguagus Bay), 340
Float plan, 7
Flowers Cove (Grand Lake), 436, 437
Flying Dragon, 112
Fog, 15, 16, 343; at Petit Manan, 338; at Machias Seal Island, 361; in Bay of Fundy, 404; navigating in, 17, 22
Folly Island (Bartlett Narrows), 300, 302
Folly Island (Cape Porpoise), 51
Foraging, 10
Fore River (Casco Bay), 65, 70
Fore River (Portland), 65
Forresters Cove (Kennebecasis Bay), 423
Fort Baldwin (Popham Beach), 110
Fort Charles (Pemaquid Harbor), 151
Fort Edgecomb (Sheepscot River), 133
Fort Foster (Gerrish Island), 37
Fort Frederick, 151
Fort Gorges (Portland Harbor), 64, **70**
Fort Island (Damariscotta River), 141, 146
Fort Island (Stage Island Harbor), 53
Fort Kent, 243
Fort Knox (Penobscot River), 181
Fort Madison (Castine), 247, 249
Fort McClary (Pepperrell Cove), 34, 36, 37
Fort McKinley (Great Diamond Island), 73, 74
Fort Pemaquid (Pemaquid Harbor), 151
Fort Pentagoet (Castine), 247
Fort Point Cove (Penobscot River), 242-243
Fort Popham (Kennebec River), **110**; tidal range, 108
Fort Pownall (Penobscot River), 243
Fort Preble (South Portland), 70

Fort St. George (Kennebec River), 110
Fort Scammel (Casco Bay), 243
Fort William and Mary (Portsmouth Harbor), 33
Fort William Henry (Pemaquid Harbor), 151
Foster Island (Johns River), 153
Fox Island (Cobscook Bay), 391
Fox Islands (North Haven and Vinalhaven), 182
Fox Islands Thorofare (Penobscot Bay), 182, 210, 211, **213-214**, 215-219, 270; currents, 22
Francis Todd, 325
Franklin Island (Muscongus Bay), 154, 163, 164
Fredericton, N.B. (Saint John River), 405, 412, 438, **439–440**
Fredericton Yacht Club, 440
Freeport, 85
Freese Island (Pickering Cove), 280
French Lake (Grand Lake), 412, 433, **438**
Frenchboro (Long Island), 269, **291-293**, 303
Frenchman Bay, 6, 269, 293, 323, 327, 328, 332
Friar Roads, 380, 387, 392
Friars Bay (Campobello Island), 378, 381, 382
Friendship Harbor (Muscongus Bay), 154, **162–163**
Friendship Long Island (Muscongus Bay), 162, 164
Friendship sloops, 119, 162, **163**, 176, 284, 348; illus., 163
Friends of Casco Bay, 66
Friends of Nature, 277
Frye Island, N.B. (Bay of Fundy), 404, 406, 407
Fundy Traffic, 24, 365, 372, 405, 410, 416

Gage, Colonel Thomas, 432
Gagetown, N.B. (Saint John River), 412, 429, **432**, 438
Gagetown Island (Saint John River), 432, 433, 438
Galley Cove (Bartlett Island), 300
Gardiner, Sylvester, 117
Gardiner (Kennebec River), 108, **116–117**; tidal range, 108
Garrison Island (Friendship Harbor), 162
Gay Island (Muscongus Bay), 154, 162, 164, 173
George Island (Ridley Cove), 95, 96
Georges Harbor (Muscongus Bay), 154, 171
Georges Islands (Muscongus Bay), 171
Georgetown Island (Sheepscot River), 120
Gerrish Island (Pepperrell Cove), 37
Ghost Island (Belleisle Bay), 428
Gifte of God, 110
Gilbert, Captain Raleigh, 110
Gilbert Island (Saint John River), 438
Gilpatrick Cove (Northeast Harbor), 309
Goat Island (Cape Porpoise), 51, 53
Goat Island (Grand Lake), 436
Goat Island (Kennebecasis Bay), 420
Goat Island (Portsmouth, N.H.), 40
Gomez, Estavan, 166
Goose Cove (Blue Hill Bay), 303
Goose River (Rockport), 203, 310
Goose Rock Passage, 117, 119, 120, 124
Gooseberry Island (Biddeford Pool), 55
Gooseberry Island (Toothacher Bay), 288
Gorgeana, 44
Gorges, Sir Ferdinando, 44, 76, 133
Goslings, The (Casco Bay), 87–88
Gosnold, Bartholomew, 47, 156

Gosport Harbor (Isles of Shoals), 30, 31
Gouldsboro Bay, 338
Governor Curtis, 222
GPS, 17, **18**
Grace Bailey, 262
Grand Bay (Saint John River), 417, **418**, 419, 420, 421, 424
Grand Harbour (Grand Manan), 373, 376
Grand Lake (Saint John River), 405, 412, **433-437**, 438
Grand Manan Channel, 26, 367, 404; currents, 22, 370, 371
Grand Manan Island (New Brunswick), 17, 295, 358, 360, 362, 365, 366, **369–376**, 382, 384, 388, 392, 405, 407; passages to, 370
Granite, 182, 191, 198, 272, 274–278, 296; quarries, 196, 272, 273, 278, 280, 289
Grassy Island (Long Reach), 424, 426
Great Bay (Piscataqua River), 33, 39, 40, 41
Great Beach (Roque Island), **351**, 352, 353
Great Chebeague Island (Casco Bay), 63, 64, **76–77**, 81
Great Cove (Bartlett Island), 300
Great Cove (Piscataqua River), 41
Great Cranberry Island (Mount Desert), 305, 313, **316–317**
Great Diamond Island (Casco Bay), 73, 74
Great Spruce Head Island (Penobscot Bay), 182
Great Spruce Island (Roque Island), 349, 350, 352
Great Wass Island, 334, 341, 343, 344, 345, 347
Green Island (Eastern Bay), 345
Green Island (Merchants Row), 274
Green Island (Petit Manan Point), 59, 330, 338, 339
Green Islands (Ebenecook Harbor), 120, 125
Green Islands (Penobscot Bay), 181, 182, 186
Greening Island (Somes Sound), 311
Greenland Cove (Muscongus Sound), 154, 156, 157, 158
Greenlaw Cove (Deer Isle), 281
Greens Island (Vinalhaven), 226, 227
Gregorys Point (Long Reach), **424**, 425
Grimross Island (Saint John River), 438
Grindel Point (Islesboro), 230, 232, 233, **234**
Grindstone Neck (Winter Harbor), 322, 328, **329**
Grog Island (Merchants Row), 274, 280
Guillemots, 352
Gut, The (Biddeford Pool), 54, 55
Gut, The (South Bristol); east, 150, **151**; west, 143, **144**

Haddock Island (Muscongus Bay), 157
Haley, Captain Samuel, 31
Haley's Cove (Isles of Shoals), 32
Halfway Rock (Casco Bay), 16, 63, 79, 89
Halifax Island (Roque Island), 349, 351, 352
Hall, Ebenezer, 186, 187
Hall Island (Muscongus Bay), 164–165
Hall Quarry (Somes Sound), 278, 309
Ham Loaf Island (Card Cove), 95
Hampden (Penobscot River), 242, **245**
Harbor Island (Burnt Coat Harbor), 288
Harbor Island (Frenchboro), 291
Harbor Island (Merchants Row), 279
Harbor Island (Muscongus Bay), 154, **164–165**
Harbor Island (Naskeag Harbor), 285
Harbor Island (New Meadows River), 93, 98
Harbour Bridge (Saint John, N.B.), 415

Harbour de Lute (Campobello Island), 378, **381**
Hardwood Island (Bartlett Narrows), 302, 303
Hardwood Island (Merchants Row), 279
Hardwood Island (Moosabec Reach), 344
Hardwood Island (Passamaquoddy Bay), 395
Hardy, 156; *Hardy III*, 167, 168
Hardy Island (Medomak River), 160
Harkness, 184
Harmon Harbor (Sheepscot River), 121
Harpswell Harbor (Harpswell Sound), 91
Harpswell Neck (Casco Bay), 63, 87, 88, 89
Harpswell Sound (Casco Bay), 63, 91, 93
Harraseeket River (Casco Bay), 64, **84–86**
Harraseeket Yacht Club (South Freeport), 64
Harrington Bay, Harrington River, 339–341
Harris Island (York River), 42, 43
Hassam, Childe, 30
Hatchet Cove (Muscongus Bay), 161
Hatfield Point, N.B. (Belleisle Bay), 426, 428
Hawthorne, Nathaniel, 30
Hay Island (Seal Bay), 220, 221
Haycock Harbor (eastern coast), 362
Haystack Mountain School of Crafts, 280
Heads, 9. *See also* Pump-out stations
Head Harbor (Head Harbor Island), 338, **347**
Head Harbor (Isle au Haut), 261, **263**
Head Harbor Island (Eastern Bay), 342, 343, 344
Head Harbour (Campobello Island, N.B.), 367, 370, **378-379**
Head Harbour Island (Campobello Island, N.B.), 379
Head Harbour Passage, 377, 378, 379, **380**, 383, 387, 392, 393; currents, 22, 380
Heart Island (East Penobscot Bay), 259
Helen J. Steitz, 207
Hells Half Acre (Merchants Row), 274, 275, 277
Hendricks Harbor (Sheepscot River), 122
Henry Cove (Winter Harbor), 329, **331**
Henry Islands (Smith Cove), 250
Heritage, 202
Hermit Island (Cape Small Harbor), 101, 102, 104
Herons, 87, 88, 103
Herreshoff, Nathaniel, 57
Hesper, 112, 131, 133
Hewett Island (Muscle Ridge Channel), 194, 195
Hickey Island (Little Kennebec Bay), 354
High Island, High Island Harbor (Muscle Ridge), 195–196
High Island (Poorhouse Cove), 152
High Island (Tenants Harbor), 191
Hinckley, Henry R. & Co., 308
Hockomock Bay (Inside Passage), 119, 120
Hockomock Channel (Medomak River), 160
Hodgdon Cove (Townsend Gut), 127
Hodgdon Island (Back River), 128
Hodgsons Island (Damariscotta River), 146, 147
Hog Island (Cobscook Bay), 390
Hog Island (Damariscotta River), 149
Hog Island (Digdeguash River), 397
Hog Island (Marsh Cove), 219
Hog Island (Metinic Island), 183
Hog Island (Moose Snare Cove), 354
Hog Island (Muscongus Bay), 154, 158, **159**, 160, 361

Hog Island (Saint John River), 424, 426, 427, 429
Hog Island (Washademoak Lake), 430
Hog Island Passage (Belleisle Bay), 426
Holbrook Island (Cape Rosier), 250, **251-252**
Hole-in-the-Wall (Saint John River), 429, **429–430**
Holmes Bay (Machias Bay), 355
Home Harbor (Muscle Ridge Channel), 194
Homer, Winslow, 58, 68
Hontvet, Maren, 32
Hopkins, Eric, 216
Hopkins Island (New Meadows River), 99
Hornbarn Cove (Muscongus Bay), 163
Horne, Sir William Van, 398
Horse Island (Potts Harbor), 89
Horseshoe Cove (Cape Rosier), 253
Hospital Island (Smith Cove), 250
House Island (Casco Bay), 71
How to use this book, 2
Howard Cove (Machias Bay), 355, 356
Huestis Island (Saint John River), 433
Hulls Cove (Frenchman Bay), 302
Human waste. *See* Pump-out stations
Hungry Island (Medomak River), 161
Hunting Island (Sheepscot River), 121
Hupper Island (Muscongus Bay), 154, 177, 178
Hurricane Island (Hurricane Sound), 194, 213, **225-226**, 278
Hurricane Island Outward Bound School (HIOBS), 24, 172, 199, **225-226**, 355, 357
Hurricane Sound (Penobscot Bay), 182, **221**, 222-226
Hurricanes, hurricane holes, 15
Hussey Sound (Casco Bay), 63, 73
Hutchins Island (Islesboro), 231, 235

Ice, 13; harvesting, 96, 101, 144, **145**
Indian Creek (Vinalhaven), 229
Indian Island (Head Harbor Passage, N.B.), 392
Indian Island (Kennebecasis Bay), 420
Indian Island (Rockport Harbor), 170, 203
Indian Lake (Grand Lake), 412, 433, **438**
Indiana, Robert, 229
Indians, 56, 175, 198, 357, 401; Indian attacks, wars, 8, 29, 33, 44, 47, 53, 71, 116, 133, 151, 171, 186, 187; land deals, 156; middens, 128, 148, 149, 349; Passamaquoddy Tribe, 392; place names, 115, 119, 140, 183, 293, 300, 340, 355, 369, 389, 392, 408, 419, 432; taken by Waymouth expedition, 171
Indiantown Island (Townsend Gut), 128
Ingalls Head (Grand Manan), 373, **376**
Inner Duck Rock (Monhegan Island), 168
Inner Harbor (Southeast Harbor), 281
Inner Heron Island (Damariscotta River), 142, 150
Inside Passage (Bath to Boothbay Harbor), 107, 108, 111, 117–119, 120, 124, 126–178, 139; timing, 118, 120
Ironbound Island (Frenchman Bay), 269, 322
Irony Island (Casco Bay), 88
Island Institute, 188, 292, 441
Islands, general, 7
Isle au Haut (Penobscot Bay), 182, 209, **261-266**, 269, 270, 272, 273, 274, 277, 285, 295, 297, 301, 302; lighthouse, 170; *See also* Acadia National Park

Isle au Haut (village), 265-266
Isle au Haut Thorofare, 261, 263, **265-266**
Isle of Bacchus (Richmond Island), 60
Isle of Pines (Long Reach), 426
Isle of Springs (Sheepscot River), 119, 120, **128**
Isles of Shoals, 29, **30–33**, 129
Islesboro (Penobscot Bay), 182, 210, **230**, 231-232, 234-236
Islesboro Harbor (Islesboro), 236
Islesford. *See* Little Cranberry Island
Issac H. Evans, 202, 282

Jackson, President Andrew, 214
Jackson Laboratory (Bar Harbor), 301, 325
Jamaica Island (Piscataqua River), 37
Jefferson, President Thomas, 112, 132, 366
Jemseg River (Saint John River), 405, **433**
Jenkins Cove (Belleisle Bay), 426, **428**
Jenny Island (Ridley Cove), 93, 96
Jericho Bay, 269, 270, 282, 285
Jewell Island (Casco Bay), 63, 76, **79–80**
Jewett, Sarah Orne, 30, 190
Job Island (Islesboro), 231, 232
John F. Randall, 145
John Island (Bartlett Narrows), 300
Johns Bay, 143, **150–153**
Johns Island (Casco Passage), 285
Johns Island (Johns Bay), 151
Johns River (Johns Bay), 152, 153
Johnson Bay (Lubec), 383, **384**, 385
Johnson Cove (Little Kennebec Bay), 353–354
Jones, John Paul, 33
Jones Cove (Damariscotta River), 146
Jonesport (Moosabec Reach), 276, 295, 342-344, **348**, 350, 360
Jordan Island (Frenchman Bay), 322, 328
Jordan River (Mount Desert Narrows), 326
Jordans Delight (Narraguagus Bay), 339, 341
Josiah's River (Ogunquit), 45
Joy Bay (Gouldsboro Bay), 338
Julie N., 70

Katona, Steven, 295
Kench, Thomas, 286
Kennebec River, 16, 26, 29, 101, 103, *107–108*, 109-111, 114, 116, 140, 145, 152, 173, 340; currents, 108; tidal range, 21
Kennebecasis Bay (Saint John River), 412, 418, **419**, 420-423
Kennebecasis Island (Saint John River), 418, **419**, 421
Kennebecasis River (Kennebecasis Bay), 419, **423**
Kennebeccasis Yacht Club, 412, 414, 418, 419, **420**, 425
Kennebunk, Kennebunkport, 29, **47–49**, 112, 295
Kennebunk River. *See* Kennebunk, Kennebunkport
Kennebunkport Conservation Trust, 53
Kent, Rockwell, 169
Kent Cove (Fox Islands Thorofare), 217
Ketepec, N.B. (Grand Bay), 418
Kidd, Captain, 31, 79
Killick Stone Island (Muscongus Bay), 157
Kimball Island (Isle au Haut Thorofare), 264, 265
Kingston Creek (Belleisle Bay), 426, **427**
Kittery, Kittery Point, 29, 33, 36, 37, 38, 39, 112
Kittery Landing Marina, 39

Kittery Point Yacht Club, 39
Knight Island (Eastern Bay), 346
Knox, Major General Henry, 177
Knubble Bay (Inside Passage), 119, 124, 125
Kollegewidgwok Yacht Club (Blue Hill Harbor), 293, 296
Kress, Stephen, 159, 165, 361

Lakeman Island, Lakeman Harbor (Roque Island), 349, **352**
Lasell Island (Penobscot Bay), 210, **211**
Laundry Cove (Isle au Haut), 265, **266**
Laura B., 167, 169, 177, 178
Lawrys Island, Narrows (Hurricane Sound), 221, **222**, 224
Lazygut Islands (Deer Island Thorofare), 270
Leadbetter Island, Narrows (Hurricane Sound), 221-223
Leavitt Island (Ridley Cove), 96
Lee Cove (Saint John River), 417
Lem's Cove (Bucks Harbor), 254
Leonardville (Deer Island, NB), 392
Letang Harbour, River (Bay of Fundy), 406, **407-408**
Letete Passage (New Brunswick), 367, 377, 378, **394**, 404-406, 408; currents, 22
Levett, Captain Christopher, 42
Lewis Cove (Linekin Bay), 140
Lewis R. French, 112
Libby Islands (Machias Bay), 112, 355
Liberty ships, 69
Lighthouses, **170**; museums, 170, 178, 373, 440; *See also* Burgess, Abbie
Lime, 112, 175, 182, 198, 203, 204, **205**
Lime Island (Penobscot Bay), 210
Lincolnville Beach (West Penobscot Bay), 205
Lindbergh, Charles and Anne Morrow, 224, 225
Linekin Bay (Booth Bay), 140
Lines Island (Kennebec River), 114
Little Barred Island (Penobscot Bay), 257-258
Little Bay (Piscataqua River), 41
Little Birch Island (Casco Bay), 89
Little Brimstone Island (Isle au Haut Bay), 230
Little Bustins Island (Casco Bay), 64, 85
Little Chebeague Island (Casco Bay), 76
Little Cranberry Island (Mount Desert), 301, 316, **317–318**, 319
Little Cranberry Yacht Club, 318
Little Deer Isle (Penobscot Bay), 182
Little Diamond Island (Casco Bay), 73, 74
Little Eaton Island (East Penobscot Bay), 256
Little Garden Island (White Islands), 224-225
Little Green Island (Muscle Ridge Channel), 195
Little Harbor (New Hampshire), 34, 35–36
Little Hen Island (Seal Bay), 220
Little Hurricane Island (Hurricane Sound), 222
Little Jewell Island (Casco Bay), 79, 80
Little Kennebec Bay (Englishman Bay), 353, 354
Little Machias Bay, 358
Little Mark Island (Casco Bay), 81
Little Moshier Island (Casco Bay), 83
Little River. *See* Cutler
Little River (Damariscotta River), 140, **141–142**
Little River Island (Cutler), 358, 359
Little Sheepscot River, 120, **124**
Little Snow Island (Quahog Bay), 94

Little Spruce Island (Roque Island), 349, 350
Little Thorofare (Fox Islands Thorofare), 218
Little White Island (White Islands), 224-225
Lobsterboat races, 272, 328, 348
Lobster Cove (Muscle Ridge Channel), 194
Lobster Island (Southern Harbor), 214
Lobsters, Lobstering, 11, 12, **50**, 52, 369, 370; buoys, **12**, 26; lobster pounds, 104, 328, 341
Locust Island (Medomak River), 160
Longfellow, Henry Wadsworth, 30, 67
Long Cove (Hurricane Sound), 221, **223**
Long Cove (Tenants Harbor), 191–192
Long Cove (West Penobscot Bay), 240-241
Long Island (Blue Hill Bay), 293
Long Island (Casco Bay), 63, 74, 76
Long Island (Cobscook Bay), 390
Long Island (Digdeguash River), 397
Long Island (Frenchboro), 291-293
Long Island (Kennebecasis Bay), 419, 421, 422, 423
Long Island (New Meadows River), 100
Long Island (Saint John River), 429, 430
Long Point (Belleisle Bay), 428
Long Point Island (Quahog Bay), 93
Long Porcupine Island (Frenchman Bay), 322
Long Reach (Saint John River), 412, 418, **424-426**
Longfellow, Henry Wadsworth, 30, 67, 68
Loran, 17, 18, 21–24; interference, 21; in Canada, 21
Lords Cove (Deer Island, N.B.), 392, **393–394**
Louds Island (Muscongus Bay), 154, **157**, 159
Love Cove (Ebenecook Harbor), 120, 125, **126**
Lovell, General Solomon, 251; *see also* Penobscot Expedition
Lowell Cove (Wills Gut), 94
Lower Goose Island (Casco Bay), 87
Lower Hell Gate (Inside Passage), 22, 117, **119**, 124, 139
Lower Musquash Island (Saint John River), 429, 429–430
Loyalists, United Empire, 411, 429, 432, 439
Lubec, 295, 362, 365, 367, 377, 379, 380, 382, **383-385**
Lubec Narrows, 367, **377**, 378, 382, **383**, 384; currents, 22
Luckse Sound (Casco Bay), 63
Lunging Island (Isles of Shoals), 30, 32
Lunt Harbor (Frenchboro), 291-292
Luther Little, 112, 131, 133

Machiasport (Machias River), 356, 357
Machias (Machias River), 355
Machias Bay, 341, **355**, 358
Machias River (Machias Bay), 129, **355**, 357
Machias Seal Island, 165, 330, 334, 348, 358, **360-361**, 376, 400
Machiasport (Machias River), 355, 356, **357**
Mackerel Cove (Casco Bay), 89
Mackerel Cove (Swan's Island), 285, **286-287**
MacMahan Island (Sheepscot River), 119, 120, 124
MacMillan, Admiral Donald, 81, 112, 144
Mactaquac Dam (Saint John River), 405, 412, 439, 440
Maddock Cove (Ebenecook Harbor), **125**, 126
Magaguadavic River (Passamaquoddy Bay), 395, **396**
Maine Audubon Society. *See* Audubon, Maine Society
Maine Coast Heritage Trust, 11, 292, 319, 359, 441
Maine Coastal Islands National Wildlife Refuge, 8, 33, 339, 352

Maine Island Trail, Maine Island Trail Association, 9, 11, 78, 80, 99, 274, 352, 441
Maine Lights Program, 53
Maine Maritime Academy (Castine), 181
Maine Maritime Museum (Bath), 52, 111, **113-114**
Maine Seacoast Missionary Society, 186
Maine Sea Coast Mission, 292
Maine Shipyard (South Portland), 69
Maine, State of; Bureau of Public Lands, 110, 148; Dept. of Inland Fisheries and Wildlife, 115; Dept. of Marine Resources, 52, 139, 192
Maine Yankee (Montsweag Bay), 120
Malaga Island (Isles of Shoals), 30, **31**
Malaga Island (New Meadows River), 98
Malden Island (Five Islands), 123
Man Island (Head Harbor), 347
Manana Island (Monhegan), 26, 166, 168, 169
Manawagonish Island (Bay of Fundy), 410
Manset (Southwest Harbor), 308
Maple Juice Cove (St. George River), 154, 173, **174-175**
Maquapit Lake (Grand Lake), 412, 433, **438**
Maquoit Bay (Casco Bay), 63, 73, **87**
Marble Cove (Saint John River), 417
Margaretta, 355
Marine operators, 24
Mark Island (Deer Island Thorofare), 270
Mark Island (Kennebec River), 110
Mark Island (Moosabec Reach), 348
Mark Island (Penobscot Bay), 210
Mark Island (Winter Harbor), 329
Marsh Cove (North Head Island), 219
Marsh Island (Cape Split), 342
Marsh Island (Muscongus Bay), 154, 157
Marsh Island (Roque Island), 349, 352
Marshall Island (Toothacher Bay), 290
Marshall Point (Port Clyde), 107, 170, 173, 178
Mary and John, 110
Mary Barrett, 125
Mary Day, 262
Mason Bay (Chandler Bay), 353
Masts and the Broad Arrow, 67, **82**
Mather Island (Kennebecasis Bay), 423
Mathews Cove (Kennebecasis Bay), 423
Matinicus Island, Matinicus Harbor, 112, 181, 182, **183–187**
Matinicus Rock, 165, 181, 182, **188-189**, 295, 330, 361, 400; lighthouse, 26, 170, 185, 188, 189, 202; passages to, 26
Mayflower, 133
McCormack Cove (Kennebecasis Island), 418, **419**
McFarland Island (Boothbay Harbor), 136
McFarlands Cove (Johns Bay), 150
McGee Island (Muscongus Bay), 154, **172**
McGlathery Island (Merchants Row), 274, **277**
McHeard Cove (Blue Hill Bay), 298
McMaster Island (Letete Passage), 377, 394
Meadow Cove (Damariscotta River), 146
Medomak (Medomak River), 160–178
Medomak River (Muscongus Bay), 148, 154, 160, 162
Meduncook River (Muscongus Bay), 163
Meehans Cove (Kennebecasis Bay), 423
Memorial Bridge (Portsmouth-Kittery), 39

Menhaden (pogies), 92
Merchant Island, Merchant Harbor, 274, **279**
Merchant Row (passage), 269, **270**; currents, 22
Merchants Row (islands), 182, 269, 270, 272, **274–279**
Merepoint Bay (Casco Bay), 63, **87**
Merriconeag Sound (Casco Bay), 63
Merritt Island (New Meadows River), 101
Merrymeeting Bay (Kennebec River), 107, 114
Metinic Island (Penobscot Bay), 181, **183**
Metric system, conversion, 365, 377
Middle Bay (Casco Bay), 63
Middle Hardwood Island (Eastern Bay), 344
Milbridge (Narraguagus River), 340
Milkish Channel (Saint John River), 419, 421
Mill Cove (Boothbay harbor), 137, 138
Mill Cove (Grand Lake), 433, **434**
Mill Cove (Harpswell Sound), 92
Mill Cove (Long Reach), 425
Mill Cove (New Meadows River), 100
Mill Cove (Quahog Bay), 96
Mill Cove (Sheepscot River), 128-129
Mill River (Winter Harbor, Vinalhaven), 216, 219
Millay, Edna St. Vincent, 209
Million Dollar Bridge (Portland), 70
Minister Island (Passamaquoddy Bay), 398
Ministerial Island (Casco Bay), 81
Mink Island (Cross Island), 357
Mink Island (Eastern Bay), 345
Minturn (Swan's Island). *See* Swan's Island: Minturn
Mistake Cove (Long Reach), 426
Mistake Intervale (Long Reach), 424, 426
Mistake Island, Mistake Island Harbor (Eastern Bay, 335, 343, 344, 345, **346**, 347
Mohawk Island (Letete Passage), 394
Monckton, Colonel Robert, 432
Money, Canadian, 367
Monhegan Island, 107, 109, 133, 139, 154, **166–169**; excursions to, 156, 167, 177, 178; lighthouse, 26, 167, 169, 170; passages to, 26
Monroe Island (Muscle Ridge Channel), 197
Montsweag Bay (Inside Passage), 120
Moore, Ruth, 304
Moores Harbor (Isle au Haut), 263, **264**
Moosabec Reach, 310, 342, **343**, 344, 348; currents, 20
Moose Island (Blue Hill Bay), 303
Moose Island (Eastport), 385, 389, 392, 408
Moose Island (Stonington), 272
Moose Snare Cove (Little Kennebec Bay), 354
Morgan, Dodge, 66
Morgan Bay (Blue Hill Bay), 299
Morison, Samuel Eliot, 310, 349
Morse Cove (Penobscot River), 243
Morse Island (Muscongus Bay), 162, 164
Moshier Island (Casco Bay), 85
Mosquito Island (Penobscot Bay), 178
Mount Agamenticus, 29
Mount Desert Island, 6, 209, 269, 270, 277, 285, 287, 290, 293, 295, 297, **300–302**, 303, 305, 310, 311, 312, 313, 317, 322, 323, 326, 327, 328, 330, 332, 338, 347, 370; passages to, 26; visibility, 26

Mount Desert Narrows, 299, **326**
Mount Desert Rock, 269, **293**, 295, 315; lighthouse, 26, 170; passages to, 26
Mount Katahdin, 56
Mouse Island (Burnt-Pell Passage), 266
Mouse Island (Penobscot Bay), 210
Mouse Island (Perry Creek), 216
Mud Hole (Great Wass Island), 343, 344, 345
Muddy River (Merrymeeting Bay), 114
Muir, Emily, 275
Mullen Cove (North Haven Island), 218
Muscle Ridge Channel (Penobscot Bay), 182, **192–193**
Muscongus Bay, **154–178**, 361; passages through, 154
Muscongus Sound (Muscongus Bay), 154, 157, 159
Musquash Harbour, N.B. (Bay of Fundy), 409
Musquash Island, River (Musquash Harbour), 409
Mussels, **192**, 386; and red tide, 192; cultured, 191, 219

Narraguagus Bay, Narraguagus River, **339–341**, 340, 341
Narrows, The (Damariscotta River), 146, **147**, 148
Natalie Todd, 325
National Audubon Society. *See* Audubon
Nature Conservancy, The, 11, 87, 88, 134, 293, 327, 334, 339, 340, 343, 345, 346, 352, 355, 356, 357, 390, 441
Nautilus Island (Castine), 247, 251
Navigation; Bay of Fundy, Passamaquoddy Bay, 365; Canadian buoyage, 365; publications, 18, 365; Saint John River, 412; relative navigation, 18
Navy Island (St. Andrews, N.B.), 399
Negro Island (Biddeford Pool), 54
Nevers Island (Saint John River), 433
New Castle, N.H. (Portsmouth Harbor), 35
New Harbor (Muscongus Bay), 152, 154, **156**, 167, 295
New Meadows River (Casco Bay), 63, 64, 93, **96–101**
Newcastle (Damariscotta River), 141, 148, **149**
Newcastle Creek (Grand Lake), 433, **437**
Nigh Duck (Monhegan Island), 168
No Mans Island (Merchants Row), 277
No Mans Land (Matinicus), 185
North Haven (village), 182, 212, 213, 214, **215-216**
North Haven Casino (North Haven Yacht Club), 215
North Haven Island (Penobscot Bay), 182, 201, 210-216
North Head (Grand Manan), 370, **371-373**, 375
Northeast Arm (Grand Lake), 433, 435, **437**
Northeast Cove (Dyer Island), 341
Northeast Harbor (Cross Island), 357
Northeast Harbor (Mount Desert Island), 269, 295, 301, 305, 309, 310, **313-315**, 317, 319, 323
Northeast Harbor Fleet (Gilpartick Cove), 309
Northeast Marine Animal Lifeline, 20, 295
Northern Harbour (Deer Island, N.B.), 393
Northern Island (Long Cove), 191
Northwest Cove (Dyer Island), 299
Northwest Harbor (Deer Isle), 259
Northwest Harbour (Deer Island, N.B.), 392, 393
Norton, Captain Barna, 348, 360
Norton Island (Seal Harbor, Muscle Ridge), 193
Nottingham, 41
Nubble, The (Cape Neddick), 44
Nubble, The (Metinic Island), 183

Oak Bay (St. Croix River), 402
Oak Island (Muscle Ridge Channel), 195
Oak Point (Long Reach), 412, **425–426**
Oak Point (Townsend Gut), 127
Oakhurst Island (Ridley Cove), 103
Oar Island, Oar Island Cove (Hockomock Channel), 159, 160
Odiorne State Park (New Hampshire), 35
Ogilvie, Elizabeth, 187
Ogunquit, 45–46
Old Cove (Matinicus), 185
Old Harbor (Vinalhaven), 226-227
Old Man Island (Little Machias Bay), 357, 400
Old Orchard Beach (Southern Coast), 29
Old Sow (Western Passage), 367, 382, 387, **392**
Opechee Island (Casco Passage), 285
Orcutt Harbor (Cape Rosier), 254
Oromocto, N.B. (Saint John River), 438, **439**
Oromocto Boat Club (Saint John River), 439
Oromocto Island (Saint John River), 439
Oromocto River (Saint John River), 439
Orono Island (Casco Passage), 285, 287
Orrs Cove (Quahog Bay), 95–96
Orrs Island (Casco Bay), 63, 90, 94
Orrs-Bailey Yacht Club (Merriconeag Sound), 91
Ospreys, 131, **176**, 334, 339, 340, 341, 352, 355, 426
Otis Cove (St. George River), 154, 173, **175**
Otter Cove (Eastern Harbor), 342
Otter Island (Muscle Ridge Channel), 195
Otter Island (Muscongus Bay), 164
Outer Bar Island (Corea), 337, 338
Outward Bound School. *See* Hurricane Island Outward Bound School
Oven Mouth (Cross River), 22, **130**
Owls Head, Owls Head Harbor (Penobscot Bay), 193, **197**, 310; lighthouse, 170, 199; transportation museum, 202
Ox Island, Ox Island Channel (Saint John River), 438

Palmer Creek (Saint John River), 429
Palmer Island (Medomak River), 161
Parker Island (Letete Passage), 394
Partridge Island (Saint John), 170, 410
Passagassawakeag River (Belfast), 238
Passamaquoddy Bay, 365, 367, 374, 377, 378, 382, 391, 392, 394, 395, 396, 397, 398, 399, 401, 408, 414; approaches to, 376, 383; tides, 385, 395
Patten Bay (Union River Bay), 299
Pattishall, Captain Richard, 133
Peaks Island (Casco Bay), 63, 64, **71–72**, 74, 80
Peary, Admiral Robert, 81
Pemaquid Harbor (Johns Bay), 133, 150, **151-152**
Pemaquid Neck, Pemaquid Point, 107, 150, 151, 154, 157
Pemaquid River (Pemaquid Harbor), 152
Pendleton Island (Deer Island, N.B.), 396
Pennamaquan River (Cobscook Bay), 389, 390
Penobscot Bay, 6, **179–266**, 269, 276, 282, 297, 361, 370
Penobscot Expedition, 181, 242, 247, **251**
Penobscot Island (Seal Bay), 184, 220, 221
Penobscot Marine Museum (Searsport), 239, 240
Penobscot River, 16, 129, 145, 173, **242-248**; tidal range, 21
Pentecost Harbor (Georges Islands), 82, 171

Pepperrell, Sir William, 36
Pepperrell Cove (Portsmouth Harbor), 34, 35, 36–37, 38
Pepperrell Cove Yacht Club, 36
Perkins Cove (Ogunquit), 45–46
Perkins Island (Kennebec River), 110
Perry Creek (Fox Islands Thorofare), 182, **216**
Peterson, Roger Tory, 159
Petit Manan Island, 17, 332, **338**, 340, 341, 349; lighthouse, 202
Pets, 11
Pettingill Island (Casco Bay), 87
Philbrook Cove (Seven Hundred Acre Island), 234
Philip E. Lake, 96
Pickering Cove (Southeast Harbor), 280
Pickering Island (East Penobscot Bay), 182, **256**, 257
Pierce Cove (Ebenecook Harbor), 125, **126**
Pig Island (Saint John River), 426
Pigeon Hill Bay, 339
Pigs Island (Casco Bay), 87
Pine Island (Washademoak Lake), 431
Pirates, 31, 151, 349, 355. *See also* Bellamy, Captain; Blackbeard (Edward Teach); Kidd, Captain
Piscataqua River, 29, 33, 35; currents, 34
Pleasant Bay, Pleasant River, 339–341
Pleasant Cove (Damariscotta River), 141, **147**, 148
Pleasant Cove Island (Damariscotta River), 147
Pleasant Point Gut (St. George River), 154, **173**
Plymouth, Mass., Plymouth Plantation, 107, 110, 133, 175
Point Lepreau, N.B. (Bay of Fundy), 406, **408**, 409
Pole Island (Quahog Bay), 94, 95
Pond Island (Casco Passage), 285
Pond Island (Kennebec River), 108
Pond Island (Narraguagus Bay), 341
Pond Island (Penobscot Bay), 182
Pond Island (Schoodic Peninsula), 301, **332**
Pond Island Passage (Blue Hill Bay), 269
Poorhouse Cove (Johns Bay), 150, **152-153**
Popes Folly Island (Friar Roads), 384
Popes Island (Head Harbour Passage, N.B.), 392
Popham, Captain George, 110
Popham Colony, 29, 107, 110, 111, 112
Popplestone Cove (Marshall Island), 290
Porcupine Islands (Frenchman Bay), 269, 322, 323
Porcupine Island (Little Kennebec Bay), 354
Port Clyde (St. George River), 154, 168, 171, 172, **177-178**, 295
Porter, Eliot, 16, 188
Portland (Casco Bay), 24, 63, 64, **65–66**, 70, 71, 73, 74, 80, 81, 91, 148, 295
Portland Head, Portland Head Light, 65, 67, 109, 112, 170
Portland Yacht Club (Falmouth Foreside), 74, 75
Ports of Entry. *See* Canada, entering; United States, entering
Portsmouth (New Hampshire), 29, 33, **38-41**, 295
Portsmouth Yacht Club (New Castle), 31, 35, 39
Potato Island (Burnt Coat Harbor), 288
Potato Island (Merchants Row), 274
Potts Harbor (Casco Bay), 88
Pound of Tea Island (Harraseeket River), 85
Pownalborough Courthouse (Kennebec), 116
Pratts Island (Cozy Harbor), 122
Preble Island (Sorrento), 322, 327, 328
Preparations for cruising in Maine, 6

Prescott Park (Portsmouth, N.H.), 39, 40
Pretty Marsh Harbor (Blue Hill Bay), 293, **302**
Princess Margaret Bridge (Fredericton, N.B.), 439, 440
Pring, Martin, 47, 166, 211
Printz Cove (Grand Lake), 437
Private property, 8
Prospect Harbor (Schoodic Peninsula), 335, **336**
Prouts Neck (Saco Bay), 58–59
Prouts Neck Yacht Club, 59
Public and private property. *See* Private property
Public transportation, 448
Pudding Island (Ragged Island), 187, 188
Puffins, 165, 182, 188, 330, 334, 343, 348, 349, 360, **361**, 400; Puffin Project, 159, 165–166
Pulpit Harbor (North Haven Island), 50, 210, **211–212**, 213
Pump-out stations, 10, 442
Pumpkin Knob (Casco Bay), 64
Purdys Point (Long Reach), 424, 425
Putz, George, 187

Quahog Bay (Casco Bay), 63, 64, 93, **94-95**, 96
Quoddy Narrows, 383
Quoddy River (Deer Island, N.B.), 393

Rackliff Island (Seal Harbor, Muscle Ridge), 193
Radar, 17, **18**
Ragged Island (Criehaven), 181, 182, 184, 186, **187-188**
Ragged Island (Quahog Bay), 93
Ram Island (Booth Bay), 136
Ram Island (Clam Cove), 203
Ram Island (Hatchet Cove), 161, 162
Ram Island (Holbrook Island Harbor), 251, 252
Ram Island (Kennebec River), 114
Ram Island (Muscongus Sound), 158
Ram Island (Saint John River), 438
Ram Island (Tibbett Narrows, Western Bay), 343
Randolph (Kennebec River), 116–117
Ranger (J-boat), 111
Ranger (John Paul Jones), 33
Raspberry Island (Port Clyde), 178
Raspberry Island (Quahog Bay), 94
Razor Island (Cobscook Bay), 390
Razor-billed auks, razorbills, 165, 334, 343, 348, 355, 357, 360, 361, **400**
Reach, The (Vinalhaven), 211, 222, **226-227**
Red Island (Cobscook Bay), 389, 390
Red Jacket, 112, **200**
Red tide, red tide hotline, 192
Reeds Island (Little River), 141
Reindeer Cove (Greens Island), 227
Renforth, N.B. (Renforth Cove), 422
Renforth Boat Club (Renforth Cove), 422
Renforth Cove (Kennebecasis Bay), 422
Revere, Paul, 33
Reversing Falls (Saint John River), 404, 405, 411, 412, **414–417**, 418, 420, 424, 433, 437; and river level, 412; and river tide cycles, 412; approaches downriver, 415–416; approaches upriver, 415; direction of current, 405, 411; emergencies in, 416; timing passage, 406, 411; viewing, 411

Rich, Louise Dickenson, 132, 344
Rich Cove (Quahog Bay), 96
Richmond (Kennebec River), 108, 114, **115**, 116
Richmond Island, Richmond Island Harbor (Saco Bay), 29
Ridley Cove (Casco Bay), 96
Riggs Cove (Inside Passage), 124, **125**
Rivers; current, 22, 108; seamanship, 108
Roberts, Kenneth, 41
Robinhood Cove (Inside Passage), 120, 124, 125
Rockefeller family, 290, 299, 301, 319, 328
Rockland (Penobscot Bay), 170, 181, 184, 193, **198-202**, 203, 205, 213, 215, 295, 310
Rockport (Penobscot Bay), 181, **203-205**, 310
Rockport Boat Club, 203
Rodgers Island (Johnson Bay), 385
Rogues Island (New Meadows River), 93
Roosevelt, President Franklin Delano, 365, 378, 381, 382
Roque Island, Roque Island Harbor, 112, 310, 334, 338, 341, 347, 349–353, 370
Rosier, James, 166, 173, 174. *See also* Weymouth, Captain George; reports of lobsters, 50
Ross Island (Grand Manan), 370
Ross Island (Muscongus Bay), 129, 157
Rothesay, N.B. (Kennebecasis Bay), 422
Rothesay Yacht Club, 422
Round Island (Machias Bay), 355
Round Island (Merchants Row), 274, **277-278**
Round Pond (Muscongus Sound), 156, **157–158**
Royal Kennebeccasis Yacht Club, 412, 414, 418, 419, **420**, 425
Royal River (Casco Bay), 83–84
Rum Key (Frenchman Bay), 322
Rush Island (Long Reach), 426
Russ Island (Merchants Row), 274, **275**
Rutherford Island (Damariscotta River), 143

Sabbathday Harbor (Islesboro), 230, **235**, 236
Saco (Saco River), 57, 58
Saco River (Saco Bay), 29, **57–58**
Saco Yacht Club, 57
Saddle Island (Penobscot Bay), 210
Sagadahoc Bridge (Bath), 111
St. Andrews, N.B. (Passamaquoddy Bay), 365, 367, 377, 387, 392, 394, 395, **398-401**, 432
St. Andrews Yacht Club, 399
St. Croix Island (St. Croix River), 365, 401
St. Croix River, 152, 365, 392, 398, **401–402**
St. George, N.B. (Magaguadavic River), 396
St. George River (Muscongus Bay), 154, 170, 171, 172, **173–177**
St. Helena Island (Deer Island, N.B.), 393
St. Helena Island (Merchants Row), 278
Saint John, N.B. (Bay of Fundy), 360, 365, 405, 408, 409, **410-411**, 412, 414, 417, 419, 420, 422, 425, 432, 437, 439; passages to, 405
Saint John Power Boat Club (Marble Cove), 417
Saint John River (New Brunswick), 404, 405, 410, 411, **412–414**, 418, 424, 425, 426, 427, 429, 430, 433, 438, 439, 440; cable ferries, 419, 421, 424, 426, 428; Coast Guard on, 424; tides, 412, 432
St. Stephen, N.B. (St. Croix River), 401-402
Sally, 132

Sally Islands (Gouldsboro Bay), 338
Salmon, Atlantic, 108, 129, 340. *See also* Aquaculture
Salmon River (Grand Lake), 433, **437**
Salt Island (Machias Bay), 355
Salt Marsh Cove (Damariscotta River), 148
Saltonstall, Commodore Dudley, 242, 251
Samoset, Chief, 156
Sand Cove (Grand Bay), 418
Sand Cove (Marshall Island), 290
Sand Cove (Winter Harbor), 329–330, 331
Sand Cove North (Great Wass Island), 344
Sand Island (Casco Bay), 78
Sand Island (Hatchet Cove), 161
Sandy Cove (Down East), 362
Sasanoa River (Inside Passage), 108, 111, **118–119**, 139
Sawyer Cove (Bartlett Narrows), 303
Sawyer Cove (Jonesport), 348
Sawyer Island (Sheepscot River), 119, 128
Scarborough, 29
Scarborough Beach, 59
Schoodic Island (Schoodic Point), 332
Schoodic Peninsula, Schoodic Point, 301, 322, 323, 328, 329, **332**, 334, 335, 342, 349
Seagulls, **59**, 330
Seal Bay (Vinalhaven), 220-221
Seal Cove (Blue Hill Bay), 303
Seal Cove (Damariscotta River), 141, **146-147**
Seal Cove (Fox Islands Thorofare), 217
Seal Cove (Grand Manan), 360, 370, 373, **375-376**
Seal Cove (Richmond Island), 60
Seal Harbor (Islesboro), 234
Seal Harbor (Mount Desert Island), 301, 310, 319
Seal Harbor (Muscle Ridge Channel), 193
Seal Harbor Yacht Club (Mount Desert Island), 319, 322
Seal Island (Penobscot Bay), 181, 182, 295
Sealand Cove (Eastern Bay), 344
Seals, **20**; Andre, 204; captive, 401; stranding hotline, 10, 20
Search-and-Rescue. *See* **Emergency help**
Sears Island (Penobscot Bay), 181
Searsport (Penobscot Bay), 238, **239-240**
Seavey Island (Kittery), 33, 35, 37
Seavey Island (Muscongus Bay), 154
Seaveys Island (Isles of Shoals), 32
Seaweed; harvesting, 370, 372
Sebasco (New Meadows River), 98
Sebasco Harbor (New Meadows River), 64, 99, **101–102**, 103, 104
Sebascodegan Island (Casco Bay), 92, 94, 96
Sedgwick (Benjamin River), 283
Seguin, Friends of, 170
Seguin Island, 26, 101, 107, **109**, 170; lighthouse, 26, 109, 170
Seven Hundred Acre Island (Penobscot Bay), 230, 231, **232-233**, 234
Shabbit Island (Moosabec Reach), 343
Shampers Cove (Belleisle Bay), 426, **427**
Shapleigh Island (Portsmouth, N.H.), 40
Sheep Island (Blue Hill Bay), 285
Sheep Island (Gouldsboro Bay), 338
Sheep Island (Muscle Ridge Channel), 197
Sheep Island (New Meadows River), 99

Sheep Island (Smith Cove), 250
Sheep Porcupine Island (Frenchman Bay), 322, 326
Sheepscot Bay, 122
Sheepscot River, 107, **120-133**, 140, 350
Sherman Cove (Camden Harbor), 206
Sherman Zwicker, 113
Ship Island (Blue Hill Bay), 293
Shipbuilding, Casco Bay, 67, 69, 84; Down East, 358; Midcoast, 108, 110, 132, 140, 143, 144, 160, 178; Penobscot Bay, 200, 206, 207; Southern Coast, 33, 38, 47, 51. *See also* Bath Iron Works; Liberty Ships; Maine Maritime Museum; Penobscot Marine Museum
Shipstern Island (Narraguagus Bay), 339, 340, 341
Shorey Cove (Roque Island), 349, **353**
Signe, 112
Sipp Bay (Cobscook Bay), 389
Sister Island (Casco Bay), 87
Skillings River (Frenchman Bay), 326
Slins Island (Seal Harbor, Muscle Ridge), 193
Slocum, Joshua, 157
Small Point Harbor (Casco Bay), 104
Small Point Yacht Club, 103
Smith, Captain John, xi, 30, 142, 150, 166
Smith Cove (Cape Rosier), 248, **250**, 252
Smith's Isles (Isles of Shoals), 30
Smuttynose Island (Isles of Shoals), 30, 31
Smuttynose Island (Monhegan), 166, 167, 168
Snow Island (Quahog Bay), 64, **94**, 95
Somes Cove (Blue Hill Bay), 302
Somes Harbor, Somesville (Somes Sound), 309, **312**
Somes Sound (Mount Desert Island), 276, 278, 305, **309–313**, 350
Sorrento (Frenchman Bay), 269, 322, 326, **327**, 328
Sorrento Yacht Club, 327
South Addison (Cape Split), 342
South Bay (Cobscook Bay), 389, 390
South Bay (Grand Bay), 418
South Bristol (Damariscotta River), 143, 148, **151**
South Freeport (Harraseeket River), 64, 81, **84-86**
South Portland (Casco Bay), 65–66
Southeast Harbor (Deer Isle), 280, **281**
Southern Harbor (Fox Islands Thorofare), 214
Southern Island (Tenants Harbor), 190
Southern Wolf Island (The Wolves), 408
Southport Island (Sheepscot River), 120, 121, 122, 136, 139; Inside Passage, 127
Southport Yacht Club (Cozy Harbor), 122
Southwest Cove (Merchant Island), 279
Southwest Harbor (Mount Desert Island), 52, 269, 295, *305-308*, 323. *See also* Manset
Sow Island (Casco Bay), 87
Spanish-American War, 33, 70, 110
Spar Island (Mason Bay), 353
Spectacle Islands (Cape Rosier), 252
Spectacle Islands (Townsend Gut), 128
Spectacles, The (Long Cove), 191
Speedwell, 211
Spring Cove (Cushing Island), 71
Sprout Island (Merchants Row), 274
Sprucehead Island (Muscle Ridge Channel), 193, 194

Spruce Head (Muscle Ridge Channel), 194
Spruce Island (Gilkey Harbor), 231, 233
Spruce Island (Head Harbour Passage), 380
Squirrel Island (Booth Bay), **135–136**, 139, 140
Squirrel Island Boating Association, 136
Stage Island, Stage Island Harbor (Biddeford Pool), **53**, 54, 55
Stage Neck (York River), 42
Star Island (Isles of Shoals), 30, **31**
Starboard Cove, Starboard Island (Machias Bay), 355
State of Maine, 247, 249
Stave Island (Casco Bay), 78
Stave Island, Stave Island Harbor (Frenchman Bay), 328
Steele Harbor, Steel Harbor Island (Eastern Bay), 343, 346
Stephen Taber, 112
Stimpsons Island (Fox Islands Thorofare), 218
Stockman Island (Casco Bay), 78
Stockton Harbor (Penobscot Bay), 241
Stone Island (Machias Bay), 355
Stone Island (Muscongus Bay), 172
Stone Island (Starboard Cove), 355
Stonington (Deer Island Thorofare), 182, 269, 270, **272-273**, 275, 277, 278
Stratton Island (Saco Bay), 58, 59
Strawbery Banke (Portsmouth, N.H.), 40
Strout, Captain Joshua, 170
Sturgeon Island (Merrymeeting Bay), 114
Submarines, 33, 64, 70, 76, 79, 80, 358; museums, 38, 41
Sugar Loaves, The (Fox Islands Throrofare), 214
Suliote, 238
Sullivan Harbor (Frenchman Bay), 326
Sunbeam; Sunbeam IV, 313. *See also* Seacoast Missionary Society
Surry (Patten Bay), 299
Sushi, 90, 123
Sutton Island (Cranberry Islands), 176, 306, 311, 313, 316
Swan, Colonel James, 286, 290
Swan Island (Kennebec River), 108, 114, **115–116**, 340
Swan Island Yacht Club (Richmond), 115
Swan's Island, 269, 270, 278, 285, **286–289**, 290, 303, 304, 386; Atlantic, 286, 287; Minturn, 286

Tarkington, Booth, 49
Tarratine Yacht Club (Islesboro), 230, 231, **232**
Taunton Bay (Frenchman Bay), 326
Teel Island (Muscongus Bay), 172–173
Ten Pound Island (St. George River), 175
Tenants Harbor (Penobscot Bay), 168, 178, 181, 183, **190-191**, 193, 278
Tenants Island (Cape Small Harbor), 101
Tennants Cove (Saint John River), 429
Tennants Island (Cape Small Harbor), 104
Tenpound Island (Matinicus), 184, 187
Terns, 104, 188, 189, **330**, 348, 360, 400
Thatch Island (Saint John River), 438, 439
Thaxter, Celia, 30, 31, 32
Thief Island (Muscongus Bay), 157
Thomaston (St. George River), 154, 173, **175-177**, 205
Thompson Island (Muscongus Bay), 154, 178
Thorne Island (Kennebec River), 114

Thorofare, The (Roque Island), 350
Thread of Life (Johns Bay), 140, 143, **150**
Three Islands (New Meadows River), 100
Thrumcap Island (Cape Rosier), 253
Thrumcap Island (Potts Harbor), 88, 89
Thrumcap Island (Thread of Life), 150
Thumb Island (Letete Passage), 394
Tibbett Island, Tibbett Narrows (Western Bay), 343
Ticonderoga, 112
Tides, 18; anchoring with, 21; and rivers, 21; broadcasts, 19; drain, 21; neap, 19; range, 19, 365; rule of twelfths, 22; spring, 19; tables, 21; tidal currents, 19, 22
Timberwind, 203
Time zones, 367, 372
Tinker Island (Blue Hill Bay), 293, 303
'tit Manan. *See* Petit Manan Island
Tom Cod Cove (Cape Rosier), 252
Tommy Island (Narraguagus Bay), 341
Toothacher Bay (Swan's Island), 288
Toothacher Cove (Swan's Island), 386
Torrey Islands (Eggemoggin Reach), 282, 283
Torryburn Cove (Kennebecasis Bay), 422
Towing services, 65, 441
Townsend Gut (Inside Passage), 118, 119, **126–133**
Trafton Island (Narraguagus Bay), 335, 339, 340, **341**
Trash, disposing of, 9
Treasure Island (Little River), 141
Treat Island (Friar Roads), 384
Trott Island (Stage Island Harbor), 53
Trumpet Island (Blue Hill Bay), 293
Tumbler Island (Booth Bay), 136
Turkey Cove (St. George River), 173, 174
Turnip Island (Little Sheepscot River), 124
Turnip Island (Southern Harbor, North Haven I.), 214
Turnip Island (Thread of Life), 150
Turtle Head Cove (Islesboro), 230, **235**
Turtle Island (Frenchman Bay), 322
Twin Beach (Roque Island), 351
Two Bush Channel (Penobscot Bay), 192, 194
Two Bush Islands (Merchants Row), 274
Twobush Island (Muscongus Bay), 172

Union River Bay, Union River (Blue Hill Bay), 299
United States; entering, 26, **368–369**, 379; cruising permit, 369; customs, 358, 368–369, 383, 384, 387, 402; duties, 369
United States Coast Guard. *See* Coast Guard, United States
United States Fish and Wildlife Service, 330, 357
United States Navy, 74, 76, 111, 318; Portsmouth, 33, 37
Unity, 355
Upper Flag Island (Casco Bay), 81, 89
Upper Goose Island (Casco Bay), 87, 88
Upper Hell Gate (Sasanoa River), 22, 117, **118–119**, 124, 139
Urquharts Cove (Belleisle Bay), 428

Vaill Island (Casco Bay), 74
Valley Cove (Somes Sound), 301, 309, 311
Vances Beach (Kennebecasis Bay), 421
Vaughn's Island (Cape Porpoise), 53
Verrazano, Giovanni da, 23, 107, 154, 166, 300

Vessel Traffic Services. *See* Fundy Traffic
VHF radio, 19; "sécurité" calls, 65, 323; range Downeast, 335
Vinalhaven (Penobscot Bay), 182, 201, 211, 213, 218, 219, 276, 278, 386
Vines, Captain Richard, 29
Virgin Island (Sipp Bay), 389
Virginia, 110, 111, 112

Wadsworth Cove (Damariscotta River), 141, **147**
Wagner, Lewis, 32
Waldoboro (Medomak River), 112, 154, 160
War of 1812, 132, 133, 360, 400
Warren, 242
Warren Island (Gilkey Harbor), 230, 232, **233**, 234
Washademoak Creek (Saint John River), 429, **430**
Washademoak Lake (Saint John River), 412, 429, **430-431**
Washington, President George, 109, 170, 177, 286
Wasson Brook (Grand Lake), 435–436
Water, drinking, 7, 13
Water Cove (Wills Gut), 94
Waterman Cove (Fox Islands Thorofare), 217
Watts Cove (St. George River), 173, **175**
Waymouth, Captain George, 171, 175; at Monhegan, 166; at St. George River, 173, 174; capture of Indians, 171; reports of lobsters, 50; reports of masts, 82
Weather, 14; forecasts, 15; fire danger, 10; tide information, 19; patterns, 14–15
Webb Cove (Deer Island Thorofare), 270, **280**
Weirs, 203, 304, 334, 338, 339, 342, 345, 352, 362, 366, 370, 372, 375, **374**, 381, 391
Wells Harbor (southern coast), 29
Welshpool (Campobello Island), 378, **381-382**
Wentworth, Gov. John, 33
Wentworth-By-The-Sea Hotel, 35
West Bay (Gouldsboro Bay), 338
West Brown Cow (Casco Bay), 79
West Jonesport (Moosabec Reach), **343–344**, 348
West Point Harbor (New Meadows River), 102
West Quoddy Head, 16, 295, 365, 383, 384; lighthouse, 170, 366, 384
Westerly Bar Island (Roque Island), 350
Western Bay (Blue Hill Bay), **299**, 326
Western Bay (Great Wass Island), 334, **343**, 344
Western Egg Rock (Muscongus Bay), 154
Western Island (Corea), 337
Western Passage (Gouldsboro Bay), 338
Western Passage (Passamaquoddy Bay), 377, 380, 387, 391, **392**
Westfield Beach (Saint John River), 419, **424**
Whales, 209, 269, 293, **294–295**, 315, 325, 373, 380; **stranding hotline, 10, 295**
Wharton Island (Medomak River), 161
Wheat Island (Burnt-Pell Passage), 266
Wheaton Island (Matinicus), 184, 185
Whelpley Cove (Long Reach), 425
White, E.B., 283, 298
White Head (Casco Bay), 71
White Head Island (Grand Manan), 370, 373, 376
White Island (Isles of Shoals), 30, 32, 34
White Islands (Hurricane Sound), 222, **224-225**

Whites Cove (Grand Lake), 433, 434
Whites Island (Wiscasset), 133
Whitefin, 112
Whitehawk, 112
Whitehead Island (Muscle Ridge Channel), 192, 193; lighthouse, 193, 202
Whitehead Passage (Casco Bay), 64, 71
Whiting Bay (Cobscook Bay), 389
Whittier, John Greenleaf, 30, 85
Widgeon Cove (Harpswell Sound), 92
Widow Island (Fox Islands Thorofare), 214, 218
Wiggins Cove (Grand Lake), 433, **435**
William Henry Memorial Fort (Pemaquid Harbor), 151, 152
Williams Island (Merepoint Bay), 87
Williams Island (New Meadows River), 101
Williams, William, 41, 170
Wills Gut, east (Casco Bay), 94
Wilson, Jon, 284
Wilson Cove (Ragged Island), 188
Wilsons Beach (Campobello Island), 379, **380**
Windjammers, 112, 122, 136, 181, 202, 203, 206, 209, 211, **262**, 272, 277, 321
Winds, prevailing on Maine coast, 14
Winnegance Bay (New Meadows River), 100
Winter Harbor (Frenchman Bay), 269, 322, **328–331**, 335
Winter Harbor (Vinalhaven), 182, **219**
Winter Harbor Yacht Club (Sand Cove), 329, 331
Winterport (Penobscot River), 242, **245**
Wiscasset (Sheepscot River), 112, 120, **131-133**, 145

Wiscasset Yacht Club, 131
Witch Island (Johns Bay), 150, 151
Witches Island (Cape Harbor), 121
Wolves, The (Bay of Fundy), 392, 405, **408**
Wonsqueak Harbor (Schoodic Peninsula), 336
Wood Island (Biddeford Pool), 54
Wood Island (Casco Bay), 104
Wood Island (Grand Manan), 375, 376
Wood Island (Piscataqua River), 24, 33, 35
Wood Island Harbor (Biddeford Pool), 54–56
Wooden Ball Island (Penobscot Bay), 181, 182, **189**
WoodenBoat School (Eggemoggin Reach), 282, **284-285**
Woods Island (Kennebec River), 114
Working Harbors, 11, 13
World War I, 196
World War II, 8, 64, 69, 70, 73, 74, 75, 318; artifacts, 38; defenses, 33, 37, 71, 72, 76, 80; forts, 110; ships, 111
Wreck Island (Merchants Row), 274, 277, **278–279**
Wyer Island (Harpswell Sound), 91
Wyeth, Andrew, 174, 190; Jamie, 169

Yarmouth (Royal River), 83–84
Yarmouth Island (Quahog Bay), 94, 96
Yellow Head Island (Machias Bay), 355
York Beach (southern coast), 44
York Harbor, York River, York Village, 29, **42–44**, 170
York Narrows (Swan's Island), 269, **285**, 287
Youngs Cove (Grand Lake), 433, **435**

—— SIDEBARS ——

Acadia, For the Enjoyment of all People, 301
Atlantic Neptune: The Charts of DesBarres, 310
Bald Eagles, 340
Cormorants, 129
Early Explorers, 16
Granite, 276
Ice, 145
Indigenous People of the Coast, 56
Lighthouses, 170
Lime, 205
Lobster Buoys and Navigation, 26
Lobsters and Lobstering, 50
Long Islanders, 290
Masts and the Broad Arrow, 82
Matinicus, A Different Breed, 186
Mussels: The Best Things in Life are Free, 192
Norumbega, 241

Ospreys, 176
Penobscot Expedition, 251
Portland, The Beautiful Town by the Sea, 67
Puffins, 361
Razor-Billed Auks, 400
Red Jacket, 200
River Seamanship, 108
Sailing with Children, 320
Salmon Aquaculture, 386
Seagulls, Scavengers of the Coast, 59
Seals, 20
Shipbuilding in Maine, 112
Terns, 330
Weirs, 374
Whales and Whale-Watching, 294
Windjammers, 262
Wiscasset: Days of Glory, Days of Gloom, 132

Curtis Rindlaub and his wife, Carol Cartier, are long-time Maine island residents. Curtis is a writer and photographer whose work has appeared in *Nautical Quarterly*, *WoodenBoat*, *Coastal Living*, *Fine Homebuilding*, *Yankee*, and others. He is the author of the bestselling *Maine Coast Guide for Small Boats: Casco Bay*. Carol is a designer and photo stylist.

Curtis grew up sailing on Long Island Sound, in the Stockholm archipelago, and in the Oslo Fjord. After studying biology at Tufts University, he designed, built, and paddled a sea kayak alone from Glacier Bay, Alaska to Seattle, Washington. Curtis and Carol sail from Peaks Island with their two children, Nathaniel and Hannah, aboard their 35-foot Vindo ketch *Indigo*. They have been updating *A Cruising Guide to the Maine Coast* since 1994.